Fundamentals of Particle Physics

This text is a modern introduction to the Standard Model of particle physics for graduate students and advanced undergraduate students. Assuming only prior knowledge of special relativity and non-relativistic quantum mechanics, it presents all aspects of the field, including step-by-step explanations of the theory and the most recent experimental results. Taking a pedagogical, first-principles approach, it demonstrates the essential tools for students to process and analyse experimental particle physics data for themselves. While relatively short compared to some other texts, it provides enough material to be covered comfortably in a two-semester course. Some of the more technical details are given in optional supplementary boxes, while problems are provided at the end of each chapter. Written as a bridge between basic descriptive books and purely theoretical works, this text offers instructors ample flexibility to meet the needs of their courses.

Pascal Paganini is a professor at the École Polytechnique near Paris and researcher at CNRS, where he has lectured for many years on particle physics for undergraduate and graduate students. As an experimental particle physicist, he has engaged in several experiments at CERN, including the Compact Muon Solenoid experiment at the Large Hadron Collider. More recently, he has been involved in the Super-Kamiokande neutrino experiment in Japan, and designing its next-generation successor, Hyper-Kamiokande.

Fundamentals of Particle Physics

Understanding the Standard Model

PASCAL PAGANINI
École Polytechnique, CNRS, Palaiseau

CAMBRIDGE
UNIVERSITY PRESS

Shaftesbury Road, Cambridge CB2 8EA, United Kingdom

One Liberty Plaza, 20th Floor, New York, NY 10006, USA

477 Williamstown Road, Port Melbourne, VIC 3207, Australia

314–321, 3rd Floor, Plot 3, Splendor Forum, Jasola District Centre, New Delhi – 110025, India

103 Penang Road, #05–06/07, Visioncrest Commercial, Singapore 238467

Cambridge University Press is part of Cambridge University Press & Assessment, a department of the University of Cambridge.

We share the University's mission to contribute to society through the pursuit of education, learning and research at the highest international levels of excellence.

www.cambridge.org
Information on this title: www.cambridge.org/9781009171588

DOI: 10.1017/9781009171595

First published 2023

A catalogue record for this publication is available from the British Library

A Cataloging-in-Publication data record for this book is available from the Library of Congress

ISBN 978-1-009-17158-8 Hardback

Additional resources for this publication at https://www.cambridge.org/9781009171588.

To Patricia, Léo, Nathan and Mathis

Contents

Preface

Particle physics is a fascinating subject. The aim of this field is to understand the complexity of the structure of matter at the most fundamental level and its interplay with elementary forces from very simple principles. It covers subatomic and high energy physics at the smallest scales (elementary particles) and largest scales (early evolution of the Universe itself). Nothing less!

Particle physics has considerably evolved in recent decades with new experimental discoveries made possible by increasingly complex experiments, notably at the emblematic Large Hadron Collider. Its theoretical ground is now well established and described within a framework, the Standard Model (of particle physics), generally grandiosely written with capital letters.

This book attempts to present the remarkable advances of recent years and to explain the underlying theory at a level accessible to students familiar with quantum mechanics, special relativity and classical electrodynamics. It is primarily oriented towards the needs of graduate students in experimental physics (and sometimes perhaps also their supervisors) or for advanced undergraduate students. Students wishing to pursue a carrier as a theorist will need to supplement their education with more advanced books, in particular in quantum field theory. I am always surprised to meet theorists who have no idea how real detectors work. On the other hand, I am also surprised to meet experimentalists with a very rough knowledge of the underlying theory. I sincerely hope that this book can be of some help to both. With this book, students have the tools allowing them to analyse the experimental data and calculate basic formulas using quantised field theories, at least at the tree level.

I have tried to write a book as self-contained as possible. Even though quantum field theory is not the main subject of the book, it remains an essential mathematical tool for understanding particle physics. As I have always felt uneasy with presentations of basic formulas coming out of nowhere, I give a basic, but hopefully pedagogical, introduction on this complicated topic, using mostly the canonical operator formalism. In most cases, I have tried to derive everything from first principles, making the various calculations explicit. Since particle physics is not just theory, I illustrate the experimental developments with key experiments when appropriate for the understanding of the field. In this respect, a large part of the first chapter is devoted to the presentations of techniques used to detect particles in experiments. Throughout this book, I hope to have struck the right balance between theory and experiments (many good textbooks unfortunately focus on only one of the two aspects).

How to Use the Book?

This book is the result of a decade of teaching particle physics at École Polytechnique (France) to graduate students. It includes material that can be comfortably covered in two semesters. However, some suggestions for a shorter course are given in the following paragraph. For each chapter, technical details are often provided in the form of supplementary boxes. They may be ignored in a first approach. Problems are also provided at the end of the chapters.

The first chapter gives a general overview of the field. The second part of this chapter is devoted to experimental techniques, including a discussion of statistics. Students who aim at a theoretical career may be tempted to skip this chapter. I consider, however, that it would be a mistake. Particle physics as natural science is always, at the end, a question of measurements. It is, therefore, essential to understand how they are made and what their uncertainties are. Students who are comfortable with special relativity, in particular with the tensor index notation, and non-relativistic quantum mechanics can skip the second chapter, which is just a quick summary of these topics. Chapters 3 and 4 are mostly devoted to collision kinematics and conservation rules. Chapter 5 introduces relativistic quantum mechanics and begins to introduce the notion of quantum fields. The first part of Chapter 6 can be considered an introduction to quantum field theory. It is completed by the presentation of the electromagnetic interaction and the parton model in Chapter 7. This latter chapter can probably be skipped in an undergraduate course. Chapters 8 and 9 describe the strong and weak interactions. As non-Abelian interactions, several notions from group theory are used. They are explained in detail in Appendices A–D. The part of Chapter 8 devoted to non-perturbative quantum chromodynamics can be skipped as a first approach. Chapter 9 includes recent developments in neutrino physics. Chapters 10 and 11 are devoted to the electroweak model. They include recent results about the Higgs boson physics, at the core of modern particle physics. The last chapter completes the Standard Model of particle physics and presents the most important open questions that could revolutionise particle physics in the future. Some appendices, beyond those devoted to group theory, may be of interest to readers who are not comfortable with advanced mathematics (they are presented from the point of view of a physicist).

Acknowlegements

Many colleagues at École Polytechnique gave valuable assistance in writing this book. In particular, I acknowledge Matthew Nguyen, who read the entire manuscript. When he kindly accepted this task, I am not sure he was fully aware of the burden it represented. Other colleagues gave me valuable comments on parts of the manuscript depending on their expertise, especially Christoph Kopper (for most of the quantum field theory discussed in this book), Franois Gelis and Cédric Lorcé (for the quantum chromodynamics chapter), Stéphane Munier (partially for quantum chromodynamics and the appendices on group theory), Thomas Mueller (for the weak interaction chapter and neutrino physics in general) and Émilien Chapon (for the statistical part of the first chapter). This book is also the result of many discussions with colleagues over coffee (Émilie Maurice, Frédéric Fleuret, Roberto Salerno, Denis Bernard, etc.) and with former students (Matthieu Licciardi, Thomas Strebler, Guillaume Falmagne, etc.), some of whom being now also colleagues. I am grateful to the students, in general, for their excellent questions, which have contributed to improving my knowledge and, thus, indirectly, to improving this document.

Finally, I would like to thank my wife for her understanding and encouragement during this whole enterprise, which took much longer than expected.

Pascal Paganini

Notation and Constants

Generalities

- Eq. and Eqs. are abbreviations for equation(s). Similarly, Fig(s). stands for figure(s) and Ref(s). for reference(s) and p. for page.
- Spatial three-vectors are indicated by letters in boldface. For instance, $\boldsymbol{r}$.
- The symbol $\times$ used between two three-vectors denotes the cross-product of vectors. For example, $\boldsymbol{L} = \boldsymbol{r} \times \boldsymbol{p}$.
- In calculations, the symbole * is the complex conjugate operation, $\Re$ the real part and $\Im$ the imaginary part.
- The symbols $\mathbb{1}$ and $\mathbb{0}$ denote, respectively, the identity matrix and the zero matrix. Depending on the context, the dimension of these matrices may vary.
- The transpose of a matrix A is denoted as A^{T}.

Relativity

- Four-vectors are indicated in regular face except at the very beginning of Chapter 2 where boldface is used for pedagogical reasons. The four-vector position, for instance, is $x = (ct, \boldsymbol{x})$. Note that I frequently use x^μ to label the four-vector x. In other words, the distinction between the four-vector x and one of its components, x^μ, is ignored.
- Throughout this document, I use the metric tensor

$$(g^{\mu\nu}) = (g_{\mu\nu}) = \begin{pmatrix} 1 & 0 & 0 & 0 \\ 0 & -1 & 0 & 0 \\ 0 & 0 & -1 & 0 \\ 0 & 0 & 0 & -1 \end{pmatrix},$$

  with Greek indices running over 0, 1, 2 and 3. Roman indices run over the three spatial coordinate labels $1, 2$ and 3 or x, y and z depending on the context. I usually enclose the symbol for a matrix within parentheses and label the elements of that matrix without parentheses. So $(g^{\mu\nu})$ is the matrix, while $g^{\mu\nu}$ is the element, for example $g^{00} = 1$.

Quantum Mechanics and Quantum Field Theory

- The notation h.c. denotes the Hermitian conjugate, whose symbol when applied to an operator or a matrix is $\dagger$. For a matrix A, its Hermitian conjugate is $A^\dagger = A^{*\mathsf{T}}$.

- Operators acting in the Hilbert space are generally denoted with a hat symbol on top of the label, for instance $\hat{C}$. However, I do not follow this convention for quantum fields such as $\phi(x)$ and $\psi(x)$.
- The time arrow of all Feynman diagrams goes from the left to the right. The initial state is thus always on the left-hand side, and the final state on the right-hand side.

Constants

Table 1 includes the main physical constants used in this book. Their values are taken from Particle Data Group 2022. The one-standard-deviation uncertainty is indicated in parentheses and corresponds to the uncertainty in the last digits. When absent, it indicates that the value is exact.

Table 1 Main physical constants

Quantity	Symbol	Value	Units
Speed of light (vacuum)	c	299 792 458	m s^{-1}
Planck constant	h	$6.626\,070\,15 \times 10^{-34}$	J s
Reduced Planck constant	$\hbar = h/(2\pi)$	$1.054\,571\,817 \cdots \times 10^{-34}$	J s
		$6.582\,119\,569 \cdots \times 10^{-22}$	MeV s
Conversion constant	$\hbar c$	$197.326\,980 \cdots$	MeV fm
Conversion constant	$(\hbar c)^2$	$0.389\,379\,372 \cdots$	GeV^2 mb
Electron charge magnitude	e	$1.602\,176\,634 \times 10^{-19}$	C
Permittivity of free space	$\epsilon_0 = 1/(\mu_0 c^2)$	$8.854\,187\,8128(13) \times 10^{-12}$	F m^{-1}
Permeability of free space	$\mu_0/(4\pi \times 10^{-7})$	$1.000\,000\,000\,55(15)$	N A^{-2}
Fine structure constant	$\alpha = e^2/(4\pi\epsilon_0\hbar c)$	$1/137.035\,999\,084(21)^{\dagger}$	none
Electron mass	m_e	$0.510\,998\,950\,00(15)$	MeV/c^2 #
Proton mass	m_p	$938.272\,088\,16(29)$	MeV/c^2 #
Bohr magneton	$\mu_B = e\hbar/(2m_e)$	$5.788\,381\,8060(17) \times 10^{-11}$	MeV T^{-1}
Nuclear magneton	$\mu_N = e\hbar/(2m_p)$	$3.152\,451\,258\,44(96) \times 10^{-14}$	MeV T^{-1}
Boltzmann constant	k_B	$1.380\,649 \times 10^{-23}$	J K^{-1}
		$8.617\,333\,262 \times 10^{-5}$	eV K^{-1}
Gravitational constant	G_N	$6.674\,30(15) \times 10^{-11}$	m^3 kg^{-1} s^{-2}
		$6.708\,83(15) \times 10^{-39}\hbar c$	(GeV/c^2)$^{-2}$
Fermi constant	$G_F/(\hbar c)^3$	$1.166\,378\,7(6) \times 10^{-5}$	GeV^{-2}
Avogadro number	N_A	$6.022\,140\,76 \times 10^{23}$	mol^{-1}

† Value at $Q^2 = 0$, i.e. large distance.

Mass in particle physics is given in units of eV/c^2 and its multiples, with 1 eV/$c^2 \simeq 1.78 \times 10^{-36}$ kg.

1 Particle Physics Landscape

In this first chapter, a quick introduction to the Standard Model of particle physics is given. The concepts of elementary particles, interactions and fields are outlined. The experimental side of particle physics is also briefly discussed: how to produce elementary particles, observe them with detectors and make measurements with the data collected by the detectors.

1.1 Elementary Particles

Elementary particle physics, also commonly denoted as high energy physics, is the science that studies the units of matter at the most fundamental level and the nature of the fundamental interactions, the forces, governing their behaviour. Both the units of matter and the interactions are believed to be related to elementary particles. Elementary particles are the simplest objects one can think of: elementary particles have no substructure and thus are not made up of other objects. In the context of the Standard Model of particle physics, elementary particles are supposed to have no spatial extension (confirmed by experiments within the accuracy of their measurements) and are characterised by only a very few quantities: their mass (which could be zero), their spin (the intrinsic angular momentum that could also be zero) and some quantum numbers (like the electric charge) on which the forces depend.

Labelling an object as an elementary particle depends on our ability to probe its possible substructure and thus on our experiments. At the end of the nineteenth century, atomic nuclei were considered elementary. This is no longer the case, as we know that nuclei are made up of protons (discovered by E. Rutherford in the late 1910s) and neutrons (J. Chadwick in 1932), themselves made up of quarks, as revealed by experiments in the 1970s. Modern experiments have not found any substructure of quarks. They are thus considered elementary, a conclusion that could be challenged by future experiments. The electron (e^-), discovered in 1897 by J. J. Thomson, is also considered elementary. Thus, with the electrons and just two species of quarks, named *up* (denoted by the symbol u) and *down* (denoted by the symbol d), all atoms, and hence all ordinary matter observed in nature, can be described. The Standard Model manages to reduce Dmitry Mendeleev's famous periodic table of elements to just three elementary particles!

If elementary particles do not have any constituents, it does not mean that they are themselves necessarily constituents of composite structures. If electrons are the constituents of atoms, and up and down quarks are the constituents of protons and neutrons, respectively (a proton contains two up quarks and one down quark, while a neutron contains two down quarks and one up quark), then other elementary particles do not have the capability of forming larger structures. For instance, the electron neutrino (ν_e) is observed when an unstable nucleus decays by emitting an electron (nuclear beta decay). Neither ν_e nor the electron was present in the nucleus before the decay: their production is the result of the decay itself, allowed by the famous equivalence between energy and mass proposed by Einstein (in modern physics, mass is not a conserved quantity; hence, a heavy particle can potentially decay into lighter ones; only the total energy is conserved). A reason why elementary particles do not necessarily form larger structures is that most of them are very unstable, decaying promptly into lighter elementary particles. For instance, the muon (μ^-) discovered in 1937 is a replica of the electron (it carries the same quantum numbers, spin etc.), except that it is heavier. It decays into lighter particles, while the electron, being the lightest of its species, is necessarily stable. The differences of properties between muons and electrons are then entirely the consequence of the difference in their masses. Another heavy sibling of the electron is the tau (τ^-), discovered in 1975. These three elementary particles strictly carry the same quantum numbers: they all have spin 1/2 in units of $\hbar$, an electric charge $-e$, where e is the charge of the proton, etc. Thus, at a fundamental level, they all have the same interactions;[1] the μ^- and τ^- are simply about 200 and 3 500 times heavier than the electron respectively, leading to slightly different properties. In particle physics, these heavier electrons are said to belong to different *generations*. Each has its own neutrinos: the electron neutrino (ν_e) has already been introduced for the first generation, the muon neutrino (ν_μ) for the second and the tau neutrino (ν_τ) for the third. The three neutrinos share common properties with the e^-, μ^- and τ^- related to the (electro-)weak interaction, as we will discover later. Obviously, they also differ since they are the only fermions of the Standard Model that have no electric charge. The set of these six elementary particles defines the *lepton* family.

Quarks have generations too. As e^- and ν_e, up and down quarks belong to the first generation. Their heavier siblings, the *charm*-quark (c) and *strange*-quark (s), belong to the second generation, and the *top*-quark (t) and *bottom*-quark (b), also called the *truth*-quark and *beauty*-quark, respectively, belong to the third generation. All elementary fermions (spin 1/2 particles) of the Standard Model are listed in Table 1.1 with some of their properties. The six kinds of quarks and leptons are distinguished by *flavour*, i.e. species: there are six flavours of quarks (up, down, strange, charm, bottom and top) and six flavours of leptons (electron, muon, tau, electron neutrino, muon neutrino and tau neutrino).

If the ordinary matter in the universe is made with elementary particles of the first generation, particles from the two other generations have been produced in laboratories, thanks to high energy accelerators. Note that when extreme conditions are encountered in cosmic events, such as a core collapse producing a supernova, those particles must be produced, too. However, since they are highly unstable, only the first generation reaches the earth.[2] It is legitimate to ask whether there are other generations of heavier quarks or leptons. So

[1] Except with the Higgs boson. We will see in Chapter 11 that the interactions of fermions with the Higgs bosons are proportional to their mass.

[2] Muons or composite particles containing a strange quark have been observed from cosmic rays, but they are secondary particles. See Section 1.3.1.

Table 1.1 Elementary fermions of the Standard Model. Masses are indicated in brackets and expressed in MeV/c^2.

	Spin[a]	Electric charge[b]	Generation 1	Generation 2	Generation 3
Leptons	1/2	0	ν_e (~ 0)[c]	ν_μ (~ 0)[c]	ν_τ (~ 0)[c]
	1/2	-1	e^- (0.511)	μ^- (106)	τ^- (1777)
Quarks	1/2	2/3	u (~ 2)	c ($\sim 1.27 \times 10^3$)	t (172.8×10^3)
	1/2	$-1/3$	d (~ 5)	s (~ 93)	b ($\sim 4.18 \times 10^3$)

[a]In units of $\hbar$. [b]In units of proton charge. [c]See footnote 3.

far, there is no experimental evidence supporting this hypothesis. Moreover, experiments at the Large Electron-Positron collider (LEP) collider concluded in the 1990s that if there is a fourth generation, the mass of the corresponding neutrino must be larger than 45 GeV/c^2 (under some assumptions). Given that the mass of neutrinos of the first three generations is ridiculously tiny[3] compared with other elementary fermions (an upper limit of the order 1 eV/c^2), it seems unlikely to have a fourth generation with such a difference. Therefore, in the Standard Model, only three generations are assumed. At this stage, one can appreciate the similarity between leptons and quarks within a given generation: the difference in electric charge between the two quarks is always equal to one unit (of proton charge), which is also the difference in charge between the two leptons. We shall see that this is a consequence of the symmetric structure of the Standard Model in the following chapters.

One may wonder why we distinguish quarks from leptons. The reason is that both kinds of particles do not experience the same interactions, and thus they present very different properties. Whereas leptons can be observed in their free states (i.e., propagating freely), quarks cannot. Quarks are always confined to bound states that are generically called *hadrons*. Protons or neutrons are two examples of hadrons, but there are hundreds of others. Table 1.2 gives the most common hadrons. One can notice in the table that there are two kinds of hadrons: those containing three quarks form the family of *baryons*, and those containing a quark and an antiquark (antiquarks are denoted by a symbol with a bar over the quark symbol, i.e., $\bar{u}$) called *mesons*. Since this is the first time we encounter an *antiparticle*, it is worth introducing them. A particle (elementary or not) has a corresponding antiparticle with the same mass, lifetime and spin, but opposite internal quantum numbers (electric charge is an internal quantum number, and we shall see that there are others). For instance, the positron (e^+) is the antiparticle of the electron: it is stable, has the same mass as the electron (511 keV/c^2), but a positive electric charge. Some particles are their own antiparticles, for example the π^0 that contains as many quarks as antiquarks of a given flavour (see Table 1.2). The photon (γ) is an example of an elementary particle

[3] In Table 1.1, I oversimplify the notion of neutrino mass, suggesting that ν_e, ν_μ and ν_τ have a definite mass close to zero. Actually, the neutrinos that have this tiny mass are a quantum mechanical superposition of the listed neutrinos, with the latter not having, strictly speaking, a definite mass.

Table 1.2 Example of hadrons.

	Baryons			Mesons	
Name	Proton	Neutron	Lambda	Pions	Kaons
Symbol	p	n	Λ	π^+, π^0, π^-	$K^+, K^0, \overline{K^0}, K^-$
Quark content	uud	udd	uds	$u\bar{d}$, $\frac{u\bar{u}-d\bar{d}}{\sqrt{2}}$, $\bar{u}d$	$u\bar{s}$, $d\bar{s}$, $\bar{d}s$, $\bar{u}s$
Mass (MeV/c^2)	938	940	1116	140, 135, 140	494, 498, 498, 494

being its own antiparticle. Obviously, only particles having all their internal quantum numbers equal to zero can share this property. When a particle and its antiparticle meet, they annihilate, reduced to pure energy with no residual quantum numbers, from which another pair of particle–antiparticle can emerge. An example is the reaction $e^- + e^+ \to u + \bar{u}$.

1.2 Fundamental Interactions

1.2.1 Quick Overview

Four fundamental interactions (or forces) are known in nature: gravitational, electromagnetic, weak and strong interactions.

In classical physics, gravitation is the attractive force felt by massive objects and is described by Newton's well-known law. In the context of general relativity, gravity becomes a geometric property of spacetime, which is shaped by the energy and momentum of all possible objects, not only matter but also radiation. As we shall see in this section, gravitation does not have any impact on particles, at least when their energy is far from the Planck scale (defined below). Hence, this book mostly ignores it.

Electromagnetism is the attractive (or repulsive) force felt by objects having opposite (or the same) electric charges and nicely described at the classical level by Maxwell's equations. Its generalisation at the quantum level, the quantum electrodynamics (QED) presented in this book, describes the interaction as an exchange of photons between charged particles. Therefore, all charged particles experience the electromagnetic interactions.

The weak interaction is responsible for many decays of unstable particles. It is the interaction that explains beta radioactivity, where there is the emission of an electron and an anti-neutrino, when a neutron decays into a proton, $n \to p + e^- + \bar{\nu}_e$. Hence, the weak interaction acts on both quarks and leptons. We shall see that despite their very different manifestations, the electromagnetic and the weak interactions appear as two aspects of a more basic interaction called the electroweak interaction.

Finally, the strong interaction is the force binding quarks in hadrons. Among elementary fermions, only quarks experience this interaction. Its description is quite close in spirit to electromagnetism since in quantum chromodynamics, the theory describing the strong interaction at the quantum level, it arises through an exchange of a massless particle called the gluon.

Table 1.3 Interactions of elementary fermions.

	Weak	Electromagnetic	Strong
Quarks	□	□	□
$e^\pm, \mu^\pm, \tau^\pm$	□	□	□
ν_e, ν_μ, ν_τ	□	□	□

Table 1.3 summarises the interactions to which elementary fermions of the Standard Model are subjected.

Whereas the electromagnetic interaction (and the gravity obviously) is familiar to us in our everyday life, it is not the case of weak and strong interactions. Electromagnetic interaction and gravity are long-range interactions, with the strength of the force decreasing inversely proportional to the distance squared (e.g., in Problem 1.1, Coulomb's law is deduced from Maxwell's equations). Macroscopic objects can then experience these interactions. However, both weak and strong interactions turn out to be only short-range interactions (at the nucleon scale or even less) as we shall see in Sections 1.2.3 and 1.2.4.

In the subatomic world, the gravitational force can be safely ignored compared to the three others for two reasons: first, particles are very light; and second, the Newton constant is also extremely small. For instance, the strength of the gravitational force between two protons, a distance r apart, is

$$f_G = \frac{G_N m_p^2}{r^2} = 5.9 \times 10^{-39} \frac{\hbar c}{r^2},$$

with $G_N \sim 6.7 \times 10^{-39} \hbar c\ (\mathrm{GeV}/c^2)^{-2}$. On the other hand, the strength of the electromagnetic force between those protons is

$$f_{\mathrm{EM}} = \frac{e^2}{4\pi\epsilon_0 r^2} = \alpha \frac{\hbar c}{r^2} = \frac{1}{137} \frac{\hbar c}{r^2},$$

with $\alpha \sim 1/137$. (See Table 1, p. xix for the expressions and numerical values of the constants). The ratio of the two shows that the gravitational force is about 10^{36} times weaker than the electromagnetic force! Notice that both interactions would be of the same order of magnitude, i.e. $\hbar c/r^2$, for masses about the Planck mass $m_{Pl} = \sqrt{\hbar c/G_N} \sim 1.2 \times 10^{19}\ \mathrm{GeV}/c^2$. Such masses or equivalently such energies[4] are far beyond the reach of any modern accelerator or known cosmic events (the cosmic ray with the highest energy ever seen on earth is about 3×10^{11} GeV and the most powerful accelerator barely boosts protons to 7×10^3 GeV). If the electromagnetic force is so much stronger than gravity, then it might be surprising that, in everyday life, the effects of gravity seem to be rather dominant. The reason for this is simply that in electromagnetism, positive and negative electric charges compensate each other, with the matter being globally neutral, whereas the gravitational force increases with mass without any such compensation.

There is another more worrying reason why the gravitational force does not belong to the corpus of the Standard Model of particle physics. As we will see in the next three sections,

[4] In general relativity, the gravity's source is the energy–momentum tensor, affecting the spacetime curvature, not only the mass, as in Newtonian physics.

interactions are described in terms of the exchange of elementary particles using quantum field theory as a theoretical framework. Unfortunately, there is not yet a well-established quantum theory of gravitation! Candidates for such a theory to reconcile quantum mechanics and general relativity, such as string theory or loop quantum gravity, are still unproven. Consequently, only the electromagnetic, weak and strong interactions are described by the Standard Model of particle physics.

1.2.2 Need for Fields

Modern physics introduces the notion of fields to avoid the issues of forces capable of acting at a distance instantly. As an example, imagine that suddenly the sun vanishes. Since in modern physics, nothing can propagate faster than light, the effect of the sun disappearance must take at least eight minutes before reaching the earth, and thus the gravitational force due to the sun must continue to act during this interval even if the sun is no longer present. Hence, the sun itself is not enough to explain the transmission of the gravitational force. This role is played by the gravitational field. Similarly, there is an electromagnetic field, a strong field and a weak field associated with these interactions. These fields permeate all of space. They respond locally to sources (masses for classical gravitation, electric charges for electromagnetism, etc.) and act on another distant point x, propagating the action to it at a finite velocity (at most the speed of light). The force felt at x thus results from the state of the field *locally* at that point. The action at a distance of classical physics is then avoided.

If fields are the natural consequence of the principle of relativity in modern physics, modern physics also includes another key ingredient, quantum mechanics, which implies that fields can only exist in well-defined states of definite quantised energy. This is very similar to the harmonic oscillator quantisation with which the reader should already be familiar. The next conceptual step is to identify those quanta as particles. At the beginning of the twentieth century, Einstein was the first to realise that the photon was the quantum of the electromagnetic field, manifesting a particle-like aspect in the photoelectric effect. It took about 80 more years to identify the quanta of the weak and strong interactions – the $W^{\pm}$ and Z^0 bosons for the weak interaction and the gluons for the strong interaction. These bosons (they are spin 1 particles) are the force carriers of the interactions, with the action of a force being the result of the exchange of quanta (i.e., the force-carrier particles) of the associated field. The following chapters will be mostly devoted to their presentation: quantum numbers carried by the fields, equation of propagation of the field, etc. Table 1.4 summarises some of their properties.

Notice that because of quantum mechanics, even elementary fermions have a wave-like or field aspect. Therefore, the classical conceptual distinction between matter and forces based on the notion of particles on the one side and fields on the other is no longer relevant. All particles, elementary fermions or force carriers, must be described in terms of quanta of fields. The appropriate theoretical framework is then the relativistic quantum field theory, briefly introduced in Chapters 5 and 6. It turns out that the force fields (and their quanta) emerge naturally from the necessity to respect some particular symmetries at the *local* level: physics must be invariant under such transformations that depend on the spacetime point (as opposed to a *global* symmetry transformation that acts on every spacetime point

Table 1.4 Elementary bosons of the Standard Model.

		Spin[a]	Electric charge[b]	Multiplicity	Mass (GeV/c^2)	Force carrier of
Photon	γ	1	0	1	0	Electromagnetism (QED)
W bosons	$W^{\pm}$	1	± 1	2	80.4	Weak force
Z boson	Z^0	1	0	1	91.2	Weak force
Gluons	g	1	0	8	0	Strong force (QCD)
Higgs	H	0	0	1	125.2	None

[a]In units of $\hbar$. [b]In units of proton charge.

in the same manner). Then, the quanta (the force carriers) of the force fields described by the Standard Model of particle physics must be spin 1 particles (in units of $\hbar$). For instance, anticipating Chapter 6, we will see that the gauge invariance of Maxwell's equations, probably already known by the reader, is connected to the local phase invariance of the fermion fields. Indeed, the local phase invariance imposes a specific transformation of the electromagnetic field (whose quantum is the photon, a spin 1 boson) that precisely leaves invariant Maxwell's equations. This is an example of a gauge theory that will be further generalised to more complicated symmetries for the other interactions (Chapters 8 and 10 of this book).

1.2.3 Yukawa's Theory of Short-Range Interactions

Historically, the strong force was first discovered as the nuclear force binding protons and neutrons in the atomic nucleus. A model due to the Japanese physicist Hideki Yukawa (1935) assumed a potential of the form

$$\varphi(\boldsymbol{r}) = \frac{g}{4\pi}\frac{e^{-r/r_0}}{r}, \tag{1.1}$$

where $r = |\boldsymbol{r}|$, r_0 is a representative parameter of the range of the interaction, and g is a constant analogous to the electric charge, representative of the strength of the interaction. For $r > r_0$, the potential becomes rapidly negligible, leading to an interaction range of the order of r_0. Since the range of the nuclear force is at most the size of a few protons (the nuclear force is responsible for the cohesion of nuclei but not for the cohesion of larger structures), r_0 must be of the order of a few fm. Let us emphasise the difference between Eq. (1.1) and the familiar electrostatic scalar potential due to a point-like charge, e. Since in electrostatics, the electric field simply satisfies $\boldsymbol{E} = -\nabla V$, the Maxwell equation $\nabla \cdot \boldsymbol{E} = \rho/\epsilon_0$ with $\rho(\boldsymbol{r}) = e\delta(\boldsymbol{r})$ implies Poisson's equation

$$\nabla^2 V = -\frac{e}{\epsilon_0}\delta(\boldsymbol{r}),$$

whose solution (Problem 1.2) is the electrostatic potential

$$V(\boldsymbol{r}) = \frac{e}{4\pi\epsilon_0}\frac{1}{r}. \tag{1.2}$$

It depends only on $1/r$ and thus allows a long-range interaction. In addition, unlike $V(\boldsymbol{r})$, $\varphi(\boldsymbol{r})$ in Eq. (1.1) is a solution (Problem 1.2) of

$$\left(\nabla^2 - \frac{1}{r_0^2}\right)\varphi = -g\delta(\boldsymbol{r}).$$

The parameter r_0 is naturally related to the characteristic mass

$$m_0 = \frac{\hbar}{r_0 c} \tag{1.3}$$

(check the consistency of the units), so that

$$\left(\nabla^2 - \left(\frac{m_0 c^2}{\hbar c}\right)^2\right)\varphi = -g\delta(\boldsymbol{r}).$$

Yukawa then extended this equation to the non-static case and interpreted $\varphi(\boldsymbol{r}, t)$ as a quantised field satisfying in the vacuum

$$\left(\nabla^2 - \frac{\partial^2}{c^2\partial t^2} - \left(\frac{m_0 c^2}{\hbar c}\right)^2\right)\varphi(\boldsymbol{r}, t) = 0. \tag{1.4}$$

The parameter m_0 is then the mass associated with the quanta of the field $\varphi(\boldsymbol{r}, t)$. The reader might recognise the relativistic Klein–Gordon equation; more details about this equation and its quantisation are given in Chapters 5 and 6. Hence, in this model, a short-range interaction implies a massive field and, thus, a massive carrier of the interaction. In his publication (Yukawa, 1935), Yukawa set the value $r_0 = 2$ fm and hence predicted, according to Eq. (1.3), that the quantum of the field φ is a new particle with a mass $m_\varphi = m_0 = 100\ \text{MeV}/c^2$. He observed that the scattering of a neutron by a proton $n + p \to n + p$ could be described by the combination of two elementary processes: the incident neutron emits the quantum particle φ and becomes a proton, $n \to p + \varphi$, while the incident proton absorbs the quantum-particle and becomes a neutron $p + \varphi \to n$. He stated that the exchanged quantum particle 'cannot be emitted into the outer space' since[5] $m_n \ll m_p + m_\varphi$. In modern language, in this scattering, Yukawa stated that the intermediate particle carrying the interaction must be a *virtual particle*. Virtuality is allowed by the uncertainty principle of quantum mechanics

$$\Delta E \Delta t \gtrsim \hbar,$$

whose interpretation is not straightforward [see, e.g., (Aharonov and Bohm, 1961) for an interesting discussion]. Here, if the intermediate state $p + \varphi$ has a lifetime Δt (the time interval between the emission of φ and its absorption), its energy uncertainty is of the order of ΔE given by the above relation. For two nucleons separated by a distance r, and assuming a velocity of φ of the order of c (which is, of course, an upper bound), $\Delta t \sim r/c$,

[5] In the rest frame of φ, if the proton has a momentum $\boldsymbol{p}$, the neutron has $\boldsymbol{p}$ and denoting by p_n and p_p the 4-momentum of the neutron and the proton, respectively, the conservation of the energy–momentum requires

$$m_\varphi^2 = (p_n - p_p)^2 = m_n^2 + m_p^2 - 2\left(\sqrt{m_n^2 + |\boldsymbol{p}|^2}\sqrt{m_p^2 + |\boldsymbol{p}|^2} + |\boldsymbol{p}|^2\right) \leq (m_n - m_p)^2,$$

which is not possible for $m_\varphi = 100\ \text{MeV}/c^2$, given that $m_n \sim m_p \sim 940\ \text{MeV}/c^2$.

and thus $\Delta E \gtrsim \hbar c/r$. When r is small enough, ΔE can reach $m_\varphi c^2$ without contradicting the energy conservation principle. This implies that r is at most $\hbar c/m_\varphi c^2 = r_0$. The parameter r_0 can then be seen as the range of the interaction. While the presence of the quantum particle φ cannot be observed directly in this scattering, Yukawa's theory implies its existence, and some kinematically possible processes should allow its production and detection. Two years after Yukawa's paper, while analysing cosmic rays, Anderson and Neddermeyer (1937, 1938), followed shortly by Street and Stevenson (1937), discovered a new charged particle whose mass was around 100 MeV/c^2. It was a perfect candidate for φ. However, a few years later, experiments showed that these cosmic ray particles barely interacted with nuclei, contrary to expectation. It turns out that this new particle was actually the muon. It was only in 1947 that, while analysing cosmic rays, Lattes et al. (1947) found a heavier particle, with a mass of about 140 MeV/c^2, interacting with nuclei in a manner consistent with Yukawa's prediction. This particle is known today as the pion, with its three charge states denoted as π^+, π^- and π^0.

1.2.4 A Glimpse of the Weak and Strong Interactions

Yukawa's theory relates the range of interactions with the mass of the force carrier via Eq. (1.3). With masses of the order of 100 GeV/c^2 (see Table 1.4), the carriers of the weak interaction, $W^\pm$ and Z^0, imply an interaction range of about 10^{-18} m. The pions, being 1 000 times lighter, imply a nuclear interaction range of about 10^{-15} m. They can be considered as the carriers of the effective strong force at the scale of the nucleon. Notice that pions are spin 0 bosons and not elementary particles since they are mesons. Therefore, they cannot be the carriers of the strong interaction at the most fundamental level, i.e. at the scale of the quarks. Those are the gluons, and as expected they are spin 1 elementary particles. However, the attentive reader will have remarked that they are massless (cf. Table 1.4). When the force carrier of an interaction is a massless particle, according to Eq. (1.3), we would expect an infinite range. The electromagnetic interaction is clearly active over a very large range since the light of distant galaxies can be observed on earth. However, massless gluons seem to contradict the apparent short-range behaviour of the strong interaction. In order to resolve this contradiction, one should dive deep into the theoretical framework of the strong interaction, quantum chromo dynamics (QCD), a theory presented in Chapter 8. A simplified summary is given here.

When two quarks strongly interact, they exchange gluons as they similarly exchange photons in electromagnetic interactions (QED). The role played by the electric charge in QED is replaced by another kind of charge, the *colour*. If there is only one elementary electric charge e (all charges of particles are just a – possibly fractional – multiple of e), quarks can have three 'colour charges', arbitrarily called *red*, *green* and *blue*. Hence, a quark is labelled by its flavour but also its colour, for instance, u_r for a red up quark. Since the gluons are massless particles, there is intrinsically no theoretical limit to the range of the interaction. However, there is a specific property of QCD, the *colour confinement*, that prevents the observation of quarks as free isolated particles. Experimentally, even by smashing two hadrons at very high energy, detectors observe hadrons again. The higher the energy, the greater the number of hadrons that is detected, without producing isolated quarks. QCD asserts that all particle states observed in nature must be 'colourless' (the appropriate term is colour singlet coming from group theory, meaning that a rotation in the colour space leaves the state invariant, but at this stage of the book, colourless is probably more suitable). Of course, a single quark carrying a colour cannot be colourless. But mesons made of a quark and an antiquark can: if the quark has a colour, the antiquark has an anti-colour. For example, if the quark is red, it is sufficient that the antiquark is anti-red to cancel the overall colour of the quark–antiquark system. Baryons are also colourless because their three quarks have three different colours: inspired by the colour theory of visual arts,[6] the superposition of red, green and blue produces the achromatic white. (From this perspective, an anti-colour can be seen as the complementary colour.) How can the colour confinement be qualitatively explained, and why does it have an impact on the range of the strong interaction? We saw that the exchange of a massless boson leads naturally to a Coulomb-like potential with a $1/r$ dependence, incompatible with confinement. At very short distances ($< 10^{-15}$ m), the QCD colour potential between a quark q_1 and an antiquark $\bar{q}_2$, separated by the distance r, is indeed in $1/r$, adequately described by the exchange of a gluon. However, for longer distances, there is a linear behaviour, leading to the phenomenological potential

$$V(r) = -\frac{A}{r} + \sigma r, \tag{1.5}$$

where A and σ are two positive constants. Therefore, the stored potential energy between the $q_1\bar{q}_2$ pair increases with distance and can become so high that it exceeds the energy required to produce a new pair of $q_3\bar{q}_4$. It costs less energy to form the two pairs $q_1\bar{q}_4$ and $q_3\bar{q}_2$ because the distance between the quark and the antiquark of these pairs is reduced with respect to the initial distance between q_1 and $\bar{q}_2$. Hence, quark–antiquark systems are necessarily confined. This mechanism is, however, not yet quantitatively understood in detail. At a larger scale, the residual nuclear force between hadrons can be understood as follows: far from a hadron, the colour of the quarks constituent of the hadron is not seen, since the quark configuration is such that the hadron is colourless. Therefore, the strong interaction between hadrons rapidly vanishes, producing a short-range nuclear force. In Chapter 8, when we are more familiar with quantum chromodynamics and its Feynman diagrams, we will see how we can recover the pion model exchange from gluon emissions. It is often said that nuclear force is analogous to the Van der Waals forces, where electric dipoles provide intermolecular bondings. Actually, if a similar effect could exist with colour charge instead of electric charge, its magnitude would be too weak to explain the bonding between nucleons (Povh et al., 2008, Chapter 16). The mediation via pions (and possibly heavier mesons) is really required to describe the nuclear force.

After its range, another aspect characterising an interaction is its strength. If it was easy to define the strength of the gravitation and electromagnetic interactions in Section 1.2.1, it is less obvious for the weak and strong interactions. The weak interaction is called 'weak' because its strength appears to be the weakest of the three interactions described by the Standard Model. However, we shall see in Chapter 9 that intrinsically, the interaction is

[6] The visual arts provide a simple analogy for colour combination, but the exact rules are governed by group theory.

Table 1.5 Decays of Λ and Σ baryons.

Baryons	Quark contents	Mass (GeV/c^2)	Lifetime (s)	Decay channels
Λ^0	*uds*	1 116	2.6×10^{-10}	$p + \pi^-, n + \pi^0$
Σ^0	*uds*	1 193	7.4×10^{-20} † #	$\Lambda^0 + \gamma$
$\Sigma^0(1\,385)$	*uds*	1 384	$\sim 2 \times 10^{-23}$ † ♭	$\Lambda^0 + \pi^0$, $\Sigma^0 + \pi^0$

† Such a short lifetime cannot be measured from the decay length distribution in detectors.

The Σ^0 lifetime can be indirectly measured via the cross-section $\Lambda + \gamma \to \Sigma^0 + \gamma$, where the γs come from the Coulomb field of a nucleus (Dydak et al., 1977).

♭ The $\Sigma^0(1\,385)$ lifetime can only be estimated from the intrinsic energy uncertainty of unstable states, via $\tau \sim \hbar/\Delta E$ (see Section 3.4).

not that weak, in particular at high energy. Similarly, the strength of the strong interaction (the strongest of the three) depends on the energy scale: we will see in Chapter 8 that at high energy, i.e. at very small distances, its strength is rather weak, while for distances of the order of the size of the hadrons, its strength is extremely powerful. However, even if the notion of interaction strength becomes complicated, a crude estimate can be inferred at low energy by comparing the lifetimes of similar particles that decay mostly in a decay channel involving a specific interaction. Intuitively, the stronger the interaction, the higher the probability of decay, and therefore the shorter the particle's lifetime. Let us introduce two dimensionless constants, α_w and α_s, for the weak and strong interactions, respectively. They are analogous to α, the fine structure constant of electromagnetism. We will learn how to calculate the lifetimes of particles in the following chapters, but at this stage, let us admit that if the boson, carrier of the interaction, is emitted or absorbed, the lifetime formula gets a factor proportional to the inverse of the constant. In Table 1.5, three baryons containing the same quark content (*uds*) are compared. They differ by their quantum numbers. Both Λ^0 and Σ^0 are spin 1/2 particles, while $\Sigma^0(1\,385)$ has a spin 3/2. The difference between Λ^0 and Σ^0 is the isospin quantum number[7] since $I = 0$ for Λ^0 and 1 for Σ^0. Their lifetime and their most frequent decay channels, accounting for almost 100% of the cases, are listed in the table. The only interaction allowing a change of generation among the elementary fermions is the weak interaction. The decays $\Lambda^0 \to p + \pi^-$ and $\Lambda^0 \to n + \pi^0$ proceed then via this interaction. At the quark level, the strange quark of the Λ^0 decays as $s \to u + W^-$, followed by $W^- \to d + \bar{u}$. The virtual W^-, carrier of the weak interaction, is then emitted and absorbed, yielding a factor α_w^{-2} in the Λ^0 lifetime formula. On the other hand, the decay $\Sigma^0 \to \Lambda^0 + \gamma$ involves the electromagnetic interaction since a real photon is emitted, but not absorbed. It contributes to a factor α^{-1} to the Σ^0 lifetime. Hence, the ratio of those two lifetimes satisfies

$$\frac{\tau_{\Lambda^0}}{\tau_{\Sigma^0}} \propto \frac{\alpha}{\alpha_w^2}.$$

[7] Isospin symmetry (introduced in Section 8.1.1) assumes that the up and down quarks belong to a two-component vector of isospin value $I = 1/2$. The up quark is the upper component with $I_z = +1/2$, and the down quark is the lower component with $I_z = -1/2$. The maths of the isospin is the same as that of the usual spin. The up and down quarks were so named because of that symmetry.

In order to get a crude estimate of the constants, we will assume that the previous formula is actually an equality, neglecting all other differences in the lifetime formula (like the phase space). We conclude, using the value $\alpha \sim 1/137$, that

$$\alpha_W \sim \sqrt{\alpha \frac{\tau_{\Sigma^0}}{\tau_{\Lambda^0}}} \sim 10^{-6}.$$

The weak interaction is thus about 10^4 weaker than the electromagnetism interaction. Finally, the decay of $\Sigma^0(1\,385)$ is a strong decay that conserves the isospin numbers (the π^0 has $I = 1$ and $I_z = 0$). One of the quarks of the $\Sigma^0(1\,385)$ emits a virtual gluon that converts into a pair $u\bar{u}$ or $d\bar{d}$, becoming the π^0. Hence, we expect $\tau_{\Sigma^0(1\,385)} \propto \alpha_s^{-2}$, allowing us to conclude

$$\alpha_s \sim \sqrt{\alpha \frac{\tau_{\Sigma^0}}{\tau_{\Sigma^0(1\,385)}}} \sim 1\text{–}10.$$

In summary, at low energy, the strength of the interactions of the Standard Model is roughly

$$\begin{array}{ll} \text{strong:} & \alpha_s \sim 1\text{–}10, \\ \text{electromagnetism:} & \alpha = 1/137 \sim 10^{-2}, \\ \text{weak:} & \alpha_W \sim 10^{-6}. \end{array} \tag{1.6}$$

Do not take those numbers too seriously: other decay channels may lead to different estimates. They are just indicative of the order of magnitude of the interaction strengths. We can, however, keep in mind that the lifetime of particles whose decay is dominated by a single interaction is typically

$$\begin{array}{ll} \text{strong:} & \tau_s \sim 10^{-24}\text{–}10^{-21}\ \text{s}, \\ \text{electromagnetism:} & \tau_{\text{E.M.}} \sim 10^{-20}\text{–}10^{-16}\ \text{s}, \\ \text{weak:} & \tau_W \sim 10^{-13}\text{–}10^{3}\ \text{s}. \end{array} \tag{1.7}$$

1.3 Production of Particles

To study particles, the first step is to produce them. Two mechanisms can be identified: natural production originating from cosmic rays or artificial production requiring an accelerator.

1.3.1 Cosmic Rays

Until the early 1950s, the only source of high energy particles was the interaction of primary cosmic rays with the atmosphere. Primary cosmic rays are charged particles, whose lifetime is long enough to be produced by astrophysical sources and reach the earth. They are mostly protons, and nuclei, whose relative abondance depends on the energy range considered. For example, at about 10 GeV per nucleon, 94% are protons, 6% helium nuclei (alpha particle) and the remaining 1% are stable nuclei synthesised in stars (Particle Data Group, 2022).

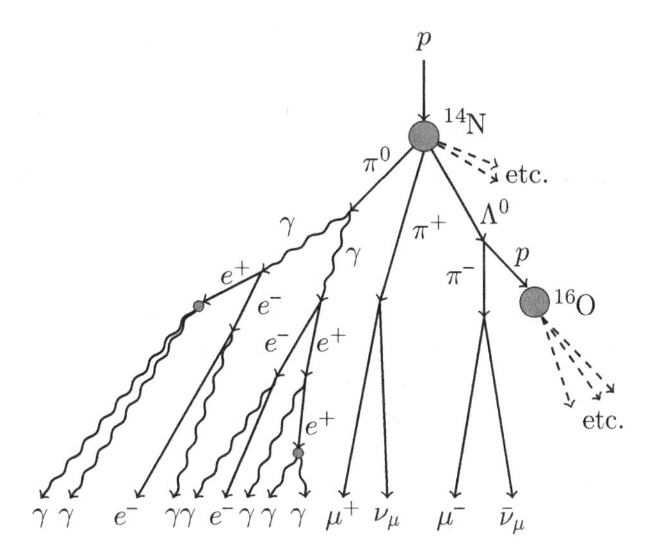

Fig. 1.1 Development of cosmic rays air shower. The small grey circles are electrons in the atmosphere.

By entering the stratosphere, primary cosmic rays cause nuclear collisions with mostly oxygen and nitrogen atoms. These collisions produce secondary particles, including protons, neutrons and charged mesons, such as pions ($\pi^{\pm}$) and kaons ($K^{\pm}$), which typically decay into muons and neutrinos. Since the lifetimes of charged pions and kaons are large enough to travel over significant distances, some may collide with another nucleus in the air before decaying. Neutral pions (π^0) are also produced from the primary interaction. They typically decay almost immediately into a pair of high energy photons that interact electromagnetically with the Coulomb field of atoms. This results in pairs of electron and positron, which can further radiate photons (gamma rays). This process continues, creating a cascade of electrons, positrons and photons. Figure 1.1 illustrates the interaction of a primary cosmic rays in the atmosphere. The number of particles increases rapidly as the cascade of particles, also called *shower*, moves downwards. In each interaction, the particles lose energy, and eventually will not be able to create new particles. After some point, more particles are stopped than created, and the number of shower particles declines.

Energetic secondary particles reaching sea level are dominated by neutrinos and muons. To give an order of magnitude, the neutrino flux around 1 GeV (where the probability of interaction is maximum) is about 1 cm^{-2} s^{-1} from all directions (Gaisser, 1990). For muons with mean energy above 1 GeV, the intensity of the vertical flux is $I_v \simeq 7 \times 10^{-3}$ cm^{-2} s^{-1} sr^{-1} (Grieder, 2001). Given that the muon flux intensity varies with the zenith angle, θ, as $I(\theta) \simeq I_v \cos^2\theta$ (empirical relationship), the muon flux coming from the sky collected by a horizontal detector is then of the order of 1 cm^{-2} mn^{-1} (Problem 1.3).

Before the 1950s, it was the analysis of the secondary particles recorded in cloud chambers at the ground level or detected in photographic emulsions flown in a balloon in the upper atmosphere that led to the discoveries of the positron (1932), muon (1937), pion (1947), kaon (1947) and Λ^0 (1950).

1.3.2 Accelerators

After the 1950s came the era of particle accelerators. They have the great advantage to provide controlled collisions: the energy of the incident particle and the projectile is reasonably determined, as well as their species, and high energy collisions can be reliably repeated. The technology of accelerators is complex and deserves a book on its own. See, for example, Wiedemann (2007). I limit myself here to very general considerations.

Cyclotrons were first built: they consist of two D-shaped cavities, called dees, facing each other (ᗡD) in which a magnetic field bends the trajectory of a charged particle on a circular path. The dees are kept at different electrical potentials to accelerate the particle as it passes from one dee to the other. The process is repeated twice per turn and therefore requires swapping the potentials of the two dees to maintain the acceleration in the appropriate direction. At each turn, the energy of the particle increases, and thus, the radius of the orbit gets larger and larger. After several turns, the particle reaches the rim of the dees and can hit a target where the collision occurs. Cyclotrons had limitations, for instance, the velocity of the accelerated particles could not exceed about $0.1c$ because the circular motion at high speed could not be maintained synchronous with the accelerating field.

Cyclotrons are now replaced by *synchrotrons*, where particles orbit in a fixed circular ring thanks to an adjustable magnetic field. The acceleration is realised thanks to a radiofrequency (RF) cavity located along the ring. RF cavities are metallic chambers that contain an electromagnetic field generated by an RF power generator. They are shaped so that electromagnetic waves become resonant inside the cavity. Their frequency is synchronised to accelerate the particles every time they pass the RF cavity. The Large Hadron Collider (LHC) at CERN (European Organisation for Nuclear Research) near Geneva, Switzerland, is nowadays the most emblematic and powerful synchrotron, colliding two beams of protons, each at an energy of 6.5 TeV, or for one month a year, beams of heavy ions with 2.56 TeV per nucleons. In the 2020s, it is expected that the proton beam energy would be upgraded to 13.6 or even 7 TeV (and the heavy ions beam to 2.76 TeV per nucleons). To accelerate particles to very high energy, synchrotrons require strong magnetic fields to maintain the particles on their circular path. At the LHC, magnetic fields up to 7.74 T are generated by 1 232 superconducting dipoles operating at 1.9 K, and eight superconducting RF cavities delivering 5 MV/m are used for the acceleration. Synchrotrons are limited by two main factors: the maximum magnetic field strength of the bending magnets and the *synchrotron radiation*. A charged particle in a circular orbit undergoes an acceleration and thus emits electromagnetic radiation, the synchrotron radiation, whose radiated power is proportional to $E^4/(m^4R^2)$, with E and m being the energy and mass of the particle, respectively, and R the radius of the orbit. To limit this radiation, it is thus better to use heavy particles and very large synchrotrons. That is why the LHC accelerates protons to 6.5 TeV in a ring about 27 km in circumference, whereas its predecessor, the Large Electron–Positron collider (LEP), accelerated electrons at most to 104.5 GeV in that same ring.

In order to circumvent synchrotron radiation, another strategy is to use *linear accelerators*. However, unlike circular accelerators, an RF cavity can accelerate particles only once on a linear trajectory. The acceleration voltage per unit length is thus the limiting factor in the energy to which a linear accelerator can boost particles. In addition, beams can only collide once, limiting the probability of interaction. The Stanford Linear Collider (SLC), operating until 1998 at Stanford (CA, USA), was the longest linear accelerator (3.2 km), accelerating electrons and positrons to an energy of 50 GeV. It is, to date (and will stay even in the near future), the only example of a linear collider that has produced high energy particle collisions (i.e., tens of GeV) at a high rate.

In summary, particles can be produced from collisions between cosmic rays with the atmosphere or in a controlled environment using particle accelerators. Although accelerators are the primary tool of the high energy physicist, cosmic rays are also studied, particularly to understand the astrophysical phenomena that accelerate particles to such high energies, and for neutrino physics (see Chapter 9). In the history of particle physics, probing collisions of increasingly high energy has led to many discoveries. The higher the energy, the smaller the spatial extent of the object that can be probed. From this perspective, the key parameter of accelerators is the maximum energy reached by the accelerated particles. The other important parameter is the *luminosity* (defined in Section 3.5). The luminosity, $\mathcal{L}$, has the dimension of events per unit time per unit area and determines, for a given reaction, the number of potential collisions that can be produced per second, dN/dt. More specifically,

$$\frac{\mathrm{d}N}{\mathrm{d}t} = \mathcal{L} \times \sigma, \tag{1.8}$$

where σ is the cross section of the reaction. The cross section is defined in Section 3.5, but at this stage, let us say that it is a quantity representative of the probability of the interaction. Therefore, the higher the luminosity, the smaller the cross section to be studied can be, revealing rare processes. With a luminosity of the order of 10^{34} cm^{-2} s^{-1}, the LHC yields about one billion proton–proton collisions per second at a centre-of-mass energy of 13 TeV. The integrated luminosity, i.e. the integral of the luminosity over the operating time of the collider, is then directly related to the total amount of collisions produced by the collider. It is, therefore, a metric of its performance.

1.4 Detection of Particles

The complexity of detectors has considerably increased over time, in particular, because of the ever higher particle energies involved and the very large number of particles produced in the collisions generated by modern colliders. Detectors now include devices able to identify particles and measure with a high precision their positions, momenta, energies and possibly their lifetimes. The precise description of the technology used in detectors goes beyond the scope of this book. There are dedicated books about this topic, such as those authored by Rossi (1952), Leo (1994) and Grupen and Shwartz (2011). This section is more focused on the interaction mechanisms exploited by particle physics detectors.

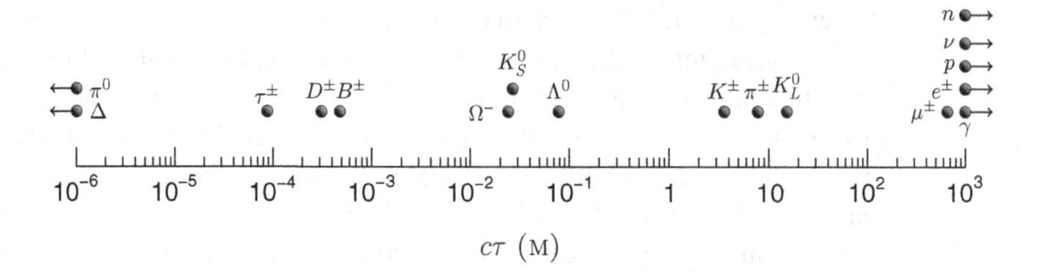

Fig. 1.2 Lifetime of particles multiplied by the speed of light. The arrows indicate that the actual value of $c\tau$ exceeds the scale in the figure.

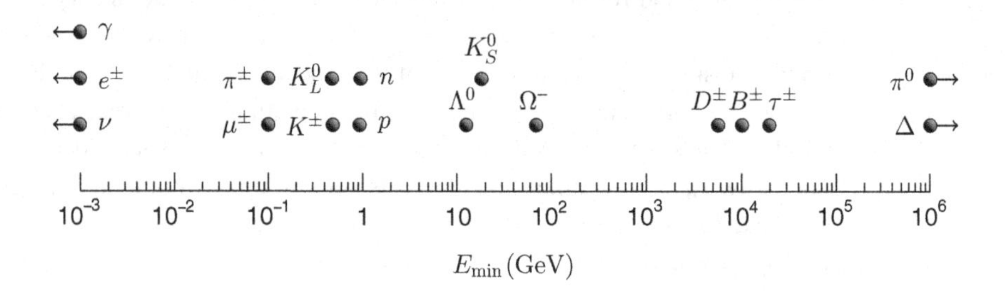

Fig. 1.3 Required minimum particle energy, $E_{\rm min}$ from Eq. (1.9) in GeV, to travel over a distance of 1 m. The arrows indicate that the actual value of $E_{\rm min}$ exceeds the scale in the figure.

1.4.1 General Considerations

Most of the particles are highly unstable with a lifetime so short that they almost immediately decay after their production. Hence, they cannot be observed directly in the detectors, and only their long-lived decay products can reach detectors or interact long enough with detectors. Figure 1.2 shows the average distance travelled in the vacuum by a representative sample of particles, assuming that their velocity is equivalent to the speed of light (ultra-relativistic approximation). Only a subset of unstable particles can travel over macroscopic distances, let us say at least a few millimetres, before decaying. All such particles decay by the weak interaction. If we require that a particle of mass m and lifetime τ travels on average over a distance l in the vacuum, the minimum energy is constrained to be (see Problem 1.4)

$$E_{\rm min} = mc^2 \sqrt{1 + \left(\frac{l}{c\tau}\right)^2}. \tag{1.9}$$

To give an order of magnitude, let us choose $l = 1$ m (the typical size of modern detectors is a few tens of metres). Figure 1.3 shows the corresponding values of $E_{\rm min}$ for various particles. Modern accelerators are powerful enough to produce collisions from which particles with energies of the order of a few GeV emerge. Therefore, we conclude from Fig. 1.3 that, statistically, the only particles that can possibly interact with detectors over a significant distance are:

$$\gamma, e^{\pm}, \nu\ (\bar{\nu}), \pi^{\pm}, \mu^{\pm}, K^{\pm}, K_S^0, K_L^0, p\ (\bar{p}), n\ (\bar{n}), \Lambda^0\ (\overline{\Lambda^0}). \tag{1.10}$$

Particles of this list can be classified into two categories: charged and uncharged particles. Since they experience different interactions with matter, different devices are needed for their detection.

1.4.2 Interaction of Charged Particles with Matter

Energy Loss by Ionisation

When a charged particle passes through a medium (called below *absorber*), it continuously interacts with the electrons of its atoms, or classically speaking, it undergoes inelastic collisions with the electrons. Thus, it transfers a fraction of its energy to the electrons of the absorber, ionising the atoms when the transfer is large enough or exciting them (raising the electrons to a higher lying shell) otherwise. The energy transfer causes an energy loss of the incident particle, thus reducing its velocity. The energy transfer to the electrons may be so large that these electrons may cause further ionisation of the atoms. Such electrons are called δ-rays. Scattering from nuclei also occurs. Since, in general, the mass of the incident particle is significantly smaller than that of the nucleus of the absorber (almost no recoil), very little energy is transferred by this mechanism. In this case, only elastic scattering occurs.[8]

The energy loss by ionisation or excitation is statistical in nature, governed by quantum mechanical probabilities. For a macroscopic path length, it occurs many times, and thus the fluctuations in the energy loss are small. As a first approximation, one can consider the average energy loss. For an incident particle with an electric charge ze, the mean energy loss by ionisation or excitation per unit length, $-\,\mathrm{d}E/\,\mathrm{d}x$, normalised to the absorber density ρ (g/cm^3) is given by the Bethe–Block formula (Bethe, 1930, 1932; Bloch, 1933)

$$\frac{1}{\rho}\left(-\frac{\mathrm{d}E}{\mathrm{d}x}\right)_{\text{ion.}} \simeq Kz^2\frac{Z}{A}\frac{1}{\beta^2}\left(\ln\frac{2m_ec^2\gamma^2\beta^2}{I} - \beta^2 - \cdots\right). \tag{1.11}$$

This formula is an approximation valid for incident charged particles heavier than the electron, and for moderately relativistic velocities, $\beta\gamma \sim 0.1-1{,}000$, with $\beta = v/c$ and $\gamma = 1/\sqrt{1-\beta^2}$. In Eq. (1.11), $K = 0.3071$ MeV mol^{-1} cm^2 is a constant, Z is the atomic number of the absorber, A is its atomic mass in g mol^{-1} and m_e is the electron mass. The parameter I is called the mean excitation energy: it is an effective ionisation potential, averaged over all electrons of the absorber. It depends only on the absorber, and its value for various absorbers can be found in the literature (e.g., in Particle Data Group, 2022). To give an order of magnitude, a usual parametrisation for $z > 1$ is $I \sim 16\ Z^{0.9}$ eV (Grupen and Shwartz, 2011). The dots in Eq. (1.11) represent a set of small corrections. They concern incident particles moving at low velocities that are comparable to or smaller than those of atomic electrons but also high energy particles whose electric field is screened by the electric polarisation of the medium. They can be found in Particle Data Group (2022). The quantity $2m_ec^2\gamma^2\beta^2$ in the numerator of the logarithm in Eq. (1.11) is an

[8] By definition, in elastic scattering, the total kinetic energy is conserved.

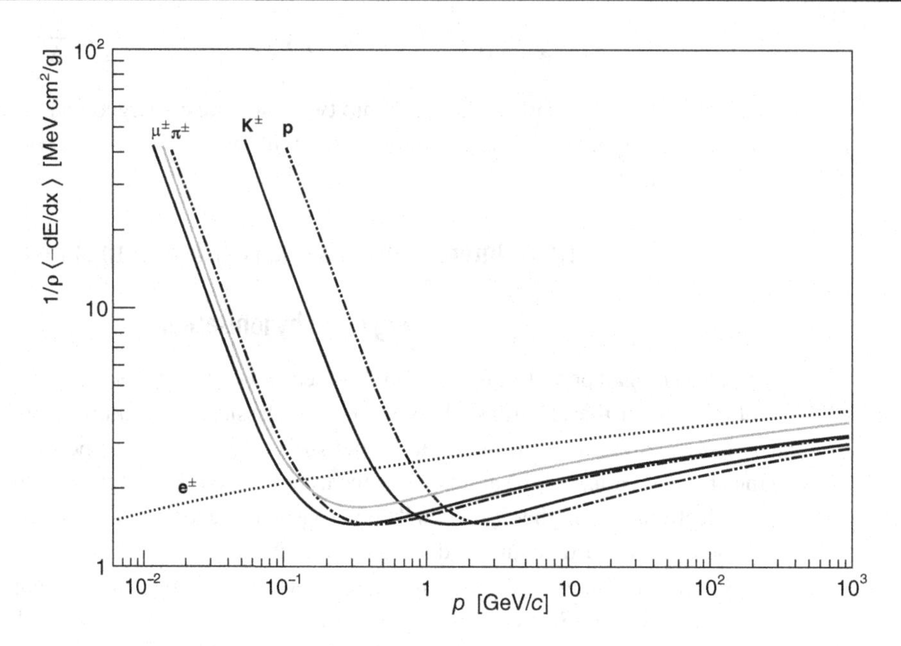

Fig. 1.4 The energy loss by ionisation or excitation (stopping power), $\langle dE/dx\rangle/\rho$, as a function of the momentum of different particles. The absorber here is copper ($Z = 29$, $A = 63.5$ g/mol, $I = 322$ eV), except for the grey curve, which represents the stopping power of muons in silicon ($Z = 14$, $A = 28.1$ g/mol, $I = 173$ eV).

approximation, valid at moderately relativistic velocities, of the maximum kinetic energy that can be transferred to an electron.[9]

For a given absorber, the energy loss in Eq. (1.11), also called the *stopping power*, depends only on the velocity (β) and the charge of the incident particle. It does not depend on its mass, to first order. It is shown in Fig. 1.4 as a function of the momentum $p = \gamma m\beta c$, for different particles passing through copper. The energy loss is maximal at low velocity since slower particles experience the electron field for a longer time. In this regime, dE/dx is dominated by $1/\beta^2$ and decreases with increasing velocity. It reaches a minimum when $\gamma\beta = p/(mc) \sim 3$–$4$ or, equivalently, when $v \sim 0.95c$–$0.96c$. Particles at this point are called MIP or *Minimum Ionising Particles*. According to Fig. 1.4, an MIP loses about 1.4 MeV cm^2/g. This value varies little over a wide range of absorbers. At first order, the incidence of the absorber in Eq. (1.11) comes from the factor Z/A, which changes little for comparable Z. Hence, when dE/dx is normalised to the density, as in Eq. (1.11), it is almost independent of the absorber. At higher energy than that of the MIP, the factor β becomes almost constant (equal to 1), and the factor in $1/\beta^2$ becomes irrelevant. The energy loss rises though, because of the term in $\ln(\gamma^2\beta^2)$, an effect called the relativistic rise. The previously mentioned corrections become significant, however, and tend to

[9] Without this simplification, the correct expression of the numerator in Eq. (1.11) is $\sqrt{2m_ec^2\gamma^2\beta^2T_{\max}}$, with $T_{\max} = 2m_ec^2\gamma^2\beta^2/(1 + 2\gamma m_e/m + (m_e/m)^2)$, where m is the mass of the incident particle. Equation (1.11) is recovered when $2\gamma m_e/m \ll 1$ and $m \gg m_e$.

moderate the rise. Hence, in practice, most relativistic particles have a mean energy loss by ionisation that is close to the minimum.

Formula (1.11) is valid for charged particles heavier than the electron. For electrons or positrons, one has to take into account that the mass of the incident particle and that of the target electron is the same [the derivation of Eq. (1.11) assumes that the incident particle remains undeflected during the collision]. In addition, incident electrons cannot be distinguished from those of the target, requiring an appropriate treatment in the calculation. The expression of the energy loss for electrons or positrons due to collisions on electrons can be found in Leo (1994) and Particle Data Group (2022). They are represented with the dashed line in Fig. 1.4, where they cannot be distinguished from one another in this momentum region. At relativistic energies, above tens of MeV, on the other hand, the energy loss of positrons and electrons is dominated by another source of energy loss: *bremsstrahlung*, i.e. the energy loss by radiation (see Section 'Energy Loss by Radiation: The Bremsstrahlung').

The $\mathrm{d}E/\mathrm{d}x$ curves allow the identification of charged particles. If both the momentum $p = \gamma\beta mc$ and the energy loss $\mathrm{d}E/\mathrm{d}x$ are measured with sufficient accuracy, it gives a point in the $(\mathrm{d}E/\mathrm{d}x, p)$ plane that can be used to determine the species of the particle.

Energy Loss by Radiation: The Bremsstrahlung

Charged particles interact not only with the electrons of atoms (contributing to the energy loss by ionisation and excitation of the atoms) but also with the Coulomb field of the nuclei. The incident particle is then deviated from its straight-line course by the electrical interaction with the nuclei. It contributes to *multiple scattering* but also to the emission of electromagnetic radiation due to the acceleration of the charged particle. As synchrotron radiation is emitted because of the acceleration in a magnetic field, *bremsstrahlung* is radiation emitted because of the acceleration (or rather a deceleration) in the electric Coulomb field (the German word bremsstrahlung means 'deceleration radiation').

For high energies, the energy loss by bremsstrahlung can be described by Rossi (1952)

$$\left\langle -\frac{\mathrm{d}E}{\mathrm{d}x} \right\rangle_{\text{brem.}} \simeq 4\alpha N_A \frac{Z^2}{A} \left(\frac{1}{4\pi\epsilon_0} \frac{(ze)^2}{mc^2} \right) E \ln\left(183\, Z^{-1/3}\right), \tag{1.12}$$

where ze, m and E are the charge, mass and energy of the incident charged particle, respectively, and Z and A are, respectively, the atomic number and atomic mass of the absorber. Whereas the energy loss by ionisation varies logarithmically with energy (Fig. 1.4) and linearly with Z [Eq. (1.11)], the energy loss by bremsstrahlung increases linearly with E and quadratically with Z. In addition, it is proportional to the inverse of the particle mass squared. Hence, at high energy, only low mass particles, namely positrons and electrons, lose significant energy by bremsstrahlung. The *critical energy*, E_c, is defined as

$$\left\langle -\frac{\mathrm{d}E}{\mathrm{d}x} \right\rangle_{\text{brem.}} = \left\langle -\frac{\mathrm{d}E}{\mathrm{d}x} \right\rangle_{\text{ion.}} \quad \text{for } E = E_c. \tag{1.13}$$

For electrons and positrons, it is about a few tens MeV in most solid absorbers, with an approximate formula being $E_c \simeq 1600\, m_e c^2/Z$ (Leo, 1994). For the next lightest charged particle, the muon, the energy loss by bremsstrahlung dominates over that by ionisation for muon energies of a few hundred GeV (Lohmann et al., 1985).[10] With such a high value, we will restrict ourselves to the bremsstrahlung of electrons and positrons.

Since the energy loss by bremsstrahlung is proportional to energy, Eq. (1.12) can be rewritten (forgetting the mean value) as

$$-\frac{dE}{dx} = \frac{E}{X_0},$$

whose solution is

$$E(x) = E_0 \exp(-x/X_0). \tag{1.14}$$

Here, the quantity X_0 is called the *radiation length.* It corresponds to the average length over which the energy is reduced by a factor $1/e$, due to bremsstrahlung radiation. According to Eq. (1.12), for electrons and positrons $z^2 = 1$ and $m = m_e$, such that introducing the classical radius of the electron $r_e = e^2/(4\pi\epsilon_0 m_e c^2) = 2.82 \times 10^{-13}$ cm, the radiation length in g/cm^2 reads

$$X_0 = \frac{A}{4\alpha N_A Z^2 r_e^2 \ln\left(183\, Z^{-1/3}\right)}. \tag{1.15}$$

Table 1.6 gives the values of the critical energy and the radiation length for a few common absorbers. Formula (1.15) is approximate: it neglects the bremsstrahlung due to the electron cloud and possible screening of the Coulomb field of nuclei by electrons. Therefore, the values in Table 1.6 differ from the results of Eq. (1.15) by about 10–20%. A better approximate formula is (Grupen and Shwartz, 2011)

$$X_0 = \frac{716.4\, A\,[\mathrm{g/mol}]}{Z(Z+1)\ln\left(287\sqrt{Z}\right)}\ \mathrm{g/cm^2}. \tag{1.16}$$

Since the energy loss by bremsstrahlung is proportional to Z^2 and thus X_0 in $1/Z^2$ at first order, if one wants to stop a high energy electron, it is better to use absorbers with a high atomic number. This determines the choice of the materials used in electromagnetic calorimeters, as we will see in Section 1.4.4.

Emission of Cherenkov Radiation

Another kind of radiation occurs (but not in the same wavelength spectrum) when a charged particle traverses a medium at a velocity larger than the phase velocity of light in that medium, i.e. when the velocity of the particle satisfies

$$v > c/n,$$

where n is the index of refraction. The passage of the charged particle transiently polarises the atoms or the molecules of the medium, generating electric dipoles. Since the polarisation varies with time, it is well known from classical electrodynamics that the dipoles emit

[10] Actually, for high energy muons, another radiation process called *pair production* dominates, where the interaction of a muon with a nucleus produces $\mu^\pm + e^+ + e^- + X$, with X being any other decay product. It is about 1.5 times larger than bremsstrahlung at energies above a few hundred GeV (Lohmann et al., 1985). Hence, the energy loss by ionisation becomes negligible with respect to all other contributions for $E \gtrsim 350$ GeV in copper (Particle Data Group, 2022).

Table 1.6 Critical energies (electron) and radiation lengths for various absorbers.

Absorber	Z	A	E_c	X_0	
		(g/mol)	(MeV)	(g/cm^2)	(cm)
Air (dry, 1 atm)	–	–	87.92	36.62	30390
H_2O	–	–	78.33	36.08	36.08
Pb	82	207.2	7.43	6.37	0.56
Cu	29	63.55	19.42	12.86	1.44
Fe	26	55.85	21.68	13.84	1.76

Source: From Particle Data Group (2022, Section: 'Atomic and Nuclear Properties of Materials').

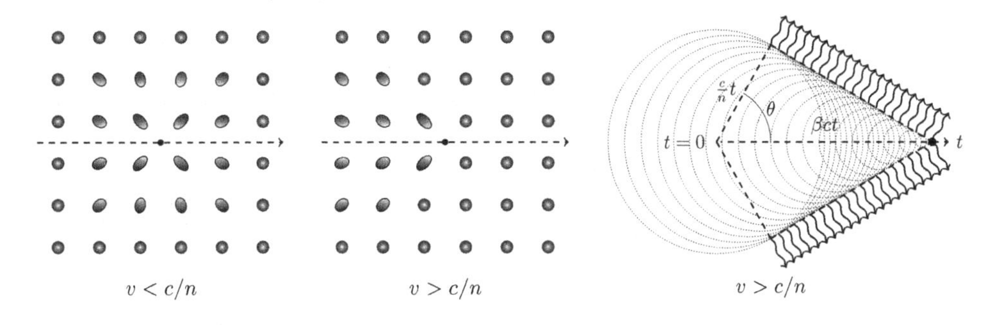

Fig. 1.5 Illustration of the Cherenkov effect when an electron (black circle) passes through a medium. Left-hand and middle diagrams: polarisation of the medium when $v < c/n$ and $v > c/n$, respectively. Atoms are represented by shaded circles or ellipses. The shading is white for positive charges and black for negative charges. Right-hand side: emission of Cherenkov light (wavy lines) with its characteristic angle θ. The dotted circles represent the wavefront of the light at different times.

electromagnetic radiation (see, e.g., Jackson, 1998). The emission stops when the atoms or the molecules return to their initial state without polarisation. However, when dipoles are symmetrically arranged with respect to the track of the charged particle, the resulting electric field of all dipoles vanishes, preventing the emission of the radiation. This is only possible when the particle travels slower than light in the medium, $v < c/n$, as illustrated in the left-hand diagram of Fig. 1.5. In contrast, for velocities greater than c/n (middle diagram), there is a resultant dipole field along the axis of the particle track (and not elsewhere), varying with time. Hence, radiation is emitted at each point along the axis. The wavelets from all portions of the axis are then in phase with one another (right-hand diagram of Fig. 1.5), such that a distant observer can see the Cherenkov light only at a single angle, θ, with respect to the track of the particle. By symmetry about the axis of the particle, the light emitted on each point of the axis propagates along the surface of a cone. Elementary geometry allows us to determine the opening angle of the cone. Between $t = 0$ and t, the charged particle travels over a distance $vt = \beta ct$, while the light emitted at $t = 0$ defines a wavefront of radius $c/n \times t$. Thus, the cosine of the angle, the ratio of these two quantities, satisfies

$$\cos\theta = \frac{1}{\beta n}. \tag{1.17}$$

For example, in water where $n = 1.33$, the maximum Cherenkov angle is reached when $\beta = 1$, giving $\theta = 41.4^\circ$, while in air ($n = 1.000293$), it is only $\theta = 1.39^\circ$.

One can show that the number of photons radiated per unit path per unit wavelength is [see Jelley (1958), an excellent book on the Cherenkov effect and its applications]

$$\frac{\mathrm{d}^2N}{\mathrm{d}x\,\mathrm{d}\lambda} = \frac{2\pi\alpha z^2}{\lambda^2}\sin^2\theta = \frac{2\pi\alpha z^2}{\lambda^2}\left(1 - \frac{1}{\beta^2 n^2}\right), \tag{1.18}$$

where the charge of the particle is ze. A naive interpretation of Eq. (1.18) would lead us to conclude that very short wavelengths should be favoured. However, all media are more or less dispersive (except the vacuum), meaning that the refractive index actually depends on the wavelength of the photon (or its energy), $n = n(\lambda)$. In the X-ray region and at higher energy, $n(\lambda)$ is always less than 1,[11] forbidding the emission of radiation since $\beta = v/c > n > 1$ is impossible. Hence, ultraviolet and visible photons are the most numerous. This imposes the use of a transparent medium for the construction of detectors using the Cherenkov effect.

The energy loss due to the emission of Cherenkov photons is negligible compared with the ionisation or bremsstrahlung loss. Since $E_\gamma = hc/\lambda$, it follows from Eq. (1.18),

$$\frac{\mathrm{d}^2N}{\mathrm{d}x\,\mathrm{d}E_\gamma} = \frac{\alpha z^2}{\hbar c}\sin^2\theta \simeq 370\, z^2 \sin^2\theta \text{ eV}^{-1}\text{cm}^{-1}. \tag{1.19}$$

Thus, the energy lost by the charged particle is $-d^2E = E_\gamma d^2N$, yielding

$$-\left.\frac{\mathrm{d}E}{\mathrm{d}x}\right|_{\text{Cher.}} = \frac{\alpha z^2}{\hbar c}\int_{\beta n>1}\sin^2\theta\, E_\gamma\, \mathrm{d}E_\gamma \simeq 370\, z^2 \int_{\beta n>1}\sin^2\theta\, E_\gamma\, \mathrm{d}E_\gamma. \tag{1.20}$$

Assuming that the medium is reasonably not dispersive between 0 and $E_\gamma^{\max}$, the integration of Eq. (1.20) gives $370\, z^2 \sin^2\theta\, (E_\gamma^{\max})^2/2$. For instance, with $E_\gamma^{\max} = 6.6$ eV (ultraviolet photon, $\omega \sim 10^{16}$ Hz), a 100 MeV electron ($z^2 = 1, \beta \sim 1$) passing through 1 cm of water ($\sin^2\theta = 0.25$) loses about 2×10^{-3} MeV energy. That same electron, according to Eq. (1.14) and Table 1.6, would lose about 2.7 MeV energy by bremsstrahlung– three orders of magnitude larger. Note that the energy loss due to Cherenkov radiation is already included in Eq. (1.11) via the correction factor at high energy that takes into account the polarisation of the medium. This is the part preventing $-\,\mathrm{d}E/\mathrm{d}x$ to rise as $\log(\gamma\beta)$ at high energy (Fermi plateau). It is important to realise that both the Cherenkov radiation and bremsstrahlung radiation are of different nature: the Cherenkov radiation arises only from the macroscopic properties of the medium, whereas in the case of bremsstrahlung radiation, it is the individual interaction of the charged particle with the Coulomb field of atoms that matters.

The measurement of the Cherenkov angle provides another way of identifying particles, as soon as the momentum p is known, since from Eq. (1.17),

$$\theta = \arccos\left(\frac{E}{npc}\right) = \arccos\left(\frac{\sqrt{p^2c^2 + m^2c^4}}{npc}\right). \tag{1.21}$$

[11] The phase velocity of light can be greater than c.

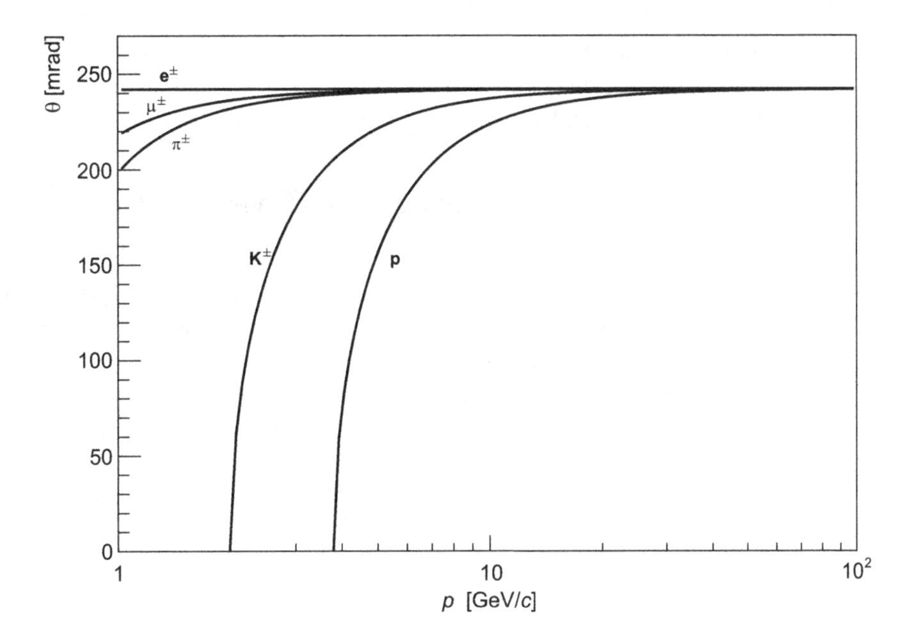

Fig. 1.6 Cherenkov angle as a function of the momentum of various particles. The medium here is the aerogel, with a reflective index $n = 1.03$.

Identification of particles is only possible when the Cherenkov effect is present, i.e. when $\beta > 1/n$. It imposes the following constraints on the energy and momentum of the particle:

$$E > \frac{n}{\sqrt{n^2 - 1}} mc^2, \quad p > \frac{1}{\sqrt{n^2 - 1}} mc. \tag{1.22}$$

Figure 1.6 shows the Cherenkov angles of various particles as a function of their momentum in the aerogel, a mixture of SiO_2 and H_2O. The region of momenta where identification is possible is narrow; thus, detectors frequently use several media to extend it.

For completeness, radiation similar to the Cherenkov effect exists when a charged particle crosses from one medium to another with different dielectric properties, for example, from the vacuum to a pure dielectric or a perfect conductor. This *transition radiation* is emitted at the interface between the two media. Unlike Cherenkov radiation, it occurs at any velocity of the particle. However, the probability increases with γ, so in practice, only highly relativistic particles ($\gamma \gtrsim 10^3$) emit substantial radiation. This characteristic is sometimes used in detectors to discriminate heavy particles, such as hadrons from lighter ones, typically electrons, since for a given energy $E = \gamma mc^2$, they have very different γ factors. The interested reader can consult Grupen and Shwartz (2011) where many examples of transition radiation detectors are given.

1.4.3 Interaction of Neutral Particles with Matter

Uncharged particles do not interact through the Coulomb force, and therefore, while passing through a medium, they must first undergo a strong, weak or, in the case of photons, electromagnetic interaction. The first interaction of the particle with the medium often involves the nucleus of constituent atoms of the medium and produces charged particles that are then detected.

Specific Interactions of High-Energy Photons

Interactions of high energy photon (X-ray or γ-ray) produce electrons through three main modalities: the *photoelectric effect*, *Compton scattering* and *pair production* of electrons.

In the photoelectric effect, a photon is absorbed by an atomic electron, leading to the ejection of the electron from the atom: $\gamma + \text{atom} \rightarrow \text{atom}^+ + e^-$. This process is the predominant mode of interaction for γ-rays or X-rays of relatively low energy, typically below 0.5 MeV. The ejected electron, called *photoelectron*, has energy given by $E_{\text{p.e}} = h\nu - E_b$, where $h\nu$ is the energy of the incident photon and E_b is the binding energy of the photoelectron in its original atomic shell (the recoil of the atoms can be safely ignored). Since E_b is small compared with γ-ray energy, the photoelectron carries most of the photon energy. Hence, for an evaluation of the photon energy, this process is ideal.

In Compton scattering, the incoming photon is not absorbed but deflected by an electron, transferring a portion of its energy to the electron that recoils (to conserve the energy and momentum): $\gamma + e^- \rightarrow \gamma + e^-$. The energy spectrum of the recoil electron is then a continuum, whose maximum energy is $m_e c^2 + E_\gamma \times \epsilon/(1+\epsilon)$, with $\epsilon = 2E_\gamma/(m_e c^2)$. In Chapter 6, we will learn how to calculate the cross section of this process, assuming that the electron is free. However, as soon as the photon energy is much larger than the binding energy of the electron, which is the case for γ-ray photons, the results obtained in Chapter 6 remain valid (including the maximum value given above). Compton scattering is the dominant process for photon energy between 0.5 and 10 MeV.

Finally, at energies higher than 10 MeV, pair production dominates. It mostly results from the interaction of a photon with the Coulomb field of the nucleus: $\gamma + \text{nucleus} \rightarrow \text{nucleus} + e^+ + e^-$. In this process, the photon is thus absorbed as in the photoelectric effect. Pair production can be viewed as a process similar to bremsstrahlung since for the latter, $e^- + \text{nucleus} \rightarrow \text{nucleus} + e^- + \gamma$. Both bremsstrahlung and pair production involve the same set of particles, provided an electron is exchanged for a positron. We will learn in Chapter 6 that, because of quantum field theory properties, the two processes can then be calculated from one another. Consequently, just as the energy of electrons decreases exponentially in matter by bremsstrahlung emission as a function of the radiation length, X_0 [cf. Eq. (1.14)], for pair production, the intensity of high energy photons behaves the same way, i.e.

$$I(x) = I_0\, e^{-x/\lambda}, \tag{1.23}$$

where, for photon energy greater than 1 GeV, the *attenuation length* or *mean free path* λ is given by Particle Data Group (2022)

$$\lambda \simeq \frac{9}{7} X_0. \tag{1.24}$$

The parameter $\mu = 1/\lambda$ is thus an absorption coefficient. When it is normalised to the density of the medium $\mu' = \mu/\rho$ (like the radiation length), it is called the *mass absorption coefficient*.

Interaction of Hadrons

Hadrons experience the strong interaction and can interact with the nuclei of the medium. For neutral hadrons, such as neutrons, this is the main interaction in matter. Since the strong force is a short-range interaction, the probability that the hadron is close enough to a given nucleus (about 1 fm) is low, and hence neutral hadrons are penetrating particles. Charged hadrons can also have electromagnetic interactions, leading to the mechanisms of energy loss already presented.

The interactions of neutrons depend strongly on their energy. At low kinetic energy, in the MeV region and below, elastic scattering of neutrons with nuclei contributes to their energy loss. Inelastic scattering is also possible, leaving the nucleus in an excited state that will further de-excite emitting photons or charged particles. When neutrons are sufficiently slowed down, they may be captured by a nucleus that emits photons, or they may induce a nuclear reaction (fission etc.). In particle physics, those low-energy neutrons are generally identified via those secondary particles (photons or charged particles). On the other hand, neutrons or any hadrons with high energy (kinetic energy above 100 MeV, so most of the hadrons produced with accelerators) undergo nuclear reactions that produce secondary hadrons, that themselves generate nuclear reactions and so on. This process produces a *hadronic shower* that can be captured by a hadronic calorimeter (see Section 'Detection of Hadrons: Hadronic Calorimeters').

1.4.4 Detectors

Measurement of the Momentum and dE/dx: Tracker Detectors

Track detectors, or *trackers* for short, are detectors that measure the particle tracks (i.e., the trajectory of the particle and the energy deposited along the track). Because of the large number of particles produced by collisions at modern colliders, most detectors measuring p and dE/dx are nowadays based on semiconductors devices (mainly silicon). The previous generation of detectors was often gaseous detectors like *time projection chambers*, where the passage of the ionising particle created electron–ion pairs collected by an electric field. In the case of semiconductors, the electric charge is carried by electrons that have been excited to the conduction band from the valence band and by the corresponding holes created in the valence band.[12] The crystal lattice of semiconductors is generally doped with elements that can easily provide an electron (donor) or capture it (acceptor). A few impurity atoms per billion semiconductor atoms are sufficient. Silicon has four valence electrons. Donors then have five valence electrons, with the excess electron becoming easily a free electron (its energy level is very close to the lower bound of the conduction band, about

[12] The reader not familiar with solid state physics can consult, for example, Kittel (2005).

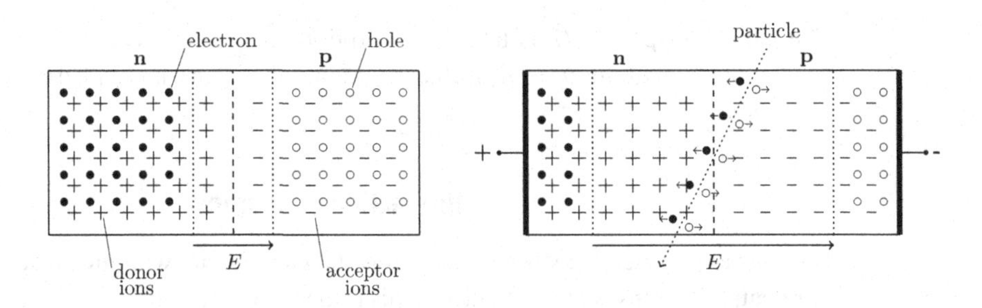

Fig. 1.7 Working principle of a p–n junction. On the left-hand side, no voltage is applied. On the right-hand side, a reverse bias is applied to the junction i.e. a positive voltage on the n-side and a negative voltage on the p-side. The effect of the passage of a particle is illustrated with the drift of the charge carriers.

0.05 eV in silicon). Silicon doped in this way is called n-doped silicon (n for negative). If acceptors with three valence electrons are added to the crystal lattice, one electron is missing for the covalent bond, creating a hole (and a new energy level in the energy gap, very close to the upper bound of the valence band). It forms p-doped silicon. Silicon detectors are based on p–n junctions, where p-doped silicon and n-doped silicon are in contact, as illustrated in the left-hand sketch in Fig. 1.7. In the contact area, there is a natural migration of electrons from the n-region to the p-region to annihilate the holes. Similarly, holes from the p-region diffuse towards the n-region. Hence, both sides of the junction are filled with ionised donors and acceptors, which are immobile, as they are bound in the crystal lattice. This creates an electric field, which balances the diffusion force of electrons and holes, generating a depleted region of free charges (electrons or holes). When an ionising particle enters the depleted region, electron–hole pairs are created. Both charge carriers then feel electrostatic forces and drift in opposite directions. These charges can then be collected by electrodes, inducing a current pulse. However, the intrinsic electric field is not intense enough to collect enough charges. Moreover, the depletion region is too thin to allow high energy particles to lose enough energy. Therefore, a voltage is applied to the junction (right-hand sketch in Fig. 1.7) such that it increases the electric field and the thickness of the depleted region. The total charge collected is proportional to the energy deposited in the depletion layer. The average energy required for the creation of an electron–hole pair is typically about 3 eV, which is an order of magnitude less than the energy required for the creation of electron–ion pairs in a gaseous detector.[13] Hence, for a given energy, semiconductors produce about 10 times more electron–hole pairs than electron–ion pairs. Moreover, they have a greater density and thus a larger stopping power. This explains their popularity despite their relatively high cost.

Trackers based on silicon use a very large number of silicon sensors with p–n junctions (some have nearly 100 million) that can be shaped into thin strips or pixels. In its simplest form, a silicon strip detector is constructed as an array of p-type junctions on a single n-type

[13] More specifically, for silicon, it is 3.62 eV at room temperature. Given that the energy gap is 1.14 eV, only one-third of the energy is spent on the production of the electron–hole pair. The other two-thirds excite the vibration states of the lattice.

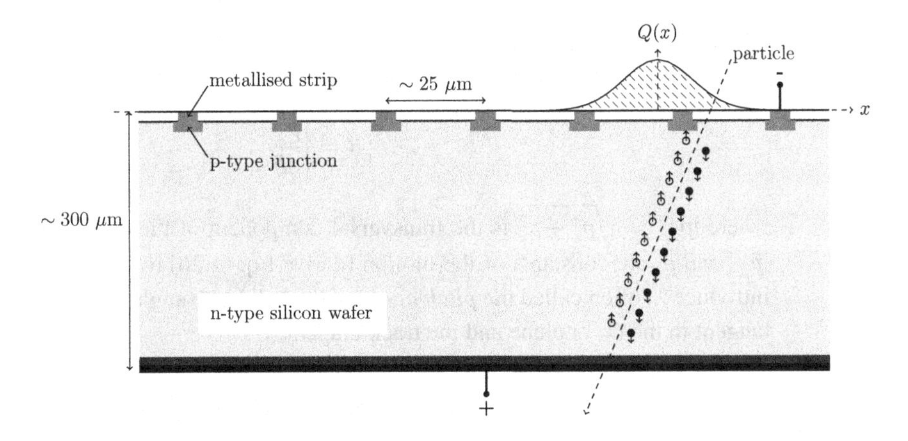

Fig. 1.8 Simplified schematic layout of a silicon strip detector. Electrons are represented by the black circles and holes by the open circles. The quantity $Q(x)$ represents the charge collected over several strips. Weighting the position of the strips by their charge gives interpolation of the particle position x.

silicon wafer (Fig. 1.8). The typical strip width is 10 μm on an about 25 μm pitch. When a particle passes through the depleted region (most of the n-type region is usually depleted), electron–hole pairs are produced, inducing a current in the strips, collected by readout amplifiers (one per strip). Given that an MIP leaves about 1.66 MeV/cm^2 (see the grey curve in Fig. 1.4), when a particle crosses 300 μm of silicon (density $\rho = 2.33$ g/cm^3), it loses about 116 keV. With 3.62 eV required to create a pair, we conclude that about 30 000 pairs are produced, giving a clear signal. The distribution of the collected charges over multiple strips allows the position interpolation of the particle in one dimension. To give an order of magnitude, the position resolution is about a few micrometres.[14] Double-sided strips with two orthogonal planes of silicon strips as well as an array of silicon pixels give position information in two dimensions. Detectors installed at accelerators have several concentric layers of silicon strips or pixels surrounding the collision region. They provide the third dimension of position reconstruction.

Track detectors frequently operate in a strong magnetic field $\boldsymbol{B}$ that deflects the charged particle trajectory by virtue of the Lorentz force,

$$\frac{\mathrm{d}\boldsymbol{p}}{\mathrm{d}t} = q\boldsymbol{v} \times \boldsymbol{B}, \tag{1.25}$$

where $\boldsymbol{p} = \gamma m \boldsymbol{v}$ and q is the charge of the particle. Solving this equation (see Problem 1.5) shows that the trajectory is that of non-relativistic mechanics, i.e. a helical trajectory, except that the mass of the particle, m, is substituted with γm,

$$\begin{cases} x(t) = & x(0) + [p_y(0) + p_x(0)\sin(\omega t) - p_y(0)\cos(\omega t)]/(q|\boldsymbol{B}|) \\ y(t) = & y(0) + [-p_x(0) + p_y(0)\sin(\omega t) + p_x(0)\cos(\omega t)]/(q|\boldsymbol{B}|) \\ z(t) = & z(0) + p_z(0)\,\omega t/(q|\boldsymbol{B}|), \end{cases} \tag{1.26}$$

[14] If we assume that a single strip is read out with a pitch p between strips, and that the track could be anywhere between $-p/2$ and $p/2$ with a uniform distribution, simple statistics show that the position resolution is $p/\sqrt{12}$. With $p = 25$ μm, we obtain 7 μm.

where $\omega = q|\boldsymbol{B}|/(\gamma m)$ and $\boldsymbol{B}$ is chosen along z. Projected on the plane orthogonal to the magnetic field, (x, y), the trajectory is a circle with a bending radius R satisfying

$$R = \left| \frac{\boldsymbol{p}_\perp}{q\boldsymbol{B}} \right|, \tag{1.27}$$

where $|\boldsymbol{p}_\perp| = \sqrt{p_x^2 + p_y^2}$ is the transversal component of the momentum. Notice that $|\boldsymbol{p}|$, $|\boldsymbol{p}_\perp|$ and p_z are constants of the motion [derive Eq. (1.26) to check it]. Therefore, we can introduce λ, often called the *pitch angle* of the helix, the angle between the projected track tangent in the (x, y) plane and the track tangent,

$$\cos\lambda = \left| \frac{\boldsymbol{p}_\perp}{\boldsymbol{p}} \right|, \quad \tan\lambda = \frac{p_z}{|\boldsymbol{p}_\perp|}. \tag{1.28}$$

Similarly, let us introduce ϕ, the angle between the x-axis and the projected track tangent at $t = 0$, i.e.

$$\cos\phi = \frac{p_x(0)}{|\boldsymbol{p}_\perp|}, \quad \sin\phi = \frac{p_y(0)}{|\boldsymbol{p}_\perp|}.$$

Using these two angles and Eq. (1.27) yields the simple trajectory equations

$$\begin{cases} x(t) = & x(0) + R\sin\phi + R\sin(\omega t - \phi) \\ y(t) = & y(0) - R\cos\phi + R\cos(\omega t - \phi) \\ z(t) = & z(0) + R\,\omega t\tan\lambda. \end{cases} \tag{1.29}$$

The quantity $R\omega t$ represents the arc length of the track circle in the transverse plane. Figure 1.9 illustrates how the measurement of the signals in several layers of the tracker gives access to the bending radius in the transverse plane and the arc length (left-hand schema).

The angle λ can then be evaluated by fitting the slope in the plane (arc length, z) as shown in the right-hand schema. Having R and λ, we can then access the total momentum thanks to Eqs. (1.27) and (1.28). Using units where the momentum is in GeV$/c$ and the charge is in units of the elementary charge e yields

$$|\boldsymbol{p}|\cos\lambda = \delta\,|q\boldsymbol{B}|\,R, \tag{1.30}$$

where $\delta = 10^{-9}c \simeq 0.3$. One can show [see Grupen and Shwartz (2011), for instance] that the relative resolution of the momentum measurement can be parametrised as

$$\frac{\sigma(|\boldsymbol{p}|)}{|\boldsymbol{p}|} = \sqrt{a^2|\boldsymbol{p}|^2 + b^2}, \tag{1.31}$$

where a and b are two constants encoding the intrinsic resolution coming from the measurement points along the track and the effect of multiple scattering, respectively. Qualitatively, the linear dependence of the first term with the momentum can be understood as follows: when the momentum is very high, the bending radius becomes almost a straight line, and thus, the uncertainty on the momentum gets very large. This follows from the fact that what is measured is not directly the bending radius but rather its inverse, the curvature $\kappa = 1/R$ (see Problem 1.6). Since according to Eq. (1.30) the uncertainty of κ

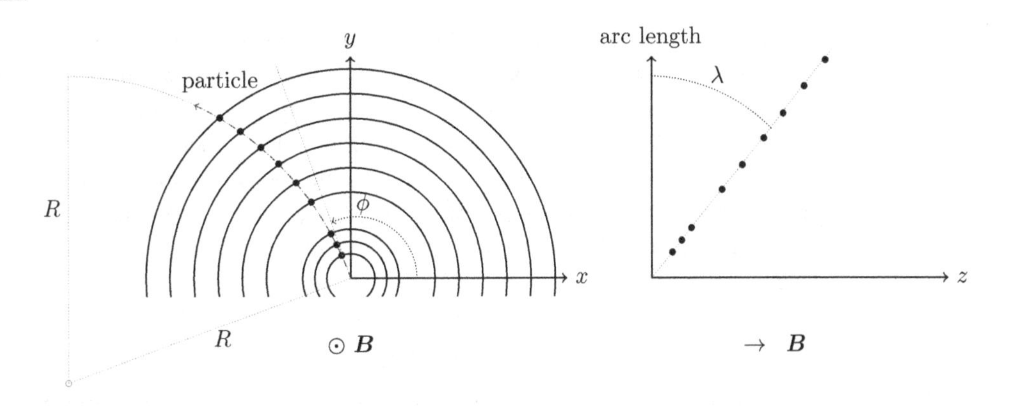

Fig. 1.9 A tracker with nine detecting layers. The trajectory, given by Eq. (1.29), is depicted with the dashed line. The position measurements on layers are represented by black circles. They enable the measurement of R, ϕ and λ and thus the momentum. In this example, the particle is produced in the centre of the detector. Otherwise, the distance of the closest approach to the origin in the (x, y) plane would have to be determined.

is $\sigma(\kappa) \propto \sigma(1/|\boldsymbol{p}|) = \sigma(|\boldsymbol{p}|)/|\boldsymbol{p}|^2$, it follows that $\sigma(|\boldsymbol{p}|)/|\boldsymbol{p}| \propto |\boldsymbol{p}|$. A typical order of magnitude of the constant a in Eq. (1.31) is $10^{-3} - 10^{-4}$, when $|\boldsymbol{p}|$ is expressed in GeV/c.

Detection of High-Energy Electrons and Photons: Electromagnetic Calorimeters

Calorimeters are detectors that aim to measure the energy of a particle, neutral or charged, by converting its energy into a measurable signal. Ideally, a calorimeter absorbs the whole energy of a particle. Their role is complementary to that of trackers that measure the momentum of a particle, ideally without perturbing the particle track much.

Electromagnetic calorimeters are dedicated to the energy measurement of photons and electrons. They are optimised to exploit the energy-loss mechanisms of these particles at high energies, in particular, bremsstrahlung emission for electrons and pair production for photons (see Sections 1.4.2 and 1.4.3). Via the interaction with the absorber of the calorimeter, a high energy electron emits a bremsstrahlung photon, which creates an electron–positron pair. The members of the pair can in turn emit photons, and the process continues until the produced particles have energies below the critical energy, at which point they lose their energy by ionisation or Compton scattering, for electrons and photons, respectively. This process creates a cascade of electrons, positrons and photons, called an *electromagnetic shower*. Similarly, when a high energy photon enters the calorimeter, it converts first into an electron–positron pair that generates the electromagnetic shower. An illustration is given in Fig. 1.1, p. 13, where the two photons that are the decay products of the π^0 on the left-hand side of the sketch are the source of the electromagnetic shower in the air. Air is not a good calorimeter, with its radiation length, X_0, being too large (see Table 1.6). It is much better to use dense materials with high atomic numbers, favouring large energy loss, such that the value of X_0 is small [cf. Eq. (1.15)].

Although the development of the electromagnetic shower is a statistical process, its main characteristics can be inferred from a simple model (proposed by Heitler), where after each radiation length, X_0, the number of particles is doubled and their energy is divided by 2. In other words, in the processes $\gamma \to e^+ + e^-$ or $e^\pm \to \gamma + e^\pm$, the energy of the initial particle is assumed to be symmetrically shared between the particles of the final state. At a depth of t radiation lengths, the number of particles is then $N(t) = 2^t$, and their mean energy $E(t) = E_0/2^t$, where E_0 is the energy of the incident particle in the calorimeter. The shower should stop its development when the energy of the particles is below the critical energy, $E(t) < E_c$. The maximum number of particles is then reached after $t_{\max}$ radiation lengths, with $E(t_{\max}) = E_c$, i.e.

$$t_{\max} = \frac{\ln(E_0/E_c)}{\ln(2)}, \quad N(t_{\max}) = \frac{E_0}{E_c}. \tag{1.32}$$

Hence, after an exponential rise of the number of particles, the development of an electromagnetic shower reaches a maximum at a depth (in units of the radiation length) that scales logarithmically with the initial energy E_0, while the number of particles is proportional to E_0. For instance, according to Table 1.6, a 100 GeV electromagnetic shower reaches its maximum at $t_{\max}X_0 \simeq 12 \times 1.76 = 21$ cm in iron and about $14 \times 0.56 = 8$ cm in lead. A calorimeter built with a lead absorber is thus much more compact.

Equation (1.32) gives information only about the longitudinal profile (i.e., along the direction of the incident particle) of the shower. However, the shower also expands transversally, in particular, because multiple scattering of electrons becomes significant below the critical energy. The transversal profile is characterised by a lateral width, R_M, called the Molière *radius*, which is given by Particle Data Group (2022)

$$R_M = \frac{\sqrt{4\pi\alpha}m_ec^2}{E_c}X_0 \simeq \frac{21.2\ [\text{MeV}]}{E_c}X_0. \tag{1.33}$$

The Molière radius is such that, on average, 90% of the total energy of the shower (and thus, of the incident particle, approximately) is contained in a cylinder around the shower axis whose radius is $r = R_M$, and 95% for $r = 2R_M$.

An accurate description of showers based on analytical formulas is complicated. They are, however, rather well described by Monte-Carlo simulation.[15] A simulation of an electromagnetic shower created by a 100 GeV photon in the air is shown in Fig. 1.10a. The shape in denser materials is similar, with a much smaller longitudinal and lateral extension.

Calorimeters can be classified into two categories: homogeneous calorimeters and sampling calorimeters. In the first one, the same material combines the properties of an absorber and the ability to produce a measurable signal. An example is the Compact Muon Solenoid (CMS) electromagnetic calorimeter, made of 61 200 lead tungstate ($PbWO_4$) crystals in the central region of the detector (the so-called barrel). These crystals are inorganic scintillators,[16] with a length of 25 radiation lengths and a transverse size at the front face of about $R_M \times R_M$, ($X_0 = 0.85$ cm and $R_M = 2.19$ cm in $PbWO_4$ crystals). An electromagnetic shower is then fully contained in length and spreads over a few crystals

[15] Monte carlo simulations use random numbers to solve problems with many degrees of freedom.

[16] A scintillator is a medium that converts the excitation of its constituents (for example, the crystal lattice), due to the energy loss of charged particles, into visible (or often ultraviolet) light. There are many types of scintillators: organic crystals, glass, or various other liquid, plastic or gaseous materials etc. Plastic scintillators are often used because they can be easily shaped. See Leo (1994) for more details.

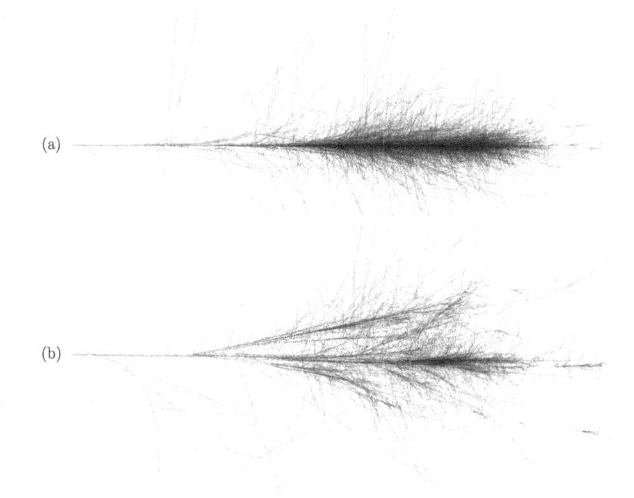

Fig. 1.10 Simulation of showers produced by a 100 GeV photon (a) and a 100 GeV proton (b) in the air. The longitudinal size is 30.1 km and the transversal size, 10 km. Images from F. Schmidt, J. Knapp, CORSIKA Shower Images, 2005, www-zeuthen.desy.de/jknapp/fs/showerimages.html.

transversally. Due to the large magnetic field (4 T), conventional photomultipliers (presented in the next section) cannot be used to detect the scintillation photons (their gain and linearity are affected by the field). Instead, avalanche photodiodes (denoted APD) are used to read the scintillation light. They are semiconductor devices (cf. Fig. 1.7) configured with a high reverse bias voltage and optimised in terms of doping layers to reproduce the avalanche of conventional photomultipliers when a photon enters the device (see the next section).

The second category of calorimeters are sampling calorimeters that alternate the layers of the absorber material with the active media that provide the measurable signal. Hence, only a sample of the total energy deposition is measured, limiting their energy resolution. However, they generally have the advantage of being more economical.

The relative energy resolution $\sigma(E)/E$ of calorimeters improves with the energy of the incident particle because the higher the energy, the larger the number of particles that are statistically produced in the shower. A standard parametrisation is

$$\frac{\sigma(E)}{E} = \sqrt{\left(\frac{a}{\sqrt{E}}\right)^2 + \left(\frac{b}{E}\right)^2 + c^2}, \tag{1.34}$$

where a, b and c are constants reflecting the different sources contributing to the resolution. For showers with energies a few tens of GeV, the dominating term is the stochastic term, $a/\sqrt{E}$, due to the fluctuations related to the physical development of the shower. At very high energy (hundreds of GeV and beyond), the limiting factor is the constant term, c (generally a fraction of percent), which includes instrumental contributions that do not

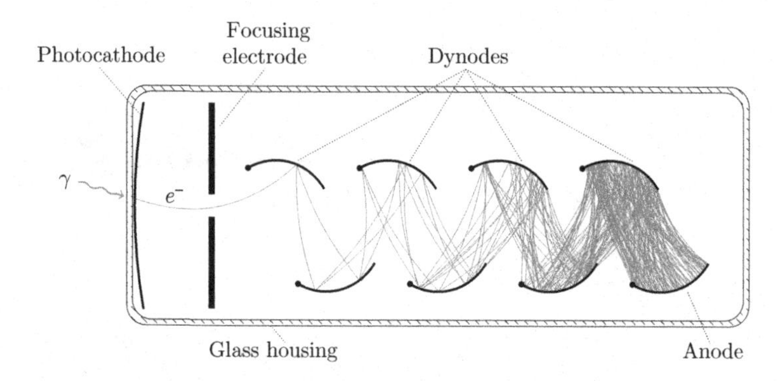

Fig. 1.11 Schematic diagram of a photomultiplier tube. The incident photon is represented by the grey wavy line and electrons by the grey curves.

depend on the energy of the particle (for instance, inhomogeneity of the calorimeter). Typical energy resolutions are $(1-5\%)/\sqrt{E[\text{GeV}]}$ and $(5-20\%)/\sqrt{E[\text{GeV}]}$, respectively, for homogeneous and sampling electromagnetic calorimeters (Fabjan and Gianotti, 2003).

Detection of Visible and Ultraviolet Photons: Photomultipliers

Photomultipliers convert light into a measurable electric current. They are used in many detectors to convert scintillation or Cherenkov light, for instance. Figure 1.11 shows a schematic diagram of a photomultiplier tube (denoted PMT).

When a photon impinges upon a photocathode, an electron, called a *photoelectron*, is extracted by the photoelectric effect. It is focused by an electric guiding field onto the first dynode, where it transfers some of its energy to the electrons in the dynode, producing secondary electrons, which are in turn accelerated toward the next dynode. The process continues by creating an 'avalanche' of electrons, eventually collected by the anode. The voltage between the photocathode and the anode is subdivided by a chain of resistors such that it increases gradually from one dynode to the next. Voltages of the order of 1 000–2 000 volts are typically used to accelerate electrons within the chain of dynodes, with the most negative voltage connected to the cathode and the most positive voltage connected to the anode. The current collected at the anode is then proportional to the number of incident photons. Typically three to five electrons are emitted by a dynode, for each incident electron. Most of the photomultipliers contain 10–14 stages of dynodes. Hence, gains ranging from 10^6 to 10^8 are achievable, allowing the detection of a single photon. A photomultiplier, consisting of N dynodes and producing, on average, δ secondary electrons per dynode, generates a total charge of $\delta^N e$ Coulombs per photoelectron. Because electrons follow different paths, there is a natural variation in their transit time (denoted TTS) up to the anode. Very large photomultipliers can have TTS $\sim$ 5 ns. It thus takes about 5 ns to collect the total charge by the anode. Hence, such a photomultiplier, with $N = 12$ and $\delta = 4$ (these are typical numbers), would produce a current of about 0.5 mA per photoelectron, an amply measurable current.

Photomultipliers are characterised by their TTS but also by their quantum efficiency, i.e. the number of photoelectrons released by the photocathode divided by the number of incident photons. Depending on the material of the photocathode, the photomultiplier can be sensitive to photons in the infrared to ultraviolet spectrum. Most of the photocathodes are made of semiconductor materials because of their good quantum efficiency, ranging typically from 10% to 30%.

In recent years, silicon photomultipliers (SiPM)[17] have become very popular in particle physics detectors (in Simon, 2019, many recent applications are given). A SiPM is a set of pixels where each pixel (whose typical size is a few tens of micrometres and a SiPM can have over a thousand pixels per mm^2) is a silicon photodiode. However, all pixels are built on a common silicon substrate, and the photodiodes are configured in Geiger mode (i.e., producing a large amplitude in response to the detection of even a single photon), by means of a very strong bias voltage above the breakdown voltage, $V_{\text{bkd.}}$. Instead of producing an avalanche (i.e., a charge carrier multiplication) that leads to a signal proportional to the optical input signal as in the APD, the avalanche is self-sustaining, producing a signal pulse that is not dependent on the intensity of the optical input signal. It is, therefore, an all-or-nothing mode. Once there is an avalanche, the bias voltage automatically reduces to below $V_{\text{bkd.}}$, resulting in the quenching of the avalanche, and quickly returns above that value after a recovery time, in order to be ready to sense new optical input signals [see Buzhana et al. (2003) for more technical details]. Hence, if a SiPM pixel operates as a binary device, the SiPM as a whole is an analogue detector, which can measure the light intensity by counting the numbers of illuminated pixels, and, if the recovery time of each pixel is short enough compared to the duration of the light source (recovery times of the order of a few nanoseconds are possible), by counting the number of times a given pixel is fired. There are many advantages of SiPMs for detectors; in particular, their small size greatly improves the detector granularity compared with standard photomultipliers, while still allowing a single photon to be counted. In addition, they are very fast and insensitive to magnetic fields. Their dynamic range is, however, intrinsically limited by the number of pixels.

Detection of Hadrons: Hadronic Calorimeters

Charged or neutral hadrons with high energy (kinetic energy above 100 MeV) undergo inelastic nuclear reactions with matter that produce secondary hadrons, which in turn generate nuclear reactions and so on. This process produces a *hadronic shower*. The quantity that characterises the development of a hadronic shower is the inelastic *nuclear interaction length* λ_I that is analogous to the electromagnetic radiation length, X_0, for electromagnetic showers. It governs the absorption of hadrons by the matter:

$$N(x) = N_0\, e^{-x/\lambda_I},$$

where $N(x)$ is the number of hadrons surviving after passing through a length x in the matter. Generally, λ_I is much larger than X_0. For instance, in iron $\lambda_I = 16.8$ cm, whereas

[17] SiPM are also called Multi-Pixel Photon Counter and Geiger-mode avalanche photo diode (MPPC and GAPD), and by many other names depending on the manufacturer.

$X_0 = 1.76$ cm. Hence, a hadronic shower tends to develop over a larger length scale than an electromagnetic one, and hadronic calorimeters must then be thicker than electromagnetic calorimeters to contain the shower. In practice, the size of hadronic calorimeters with respect to that of electromagnetic ones cannot scale as λ_I/X_0. (If the electromagnetic calorimeter is 1 m thick, the hadronic one should be almost 10 m thick!) Consequently, only a fraction of the longitudinal development of the shower may be contained, with the thickness of hadronic calorimeters being typically of the order of 5–10 interaction lengths (while that of electromagnetic calorimeters ranges from 20 to 30 X_0). Because of large transverse momentum transfers in nuclear interactions, the lateral extension of hadronic showers is also larger than that of electromagnetic showers. Figure 1.10 illustrates the difference between a shower initiated by a photon and a proton in the air (for non-dense media such as air, λ_I and X_0 can be of the same order of magnitude, explaining why the two showers have a similar longitudinal extension in the figure). The dynamics of hadronic showers are complicated. First, about 40% of the energy of the incident energy is dissipated and does not contribute to the production of secondary hadrons (Grupen and Shwartz, 2011). It can go to binding energy, can be transferred to recoils of nuclei or can simply escape detection through the production of neutrinos. Second, nuclear reactions produce many pions, of which about one-third are π^0 that decay into two photons. They induce an electromagnetic shower, mainly concentrated around the hadronic shower core since electromagnetic showers are narrower than hadronic ones. The π^0 production, however, is subject to large fluctuations. Hence, from one event to another, the shower can contain a large or negligible fraction of electromagnetic contribution. Since, in the end, the measured energy is coming from low- energy charged particles (electrons, pions, protons, etc.) interacting with the active material, it is then difficult to optimise a calorimeter to respond similarly to electrons and hadrons. All these complications imply that the energy resolution of hadronic calorimeters is worse than that of pure electromagnetic ones, typically,

$$\frac{\sigma(E)}{E} \sim \frac{(30\% \text{ to } 120\%)}{\sqrt{E[\text{GeV}]}}.$$

The best resolution is generally obtained with calorimeters that use uranium as an absorber, because in ^{238}U, many neutrons and energetic photons are produced by its fission that can eventually generate a visible signal.

Most of hadronic calorimeters are sampling calorimeters. The nuclear interaction length of the active material that produces the visible signal is so large that it cannot contain a significant fraction of the shower. Therefore, layers of absorber material are required. Iron or copper are very common absorbers, while the active material is often composed of plastic scintillator tiles.

Detection of Muons

Muons mainly lose their energy by ionisation. To give an order of magnitude, in copper ($\rho = 8.96$ g/cm^3), an MIP loses about $1.4 \times 8.96 \sim 12.5$ MeV/cm (Fig. 1.4). Hence, to first approximation, a 12.5 GeV muon would travel over 10 m (although the actual range is shorter because muons at that energy are not MIPs, and therefore they lose slightly more energy). Muons are thus highly penetrating particles – a unique feature among charged elementary particles. In other words, they are the only surviving charged particles after travelling a few metres of matter.

Particle detectors at accelerators use this property extensively (in addition to possible identification from the $\mathrm{d}E/\mathrm{d}x$ or the Cherenkov effect). In the outermost parts of the detector, beyond the absorbers of the calorimeters, tracking detectors are installed to detect the passage of muons. Since the flux of particles in these regions is strongly reduced compared to the region close to the collision point, gaseous detectors are often used. In those detectors, the muon ionises the atoms of the gas, creating electron–ion pairs. The average energy needed to produce a pair is about 30 eV, with a weak dependence on the gas (compared to the 3 eV in semiconductor devices for electron–hole pairs). A system of anodes and cathodes with a large potential difference amplifies the number of primary electron–ion pairs, creating an avalanche proportional to the initial signal. Generally, a noble gas, such as argon, is chosen since it requires a lower electric field to produce the avalanche. Many muon detectors work on this principle: *cathode strip chambers*, *drift tubes*, *resistive plate chambers*, *gas electron multipliers*, etc. They mostly differ by the structure of their system of anodes or cathodes and the shapes of the latter (wires, strips, plates, etc.), leading to different performance in terms of timing response and accuracy of the position measurement.

Particle detectors for non-accelerator physics generally use other techniques to identify muons. For instance, in water Cherenkov detectors, such as Super-Kamiokande in Japan, muons produce sharp Cherenkov rings (resulting from the projection of the Cherenkov light-cone onto photomultipliers), whereas rings originating from electrons are fuzzier because of the multiple scattering that affects electrons. The analysis of the rings thus makes it possible to identify the particles.

Notice that for very high energy muons (several hundreds of GeV), the main source of energy loss is not ionisation but radiative processes (e^+e^- pair production, bremsstrahlung, and photonuclear interactions).[18] Hence, they can also produce electromagnetic showers (due to e^+e^- pair production and bremsstrahlung) or hadronic showers (due to photonuclear interaction) in calorimeters.

Detection of Neutrinos

Neutrinos are stable particles and light enough to travel over large distances at (almost) the speed of light, even at low energy (cf., Fig. 1.3). However, since they only experience the weak interaction, the probability that they interact with matter turns out to be ridiculously weak. Despite this, there are specific detectors dedicated to their detection, as mentioned in Chapter 9. These detectors usually exploit the production of a charged lepton, coming from the interaction of a neutrino or anti-neutrino with a nucleon. To give an order of magnitude, the cross section (whose exact definition is given in Section 3.5) of the reaction $\nu_e + n \to e^- + p$ is about 0.9×10^{-38} cm^2, for a neutrino energy of 1 GeV (Formaggio and Zeller, 2012). As a typical example, given that the flux of atmospheric neutrinos at that energy is

[18] Photonuclear production of muons is the process $\mu^\pm$ + nucleon $\to \mu^\pm$ + hadrons, where a virtual photon is exchanged between the muon and the nucleon.

about 1 $\text{cm}^{-2}\,\text{s}^{-1}$, we obtain an interaction rate per nucleon of $0.9 \times 10^{-38}\,\text{s}^{-1}$. Hence, only detectors containing a huge amount of matter can expect to observe these neutrinos via the detection of the electron or the proton. Since 1 g of matter contains about $N_A = 6 \times 10^{23}$ nucleons (the Avogadro number), a one kiloton detector running one year (3.15×10^7 s) would detect at most 170 neutrinos. This is why modern neutrino detectors, aiming to observing the atmospheric flux, are usually tens of kilotons. The situation is even worse for lower energy neutrinos. At about 1 MeV, the cross section of the neutrino/antineutrino-nucleon inelastic scattering is about $\sigma = 10^{-43}\,\text{cm}^2$ (Giunti and Kim, 2007). The mean free path (introduced in Section 3.5), i.e. the average distance between two interactions, is given by

$$\lambda = \frac{1}{\rho\sigma},$$

where ρ is the density of nucleons. For example, in lead, $\rho = N_A \times 11.3$ (density) $\simeq 7 \times 10^{24}$ nucleons/cm^3, and hence, $\lambda \sim 1.5 \times 10^{18}$ cm. More than one lightyear in lead without an interaction!

Detectors operating at colliders cannot afford to have so much material dedicated to the detection of neutrinos. Therefore, neutrinos usually simply escape those detectors without leaving signals of their presence. Nevertheless, in the centre-of-mass frame of the two incident particles from the collision of the beams, the total momentum should be zero by definition. Hence, if all particle momenta can be observed except those of neutrinos, the missing momentum can be attributed to the presence of neutrinos (assuming, of course, that no other unknown particles escape the detector), i.e.

$$\boldsymbol{p}_\nu \simeq \boldsymbol{p}_{\text{miss}} = -\sum_i \boldsymbol{p}_i,$$

where i runs over the momenta of the observed particles. However, this method gives only a crude estimate of the neutrino momentum, limited by the momentum resolution of the measured particles and the hermiticity of detectors (i.e., their coverage in solid angle, ideally, 4π).

Example of a General-Purpose Detector at a Collider

General-purpose detectors aim to cover the full diversity of reactions issued from beam collisions. They must be able to identify all long-lived particles reaching the detector after a collision (cf. list 1.10). Detectors at accelerators consist of a succession of concentric sub-detectors surrounding the collision point, each specialised in a given task: track determination, momentum measurement, energy measurement, and particle identification. Figure 1.12 shows the layout of a typical detector in the transverse view, i.e. in the view perpendicular to the beam direction. The concentric sub-detectors form the central part of the detector called the *barrel*. The barrel is completed by two *end-caps* closing the barrel from each side to detect particles emitted from all angles to the beam axis (end-caps are not represented in Fig. 1.12). Such a detector is referred to as *hermetic*. It covers nearly a solid angle of 4π around the collision point, the only particles not crossing active elements of the detector being those emitted into the beam pipe (with angles of less than about 1°

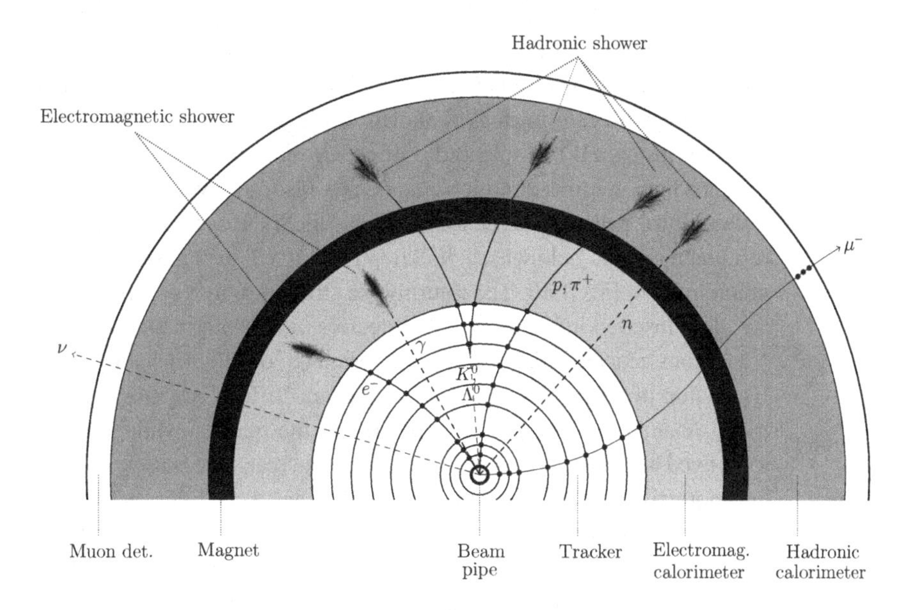

Fig. 1.12 Layout (transverse view) of a general-purpose detector at colliders. The beams circulate in the dimension perpendicular to the page and collide in the centre of the detector. The position of the magnet (a solenoid) in between the two calorimeters is purely indicative; in some detectors, it surrounds the tracker, while in others, it is outside the two calorimeters. The magnetic field is perpendicular to the page, towards the reader. Charged particle trajectories are represented by solid curves, while those of neutral particles are shown by dashed lines.

to the beam axis). Recent examples of such detectors are the ATLAS and CMS detectors operating at the LHC at CERN.

At the very core of the detector, a tracker is in charge of the momentum measurement of charged particles. The tracker determines the curvature of the trajectories of the charged particles in a powerful magnetic field. The innermost layers of the sensitive material (very often silicon pixels) are located radially at a few centimetres from the interaction point, which is, to a first approximation, in the centre of the detector in transverse view. They have the finest granularity of all sub-detectors in order to cope with the high density of particles, and they are crucial to the reconstruction of tracks from short-lived particles. By interpolating tracks towards the centre of the detector, the actual position of the interaction point, usually called the *primary vertex*, can be estimated. Particles such as B hadrons containing a b-quark, which have a significant lifetime of the order of 10^{-12} s, cannot reach those first layers; however, they travel over a few millimetres before decaying. The interpolation of the charged tracks of their decay products allows the identification of a secondary vertex that is displaced with respect to the primary vertex. The distance between the secondary and primary vertices thus gives a measurement of the distance travelled by the meson and indirectly its lifetime.

The next sub-detector encountered by long-lived particles is the electromagnetic calorimeter. It stops electrons or photons by absorbing (ideally all) their energy through the development of electromagnetic showers. An electron is thus distinguished from a photon by the presence of a track in the tracker.

Depending on the experiment, a solenoid providing a powerful magnetic field along the beam axis (typically between 1 and 4 T) can be found surrounding the tracker, the electromagnetic calorimeter (as shown in Fig. 1.12) or even the hadronic calorimeter (in CMS for instance). The solenoid is often surrounded by steel or iron structures, the 'return yoke', which confines the high magnetic field to the volume of the detector (it provides a flux return path for the central solenoid field). Hence, outside the solenoid, there is a residual magnetic field whose direction is reversed, explaining the strange trajectory of the muon track in Fig. 1.12. The return yoke structures are not represented in Fig. 1.12. They are often inserted in the muon sub-detectors or the calorimeters.

Hadrons reaching the hadronic calorimeter develop a hadronic shower (see Section 'Detection of Hadrons: Hadronic Calormeters'). They are mostly pions, protons and neutrons. A neutron does not leave a track in the tracker, while charged hadrons do. Very long-lived hadrons, such as the K^0 and Λ^0, can reach the tracker when sufficiently boosted, but more likely they decay in the tracker after travelling a significant distance (several tens of centimetres). Since they often decay into two charged hadrons, their signature in the detector is of V-shape, with the vertex clearly separated from the primary vertex (see Fig. 1.12). Incidently, in the early days of particle physics (mid-twentieth century), those particles were called V-particles.

Finally, the outermost sub-detector is the muon sub-detector. Usually, muons are the only charged particles that can pass through the rest of the detector without being stopped (see Section 'Detection of Muons'). Note, however, that charged pions produced in a hadronic shower in the hadronic calorimeter can sometimes leak into the muon detector, mimicking a muon signal.

1.5 What Is a Measurement?

Physics is an experimental science. Even the most beautiful theories must be falsifiable (to claim to be science) and hence should be confronted, at least conceptually, with experimental measurements. Experimental measurements can be used to estimate a quantity from some data, and the question then arises of the best estimate of that quantity and its uncertainty. Measurements can also be used to test a hypothesis that is based on a particular model and to test the consistency of the data with that model. This section assumes that the reader already has an elementary knowledge of probability theory and statistics. Only a few topics relevant to particle physics are covered in this section, without proving them in depth. The reader can (and should) consult statistics textbooks, such as Barlow (1989), James (2006), Bohm and Zech (2010) and Lyons (1986) to fill in the gaps.

1.5.1 Generalities

Randomness is inherent in particle physics because it is built upon quantum mechanics. For instance, a muon decays *on average* in 2.2 microseconds, but sometimes it takes less and sometimes more. We can calculate the probability that it decays at a given time, but we cannot say precisely when it will happen. Even the signal collected by a detector has a degree of randomness: a charged particle passing through a medium has a certain probability of ionising the atoms of that medium, which induces intrinsic fluctuations in the collected current.

Continuous random processes are described by *probability density functions* (p.d.f.). If x is a random variable and $p(x)$ is its p.d.f. normalised to unity, then $p(x)\ \mathrm{d}x$ is the probability of finding x between x and $x + \mathrm{d}x$. The first two moments of the distribution, the *mean* μ and *variance* σ^2, are given by

$$\mu = E[x] = \int x p(x)\ \mathrm{d}x, \tag{1.35}$$

where $E[x]$ is the expectation value and

$$\sigma^2 = E[(x-\mu)^2] = E[x^2] - \mu^2. \tag{1.36}$$

The square root of the variance, σ, is the *standard deviation*, and it represents a measure of the width of the distribution. For outcomes of a process depending on several random variables, let us say x and y, the *covariance* is

$$\mathrm{cov}[x,y] = E[xy] - E[x]E[y] = E[(x-\mu_x)(y-\mu_y)], \tag{1.37}$$

where μ_x and μ_y are the means of x and y, respectively. These mean values are given by Eq. (1.35), inserting a multivariable p.d.f., $p(x, y)$, and performing a double integral over x and y. The variance of x is trivially deduced from the covariance, with $\sigma_x^2 = \mathrm{cov}[x, x]$. It is convenient to introduce the dimensionless *correlation coefficient*

$$\rho_{xy} = \frac{\mathrm{cov}[x,y]}{\sigma_x \sigma_y}, \tag{1.38}$$

which varies between -1 and $+1$. When x and y are independent, namely when the p.d.f. is factorisable, i.e. $p(x, y) = p_x(x)p_y(y)$, the covariance and correlation coefficient are zero.[19] It is standard to have random variables that depend on other random variables. If x and y are two random variables, and $x'(x, y)$ and $y'(x, y)$ are two variables depending on x and y, the covariance of x' and y' is related to that of x and y via the formula

$$\begin{aligned}\mathrm{cov}[x',y'] = & \tfrac{\partial x'}{\partial x}\tfrac{\partial y'}{\partial x}\mathrm{cov}[x,x] + \tfrac{\partial x'}{\partial y}\tfrac{\partial y'}{\partial y}\mathrm{cov}[y,y] \\ & + \tfrac{\partial x'}{\partial x}\tfrac{\partial y'}{\partial y}\mathrm{cov}[x,y] + \tfrac{\partial x'}{\partial y}\tfrac{\partial y'}{\partial x}\mathrm{cov}[y,x],\end{aligned} \tag{1.39}$$

where all derivatives are evaluated with $x = \mu_x, y = \mu_y$. The variance, $\sigma_{x'}^2 = \mathrm{cov}[x', x']$, is then

$$\sigma_{x'}^2 = \left(\frac{\partial x'}{\partial x}\right)^2 \sigma_x^2 + \left(\frac{\partial x'}{\partial y}\right)^2 \sigma_y^2 + 2\frac{\partial x'}{\partial x}\frac{\partial x'}{\partial y}\rho_{xy}\sigma_x\sigma_y. \tag{1.40}$$

The extension of Formulas (1.39) and (1.40) to more variables, $x_1, \ldots, x_n$, is straightforward using the covariance matrix

$$V_{ij} = \mathrm{cov}[x_i, x_j] = \rho_{ij}\sigma_i\sigma_j. \tag{1.41}$$

[19] The converse is not necessarily true: the covariance can be 0 when x and y depend on each other. This is true, for example, if x is uniformly distributed between 0 and 1 and $y = \pm x$, where the sign is randomly chosen.

For instance, Eq. (1.40) becomes

$$\sigma_{x'_i}^2 = \sum_{k,l} \frac{\partial x'_i}{\partial x_k} \frac{\partial x'_i}{\partial x_l} V_{kl} = \sum_k \left(\frac{\partial x'_i}{\partial x_k} \right)^2 \sigma_k^2 + 2 \sum_{k>l} \frac{\partial x'_i}{\partial x_k} \frac{\partial x'_i}{\partial x_l} \rho_{kl} \sigma_k \sigma_l. \tag{1.42}$$

Notice that these formulas are an approximation when the variables x' and y' are non-linear combinations of x and y.

Among the infinite possibilities of p.d.f., the Gaussian (or normal) distribution and the Poisson distribution are the most often encountered. The Gaussian distribution is of the utmost importance because of the *central limit theorem*, which states that the sum of n random variables distributed according to any p.d.f. with finite mean and variance has a p.d.f. approaching the Gaussian distribution for large n. As many measurements are based on averaging samples of data, Gaussian distributions are often observed. The well-known formula of the Gaussian density function with mean μ and variance σ^2 (and thus standard deviation, σ) is

$$G(x\,;\mu,\sigma) = \frac{1}{\sqrt{2\pi}\sigma} \exp\left[-\frac{1}{2} \left(\frac{x-\mu}{\sigma} \right)^2 \right]. \tag{1.43}$$

The Gaussian density function is represented in Fig. 1.13. The probability that the random variable x lies in the interval $[a, b]$ is given by

$$p_{a,b} = \int_a^b G(x\,;\mu,\sigma)\, \mathrm{d}x = F_G(b) - F_G(a), \tag{1.44}$$

where $F_G(x) = \int_{-\infty}^{x} G(x'\,;\mu,\sigma)\, \mathrm{d}x'$ is the *cumulative function* of the Gaussian. Conversely, the probability of being outside this interval is $1 - p_{a,b}$. The canonical values in terms of numbers of standard deviations from the mean value, $\mu \pm n\sigma$, are given in Fig. 1.13. With the change of variable $x \to t = (x-\mu)/(\sqrt{2}\sigma)$, $p_{a,b}$, with $a = \mu - n\sigma$ and $b = \mu + n\sigma$, takes the expression $p_{a,b} = \mathrm{erf}(n/\sqrt{2})$, where erf($x$) is the error function $\mathrm{erf}(x) = 1/\sqrt{\pi} \times \int_0^x \exp(-t^2)\, \mathrm{d}t$.

Whereas the central limit theorem states that a variable that is the *sum* of a large number of random variables is described by the Gaussian (or normal) density function, a variable that is the *product* of random variables is described by the log-normal density function. The logarithm of such a variable would follow a Gaussian distribution $G(\ln(x)\,;\mu,\sigma)$, and thus the p.d.f. of x satisfies $f(x\,;\mu,\sigma)\, \mathrm{d}x = G(\ln(x)\,;\mu,\sigma)\, \mathrm{d}(\ln(x))$.[20] It follows that the log-normal density function is

$$f(x\,;\mu,\sigma) = \frac{1}{x\sqrt{2\pi}\sigma} \exp\left[-\frac{1}{2} \left(\frac{\ln(x)-\mu}{\sigma} \right)^2 \right], \tag{1.45}$$

with $x > 0$. The mean and standard deviation are $e^{\mu+\sigma^2/2}$ and $e^{\mu+\sigma^2/2}\sqrt{e^{\sigma^2}-1}$, respectively. The log-normal distribution is encountered in processes with many small multiplicative variations. A typical example is shower development. An electromagnetic shower

[20] In general, if $\boldsymbol{x}$ is a vector of random variables described by the p.d.f. $p_x(\boldsymbol{x})$ and $\boldsymbol{y}$ is a vector of random variables obtained by the one-to-one correspondence $\boldsymbol{y} = f(\boldsymbol{x})$, then the p.d.f. of $\boldsymbol{y}$ is $p_y(\boldsymbol{y}) = p_x(\boldsymbol{x})|J| = p_x(f^{-1}(\boldsymbol{y}))|J|$, where $|J|$ is the determinant of the Jacobian transformation, i.e. $J_{ij} = \partial x_i/\partial y_j$. In the case of a single random variable, the relation is equivalent to $p_y(y)\, \mathrm{d}y = p_x(x)\, \mathrm{d}x$.

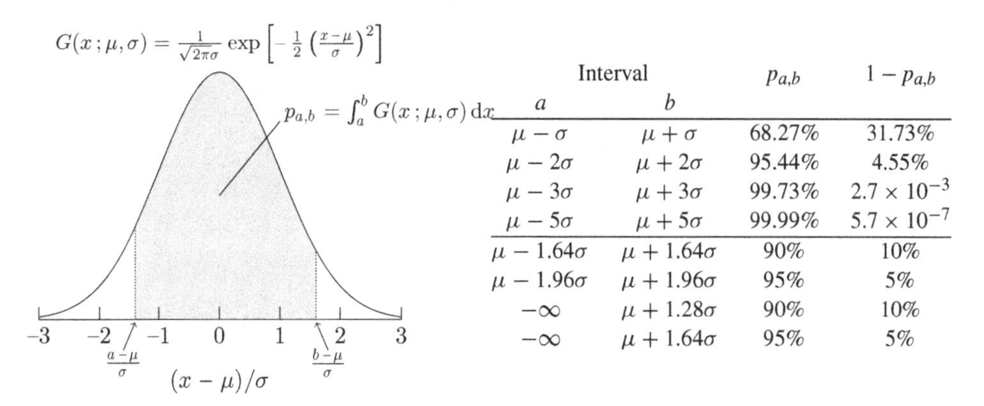

Interval a	Interval b	$p_{a,b}$	$1 - p_{a,b}$
$\mu - \sigma$	$\mu + \sigma$	68.27%	31.73%
$\mu - 2\sigma$	$\mu + 2\sigma$	95.44%	4.55%
$\mu - 3\sigma$	$\mu + 3\sigma$	99.73%	2.7×10^{-3}
$\mu - 5\sigma$	$\mu + 5\sigma$	99.99%	5.7×10^{-7}
$\mu - 1.64\sigma$	$\mu + 1.64\sigma$	90%	10%
$\mu - 1.96\sigma$	$\mu + 1.96\sigma$	95%	5%
$-\infty$	$\mu + 1.28\sigma$	90%	10%
$-\infty$	$\mu + 1.64\sigma$	95%	5%

Fig. 1.13 The Gaussian density function and the fraction of values, with $p_{a,b}$, lying within an interval $[a, b]$.

results mainly from the multiplication of the pair-production and bremsstrahlung processes (see Section 'Detection of High-Energy Electrons and Photons: Electromagnetic Calorimeters'), each with their corresponding energy loss. Hence, fluctuations in shower size at a given depth, in a sample of showers of the same energy, are approximatively log normal (Gaisser, 1990, chapter 15).

The Poisson density function is also very often encountered in physics experiments. It describes the probability of n events occurring (n, an integer number) given that the mean expected number is μ (a real positive number). The Poisson p.d.f.[21] is

$$P(n\,;\mu) = \frac{\mu^n}{n!}e^{-\mu}. \tag{1.46}$$

Each event is understood to occur independently of one another. It can be shown that the Poisson distribution is a limiting case of the binomial distribution.[22] By construction, not only the mean of the Poisson distribution is given by μ but also its variance (the standard deviation is thus $\sqrt{\mu}$). The Poisson distribution then depends on a single number. Although the Poisson distribution is discrete, it is normalised to unity in the sense that $\sum_{n=0}^{\infty} P(n\,;\mu) = 1$. A frequent application is when the mean rate of a process (mean number of reactions per second) is known, for example, the decay rate of a particle, but one wishes to evaluate the probability of observing n events during the time t. If the rate is λ, then the mean number of events during t is $\mu = \lambda t$, and the probability of observing n events is thus given by $P(n\,;\lambda t)$. An even more frequent application is when data are sampled, i.e. when data are analysed in discrete intervals (called bins), for example, bins in the energy channel of a detector, bins in time of arrival of a particle, etc. One then counts the number of events in each bin. In most of these cases, this number is assumed to follow

[21] It is formally not a p.d.f. since the Poisson distribution is not continuous. For discrete random variables, it is called a probability mass function (p.m.f.), and it directly gives a probability.

[22] Probably, the binomial distribution is already well known by the reader. It describes cases where there are only two possible outcomes (such as heads and tails in a coin toss). If one calls one of the two outcomes 'success', then the probability of obtaining r success in N independent trials is $\frac{N!}{r!(N-r)!}p^r(1-p)^{n-r}$, where p is the intrinsic probability of success. The Poisson distribution is obtained when $p \to 0$, $N \to \infty$ and $N \times p = \mu$.

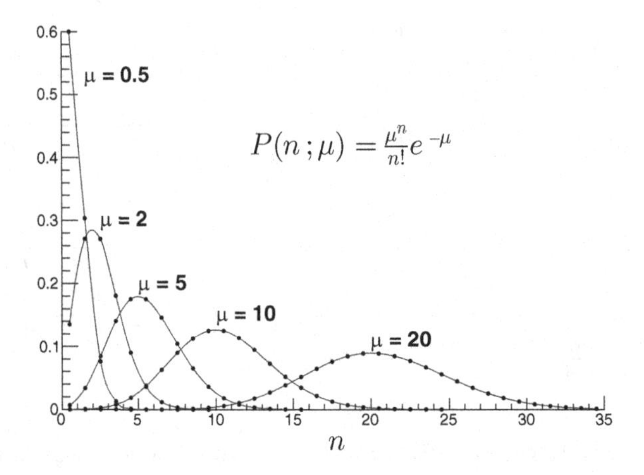

Fig. 1.14 The Poisson distribution. The discrete points for each value of μ are connected by smooth curves for clarity.

a Poisson distribution (counting can never be negative). If n_i events are observed in bin i, assuming that n_i represents the mean number of events of the Poisson distribution of that bin, the uncertainty is then $\sqrt{n_i}$, according to the standard deviation of the Poisson distribution. Notice that it is the estimated uncertainty on the mean of the underlying Poisson distribution. The quantity $\sqrt{n_i}$ is therefore not the error on the number of observed events, which is assumed to be an error-free count. Figure 1.14 presents the Poisson distribution for various values of μ. Notice that the distribution is not symmetric around μ. However, it becomes more symmetric as μ increases. In fact, when μ is large (in practice, above 20), the Poisson distribution $P(n\,;\mu)$ approaches the Gaussian distribution $G(n\,;\mu, \sigma = \sqrt{\mu})$.

1.5.2 Measurement Uncertainties

Measurement uncertainties (or errors, both words are used equivalently) are classified into two categories: *statistical uncertainties* often called random uncertainties and *systematic uncertainties* (in short systematics). The former arise, for example, from the inherent statistical nature of the phenomena studied in particle physics. The limiting ability of any device to give measurement with an infinite accuracy also contributes to these kinds of errors. In general, statistical uncertainties are supposed to follow a known statistical distribution. They all share the characteristics that sequential measurements are statistically uncorrelated. The precision is thus improved by combining several measurements. Indeed, if r_i is the result of the ith measurement and σ_{r_i}, its uncertainty, the error resulting from the average of N measurements is, according to Eq. (1.42),

$$\sigma^2 = \frac{1}{N^2}\sum_{i=1}^{N}\sigma_{r_i}^2.$$

Assigning to all measurements the same error σ_r, it follows that $\sigma = \sigma_r/\sqrt{N}$, scaling with the inverse of the square root of the number of measurements. Notice that, in general, when each measurement has a different uncertainty, it is more appropriate to perform a weighted average

$$r = \frac{1}{\sum_i w_i} \sum_{i=1}^{N} w_i r_i,$$

where $w_i = 1/\sigma_{r_i}^2$ (justified in Problem 1.7). Assuming no correlation in Eq. (1.42), the uncertainty of r is then

$$\sigma^2 = \frac{1}{\left(\sum_i w_i\right)^2} \sum_{i=1}^{N} w_i^2 \sigma_{r_i}^2 = \frac{1}{\sum_i w_i}. \tag{1.47}$$

Unlike statistical errors, systematic uncertainties do not necessarily improve with more data. They generally represent a possible bias, mistake, etc. They are not directly due to the statistic of the data. A simple example is a thermometer, whose 0° is actually shifted by a constant value. All measurements will be affected by that value. A more realistic example in the context of particle physics is a calibration error of a calorimeter that affects the energy scale, i.e. the response of the calorimeter to the energy of the incident particle. Another example is a measurement using inputs from a theoretical model that is actually wrong or inputs from a simulation that does not describe properly the real data. Systematic errors affect different measurements made in identical conditions in the same manner. Hence, repeated measurements do not reduce systematic errors. It is always delicate to estimate systematic errors. There is no universal recipe, and it often relies on the experience of the experimentalist who can guess what parameter has a significant effect on the final result of an analysis. A case often encountered is an external input parameter with known uncertainty σ. For example, the luminosity of an accelerator. The resulting systematic uncertainty on the analysis is then obtained by varying the input parameter by $\pm\sigma$. The deviation from the initial analysis result is then considered as a systematic uncertainty for this parameter.

At the end of the analysis, all sources of systematic uncertainties have to be combined with the statistical error. Implicitly we assume that the value x that is measured results from the summation, $x = x_{\text{true}} + \Delta x_{\text{stat}} + \sum_i \Delta x_{i,\text{syst}}$, where Δx_{stat} and $\Delta x_{i,\text{syst}}$ have the respective errors σ_{stat} and $\sigma_{i,\text{syst}}$. No correlation is assumed between the statistical uncertainty and the sources of systematic uncertainties. Formula (1.42) then yields

$$\sigma_x^2 = \sigma_{\text{stat}}^2 + \underbrace{\sum_{k,l} \rho_{kl}\, \sigma_{k,\text{syst}} \sigma_{l,\text{syst}}}_{\equiv \sigma_{\text{syst}}^2} .$$

The final result is then generally reported as $r = x \pm \sigma_{\text{stat}} \pm \sigma_{\text{syst}}$. For example, when the Higgs boson was discovered by the ATLAS and CMS experiments in 2012, the publication CMS Collaboration (2012b) reported the mass of the boson as $m_H = 125.3 \pm 0.4$ (stat.) $\pm$ 0.5 (syst.) GeV$/c^2$. Separately quoting the statistical and systematic uncertainties has the advantage of indicating whether taking more data would significantly reduce the global uncertainty.

1.5.3 Parameter Estimation: Maximum Likelihood and Least-Squares Methods

As physicists, we very often measure quantities that we want to compare with a model. The model may depend on parameters that we would like to evaluate. Two general methods are briefly described in this section: the maximum likelihood and least-squares methods.

The Maximum Likelihood Method

Let $\boldsymbol{x} = x_1, \ldots, x_n$ be n independent observations. We assume that they all follow the same p.d.f., $f(x\,;\theta)$, which depends on a parameter θ that we wish to evaluate. The joint probability for obtaining $\boldsymbol{x}$ is the product of the p.d.f. (there should be an extra $n!$ if the order of observations does not matter, but we will see below that this factor will not play any role). When this joint probability is interpreted as a function of the parameter of interest, it is called the *likelihood function*,

$$L(\theta) \equiv L(\boldsymbol{x}\,;\theta) = \prod_{i=1}^{n} f(x_i\,;\theta). \tag{1.48}$$

Notice that $L(\theta)$ is not the p.d.f. of θ. Let us call $\hat{\theta}$ an estimator of θ. One can show that when n becomes infinite, then $\hat{\theta}$, given by the global maximum of the likelihood, is a *consistent* estimator, i.e. it converges to the true value of θ. It is *unbiased*, i.e. its expectation value is θ, and its variance converges to the minimum possible variance for an unbiased estimator (it is then qualified as *efficient*). Beware that these nice properties are only true in the asymptotic limit, i.e. when n becomes infinite. For small n, the estimator is usually biased. The name of the method, the maximum likelihood, may suggest that the estimate is the most likely value, whereas it is actually the estimate that makes the data most likely. In practice, it is more convenient to find the global minimum of the negative log-likelihood function (because it converts a product into a sum, which is easier to handle in terms of numerical accuracy),

$$\mathcal{L}(\theta) = -\ln\left(L(\theta)\right) = -\sum_{i=1}^{n} \ln\left(f(x_i\,;\theta)\right). \tag{1.49}$$

The estimator $\hat{\theta}$ is then a solution of

$$\left.\frac{\partial}{\partial\theta}\mathcal{L}(\theta)\right|_{\theta=\hat{\theta}} = -\sum_{i=1}^{n} \left.\frac{\partial}{\partial\theta}\ln\left(f(x_i\,;\theta)\right)\right|_{\theta=\hat{\theta}} = 0. \tag{1.50}$$

In general, there is no analytic solution to this equation, and the solution must be found numerically. Once $\hat{\theta}$ is found, we can expand $\mathcal{L}(\theta)$ around this value and obtain

$$\mathcal{L}(\theta) = \underbrace{\mathcal{L}(\hat{\theta})}_{\mathcal{L}_{\min}} + \frac{1}{2!}\left.\frac{\partial^2\mathcal{L}(\theta)}{\partial\theta^2}\right|_{\theta=\hat{\theta}} (\theta-\hat{\theta})^2 + O(\theta-\hat{\theta})^3. \tag{1.51}$$

Let us set

$$\sigma_{\hat{\theta}} = \left.\left(\frac{\partial^2\mathcal{L}(\theta)}{\partial\theta^2}\right)^{-1/2}\right|_{\theta=\hat{\theta}}. \tag{1.52}$$

Equation (1.51) then becomes

$$\mathcal{L}(\theta) = \mathcal{L}_{\text{min}} + \frac{1}{2}\left(\frac{\theta - \hat{\theta}}{\sigma_{\hat{\theta}}}\right)^2 + O(\theta - \hat{\theta})^3. \tag{1.53}$$

Or equivalently,

$$L(\theta) \simeq L_{\text{max}} \times \exp\left(-\frac{1}{2}\left(\frac{\theta - \hat{\theta}}{\sigma_{\hat{\theta}}}\right)^2\right).$$

The quantity $\sigma_{\hat{\theta}}$ in Eq. (1.52) represents the standard deviation of $\hat{\theta}$ only when $\mathcal{L}$ is reasonably parabolic near $\hat{\theta}$. Equivalently, $\sigma_{\hat{\theta}}$ can be obtained from Eq. (1.53) since for $\theta = \hat{\theta} \pm \sigma_{\hat{\theta}}$,

$$\mathcal{L}(\hat{\theta} \pm \sigma_{\hat{\theta}}) = \mathcal{L}_{\text{min}} + 1/2 \;\text{ or }\; L(\hat{\theta} \pm \sigma_{\hat{\theta}})/L_{\text{max}} = \exp(-1/2). \tag{1.54}$$

The standard deviation is then obtained when $\mathcal{L}$ increases by 1/2 from its minimum value. Similarly, the n standard deviations are obtained when $\mathcal{L}$ increases by $n^2/2$ from its minimum value. Even for non-parabolic forms, Eq. (1.54) is used to derive the error on $\hat{\theta}$. In such a case, the error may not be symmetric around $\hat{\theta}$. We can do so because estimators obtained with the likelihood method are *invariant* under parameter transformations: if $\alpha = g(\theta)$, with g being a one-to-one transformation function, then the estimator of α that maximises the likelihood is just $\hat{\alpha} = g(\hat{\theta})$. Hence, if $\mathcal{L}(\theta)$ is not parabolic around $\hat{\theta}$, we can perform the appropriate transformation $\alpha = g(\theta)$, such that $\mathcal{L}(\alpha)$ becomes parabolic. The invariance property not only applies to the maximum of the likelihood L but also to its relative values (James, 2006) and thus to the differences of its logarithm $\mathcal{L}$. Hence, the values of α satisfying $\mathcal{L}(\alpha) = \mathcal{L}(\hat{\alpha}) + 1/2$ correspond to the values of θ satisfying $\mathcal{L}(\theta) = \mathcal{L}(\hat{\theta}) + 1/2$, and the 1σ domain (or any confidence interval) can be found without explicitly finding the transformation function g. Figure 1.15 illustrates the situation for a non-parabolic shape, where the 68.3% confidence interval $[\theta_-, \theta_+]$ containing the true value of θ (cf. table in Fig. 1.13) is not symmetric around $\hat{\theta}$. The measured value of θ is then noted with its errors as $\hat{\theta}^{+\sigma_+}_{-\sigma_-}$ instead of $\hat{\theta} \pm \sigma_{\hat{\theta}}$. For instance, the mass of the u quark in Particle Data Group (2022) is noted $m_u = 2.16^{+0.49}_{-0.26}$ MeV. When the number of observations is large, $\mathcal{L}$ gets closer to a parabola (a consequence of the central limit theorem), and σ_-, σ_+ in Fig. 1.15 converge to $\sigma_{\hat{\theta}}$, the standard deviation of $\hat{\theta}$ in the parabolic case.

For a set of parameters $\boldsymbol{\theta} = \theta_1, \ldots, \theta_m$, there are m equations such as Eq. (1.50) to solve, i.e.

$$\frac{\partial}{\partial \theta_k}\mathcal{L}(\boldsymbol{\theta})\bigg|_{\boldsymbol{\theta}=\hat{\boldsymbol{\theta}}} = -\sum_{i=1}^{n} \frac{\partial}{\partial \theta_k} \ln\left(f(x_i\,;\boldsymbol{\theta})\right)\bigg|_{\boldsymbol{\theta}=\hat{\boldsymbol{\theta}}} = 0, \;\; k = 1, \ldots, m.$$

The standard deviation of the parameters is obtained for the parabolic case by finding the covariance matrix where the elements of its inverse are given by

$$\left(V^{-1}\right)_{ij} = \frac{\partial^2 \mathcal{L}(\boldsymbol{\theta})}{\partial \theta_i \partial \theta_j}\bigg|_{\boldsymbol{\theta}=\hat{\boldsymbol{\theta}}}. \tag{1.55}$$

For non-parabolic cases, Eq. (1.54) extended to the m-dimension space must be applied.

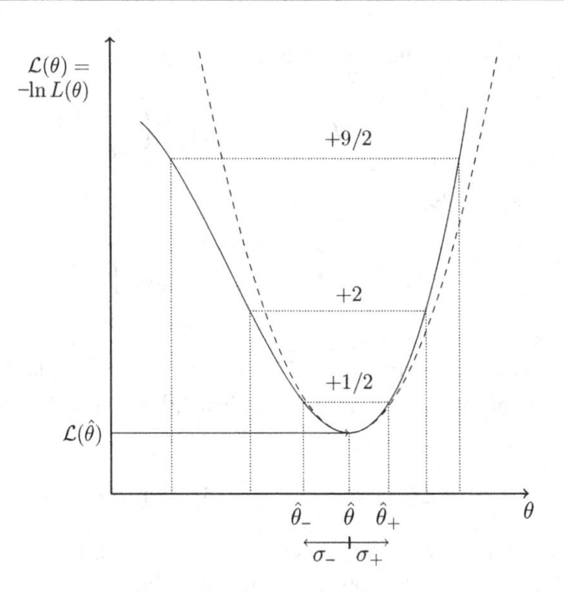

Fig. 1.15 The negative log-likelihood function (solid curve) and the parabolic approximation around the minimum value (dashed line). The asymmetric errors are σ_+ and σ_-, corresponding to the '1σ domain' (68.3%), obtained with $\mathcal{L}(\hat{\theta}) + 1/2$. The '$2\sigma$ and 3σ domains' (95.4% and 99.7%, respectively) are obtained with $\mathcal{L}(\hat{\theta}) + 2$ and $\mathcal{L}(\hat{\theta}) + 4.5$, respectively.

For processes where the number of events is not fixed, if one wants to compare the observed number of events n to a theory prediction ν that might partially depend on the other parameters $\boldsymbol{\theta}$, an additional Poisson term can be incorporated in the likelihood function. Equation (1.48) then becomes

$$L(\nu, \boldsymbol{\theta}) \equiv L(\boldsymbol{x}\,;\nu, \boldsymbol{\theta}) = e^{-\nu}\frac{\nu^n}{n!}\prod_{i=1}^{n} f(x_i\,;\boldsymbol{\theta}). \tag{1.56}$$

This variation of the maximum likelihood method is called the *extended maximum likelihood.* The unknown parameters $\boldsymbol{\theta}$ are now not only encoded in the shape of the data distribution but also in the number of events. The estimate of the expected number of events $\hat{\nu}$ is then obtained by the minimisation of

$$\mathcal{L}(\nu, \hat{\boldsymbol{\theta}}) = -\ln(L(\nu, \hat{\boldsymbol{\theta}})) = \nu - n\ln n - \sum_{i=1}^{n} \ln\left(f(x_i\,;\theta)\right) + \text{constant}$$

with respect to all free parameters. If ν does not depend on $\hat{\boldsymbol{\theta}}$, the constraint $\partial\mathcal{L}/\partial\nu = 0$ yields $\hat{\nu} = n$, and the values $\hat{\boldsymbol{\theta}}$ are the same as those obtained from the standard maximum likelihood method. Otherwise, the additional constraint improves the accuracy on $\hat{\boldsymbol{\theta}}$.

The Least-Squares Method

Let us suppose that we have n observations of two variables, x and y. The x_i variables ($i \in [1, n]$) are assumed to have negligible uncertainties, whereas y_i has an uncertainty

σ_i. A situation often encountered is the case of binned data, where x is displayed in a histogram with n bins, which represents the observed distribution of x. For bin i, $x = x_i$ and y_i would be the bin content, i.e. the number of events corresponding to the value x_i. Let us imagine that the relation between x_i and y_i is given by theoretical supposition, $s(x_i\,;\theta)$, which depends on a parameter θ that we wish to estimate. For example, x represents the set of decay times of many identical particles, and $s(x_i\,;\theta)$ would be the exponential function, with θ being the lifetime of the particle. If $s(x_i\,;\theta)$ describes reasonably well the data, the variable y_i should be very close to $s(x_i\,;\theta)$. The least-squares method consists in minimising the squares of the residuals,

$$\chi^2 = \sum_{i=1}^{n} \left[\frac{y_i - s(x_i\,;\theta)}{\sigma_i} \right]^2, \tag{1.57}$$

with respect to the parameter of interest, i.e.

$$\left. \frac{\partial \chi^2}{\partial \theta} \right|_{\theta=\hat{\theta}} = 0. \tag{1.58}$$

Formula (1.57) assumes that the y_i variables are uncorrelated. If it is not the case, one has to take into account their correlation, using the correlation matrix V, and thus Eq. (1.57) becomes

$$\chi^2 = \sum_{i,j=1}^{n} [y_i - s(x_i\,;\theta)] \left(V^{-1}\right)_{ij} [y_j - s(x_j\,;\theta)]. \tag{1.59}$$

Strictly speaking, χ^2 in Eq. (1.57) or (1.59) is a true *chi-square* only if y_i is Gaussian distributed with mean $s(x_i\,;\theta)$ and standard deviation σ_i. It is generally a reasonable assumption since we saw previously that many distributions with large statistics converge to a Gaussian distribution. In such a case, χ^2 follows a distribution called the *chi-square distribution* given by the p.d.f.

$$f_\chi(\chi^2\,;n) = \frac{2^{-\frac{n}{2}}}{\Gamma\left(\frac{n}{2}\right)} \left(\chi^2\right)^{\frac{n-2}{2}} \exp\left(-\frac{\chi^2}{2}\right), \tag{1.60}$$

which has a mean value n and a variance $2n$. The Gamma function is defined as $\Gamma(x) = \int_0^\infty t^{x-1} e^{-t}\, dt$ and corresponds to $(x-1)!$ when x is an integer. For large n (typically above 30), $f_\chi(\chi^2\,;n)$ approaches a Gaussian p.d.f. Notice that because of the constraint (1.58), we adjust one parameter, reducing the number of degrees of freedom (n.d.f.) by one unit. Indeed, θ could be chosen to perfectly match one of the y_i, and hence the variability of χ^2 will not be due to n independent random variables but to $n-1$. Hence, if $\chi^2(\theta)$ in Eq. (1.57) follows the chi-square distribution $f_\chi(\chi^2\,;n)$, $\chi^2(\hat{\theta})$ actually follows $f_\chi(\chi^2\,;n-1)$. Similarly, if m parameters $\boldsymbol{\theta} = \theta_1, \ldots, \theta_m$ are estimated, the chi-square distribution is given by $f_\chi(\chi^2\,;n-m)$.

The least-squares method is rather general since knowledge of the actual distribution of y_i is not required, only of its variance. It is thus simple to implement. (An example of its application to estimate parameters is proposed in Problem 1.8.) However, if the distribution of y_i is known, it is better to use the maximum likelihood method since it will be more accurate. Notice that both give the same estimator if y_i is Gaussian distributed. Indeed, in such a case, the p.d.f. of the y_i (uncorrelated) measurements is

$$f(y\,;\theta) = \prod_{i=1}^{n} \frac{1}{\sqrt{2\pi}\sigma_i} \exp\left[-\frac{1}{2}\left(\frac{y_i - s(x_i\,;\theta)}{\sigma_i}\right)^2\right],$$

and thus the negative log-likelihood reads

$$\mathcal{L}(y\,;\theta) = -\ln\left(f(y\,;\theta)\right) = \chi^2/2 + \sum_{i=0}^{n} \ln\left(\sqrt{2\pi}\sigma_i\right). \tag{1.61}$$

Hence, minimising $\mathcal{L}$ is equivalent to minimising χ^2 (since only χ^2 depends on θ). We can then take advantage of what we have learned with the likelihood to deduce that, due to Eq.(1.54), the standard deviation error of the estimated parameter θ in the least-squares method is obtained when $\chi^2 = \chi^2_{\text{min}} + 1$, while the domain at n-sigma is obtained with $\chi^2 = \chi^2_{\text{min}} + n^2$. Moreover, in the specific case of Gaussian distributed variables, the transposition of Eq. (1.55) into the least-squares language is

$$\left(V^{-1}\right)_{ij} = \frac{1}{2} \frac{\partial^2 \chi^2}{\partial\theta_i \partial\theta_j}\bigg|_{\theta=\hat{\theta}}, \tag{1.62}$$

and the standard deviation of the parameters can be recovered directly from the reading of the covariance matrix (at the cost of a matrix inversion). Both methods require the minimisation (or maximisation) of quantities, which generally cannot be performed analytically and require numeric tools. In the high energy physics community, a software package called ROOT (Brun and Rademakers, 1996) is extensively used. It provides many tools and has the great advantage to be free and open source.[23]

The maximum likelihood and the least-squares methods are extensively used to estimate parameters with their uncertainties. One may wonder what is the meaning of the parameter uncertainty or its confidence intervals since, in principle, a parameter (even if unknown) is not expected to be a random variable. When probabilities are interpreted from a frequentist point of view,[24] quoting that $\theta = \hat{\theta} \pm \sigma_\theta$ means that out of N measurements of the parameter θ, one should expect that on average about 68.3% contain the true value within their error intervals. For Bayesians, the parameter θ has a p.d.f. and is therefore seen as a random variable for which σ_θ has a clear interpretation.

1.5.4 Model and Hypothesis Testing

Very often, one wants to test hypotheses such as 'Is there an unknown particle in my data?', 'Is this track a muon or an electron?' and 'Does the energy loss increase logarithmically

[23] See https://root.cern.

[24] There are two schools of thought among statisticians: frequentist and Bayesian. For frequentists, if an experiment is reproduced N times in the same condition, then the probability of a given outcome is the number of times it is observed divided by N when N becomes infinite. It is intuitive, but, in practice, N can never be infinite. Moreover, how should one define a probability, when an experiment can only be realised once, as, for example, the big bang? For Bayesians, probabilities are interpreted as a degree of belief that something will happen. It could be measured, for instance, by considering the odds offered for a bet. This approach extends to more conceptual objects, such as the probability of a theory. Hence, there is a degree of subjectivity, but the definition of probabilities does not suffer from frequentist limitations. Both approaches are complementary.

with the velocity of the particle?'. In the first two examples, we compare two hypotheses: physics beyond the Standard Model against the physics of the Standard Model and track left by a muon against track left by an electron. The question arises as to which hypothesis is more likely, and one should have clear criteria to claim a discovery or refute it. In the third example, there is no unique alternative hypothesis to which we can compare the hypothesis 'the energy loss increases logarithmically with the velocity of the particle'. Actually, there are even an infinite number of alternative hypotheses. This last case is known as the *goodness of fit*. In all cases, one then needs to define criteria, i.e. a test statistic, to determine the level of agreement of a hypothesis with the observation.

Goodness of Fit

When a fit is performed, i.e. the adjustment of parameters $\hat{\boldsymbol{\theta}}$ of a model using a functional form $s(\boldsymbol{x}, \boldsymbol{\theta})$ to the data $\boldsymbol{x}$, the hypothesis we usually want to test is whether the model is consistent with the data. The χ^2-test is the most popular test statistic [but there are others, see the textbooks James (2006) and Barlow (1989), for instance]. It consists in calculating χ^2 with Eq. (1.59) and evaluating the p-value, i.e. the probability to get χ^2 values that are equal to or greater than the actual value observed in the data, χ^2_{obs}. Since χ^2 is supposed to follow the chi-square probability function (1.60), the p-value is defined as

$$p\text{-value} = \int_{\chi^2_{\text{obs}}}^{\infty} f_\chi(\chi^2; n_{\text{dof}})\, \mathrm{d}\chi^2, \tag{1.63}$$

where n_{dof} is the number of degrees of freedom. `ROOT` implements this calculation for us with the function `TMath::Prob(Double_t chi2, Int_t ndf)`. If m parameters $\hat{\boldsymbol{\theta}}$ were previously adjusted with the same n data points $\boldsymbol{x}$, then $n_{\text{dof}} = n - m$. Moreover, if the least-squares method was used, then $\chi^2_{\text{obs}} = \chi^2_{\text{min}}$. The p-value, being built with the data, is itself a random variable. By construction, it lies between 0 and 1 and is uniformly distributed (if the errors are really Gaussian distributed). A large value of χ^2_{obs}, thus of the squares of the residuals, should indicate a poor fit of the model to the data and leads to a small p-value. It could also be due to an underestimation of errors. Conversely, a very low χ^2_{obs} value is probably due to an overestimation of errors. Assuming that the errors are correctly estimated, the model described by the function $s(\boldsymbol{x}, \boldsymbol{\theta})$ can then be rejected if

$$p\text{-value} < \alpha, \tag{1.64}$$

where $1 - \alpha$ is called the *confidence level*. The choice of α is subjective, but the standard values of α are 1%, 5% or 10%. Instead of the p-value, one can use the following rule of thumb as a quick check: since according to the chi-square p.d.f., Eq. (1.60), the mean value of χ^2 is n_{dof} and the standard deviation is $\sqrt{2n_{\text{dof}}}$, one expects χ^2_{obs} to be reasonably close to n_{dof}. More specifically, the rule of thumb consists in checking that $|\chi^2_{\text{obs}} - n_{\text{dof}}|/\sqrt{2n_{\text{dof}}}$ is less than or similar to 1.

When fit parameters are determined with the least-squares method, it is natural to use the χ^2-test to check the goodness of fit. But if the maximum likelihood method is used with unbinned data, how can we check the quality of the description of the data by the model? A possible solution is to bin the data and perform a χ^2-test using the best-fit parameters

$\hat{\boldsymbol{\theta}}$ obtained with the first method to predict the bin content. Let us denote by n the number of bins, by n_i the number of events in bin i and by $N = \sum_i n_i$ the total number of events. The binning is chosen such that n_i is large enough to consider that the Poisson distribution approaches a Gaussian (in practice, $n_i \geq 5$ is acceptable). If the p.d.f. of x is $f(x;\boldsymbol{\theta})$, the expected probability at the best-fit parameter $\boldsymbol{\theta} = \hat{\boldsymbol{\theta}}$, for an event to appear in bin i, is

$$p_i(\hat{\boldsymbol{\theta}}) = \frac{1}{C} \int_{x_i^{\text{low}}}^{x_i^{\text{up}}} f(x;\hat{\boldsymbol{\theta}})\, dx, \tag{1.65}$$

where x_i^{low} and x_i^{up} are the bin limits and C is a normalisation constant. When the histogram range covers all possible values of x, $C = 1$ since the p.d.f. is normalised; otherwise, $C = \sum_i \int_{x_i^{\text{low}}}^{x_i^{\text{up}}} f(x;\hat{\boldsymbol{\theta}})\, dx$. In practice, the integral in Eq. (1.65) is often approximated with $\Delta x_i f(x_i^c;\hat{\boldsymbol{\theta}})$, where $\Delta x_i = x_i^{\text{up}} - x_i^{\text{low}}$ and x_i^c is the value at the bin centre. When the total number of events is fixed to N (thus, we only check the shape of the distribution, not the normalisation), we can calculate the following χ^2:

$$\chi^2 = \sum_{i=1}^{n} \frac{\left[n_i - Np_i(\hat{\boldsymbol{\theta}})\right]^2}{Np_i(\hat{\boldsymbol{\theta}})}. \tag{1.66}$$

It follows a chi-square probability function (1.60), with $(n - m - 1)$ degrees of freedom (the -1 is coming from the constraint on the total number of events). Then, the p-value (1.63) provides a criterion to evaluate the quality of the fit. Alternative χ^2 can be built to test the goodness of fit. They are described extensively in Baker and Cousins (1984).

The χ^2-tests that use the p.d.f. (1.60) are limited by the underlying assumption that the measurements have Gaussian errors. It is generally a good assumption, but not always. For instance, binned data can contain nearly empty bins, where the Poisson distribution with a low mean cannot be approximated by a Gaussian. Also, the real distribution of errors could look more or less Gaussian, but it actually has long tails that cannot be described by a Gaussian. A possible workaround is to find a function that transforms the data into another variable $\boldsymbol{x}' = f(\boldsymbol{x})$ whose errors are more Gaussian-like. The χ^2-test is then performed on $\boldsymbol{x}'$. If it is not possible, one can always generate the expected χ^2 distribution with simulated pseudo-experiments from which a p-value is calculated.

Claiming a Discovery and Setting Limits

Let us imagine that we observe a bump in the distribution of some data, which might be a hint of a new particle. To quantify whether the bump is significant, we introduce two hypotheses: the null hypothesis, H_0, corresponding to the absence of new physics (i.e., the known background), considered to be true by default; and the alternative hypothesis, H_1, where a new signal is allowed on top of the usual background. The joint p.d.f., $f(\boldsymbol{x}|H)$, of the measured variables $\boldsymbol{x} = x_1, \ldots, x_n$ depends on the hypothesis. Let us assume that a test statistic (which depends on the data), $t(\boldsymbol{x})$, that discriminates H_0 from H_1 can be found. It depends on $\boldsymbol{x}$, and so is itself a random variable described by its own p.d.f., $f_t(t|H)$. The two hypotheses H_0 and H_1 can be discriminated if $f_t(t|H_0)$ and $f_t(t|H_1)$ are significantly different. Suppose that by convention, $t(\boldsymbol{x})$ tends to have larger values when H_1 is true. For

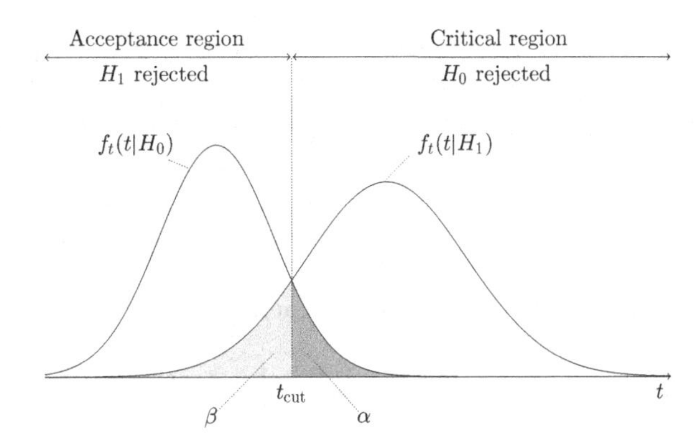

Fig. 1.16 Probability distribution of the test statistic t for the null hypothesis H_0 and the alternative H_1. The critical and acceptance regions are delimited by t_{cut}, which determines the values of the significance level, α, and the power of test $1-\beta$ (see text).

this purpose, a value t_{cut} is chosen that delimits two regions: the so-called *critical region*, when $t > t_{cut}$, and the *acceptance region*, when $t < t_{cut}$. If the H_0 hypothesis is true, the probability to get a value t in the critical region is

$$\int_{t_{cut}}^{\infty} f_t(t|H_0)\, dt = \alpha, \tag{1.67}$$

while the corresponding probability if H_1 is true is

$$\int_{t_{cut}}^{\infty} f_t(t|H_1)\, dt = 1 - \int_{-\infty}^{t_{cut}} f_t(t|H_1)\, dt = 1 - \beta. \tag{1.68}$$

The quantities α and β are represented in Fig 1.16 along the different regions. Testing the null hypothesis H_0 against the alternative H_1 consists in setting a significance level α (for instance, $\alpha = 1\%$), which determines the value of t_{cut}, and thus of β. The quantity $1-\alpha$ is usually called the *confidence level* and is denoted CL, while $1-\beta$ is called the power of the test because if $1-\beta$ is large, the alternative hypothesis is well separated from the null hypothesis. The lower the values of α and β, or equivalently the larger the confidence level and the power, the better the test statistic is at discriminating between the two hypotheses. If the observed value of the statistic test, t_{obs}, falls in the critical region, i.e. $t_{obs} > t_{cut}$, H_0 is considered rejected. Hence, α is the probability of rejecting H_0, despite being true: it is the probability of a false discovery claim. On the other hand, β is the probability of rejecting H_1, although it is true (and thus of not rejecting H_0, despite being false): it is the probability of missing a discovery when there should be one. In the rare cases where H_0 and H_1 are the only possible hypotheses that could lead to the same data, rejecting one hypothesis is equivalent to accepting the other. In most cases, it is not equivalent because of possible alternative hypotheses not considered.

In order to claim a discovery of a new signal on top of the known background, the two hypotheses $H_0 = H_{bkg}$ and $H_1 = H_{sig+bkg}$ are then tested. The hypothesis $H_{sig+bkg}$ usually depends on a parameter of interest that we wish to measure. It is typically the cross section, σ_{sig}, of the new signal. In high energy physics, the convention is to claim a discovery, or *observation*, if H_{bkg} is rejected with a significance level as low as $\alpha = 2.87 \times 10^{-7}$, equivalent to the one-sided tail probability of $Z = 5$ standard deviations of the unit Gaussian.[25] One simply says that the significance level is 5σ. As before the p-value (the observed level of significance) of the hypothesis H is

$$p\text{-value} = \int_{t_{obs}}^{\infty} f_t(t|H)\ \mathrm{d}t. \tag{1.69}$$

Hence, when $H_0 = H_{bkg}$ is rejected, it implies

$$p_{bkg} = \int_{t_{obs}}^{\infty} f_t(t|H_{bkg})\ \mathrm{d}t < \alpha. \tag{1.70}$$

If only a significance level of 3σ (i.e., $Z = 3$ and $\alpha = 1.35 \times 10^{-3}$) can be reached, it is not considered a discovery, but rather *evidence*, in the accepted jargon. Discoveries are rare, but evidence regularly appears before vanishing with the acquisition of more data. One of the reasons for this is the difficulty of correctly assessing the measurement errors, which can result from poorly modelled long tails, and thus a poor description of the tail of the probability distribution of the test statistic. So before getting too excited by an evidence, it is wiser to keep this in mind. In order to convince oneself of the discovery or the evidence, many additional elements are usually provided. For instance, one gives a confidence interval estimate of σ_{sig}, the parameter of interest under $H_{sig+bkg}$. Also, a simulation of the signal is often used to evaluate the expected p-value under the H_{bkg} hypothesis. A reasonable agreement between the expected p-value and the measured one in Eq. (1.70) strengthens the confidence in the result.

When Eq. (1.70) is not satisfied, i.e. when $p_{bkg} > \alpha$, we fail to reject H_{bkg} at the significance level, α. However, it does not imply that we have no sensitivity at all to values of σ_{sig} under $H_{sig+bkg}$. In other words, some values may be excluded and others may not. Another test is then needed to find the excluded values. It consists in testing the hypotheses $H_{sig+bkg}$ for a given value, σ^0_{sig}, against H_{bkg}. The hypothesis $H_0 = H_{sig+bkg}{:}\sigma^0_{sig}$ is excluded at the significance level, α, if its p-value satisfies

$$p_{sig+bkg} = \int_{t_{obs}}^{\infty} f_t(t|H_{sig+bkg}{:}\sigma^0_{sig})\ \mathrm{d}t < \alpha.$$

One varies the value of σ^0_{sig} and performs the test for each value to obtain the region of excluded σ_{sig}. For a particle-production rate or a cross section, since it is assumed that a new production mechanism can only add more events than what is expected with the background only, the excluded region is an interval of the form $[\sigma^{lim}_{sig}, +\infty]$. The standard p-value threshold for exclusion in high energy physics is $\alpha = 0.05$, i.e. 95% confidence level.

Sometimes, the number of background events may fluctuate downwards, leading to the exclusion of $H_{sig+bkg}$, although there is no sensitivity to distinguish $H_{sig+bkg}$ from H_{bkg}. In

[25] if $p = \int_Z^{\infty} \frac{1}{\sqrt{2\pi}} e^{-x^2/2}\ \mathrm{d}x$, then $Z = \Phi^{-1}(1-p)$, where $\Phi^{-1}(x)$ is the inverse function of the cumulative distribution of the unit Gaussian, i.e. $\Phi(x) = \int_{-\infty}^{x} \frac{1}{\sqrt{2\pi}} e^{-x'^2/2}\ \mathrm{d}x'$. Hence, $\alpha = 2.87 \times 10^{-7} = \int_5^{\infty} \frac{1}{\sqrt{2\pi}} e^{-x^2/2}\ \mathrm{d}x = 1 - \Phi(5)$.

extreme cases, one may even obtain unphysical $\sigma_{\text{sig}}^{\text{lim}} < 0$. To prevent those cases, instead of using the p-value, $p_{\text{sig+bkg}}$, the high energy physics community often uses the ratio

$$\text{CL}_s = \frac{p_{\text{sig+bkg}}}{1 - p_{\text{bkg}}}$$

to set exclusion limits. Requiring $\text{CL}_s < \alpha$ is more stringent than $p_{\text{sig+bkg}} < \alpha$ since CL_s is necessarily larger than $p_{\text{sig+bkg}}$. It follows that CL_s is more robust against unphysical exclusions.

One may wonder how to choose the test statistic $t(\boldsymbol{x})$. In simple analyses, it is natural to use, for instance, the number of events for counting analyses, or the distribution of an observable such as the reconstructed energy of a particle. In more complicated analyses, when the likelihood function is known, a widely used test statistic to establish discovery or exclusion limits is based on the *likelihood ratio*.[26] With $\hat{\sigma}_{\text{sig}} > 0$ denoting the best fit value of the cross section of the signal hypothesis $H_{\text{sig+bkg}}$ (the one maximising the likelihood), this ratio inspired from Eq. (1.54) is defined as

$$q(\sigma_{\text{sig}}) = \frac{L(\sigma_{\text{sig}})}{L(\hat{\sigma}_{\text{sig}})}. \tag{1.71}$$

Notice that the value $\sigma_{\text{sig}} = 0$ corresponds to hypothesis H_{bkg}. For large samples, the quantity $-2\ln q(\sigma_{\text{sig}})$ is asymptotically distributed as χ^2 with, in our example of Eq. (1.71), one degree of freedom. It thus becomes easy to construct confidence intervals, as suggested by Feldman and Cousins (1998). The likelihood ratio also allows the inclusion of systematic errors as nuisance parameters that must be fitted from the data. This approach (called profile likelihood ratio) is beyond the scope of this introduction. It is detailed in Cowan et al. (2011).

Problems

1.1 **Coulomb's law.** Consider an electric charge density distribution $\rho(\boldsymbol{r})$ creating the electric field $\boldsymbol{E}$. Starting with Maxwell's equation $\boldsymbol{\nabla} \cdot \boldsymbol{E} = \rho/\epsilon_0$, integrate the equation over a volume V, and show that

$$\oint_S \boldsymbol{E} \cdot \mathrm{d}S = Q/\epsilon_0,$$

where S is the surface encapsulating the volume V and Q is the charge contained in V. Assuming a static point-like particle with a charge q_0 located at $\boldsymbol{0}$, show that the electric field at $\boldsymbol{r}$ is

$$\boldsymbol{E} = \frac{q_0}{4\pi\epsilon_0|\boldsymbol{r}|^2}\frac{\boldsymbol{r}}{|\boldsymbol{r}|}.$$

Conclude that the force felt by a charge q at $\boldsymbol{r}$ is given by Coulomb's law,

$$\boldsymbol{f} = \frac{q\,q_0}{4\pi\epsilon_0|\boldsymbol{r}|^2}\frac{\boldsymbol{r}}{|\boldsymbol{r}|}.$$

[26] The popularity of the likelihood ratio is coming from the Neyman–Pearson lemma that states that the likelihood ratio between two hypotheses represents the optimal test statistic.

1.2 **Yukawa potential.** Consider the general equation

$$(\nabla^2 - \lambda^2)f(\boldsymbol{r}) = -\delta(\boldsymbol{r}),$$

where λ is a real number. Using the Fourier transform $\tilde{f}(\boldsymbol{k}) = \iiint f(\boldsymbol{r})e^{-i\boldsymbol{k}\cdot\boldsymbol{r}}\,\mathrm{d}\boldsymbol{r}$ and its inverse transform $f(\boldsymbol{r}) = 1/(2\pi)^3 \iiint \tilde{f}(\boldsymbol{k})e^{i\boldsymbol{k}\cdot\boldsymbol{r}}\,\mathrm{d}\boldsymbol{k}$, show that the solution of the equation is

$$f(\boldsymbol{r}) = \frac{1}{(2\pi)^3}\iiint \frac{e^{i\boldsymbol{k}\cdot\boldsymbol{r}}}{|\boldsymbol{k}|^2 + \lambda^2}\,\mathrm{d}\boldsymbol{k}.$$

Using polar coordinates and orienting the z-axis of the reference frame along the $\boldsymbol{r}$ direction, check that

$$f(\boldsymbol{r}) = \frac{1}{2\pi^2}\frac{1}{|\boldsymbol{r}|}\int_0^\infty \frac{\sin(|\boldsymbol{k}||\boldsymbol{r}|)}{|\boldsymbol{k}|^2 + \lambda^2}|\boldsymbol{k}|\,d|\boldsymbol{k}|.$$

Using a contour integration in the complex plane (if you are not familiar with this technique, read Appendix F), show that

$$f(\boldsymbol{r}) = \frac{e^{-\lambda|\boldsymbol{r}|}}{4\pi|\boldsymbol{r}|}.$$

Deduce the expression of the electrostatic potential, Eq. (1.2), and the Yukawa potential, Eq. (1.1).

1.3 Let I_v be the vertical flux of muons. Given that the flux intensities coming from the zenith angle, θ, is given by $I(\theta) = I_v \cos^2\theta$, show that the muon flux from the sky collected by a horizontal detector should be $I_v\pi/2$.

1.4 Playing with relativistic formulas that are recalled in Section 2.2, check Eq. (1.9).

1.5 Without loss of generality, let $\boldsymbol{B}$ be a static magnetic field along the z-axis. Check that the power from the Lorentz force is zero and deduce that γ is a constant of the motion. Conclude that

$$\gamma m \frac{\mathrm{d}\boldsymbol{v}}{\mathrm{d}t} = q\boldsymbol{v} \times \boldsymbol{B}.$$

Show that the solution of this equation yields the following coordinates of the trajectory:

$$\begin{aligned}
x(t) &= x(0) + v_y(0)/\omega + [v_x(0)\sin(\omega t) - v_y(0)\cos(\omega t)]/\omega \\
y(t) &= y(0) - v_x(0)/\omega + [v_y(0)\sin(\omega t) + v_x(0)\cos(\omega t)]/\omega \\
z(t) &= z(0) + v_z(0) \times t,
\end{aligned}$$

with $\omega = q|\boldsymbol{B}|/(\gamma m)$. One could introduce the variable $u = x + iy$ to ease the calculation. Finally, verify that the bending radius in the (x, y) plane satisfies Eq. (1.27).

1.6 Sagitta determination.

The momentum measurement is related to the determination of the sagitta, s (image on the right). Imagine that three points are used to determine the particle track. In the approximation where the angle ϕ is small (or equivalently $s \ll L$), show that

$$s \simeq R\frac{\phi^2}{8} = \frac{L^2}{8R}.$$

Using Eq. (1.30) with $\delta = 0.3$ and $|q| = 1$, deduce that the transverse momentum is

$$|\boldsymbol{p}_\perp| = \frac{0.3|\boldsymbol{B}|L^2}{8s}.$$

Hence, conclude that the uncertainty is given by $\sigma(|\boldsymbol{p}_\perp|)/|\boldsymbol{p}_\perp| = \sigma(s)\, 8|\boldsymbol{p}_\perp|/(0.3BL^2)$.

1.7 Weighted sum and likelihood. Let r_i be the result of the ith measurement and σ_{r_i} its uncertainty. The r_i random variables, $i = 1, \ldots, n$, are assumed to follow a Gaussian p.d.f. $G(r_i\,; \mu, \sigma_{r_i})$. Apply the maximum likelihood method to evaluate the parameter μ. Show that the minimisation of the negative log-likelihood yields the estimator

$$\hat{\mu} = \frac{1}{\sum_i w_i} \sum_{i=1}^{N} w_i r_i,$$

with $w_i = 1/\sigma_{r_i}^2$.

1.8 Particle lifetime and least squares. The least-squares method can be solved analytically when the data are described by a linear function in the parameters to estimate. We will adapt it to the determination of the lifetime τ of the nuclei constituting a radioactive source. A detector records its activity during consecutive intervals of $\Delta t = 20$ ns:

Interval, k	1	2	3	4	5
Number of counts, n_i	2 797	1 241	570	264	128

The number of atoms of the source at a time t is given by $N(t) = N_0 \exp(-\gamma t)$, with $\gamma = 1/\tau$ ($t = 0$ is the beginning of data acquisition).

1. Show that the number of decays (i.e., counts) between t and $t + \Delta t$ satisfies $\ln[N_{\text{decay}}(t)] = -\gamma t + \alpha$, with $\alpha = \ln[N_0(1 - \exp(-\gamma \Delta t))]$.
2. Assuming that the number of counts n_k is Gaussian distributed, with a standard deviation $\sigma_k = \sqrt{n_k}$ (justified by their large values), what standard deviation should be used for $\ln(n_k)$?
3. Apply the least-squares method to the variable $\ln(n_k)$ and show that the estimator of γ is

$$\hat{\gamma} = \frac{1}{\Delta t} \frac{\left(\sum_k n_k \ln n_k\right)\left(\sum_k n_k (k-1)\right) - \left(\sum_k n_k\right)\left(\sum_k n_k \ln n_k (k-1)\right)}{\left(\sum_k n_k\right)\left(\sum_k n_k (k-1)^2\right) - \left(\sum_k n_k (k-1)\right)^2},$$

where k varies from 1 to 5. Deduce the value of τ and compare it to the true value $\tau = 25$ ns.

2 Preliminary Concepts: Special Relativity and Quantum Mechanics

Elementary particles are almost massless, travelling at a significant fraction of the speed of light, and are governed by the laws of quantum mechanics. Hence, a good knowledge of special relativity and quantum mechanics is essential for studying particle physics. The reader is assumed to be already familiar with these two theories. However, in the first part of the chapter, a brief review of special relativity is given with emphasis on the covariant and contravariant notations, which may be less well mastered but are very useful in particle physics. The second part of the chapter deals with an important aspect of quantum mechanics for particle physics: angular momentum. Other aspects will be addressed in the next two chapters, as appropriate.

2.1 Covariant and Contravariant Coordinates, and the Scalar Product

In special relativity, the position in spacetime requires four numbers (ct, x, y, z), where c is the velocity of light. Given a basis of a reference frame, $\{\boldsymbol{e}_\mu\}$, with $\mu = 0 - 3$, the vector position $\boldsymbol{X}$ is then

$$\boldsymbol{X} = \sum_{\mu=0}^{3} x^\mu \boldsymbol{e}_\mu = x^\mu \boldsymbol{e}_\mu,$$

where in the last term, the summation over μ is implicit (this is the Einstein notation for repeated indices). Since a vector exists independently of any basis, a change of reference frame should not modify the vector. As such a change affects the basis, the coordinates of the vector must be modified to compensate for the change in basis. For instance, if the basis is scaled up, the coordinates, x^μ, must be scaled down. Those coordinates, countering the change of basis elements, are then called the *contravariant* coordinates and are denoted as x^μ with Lorentz indices, μ, in the upper position. For the 4-vector position, the contravariant (cartesian) coordinates are

$$x^0 = ct,\ x^1 = x,\ x^2 = y,\ x^3 = z.$$

A common abuse of notations consists in denoting the 4-vector by its coordinates, such as

$$x^\mu = (ct, \boldsymbol{x}) = (ct, x, y, z). \tag{2.1}$$

The scalar product of two 4-vectors, $\boldsymbol{X} = x^\mu \boldsymbol{e}_\mu$ and $\boldsymbol{Y} = y^\nu \boldsymbol{e}_\nu$, is defined by

$$\boldsymbol{X} \cdot \boldsymbol{Y} = x^\mu y^\nu \boldsymbol{e}_\mu \cdot \boldsymbol{e}_\nu = g_{\mu\nu} x^\mu y^\nu, \tag{2.2}$$

with $g_{\mu\nu} = \boldsymbol{e_\mu} \cdot \boldsymbol{e_\nu}$. The scalar product must be consistent with the definition of the spacetime interval of special relativity. Recall that the spacetime interval between an event occurring at $\boldsymbol{X_1}$ and another event at $\boldsymbol{X_2}$ is

$$\Delta s^2 = (\boldsymbol{X_1} - \boldsymbol{X_2}) \cdot (\boldsymbol{X_1} - \boldsymbol{X_2}) = c^2(t_1 - t_2)^2 - (x_1 - x_2)^2 - (y_1 - y_2)^2 - (z_1 - z_2)^2. \quad (2.3)$$

It is the relativistic equivalent of the (squared) Euclidean distance. It can be positive, negative or null. Since the speed of light is assumed to be constant in any frame, $\Delta s^2 = 0$ for a light-like interval. When two events can be causally related, $\Delta s^2 > 0$, and the interval is referred to as a time-like interval. Otherwise, for $\Delta s^2 < 0$, it is a space-like interval (the two events being causally separated). According to Eq. (2.2), the scalar product $(\boldsymbol{X_1} - \boldsymbol{X_2}) \cdot (\boldsymbol{X_1} - \boldsymbol{X_2})$ matches the definition of the spacetime interval (2.3) if

$$(g_{\mu\nu}) = \begin{pmatrix} g_{00} & g_{01} & g_{02} & g_{03} \\ g_{10} & g_{11} & g_{12} & g_{13} \\ g_{20} & g_{21} & g_{22} & g_{23} \\ g_{30} & g_{31} & g_{32} & g_{33} \end{pmatrix} = \begin{pmatrix} 1 & 0 & 0 & 0 \\ 0 & -1 & 0 & 0 \\ 0 & 0 & -1 & 0 \\ 0 & 0 & 0 & -1 \end{pmatrix}. \quad (2.4)$$

The tensor, $g_{\mu\nu}$, is the metric tensor of the Minkowski space. A new set of coordinates, called *covariant*, can then be defined as

$$x_\mu = g_{\mu\nu} x^\nu, \quad (2.5)$$

with $x_\mu = (ct, -\boldsymbol{x})$. Lower and upper indices always indicate covariant and contravariant coordinates, respectively. The contravariant coordinates are obtained from x_μ as

$$x^\mu = g^{\mu\nu} x_\nu, \quad (2.6)$$

where $(g^{\mu\nu})$ is the inverse matrix of $(g_{\mu\nu})$ so that

$$g^{\mu\nu} = g_{\mu\nu}. \quad (2.7)$$

The tensor $g^{\mu\nu}$ is thus the inverse of $g_{\mu\nu}$. Indeed, the combination of Eqs. (2.5) and (2.6) implies

$$g_{\mu\rho} g^{\rho\nu} = \delta_\mu^\nu = \begin{cases} 1 & \text{if } \mu = \nu, \\ 0 & \text{otherwise,} \end{cases} \quad (2.8)$$

where δ_μ^ν is the Kronecker symbol,[1] whose action on a tensor A is to replace an index, i.e.

$$\delta_\mu^\nu A_{\nu\sigma} = A_{\mu\sigma}, \quad \delta_\mu^\nu A^{\mu\sigma} = A^{\nu\sigma}. \quad (2.9)$$

Beware that if $g_{0\rho} g^{\rho 0} = \delta_0^0 = 1$, and similarly $g_{1\rho} g^{\rho 1} = g_{2\rho} g^{\rho 2} = g_{3\rho} g^{\rho 3} = 1$, then,

$$g_{\mu\rho} g^{\rho\mu} = 4, \quad (2.10)$$

because the symbol δ_μ^μ implies an implicit summation ($\delta_\mu^\mu = \delta_0^0 + \delta_1^1 + \delta_2^2 + \delta_3^3 = 4$). The metric tensor being symmetric, the order of indices does not matter ($g_{\mu\nu} = g_{\nu\mu}$).

[1] Because of the symmetry of the metric tensor, there is no need to keep track of the position of the indices in the Kronecker symbol, i.e. $\delta_\mu^\nu = \delta^\nu{}_\mu = \delta_\mu{}^\nu$.

Given the relation (2.6), the scalar product of two 4-vectors has then the following equivalent definitions:

$$\boldsymbol{X} \cdot \boldsymbol{Y} = g_{\mu\nu} x^{\mu} y^{\nu} = g^{\mu\nu} x_{\mu} y_{\nu} = x^{\mu} y_{\mu} = x_{\mu} y^{\mu}. \tag{2.11}$$

From now on, 4-vectors will be written with a regular font using indifferently, as a symbol, the label of the 4-vector or its components (a common abuse of notation). For example, the 4-vector position can be simply written as x or x^{μ}, and the scalar product above is just $x \cdot y$.

2.2 Lorentz Transformations

2.2.1 Definition and Properties

A Lorentz transformation is a linear transformation that conserves the scalar product and hence the spacetime interval between two events. It is thus a linear map from x to x' of the form

$$x'^{\mu} = \Lambda^{\mu}{}_{\nu} x^{\nu}, \tag{2.12}$$

where Λ is the Lorentz transformation matrix [again, the implicit summation is used in Eq. (2.12)]. The corresponding transformation of the covariant coordinates is

$$x'_{\mu} = g_{\mu\rho} x'^{\rho} = g_{\mu\rho} \Lambda^{\rho}{}_{\sigma} x^{\sigma} = g_{\mu\rho} \Lambda^{\rho}{}_{\sigma} g^{\sigma\nu} x_{\nu} = \Lambda_{\mu}{}^{\nu} x_{\nu}, \tag{2.13}$$

where in the last step, the usual rule for lowering or raising the indices is applied to the Lorentz matrix element. As the scalar product is conserved, $x'^2 = x^2$. It follows, using matrix notation[2] $g \equiv (g_{\mu\nu})$ and $\Lambda \equiv (\Lambda^{\mu}{}_{\nu})$,

$$x^{\mathsf{T}} \Lambda^{\mathsf{T}} g \Lambda x = x^{\mathsf{T}} g x,$$

with the symbol ^{T} denoting the transpose matrix. The corresponding equality with the tensorial notation (2.12) is (see Problem 2.1)

$$g_{\rho\sigma} \Lambda^{\rho}{}_{\mu} x^{\mu} \Lambda^{\sigma}{}_{\nu} x^{\nu} = g_{\mu\nu} x^{\mu} x^{\nu}.$$

Since the last two equalities are valid for any x, necessarily

$$\Lambda^{\mathsf{T}} g \Lambda = g \quad \text{and} \quad g_{\rho\sigma} \Lambda^{\rho}{}_{\mu} \Lambda^{\sigma}{}_{\nu} = g_{\mu\nu}. \tag{2.14}$$

Equation (2.14) defines the conditions that any Lorentz transformation must satisfy. Incidentally, the two equations in Eq. (2.14) are striclty equivalent. Indeed, by definition of the transposition, $\Lambda_{\eta\mu} = \Lambda^{\mathsf{T}}{}_{\mu\eta}$. It follows that $\Lambda^{\rho}{}_{\mu}$ in Eq. (2.14) is

$$\Lambda^{\rho}{}_{\mu} = g^{\rho\eta} \Lambda_{\eta\mu} = g^{\rho\eta} \Lambda^{\mathsf{T}}{}_{\mu\eta} = \Lambda^{\mathsf{T}}{}_{\mu}{}^{\rho}. \tag{2.15}$$

Hence, the equation on the right-hand side of (2.14) is equivalent to

$$\Lambda^{\mathsf{T}}{}_{\mu}{}^{\rho} g_{\rho\sigma} \Lambda^{\sigma}{}_{\nu} = g_{\mu\nu}.$$

[2] I usually enclose the symbol for a matrix within parentheses and label the elements of that matrix without parentheses.

The indices μ, ρ, σ, ν are thus in the correct order to be consistent with the matrix notation[3] $\Lambda^\mathsf{T} g \Lambda$.

Formula (2.14) allows us to deduce several properties of the Lorentz matrix. Taking the determinant, it follows that $\det(\Lambda^\mathsf{T})\det(\Lambda) = 1$, which implies

$$\det{}^2(\Lambda) = 1. \tag{2.16}$$

Consequently, Λ has determinant ± 1, and, therefore, the inverse matrix, Λ^{-1}, necessarily exists. Let us multiply $\Lambda^\mathsf{T} g \Lambda = g$ by Λ^{-1} from the right and $\left(\Lambda^{-1}\right)^\mathsf{T}$ from the left. Given the usual matrix property $\left(\Lambda^{-1}\right)^\mathsf{T} = (\Lambda^\mathsf{T})^{-1}$, it follows that

$$g = \left(\Lambda^{-1}\right)^\mathsf{T} g \Lambda^{-1}.$$

This shows that Λ^{-1} is also a Lorentz transformation since it satisfies the general definition (2.14). Problem 2.2 shows that its components are given by

$$\left(\Lambda^{-1}\right)^{\sigma}{}_{\rho} = \Lambda_{\rho}{}^{\sigma}. \tag{2.17}$$

Pay attention to the order and position of indices. Consequently, the transformation of the covariant coordinates (2.13) can be rewritten as

$$x'_\mu = \Lambda_\mu{}^\nu x_\nu = \left(\Lambda^{-1}\right)^\nu{}_\mu x_\nu = \left(\left(\Lambda^{-1}\right)^\mathsf{T}\right)_\mu{}^\nu x_\nu, \tag{2.18}$$

where the transpose matrix is introduced in order to have the appropriate order of the indices.[4] Equation (2.18) gives us the transformation of the covariant coordinates when the contravariant coordinates transform as Eq. (2.12).

We saw previously that the inverse of the Lorentz matrix is defined. In addition, the product of two Lorentz transformations satisfies

$$(\Lambda_1\Lambda_2)^\mathsf{T} g\, (\Lambda_1\Lambda_2) = \Lambda_2^\mathsf{T} \underbrace{\Lambda_1^\mathsf{T} g \Lambda_1}_{g} \Lambda_2 = g,$$

proving that $\Lambda_1\Lambda_2$ is a Lorentz transformation, too. Since the identity matrix satisfies (2.14), we have all the properties of a symmetry group (see Section A.1). When $\det(\Lambda) = 1$, the Lorentz transformation conserves the spatial orientation and is called the proper Lorentz transformation. When $\det(\Lambda) = -1$, the transformation is called improper and may change the spatial or time orientation. A simple example is $\Lambda = g$, which reverses the space coordinates and thus corresponds to the parity transformation (see Section 4.3.3). The component $\mu = \nu = 0$ in Eq. (2.14) reads $g_{\rho\sigma}\Lambda^\rho{}_0\Lambda^\sigma{}_0 = g_{00} = 1$. But it is also true that

$$1 = g_{\rho\sigma}\Lambda^\rho{}_0\Lambda^\sigma{}_0 = (\Lambda^0{}_0)^2 - \sum_{i=1}^{3}(\Lambda^i{}_0)^2,$$

implying that $(\Lambda^0{}_0)^2 \geq 1$. When $\Lambda^0{}_0 \geq 1$, the transformation is orthochronous, meaning that the time direction is conserved. If $\Lambda^0{}_0 \leq -1$, future and past are swapped, defining an antichronous transformation. A trivial example is $\Lambda = -g$, which defines the time-reversal

[3] The product of the three matrices A, B and C defines a new matrix D whose elements are $D_{ij} = \sum_{k,l} A_{ik}B_{kl}C_{lj}$.

[4] If the covariant coordinates x_μ are put in a column vector $\tilde{x}$, then Eq. (2.18) reads $\tilde{x'} = \left(\Lambda^{-1}\right)^\mathsf{T}\tilde{x}$.

transformation. The transformations having both $\det(\Lambda) = 1$ and $\Lambda^0_{\ 0} \geq 1$ form a subgroup of the Lorentz transformation group called the restricted Lorentz group. I give some details in Appendix D.1. The set of proper orthochronous transformations consist of boosts and rotations in the three-dimensional space.

2.2.2 Special Lorentz Transformations

In the framework of special relativity, the Lorentz transformation corresponds to the law that governs changes of inertial (or Galilean) reference frames, for which the physics equations must be preserved (i.e., they must have the same mathematical form whatever the inertial frame). Moreover, the speed of light, as well as the orientation of space and time, must be the same for any Galilean frame. To have equations that are consistent with these principles, both sides of the equations must transform in the same way under Lorentz transformations.[5] It ensures that the equations will hold in any inertial frame. Such equations are called covariant. With these assumptions, one can derive that, for a frame $\Re'$ moving at a velocity v along the x-axis with respect to another frame $\Re$ (with x and x' being aligned), both sets of coordinates are related by the Lorentz transformation $x'^{\mu} = \Lambda^{\mu}_{\ \nu} x^{\nu}$ with (see any special relativity course)

$$(\Lambda^{\mu}_{\ \nu}) = \begin{pmatrix} \Lambda^0_{\ 0} & \Lambda^0_{\ 1} & \Lambda^0_{\ 2} & \Lambda^0_{\ 3} \\ \Lambda^1_{\ 0} & \Lambda^1_{\ 1} & \Lambda^1_{\ 2} & \Lambda^1_{\ 3} \\ \Lambda^2_{\ 0} & \Lambda^2_{\ 1} & \Lambda^2_{\ 2} & \Lambda^2_{\ 3} \\ \Lambda^3_{\ 0} & \Lambda^3_{\ 1} & \Lambda^3_{\ 2} & \Lambda^3_{\ 3} \end{pmatrix} = \begin{pmatrix} \gamma & -\beta\gamma & 0 & 0 \\ -\beta\gamma & \gamma & 0 & 0 \\ 0 & 0 & 1 & 0 \\ 0 & 0 & 0 & 1 \end{pmatrix}, \tag{2.19}$$

and

$$\boxed{\beta = \frac{\mathrm{v}}{c},\ \gamma = \frac{1}{\sqrt{1-\beta^2}}}. \tag{2.20}$$

This rotation-free Lorentz transformation is called a Lorentz *boost*, and the quantity γ is the Lorentz factor. The coordinates x^{μ} are deduced from x'^{μ} by replacing β with $-\beta$. Therefore, the Lorentz transformation describing the change from the particle rest frame to the lab frame is

$$\Lambda_{\text{particle to lab}} = \begin{pmatrix} \gamma & \beta\gamma & 0 & 0 \\ \beta\gamma & \gamma & 0 & 0 \\ 0 & 0 & 1 & 0 \\ 0 & 0 & 0 & 1 \end{pmatrix}. \tag{2.21}$$

For an arbitrary direction of the $\Re'$ frame, $\boldsymbol{\beta} = \mathbf{v}/c$, the transformation (2.19) becomes

$$\begin{aligned} x'^0 &= \gamma(x^0 - \boldsymbol{\beta}\cdot\boldsymbol{x}), \\ \boldsymbol{x}' &= \boldsymbol{x} + (\gamma-1)(\boldsymbol{\beta}\cdot\boldsymbol{x})\frac{\boldsymbol{\beta}}{\beta^2} - \boldsymbol{\beta}\gamma x^0, \end{aligned} \tag{2.22}$$

[5] Scalars are unaffected, 4-vectors transform as in Eq. (2.12), two-dimensions tensors $F^{\mu\nu}$ as $\Lambda^{\mu}_{\ \rho}\Lambda^{\nu}_{\ \sigma}F^{\rho\sigma}$, etc.

where $\boldsymbol{x} = (x^1, x^2, x^3)$. In terms of matrix representation, it corresponds to

$$\begin{pmatrix} x'^0 \\ x'^1 \\ x'^2 \\ x'^3 \end{pmatrix} = \begin{pmatrix} \gamma & -\beta_x\gamma & -\beta_y\gamma & -\beta_z\gamma \\ -\beta_x\gamma & 1+(\gamma-1)\frac{\beta_x^2}{\beta^2} & (\gamma-1)\frac{\beta_x\beta_y}{\beta^2} & (\gamma-1)\frac{\beta_x\beta_z}{\beta^2} \\ -\beta_y\gamma & (\gamma-1)\frac{\beta_y\beta_x}{\beta^2} & 1+(\gamma-1)\frac{\beta_y^2}{\beta^2} & (\gamma-1)\frac{\beta_y\beta_z}{\beta^2} \\ -\beta_z\gamma & (\gamma-1)\frac{\beta_z\beta_x}{\beta^2} & (\gamma-1)\frac{\beta_z\beta_y}{\beta^2} & 1+(\gamma-1)\frac{\beta_z^2}{\beta^2} \end{pmatrix} \begin{pmatrix} x^0 = ct \\ x^1 = x \\ x^2 = y \\ x^3 = z \end{pmatrix}. \tag{2.23}$$

For completeness, the law of transformation of velocities can be easily deduced from Eq. (2.22), using its differential form. Denoting by $\boldsymbol{w} = \mathrm{d}\boldsymbol{x}/\mathrm{d}t$ and $\boldsymbol{w}' = \mathrm{d}\boldsymbol{x}'/\mathrm{d}t$, the velocities in the frames $\mathfrak{R}$ and $\mathfrak{R}'$, respectively, it follows that

$$\boldsymbol{w}' = \frac{\boldsymbol{w} + (\gamma-1)(\boldsymbol{\beta}\cdot\boldsymbol{w})\frac{\boldsymbol{\beta}}{\beta^2} - \boldsymbol{\beta}c\gamma}{\gamma\left(1 - \frac{\boldsymbol{\beta}\cdot\boldsymbol{w}}{c}\right)}. \tag{2.24}$$

For instance, if the frame $\mathfrak{R}'$ is moving along the z-axis $\boldsymbol{\beta} = \beta\boldsymbol{u}_z$, the three components of the velocity in that frame are then reduced to

$$w'_x = \frac{w_x}{\gamma\left(1 - \frac{\beta w_z}{c}\right)}, \quad w'_y = \frac{w_y}{\gamma\left(1 - \frac{\beta w_z}{c}\right)}, \quad w'_z = \frac{w_z - \beta c}{1 - \frac{\beta w_z}{c}}.$$

2.2.3 Proper Time

In the rest frame of a particle, the infinitesimal spacetime interval $\mathrm{d}s^2$ is simply reduced to its proper time contribution $c^2\,\mathrm{d}\tau^2$, while in the lab frame, the interval is $c^2\,\mathrm{d}t^2 - \mathrm{d}x^2 - \mathrm{d}y^2 - \mathrm{d}z^2 = c^2\,\mathrm{d}t^2(1 - v^2/c^2)$. Hence, $\mathrm{d}\tau = \sqrt{(1 - v^2/c^2)}\,\mathrm{d}t$, i.e.

$$\boxed{\mathrm{d}\tau = \frac{\mathrm{d}t}{\gamma}}. \tag{2.25}$$

Consequently, for an unstable particle, its lifetime in the particle rest frame is always shorter than its lifetime measured in another frame.

2.2.4 Example of 4-Vectors

So far, only the position 4-vector has been explicitly considered. However, any four numbers that follow the same transformation under a change of frame as the position 4-vector form by definition a 4-vector. In the following, some examples of 4-vectors of prime importance in particles physics are provided.

Derivative

If F is a scalar function, then the quantity

$$\mathrm{d}F = \frac{\partial F}{\partial x^\mu}\,\mathrm{d}x^\mu$$

is a scalar as well. Since $\mathrm{d}x^\mu$ is contravariant, the four components

$$\partial_\mu = \frac{\partial}{\partial x^\mu} = \left(\frac{1}{c}\frac{\partial}{\partial t}, \frac{\partial}{\partial x}, \frac{\partial}{\partial y}, \frac{\partial}{\partial z}\right) \tag{2.26}$$

are the covariant components of the derivative 4-vector. Equivalently,

$$\partial^\mu = \frac{\partial}{\partial x_\mu} = \left(\frac{1}{c}\frac{\partial}{\partial t}, -\frac{\partial}{\partial x}, -\frac{\partial}{\partial y}, -\frac{\partial}{\partial z}\right). \tag{2.27}$$

The Lorentz square of the derivative is a 4-scalar, called the d'Alembertian operator, given by

$$\square \equiv \partial^\mu \partial_\mu = \partial_\mu \partial^\mu = \frac{1}{c^2}\frac{\partial^2}{\partial t^2} - \frac{\partial^2}{\partial x^2} - \frac{\partial^2}{\partial y^2} - \frac{\partial^2}{\partial z^2}. \tag{2.28}$$

4-Velocity

In order to define the 4-velocity, u^μ, one has to divide the variation of the 4-vector position by a time interval. The obvious choice is to use the proper time since it is independent of the choice of the reference frame. Hence,

$$u^\mu = \frac{\mathrm{d}x^\mu}{\mathrm{d}\tau} = \gamma \frac{\mathrm{d}x^\mu}{\mathrm{d}t} = \gamma(c, \mathbf{v}). \tag{2.29}$$

One can easily check that $u^2 = c^2$.

4-Momentum

Inspired by classical mechanics, the 4-momentum is defined by the product of the mass with the 4-velocity

$$p^\mu = m\, u^\mu = (\gamma m c, \gamma m \mathbf{v}) = (E/c, \boldsymbol{p}). \tag{2.30}$$

It is clearly a 4-vector since the mass, m, is a 4-scalar independent of the frame,[6] whereas u^μ is a 4-vector. Its norm is $p^2 = m^2 u^2 = m^2 c^2$. Therefore, defining the classical 3-momentum vector as $\boldsymbol{p} = m\boldsymbol{u} = m\gamma\mathbf{v}$ ensures that the law of momentum conservation is consistent with the principle of (special) relativity. In addition, since p^2 is also $E^2/c^2 - |\boldsymbol{p}|^2$, one has the important relation

$$\boxed{E^2 = |\boldsymbol{p}|^2 c^2 + m^2 c^4}. \tag{2.31}$$

A particle satisfying $p^2 = m^2$ is said to be *on the mass shell*. Equation (2.31) is usually called the mass-shell condition. Incidentally, Eq. (2.30) implies

$$|\boldsymbol{p}| = \frac{\beta}{c} E. \tag{2.32}$$

Notice that when $\beta = 1$, $|\boldsymbol{p}| = E/c$, which, inserted in Eq. (2.31), implies that $m = 0$.

[6] There is a source of confusion in some old textbooks (a historical artefact of special relativity), where it is sometimes said that the mass increases with the velocity of the particle. This approach is now completely obsolete and better to avoid. The mass is invariant but not the energy and the momentum that do increase with the velocity.

4-Acceleration

Following the same argument as for the 4-velocity, the 4-acceleration is defined as the variation of the 4-velocity with respect to the proper time, i.e.

$$\Gamma^{\mu} = \frac{\mathrm{d}u^{\mu}}{\mathrm{d}\tau} = \gamma\left(\frac{\mathrm{d}\gamma}{\mathrm{d}t}c, \frac{\mathrm{d}\gamma}{\mathrm{d}t}\mathbf{v} + \gamma \boldsymbol{a}\right), \tag{2.33}$$

where $\boldsymbol{a} = \mathrm{d}^2\boldsymbol{x}/\,\mathrm{d}t^2$ is the classical 3-vector acceleration.

4-Current

Let us consider an electric charge with a constant density ρ in the small element of volume $\mathrm{d}V$ and moving at velocity $\mathbf{v}$. The current density is thus $\boldsymbol{j} = \rho\mathbf{v}$, while the charge is $q = \rho\,\mathrm{d}V$. It would be tempting to define the 4-current as $(q,\boldsymbol{j})$, but such an object is not a 4-vector. Indeed, the charge being an intrinsic property of the particle cannot depend on the reference frame. Therefore, it is invariant and cannot transform as the time component of a 4-vector. However, we must have $\rho\,\mathrm{d}V = \rho'\,\mathrm{d}V'$. In a frame $\mathfrak{R}$, the variation of the spacetime position of the charge during $\mathrm{d}t$ is $\mathrm{d}x^{\mu} = (c\,\mathrm{d}t, \mathrm{d}\boldsymbol{x})$, which is obviously a 4-vector. So the quantity

$$\rho\,\mathrm{d}V\,\mathrm{d}x^{\mu} = \rho\,\mathrm{d}t\,\mathrm{d}V(\mathrm{d}x^{\mu}/\,\mathrm{d}t) = \rho(\mathrm{d}\Omega/c)(\mathrm{d}x^{\mu}/\,\mathrm{d}t)$$

is a 4-vector as well, where $\mathrm{d}\Omega = c\,\mathrm{d}t\,\mathrm{d}V$ is the elementary 'volume' of the spacetime for an infinitesimal time interval $\mathrm{d}t$. With the quantity $\mathrm{d}\Omega$ being a 4-scalar (see Problem 2.3), $\rho(\mathrm{d}x^{\mu}/\,\mathrm{d}t)$ must be a 4-vector. The 4-current is then defined as

$$j^{\mu} = \rho\frac{\mathrm{d}x^{\mu}}{\mathrm{d}t} = (c\rho,\boldsymbol{j}). \tag{2.34}$$

We shall see in Chapter 6 the importance of the 4-current for the description of the interaction between charged particles and photons.

Electromagnetic 4-Potential

In classical electrodynamics, the electric field, $\boldsymbol{E}$, and the magnetic field, $\boldsymbol{B}$, are the solution of Maxwell's equations,

$$\begin{array}{ll} (a)\ \nabla\cdot\boldsymbol{E} = \rho/\epsilon_0 & (c)\ \nabla\cdot\boldsymbol{B} = 0 \\ (b)\ \frac{\partial\boldsymbol{B}}{\partial t} + \nabla\times\boldsymbol{E} = 0 & (d)\ \nabla\times\boldsymbol{B} - \frac{1}{c^2}\frac{\partial\boldsymbol{E}}{\partial t} = \mu_0\boldsymbol{j}, \end{array} \tag{2.35}$$

where ρ and $\boldsymbol{j}$ are the charge density and current density, respectively. Let us postulate the 4-potential from the scalar potential, V, and the vector potential, $\boldsymbol{A}$,

$$A^{\mu} = (V/c, \boldsymbol{A}), \tag{2.36}$$

and the rank 2 electromagnetic tensor with

$$F^{\mu\nu} = \partial^{\mu}A^{\nu} - \partial^{\nu}A^{\mu}. \tag{2.37}$$

We shall see that we can easily recover Maxwell's equations from Eqs. (2.36) and (2.37). Such a reformulation will be explicitly covariant. First, note that $F^{\mu\nu}$ is antisymmetric and, therefore, it depends on only six independent parameters. Making explicit the quantities

$$\begin{aligned} F^{01} &= \partial^0 A^1 - \partial^1 A^0 = 1/c(\partial A_x/\partial t + \partial V/\partial x) \\ F^{02} &= \partial^0 A^2 - \partial^2 A^0 = 1/c(\partial A_y/\partial t + \partial V/\partial y) \\ F^{03} &= \partial^0 A^3 - \partial^3 A^0 = 1/c(\partial A_z/\partial t + \partial V/\partial z), \end{aligned}$$

we can set $F^{01} = 1/c(-E_x)$, $F^{02} = 1/c(-E_y)$ and $F^{03} = 1/c(-E_z)$, such as $\boldsymbol{E}$ satisfies the electric field relation

$$\boldsymbol{E} = -\frac{\partial \boldsymbol{A}}{\partial t} - \boldsymbol{\nabla} V. \tag{2.38}$$

We can set the three other independent parameters to define the magnetic field

$$\begin{aligned} F^{12} &= -(\partial A_x/\partial y - \partial A_y/\partial x) \equiv -B_z \\ F^{13} &= -(\partial A_x/\partial z - \partial A_z/\partial x) \equiv B_y \\ F^{23} &= -(\partial A_y/\partial z - \partial A_z/\partial y) \equiv -B_x. \end{aligned}$$

These three equations are equivalent to

$$\boldsymbol{B} = \boldsymbol{\nabla} \times \boldsymbol{A}. \tag{2.39}$$

Consequently, $F^{\mu\nu}$ reads

$$(F^{\mu\nu}) = \begin{pmatrix} 0 & -E_x/c & -E_y/c & -E_z/c \\ E_x/c & 0 & -B_z & B_y \\ E_y/c & B_z & 0 & -B_x \\ E_z/c & -B_y & B_x & 0 \end{pmatrix}. \tag{2.40}$$

The combination of Eqs. (2.38) and (2.39) leads to Maxwell's Equation (2.35-b), $\frac{\partial \boldsymbol{B}}{\partial t} + \boldsymbol{\nabla} \times \boldsymbol{E} = 0$, while Eq. (2.39) implies Eq. (2.35-c), $\boldsymbol{\nabla} \cdot \boldsymbol{B} = 0$. Actually, both Eqs. (2.35-b) and (2.35-c) can be recast into the single equation

$$\partial^\mu F^{\nu\rho} + \partial^\nu F^{\rho\mu} + \partial^\rho F^{\mu\nu} = 0. \tag{2.41}$$

For example, the x-component of Eq. (2.35-b), $\partial B_x/\partial t + \partial E_z/\partial y - \partial E_y/\partial z = 0$, corresponds to the choice $\mu = 0, \nu = 3, \rho = 2$, while Eq. (2.35-c) is obtained with $\mu = 1, \nu = 3, \rho = 2$. To find Eqs. (2.35-a) and (2.35-d), we just have to write

$$\partial_\mu F^{\mu\nu} = \mu_0 j^\nu, \tag{2.42}$$

with j^ν being the 4-current. Let us check the component $\nu = 0$ for which $j^0 = c\rho$. The left-hand side of Eq. (2.42) reads

$$\partial_\mu F^{\mu 0} = \frac{1}{c}\frac{\partial E_x}{\partial x} + \frac{1}{c}\frac{\partial E_y}{\partial y} + \frac{1}{c}\frac{\partial E_z}{\partial z} = \frac{1}{c}\boldsymbol{\nabla} \cdot \boldsymbol{E}.$$

Recalling the formula $\epsilon_0 \mu_0 = 1/c^2$, it yields $\boldsymbol{\nabla} \cdot \boldsymbol{E} = \rho/\epsilon_0$, i.e. Maxwell's Equation (2.35-a). Similarly, the three components of Maxwell's Equation (2.35-d) are obtained by setting $\nu = 1, 2$ and 3 in Eq. (2.42).

Before concluding this section, it is worth recalling how the electric and magnetic fields are affected by a change of reference frame. Both are entangled, and, for instance, a pure electric field in one frame may be seen as a magnetic field (or a mixture of electric and magnetic fields) in another frame and vice versa. Technically, the transformation of the fields can be obtained by applying the Lorentz transformation to the electromagnetic tensor, i.e.

$$F'^{\mu\nu} = \Lambda^{\mu}{}_{\rho}\Lambda^{\nu}{}_{\sigma}F^{\rho\sigma} = \Lambda^{\mu}{}_{\rho}\Lambda^{\intercal}{}_{\sigma}{}^{\nu}F^{\rho\sigma} = \Lambda^{\mu}{}_{\rho}F^{\rho\sigma}\Lambda^{\intercal}{}_{\sigma}{}^{\nu},$$

which, in matrix notation, reads $F' = \Lambda F \Lambda^{\intercal}$. The components of $\boldsymbol{E}'$ and $\boldsymbol{B}'$ are then read from F' with Eq. (2.40). An alternative is to transform the 4-vector, A^{μ}, with $A'^{\mu} = \Lambda^{\mu}{}_{\nu}A^{\nu}$, and apply Eqs. (2.38) and (2.39). Beware that the derivatives present in those equations are then taken with respect to the new coordinates. It thus requires another Lorentz transformation to express the derivatives in the original frame.

2.2.5 Rapidity

In special relativity, the rapidity ζ is a hyperbolic angle defined by

$$\zeta = \tanh^{-1}\beta, \tag{2.43}$$

which implies

$$\beta = \tanh\zeta, \quad \gamma = \cosh\zeta, \quad \beta\gamma = \sinh\zeta. \tag{2.44}$$

Unlike the velocities, when two frames $\mathfrak{R}'$ and $\mathfrak{R}''$ are moving in the same direction with respect to $\mathfrak{R}$, the rapidities of two Lorentz transformations just add up,

$$\zeta_{\mathfrak{R}''/\mathfrak{R}} = \zeta_{\mathfrak{R}''/\mathfrak{R}'} + \zeta_{\mathfrak{R}'/\mathfrak{R}}. \tag{2.45}$$

Incidentally, the addition of rapidities provides an easy way to remember the addition of relativistic velocities,[7]

$$\beta_{\mathfrak{R}''/\mathfrak{R}} = \tanh(\tanh^{-1}\beta_{\mathfrak{R}''/\mathfrak{R}'} + \tanh^{-1}\beta_{\mathfrak{R}'/\mathfrak{R}}) = \frac{\beta_{\mathfrak{R}''/\mathfrak{R}'} + \beta_{\mathfrak{R}'/\mathfrak{R}}}{1 + \beta_{\mathfrak{R}''/\mathfrak{R}'}\beta_{\mathfrak{R}'/\mathfrak{R}}}. \tag{2.46}$$

As $\tanh\zeta = (e^{\zeta} - e^{-\zeta})/(e^{\zeta} + e^{-\zeta})$, the rapidity in Eq. (2.43) can be expressed as a function of β with

$$\zeta = \frac{1}{2}\ln\left(\frac{1+\beta}{1-\beta}\right). \tag{2.47}$$

Hence, at a low velocity, $\beta \ll 1$, the rapidity,

$$\zeta = \frac{1}{2}\left[\ln(1+\beta) - \ln(1-\beta)\right] \approx \beta,$$

converges to the velocity. According to the addition of rapidities (2.46), we thus recover the usual Galilean addition of velocities at low velocity.

We can apply the addition of rapidities to a frequent case in particle physics. Imagine that we have a beam of incident particles that strikes a target. The collision generally produces

[7] Recall the trigonometric relation, $\tanh(a+b) = [\tanh(a) + \tanh(b)]/[1 + \tanh(a)\tanh(b)]$.

many particles. Assume that the rapidity ζ of one of those particles is measured in the laboratory frame. We might be interested in the value of the rapidity ζ' in the centre-of-mass frame of the collision (simply because theoretical calculations are generally performed in that frame). As the target is at rest, the centre-of-mass frame necessarily moves along the beam axis, let us say at a velocity β. According to Eq. (2.45), the rapidity ζ' satisfies

$$\zeta = \zeta' + \tanh^{-1} \beta, \tag{2.48}$$

if and only if the particle travels along the beam axis, which is restrictive. To circumvent this constraint, experimental particle physicists use a modified definition of rapidity by considering only the longitudinal component of the particle momentum along the beam axis, p_L. Let us see how. Using Eq. (2.32) to express the particle velocity as a function of its energy and momentum, the rapidity (2.47) reads

$$\zeta = \frac{1}{2} \ln \left(\frac{E + |\boldsymbol{p}|c}{E - |\boldsymbol{p}|c} \right).$$

Now, if instead of $|\boldsymbol{p}|$, we use the longitudinal component of the momentum along the beam axis p_L, the modified rapidity is

$$\zeta_L = \frac{1}{2} \ln \left(\frac{E + p_L c}{E - p_L c} \right). \tag{2.49}$$

This rapidity would by definition correspond to that of a particle moving along the same axis as the centre-of-mass frame. Therefore, the additivity of rapidities remains valid, and Eq. (2.48) can be used. Since in Eq. (2.48), β is a constant that only depends on the characteristic of the beam and the target, we conclude that $\mathrm{d}\zeta'_L = \mathrm{d}\zeta_L$, i.e. the variation of rapidities is Lorentz invariant. Consequently, the rapidity distribution, i.e. the number of particles in a given rapidity interval, $\mathrm{d}N/\,\mathrm{d}\zeta_L$, is thus Lorentz invariant too and can eventually be compared directly to the predictions of a theoretical model.

2.3 Angular Momentum of Particles

The purpose of this section is to remind the reader of important properties of the angular momentum in quantum mechanics that are often used in particle physics. More thorough derivations can be found in quantum mechanics textbooks, such as Griffiths (1995) or Cohen-Tannoudji et al. (1977).

2.3.1 Orbital Angular Momentum in Quantum Mechanics

In classical mechanics, the angular momentum is defined as $\boldsymbol{L} = \boldsymbol{r} \times \boldsymbol{p}$. It depends on six numbers r_x, r_y, r_z, p_x, p_y and p_z. In quantum mechanics, due to the Heisenberg uncertainty principle, it is not possible to simultaneously measure these six numbers with arbitrary

Supplement 2.1. Pseudorapidity

In Formula (2.49), the mass and the momentum of the particle have to be known (or equivalently, both E and p_L have to be measured) independently. However, as soon as $E \gg mc^2, E \simeq |\boldsymbol{p}|c$ and

$$p_L = |\boldsymbol{p}| \cos\theta \simeq \frac{E}{c} \cos\theta,$$

with the angle θ being the angle between the particle momentum and the beam axis. The rapidity (2.49) then approaches the pseudorapidity, η, defined as

$$\eta = \frac{1}{2} \ln\left(\frac{1+\cos\theta}{1-\cos\theta}\right) = -\ln\left(\tan\frac{\theta}{2}\right). \tag{2.50}$$

When θ is close to zero, the particle is on the beam axis, and η diverges to infinity. By construction, η and ζ are very close as soon as $E \gg mc^2$, and the angle of the particle with respect to the beam axis is large enough. The pseudorapidity has the advantage of not depending on the mass and thus on the species of the particle, which is not necessarily known in experiments. Only the angle θ has to be measured. However, whereas the behaviour of the rapidity is clear under a Lorentz boost, $\zeta_2' - \zeta_1' = \zeta_2 - \zeta_1$, it is not the case for the pseudorapidity, where $\eta_2' - \eta_1' \neq \eta_2 - \eta_1$.

precision. It turns out that the best that one can do is to simultaneously measure both the magnitude of the angular momentum vector and its component along one axis. The angular momentum is quantised and defined as an operator acting on the wave function. Its expression is

$$\hat{\boldsymbol{L}} = \hat{\boldsymbol{r}} \times \hat{\boldsymbol{p}} = -i\hbar\, \hat{\boldsymbol{r}} \times \boldsymbol{\nabla}, \tag{2.51}$$

where the correspondence principle stating that $\hat{\boldsymbol{p}} = -i\hbar\boldsymbol{\nabla}$ (see Chapter 5) is used to express the angular momentum operator. It satisfies the following canonical commutation relations expressing the impossibility to simultaneously measure all the angular momentum components:

$$[\hat{L}_x, \hat{L}_y] = i\hbar\hat{L}_z,\ [\hat{L}_y, \hat{L}_z] = i\hbar\hat{L}_x,\ [\hat{L}_z, \hat{L}_x] = i\hbar\hat{L}_y. \tag{2.52}$$

These three commutation relations can simply be grouped into

$$[\hat{L}_i, \hat{L}_j] = i\hbar\epsilon_{ijk}\hat{L}_k,$$

with the symbol ϵ_{ijk} denoting the spatial antisymmetric tensor, satisfying

$$\begin{aligned} \epsilon_{123} &= \epsilon_{231} = \epsilon_{312} = 1; \\ \epsilon_{213} &= \epsilon_{132} = \epsilon_{321} = -1; \\ \epsilon_{ijk} &= 0 \text{ if any two indices are equal.} \end{aligned} \tag{2.53}$$

It follows that $\hat{L}^2 = \hat{L}_x^2 + \hat{L}_y^2 + \hat{L}_z^2$ satisfies

$$[\hat{L}^2, \hat{L}_{x,y,z}] = 0. \tag{2.54}$$

Consequently, one can define a basis of eigenvectors of both $\hat{L}^2$ and $\hat{L}_z$. These eigenvectors, $|l, m\rangle$, labelled by the quantum number of the angular momentum magnitude, l, and the

Table 2.1 Spherical harmonics Y_l^m.

$m\backslash l$	0	1	2
0	$\sqrt{\frac{1}{4\pi}}$	$\sqrt{\frac{3}{4\pi}}\cos\theta$	$\sqrt{\frac{5}{4\pi}}\left(\frac{3}{2}\cos^2\theta-\frac{1}{2}\right)$
1		$\sqrt{\frac{3}{8\pi}}\sin\theta e^{i\phi}$	$-\sqrt{\frac{15}{8\pi}}\sin\theta\cos\theta e^{i\phi}$
2			$\frac{1}{4}\sqrt{\frac{15}{2\pi}}\sin^2\theta e^{2i\phi}$.

quantum number of the projection of the angular momentum on the z-axis,[8] m, satisfy

$$\begin{aligned}\hat{L}^2\,|l,m\rangle &= l(l+1)\hbar^2\,|l,m\rangle\,,\\ \hat{L}_z\,|l,m\rangle &= m\hbar\,|l,m\rangle\,,\ m\in[-l,l].\end{aligned} \tag{2.55}$$

For the orbital angular momentum, l and m are two integers. Using the spherical coordinates

$$x=r\sin\theta\cos\phi,\ y=r\sin\theta\sin\phi,\ z=r\cos\theta,$$

with θ and ϕ being the polar and azimuthal angles, respectively; it follows from the expression of the operator of the orbital angular momentum (2.51) that

$$\hat{L}^2=-\hbar^2\left(\frac{\partial^2}{\partial\theta^2}+\frac{1}{\tan\theta}\frac{\partial}{\partial\theta}+\frac{1}{\sin^2\theta}\frac{\partial^2}{\partial\phi^2}\right),\ \ \hat{L}_z=-i\hbar\frac{\partial}{\partial\phi}. \tag{2.56}$$

The spherical harmonics $Y_l^m(\theta,\phi)=\langle\theta,\phi|l,m\rangle$ are the eigenfunctions of the orbital angular momentum and thus satisfy Eq. (2.55) with the operators (2.56). The expressions of the most useful spherical harmonics for particles physics are given in Table 2.1. In general, the spherical harmonics can be expressed in terms of the associated Legendre polynomials $P_l^m(x)$,

$$Y_l^m(\theta,\phi)=(-1)^m\sqrt{\frac{(2l+1)}{4\pi}\frac{(l-|m|)!}{l+|m|)!}}P_l^{|m|}(\cos\theta)e^{im\phi}, \tag{2.57}$$

where

$$P_l^m(x)=\frac{1}{2^l l!}(1-x^2)^{m/2}\frac{\mathrm{d}^{(l+m)}}{\mathrm{d}x^{l+m}}(x^2-1)^l. \tag{2.58}$$

Using the two previous formulas, one can easily check that

$$Y_l^{-m}(\theta,\phi)=(-1)^m\left[Y_l^m(\theta,\phi)\right]^*,\quad m\geq 0.$$

This allows us to deduce the expression of the spherical harmonics for $m<0$.

[8] The z-axis is chosen by convention.

2.3.2 Spin Angular Momentum

As the name suggests, spin was originally conceived as the rotation of a particle about its axis. This name is unfortunate: elementary particles are described as point-like objects for which 'rotation about its axis' is meaningless. One has to give up the classical representation and just admit that spin is a kind of angular momentum in the sense that it obeys the same mathematical laws that quantised orbital angular momentum, i.e.

$$\begin{aligned} \hat{S}^2 \left|s, m\right\rangle &= s(s+1)\hbar^2 \left|s, m\right\rangle , \\ \hat{S}_z \left|s, m\right\rangle &= m\hbar \left|s, m\right\rangle , \ m \in [-s, s]. \end{aligned} \tag{2.59}$$

However, spin has peculiar properties that distinguish it from orbital angular momenta. The spin quantum number, s, may take positive half-integer values (for fermions). Spin is often presented as the *intrinsic* angular momentum, suggesting that it is a residual angular momentum even if the particle is at rest (and hence, $p = 0$). This image is not appropriate for massless particles that always propagate at the velocity of light.

A case of utmost importance is the spin 1/2 (all elementary fermions of the Standard Model have spin 1/2). Such a particle has two possible projections, $|\frac{1}{2}, +\frac{1}{2}\rangle$ and $|\frac{1}{2}, -\frac{1}{2}\rangle$, which can be gathered into a two-component notation named the Pauli spinor,

$$\left|\frac{1}{2}, +\frac{1}{2}\right\rangle = \begin{pmatrix} 1 \\ 0 \end{pmatrix} \equiv |\uparrow\rangle_z , \quad \left|\frac{1}{2}, -\frac{1}{2}\right\rangle = \begin{pmatrix} 0 \\ 1 \end{pmatrix} \equiv |\downarrow\rangle_z ,$$

with

$$\hat{S}_z |\uparrow\rangle_z = \frac{\hbar}{2} |\uparrow\rangle_z , \quad \hat{S}_z |\downarrow\rangle_z = -\frac{\hbar}{2} |\downarrow\rangle_z .$$

The states $|\uparrow\rangle_z$ and $|\downarrow\rangle_z$ are thus the eigenstates of the $\hat{S}_z$ operator for which a measurement of the spin along the z-axis always gives the same value (always $+\hbar/2$ or always $-\hbar/2$). However, these states are not eigenstates of the spin operators $\hat{S}_x$ or $\hat{S}_y$, corresponding respectively to a spin measurement along the x- or y-axis [see, for instance, Cohen-Tannoudji et al. (1977) or Appendix B]. More specifically,

$$\hat{S}_x |\uparrow\rangle_z = \frac{\hbar}{2} |\downarrow\rangle_z , \quad \hat{S}_x |\downarrow\rangle_z = \frac{\hbar}{2} |\uparrow\rangle_z , \quad \hat{S}_y |\uparrow\rangle_z = i\frac{\hbar}{2} |\downarrow\rangle_z , \quad \hat{S}_y |\downarrow\rangle_z = -i\frac{\hbar}{2} |\uparrow\rangle_z . \tag{2.60}$$

Then, the mean value of the spin on the x- and y-axes for a system prepared in $|\uparrow\rangle_z$ or $|\downarrow\rangle_z$ gives 0 (i.e., ${}_z\langle\uparrow |\hat{S}_x| \uparrow\rangle_z = {}_z\langle\downarrow |\hat{S}_x| \downarrow\rangle_z = 0$). The eigenstates of the spin operators $\hat{S}_x$ and $\hat{S}_y$ instead are the linear combinations

$$\begin{aligned} |\uparrow\rangle_x &= (|\uparrow\rangle_z + |\downarrow\rangle_z)/\sqrt{2}, \qquad & |\downarrow\rangle_x &= (|\uparrow\rangle_z - |\downarrow\rangle_z)/\sqrt{2}, \\ |\uparrow\rangle_y &= (|\uparrow\rangle_z + i\,|\downarrow\rangle_z)/\sqrt{2}, \qquad & |\downarrow\rangle_y &= (|\uparrow\rangle_z - i\,|\downarrow\rangle_z)/\sqrt{2}. \end{aligned}$$

The projection on the three different axes is obtained from Eq. (2.60), using the spinor notation, with 2×2 matrices, called the Pauli matrices (see Appendix B), where

$$\hat{\mathbf{S}} = \frac{\hbar}{2}\boldsymbol{\sigma}. \tag{2.61}$$

This vectorial notation actually means

$$\hat{S}_x = \frac{\hbar}{2}\sigma^1, \quad \hat{S}_y = \frac{\hbar}{2}\sigma^2, \quad \hat{S}_z = \frac{\hbar}{2}\sigma^3,$$

where $\sigma^{i=1,2,3}$ are given by[9]

$$\sigma^1 = \begin{pmatrix} 0 & 1 \\ 1 & 0 \end{pmatrix}, \quad \sigma^2 = \begin{pmatrix} 0 & -i \\ i & 0 \end{pmatrix}, \quad \sigma^3 = \begin{pmatrix} 1 & 0 \\ 0 & -1 \end{pmatrix}. \tag{2.62}$$

If instead of the x-, y- or z-axis, the spin measurement is performed along an arbitrary axis $\boldsymbol{u} = (\cos\varphi \sin\theta, \sin\varphi \sin\theta, \cos\theta)^{\mathsf{T}}$, then the corresponding spin operator is simply

$$\hat{S}_{\boldsymbol{u}} = \hat{\boldsymbol{S}} \cdot \boldsymbol{u} = \hat{S}_x u_x + \hat{S}_y u_y + \hat{S}_z u_z = \frac{\hbar}{2} \begin{pmatrix} \cos\theta & e^{-i\varphi} \sin\theta \\ e^{i\varphi} \sin\theta & -\cos\theta \end{pmatrix}. \tag{2.63}$$

It follows that the eigenstates of $\hat{S}_{\boldsymbol{u}}$ are then given by

$$\begin{aligned} |\uparrow\rangle_{\boldsymbol{u}} &= e^{-i\frac{\varphi}{2}} \cos\tfrac{\theta}{2} |\uparrow\rangle_z + e^{+i\frac{\varphi}{2}} \sin\tfrac{\theta}{2} |\downarrow\rangle_z, \\ |\downarrow\rangle_{\boldsymbol{u}} &= -e^{-i\frac{\varphi}{2}} \sin\tfrac{\theta}{2} |\uparrow\rangle_z + e^{+i\frac{\varphi}{2}} \cos\tfrac{\theta}{2} |\downarrow\rangle_z. \end{aligned} \tag{2.64}$$

The reader can easily check that the $\hat{\mathbf{S}}$ operators in Eq. (2.61) satisfy all the conditions of the angular momentum operators. They obey the commutation relation $[\hat{S}_x, \hat{S}_y] = i\hbar\hat{S}_z$, and the eigenstates of $\hat{S}^2 = \hat{S}_x^2 + \hat{S}_y^2 + \hat{S}_z^2$ and $\hat{S}_z$ are the spin-up and spin-down states with respective eigenvalues $s(s+1)\hbar^2$ and $\pm s\hbar$, where $s = 1/2$. They represent the spin operators for spin-1/2 states.

2.3.3 Total Angular Momentum and Addition of Angular Momenta

The total angular momentum is the sum of the orbital angular momentum and spin

$$\hat{\boldsymbol{J}} = \hat{\boldsymbol{L}} + \hat{\boldsymbol{S}},$$

with the operator $\hat{J}$ satisfying

$$\begin{aligned} \hat{J}^2 |j,m\rangle &= j(j+1)\hbar^2 |j,m\rangle, \\ \hat{J}_z |j,m\rangle &= m\hbar |j,m\rangle, \quad m \in [-j,j]. \end{aligned} \tag{2.65}$$

The quantum numbers j and m are integers or half-integers. The dimension of the Hilbert space $\mathcal{H}(j)$ generated by the states of the total angular momentum is $2j+1$.

When two angular momenta $\boldsymbol{J_1}$ and $\boldsymbol{J_2}$ are added, the new space is the tensorial product $\mathcal{H}_1(j_1) \otimes \mathcal{H}_2(j_2)$, of dimension $(2j_1+1) \times (2j_2+1)$, for which a complete basis can be denoted by the eigenvectors

$$|j_1, j_2, m_1, m_2\rangle \equiv |j_1, m_1\rangle \otimes |j_2, m_2\rangle.$$

By construction, they satisfy

$$\hat{J}_i^2 |j_1, j_2, m_1, m_2\rangle = j_i(j_i+1)\hbar^2 |j_1, j_2, m_1, m_2\rangle, \quad \hat{J}_{i_z} |j_1, j_2, m_1, m_2\rangle = m_i\hbar |j_1, j_2, m_1, m_2\rangle,$$

with $i = 1$ or 2. Another complete basis of this space exists, which corresponds to the total angular momentum $\boldsymbol{J} = \boldsymbol{J_1} + \boldsymbol{J_2}$. This basis is generated by the eigenvectors of $\hat{J}_1^2, \hat{J}_2^2, \hat{J}^2$ and $\hat{J}_z$ that commute among each other. The eigenstates are denoted as

$$|j_1, j_2, j, m\rangle, \text{ with } j \in j_1 + j_2,\ j_1 + j_2 - 1,\ \ldots,\ |j_1 - j_2| \text{ and } m = m_1 + m_2 \in [-j, j].$$

[9] Pauli matrices are often denoted as σ_x, σ_y and σ_z.

It follows that

$$\hat{J}_1^2 \, |j_1,j_2,j,m\rangle = j_1(j_1+1)\hbar^2 \, |j_1,j_2,j,m\rangle\,, \quad \hat{J}_2^2 \, |j_1,j_2,j,m\rangle = j_2(j_2+1)\hbar^2 \, |j_1,j_2,j,m\rangle\,,$$
$$\hat{J}^2 \, |j_1,j_2,j,m\rangle = j(j+1)\hbar^2 \, |j_1,j_2,j,m\rangle\,, \qquad \hat{J}_z \, |j_1,j_2,j,m\rangle = m\hbar \, |j_1,j_2,j,m\rangle\,.$$

In other words, the resulting space is decomposed as

$$\mathcal{H}_1(j_1) \bigotimes \mathcal{H}_2(j_2) = \bigoplus_{j=|j_1-j_2|}^{j_1+j_2} \mathcal{H}(j).$$

The relation between the basis $|j_1,j_2,j,m\rangle$ and $|j_1,j_2,m_1,m_2\rangle$ is given by the *Clebsch-Gordan* coefficients, chosen as real numbers and denoted as $\langle j_1,j_2,m_1,m_2|j_1,j_2,j,m\rangle$

$$|j_1,j_2,j,m\rangle = \sum_{m_1=-j_1}^{j_1} \sum_{m_2=-j_2}^{j_2} \langle j_1,j_2,m_1,m_2|j_1,j_2,j,m\rangle \, |j_1,j_2,m_1,m_2\rangle\,,$$

where the non-zero coefficients are those for which $m = m_1 + m_2$. Starting from the state with the highest weight, i.e. $|j_1,j_2,j,\ m=j\rangle$ and $|j_1,j_2,\ m_1=j_1,\ m_2=j_2\rangle$, the coefficients are obtained by applying as many times as necessary the ladder operator $\hat{J}_- = \hat{J}_{1-} + \hat{J}_{2-}$, where $\hat{J}_{1-} = \hat{J}_{1x} - i\hat{J}_{1y}$ (and similarly for $\hat{J}_{2-}$), and using orthogonality between the states. The most useful coefficients for particle physics are given concisely in Fig. 2.1. Looking, for example, at the table labelled $1/2 \times 1/2$, it is easy to check that the addition of two spins 1/2 ($j_1 = j_2 = 1/2$) gives the following well-known expression of $|j,m\rangle$ as a function of $|m_1,m_2\rangle$:

$$|1,1\rangle = |+\tfrac{1}{2},+\tfrac{1}{2}\rangle, \qquad |1,-1\rangle = |-\tfrac{1}{2},-\tfrac{1}{2}\rangle,$$
$$|1,0\rangle = \tfrac{1}{\sqrt{2}}\,|+\tfrac{1}{2},-\tfrac{1}{2}\rangle + \tfrac{1}{\sqrt{2}}\,|-\tfrac{1}{2},+\tfrac{1}{2}\rangle, \qquad |0,0\rangle = \tfrac{1}{\sqrt{2}}\,|+\tfrac{1}{2},-\tfrac{1}{2}\rangle - \tfrac{1}{\sqrt{2}}\,|-\tfrac{1}{2},+\tfrac{1}{2}\rangle\,.$$

Supplement 2.2. Spin of Composite Particles

The spin of a composite particle is the sum of the spins and orbital angular momenta of its constituents. Here, the designation 'constituent' is general. For instance, for a hadron, it means its valence quarks but also quarks or gluons from quantum fluctuations. It is known nowadays that the spin of hadrons cannot be explained only by the contribution of the valence quarks, as we shall see in Chapter 7.

2.3.4 Effect of Rotations on Angular Momentum States

Let us start with the rotations in the usual $\mathbb{R}^3$ Euclidian space. In general, a rotation $\mathcal{R}(\boldsymbol{n},\Theta)$ can be defined by two quantities: the rotation axis $\boldsymbol{n}$ and the rotation angle Θ. Three real numbers are thus needed: the two polar coordinates (φ,θ) of the rotation axis and Θ. Two viewpoints can be adopted: active or passive. An active rotation, denoted by the label (a) in the following equation, is a vector rotation in a fixed coordinate frame $(\boldsymbol{e}_x,\boldsymbol{e}_y,\boldsymbol{e}_z)$. In cartesian coordinates, a vector $\boldsymbol{r_1}$ becomes $\boldsymbol{r_2}$, with

$$\boldsymbol{r_1} = x_1\boldsymbol{e}_x + y_1\boldsymbol{e}_y + z_1\boldsymbol{e}_z \rightarrow \boldsymbol{r_2} = \mathcal{R}^{(a)}(\boldsymbol{n},\Theta)\boldsymbol{r_1} = x_2\boldsymbol{e}_x + y_2\boldsymbol{e}_y + z_2\boldsymbol{e}_z.$$

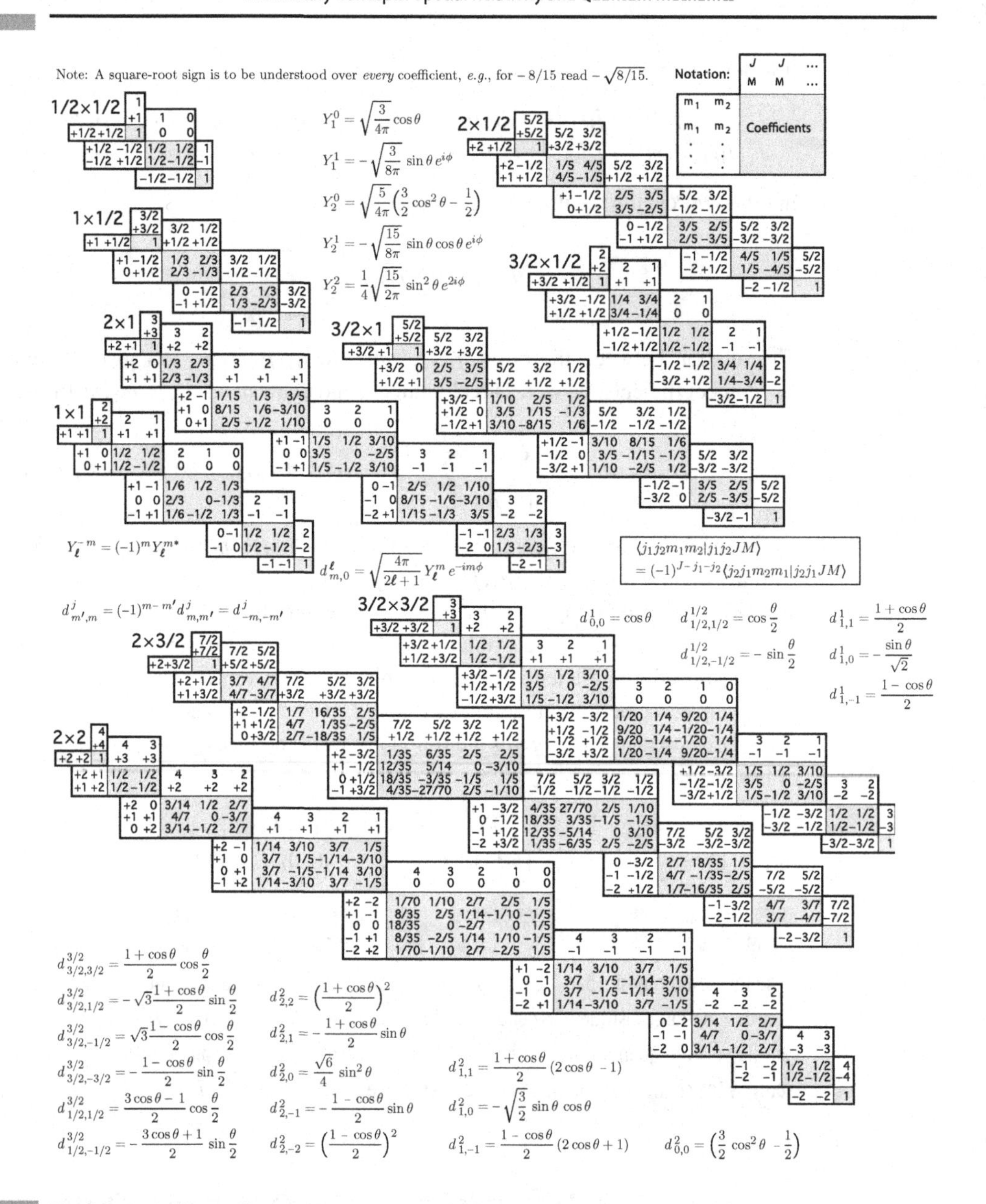

Fig. 2.1 The Clebsch–Gordan coefficients, spherical harmonics and d-functions. Image from R. L. Workman et al. (Particle Data Group, 2022).

In contrast, a passive rotation denoted by the label (p) is a rotation of the coordinate system itself, whereas the vector remains fixed. Its coordinates are changed in the new frame. Denoting the rotated frame by (e'_x, e'_y, e'_z), the vector $\boldsymbol{r}_1$ becomes $\boldsymbol{r}'_1$, with

$$\boldsymbol{r}_1 = x_1\boldsymbol{e}_x + y_1\boldsymbol{e}_y + z_1\boldsymbol{e}_z \rightarrow \boldsymbol{r}'_1 = \mathcal{R}^{(p)}(\boldsymbol{n}, \Theta)\boldsymbol{r}_1 = x'_1\boldsymbol{e}'_x + y'_1\boldsymbol{e}'_y + z'_1\boldsymbol{e}'_z.$$

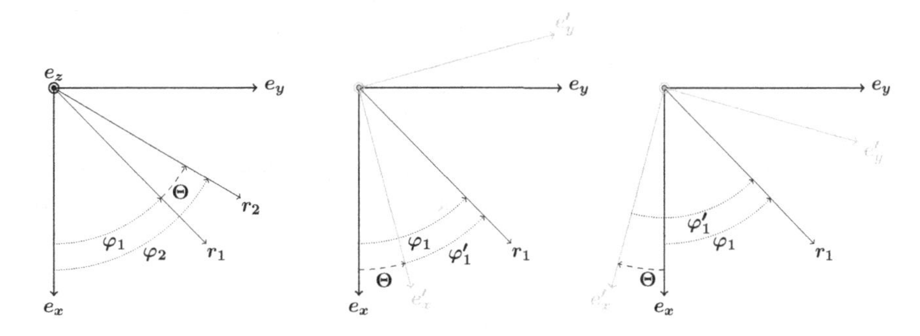

Fig. 2.2 A rotation with an angle Θ around the z-axis (perpendicular to the figure plane). Left: active rotation with an angle Θ applied on the vector $\boldsymbol{r}_1$. Middle: passive rotation with an angle Θ applied on the frame in black. Right: passive rotation with an angle $-\Theta$.

Note that the geometrical rotation described by $\mathcal{R}^{(a)}(\boldsymbol{n},\Theta)$ is not equivalent to that of $\mathcal{R}^{(p)}(\boldsymbol{n},\Theta)$. Indeed, as illustrated in Fig. 2.2, both geometrical rotations are equivalent provided we change the sign of the rotation angle, i.e.

$$\mathcal{R}^{(a)}(\boldsymbol{n},\Theta) \Leftrightarrow \mathcal{R}^{(p)}(\boldsymbol{n},-\Theta). \tag{2.66}$$

The coordinates in the left-hand sketch are equal to those of the right-hand sketch. For example, the angle φ'_1 of $\boldsymbol{r_1}$ in the rotated frame (grey) is the angle φ_2 of the rotated vector $\boldsymbol{r_2}$ in the fixed frame (black). Therefore, both rotations $\mathcal{R}^{(a)}(\boldsymbol{n},\Theta)$ and $\mathcal{R}^{(p)}(\boldsymbol{n},-\Theta)$ are represented by the same orthogonal matrix, $R(\boldsymbol{n},\Theta)$, acting on the components of the vectors. The expression of the matrix depends on the parametrisation used to describe the rotation. A frequent parametrisation uses the Euler angles. In this formulation, an arbitrary rotation is carried out via three successive rotations about prescribed axes. Several equivalent possibilities exist. For example: rotate first through an angle $\psi > 0$ about the fixed axis $\boldsymbol{e}_z$, then rotate through $\theta > 0$ about the fixed axis $\boldsymbol{e}_y$ and finally rotate again about the fixed axis $\boldsymbol{e}_z$ through an angle $\varphi > 0$. The final rotation matrix is then

$$\begin{aligned} R(\varphi,\theta,\psi) &= R(\boldsymbol{e}_z,\varphi) \times R(\boldsymbol{e}_y,\theta) \times R(\boldsymbol{e}_z,\psi) \\ &= \begin{pmatrix} \cos\varphi & -\sin\varphi & 0 \\ \sin\varphi & \cos\varphi & 0 \\ 0 & 0 & 1 \end{pmatrix} \begin{pmatrix} \cos\theta & 0 & \sin\theta \\ 0 & 1 & 0 \\ -\sin\theta & 0 & \cos\theta \end{pmatrix} \begin{pmatrix} \cos\psi & -\sin\psi & 0 \\ \sin\psi & \cos\psi & 0 \\ 0 & 0 & 1 \end{pmatrix}. \end{aligned} \tag{2.67}$$

Let us now move to the rotation of a quantum mechanics system. The rotation of a system in the $\mathbb{R}^3$ Euclidian space must have a counterpart in the Hilbert space of quantum states. We focus here on angular momentum states. Consider an angular momentum state, $|j,m\rangle_z$, where the reference axis is $\boldsymbol{e}_z$. This state satisfies the relations $\hat{J}^2|j,m\rangle_z = j(j+1)\hbar^2|j,m\rangle_z$ and $\hat{J}_z|j,m\rangle_z = m\hbar|j,m\rangle_z$. After a rotation performed in $\mathbb{R}^3$, the state must still have the projection $m\hbar$, but this time with respect to the rotated axis $\boldsymbol{e}'_z = \mathcal{R}\boldsymbol{e}_z$. Let us denote the state after rotation by $|j,m\rangle_{z'}$. It is no longer an eigenstate of $\hat{J}_z$, but rather an eigenstate of $\hat{J}_{z'}$, with $\hat{J}_{z'}|j,m\rangle_{z'} = m\hbar|j,m\rangle_{z'}$. We do not expect the rotation to change the norm of the angular momentum, so $|j,m\rangle_{z'}$ is still an eigenstate of $\hat{J}^2$, with $\hat{J}^2|j,m\rangle_{z'} = j(j+1)\hbar^2|j,m\rangle_{z'}$.

How is $|j,m\rangle_{z'}$ related to $|j,m\rangle_z$? There must be a linear operator, $\hat{R}$, describing the action of the rotation in the Hilbert space, i.e.

$$|j,m\rangle_{z'} = \hat{R}\,|j,m\rangle_z\,. \tag{2.68}$$

One can show [see Appendix A for an explanation with a group theory or a quantum mechanics textbook, such as Cohen-Tannoudji et al. (1977)] that for a rotation through an angle Θ about the axis $\boldsymbol{n}$, the operator takes the form[10]

$$\hat{R}(\boldsymbol{n},\Theta) = e^{-i\frac{\Theta}{\hbar}\boldsymbol{n}\cdot\hat{\boldsymbol{J}}}. \tag{2.69}$$

Note that this is the operator corresponding to an active rotation of angle Θ. Due to the equivalence given in Eq. (2.66), it is also the operator for a passive rotation of angle $-\Theta$. A passive rotation with an angle Θ would thus have a plus sign in the exponential, and consequently, its operator would be $\hat{R}^\dagger$ (since $\hat{J}$ is Hermitian). Parametrising the (active) rotation with the Euler angles, the operator is then

$$\hat{R}(\varphi,\theta,\psi) = e^{-i\frac{\varphi}{\hbar}\hat{J}_z}e^{-i\frac{\theta}{\hbar}\hat{J}_y}e^{-i\frac{\psi}{\hbar}\hat{J}_z}. \tag{2.70}$$

This formula is just the counterpart in the Hilbert space of Formula (2.67) used in $\mathbb{R}^3$. Using the completeness of the angular momentum basis $\{|j,m\rangle_z\}$, Eq. (2.68) becomes

$$|j,m\rangle_{z'} = \hat{R}\,|j,m\rangle_z = \sum_{m'=-j}^{j} |j,m'\rangle_{zz}\langle j,m'|\hat{R}\,|j,m\rangle_z = \sum_{m'=-j}^{j} |j,m'\rangle_z D^j_{m',m}, \tag{2.71}$$

where the quantities

$$D^j_{m',m} = {}_z\langle j,m'|\hat{R}\,|j,m\rangle_z$$

are called the *Wigner* rotation D-functions.[11] They depend on the Euler angles, and their interpretation is the following: the left-hand side of Eq. (2.71) describes a rotated ket, which is an eigenvector of $J_{z'}$, where the axis is the rotated axis, z'. On this axis, its eigenvalue is $m\hbar$. The kets on the right-hand side are the eigenvectors of J_z, with eigenvalue $m'\hbar$ on the initial z-axis. Thus, $D^j_{m',m}$ is simply the amplitude to measure the value $m'\hbar$ along the z-axis when the angular momentum points in the direction of the rotated axis with component $m\hbar$. Using the Euler angle rotation operator (2.70), the expression of $D^j_{m',m}$ can be simplified since $|j,m\rangle_z$ and $|j,m'\rangle_z$ are the eigenvectors of J_z, yielding

$$D^j_{m',m}(\varphi,\theta,\psi) = e^{-i\varphi m'}\, d^j_{m',m}(\theta)\, e^{-i\psi m}, \tag{2.72}$$

where the d-functions

$$d^j_{m',m}(\theta) = {}_z\langle j,m'|e^{-i\frac{\theta}{\hbar}\hat{J}_y}\,|j,m\rangle_z \tag{2.73}$$

are the *reduced rotation matrices* for which the most useful expressions are given in Fig. 2.1. Since we are interested in the angular momentum measured along the z-axis, only rotations that affect the z-axis matter. Therefore, we can simply ignore the first rotation with the ψ angle in Eq. (2.72) and conventionally choose $\psi = 0$. Such a rotation $\hat{R}(\varphi,\theta,0)$

[10] The angular momentum operators are the (infinitesimal) generators of rotations, as shown in Section A.4.5.

[11] For passive rotation we would have ${}_z\langle j,m'|\hat{R}^\dagger\,|j,m\rangle_z = \left({}_z\langle j,m|\hat{R}\,|j,m'\rangle_z\right)^* = \left(D^j_{m,m'}\right)^*$.

would bring the z-axis in the same position as $\hat{R}(\varphi,\theta,\psi)$ (only the x- and y-axes would differ). Consequently, Eq. (2.72) finally reduces to

$$D^{j}_{m',m}(\varphi,\theta,0) = e^{-i\varphi m'} d^{j}_{m',m}(\theta). \tag{2.74}$$

As an example, it is interesting to consider a spin 1/2. Given that $\hat{J}_y = \hat{S}_y = \hbar\sigma_y/2$, the reduced rotation matrix is

$$d^{\frac{1}{2}}_{m',m}(\theta) = \left\langle \frac{1}{2}, m' \right| \exp\left(-i\frac{\theta}{2}\sigma_y\right) \left| \frac{1}{2}, m \right\rangle = \cos\frac{\theta}{2}\delta_{m',m} - i\sin\frac{\theta}{2}\left\langle \frac{1}{2}, m' \right| \sigma_y \left| \frac{1}{2}, m \right\rangle,$$

where the exponential is decomposed as $\exp\left(-i\theta/2\sigma_y\right) = \cos(\theta/2)\mathbb{1} - i\sigma_y\sin(\theta/2)$ (use the Taylor series to find this identity using $\sigma_y^2 = \mathbb{1}$). Therefore, using the spinor notation introduced in Section 2.3.2, it follows

$$d^{\frac{1}{2}}_{\frac{1}{2},\frac{1}{2}} = d^{\frac{1}{2}}_{-\frac{1}{2},-\frac{1}{2}} = \cos\frac{\theta}{2},\quad d^{\frac{1}{2}}_{\frac{1}{2},-\frac{1}{2}} = -\sin\frac{\theta}{2},\quad d^{\frac{1}{2}}_{-\frac{1}{2},\frac{1}{2}} = \sin\frac{\theta}{2},$$

and thus,

$$D^{\frac{1}{2}}_{\frac{1}{2},\frac{1}{2}} = e^{-i\frac{\varphi}{2}}\cos\tfrac{\theta}{2},\quad D^{\frac{1}{2}}_{-\frac{1}{2},\frac{1}{2}} = e^{+i\frac{\varphi}{2}}\sin\tfrac{\theta}{2},\quad D^{\frac{1}{2}}_{\frac{1}{2},-\frac{1}{2}} = -e^{-i\frac{\varphi}{2}}\sin\tfrac{\theta}{2},\quad D^{\frac{1}{2}}_{-\frac{1}{2},-\frac{1}{2}} = e^{+i\frac{\varphi}{2}}\cos\tfrac{\theta}{2}.$$

These results correspond to the formulas given in Fig. 2.1. It follows from Eq. (2.71) that

$$\begin{aligned} \hat{R}\,|\tfrac{1}{2},\tfrac{1}{2}\rangle &= |\tfrac{1}{2},-\tfrac{1}{2}\rangle\, e^{+i\frac{\varphi}{2}}\sin\tfrac{\theta}{2} + |\tfrac{1}{2},\tfrac{1}{2}\rangle\, e^{-i\frac{\varphi}{2}}\cos\tfrac{\theta}{2}; \\ \hat{R}\,|\tfrac{1}{2},-\tfrac{1}{2}\rangle &= |\tfrac{1}{2},-\tfrac{1}{2}\rangle\, e^{+i\frac{\varphi}{2}}\cos\tfrac{\theta}{2} - |\tfrac{1}{2},\tfrac{1}{2}\rangle\, e^{-i\frac{\varphi}{2}}\sin\tfrac{\theta}{2}, \end{aligned}$$

or in matrix notation in the basis $\{|1/2,\pm 1/2\rangle\}$[12],

$$\hat{R}(\theta,\varphi) = \begin{pmatrix} e^{-i\frac{\varphi}{2}}\cos\frac{\theta}{2} & -e^{-i\frac{\varphi}{2}}\sin\frac{\theta}{2} \\ e^{+i\frac{\varphi}{2}}\sin\frac{\theta}{2} & e^{+i\frac{\varphi}{2}}\cos\frac{\theta}{2} \end{pmatrix}. \tag{2.75}$$

The method exposed here is general once the expression of $\hat{J}_y$ is known. For spin-1/2 states, these results could have been obtained more easily. We said previously that $D^{j}_{m',m}$ is the amplitude to measure the value $m'\hbar$ along the z-axis when the angular momentum points in the direction of the rotated axis with component $m\hbar$. In other words, $D^{j}_{m',m}$ is the projection of the final spin state on the initial state. Since the eigenstates for spin 1/2 in an arbitrary direction $\boldsymbol{u}$ are already known, see Eq. (2.64), $D^{1/2}_{-1/2,1/2}$, for instance, is simply ${}_z\langle\downarrow\,|\,\uparrow\rangle_{\boldsymbol{u}}$.

Let us see the effect of the rotation (2.75) when a system $|\psi\rangle = \alpha\,|\uparrow\rangle + \beta\,|\downarrow\rangle$ is brought back in its original position, i.e. with $\theta = 2\pi$ and $\varphi = 0$. We obtain the famous result $\hat{R}(2\pi,0)\,|\psi\rangle = -\,|\psi\rangle$. For spin-1/2 particles, a rotation of 2π is not enough to recover the initial state, but two turns (4π) are actually needed. There is a direct connection with the *spin-statistics theorem* that states that the wave function of a system of identical particles with half-integer spin changes sign when two particles are swapped. Indeed, exchanging two particles is topologically equivalent to the rotation of either one of them by an angle 2π, as shown in Fig. 2.3. Thus, the interchange of two spin-1/2 particles generates a minus sign. For example, if $\psi(1)\chi(2)$ describes particle 1 in state ψ and particle 2 in state χ,

[12] The reader could have directly found this result by expanding the formula $e^{-i\frac{\varphi}{\hbar}\hat{S}_z}e^{-i\frac{\theta}{\hbar}\hat{S}_y}$, with S_z and S_y given by the Pauli matrices [Eq. (2.61)] .

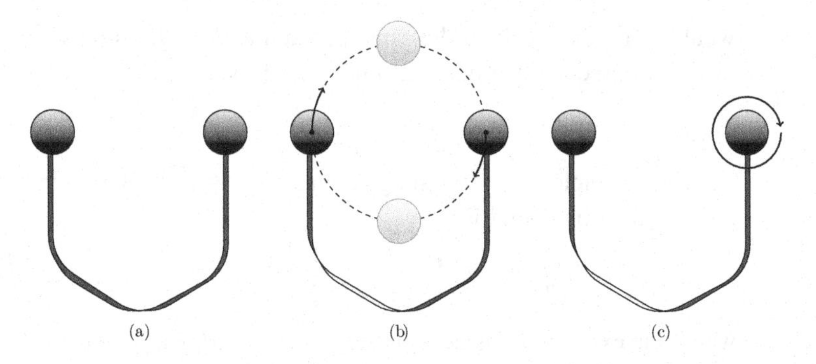

Fig. 2.3 Illustration of the topological equivalence between the exchange of two particles (balls) and the rotation by an angle 2π of either one of them. To check the equivalence, a ribbon connects two particles in (a). One of its sides is dark, whereas the other is bright. In (b), the particles are exchanged. The ribbon is then twisted, as illustrated by the visible bright side. An object rolling on the dark side from one particle to the other would undergo a complete turn on itself. In (c), the particle on the right-hand side is rotated by 2π. The ribbon is also twisted as in (b). The sequence leading to (c) is thus equivalent to that leading to (b).

the interchange of the particles must be described by $-\psi(2)\chi(1)$, and since both particles are identical, the correct wave function is the linear superposition $\psi(1)\chi(2) - \psi(2)\chi(1)$, which is antisymmetric, as required for fermions. The reader can check that with spin-1 particles, one single turn is enough to recover the initial state.

2.4 Units

Elementary particle energies are usually very low compared to macroscopic objects. Therefore, instead of using the Joule, it is more convenient to use the electron volt (eV) or its multiples (keV, MeV, GeV, TeV, etc.), where

$$1 \text{ eV} = 1.6 \times 10^{-19} \text{ J}. \tag{2.76}$$

For instance, the tiny rest energy of the electron then becomes a significant number $m_e c^2 = 8.19 \times 10^{-14}$ J $= 0.511$ MeV. Moreover, elementary particles are quantum objects governed by the Planck constant $\hbar = h/(2\pi) = 1.0546 \times 10^{-34}$ J $\cdot$ s $= 6.582 \times 10^{-22}$ GeV $\cdot$ s. They usually travel at a significant fraction of the speed of light $c = 2.998 \times 10^8$ ms^{-1}. Hence, many formulas encountered in particle physics are functions of these two fundamental constants. Instead of using the International System of Units (S.I.) with m, s, kg, for the sake of simplicity, it is more appropriate to use the so-called *natural units*, where

$$c = \hbar = 1 \text{ (unitless)}. \tag{2.77}$$

Therefore, length units become equivalent to time units, and energy units are equivalent to mass units or the inverse of time units. More specifically, $\hbar = 1$ implies

$$1 \text{ GeV}^{-1} \equiv 6.582 \times 10^{-25} \text{ s}, \tag{2.78}$$

while $\hbar c = 1$ yields

$$1 \text{ fm} \equiv \frac{1}{0.197} \text{ GeV}^{-1} = (197 \text{ MeV})^{-1}. \tag{2.79}$$

So, in natural units, a length is expressed in GeV^{-1}, an area (and thus a cross-section) in GeV^{-2}, a time in GeV^{-1}, and energy, mass and momentum in GeV.

In the same spirit, electromagnetic equations are simplified by adopting Heaviside–Lorentz units, where the permittivity of free space ϵ_0 and the permeability μ_0 are set to 1. For example, consider the Coulomb electromagnetic energy potential between two particles of the elementary electric charge e (or $-e$). In S.I. units, it is

$$V(r) = \frac{e^2}{4\pi\epsilon_0 r} = \hbar c \frac{\alpha}{r},$$

with α being a unitless number called the fine structure constant,

$$\alpha = \frac{e^2}{4\pi\epsilon_0 \hbar c} \simeq (137.036)^{-1}. \tag{2.80}$$

The combination of natural and Heaviside–Lorentz units leads to the potential energy $V(r) = \alpha/r$. As $1/r$ is in energy units [Eq. (2.79)], $V(r)$ has the correct units. In natural units, the value of the elementary electric charge is

$$e = \sqrt{4\pi\alpha} \simeq 0.3.$$

In the rest of the book, natural units will be used.

Problems

2.1 Check that if $x'^{\mu} = \Lambda^{\mu}{}_{\nu} x^{\nu}$, then $x' \cdot x' = x \cdot x$ implies $g_{\rho\sigma} \Lambda^{\rho}{}_{\mu} x^{\mu} \Lambda^{\sigma}{}_{\nu} x^{\nu} = g_{\mu\nu} x^{\mu} x^{\nu}$.

2.2 Starting from Eq. (2.14), show that $\Lambda_{\rho}{}^{\sigma} \Lambda^{\rho}{}_{\mu} = \delta^{\sigma}_{\mu}$. Conclude that $\left(\Lambda^{-1}\right)^{\sigma}{}_{\rho} = \Lambda_{\rho}{}^{\sigma}$.

2.3 Using Lorentz transformation, show that $\mathrm{d}\Omega = c\,\mathrm{d}t\,\mathrm{d}V$ is a 4-scalar. Hint: do not forget the Jacobian of the transformation!

2.4 What is the appropriate factor to convert cross sections expressed in natural units (GeV^{-2}) to μb? (1 barn (b) $= 10^{-24}$ cm^2).

2.5 Relative velocity in special relativity. Consider two particles (1) and (2) moving at $\boldsymbol{v}_1$ and $\boldsymbol{v}_2$ in the lab frame. Using Eq. (2.24), show that the velocity $\boldsymbol{v}_1^{(2)}$ of the particle (1) in the rest frame of (2) satisfies

$$\boldsymbol{v}_1{}^{(2)} \cdot \boldsymbol{v}_2 = \frac{(\boldsymbol{v}_1 - \boldsymbol{v}_2) \cdot \boldsymbol{v}_2}{1 - \frac{\boldsymbol{v}_1 \cdot \boldsymbol{v}_2}{c^2}}, \quad \boldsymbol{v}_1{}^{(2)} \times \boldsymbol{v}_2 = \frac{\boldsymbol{v}_1 \times \boldsymbol{v}_2}{\gamma\left(1 - \frac{\boldsymbol{v}_1 \cdot \boldsymbol{v}_2}{c^2}\right)}.$$

Using the identity $(\boldsymbol{a} \times \boldsymbol{b})^2 = |\boldsymbol{a}|^2 |\boldsymbol{b}|^2 - (\boldsymbol{a} \cdot \boldsymbol{b})^2$, deduce that

$$|\boldsymbol{v}_1{}^{(2)}| = \frac{\sqrt{(\boldsymbol{v}_1 - \boldsymbol{v}_2)^2 - \left(\frac{\boldsymbol{v}_1 \times \boldsymbol{v}_2}{c}\right)^2}}{1 - \frac{\boldsymbol{v}_1 \cdot \boldsymbol{v}_2}{c^2}}.$$

2.6 Addition of two spin 1. For this problem, use the tables of Fig. 2.1.

1. How can a spin-2 state with $s_z = 0$ be formed by adding two spin-1 particles?
2. A spin-1 particle with $s_z = -1$ and a second with $s_z = 1$ are prepared. What spin configuration can they form?
3. Two particles of spin-1 form a spin 1 system. How many different states are possible? Give all combinations.

3 Collisions and Decays

The determination of particle properties relies mostly on experimental measurements based on their collisions and decays. This chapter introduces the concepts of the reaction cross section and the particle decay rate. The relevant formulas are derived in detail.

3.1 Generalities

There are two kinds of measurements that are typically made with particles: you can smash them together or observe their spontaneous decay for the unstable ones. In the former case, you should then be interested in studying the collisions by quantifying the probability that a given reaction occurred, observing how particles are scattered or observing new particles emerging from the collisions. These types of measurements are related to the notion of the cross section. In the latter case, you will try to measure the lifetime of unstable particles and the probability that a given decay channel occurs. The cross section and the lifetime depend on both the dynamics of the process (the way the interactions act on particles) and on the kinematics, which in turn depend on the particle 4-momenta. The dynamics are encoded in the amplitude, which can be calculated with the use of Feynman diagrams, which are the subject of the following chapters. The kinematics are contained in the phase space and are described in this chapter.

3.2 Phase Space

Let us consider a reaction with $n_i \geq 1$ particles in the initial state and $n_f \geq 1$ in the final state (with n_f not necessarily equal to n_i). Ideally, in an experiment, the incident particles are prepared in a known state (a given energy and momentum typically), and the degrees of freedom of the reaction are thus related to the particles of the final state. Let us consider here only the degrees of freedom due to the kinematics; other degrees of freedom, such as spin and colour of quarks, will be treated later. Each 4-momentum has four degrees of freedom, and hence, in principle, the final state is characterised by $4n_f$ degrees of freedom. However, particles in the final (and initial) states are real particles, considered to be on the mass shell (see Section 2.2.4), and therefore for a given set of outgoing particles (whose masses are known), the energy dispersion relation $E^2 = |\boldsymbol{p}|^2 + m^2$ reduces the final state to $4n_f - n_f = 3n_f$ free parameters. Moreover, because of the conservation of the total 4-momentum between the initial and the final states, four additional constraints are added leaving only

$$N^{\text{d.o.f}}_{\text{phase space}} = 3n_f - 4 \tag{3.1}$$

independent parameters. A reaction thus corresponds to a single point on a hyper-surface of dimension $3n_f - 4$ within a space of dimension $4n_f$. This hyper-surface is called the *phase space*: it corresponds to all values, kinematically allowed, of 4-momentum components of the particles in the final state of the reaction. If the total energy of the initial state is E_{in}, the components of the 4-momentum of a particle in the final state necessarily satisfy $|p^\mu| \le E_{\text{in}}$, and thus the 'volume' of the phase space is contained within a hyper-sphere of radius E_{in}. The larger the volume, the more states are accessible in the final state.

3.3 Transition Rate

In Chapter 6, we will show that the general formula of the elementary transition rate (probability per unit time) from an initial state $|i\rangle$ made of n_i particles to a final state $|f\rangle$ made of n_f particles is given by[1]

$$\boxed{\mathrm{d}\Gamma_{i\to f} = \mathcal{V}^{1-n_i}(2\pi)^4\delta^{(4)}(p'_1 + \cdots + p'_{n_f} - p_1 - \cdots - p_{n_i})|\mathcal{M}_{fi}|^2 \prod_{k=1}^{n_i}\frac{1}{2E_k}\prod_{k=1}^{n_f}\frac{\mathrm{d}^3\boldsymbol{p}'_k}{(2\pi)^3 2E'_k}}, \tag{3.2}$$

where the prime symbols denote final-state quantities and $\mathcal{M}_{fi}$ is a complex number characterising the probability amplitude of the process called the *invariant scattering amplitude*, often referred to as the *matrix element*. The dynamics of the scattering process are encoded in the amplitude, which can be computed with Feynman diagrams, as we will learn in Chapter 6. The amplitude is squared, and various degrees of freedom that are not observed, such as spin or colour, are summed or averaged. The $|\mathcal{M}_{fi}|^2$ appearing here implicitly includes these factors. The symbol $\mathcal{V}$ in Eq. (3.2) is the volume of the 3-space used to perform the integrations over the space. This arbitrary parameter appears here to avoid diverging quantities, but as we shall see in the next sections, all physical quantities, such as decay rates (and thus the lifetime of a particle) and cross sections, do not depend on it. Therefore, the limit to infinite space volume will be safely taken. The other parameters of Formula (3.2) depend on the kinematics of the transition through the momenta and the energies of particles. The infinitesimal quantity

$$\mathrm{d}\Phi = (2\pi)^4\delta^{(4)}(p'_1 + \cdots + p'_{n_f} - p_1 - \cdots - p_{n_i})\prod_{k=1}^{n_f}\frac{\mathrm{d}^3\boldsymbol{p}'_k}{(2\pi)^3 2E'_k} \tag{3.3}$$

is called the Lorentz invariant phase space, often denoted dLIPS, and as its name suggests, this quantity is Lorentz invariant (this is shown in Problem 3.1). It corresponds to particles

[1] We also provide in Appendix H a 'classical' justification of this formula based on perturbation theory in quantum mechanics, leading to Fermi's golden rule.

in the final state whose momentum lies in the range $\mathrm{d}^3\boldsymbol{p}'_k$ about $\boldsymbol{p}'_k$. The infinitesimal transition rate $\mathrm{d}\Gamma_{i\to f}$ in Eq. (3.2) is proportional to $\mathrm{d}\Phi$. The actual transition rate is obtained by integrating $\mathrm{d}\Gamma_{i\to f}$ over all possible values of the momenta of the outgoing particles. When the amplitudes of two reactions depend in the same way on the momenta, the difference between their transition rates becomes dominated by the integral of $\mathrm{d}\phi$, which is proportional to the volume of the phase space introduced in Section 3.2. Reactions with a larger phase space (volume) are then more likely to occur. The volume of the phase space depends mostly on the energy excess between the initial state and the final state and thus approximately on their mass difference. For instance, the mass difference between a neutron and a proton is small compared with their mass, about 1.3 MeV. Therefore, the decay rate of $n \to p+e^-+\bar{\nu}_e$ is very small, explaining why (free-)neutrons have a large lifetime.[2]

3.4 Decay Width and Lifetime

Let us apply the elementary transition rate Formula (3.2) to the case of one unstable particle decaying into n particles: $n_i = 1$ and $n_f = n$,

$$\mathrm{d}\Gamma_{1\to n} = (2\pi)^4\delta^{(4)}(p'_1 + \cdots + p'_n - p_1)|\mathcal{M}|^2\frac{1}{2E_1}\prod_{k=1}^{n}\frac{\mathrm{d}^3\boldsymbol{p}'_k}{(2\pi)^3 2E'_k}.$$

In the rest frame of the decaying particle, E_1 is reduced to the mass of the particle, $E_1 = m$. The finite transition rate is obtained in that frame by integrating over the phase space,

$$\boxed{\Gamma_{1\to n} = (2\pi)^4\frac{1}{2m}\frac{1}{\mathcal{S}}\int \delta^{(4)}(p'_1 + \cdots + p'_n - p_1)|\mathcal{M}|^2\prod_{k=1}^{n}\frac{\mathrm{d}^3\boldsymbol{p}'_k}{(2\pi)^3 2E'_k}}. \tag{3.4}$$

The factor $\mathcal{S}$ is a combinatorial factor that avoids overcounting identical configurations whenever there are identical particles in the final state.[3] It is given by

$$\mathcal{S} = \prod_a n_a!\,, \tag{3.5}$$

where n_a is the number of identical particles of type a. For distinct particles in the final state, $\mathcal{S} = 1$. The rate is in units of s^{-1}, equivalent to GeV in natural units. It is also called the *partial decay width*. The *total decay width* or *total transition rate* is obtained by summing up all possible decays,

$$\boxed{\Gamma_{\mathrm{tot}} = \sum_{\text{final states } \mathrm{f_i}} \Gamma_{1\to f_i}}. \tag{3.6}$$

[2] The decay rate is also small because the involved interaction is the weak interaction.

[3] For instance, when there are two identical particles with momenta $\boldsymbol{p}'_1$ and $\boldsymbol{p}'_2$, the couple $(\boldsymbol{p}'_1, \boldsymbol{p}'_2)$ and $(\boldsymbol{p}'_2, \boldsymbol{p}'_1)$ describes the same physical configuration. Since the integration over the whole phase space contains both configurations, which are physically identical, we have to divide it by a factor $\mathcal{S} = 2$.

The fraction of decays into a given final state is called the *branching ratio*, given by

$$\boxed{\mathrm{BR}_{1\to f_i} = \frac{\Gamma_{1\to f_i}}{\Gamma_{\mathrm{tot}}}}. \tag{3.7}$$

When a particle is stable, the time evolution of the wave function eigenstate of the Hamiltonian is simply given from quantum mechanics by

$$\psi(t) = \psi(0)e^{-iE_0 t}, \tag{3.8}$$

with E_0 being the energy, a real number, such that $|\psi(t)|^2 = |\psi(0)|^2$ does not vary with time. When the particle is unstable, its state cannot be a stationary state with a fixed energy. If we assume that the probability for decaying during a time δt is just proportional to δt (regardless of the previous history of the particle) and suppose that the proportionality coefficient is given by the total transition rate, then the survival probability at $t + \delta t$ is

$$P_s(t+\delta t) = P_s(t) \times (1 - \delta t \Gamma_{\mathrm{tot}}).$$

In other words, the particle is alive at $t + \delta t$ if it was alive at t and did not decay during $[t, t + \delta t]$. Taking the limit $\delta t \to 0$, it follows that

$$\frac{\mathrm{d}P_s(t)}{\mathrm{d}t} = -P_s(t)\Gamma_{\mathrm{tot}},$$

and hence, the survival probability follows an exponential law,

$$P_s(t) = e^{-\Gamma_{\mathrm{tot}} t}.$$

The probability density function of the decay time is then

$$p(t) = -\frac{\mathrm{d}P_s(t)}{\mathrm{d}t} = \Gamma_{\mathrm{tot}}\, e^{-\Gamma_{\mathrm{tot}} t},$$

leading to the average decay time

$$\langle t \rangle = \int_0^\infty \mathrm{d}t\, t\, p(t) = \Big[-te^{-\Gamma_{\mathrm{tot}} t} \Big]_0^\infty + \int_0^\infty \mathrm{d}t\, e^{-\Gamma_{\mathrm{tot}} t} = \frac{1}{\Gamma_{\mathrm{tot}}}.$$

This shows that the particle lifetime, in natural units, is identified as the inverse of the total decay width,

$$\boxed{\tau = \frac{1}{\Gamma_{\mathrm{tot}}}}. \tag{3.9}$$

Therefore, the wave function of an unstable particle satisfies

$$|\psi(t)|^2 = |\psi(0)|^2 e^{-t/\tau}. \tag{3.10}$$

Comparing expressions (3.8) and (3.10), which are valid respectively for stable and unstable particles, we see that substituting E_0 in (3.8) by $E_0 - i\Gamma_{\mathrm{tot}}/2 = E_0 - i/(2\tau)$ leads to Eq. (3.10). In other words, a phenomenological description of the wave function of unstable particles can be written as

$$\psi(t) = \psi(0)e^{-iE_0 t - \frac{\Gamma_{\mathrm{tot}}}{2} t}. \tag{3.11}$$

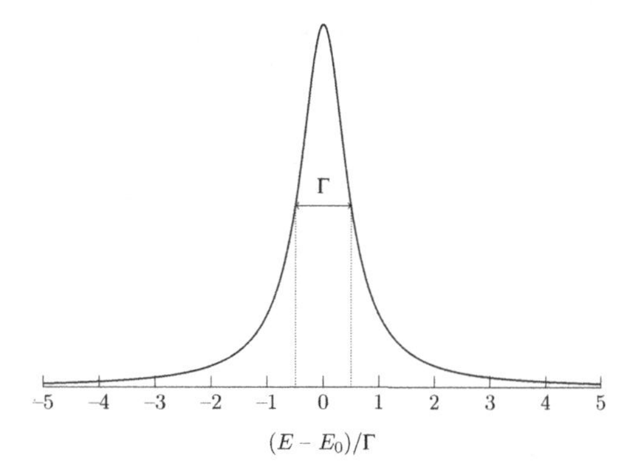

Fig. 3.1 The Breit–Wigner distribution. When $E_0 \gg \Gamma_{\text{tot}}$ (narrow width approximation), the shape of $f_{\text{BW}}(E)$ in Eq. (3.12) is almost identical to that of $f^{\text{rel}}_{\text{BW}}(E)$ in Eq. (3.13).

This has an interesting consequence on the energy distribution. The energy distribution is related to the square of the module of the Fourier transform of the single-particle wave-function, $\psi(t)$.[4] The Fourier transform reads

$$\tilde{\psi}(E) = \frac{1}{2\pi}\int_{-\infty}^{+\infty} \mathrm{d}t\, e^{iEt}\psi(t) = \frac{i\psi(0)}{2\pi}\frac{1}{(E-E_0)+i\frac{\Gamma_{\text{tot}}}{2}},$$

where $\psi(t<0)=0$ (the particle is taken to be created at $t=0$ for convenience) and $\Gamma_{\text{tot}} \neq 0$. Consequently, the probability of finding the energy E is

$$|\tilde{\psi}(E)|^2 = \frac{|\psi(0)|^2}{(2\pi)^2}\frac{1}{(E-E_0)^2+\frac{\Gamma^2_{\text{tot}}}{4}}.$$

The distribution of the energy, or equivalently the mass in the rest frame of the decaying particle, follows a Breit–Wigner law:

$$f_{\text{BW}}(E) = \frac{1}{2\pi}\frac{\Gamma_{\text{tot}}}{(E-E_0)^2+\frac{\Gamma^2_{\text{tot}}}{4}}, \tag{3.12}$$

where $f_{\text{BW}}(E)$ is normalised to 1. The energy distribution (or mass) peaks at E_0 and falls to half its maximum when $E = E_0 \pm \Gamma_{\text{tot}}/2$. It is depicted in Fig. 3.1. Formula (3.12) is well adapted to describe unstable states in nuclear physics. It is actually an approximation of the more general *relativistic Breit–Wigner distribution*, which is valid for relativistic particles,

$$f^{\text{rel}}_{\text{BW}}(E) = \frac{2E}{\pi}\frac{E_0\Gamma_{\text{tot}}}{(E^2-E_0^2)^2+(E_0\Gamma_{\text{tot}})^2}. \tag{3.13}$$

For non-relativistic particles, for which $E \simeq E_0 = m$, one recovers Eq. (3.12). We shall see in the coming chapters that the denominator of Eq. (3.13) is related to the propagator of particles originating in quantum field theory. As soon as the particle is unstable, provided

[4] Recall that in quantum mechanics, the Fourier transform of the wave function in position space gives the wave function in momentum space. Similarly, the Fourier transform of the wave function in time space gives the wave function in energy space.

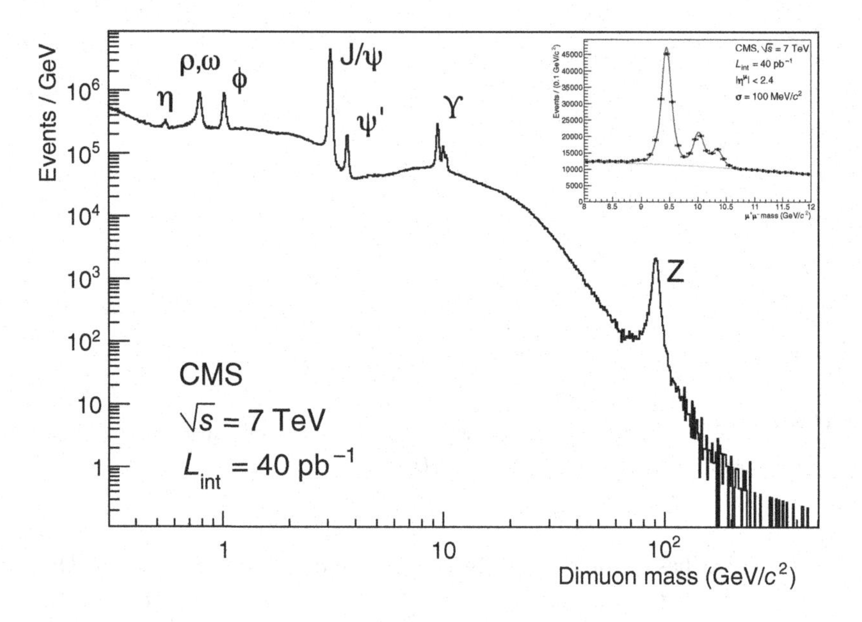

Fig. 3.2 Invariant mass spectrum of di-muons in events collected by the CMS experiment in 2010. The inset is a zoom of the Υ mass region. Image from CMS coll (2012a), licensed under Creative Commons CC BY 3.0.

that the total decay width is greater than the experimental energy resolution, the shape of its energy distribution observed in an experiment can be reasonably approximated by the Breit–Wigner parametrisation. For instance, Fig. 3.2 shows the reconstructed mass of particles decaying into two muons in the CMS detector at LHC (issued from proton–proton collisions). The presence of the peaks is the sign of the intermediate unstable particles that are called *resonances*. When the resonance width is narrow enough without other resonances nearby and the energy of the reaction is far from production thresholds of resonances, the shape of the peaks is well described by a Breit–Wigner convoluted by a function (typically a Gaussian function) representing the experimental uncertainty of the mass. From the Breit–Wigner function, one can access the nominal mass of the particle and its total decay width and thus its lifetime. For particles having a very long lifetime such as muons, the total decay width is too small to be measured. However, the lifetime is large enough so that the particle can travel over distances that are measurable by the detectors, giving a direct measurement of the lifetime.

The relationship between the lifetime and the (total) decay width given by Eq. (3.9), $\Gamma_{\text{tot}} \times \tau = 1$, can be interpreted via Heisenberg's uncertainty principle. The uncertainty ΔE on the energy measurement must be intrinsically of the order Γ_{tot}, while Δt is about τ, leading to $\Delta t \Delta E \simeq \hbar$ in S.I. units. The greater the mass (or energy) shifted from its nominal value, the shorter the particle lives.

Note that since the partial decay in Eq. (3.4) is evaluated in the rest frame of the unstable particle, its lifetime is by definition its proper time.

3.5 Cross Sections

The interaction *cross section* is a number representative of the probability of the interaction. In particle physics, cross sections are measured in barns $= 10^{-24}$ cm^2, a surface unit, by reference to the classical image of a spherical particle **1**, of radius r_1, impinging on particle **2**, of radius r_2, which is part of a set of particles referred to as the target of volume $\mathcal{V}$. Two particles can collide only if the distance between their centres is less than $r_1 + r_2$. Let us imagine that the particle **1** randomly hits the surface S of the target material, which contains ρ_2 particles **2** per unit of volume, as illustrated in Fig. 3.3. If around the centre of each target particle, we draw a disc of area $\sigma = \pi(r_1 + r_2)^2$, the probability of interaction is given by the number of particles **2** seen by **1** times the ratio of σ and the surface of the target. If the particle **1** travels over a distance dx, **1** crosses a region of the target containing $\rho_2 S\,\mathrm{d}x$ particles of the target so that the probability of interaction is

$$\mathrm{d}P = \frac{\sigma}{S}\,\rho_2\,S\,\mathrm{d}x = \sigma\rho_2\,\mathrm{d}x. \tag{3.14}$$

Now, instead of a single particle **1**, let us consider N_1 particles. The number of particles without interaction at $x + \mathrm{d}x$ is then

$$N_1(x + \mathrm{d}x) = N_1(x) - \mathrm{d}P \times N_1(x) = N_1(x) - \sigma\rho_2\,\mathrm{d}x\,N_1(x),$$

whose solution is

$$N_1(x) = N_1(0)e^{-\sigma\rho_2 x} = N_1(0)e^{-x/\lambda}.$$

The characteristic distance

$$\lambda = \frac{1}{\rho_2\sigma}. \tag{3.15}$$

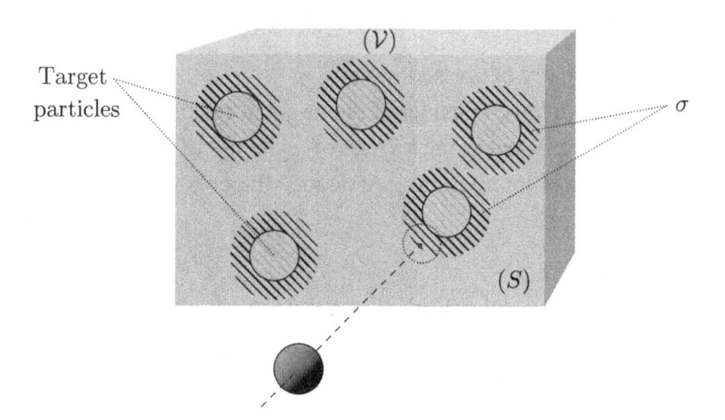

Fig. 3.3 Classical view of the cross section: the shaded area, σ, is the cross section depending on the size of the particles (see text). Target particles are drawn with plain circles, and the impact of the incident particle is depicted as the dashed circle.

is called the *mean free path*. The expectation value of an exponential probability density function is its slope. Therefore, the mean free path is interpreted at the mean distance after which one interaction occurs.

The incident particles may belong to a particle beam produced by an accelerator. Let us assume that each particle travels at a velocity $\boldsymbol{v_1}$ and that the beam has a density of ρ_1 particles per unit volume. The beam size is supposed larger than the surface S of the target. Considering an arbitrary volume $\mathcal{V}$ of the target, the number of interactions during $\mathrm{d}t$ is given by the number of particles of the beam that can interact in that volume (i.e., $\rho_1 \mathcal{V}$) times the elementary probability (3.14), with $\mathrm{d}x = |\boldsymbol{v_1}|\,\mathrm{d}t$. It follows that

$$\mathrm{d}N = \sigma \rho_1 \rho_2 |\boldsymbol{v_1}|\, \mathcal{V}\, \mathrm{d}t = \sigma\, I\, \mathcal{V}\, \mathrm{d}t, \tag{3.16}$$

where

$$I = \rho_2 \phi \tag{3.17}$$

is proportional to the flux of incident particles $\phi = \rho_1 |\boldsymbol{v_1}|$. Given that $\rho_2 \mathcal{V} = N_2$ is just the number of particles of the target contained in the volume $\mathcal{V}$, the rate Γ, defined as the number of interactions per unit time normalised to the number of particles of the target, is then

$$\Gamma = \frac{1}{N_2} \frac{\mathrm{d}N}{\mathrm{d}t} = \sigma \phi.$$

Thus, the cross section is defined as

$$\boxed{\sigma = \frac{\text{Number of interactions per unit time per target particle}}{\text{Flux of incident particles}}} = \frac{\Gamma}{\phi}. \tag{3.18}$$

It has the dimension of an area but unlike the classical case, in quantum mechanics (or quantum field theory), it should be interpreted as an effective area. The proportionality factor between the cross section and $\mathrm{d}N/\,\mathrm{d}t$ is called the (instantaneous) *luminosity*,

$$\mathcal{L} = N_2 \phi = \rho_2 \mathcal{V} \phi = I \mathcal{V}, \tag{3.19}$$

and it is usually expressed in $\mathrm{cm}^{-2}\ \mathrm{s}^{-1}$.

So far, we have considered the target at rest, so our calculation is valid in the rest frame of the target. What is the impact of a change of frame? Coming back to Formula (3.16) of the number of interactions, we deduce the cross section

$$\sigma = \frac{1}{I} \frac{1}{\mathcal{V}} \frac{\mathrm{d}N}{\mathrm{d}t}. \tag{3.20}$$

In this formula, $\mathrm{d}N/(\mathcal{V} dt)$ does not depend on the frame. Indeed, if there are $\mathrm{d}N$ reactions during $\mathrm{d}t$ in a box of volume $\mathcal{V}$ since the 4-D volume $\mathcal{V}\,\mathrm{d}t$ is Lorentz invariant (see Problem 2.3), there would still be $\mathrm{d}N$ reactions in any frame. If we succeed in showing that I is Lorentz invariant, then the cross section will be Lorentz invariant as well. Let us see how I in Eq. (3.17) can be expressed in any frame. We saw that in the rest frame of the target we have

$$I = \rho_1^{(2)} \rho_2^{(2)} |\boldsymbol{v}_1^{(2)}|, \tag{3.21}$$

where the upper index (2) is added to specify that the quantities were calculated in the rest frame of the particle **2**. Notice that $\boldsymbol{v}_1^{(2)}$, which is the velocity of particles **1** as viewed from particles **2**, is then the relative velocity $\boldsymbol{v}_{\mathbf{rel}}$ between **1** and **2**. Now, if particles **2** have a velocity $\boldsymbol{v}_2$ in the lab frame, in classical physics (non-relativistic), $\boldsymbol{v}_1^{(2)} = \boldsymbol{v}_{\mathbf{rel}} = \boldsymbol{v}_1 - \boldsymbol{v}_2$, and thus, the expression of I with quantities calculated in the lab frame would be

$$I = \rho_1 \rho_2 |\boldsymbol{v}_1 - \boldsymbol{v}_2|. \tag{3.22}$$

However, this formula does not hold for the relativistic case. First, one has to take into account the appropriate formulation for the addition of relativistic velocities. It yields (see Problem 2.5), using as usual $c = 1$,

$$|\boldsymbol{v}_1^{(2)}| = |\boldsymbol{v}_{\mathbf{rel}}| = \frac{\sqrt{(\boldsymbol{v}_1 - \boldsymbol{v}_2)^2 - (\boldsymbol{v}_1 \times \boldsymbol{v}_2)^2}}{1 - \boldsymbol{v}_1 \cdot \boldsymbol{v}_2}. \tag{3.23}$$

Second, the particle densities depend on the frame since the density is the time-like component of the 4-current density $j = (\rho, \rho \boldsymbol{v})$, as we saw in Section 2.2.4. For example, moving from the rest frame of the particles to the lab frame, their densities become

$$\rho_1 = \gamma_1 \rho_1^{(1)}, \quad \rho_2 = \gamma_2 \rho_2^{(2)},$$

while moving from the rest frame of particle **1** to the rest frame of particle **2**, the density of particle **1** is modified in

$$\rho_1^{(2)} = \gamma_{\mathrm{rel}}\, \rho_1^{(1)} = \gamma_{\mathrm{rel}} \frac{\rho_1}{\gamma_1},$$

with

$$\gamma_{\mathrm{rel}} = \frac{1}{\sqrt{1 - |\boldsymbol{v}_{\mathbf{rel}}|^2}}.$$

It follows that I in Eq. (3.21) can be expressed with $|\boldsymbol{v}_{\mathbf{rel}}|$ and quantities in the lab frame, yielding

$$I = \frac{\rho_1}{\gamma_1} \frac{\rho_2}{\gamma_2} \frac{|\boldsymbol{v}_{\mathbf{rel}}|}{\sqrt{1 - |\boldsymbol{v}_{\mathbf{rel}}|^2}}. \tag{3.24}$$

Now, according to the Formula (3.23) of $|\boldsymbol{v}_{\mathbf{rel}}|$,

$$1 - |\boldsymbol{v}_{\mathbf{rel}}|^2 = \frac{(1 - \boldsymbol{v}_1 \cdot \boldsymbol{v}_2)^2 - (\boldsymbol{v}_1 - \boldsymbol{v}_2)^2 + (\boldsymbol{v}_1 \times \boldsymbol{v}_2)^2}{(1 - \boldsymbol{v}_1 \cdot \boldsymbol{v}_2)^2}.$$

This formula can be rearranged using the relation

$$(\boldsymbol{v}_1 \times \boldsymbol{v}_2)^2 = |\boldsymbol{v}_1|^2 |\boldsymbol{v}_2|^2 - (\boldsymbol{v}_1 \cdot \boldsymbol{v}_2)^2, \tag{3.25}$$

giving

$$1 - |\boldsymbol{v}_{\mathbf{rel}}|^2 = \frac{\left(1 - |\boldsymbol{v}_1|^2\right)\left(1 - |\boldsymbol{v}_2|^2\right)}{(1 - \boldsymbol{v}_1 \cdot \boldsymbol{v}_2)^2} = \frac{1}{\left[\gamma_1 \gamma_2 (1 - \boldsymbol{v}_1 \cdot \boldsymbol{v}_2)\right]^2}. \tag{3.26}$$

Inserting Eqs. (3.23) and (3.26) into the expression (3.24) of I yields

$$I = \rho_1 \rho_2 \sqrt{(\boldsymbol{v}_1 - \boldsymbol{v}_2)^2 - (\boldsymbol{v}_1 \times \boldsymbol{v}_2)^2}. \tag{3.27}$$

Notice that when $\mathbf{v_1} = \mathbf{0}$ or $\mathbf{v_2} = \mathbf{0}$ or $\mathbf{v_1}$ and $\mathbf{v_2}$ are collinear, we recover Formula (3.22) valid for non-relativistic collisions. We can go one step further and express I in terms of manifestly covariant quantities. Since $\gamma_1\gamma_2 = (E_1E_2)/(m_1m_2)$ and $\gamma_1\gamma_2\,\mathbf{v_1}\cdot\mathbf{v_2} = \mathbf{p_1}\cdot\mathbf{p_2}$, we deduce from Eq. (3.26) the two equalities

$$\sqrt{1-|\mathbf{v}_{\mathbf{rel}}|^2} = \frac{m_1m_2}{p_1\cdot p_2}, \quad |\mathbf{v}_{\mathbf{rel}}| = \frac{\sqrt{(p_1\cdot p_2)^2-(m_1m_2)^2}}{p_1\cdot p_2}.$$

It follows from Eq. (3.24) that

$$I = \frac{\rho_1}{\gamma_1 m_1}\frac{\rho_2}{\gamma_2 m_2}\sqrt{(p_1\cdot p_2)^2-(m_1m_2)^2} = \frac{\rho_1\rho_2}{E_1E_2}\sqrt{(p_1\cdot p_2)^2-(m_1m_2)^2}. \tag{3.28}$$

This expression is clearly Lorentz invariant [since both the density and the energy are, respectively, the time-like components of the 4-vectors, $j = \rho(1,\mathbf{v})$ and $p = \gamma m(1,\mathbf{v})$], and consequently, as asserted, the cross section in Eq. (3.20) is invariant. Finally, inserting the expression (3.28) of I in Eq. (3.20) yields

$$\sigma = \frac{E_1E_2}{\rho_1\rho_2\sqrt{(p_1\cdot p_2)^2-(m_1m_2)^2}}\frac{1}{\mathcal{V}}\frac{\mathrm{d}N}{\mathrm{d}t} = \frac{E_1E_2\mathcal{V}}{\sqrt{(p_1\cdot p_2)^2-(m_1m_2)^2}}\frac{1}{N_1N_2}\frac{\mathrm{d}N}{\mathrm{d}t},$$

where $N_1 = \rho_1\mathcal{V}$ is the number of incident particles interacting in the arbitrary volume $\mathcal{V}$ with the $N_2 = \rho_2\mathcal{V}$ particles of the target. In general, we are interested in the collision of one single particle with another single particle. Hence, $N_1 = N_2 = 1$. The elementary cross section is thus

$$\mathrm{d}\sigma = \frac{E_1E_2\mathcal{V}}{\sqrt{(p_1\cdot p_2)^2-(m_1m_2)^2}}\,\mathrm{d}\Gamma_{2\to n},$$

where $\mathrm{d}\Gamma_{2\to n}$ is given by Eq. (3.2) with $n_i = 2$ particles in the initial state and $n_f = n$ particles in the final state, i.e.

$$\mathrm{d}\Gamma_{2\to n} = \frac{1}{\mathcal{V}}(2\pi)^4\delta^{(4)}(p'_1+\cdots+p'_n-p_1-p_2)|\mathcal{M}|^2\frac{1}{2E_1}\frac{1}{2E_2}\prod_{k=1}^{n}\frac{\mathrm{d}^3\mathbf{p}'_k}{(2\pi)^3 2E'_k}.$$

It follows that the differential cross section is

$$\boxed{\mathrm{d}\sigma_{2\to n} = \frac{1}{4\sqrt{(p_1\cdot p_2)^2-(m_1m_2)^2}}(2\pi)^4\delta^{(4)}(p'_1+\cdots+p'_n-p_1-p_2)|\mathcal{M}|^2\prod_{k=1}^{n}\frac{\mathrm{d}^3\mathbf{p}'_k}{(2\pi)^3 2E'_k}}. \tag{3.29}$$

The factor

$$F = \sqrt{(p_1\cdot p_2)^2-(m_1m_2)^2} = E_1E_2\sqrt{(\mathbf{v_1}-\mathbf{v_2})^2-(\mathbf{v_1}\times\mathbf{v_2})^2}, \tag{3.30}$$

is called the *flux factor*. The total cross section is obtained by integrating over the whole phase space and by taking into account the combinatorial factor $\mathcal{S}$ of Eq. (3.5) from identical particles, i.e.

$$\sigma_{2\to n} = \frac{1}{\mathcal{S}}\int \mathrm{d}\sigma_{2\to n}. \tag{3.31}$$

For distinct particles in the final state, $\mathcal{S} = 1$.

3.6 Two-Body States

3.6.1 Centre-of-Mass Frame Variables

Decays into two particles or interactions between two particles are very often encountered in particle physics. Besides, more complex states can always be decomposed into two-body systems. It is then very useful to establish the kinematics of such a system in the centre-of-mass frame of the two bodies (labelled CM in what follows). In the lab frame, the two particles have energies, momenta and masses denoted by E_i, $\boldsymbol{p}_i$ and m_i, with $i = 1, 2$, respectively. In the CM, these quantities are labelled with a star. Since by definition of the CM, $\boldsymbol{p}_1^* + \boldsymbol{p}_2^* = \boldsymbol{0}$, the norm of the two momenta is equal to a common value, $|\boldsymbol{p}^*|$. The total energy[5] in the CM is

$$\sqrt{s} = E_1^* + E_2^* = E_1^* + \sqrt{m_2^2 + |\boldsymbol{p}^*|^2} = E_1^* + \sqrt{m_2^2 + E_1^{*2} - m_1^2}.$$

It follows that

$$E_1^* = \frac{s + m_1^2 - m_2^2}{2\sqrt{s}}, \tag{3.32}$$

and since $E_2^* = \sqrt{s} - E_1^*$,

$$E_2^* = \frac{s + m_2^2 - m_1^2}{2\sqrt{s}}. \tag{3.33}$$

The momentum is then

$$|\boldsymbol{p}^*|^2 = E_1^{*2} - m_1^2 = \frac{s^2 + m_1^4 + m_2^4 - 2m_1^2 m_2^2 - 2sm_1^2 - 2sm_2^2}{4s},$$

or equivalently,

$$|\boldsymbol{p}^*|^2 = \frac{\lambda(s, m_1^2, m_2^2)}{4s} = \frac{[s - (m_1 + m_2)^2][s - (m_1 - m_2)^2]}{4s}. \tag{3.34}$$

The function $\lambda(x, y, z)$ is the triangle function, also known as the Källen function, and is defined as

$$\lambda(x, y, z) = x^2 + y^2 + z^2 - 2xy - 2xz - 2yz = [x - (\sqrt{y} + \sqrt{z})^2][x - (\sqrt{y} - \sqrt{z})^2]. \tag{3.35}$$

We notice that the energies and the norm of the momenta are fully determined by $\sqrt{s}$ and the masses. The only degrees of freedom are related to the direction of the momentum $\boldsymbol{p}^*$, i.e. two parameters. This is consistent with the dimension of the phase space given by Eq. (3.1) in Section 3.2 for a two-body system: $3 \times 2 - 4 = 2$.

[5] The exact definition of the Mandelstam variable s is clarified in Section 3.6.3.

3.6.2 Decay into Two Particles

Decays into two bodies being quite common, it is interesting to establish the corresponding partial decay width. Let us denote by $p = (E,\boldsymbol{p})$ the initial particle 4-momentum and $p_f = (E_f,\boldsymbol{p_f})$, with $f = 1,2$, the two 4-momenta of the decay products. Applying the general formula of the decay width, Eq. (3.4) yields the two-body partial decay width

$$\Gamma_{1\to 2} = \frac{(2\pi)^4}{S}\frac{1}{2E}\int \delta^{(4)}(p_1+p_2-p)|\mathcal{M}|^2\,\frac{\mathrm{d}^3\boldsymbol{p_1}}{(2\pi)^3 2E_1}\frac{\mathrm{d}^3\boldsymbol{p_2}}{(2\pi)^3 2E_2}.$$

In the rest frame of the decaying particle (or equivalently the CM frame of the two bodies), the 4-momenta are

$$p = (m,0),\quad p_1 = \left(\sqrt{|\boldsymbol{p}^*|^2+m_1^2},\boldsymbol{p}^*\right),\quad p_2 = \left(\sqrt{|\boldsymbol{p}^*|^2+m_2^2},-\boldsymbol{p}^*\right).$$

Integrating over $\boldsymbol{p_2}$, it follows that

$$\Gamma_{1\to 2} = \frac{(2\pi)^{-2}}{S}\frac{1}{2m}\int \delta\left(\sqrt{|\boldsymbol{p}^*|^2+m_1^2}+\sqrt{|\boldsymbol{p}^*|^2+m_2^2}-m\right)|\mathcal{M}|^2$$
$$\frac{\mathrm{d}^3\boldsymbol{p}^*}{2\sqrt{|\boldsymbol{p}^*|^2+m_1^2}}\frac{1}{2\sqrt{|\boldsymbol{p}^*|^2+m_2^2}},$$

where now $|\mathcal{M}|$ depends on $\boldsymbol{p}^*$. The integration is made easier by using spherical coordinates $\mathrm{d}^3\boldsymbol{p}^* = |\boldsymbol{p}^*|^2\ \mathrm{d}|\boldsymbol{p}^*|\ \mathrm{d}\Omega$, where $\mathrm{d}\Omega = \sin\theta\ \mathrm{d}\theta\ \mathrm{d}\phi$ is the infinitesimal solid angle. The decay width is then

$$\Gamma_{1\to 2} = \frac{1}{32\pi^2 mS}\int \mathrm{d}\Omega\int_0^{+\infty}\mathrm{d}|\boldsymbol{p}^*|\,\frac{\delta\left(\sqrt{|\boldsymbol{p}^*|^2+m_1^2}+\sqrt{|\boldsymbol{p}^*|^2+m_2^2}-m\right)}{\sqrt{|\boldsymbol{p}^*|^2+m_1^2}\sqrt{|\boldsymbol{p}^*|^2+m_2^2}}|\boldsymbol{p}^*|^2|\mathcal{M}|^2.$$

We could use Formula (E.4) (see the appendix for reminders about the delta functions) to perform the integration over $|\boldsymbol{p}^*|$, but it is simpler to change the variable of integration to the total energy,

$$E = \sqrt{|\boldsymbol{p}^*|^2+m_1^2}+\sqrt{|\boldsymbol{p}^*|^2+m_2^2},$$

since

$$\mathrm{d}E = \frac{E|\boldsymbol{p}^*|}{\sqrt{|\boldsymbol{p}^*|^2+m_1^2}\sqrt{|\boldsymbol{p}^*|^2+m_2^2}}\,\mathrm{d}|\boldsymbol{p}^*|.$$

It follows then

$$\Gamma_{1\to 2} = \frac{1}{32\pi^2 mS}\int \mathrm{d}\Omega\int_{m_1+m_2}^{+\infty}\frac{\mathrm{d}E}{E}\delta(E-m)|\boldsymbol{p}^*||\mathcal{M}|^2,$$

and finally the two-body decay width is

$$\boxed{\Gamma_{1\to 2} = \frac{|\boldsymbol{p}^*|}{32\pi^2 m^2 S}\int \mathrm{d}\Omega|\mathcal{M}|^2\begin{cases}\text{if } m > m_1+m_2,\\ 0 \text{ otherwise.}\end{cases}}\qquad (3.36)$$

The momentum in the rest frame, $|\boldsymbol{p}^*|$, satisfies $m = \sqrt{|\boldsymbol{p}^*|^2 + m_1^2} + \sqrt{|\boldsymbol{p}^*|^2 + m_2^2}$. Its expression has been calculated previously in Eq. (3.34). The amplitude $\mathcal{M}$ must be evaluated at $\boldsymbol{p}_1 = \boldsymbol{p}^*$ and $\boldsymbol{p}_2 = -\boldsymbol{p}^*$. This result is general; no assumptions were made on the detailed shape of $\mathcal{M}$. If we also assume that the initial particle is spinless, then there is no privileged direction and, by symmetry, the amplitude $\mathcal{M}$ cannot depend on the solid angle. With these assumptions, knowing $\int \mathrm{d}\Omega = \int \sin\theta \,\mathrm{d}\theta \;\mathrm{d}\phi = 4\pi$, we obtain the two-body decay width,

$$\Gamma_{1\to 2} = \frac{|\boldsymbol{p}^*|}{8\pi m^2 \mathcal{S}} |\mathcal{M}|^2 \begin{cases} \text{if } m > m_1 + m_2, \\ 0 \text{ otherwise.} \end{cases}$$

3.6.3 Two-Particle Scattering and the Mandelstam Variables

Let us consider the collision sketched in Fig. 3.4. The particles labelled (1) and (2) scatter and become particles (1′) and (2′). The amplitude (probability) of the process must not depend on the frame, so only 4-scalars must be involved. With momenta of particles, only 4-momenta scalar products can then be present. There are 10 possible combinations, but because of momentum conservation and the mass-shell constraint, only two independent combinations remain. It is often convenient to express the amplitude as a function of other Lorentz-invariant variables called Mandelstam variables:

$$\boxed{\begin{aligned} s &= (p_1 + p_2)^2 = (p_1' + p_2')^2, \\ t &= (p_1 - p_1')^2 = (p_2' - p_2)^2, \\ u &= (p_1 - p_2')^2 = (p_1' - p_2)^2. \end{aligned}} \tag{3.37}$$

In the CM, $\boldsymbol{p}_1^* + \boldsymbol{p}_2^* = \boldsymbol{0}$, so $(p_1^* + p_2^*)^2 = (E_1^* + E_2^*)^2$. The variable s is thus the square of the total energy in this frame. If we suppose that in Fig. 3.4, the particle (1) becomes (1′) after the scattering, then the variable t is the square of the 4-momentum transferred to the particle (2). For this reason, t is usually called the 4-momentum transfer squared. Note that in the case of a real scattering where the incoming particles are the same as the outgoing particles, $m_1 = m_1'$ and $m_2 = m_2'$, and thus according to Eqs. (3.32) and (3.33), E_1^* and E_2^* are necessarily equal to $E_1'^*$ and $E_2'^*$, respectively. Therefore, $t = (p_1 - p_1')^2 = -|\boldsymbol{p}_1^* - \boldsymbol{p}_1'^*|^2$ becomes a simple momentum transfer squared in the CM frame. The variable u plays a role similar to t: it is a 4-momentum transfer between particles (1) and (2′) and is sometimes called the crossed 4-momentum transfer squared. Since a two-particle scattering involves

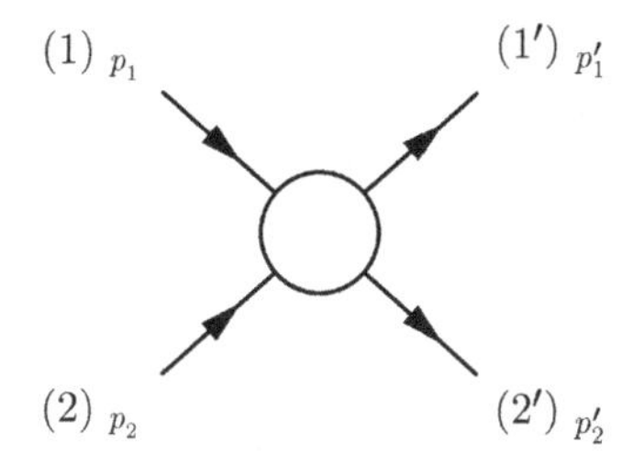

Fig. 3.4 A scattering with two particles.

only two independent 4-scalars, the three Mandelstam variables cannot be independent. Expanding the expressions of Eq. (3.37), and using the relation $p_1 + p_2 = p'_1 + p'_2$, it follows that

$$s + t + u = \sum_{i=1}^{4} m_i^2. \tag{3.38}$$

In a collision, the CM energy, $\sqrt{s}$, determines the energy available for the production of new particles. For relativistic head-on beams, the two particles have $|\boldsymbol{p_1}| \simeq E_1$ and $|\boldsymbol{p_2}| \simeq E_2$, so that, according to Eq. (3.37),

$$s \simeq 4E_1 E_2.$$

Assuming $E_1 = E_2 = E_{\text{beam}}$ (which is often the case in colliders), $\sqrt{s} = 2E_{\text{beam}}$. In the case of fixed target experiments, the particle in the target, let us say particle (2), is at rest (in the lab frame), and thus $E_2 = m_2$. The other particle has the energy of the beam $E_1 = E_{\text{beam}}$. Therefore, s becomes

$$s = m_1^2 + m_2^2 + 2E_{\text{beam}} m_2 \simeq 2E_{\text{beam}} m_2,$$

giving $\sqrt{s} = \sqrt{2m_2}\sqrt{E_{\text{beam}}}$. If the goal is to reach the highest $\sqrt{s}$, it is thus obvious that colliding beams are more suitable, as the CM energy grows linearly with the beam energy. Colliders are, however, more complex to operate, whereas fixed target experiments offer other advantages such as a high luminosity (due to the number of nuclei in the target) and the flexibility to easily change the nature of the target.

3.6.4 Two-Particle Scattering Cross Section

Let us apply the infinitesimal cross-section Formula (3.29) with two particles in the initial state and two in the final state, $n_i = n_f = 2$. It reads

$$d\sigma_{2\to 2} = \frac{(2\pi)^4}{4\sqrt{(p_1 \cdot p_2)^2 - (m_1 m_2)^2}} \delta^{(4)}(p'_1 + p'_2 - p_1 - p_2)|\mathcal{M}|^2 \frac{d^3\boldsymbol{p}'_1}{(2\pi)^3 2E'_1} \frac{d^3\boldsymbol{p}'_2}{(2\pi)^3 2E'_2}. \tag{3.39}$$

In the CM frame, the two incoming 4-momenta are

$$p_1 = \left(E_1^* = \sqrt{|\boldsymbol{p}^*| + m_1^2}, \boldsymbol{p}^*\right), \quad p_2 = \left(E_2^* = \sqrt{|\boldsymbol{p}^*| + m_2^2}, -\boldsymbol{p}^*\right),$$

with $E_1^* + E_2^* = \sqrt{s}$. Hence, the flux factor F, Eq. (3.30), simplifies to $F = (E_1^* + E_2^*)|\boldsymbol{p}^*|$, yielding

$$d\sigma_{2\to 2} = \frac{1}{4\sqrt{s}|\boldsymbol{p}^*|(2\pi)^2} \delta(E_1'^* + E_2'^* - \sqrt{s})\delta^{(3)}(\boldsymbol{p}_1'^* + \boldsymbol{p}_2'^*)|\mathcal{M}|^2 \frac{d^3\boldsymbol{p}_1'^*}{2E_1'^*} \frac{d^3\boldsymbol{p}_2'^*}{2E_2'^*}, \tag{3.40}$$

which gives the cross section for a process in which $\boldsymbol{p}_{1,2}'^*$ lies in the range $d^3\boldsymbol{p}_{1,2}'^*$. In general, one is interested in the probability of observing one of the final products in a direction within an element of the solid angle $d\Omega^* = \sin\theta^*\, d\theta^*\, d\phi^*$. Let us integrate over the other decay product, let us say $\boldsymbol{p}_2'^*$, and writing $d^3\boldsymbol{p}_1'^* = |\boldsymbol{p}'^*|^2\, d|\boldsymbol{p}'^*|\, d\Omega^*$, it follows[6]

$$d\sigma_{2\to 2} = \frac{1}{4\sqrt{s}|\boldsymbol{p}^*|(2\pi)^2}\delta(E_1'^* + E_2'^* - \sqrt{s})|\mathcal{M}|^2\, \frac{|\boldsymbol{p}'^*|^2\, d|\boldsymbol{p}'^*|\, d\Omega^*}{2E_1'^*}\frac{1}{2E_2'^*}. \tag{3.41}$$

Replacing $E_i'^*$ by $\sqrt{|\boldsymbol{p}'^*|^2 + m_i'^2}$, with $i = 1$ or 2, the *differential cross section* becomes

$$\frac{d\sigma_{2\to 2}}{d\Omega^*} = \frac{1}{16\sqrt{s}|\boldsymbol{p}^*|(2\pi)^2}\int_0^{+\infty} d|\boldsymbol{p}'^*|\delta\left(\sqrt{|\boldsymbol{p}'^*|^2 + m_1'^2} + \sqrt{|\boldsymbol{p}'^*|^2 + m_2'^2} - \sqrt{s}\right)|\mathcal{M}|^2$$
$$\times \frac{|\boldsymbol{p}'^*|^2}{\sqrt{|\boldsymbol{p}'^*|^2 + m_1'^2}\sqrt{|\boldsymbol{p}'^*|^2 + m_2'^2}}.$$

Using the same change of variable as in Section 3.6.2, $E'^* = \sqrt{|\boldsymbol{p}'^*|^2 + m_1'^2} + \sqrt{|\boldsymbol{p}'^*|^2 + m_2'^2}$, it follows that

$$\frac{d\sigma_{2\to 2}}{d\Omega^*} = \frac{1}{16\sqrt{s}|\boldsymbol{p}^*|(2\pi)^2}\int_{m_1'+m_2'}^{+\infty} \delta\left(E'^* - \sqrt{s}\right)|\mathcal{M}|^2\frac{dE'^*}{E'^*}|\boldsymbol{p}'^*|.$$

Hence, the *differential cross section* is

$$\boxed{\frac{d\sigma_{2\to 2}}{d\Omega^*} = \frac{1}{64\pi^2 s}|\mathcal{M}|^2\frac{|\boldsymbol{p}'^*|}{|\boldsymbol{p}^*|}\begin{cases}\text{if } \sqrt{s} > m_1' + m_2',\\ 0 \text{ otherwise.}\end{cases}} \tag{3.42}$$

The momenta in the CM, $|\boldsymbol{p}'^*|$ and $|\boldsymbol{p}^*|$, were calculated using Eq. (3.34). In order to calculate the total cross section, it is sufficient to integrate the previous expression. However, if we are interested in the differential cross section, Formula (3.42) is not so useful because it is only valid in the CM. We can go a bit further by expressing the infinitesimal solid angle, $d\Omega^*$, as a function of the Mandelstam variable

$$t = (p_1 - p_1')^2 = m_1^2 + m_1'^2 - 2E_1^* E_1'^* + 2\boldsymbol{p}^* \cdot \boldsymbol{p}'^*.$$

Choosing the orientation of the reference frame such that the momenta are $\boldsymbol{p}^* = (0, 0, |\boldsymbol{p}^*|)$ and $\boldsymbol{p}'^* = (|\boldsymbol{p}'^*|\sin\theta^*, 0, |\boldsymbol{p}'^*|\cos\theta^*)$ yields

$$t = m_1^2 + m_1'^2 - 2E_1^* E_1'^* + 2|\boldsymbol{p}^*||\boldsymbol{p}'^*|\cos\theta^*.$$

The only degree of freedom is the angle, as the value of the energies and the magnitude of the momenta are fixed in the CM frame for a two-body scattering according to Eqs. (3.32)–(3.34). It follows that

$$\frac{d(\cos\theta^*)}{dt} = \frac{1}{2|\boldsymbol{p}^*||\boldsymbol{p}'^*|}.$$

[6] I use a sloppy but typical notation here, keeping the same symbol, $d\sigma_{2\to 2}$, before and after the integration over the momentum.

Replacing $\mathrm{d}\Omega^*$ by $\mathrm{d}(\cos\theta^*)\,\mathrm{d}\phi^*$ in Eq. (3.42) yields[7]

$$\frac{\mathrm{d}\sigma_{2\to 2}}{\mathrm{d}t} = \frac{\mathrm{d}\sigma_{2\to 2}}{\mathrm{d}\Omega^*}\frac{\mathrm{d}\Omega^*}{\mathrm{d}t} = \frac{1}{128\pi^2 s}|\mathcal{M}|^2\frac{\mathrm{d}\phi^*}{|\boldsymbol{p}^*|^2}. \tag{3.43}$$

Let us assume that $\mathcal{M}$ does not depend on ϕ^*, although this is not necessarily the case, because the symmetry is broken by the beam axis, in particular, if the beam is polarised. After integrating over the angle,

$$\frac{\mathrm{d}\sigma_{2\to 2}}{\mathrm{d}t} = \frac{1}{64\pi s}\frac{|\mathcal{M}|^2}{|\boldsymbol{p}^*|^2}. \tag{3.44}$$

This formula is now valid in all frames, with $|\boldsymbol{p}^*|$ being given by the expression (3.34).

3.7 Crossed Reactions

So far, the three Mandelstam variables have been simply presented as a combination of 4-momenta, with s being the energy squared in the CM, t the 4-momentum transfer squared and u the cross 4-momentum transfer squared. However, we can go further by interpreting the negative signs in the definition of the Mandelstam variables (3.37). We shall see in Chapters 5 and 6 that a particle with 4-momentum $-p^\mu$ may be interpreted as an antiparticle,

$$\bar{p}^\mu \equiv -p^\mu. \tag{3.45}$$

A bar is added in the definition above to clearly state that the 4-momentum is that of an antiparticle. This approach is justified by a property of quantum field theory that states that the amplitude (a complex function describing the interaction) involving a particle in the initial state with 4-momentum p^μ is identical to the amplitude of the otherwise identical process, but with an antiparticle in the final state with momentum $-p^\mu$. This is related to the fact that the same quantum field creates the particle and destroys the antiparticle (and vice versa). Hence, the amplitudes of the following reactions can be derived from a common amplitude:

$$(1) + (2) \to (3) + (4), \tag{3.46}$$

$$(1) + (\bar{3}) \to (\bar{2}) + (4), \tag{3.47}$$

$$(1) + (\bar{4}) \to (3) + (\bar{2}), \tag{3.48}$$

and even $(1) \to (\bar{2}) + (3) + (4)$. In the previous reactions, the quantity $(\bar{k})$ denotes the antiparticle of the particle (k). However, even if a common amplitude can be used, it does not mean that all these reactions are allowed: conservation of the 4-momentum must be respected, or in other words, the kinematics must be adequate.

[7] Formally, $\int \mathrm{d}\Omega^* = \int_0^\pi \sin\theta^*\,\mathrm{d}\theta^* \int_0^{2\pi}\mathrm{d}\phi = \int_1^{-1} -\mathrm{d}(\cos\theta^*)\int_0^{2\pi}\mathrm{d}\phi^* = \int_{-1}^{1}\mathrm{d}(\cos\theta^*)\int_0^{2\pi}\mathrm{d}\phi^*$. Therefore, when $\mathrm{d}\Omega^*$ is replaced by $\mathrm{d}(\cos\theta^*)\,\mathrm{d}\phi^*$, keep in mind the appropriate order in the integral boundaries to get a positive number for the cross section.

Table 3.1 The Mandelstam variables as a function of the channels of the reaction $(1) + (2) \to (3) + (4)$.

Channel	Reaction	s-Variable	t-Variable	u-Variable
s-channel	$(1) + (2) \to (3) + (4)$	s	t	u
t-channel	$(1) + (\bar{3}) \to (\bar{2}) + (4)$	t	s	u
u-channel	$(1) + (\bar{4}) \to (3) + (\bar{2})$	u	t	s

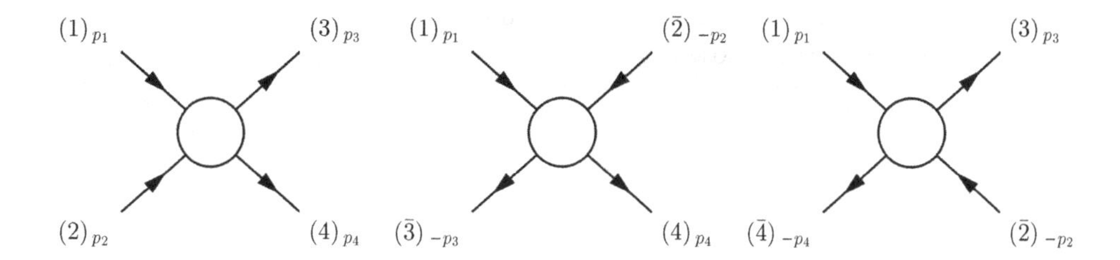

Fig. 3.5 Crossed reactions: the middle and right reactions are, respectively, the t-channel and u-channel of the reaction on the left.

The amplitude depends on the four 4-momenta or equivalently, on the Mandelstam variables. Let us denote by s, t, u, the Mandelstam variables of the reaction (3.46) and $\mathcal{M}$ its amplitude. The corresponding notation for the reaction (3.47) is s', t', u' and $\mathcal{M}'$. Thus, in terms of 4-momenta, the Mandelstam variables are

$$\begin{aligned} s &= (p_1 + p_2)^2, \quad t = (p_1 - p_3)^2, \quad u = (p_1 - p_4)^2, \\ s' &= (p_1 + p_{\bar{3}})^2, \quad t' = (p_1 - p_{\bar{2}})^2, \quad u' = (p_1 - p_4)^2. \end{aligned}$$

Reaction (3.46) is then described by $\mathcal{M}(s, t, u)$ and reaction (3.47) by $\mathcal{M}'(s', t', u')$. Applying the rule (3.45), we deduce that $s = t', t = s'$ and $u = u'$. Hence, the amplitude of reaction (3.47) is just $\mathcal{M}'(s', t', u') = \mathcal{M}(t', s', u')$. Similarly, for reaction (3.48), we would get $\mathcal{M}(u', t', s')$. These properties are very useful and avoid recalculating an amplitude from the Feynman diagrams once the amplitude of another (cross-)reaction is known. We will exploit them in Chapter 6.

The three processes in which s, t, u are the CM energy squared are called, respectively, s-, t- and u-channels and are displayed in Fig. 3.5. Reactions (3.47) and (3.48) are then, respectively, the t- and u-channel of reaction (3.46). We will see that when the interaction is modelled by the exchange of a single virtual particle, the 4-momentum squared of the virtual particle is the Mandelstam variables corresponding to the channel. A summary of those definitions is given in Table 3.1.

3.8 The Three-Particle Final State: Dalitz Plots

If the two-particle final state is very frequent, many reactions produce three particles in the final state. For example, let us consider the decay of a particle, having a mass m and

4-momentum p, into three particles with mass $m_{i=[1,3]}$ and 4-momenta $p_{i=[1,3]}$. Using the rest frame of the decaying particle, the differential decay rate, given by Eq. (3.4), reads

$$d\Gamma_{1\to 3} = (2\pi)^4 \frac{1}{2m} \delta^{(4)}(p_1 + p_2 + p_3 - p)|\mathcal{M}|^2 \frac{d^3\boldsymbol{p}_1}{(2\pi)^3 2E_1} \frac{d^3\boldsymbol{p}_2}{(2\pi)^3 2E_2} \frac{d^3\boldsymbol{p}_3}{(2\pi)^3 2E_3}.$$

Integrating over $\boldsymbol{p}_3$, it yields

$$d\Gamma_{1\to 3} = \frac{1}{16m(2\pi)^5} \delta(E_1 + E_2 + E_3 - m)|\mathcal{M}|^2 \frac{|\boldsymbol{p}_1|^2\, d|\boldsymbol{p}_1|\, d\Omega_1}{E_1} \frac{|\boldsymbol{p}_2|^2\, d|\boldsymbol{p}_2|\, d\Omega_2}{E_2} \frac{1}{E_3}.$$

Let us assume that the amplitude does not depend on the angular distributions (in other words, we suppose that there is no spin polarisation). Then, the only angular degree of freedom that is relevant is the angle θ_{12} between the two momenta $\boldsymbol{p}_1$ and $\boldsymbol{p}_2$. We can thus integrate over the other angles, with the solid angle integrations yielding

$$\iint d\Omega_1\, d\Omega_2 = 4\pi\, 2\pi\, d(\cos\theta_{12}).$$

It follows that

$$d\Gamma_{1\to 3} = \frac{1}{8m(2\pi)^3} \delta(E_1 + E_2 + E_3 - m)|\mathcal{M}|^2 \frac{|\boldsymbol{p}_1|^2 |\boldsymbol{p}_2|^2\, d|\boldsymbol{p}_1|\, d|\boldsymbol{p}_2| d(\cos\theta_{12})}{E_1 E_2 E_3}.$$

In the rest frame of the decaying particle, $|\boldsymbol{p}_3|^2 = |\boldsymbol{p}_1 + \boldsymbol{p}_2|^2 = |\boldsymbol{p}_1|^2 + |\boldsymbol{p}_2|^2 + 2|\boldsymbol{p}_1||\boldsymbol{p}_2| \cos\theta_{12}$. In order to integrate over $\cos\theta_{12}$, let us define the function[8]

$$\begin{aligned} g(\cos\theta_{12}) &= E_1 + E_2 + E_3 - m \\ &= E_1 + E_2 + \sqrt{m_3^2 + |\boldsymbol{p}_1|^2 + |\boldsymbol{p}_2|^2 + 2|\boldsymbol{p}_1||\boldsymbol{p}_2| \cos\theta_{12}} - m. \end{aligned}$$

According to the Dirac distribution property (E.4), the differential decay width reads

$$d\Gamma_{1\to 3} = \frac{1}{8m(2\pi)^3} \frac{\delta(\cos\theta_{12} - \cos\theta_{12}^0)}{\left|\frac{\partial g}{\partial \cos\theta_{12}}(\cos\theta_{12}^0)\right|} |\mathcal{M}|^2 \frac{|\boldsymbol{p}_1|^2 |\boldsymbol{p}_2|^2\, d|\boldsymbol{p}_1|\, d|\boldsymbol{p}_2| d(\cos\theta_{12})}{E_1 E_2 E_3},$$

where $\cos\theta_{12}^0$ is the solution of $g(\cos\theta_{12}^0) = 0$, i.e. $E_1 + E_2 + E_3 - m = 0$. Since

$$\frac{\partial g}{\partial \cos\theta_{12}} = \frac{|\boldsymbol{p}_1||\boldsymbol{p}_2|}{E_3},$$

the integration over $\cos\theta_{12}$ yields

$$d\Gamma_{1\to 3} = \frac{1}{8m(2\pi)^3} |\mathcal{M}|^2 \frac{|\boldsymbol{p}_1||\boldsymbol{p}_2|\, d|\boldsymbol{p}_1|\, d|\boldsymbol{p}_2|}{E_1 E_2}.$$

Given that $E_i^2 = m_i^2 + |\boldsymbol{p}_i|^2$, we have $E_i\, dE_i = |\boldsymbol{p}_i|\, d|\boldsymbol{p}_i|$ and thus, the differential decay width can be expressed as

$$d\Gamma_{1\to 3} = \frac{1}{8m(2\pi)^3} |\mathcal{M}|^2\, dE_1\, dE_2.$$

The 3-body decay is then described by two variables, here E_1 and E_2. The Dalitz plot is a plane showing the density (i.e., relative frequency) of these two variables, with one variable

[8] An alternative would be to make the change of variables: $d|\boldsymbol{p}_1|\, d|\boldsymbol{p}_2|\, d|\boldsymbol{p}_3| = |\det(J)|\, d|\boldsymbol{p}_1|\, d|\boldsymbol{p}_2|\, d(\cos\theta_{12})$, where J is the Jacobian of the transformation that is simply in this case, $\det(J) = \partial|\boldsymbol{p}_3|/\partial\cos\theta_{12} = |\boldsymbol{p}_1||\boldsymbol{p}_2|/|\boldsymbol{p}_3|$.

on the x-axis and the other on the y-axis. Traditionally, instead of energies, the squares of the masses of two pairs of the decay products are used, which has the advantage of not depending on the frame. Since $m_{13}^2 = (p_1 + p_3)^2 = (p - p_2)^2 = m^2 + m_2^2 - 2mE_2$, we have $\mathrm{d}E_2 = -\,\mathrm{d}m_{13}^2/(2m)$ and similarly $\mathrm{d}E_1 = -\,\mathrm{d}m_{23}^2/(2m)$, leading to

$$\mathrm{d}\Gamma_{1\to 3} = \frac{1}{32m^3(2\pi)^3}|\mathcal{M}|^2\,\mathrm{d}m_{13}^2\,\mathrm{d}m_{23}^2.$$

If the matrix element $|\mathcal{M}|^2$ is a constant, then the plane (m_{13}^2, m_{23}^2) should be uniformly populated. In this case, the differential decay width only depends on the phase space. The contour of the Dalitz plot is shaped by energy–momentum conservation and the masses of the particles involved in the reaction. However, as soon as there are deviations from a uniform density, this is a sign of an underlying dynamics in the process: $|\mathcal{M}|^2$ depends on m_{13}^2 or m_{23}^2. This is the case, for example, when the 3-body reaction occurs via an intermediate resonance, with $|\mathcal{M}|^2$ being larger around the resonance mass (due to an underlying Breit–Wigner shape).

As an example, in Shafer et al. (1963), the reaction $K^- + p \to \Lambda + \pi^+ + \pi^-$ was studied. The Dalitz plot is shown in Fig. 3.6. Clearly, the density is not uniform in the distorted ellipse. An accumulation of data is seen in the horizontal and vertical bands for masses $m_{\Lambda\pi^+} \approx m_{\Lambda\pi^-} \approx 1.38$ GeV. This accumulation turns out to be due to the presence of the $\Sigma^{\pm}(1385)$ resonance, having a mass of 1385 MeV. The reaction was actually: $K^- + p \to \Sigma^+ + \pi^- \to \Lambda + \pi^+ + \pi^-$ (and the C-conjugate reaction).

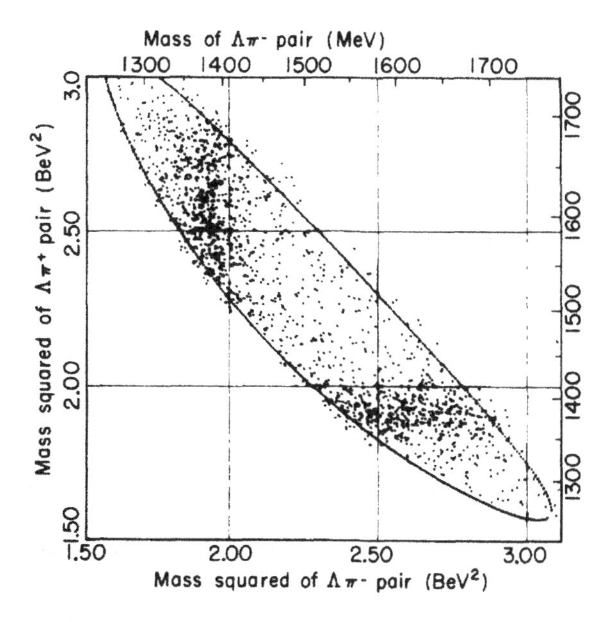

Fig. 3.6 The Dalitz plot of the reaction $K^- + p \to \Lambda + \pi^+ + \pi^-$. Source from Shafer et al. (1963), with permission from the American Physical Society.

3.9 Decomposition of the Phase Space

Once reactions involve many particles in the final state, the kinematical constraints become complicated to apply. It is then useful to simplify the calculations by decomposing a complicated phase space into smaller ones. Let us denote by P the 4-momentum of the initial state (potentially made of several particles) and $p_1, p_2, \ldots, p_n$ those of the n particles in the final state. For simplicity, we will start with $n = 2$, move to $n = 3$ and then generalise to n.

3.9.1 Two-Body LIPS

The two-body Lorentz invariant phase space deduced from Eq. (3.3) reads

$$\mathrm{d}\Phi(P \to p_1 p_2) = (2\pi)^4 \delta^{(4)}(p_1 + p_2 - P)\, \frac{\mathrm{d}^3\boldsymbol{p}_1}{(2\pi)^3 2E_1} \frac{\mathrm{d}^3\boldsymbol{p}_2}{(2\pi)^3 2E_2}.$$

We saw in Section 3.3 the relationship between the transition rate and the phase space. For two bodies, in the CM frame of the two bodies, it reads

$$\mathrm{d}\Gamma_2 = \frac{|\mathcal{M}|^2}{2m_{12}}\, \mathrm{d}\Phi(P \to p_1 p_2),$$

where $m_{12}^2 = (p_1 + p_2)^2$ is in this case the mass of the decaying particle. Actually, looking at Formula (3.36), we have already expressed $\mathrm{d}\Gamma_2$,

$$\mathrm{d}\Gamma_2 = \frac{|\boldsymbol{p}^*||\mathcal{M}|^2}{32\pi^2 m_{12}^2}\, \mathrm{d}\Omega^*.$$

Consequently, we can immediately identify

$$\mathrm{d}\Phi(P \to p_1 p_2) = \frac{|\boldsymbol{p}^*|}{16\pi^2 m_{12}}\, \mathrm{d}\Omega^*.$$

Finally, replacing $|\boldsymbol{p}^*|$ by Formula (3.34) with $\sqrt{s} = m_{12}$, it follows that

$$\mathrm{d}\Phi(p \to p_1 p_2) = \sqrt{1 - 2\frac{(m_1^2 + m_2^2)}{m_{12}^2} + \frac{(m_1^2 - m_2^2)^2}{m_{12}^4}}\, \frac{\mathrm{d}\Omega^*}{32\pi^2}. \tag{3.49}$$

As expected from Eq. (3.1), the two-body LIPS only depends on two variables: here the two angles of Ω^*.

3.9.2 Three-Body LIPS

Now, let us move to the three-body Lorentz invariant phase space,

$$\mathrm{d}\Phi(P \to p_1 p_2 p_3) = (2\pi)^4 \delta^{(4)}(p_1 + p_2 + p_3 - P)\, \frac{\mathrm{d}^3\boldsymbol{p}_1}{(2\pi)^3 2E_1} \frac{\mathrm{d}^3\boldsymbol{p}_2}{(2\pi)^3 2E_2} \frac{\mathrm{d}^3\boldsymbol{p}_3}{(2\pi)^3 2E_3}.$$

The particles in the final state are often produced via intermediate resonances and sometimes it simplifies the calculation to make this explicit. For example, let us suppose that the reaction is actually $P \to q_{12} p_3 \to p_1 p_2 p_3$, where q_{12} denotes the 4-momentum of the intermediate state that further decays in p_1p_2. One can always introduce the two identity integrals,

$$1 = \int \mathrm{d}^4 q_{12}\, \delta^{(4)}(q_{12} - p_1 - p_2)\theta(q_{12}^0)\,, \quad 1 = \int \mathrm{d}m_{12}^2\, \delta(m_{12}^2 - q_{12}^2),$$

with the quantity m_{12}^2 being the mass-squared of the particle described by the 4-momentum q_{12}. Putting together these two identity integrals and using the equivalence of Eq. (E.6), it follows that

$$1 = \iint \mathrm{d}^4 q_{12}\, \delta^{(4)}(q_{12} - p_1 - p_2)\theta(q_{12}^0)\, \mathrm{d}m_{12}^2\, \delta(m_{12}^2 - q_{12}^2) = \iint \frac{\mathrm{d}^3 \boldsymbol{q}_{12}}{2E_{12}}\, \delta^{(4)}(q_{12} - p_1 - p_2)\, \mathrm{d}m_{12}^2.$$

Consequently,

$$\begin{aligned} \mathrm{d}\Phi(P \to p_1p_2p_3) &= \frac{\mathrm{d}m_{12}^2}{2\pi}(2\pi)^4\delta^{(4)}(q_{12} + p_3 - P)\frac{\mathrm{d}^3\boldsymbol{p}_3}{(2\pi)^3 2E_3}\frac{\mathrm{d}^3\boldsymbol{q}_{12}}{(2\pi)^3 2E_{12}} \\ &\quad \times (2\pi)^4\delta^{(4)}(q_{12} - p_1 - p_2)\frac{\mathrm{d}^3\boldsymbol{p}_1}{(2\pi)^3 2E_1}\frac{\mathrm{d}^3\boldsymbol{p}_2}{(2\pi)^3 2E_2}, \end{aligned}$$

where the integral symbols are implicit. In the expression above, the three-particle phase space is decomposed into a product of smaller (two-body) phase spaces,

$$\mathrm{d}\Phi(P \to p_1p_2p_3) = \frac{\mathrm{d}m_{12}^2}{2\pi}\, \mathrm{d}\Phi(P \to q_{12}p_3)\, \mathrm{d}\Phi(q_{12} \to p_1p_2). \tag{3.50}$$

Note that the number of degrees of freedom stays the same: initially, we had $3 \times 3 - 4 = 5$ degrees of freedom [Eq. (3.1)]. Each two-body phase space has $2 \times 3 - 4 = 2$ degrees of freedom, and thus the right-hand side of the decomposition (3.50) contains $2 + 2 + 1 = 5$ degrees of freedom (the additional 1 being due to $\mathrm{d}m_{12}^2$).

3.9.3 *n*-Body LIPS

The decomposition of the three-particle final state can be easily extended to states containing n particles,

$$\mathrm{d}\Phi(P \to p_1 \cdots p_n) = \frac{\mathrm{d}m_{[1\cdots j]}^2}{2\pi}\, \mathrm{d}\Phi(P \to q_{[1\cdots j]}p_{j+1} \cdots p_n)\, \mathrm{d}\Phi(q_{[1\cdots j]} \to p_1 \cdots p_j), \tag{3.51}$$

with $m_{[1\cdots j]}^2 = (p_1 + \cdots p_j)^2 = q_{[1\cdots j]}^2$. Since each $\mathrm{d}\Phi$ is Lorentz invariant, they can be evaluated in different frames. Whether or not the particle described by q is a real resonance does not matter in this decomposition. The 4-momentum q can correspond to any sum of 4-momenta of particles belonging to the final state. However, in the case of a resonance, the q^2 distribution follows a Breit–Wigner, as seen previously.

Problems

3.1 Show that the Lorentz invariant phase space,

$$\mathrm{d}\Phi = (2\pi)^4\delta^{(4)}(p_1' + \cdots + p_{n_f}' - p_1 - \cdots - p_{n_i})\prod_{k=1}^{n_f}\frac{\mathrm{d}^3\boldsymbol{p}_k'}{(2\pi)^3 2E_k'},$$

is indeed Lorentz invariant. Hint: You can first show that $\mathrm{d}^3\boldsymbol{p}/E$ is Lorentz invariant using a Lorentz transformation and for the δ-Dirac function, use Fourier transformation.

3.2 Check that when the particle masses are negligible compared to their momenta, the three Mandelstam variables read

$$s = 4|\boldsymbol{p}^*|^2, \quad t = -2|\boldsymbol{p}^*|^2(1 - \cos\theta), \quad u = -2|\boldsymbol{p}^*|^2(1 + \cos\theta),$$

where θ is the angle between the scattered particles in the centre-of-mass frame.

3.3 Decompose the phase space of a 4-particle final state into products of 2-particle final states.

3.4 Show that for N particles on the mass shell with 4-momentum p_k and mass m_k ($k \in [1, N]$), we have the inequality:

$$\left(\sum_{k=1}^{N} p_k\right)^2 \geq \left(\sum_{k=1}^{N} m_k\right)^2.$$

3.5 π^+ decay in $\mu^+\nu_\mu$. Consider $N_0 = 600$ pions (spin 0, $m \approx 140$ MeV$/c^2$) with kinetic energy of 140 MeV. The pions decay in $\mu^+\nu_\mu$ ($m_\mu \approx 106$ MeV$/c^2$, $m_\nu \approx 0$).

1. Determine the angular distribution $\frac{\mathrm{d}N}{\mathrm{d}\cos\theta^*}$ of the μ in the π rest frame. (θ^* is the angle between the π direction and the μ direction in the π rest frame.)
2. What are the energy and momentum of the μ in the π rest frame?
3. What are the minimum energy and maximum energy of the μ in the lab frame?
4. Draw the shape of the energy distribution $\frac{\mathrm{d}N}{\mathrm{d}E_\mu}$ of the μ in the lab frame, assuming a binning of 1 MeV.

4 Conservation Rules and Symmetries

The notion of symmetry is essential in the determination of particle properties. It reveals quantities that are conserved in collisions or decays. It also constrains the mathematical formulation of theories. This chapter introduces these notions and reviews the quantities conserved in particle collisions or decays.

4.1 General Considerations

In the context of quantum physics, a symmetry corresponds to a transformation law that leaves the physics (i.e., all measurable quantities) invariant. Hence, when particles interact in a collision, some quantities must be conserved, which is a powerful tool to evaluate whether a reaction is allowed or not, or to identify a new particle produced in a reaction when all other particles involved are known. Conversely, when a quantity appears conserved in experiments, it constrains the mathematical formulation of a theory.

There are two kinds of symmetries: the ones related to the geometry of spacetime, and those, internal ones, which have no equivalence in the classical world. A further subdivision of symmetries can be distinguished, depending on the nature of the parameter(s) characterising the transformation law: it can be continuous (an angle, a phase, etc.) or discrete (a symmetry with respect to the origin of a frame, i.e. parity, or a symmetry between particles and antiparticles, etc.). Continuous symmetries are associated with quantities (quantum numbers) that are additive, while quantum numbers of discrete symmetries are multiplicative (the reason is given in Supplement 4.1). Continuous symmetries related to spacetime are normally already well known to the reader familiar with classical and quantum physics, and, therefore, in Sections 4.3.1 and 4.3.2, just basic reminders will be given. Lorentz invariance, already seen in Chapter 2, is an example of a continuous symmetry of spacetime. Boost transformations reflect the fact that the absolute velocity is not observable. Table 4.1 presents the main symmetry transformations used in particle physics. In the rest of this section, they shall be detailed. The gauge symmetries (local phase transformations) will be detailed in the next chapters.

4.2 Symmetries in Quantum Mechanics

In quantum mechanics, a system is invariant under a symmetry transformation if the absolute value of the (Hermitian) scalar product between any two physical states $|\psi_1\rangle$ and $|\psi_2\rangle$

Table 4.1 Examples of symmetries relevant for particle physics.
The symbols [+] and [×] refer to additive or multiplicative conserved quantities.

Non observable quantity	Transformation	Conserved quantity	Validity
	Continuous symmetries		
Absolute position	Space translation	Momentum [+]	Exact
Absolute time	Time translation	Energy [+]	Exact
Absolute orientation	Space rotation	Angular momentum [+][a]	Exact
Absolute velocity	Lorentz boosts	Group generators[b]	Exact
Relative phase of particle fields with different electric charge	Local phase transformation	Electric charge [+]	Exact
Relative phase between quark fields and other particle fields	Phase transformation	Baryon number [+]	Exact
Relative phase between lepton fields and other particle fields	Phase transformation	Lepton number [+]	Exact
Distinction between proton (u-quark) and neutron (d-quark)	SU(2) rotation between proton (u-quark) and neutron (d-quark)	(Strong-) Isospin [+][a]	Approx.
	Discrete symmetries		
Left–right distinction	Space inversion	Parity [×]	Approx.
Absolute time direction	Time reversal	Invariance w.r.t. time direction [×]	Approx.
Absolute particle–antiparticle distinction	Charge conjugation	Charge parity [×]	Approx.

[a] The additivity of quantum numbers for angular momentum or isospin has to be understood in terms of SU(2) addition. The projection on an axis is additive, but the quantity itself requires the use of SU(2) addition.

[b] One may wonder what conserved quantities are associated with Lorentz transformations. Since Lorentz transformations include both rotations in 3D and boosts, the conserved quantities are a kind of generalisation of the angular momentum to the Minkowski space. For instance, symmetry under rotations in the (x, y) plane corresponds to the conservation of $L_z = xp_y - yp_x$. In the same way, symmetry under hyperbolic rotation in the (t, y) plane corresponds to the conservation of $tp_y - yE$. An appropriate description of these quantities requires the identification of the six generators of the Lorentz symmetry, which is given in Appendix D.

is unchanged by the symmetry transformation, i.e.

$$|\langle(\hat{S}\psi_1)|(\hat{S}\psi_2)\rangle| = |\langle\psi_1|\psi_2\rangle|,$$

where $\hat{S}$ is an operator representing the symmetry transformation that acts on the Hilbert space of physical states. A theorem from Wigner tells us that such an operator exists for any symmetry and that it is either unitary and linear,

$$\begin{aligned}\langle(\hat{S}\psi_1)|(\hat{S}\psi_2)\rangle &= \langle\psi_1|\psi_2\rangle,\\ \hat{S}(\alpha\,|\psi_1\rangle + \beta\,|\psi_2\rangle) &= \alpha\hat{S}\,|\psi_1\rangle + \beta\hat{S}\,|\psi_2\rangle),\end{aligned} \tag{4.1}$$

or anti-unitary and anti-linear,

$$\begin{aligned}\langle(\hat{S}\psi_1)|(\hat{S}\psi_2)\rangle &= \langle\psi_1|\psi_2\rangle^* = \langle\psi_2|\psi_1\rangle,\\ \hat{S}(\alpha\,|\psi_1\rangle + \beta\,|\psi_2\rangle) &= \alpha^*\hat{S}\,|\psi_1\rangle + \beta^*\hat{S}\,|\psi_2\rangle),\end{aligned} \tag{4.2}$$

where α and β are two complex numbers. The definition of the Hermitian adjoint operator $\hat{S}^\dagger$ depends on the type of the operators. For unitary operators,

$$\langle\psi_1|(\hat{S}^\dagger\psi_2)\rangle = \langle(\hat{S}\psi_1)|\psi_2\rangle,$$

whereas for anti-unitary ones

$$\langle\psi_1|(\hat{S}^\dagger\psi_2)\rangle = \langle\psi_2|(\hat{S}\psi_1)\rangle.$$

Since these two equalities are valid for all states, we can substitute $|\psi_2\rangle$ with $\hat{S}\,|\psi_2\rangle$, yielding for unitary operators

$$\langle\psi_1|(\hat{S}^\dagger\hat{S}\psi_2)\rangle = \langle(\hat{S}\psi_1)|(\hat{S}\psi_2)\rangle\,,$$

whereas for anti-unitary operators

$$\langle\psi_1|(\hat{S}^\dagger\hat{S}\psi_2)\rangle = \langle(\hat{S}\psi_2)|(\hat{S}\psi_1)\rangle\,.$$

However, according to Eqs. (4.1) and (4.2), both previous equalities are actually equal to $\langle\psi_1|\psi_2\rangle$. Hence, with both kinds of operators, it follows necessarily that $\hat{S}^\dagger\hat{S} = \mathbb{1}$, where $\mathbb{1}$ denotes the identity operator. Similarly, $\hat{S}\hat{S}^\dagger = \mathbb{1}$. Indeed, if $|\psi_2\rangle = \hat{S}\,|\psi_2'\rangle$, then

$$\langle\psi_1|\hat{S}\hat{S}^\dagger|\psi_2\rangle = \langle\psi_1|\hat{S}\hat{S}^\dagger\hat{S}|\psi_2'\rangle = \langle\psi_1|\hat{S}\mathbb{1}|\psi_2'\rangle = \langle\psi_1|\psi_2\rangle,$$

which imposes $\hat{S}\hat{S}^\dagger = \mathbb{1}$. Therefore,

$$\hat{S}^\dagger\hat{S} = \hat{S}\hat{S}^\dagger = \mathbb{1}. \tag{4.3}$$

How are operators transformed by the symmetry? Denoting an operator by $\hat{A}$ and the transformed operator under the symmetry law by $\hat{A}'$, $\hat{A}'$ must satisfy

$$\langle\psi|\hat{A}|\psi\rangle = \langle\psi'|\hat{A}'|\psi'\rangle = \langle\psi|\hat{S}^\dagger\hat{A}'\hat{S}|\psi\rangle,$$

where $|\psi'\rangle = \hat{S}\,|\psi\rangle$. The equality being true for any $|\psi\rangle$, we conclude

$$\hat{A} = \hat{S}^\dagger\hat{A}'\hat{S} \Leftrightarrow \hat{A} \to \hat{A}' = \hat{S}\hat{A}\hat{S}^\dagger. \tag{4.4}$$

When the transformation induced by $\hat{S}$ is a true symmetry of the system, it must leave its Hamiltonian operator invariant. Hence, applying the transformation of operators (4.4) to the Hamiltonian, we deduce $\hat{H}' = \hat{S}\hat{H}\hat{S}^\dagger = \hat{H}$ or equivalently,

$$[\hat{H}, \hat{S}] = 0,$$

i.e. the Hamiltonian and the symmetry operator commute.

For continuous symmetries, infinitesimal transformations can be realised (see Appendix A about Lie groups). Denoting the infinitesimal parameter of the transformation by $\mathrm{d}\alpha \in \mathbb{R}$, we can perform this operation infinitely close to the identity,

$$\hat{S}(\mathrm{d}\alpha) = \mathbb{1} - i\,\mathrm{d}\alpha\hat{G},$$

where $\hat{G}$ is an operator called the *infinitesimal generator* of the transformation (for a reason that will be given in few lines). Exploiting the unitary of $\hat{S}$, we have

$$(\mathbb{1} + i\,\mathrm{d}\alpha\hat{G}^\dagger)(\mathbb{1} - i\,\mathrm{d}\alpha\hat{G}) = \mathbb{1},$$

which at first order in $\mathrm{d}\alpha$ imposes

$$\hat{G}^\dagger = \hat{G}.$$

The operator $\hat{G}$ is thus Hermitian, an observable, meaning that it can describe a quantity that can be measured. The Hamiltonian commuting with $\hat{S}$ then imposes that

$$[\hat{H}, \hat{S}] = [\hat{H}, \hat{G}] = 0.$$

Since $\hat{G}$ commutes with the Hamiltonian, the application of Ehrenfest's theorem[1] implies

$$\frac{\mathrm{d}\langle\hat{G}\rangle}{\mathrm{d}t} = -i\,\langle[\hat{H}, \hat{G}]\rangle = 0,$$

meaning that $\langle\hat{G}\rangle$ is a conserved quantity. This is just an illustration of Noether's famous theorem, which states that if a system has continuous symmetry properties, then there are corresponding quantities whose values are conserved in time (Noether, 1918).

We previously defined an infinitesimal transformation. The generalisation to a finite transformation of a continuous symmetry is straightforward, considering that symmetries must fulfil group properties:

1. The product of two transformations is also a transformation;
2. There is an identity transformation (which leaves physical quantities unchanged);
3. For every transformation, there is an inverse transformation;
4. Transformations are associative.

A finite transformation of parameter α is then defined as the product of N infinitesimal transformations with a parameter $\mathrm{d}\alpha = \lim_{N\to\infty} \alpha/N$, i.e.

$$\hat{S}(\alpha) = \lim_{N\to\infty} \left(\mathbb{1} - i\frac{\alpha}{N}\hat{G}\right)^N \equiv e^{-i\alpha\hat{G}}. \tag{4.5}$$

The operator $\hat{G}$ is thus the 'source' of the finite symmetry transformation, hence the name 'Generator'.

[1] We assume that the operator $\hat{G}$ and thus the symmetry $\hat{S}$ do not depend explicitly on time.

Supplement 4.1. Additive or multiplicative quantum numbers?

Let us denote by $|\psi\rangle$ the state of a system made of N subsystems, with each described by $|\psi_i\rangle$,

$$|\psi\rangle = \bigotimes_{i=1}^{N} |\psi_i\rangle,$$

where the kets $|\psi_i\rangle$ are eigenstates of the operator $\hat{S}$ associated with the symmetry transformation. Let a_i be the corresponding eigenvalues. Under the symmetry, the system is transformed as

$$\hat{S}\,|\psi\rangle \equiv \left(\bigotimes_{i=1}^{N} \hat{S}\right)\left(\bigotimes_{i=1}^{N} |\psi_i\rangle\right) = \bigotimes_{i=1}^{N} \hat{S}\,|\psi_i\rangle = \left(\prod_{i=1}^{N} a_i\right)\bigotimes_{i=1}^{N} |\psi_i\rangle.$$

Consequently, the quantum number of the system is the result of the product of the quantum numbers of the subsystems,

$$a = \prod_{i=1}^{N} a_i.$$

Thus, in general, and in particular *for discrete symmetries, quantum numbers are multiplicative*. However, in the case of continuous symmetries, the symmetry operator can be expressed as an exponential of the generator operator [Eq. (4.5)]. If the kets $|\psi_i\rangle$ are eigenstates of the generator $\hat{G}$ with eigenvalue g_i, they are clearly eigenstates of $\hat{S}(\alpha)$ with eigenvalues $a_i = \exp(-i\alpha g_i)$. Therefore, for the whole system,

$$a = \prod_{i=1}^{N} \exp(-i\alpha g_i) = \exp\left(-i\alpha \sum_{i=1}^{N} g_i\right).$$

Hence, for continuous symmetries, instead of considering $\prod_{i=1}^{N} a_i$ as the conserved quantum number, we can simply consider $g = \sum_{i=1}^{N} g_i$. That's why *continuous symmetries are associated with additive quantum numbers*. This property is clearly due to the possibility, for continuous symmetries, to 'exponentiate' the operators.

4.3 Spacetime Symmetries

4.3.1 Energy–Momentum Conservation

In fundamental physics, it is assumed that all theories are invariant under the redefinition of the origins of space and time. This implies the invariance of the equations under space translations and time translations, leading to the property that an isolated system has 4-momentum conservation. The 3-momentum operator $\hat{\boldsymbol{P}}$ is the infinitesimal generator of the space translations, and the Hamiltonian $\hat{H}$ is the infinitesimal generator of time-translations. For instance, for an infinitesimal space translation of $\delta\alpha$, we expect

$$|x + \delta\alpha\rangle = \hat{S}(\delta\alpha)\,|x\rangle = \left(\mathbb{1} - i\delta\alpha\hat{G}\right)|x\rangle.$$

Hence,

$$\langle x + \delta\alpha|\psi\rangle = \langle x|\psi\rangle + i\delta\alpha\,\langle x|\hat{G}|\psi\rangle,$$

implying

$$\langle x|\hat{G}|\psi\rangle = \frac{1}{i}\frac{\psi(x+\delta\alpha) - \psi(x)}{\delta\alpha} \xrightarrow{\delta\alpha\to 0} \frac{1}{i}\frac{\partial\psi(x)}{\partial x}.$$

As expected, we thus identify $\hat{G}$ as the operator $\frac{1}{i}\frac{\partial}{\partial x}$, the momentum operator (in one dimension in this simple example), and, therefore, the momentum $\boldsymbol{p}$ is conserved. Similarly, for a translation in time, we have

$$|\psi(t+\delta t)\rangle = \hat{S}(\delta t)\,|\psi(t)\rangle = \left(\mathbb{1} - i\delta t\hat{G}\right)|\psi(t)\rangle.$$

Therefore,

$$\hat{G}\,|\psi(t)\rangle = i\frac{|\psi(t+\delta t)\rangle - |\psi(t)\rangle}{\delta t} \xrightarrow{\delta t\to 0} i\frac{\mathrm{d}\,|\psi(t)}{\mathrm{d}t} = \hat{H}\,|\psi(t)\rangle.$$

The generator of time translation is thus the Hamiltonian, as expected, leading to the conservation of energy.

If a system consists of particles without interaction, the global momentum of the system is just the sum of the momenta of the individual particles (translation being a continuous symmetry). However, for particles in interaction with fields, the situation is more complicated, and one should take into account the momenta of the fields themselves. Fortunately, in the case of particle collisions, one assumes that the initial state is described by particles without interaction, which then interact and produce a final state in which the particles are considered again without interaction. Hence, even if one does not try to describe what happens during the collision (the dynamics of the process), one can still write that the sum of the momenta of particles in the initial state is equal to the sum of momenta of particles in the final state.

4.3.2 Total Angular Momentum Conservation

In the same spirit as for 3-momentum conservation, where the homogeneity of space is assumed, space is also considered to be isotropic. Consequently, physics should not depend on the orientation of the axes chosen to formulate the physics equations, meaning that these equations must be invariant under rotations of the whole space. Since the total angular momentum is the infinitesimal generator of the rotations [Eq. (2.69) or (2.70)], the total angular momentum is conserved in reactions. It implies that neither the spin $\boldsymbol{S}$ nor the orbital angular momentum $\boldsymbol{L}$ are conserved separately, but $\boldsymbol{J} = \boldsymbol{L} + \boldsymbol{S}$ is, i.e. both $|\boldsymbol{J}|$ and J_z are the conserved quantum numbers. As an example, consider the decay

$$\rho \to \pi^+ + \pi^-.$$

Both ρ and $\pi^\pm$ are bosons with respective spin 1 and spin 0. The sum of the two spins 0 of the pions gives obviously 0 but does not match the spin 1 of the rho. Thus, the spin number is not conserved. However, this reaction is observed because the two-pions system is produced with an orbital angular momentum $L = 1$.

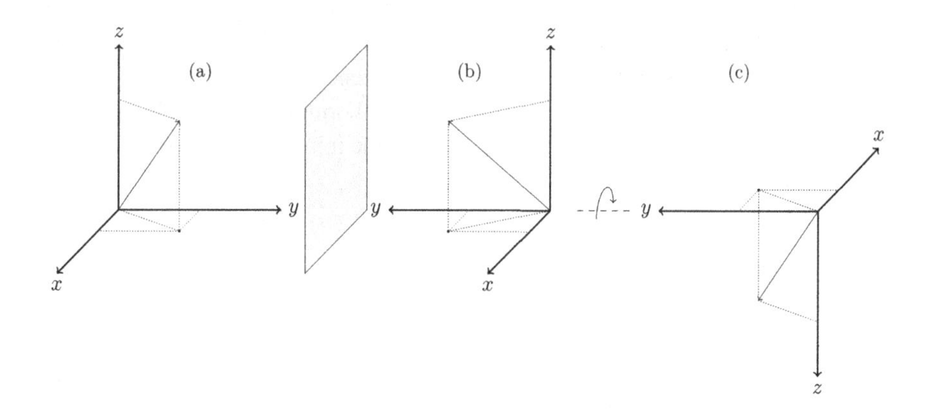

Fig. 4.1 Parity transformation. Sketch (b) is the mirror image of (a) with respect to the (x, z) plane. The mirror is in grey. Sketch (c) results from a 180° rotation of (b) about the y axis. Sketch (c) is also the parity transform of (a).

4.3.3 Parity

Parity is the transformation that performs the space inversion, i.e. that inverts the space coordinates $\boldsymbol{r} \to -\boldsymbol{r}$. It is often stated that parity transforms a system into its image in a mirror. Strictly speaking, this is not true. However, as shown in Fig. 4.1, the parity transformation is equivalent to taking the mirror image of a process, followed by a 180° rotation. Since angular momentum is conserved, rotation invariance is verified, and thus parity symmetry is equivalent to simply comparing a process with its image in a mirror (considering that at the quantum level, the parity and rotation operators commute, see Fig. 4.1).

What is the action of parity at the quantum level? Let us denote the state of the system by $|\psi\rangle$, and the parity linear operator acting in the Hilbert space by $\hat{\mathcal{P}}$. We require that the expectation values of operators describing classical quantities (position, momentum) behave like the corresponding classical variables. For instance, denoting the position operator by $\hat{\boldsymbol{x}}$, the average value of the position must satisfy

$$\boldsymbol{r} \xrightarrow{\text{parity}} -\boldsymbol{r} \Rightarrow \langle\psi'|\hat{\boldsymbol{x}}|\psi'\rangle = -\langle\psi|\hat{\boldsymbol{x}}|\psi\rangle,$$

where $|\psi'\rangle = \hat{\mathcal{P}}|\psi\rangle$. It follows that

$$\langle\psi|\hat{\mathcal{P}}^{\dagger}\hat{\boldsymbol{x}}\hat{\mathcal{P}}|\psi\rangle = -\langle\psi|\hat{\boldsymbol{x}}|\psi\rangle.$$

Since this relation holds for all $|\psi\rangle$, we conclude, given that $\hat{\mathcal{P}}$ is a unitary operator,

$$\hat{\boldsymbol{x}}\hat{\mathcal{P}} = -\hat{\mathcal{P}}\hat{\boldsymbol{x}}. \tag{4.6}$$

The operators anti-commute. We would have reached the same conclusion for the commutation of the parity operator with the momentum operator. A consequence of the relation (4.6) is that

$$\hat{\mathcal{P}}|\boldsymbol{x}\rangle = |-\boldsymbol{x}\rangle, \tag{4.7}$$

where $|\boldsymbol{x}\rangle$ is an eigenstate of the position operator. Let us check

$$\hat{\boldsymbol{x}}\hat{\mathcal{P}}|\boldsymbol{x}\rangle = -\hat{\mathcal{P}}\hat{\boldsymbol{x}}|\boldsymbol{x}\rangle = -\boldsymbol{x}\hat{\mathcal{P}}|\boldsymbol{x}\rangle,$$

which shows that $\hat{\mathcal{P}}\,|\boldsymbol{x}\rangle$ is an eigenstate of the position operator with the eigenvalue $-\boldsymbol{x}$. Hence, it is proportional to $|-\boldsymbol{x}\rangle$ (no degeneracy) up to a phase factor, which is taken to be equal to 1, by convention, leading to the Formula (4.7). Applying the parity transformation twice, we obviously expect to recover the initial state. Using twice (4.7), we obtain

$$\hat{\mathcal{P}}^2\,|\boldsymbol{x}\rangle = \hat{\mathcal{P}}\,|-\boldsymbol{x}\rangle = |\boldsymbol{x}\rangle.$$

Consequently, since this holds for all $|\boldsymbol{x}\rangle$,

$$\hat{\mathcal{P}}^2 = \mathbb{1},$$

the eigenvalues of $\hat{\mathcal{P}}$ are necessarily ± 1. Hence, if $|\psi\rangle$ is an eigenstate of $\hat{\mathcal{P}}$,

$$\hat{\mathcal{P}}\,|\psi\rangle = \pm\,|\psi\rangle.$$

Now consider a particle in an orbit of angular momentum[2] l defining a state $|l,m\rangle$. The angular part of its wave function is given by the spherical harmonics function $Y_l^m = \langle\theta,\phi|l,m\rangle$, with θ and ϕ being the angles of the spherical coordinates. The space inversion imposes $\theta \to \pi - \theta$ and $\phi \to \phi + \pi$. In the Formula (2.57) of Y_l^m, only the Legendre polynomials $P_l^m(\cos\theta)$ depend on θ, whereas only $e^{im\phi}$ depends on ϕ. According to the definition (2.58) of $P_l^m(x)$, we have $P_l^m(-x) = (-1)^{l+m}P_l^m(x)$, with the factor $(-1)^{l+m}$ being due to the derivative operator. This implies that

$$P_l^m(\cos(\pi - \theta)) = (-1)^{l+m}P_l^m(\cos\theta).$$

Moreover, when $\phi \to \phi + \pi$, then $e^{im\phi} \to (-1)^m e^{im\phi}$. Therefore, we conclude that

$$Y_l^m(\pi - \theta, \phi + \pi) = (-1)^l\, Y_l^m(\theta, \phi). \tag{4.8}$$

Property (4.8) shows that under the parity transformation $\hat{\mathcal{P}}$,

$$\hat{\mathcal{P}}\,|l,m\rangle = (-1)^l\,|l,m\rangle.$$

An eigenstate of the orbital angular momentum operator is thus also an eigenstate of the parity operator with eigenvalue $(-1)^l$. Consequently $[\hat{\boldsymbol{L}}, \hat{\mathcal{P}}] = 0$, or equivalently,

$$\hat{\mathcal{P}}\hat{\boldsymbol{L}}\hat{\mathcal{P}}^{-1} = \hat{\boldsymbol{L}}, \tag{4.9}$$

(given that $\hat{\mathcal{P}}^{-1} = \hat{\mathcal{P}}^\dagger = \hat{\mathcal{P}}$). It is not surprising that $\hat{\boldsymbol{L}}$ is not modified by the action of $\hat{\mathcal{P}}$, since classically

$$\boldsymbol{L} = \boldsymbol{r} \times \boldsymbol{p} \xrightarrow{\text{parity}} (-\boldsymbol{r}) \times (-\boldsymbol{p}) = \boldsymbol{L}. \tag{4.10}$$

[2] Orbital angular momentum is always defined relative to something. For example, in a system made of two particles, each particle has the same orbital angular momentum with respect to the CM of the system. Defining an orbital angular momentum for an isolated particle is not really meaningful.

Supplement 4.2. Pseudo-vectors and pseudo-scalars

A vector such as $\boldsymbol{L}$ in Eq. (4.10), which is invariant under parity transformation, is called a pseudo vector or equivalently an axial vector. Standard vectors that change sign under parity are polar vectors, but this denomination is rarely used, and they are simply called vectors. There are also pseudo-scalars that change sign (unlike the normal scalar). For instance, the dot-product of a pseudo-vector with a polar vector is a pseudo-scalar, while the dot-product of two pseudo-vectors or two polar-vectors is a scalar. A Lorentz 4-vector is called a vector if its space-like components form a polar vector and its time-like component a scalar. By extension, scalar mesons are particles (made of a quark and antiquark) in a configuration of total spin $J = 0$ but having even parity, while pseudo-scalar particles have instead odd parity. According to the meson parity Formula (4.17), pseudo-scalar mesons require that the internal spin configuration of the quark–antiquark pair is in a state $s = 0$ (two anti-parallel spins) without relative orbital angular momentum, i.e. a so called s-wave with state $l = 0$. Scalar mesons must have $s = 1$ (two parallel spins) and $l = 1$ (p-wave) such that $\boldsymbol{J} = \boldsymbol{L} + \boldsymbol{S}$ leads to $\mathbf{0} = \mathbf{1} + \mathbf{1}$ total angular momentum quantum number. Similarly, vector mesons are defined as having $J = 1$ and odd parity, whereas pseudo-vector mesons have even parity. Vector mesons can be obtained with two different configurations: $l = 0, s = 1$ (i.e., $\mathbf{1} = \mathbf{0} + \mathbf{1}$) or $l = 2, s = 1$ (i.e., $\mathbf{1} = \mathbf{2} + \mathbf{1}$), while pseudo-vector mesons require $l = 1, s = 0$ (i.e., $\mathbf{1} = \mathbf{1} + \mathbf{0}$) or $l = 1, s = 1$ (i.e., $\mathbf{1} = \mathbf{1} + \mathbf{1}$).

The parity quantum number $(-1)^l$ is related to the motion of particles since a particle at rest cannot have an orbital angular momentum. However, even for particles at rest, parity can be defined. It is called *intrinsic parity*. Let us see how. The relation (4.9), established for the orbital angular momentum, turns out to be also valid for the total angular momentum $\hat{\boldsymbol{J}} = \hat{\boldsymbol{L}} + \hat{\boldsymbol{S}}$,

$$\hat{\mathcal{P}}\hat{\boldsymbol{J}}\hat{\mathcal{P}}^{-1} = \hat{\boldsymbol{J}}. \tag{4.11}$$

Indeed, $\hat{\boldsymbol{J}}$ is the infinitesimal generator of rotations. Thus, if parity commutes with rotations, it necessarily commutes with their generators, and hence, the relation (4.11) holds. Now in the usual three-dimensional space, the parity operator is by definition (inversion of the axes)

$$\hat{\mathcal{P}} = -\mathbb{1}_3,$$

where $\mathbb{1}_3$ is the 3×3 identity matrix. With all matrices commuting with the identity, rotation matrices necessarily commute with $\hat{\mathcal{P}}$. We will then postulate that the commutation rules, valid in the three-dimensional space for these operators, are also valid for the expression of these operators in the Hilbert space where the particle states live. This validates the relation (4.11). Since we are interested in a particle at rest, the total angular momentum is reduced to the spin angular momentum $\hat{\boldsymbol{J}} = \hat{\boldsymbol{S}}$. Hence, Eq. (4.11) tells us that the spin is not changed by a parity transformation, meaning that an eigenstate of the spin operator stays an eigenstate with the same quantum numbers s, s_z after parity transformation. Therefore, this eigenstate can be chosen to be also an eigenstate of the parity operator (since both operators commute) with a defined eigenvalue

$$\hat{\mathcal{P}}\,|s, s_z\rangle = \eta\,|s, s_z\rangle.$$

This eigenvalue $\eta = \pm 1$ is actually the intrinsic parity. The one-particle state labelled by a given energy–momentum and spin, $|E, \boldsymbol{p}, s, s_z\rangle$, is thus for a particle at rest also the eigenstate of the parity operator, i.e.

$$\hat{\mathcal{P}}\,|m, \boldsymbol{0}, s, s_z\rangle = \eta\,|m, \boldsymbol{0}, s, s_z\rangle, \tag{4.12}$$

(with m being the mass of the particle). For moving particles, both intrinsic parity and extrinsic parity (the one coming from the orbital angular momentum that requires particle motion) have to be taken into account. Parity being a multiplicative quantum number (recall Supplement 4.1), a system made of two particles a and b, will have an overall parity[3]

$$\eta = \eta_a \eta_b (-1)^l, \tag{4.13}$$

where η_a and η_b are the intrinsic parities of particles a and b, and l is the quantum number of the orbital angular momentum of these two particles. Similarly, for systems of three particles,

$$\eta = \eta_a \eta_b \eta_c (-1)^l, \text{ with } l = l_{ab} + l_{[ab]c}, \tag{4.14}$$

where as before l_{ab} is the quantum number of the orbital angular momentum between particles a and b, and $l_{[ab]c}$ is the quantum number of the orbital angular momentum between particles c and the CM of the system made of particles a and b. The generalisation to a larger number of particles is straightforward.

The way we introduced the intrinsic parity requires that the particle can be at rest and thus seems to concern only massive particles. Actually, even for massless particles (whose velocity is always c in all reference frames), an intrinsic parity can be defined, but we have to give up with the notion of a parity-eigenstate since by definition $\hat{\mathcal{P}}\,|\boldsymbol{p}\rangle = |-\boldsymbol{p}\rangle$. Instead of relying directly on the spin, parity for massless particles relies on *helicity*. By definition, helicity is the component of the spin along the direction of motion

$$\lambda = \boldsymbol{S} \cdot \frac{\boldsymbol{p}}{|\boldsymbol{p}|}, \tag{4.15}$$

or equivalently, the component of the total angular momentum along the direction of motion since

$$\boldsymbol{J} \cdot \frac{\boldsymbol{p}}{|\boldsymbol{p}|} = \boldsymbol{S} \cdot \frac{\boldsymbol{p}}{|\boldsymbol{p}|} + (\boldsymbol{r} \times \boldsymbol{p}) \cdot \frac{\boldsymbol{p}}{|\boldsymbol{p}|} = \boldsymbol{S} \cdot \frac{\boldsymbol{p}}{|\boldsymbol{p}|} = \lambda.$$

For a particle with spin s, the helicity eigenvalues are then $-s, -(s-1), \ldots, (s-1), s$. Note that for massless particles, only the two helicities $\pm\lambda$ are allowed. Helicity is a useful concept for massless particles because it is a Lorentz-invariant quantity. For massive particles, an observer in an inertial frame travelling in the same direction as that of the particle but at a higher speed will see, from its perspective, the particle going in the opposite direction and thus would invert the sign of the helicity eigenvalue. For massless particles always travelling at c, this possibility is however forbidden, making helicity independent of the boost. In addition, helicity is invariant under rotations since it results from a scalar

[3] Strictly speaking, parity is only meaningful in the CM frame of the system, where the total momentum is null. Indeed, if the total momentum is $\boldsymbol{p}$, the application of the parity operator would change the state of the system $\hat{\mathcal{P}}\,|\boldsymbol{p}\rangle = |-\boldsymbol{p}\rangle$, excluding the possibility of becoming an eigenstate of the parity operator, unless $\boldsymbol{p} = \boldsymbol{0}$.

product. Therefore, we conclude that the helicity of massless particles is independent of the inertial frame.[4] It is thus a good quantum number to label the one-particle state of massless particles $|E = |\boldsymbol{p}|, \boldsymbol{p}, \lambda\rangle$. The intrinsic parity is then defined according to its effect on the one-particle state,

$$\hat{\mathcal{P}}\,||\boldsymbol{p}|, \boldsymbol{p}, \lambda\rangle = \eta\,||\boldsymbol{p}|, -\boldsymbol{p}, -\lambda\rangle. \tag{4.16}$$

The definition of the intrinsic parity given by (4.16) is, for massless particles, the counterpart of definition (4.12) valid for massive particles. Note that since $\boldsymbol{p}$ is reverted and $\boldsymbol{S}$ is not, helicity is reverted by the action of parity transformation. Consequently, in a theory in which the space-inversion symmetry is respected, any massless particles with non-zero helicity must also exist with the opposite helicity. We shall see respectively in Chapters 6 and 8 that both *electromagnetism and strong interactions respect the space-inversion symmetry (parity)*. Therefore, the massless boson mediators of these two interactions, respectively the photon and the gluon, have both helicities ± 1. However, *the weak interaction does not respect the space-inversion symmetry* (see Chapter 9). Consequently, the neutrinos, spin-1/2 particles that are sensitive only to the weak interaction, had only one helicity $-1/2$ (they were considered massless for decades), while the other particle with the opposite helicity was considered to be another particle,[5] the anti-neutrino.

Historically, the intrinsic parity of the proton, neutron and later on the Λ baryon (which contains an s quark) was fixed to 1. The conservation of parity by electromagnetism and strong interactions then allows one to deduce the parity of other particles. We shall see in Section 5.3 that as soon as the intrinsic parity of an elementary fermion described by the Dirac equation (spin 1/2) is fixed (equal to 1), the intrinsic parity of the anti-fermion is the opposite. We deduce that a meson made of a quark–antiquark pair has the parity,

$$\eta_{q\bar{q}} = (-1)^{l+1}. \tag{4.17}$$

In contrast, for bosons, their antiparticles have necessarily the same parity (see Section 5.2). It follows, for instance, that a $\pi^+\pi^-$ system in a relative orbital angular momentum l has parity $(-1)^l$.

4.4 Internal Symmetries

4.4.1 Charge Conjugation

The symmetries we described so far were related to the symmetries of spacetime (continuous or discrete). Charge conjugation is another kind of symmetry: it is a purely internal symmetry acting on the Hilbert space (there is no classical equivalent). It changes a particle

[4] For massive particles, since helicity does not depend on the norm of $\boldsymbol{p}$ but only on its direction, helicity does not depend on its inertial frame as soon as its velocity is smaller than that of the particle (both travelling in the same direction).

[5] The actual status of neutrinos is still an open question.

Supplement 4.3. Continuous symmetries: quantum numbers of antiparticles

We shall see in Chapter 5 that the wave function (or field) of the antiparticle is related to the complex conjugate (or the Hermitian conjugate for fields) of that of the particle. Therefore, if g is the physical quantity (quantum number) associated with the generator $\hat{G}$, then according to the expression (4.5), the wave function of the particle will get an extra phase $e^{-i\alpha g}$, while that of the antiparticle will get $e^{+i\alpha g}$, which can be written as $e^{-i\alpha(-g)}$. Therefore, the quantum number of the antiparticle will be $-g$, the opposite of that of the particle. This property is true in general for all additive quantum numbers.

state into an antiparticle state, and consequently it reverses all additive internal quantum numbers (see Supplement 4.3): electric charge (and other gauge quantum numbers such as the colour of quarks), baryon and lepton numbers, flavour numbers (isospin projection, strangeness, etc.). All quantum numbers related to spacetime symmetries are left unchanged. Let us denote the one-particle state with 4-momentum p, spin s, projection s_z and the set of additive internal quantum numbers $\{q_i\}$ by $|p, s, s_z, \{q_i\}\rangle$. Denoting the (unitary) charge conjugation operator by $\hat{C}$, one thus has

$$\hat{C}\,|p, s, s_z, \{q_i\}\rangle = \eta_C\,|p, s, s_z, \{-q_i\}\rangle\,,$$

where η_C is an arbitrary phase $|\eta_C| = 1$. The ket $|p, s, s_z, \{-q_i\}\rangle$ represents the state corresponding to the antiparticle of the state $|p, s, s_z, \{q_i\}\rangle$. When $\{q_i\} \neq 0$, one can absorb $\eta_C = e^{i\phi}$ by using $e^{-i\phi/2}\,|p, s, s_z, \{q_i\}\rangle$ to define the states of the particle and $e^{i\phi/2}\,|p, s, s_z, \{-q_i\}\rangle$ to define the states of the antiparticle. It follows that

$$\hat{C}\,|p, s, s_z, \{q_i\}\rangle = |p, s, s_z, \{-q_i\}\rangle. \tag{4.18}$$

If we apply the operator $\hat{C}$ twice, we recover the same particle and thus necessarily,

$$\hat{C}^2 = \mathbb{1}.$$

The charge conjugation operator is therefore unitary and Hermitian, just as the parity operator. When $\{q_i\} = 0$, i.e. the particle is its own antiparticle, the previous trick to absorb the phase is impossible. Consequently,

$$\hat{C}\,|p, s, s_z, \{0\}\rangle = \eta_C\,|p, s, s_z, \{0\}\rangle. \tag{4.19}$$

Since $\hat{C}^2 = \mathbb{1}$, the quantity η_C is necessarily real and $\eta_C = \pm 1$. With operator $\hat{C}$ being Hermitian, its eigenvalue η_C is observable. This multiplicative quantum number is called the *charge conjugation parity* or simply *charge parity*.

Only perfectly neutral particles (those having all $q_i = 0$) can be an eigenstate of $\hat{C}$, i.e. particles that are their own antiparticles such as the photon or neutral pion, π^0. A system made of a particle and its antiparticle is also necessarily an eigenstate of the charge conjugation operator. We are going to determine its charge conjugation parity. Let us denote the momentum of the particle in the CM frame by $\boldsymbol{p}$, its spin projection by s_z, and its additive internal quantum numbers by $\{q_i\}$. The antiparticle then has a momentum $-\boldsymbol{p}$, the additive internal quantum numbers $\{-q_i\}$ and a spin projection s'_z. The state of that system is thus described by

$$|\psi\rangle = |\boldsymbol{p}, s_z, \{q_i\}\rangle \otimes |-\boldsymbol{p}, s'_z, \{-q_i\}\rangle. \tag{4.20}$$

It follows that the application of the charge conjugation yields

$$\hat{C}\,|\psi\rangle = |\boldsymbol{p}, s_z, \{-q_i\}\rangle \otimes |-\boldsymbol{p}, s'_z, \{q_i\}\rangle. \tag{4.21}$$

In Chapter 5, we shall learn that the creation operators of two bosons commute, while the creation operators of fermions anti-commute. This property is required, in order to be consistent with the spin-statistics theorem. Consequently, if we permute the two kets on the right-hand side of (4.21), we obtain

$$\hat{C}\,|\psi\rangle = \pm\,|-\boldsymbol{p}, s'_z, \{q_i\}\rangle \otimes |\boldsymbol{p}, s_z, \{-q_i\}\rangle,$$

with a plus sign used for a system of bosons and minus sign for fermions. If the particles of the system are in a relative orbital angular momentum l, the exchange of the 2-momenta yields

$$\hat{C}\,|\psi\rangle = \pm(-1)^l\,|\boldsymbol{p}, s'_z, \{q_i\}\rangle \otimes |-\boldsymbol{p}, s_z, \{-q_i\}\rangle. \tag{4.22}$$

The last step is to exchange the two spins on the right-hand side of Eq. (4.22) in order to recover the state $|\psi\rangle$ of (4.20). If the system is a pair of scalar–antiscalar bosons, we can ignore this step since the pair will have a spin $s = 0$, as well as the individual spins $s_z = s'_z = 0$. The ket $|\psi\rangle$ would just be defined as

$$|\psi\rangle = |\boldsymbol{p}, \{q_i\}\rangle \otimes |-\boldsymbol{p}, \{-q_i\}\rangle,$$

and the Formula (4.22) would read $\hat{C}\,|\psi\rangle = (-1)^l\,|\psi\rangle$. We thus conclude

$$\eta_C(\text{scalar} - \text{antiscalar}) = (-1)^{\text{l}}. \tag{4.23}$$

Now if the system is a spin-1/2 fermion–anti-fermion pair, the effect of the exchange of the spins depends on the spin of the pair. Indeed, if the sum of the two spins forms a triplet, thus a spin 1, the swap of spins is symmetric, with the triplet being $|\uparrow\uparrow\rangle$, $(|\uparrow\downarrow\rangle + |\downarrow\uparrow\rangle)\sqrt{2}$ and $|\downarrow\downarrow\rangle$. If it forms a spin-0 singlet $(|\uparrow\downarrow\rangle - |\downarrow\uparrow\rangle)\sqrt{2}$, the swap is then antisymmetric, producing an extra -1 factor. Hence, the effect of the spin-exchange is given by the factor $(-1)^{s+1}$, where s is the spin of the system. Therefore, for a fermion–antifermion (spin 1/2) system, we conclude from Eq. (4.22),

$$\hat{C}\,|\psi\rangle = -(-1)^l(-1)^{s+1}\,|\psi\rangle,$$

It follows that

$$\eta_C(\text{fermion} - \text{anti-fermion}) = (-1)^{\text{l+s}}. \tag{4.24}$$

Charge conjugation is a useful symmetry only if it is approximatively conserved. It turns out that *both electromagnetism* and *the strong interactions respect the charge conjugation symmetry* (see Section 5.4.4 for justification). Therefore, a particle that decays via one of these two interactions can only decay into a channel having the same overall charge parity. A simple example is the π^0. Being a meson (a pair of quark–antiquark), the binding is due to the strong interaction (respecting charge conjugation symmetry), while the decay can proceed only[6] via the electromagnetic interaction. The dominant channel is $\pi^0 \to \gamma + \gamma$. Hence,

$$\eta_C(\pi^0) = \eta_C^2(\gamma) = 1.$$

Now, we shall see in Section 5.4.4 that the charge conjugation parity of the photon is $\eta_C(\gamma) = -1$. Consequently, the decay of the π^0 into an odd number of photons is forbidden. For instance, the decay into three photons has never been observed. The limit from Particle Data Group (2022) on its branching ratio is

$$\text{BR}(\pi^0 \to 3\gamma) < 3.1 \times 10^{-8}.$$

4.4.2 Baryon and Lepton Numbers

Like the electric charge, the baryon number B and the lepton number L are some examples of additive quantum numbers. *Baryon and lepton numbers are supposed to be conserved by all interactions,* but unlike the electric charge, there is no fundamental mechanism in the Standard Model justifying this assumption (while the electric charge conservation is a consequence of the gauge symmetry, as we shall see in Chapter 6).

In the late 1940s and early 1950s, in addition to the well-known electron, proton, neutron and positron, many new particles were observed, such as the π^0, $K^{0,\pm}$, Λ, Δ, and Σ. It was noticed that the number of baryons ($p, n, \Lambda, \Delta, \Sigma$), the heavy particles observed at the time with a mass larger than that of the proton, was conserved. Decays like $n \to pe^-(+\bar{\nu})$ and $\Lambda \to p\pi^-$ were observed but not $n \to e^+e^-$, $\Lambda \to \pi^+\pi^-$. A new quantum number, the baryon number (B), was then introduced. Baryons have $B \neq 0$, while the lightest ones ($m < m_p$), called mesons, have $B = 0$. Therefore, the baryon number of a state is defined as the number of baryons minus the number of anti-baryons,

$$B = N_{\text{baryons}} - N_{\text{anti-baryons}}.$$

Nowadays, we know that baryons and mesons are not elementary particles but have subconstituents, the quarks. Quarks are assumed to democratically share the baryon number, meaning that they have $B = 1/3$ since a baryon is made of three quarks (see Table 4.2). An antiquark has a baryon number $B = -1/3$, and hence, a meson made of a quark and an antiquark has baryon number zero. All other particles, including the leptons and photons, have obviously $B = 0$. With the proton being the lightest baryon, the conservation of the baryon number implies that the proton cannot decay, 'explaining' its stability. Note that the conservation of the baryon number implies that the number of baryons minus the number of anti-baryons is constant in a reaction. Consequently, if in a reaction, an anti-baryon is

[6] The π^0 is the lightest meson, so a decay via the strong interaction is kinematically impossible. Now, one cannot formally exclude a decay via the weak interaction such as $\pi^0 \to \nu\bar{\nu}$, but such a decay would be strongly suppressed by the tiny mass of the neutrinos. Indeed, with the π^0 being spinless, both ν and $\bar{\nu}$ must have the same helicity, which is almost impossible because the weak interaction depends on the chirality of the fermions.

Table 4.2 Baryon number and lepton numbers of the elementary fermions of the Standard Model.

Fermions	u,c,t d,s,b	ν_e e^-	ν_μ μ^-	ν_τ τ^-	$\bar{u},\bar{c},\bar{t}$ $\bar{d},\bar{s},\bar{b}$	$\bar{\nu}_e$ e^+	$\bar{\nu}_\mu$ μ^+	$\bar{\nu}_\tau$ τ^+
Baryon nb B	$\frac{1}{3}$	0	0	0	$-\frac{1}{3}$	0	0	0
Lepton nb L	0	1	1	1	0	-1	-1	-1
Individual lepton nb L_e	0	1	0	0	0	-1	0	0
Individual lepton nb L_μ	0	0	1	0	0	0	-1	0
Individual lepton nb L_τ	0	0	0	1	0	0	0	-1

produced while the initial state did not contain an anti-baryon, the final state must also contain a baryon to compensate for the anti-baryon production. A simple example would be $p+p \to p+p+\bar{p}+p$. At the scale of quarks, the baryon number conservation implies that the quantity $N_q - N_{\bar{q}}$ is conserved. Formally, the conservation of the baryonic number is related to the possibility of redefining the phase of quark fields. This aspect is explained in the QCD chapter, Section 8.1.1.

Similar to baryons, we observe that in experiments, the net number of leptons (i.e., the number of leptons minus the number of anti-leptons) is conserved,

$$L = N_{\text{leptons}} - N_{\text{anti-leptons}}.$$

The lepton number, L, is an additive quantum number, and the value 1 is assigned to all leptons (electron, muon, tau and the three associated neutrinos) and -1 to all anti-leptons (see Table 4.2). All other particles have a null lepton number. The conservation of the lepton number implies that in the pion decay, $\pi^- \to \mu^- + \bar{\nu}_\mu$, the neutrino produced is actually an anti-neutrino. For the same reason, the decay $\pi^+ \to \mu^+ + \nu_\mu$ implies the creation of a neutrino. As with the baryonic number, the leptonic number is also formally related to the possibility of redefining the phase of lepton fields without affecting the interactions. It is worth mentioning that *individual lepton numbers* have also been introduced: one for each generation of leptons, denoted L_e, L_μ and L_τ (cf. Table 4.2). Historically, they were introduced to explain why, experimentally, the decay $\mu^- \to e^- + \gamma$ is never observed. This decay conserves the lepton number L, but violates L_μ since $L_\mu = 1$ for the muon and zero for all other particles of the reaction. Nowadays, we know that the individual lepton numbers are not, strictly speaking, conserved. The oscillation of neutrinos makes this reaction possible in theory. We shall see in Chapter 9 that ν_μ can transform spontaneously in ν_e during its propagation. Such a transformation, experimentally observed over long distances, clearly violates the individual lepton number conservation. However, locally at the spacetime point where an interaction has occurred (the *vertex* of the interaction), the individual lepton numbers are always conserved in the Standard Model. Only during the free propagation of neutrinos, a violation can occur.

Other internal quantum numbers exist, as partially listed in Table 4.1. They will be presented in the specific chapters on interactions (mostly in Chapters 8 and 10).

Problems

4.1 Explain why the following reactions are forbidden:

(1) $p+\bar{p} \to \gamma$ (2) $n \to p+\gamma$ (3) $\Lambda^0 \to \pi^+ + e^- + \bar{\nu}_e$
(4) $K^- \to \pi^0 + e^-$ (5) $p \to n + e^+ + \nu_e$ (6) $\gamma \to e^+ + e^-$
(7) $\pi^0 \to \gamma\gamma\gamma$ (8) $p + \nu_\mu \to n + \mu^+$ (9) $p+\bar{p} \to \Lambda^0 + \Lambda^0$

For (7), assume an electromagnetic interaction (very rare decay $\pi^0 \to 3\gamma$ via a weak interaction may be possible but not yet observed). The spin J, parity P, charge conjugation parity C and mass m in MeV/c^2 of the hadrons (J^{PC}, m) are π^0:$(0^{-+}, 135)$, $\pi^\pm$:$(0^-, 140)$, $K^\pm$:$(0^-, 494)$, p:$(1/2^+, 938)$, n:$(1/2^+, 940)$, Λ^0:$(1/2^+, 1116)$. In addition, γ:$(1^{--}, 0)$, and $m_e = 0.511$, $m_\mu = 106$ and $m_\nu \sim 0$.

4.2 Consider the following reaction (where the proton is at rest): $\pi^- + p \to \Delta^0 \to \pi^0 + n$. The intermediate resonance, Δ^0, has a spin 3/2 and an intrinsic parity +1. The decay occurs via the strong interaction, which conserves the parity. The proton and the neutron have spin 1/2 and parity +1, while the pion has a spin 0 and parity −1.

1. Considering the z-axis as the direction of the π^-, show that the angular momentum conservation imposes that the projection of the Δ^0 spin is $S_z(\Delta^0) = \pm 1/2$.
2. What is the value of the orbital angular momentum l of the final system $\pi^0 n$?
3. Assuming $S_z(\Delta^0) = +1/2$, show that the neutron angular distribution is $(1 + 3\cos^2\theta)/(8\pi)$. *Hint: consider successively the neutron in the spin up and spin-down states and determine the corresponding angular distribution. Use Fig. 2.1 for the Clebsch–Gordan coefficients and the expression of the spherical harmonics.*
4. Same question if $S_z(\Delta^0) = -1/2$, can you guess the answer without any calculation?

5 From Wave Functions to Quantum Fields

The purpose of this chapter is to clearly define the mathematical objects that describe particles of various kinds: bosons (spin-0 and spin-1) or spin-1/2 fermions. This is an important step towards the calculation of Feynman diagrams.

5.1 The Schrödinger Equation

5.1.1 The Correspondence Principle

In quantum mechanics, the correspondence principle states that a quantum of momentum $\boldsymbol{p}$ is a wave with a wavevector $\boldsymbol{k}$, through the relation postulated by de Broglie, $\boldsymbol{p} = \hbar\boldsymbol{k}$. The energy of the quantum, E, is related to the angular frequency ω of the wave via the relation initially postulated by Einstein for photons, $E = \hbar\omega$. Therefore, Schrödinger expressed the phase of the quantum plane wave as $Et - \boldsymbol{p}\cdot\boldsymbol{x}$, yielding the wave function

$$\psi(\boldsymbol{x},t) = Ne^{-i(\omega t - \boldsymbol{k}\cdot\boldsymbol{x})} = Ne^{-\frac{i}{\hbar}(Et-\boldsymbol{p}\cdot\boldsymbol{x})} = Ne^{-\frac{i}{\hbar}p_\mu x^\mu} = Ne^{-\frac{i}{\hbar}p\cdot x}, \tag{5.1}$$

where N is a normalisation factor. This function is an eigenstate of the operators

$$\hat{H} = i\hbar\frac{\partial}{\partial t} \ , \quad \hat{\boldsymbol{P}} = -i\hbar\boldsymbol{\nabla}, \tag{5.2}$$

with the respective corresponding eigenvalues E and $\boldsymbol{p}$. This immediately leads to the identification of $\hat{H}$ as the energy operator and $\hat{\boldsymbol{P}}$ as the momentum operator, or in a covariant notation as the 4-momentum operator

$$\hat{P_\mu} = i\hbar\partial_\mu = i\hbar\frac{\partial}{\partial x^\mu} = \begin{pmatrix} \hat{H} \\ -\hat{\boldsymbol{P}} \end{pmatrix}. \tag{5.3}$$

Moreover, the conservation of the non-relativistic energy $E = \frac{p^2}{2m} + V$, where V is the potential energy, is translated in terms of the equality between operators

$$\hat{H} = \frac{1}{2m}\hat{\boldsymbol{P}}^2 + \hat{V}.$$

Applied to a general wave function, this implies the Schrödinger equation

$$\boxed{i\hbar\frac{\partial}{\partial t}\psi(\boldsymbol{x},t) = -\frac{\hbar^2}{2m}\boldsymbol{\nabla}^2\psi(\boldsymbol{x},t) + V\psi(\boldsymbol{x},t).} \tag{5.4}$$

5.1.2 4-Current

The notion of probability density and probability current (i.e., the flux of a probability density) is central to the interpretation of Schrödinger's equation, where $\psi(\boldsymbol{x},t)$ is the probability amplitude of measuring a particle in a given place and $\rho = |\psi(\boldsymbol{x},t)|^2$ is the probability density. In what follows, we will find this expression for the density using a method that will apply to the relativistic wave equations. The probability of measuring a particle somewhere is a conserved quantity in the sense that

$$\frac{\partial}{\partial t}\int_{\text{universe}} \rho \, \mathrm{d}^3\boldsymbol{x} = 0.$$

The conservation implies that the rate of decrease in the number of particles in a volume $\mathcal{V}$, $-\frac{\partial}{\partial t}\int_{\mathcal{V}} \rho \; \mathrm{d}^3\boldsymbol{x}$, must be equal to the total flux of particles escaping from that volume, $\oint_S \boldsymbol{j}\cdot\boldsymbol{n} \; \mathrm{d}S$, where $\boldsymbol{j}$ is the probability current and $\boldsymbol{n}$ is a unit vector normal to the element $\mathrm{d}S$ of the surface S enclosing the volume $\mathcal{V}$. Because of the Gauss theorem, the closed surface integral is equal to $\int_{\mathcal{V}} \nabla\cdot\boldsymbol{j} \; \mathrm{d}^3\boldsymbol{x}$. It follows that

$$-\frac{\partial}{\partial t}\int_{\mathcal{V}} \rho \; \mathrm{d}^3\boldsymbol{x} = \int_{\mathcal{V}} \nabla\cdot\boldsymbol{j} \; \mathrm{d}^3\boldsymbol{x}.$$

Since this equation must be true for any volume $\mathcal{V}$, we find the continuity equation,

$$\frac{\partial\rho}{\partial t} + \nabla\cdot\boldsymbol{j} = 0. \tag{5.5}$$

In order to clearly identify ρ and $\boldsymbol{j}$ from the Schrödinger equation, we first conjugate it and multiply by ψ,

$$-i\hbar\psi(\boldsymbol{x},t)\frac{\partial}{\partial t}\psi^*(\boldsymbol{x},t) = -\frac{\hbar^2}{2m}\psi(\boldsymbol{x},t)\nabla^2\psi^*(\boldsymbol{x},t) + V\psi^*(\boldsymbol{x},t)\psi(\boldsymbol{x},t),$$

while multiplying the Schrödinger equation by $\psi^*(\boldsymbol{x},t)$ yields

$$i\hbar\psi^*(\boldsymbol{x},t)\frac{\partial}{\partial t}\psi(\boldsymbol{x},t) = -\frac{\hbar^2}{2m}\psi^*(\boldsymbol{x},t)\nabla^2\psi(\boldsymbol{x},t) + V\psi^*(\boldsymbol{x},t)\psi(\boldsymbol{x},t).$$

Subtracting the two previous results leads to

$$i\hbar\frac{\partial}{\partial t}(\psi\psi^*) + \frac{\hbar^2}{2m}\nabla\cdot(\psi^*\nabla\psi - \psi\nabla\psi^*) = 0.$$

A simple comparison with the continuity equation (5.5) allows the identification

$$\rho = |\psi|^2\,, \quad \boldsymbol{j} = i\frac{\hbar}{2m}(\psi\nabla\psi^* - \psi^*\nabla\psi). \tag{5.6}$$

With the plane-wave equation (5.1), the number of particles per unit volume is $\rho = |N|^2$ and $\boldsymbol{j} = |N|^2\boldsymbol{p}/m = |N|^2\boldsymbol{v}$. As soon as the normalisation N is chosen, the number of particles per unit volume is fixed. However, if this choice is done in a particular reference frame, in another frame boosted by a γ factor, we would have a density increased by the γ factor since the volume is contracted by $1/\gamma$ in the direction of the boost. This would mean that in this other frame, one should use another $N' = \gamma N$ in order to take into account this increase. A choice depending on the frame is clearly not Lorentz invariant, which is not surprising, since we started from the non-relativistic Schrödinger equation. The adoption of relativistic wave equations will solve this issue.

5.2 Spinless Particles

5.2.1 The Klein–Gordon Equation

Let us apply the correspondence principle (5.2) to the relativistic energy, $E^2 = |\boldsymbol{p}|^2 + m^2$ (in natural units). It implies

$$-\frac{\partial^2}{\partial t^2}\phi = -\nabla^2\phi + m^2\phi,$$

which is the Klein–Gordon equation for a free particle,

$$\boxed{\left(\Box + m^2\right)\phi = 0.} \tag{5.7}$$

Notice that both the d'Alembertian $\Box = \partial^\mu\partial_\mu$ and the mass m are Lorentz scalars. If $\phi(x)$ is a (Lorentz) scalar function, as it turns out to be, then the Klein–Gordon equation is invariant under Lorentz transformations.

What are the probability density and the probability current for such an equation? Following the same procedure as for the Schrödinger equation, let us first try to determine the continuity equation induced by the Klein–Gordon solutions. Taking the complex conjugate of Eq. (5.7) and multiplying the result by ϕ yields

$$\phi\frac{\partial^2}{\partial t^2}\phi^* - \phi\nabla^2\phi^* + m^2\phi\phi^* = 0,$$

while multiplying Eq. (5.7) by ϕ^* gives

$$\phi^*\frac{\partial^2}{\partial t^2}\phi - \phi^*\nabla^2\phi + m^2\phi\phi^* = 0.$$

The subtraction of these two previous equations eliminates the mass term, yielding the continuity equation

$$\frac{\partial}{\partial t}\left(\phi^*\frac{\partial}{\partial t}\phi - \phi\frac{\partial}{\partial t}\phi^*\right) + \nabla\cdot(\phi\nabla\phi^* - \phi^*\nabla\phi) = 0.$$

The first term in the parenthesis is identified as the density, but it is purely imaginary. The equation can obviously be multiplied by a constant to get a real density. Let us simply multiply by i. It follows the identification

$$\rho = i\left(\phi^*\frac{\partial}{\partial t}\phi - \phi\frac{\partial}{\partial t}\phi^*\right), \quad \boldsymbol{j} = i(\phi\nabla\phi^* - \phi^*\nabla\phi). \tag{5.8}$$

Let us examine what we obtain with a plane wave, $\phi(x) = N\exp(-ip\cdot x)$. Inserting it in the density and current Formulas (5.8), we obtain $\rho = 2E|N|^2$ and $\boldsymbol{j} = 2\boldsymbol{p}|N|^2$, or in terms of 4-current,

$$j^\mu = \begin{pmatrix}\rho \\ \boldsymbol{j}\end{pmatrix} = 2|N|^2\begin{pmatrix}E \\ \boldsymbol{p}\end{pmatrix} = 2|N|^2 p^\mu. \tag{5.9}$$

But, by construction, the Klein–Gordon equation applied to a plane wave leads to the relativistic energy $-E^2\phi + |\boldsymbol{p}|^2\phi + m^2\phi = 0$. Consequently, two values for the energy are possible, a positive and a negative one,

$$E = \pm\sqrt{|\boldsymbol{p}|^2 + m^2}. \tag{5.10}$$

What is the problem with negative energies? Since the 3-momentum can take all positive values, the negative energies can be more and more negative, and thus there is no possible ground state corresponding to the minimum energy of the system. In addition, according to Eq. (5.9), the negative-energy solution leads to a negative density. This clearly excludes the interpretation of the density as a probability density. We shall see how to cope with this issue in the following, but at this stage, let us admit that in a given reference frame, the number of particles per unit volume is correctly given by $\rho = 2E|N|^2$, with $E > 0$. The 4-current (5.9) tells us that in another frame boosted by a γ factor, we should measure $\rho' = 2E'|N|^2$ with $E' = \gamma E$ (assuming $\boldsymbol{p} = \boldsymbol{0}$, for simplicity, in the Lorentz transformation used to calculate E'). Let us check the consistency of what we found. The measured density in the boosted frame is the number of particles contained in the original volume $\mathcal{V}$ divided by the new volume:

$$\rho' = \frac{2E|N|^2\mathcal{V}}{\mathcal{V}'} = \frac{2E|N|^2\mathcal{V}}{\mathcal{V}/\gamma} = 2E\gamma|N|^2 = 2E'|N|^2.$$

It thus validates the density[1] deduced with the Klein–Gordon equation. Consequently, unlike to what is found for the Schrödinger equation, the density deduced from the Klein–Gordon equation has the same expression in any inertial frame, i.e. the time-like component of the current (5.9). It is thus appropriate for relativistic particles.

5.2.2 Reinterpretation of the four-Current

The positive and negative energies (5.10) are associated with two solutions of the Klein–Gordon equation, respectively,

$$\phi_{1p}(x) = Ne^{-i(Et-\boldsymbol{p}\cdot\boldsymbol{x})}\ ,\ \ \phi_{2p}(x) = Ne^{-i(-Et-\boldsymbol{p}\cdot\boldsymbol{x})},$$

where the parameter $E = \sqrt{|\boldsymbol{p}|^2 + m^2}$ here is always positive. Those solutions have thus the respective energies E and $-E$ and lead to the 4-currents,

$$j^\mu(\phi_{1p}) = 2|N|^2\begin{pmatrix} E \\ \boldsymbol{p} \end{pmatrix}\ ,\ \ j^\mu(\phi_{2p}) = 2|N|^2\begin{pmatrix} -E \\ \boldsymbol{p} \end{pmatrix}.$$

With a negative density $\rho = j^0(\phi_{2p})$, the probability interpretation is impossible. However, Pauli and Weisskopf suggested (six years after the development of the Dirac equation) to interpret the 4-current as a current of charges. In order to do so, the direction of the current must logically be reversed when the charge is changed to its opposite. This is possible if the direction of the momentum is changed accordingly. Hence, let us define two new wave functions

$$\begin{aligned}
\phi_p^{(+)}(x) &= \phi_{1p}(x) = Ne^{-i(Et-\boldsymbol{p}\cdot\boldsymbol{x})} = Ne^{-ip\cdot x} \\
\phi_p^{(-)}(x) &= \phi_{2-p}(x) = Ne^{-i((-E)t-(-\boldsymbol{p})\cdot\boldsymbol{x})} = Ne^{+ip\cdot x}.
\end{aligned} \tag{5.11}$$

[1] A density has the dimensions of (volume)$^{-1}$, i.e. in natural units (energy).^{3} Hence, the constant N in $\rho = 2E|N|^2$ carries the units of energy. When $N = 1$ is chosen, we say by abuse of language that there are $2E$ particles per unit volume.

They are still the solutions of the Klein–Gordon equation with a mass m, because if $E^2 = |\boldsymbol{p}|^2 + m^2$, then $(-E)^2 = |-\boldsymbol{p}|^2 + m^2$. Using the Formula (5.8), they lead to the 4-currents

$$j^{\mu}_{\text{em}}(\phi_{\boldsymbol{p}}^{(\pm)}) = \pm 2|N|^2 \begin{pmatrix} E \\ \boldsymbol{p} \end{pmatrix}. \tag{5.12}$$

Consequently, the two wave functions $\phi_{\boldsymbol{p}}^{(\pm)}$ can now be interpreted consistently as describing states of the opposite electric charge. At the time of Pauli and Weisskopf's interpretation, Dirac had already proposed his famous equation and Carl D. Anderson, in 1932, had already discovered the anti-electron, also known as the positron. It was then logical to consider $\phi_{\boldsymbol{p}}^{(-)}$ as a function associated with the state of an antiparticle with a positive energy and a negative charge. An antiparticle is the charge conjugate state of the positive-energy state and is thus associated with the complex conjugate of the wave function of the particle, since according to the definition (5.11),

$$\phi_{\boldsymbol{p}}^{(-)}(x) = \left(\phi_{\boldsymbol{p}}^{(+)}(x)\right)^{*}.$$

With respect to the particle solution, the antiparticle has an opposite charge and a reversed momentum direction, but the same mass, for $m^2 = E^2 - |\boldsymbol{p}|^2 = (-E)^2 - |-\boldsymbol{p}|^2$. There is no place here for negative energies or propagation backwards in time. We can just say that an antiparticle can be represented by the wave function of a particle as soon as p^{μ} is reversed. This is the reason why in Feynman diagrams, an antiparticle is represented by a line with an arrow going backwards in time, as we shall see in Section 5.3.4, in the context of the Dirac equation.

5.2.3 A Few Words about the Quantised Field

So far, we have implicitly considered that the Klein–Gordon equation describes a single relativistic particle of spin 0. Indeed, if it were describing particles with non-zero spin, the solutions of the Klein–Gordon equation should be able to distinguish particles with different spin projections. However, there is no place in the two solutions (5.11) for such extra degrees of freedom. In addition, these solutions are built from 4-scalars and thus are not affected by a Lorentz transformation. Thus, the Klein–Gordon equation seems consistent with the description of a single relativistic particle of spin 0. However, as soon as the energies of the particles approach the relativistic regime, this interpretation cannot hold. Indeed, imagine a single particle of mass m in a box of size L. Because of the Heisenberg uncertainty, $\Delta p \geq \hbar/L$ (I exceptionally reintroduce the $\hbar$ and c factors in this paragraph). Since the particle is relativistic, $E \approx pc$, and hence $\Delta E \geq \hbar c/L$. But this uncertainty can potentially exceed the threshold for particle–anti-particle production from the vacuum, $\Delta E = 2mc^2$. We can then conclude that as soon as a particle is localised within a distance called the Compton wavelength, of the order of $\lambda = \hbar/mc$ (forgetting the factor 2), the probability of detecting a particle–anti-particle pair becomes significant. Thus, the notion of a single particle within a volume becomes nonsense. The formalism must be able to describe states of any number of particles. This is precisely what quantum field theory does by changing the wave functions into operators, interpreted as creation and annihilation operators, acting on a Hilbert space containing the particle states.

Let us use $N = 1$ as the normalisation of the solutions (5.11):

$$\phi_{\boldsymbol{p}}^{(\pm)}(x) = e^{-i(\pm p\cdot x)}. \tag{5.13}$$

The wave functions $\phi_{\boldsymbol{p}}^{(\pm)}(x)$ constitute a basis for a particle with momentum $\boldsymbol{p}$, and hence any wave function can be expressed as

$$\phi(x) = \int \frac{d^3\boldsymbol{p}}{(2\pi)^3 2E_p} \left[a_p \phi_{\boldsymbol{p}}^{(+)}(x) + b_p^* \phi_{\boldsymbol{p}}^{(-)}(x)\right], \text{ with } E_p = \sqrt{|\boldsymbol{p}|^2 + m^2}. \tag{5.14}$$

Note that an integration is used and not a simple summation. Implicitly, the volume in which the particle evolves is large enough (it could be the universe) to consider continuous values of the momenta. The complex numbers a_p and b_p^* are the amplitudes of the eigenmodes of ϕ with momentum $\boldsymbol{p}$, and they correspond to the coefficients of the Fourier transform of the wave function basis. The relativistic normalisation is explicit with the coefficient $1/E_p$. Alternatively, it could have been absorbed in the definition of the coefficients a_p and b_p^*. At the stage of Eq. (5.14), $\phi(x)$ is simply a classical field (or equivalently a wave function), a complex function spreading over the whole spacetime. The next stage is to quantise the field itself: $\phi(x)$ is going to be an operator acting on the Hilbert space of state vectors, the states corresponding to a certain number of particles with a definite momentum. The field still obeys the same dynamic equation (Klein–Gordon), but a_p and b_p^* are now operators. Now, b_p^* is interpreted as the Hermitian conjugate of the operator b_p. It is thus written $b_p^\dagger$. The quantum field and its Hermitian adjoint are then

$$\begin{aligned}
\phi(x) &= \int \frac{d^3\boldsymbol{p}}{(2\pi)^3 2E_p} \left[a_p \phi_{\boldsymbol{p}}^{(+)}(x) + b_p^\dagger \phi_{\boldsymbol{p}}^{(-)}(x)\right], \\
\phi^\dagger(x) &= \int \frac{d^3\boldsymbol{p}}{(2\pi)^3 2E_p} \left[a_p^\dagger \phi_{\boldsymbol{p}}^{(+)*}(x) + b_p \phi_{\boldsymbol{p}}^{(-)*}(x)\right].
\end{aligned} \tag{5.15}$$

One can show that a_p and $a_p^\dagger$ (as well as b_p and $b_p^\dagger$) obey commutation rules similar to those of the harmonic oscillator operators[2]: they are respectively annihilation and creation operators. So, $a_p^\dagger$ creates a quantum of excitation associated with the plane wave $\phi_{\boldsymbol{p}}^{(+)}$ that plays the role of a propagation mode with vector $\boldsymbol{p}$. This quantum is interpreted as a particle of mass m propagating with momentum $\boldsymbol{p}$. The operators b_p and $b_p^\dagger$ form another set of annihilation and creation operators associated with the plane wave $\phi_{\boldsymbol{p}}^{(-)}$ and similarly, $b_p^\dagger$ creates a particle with the same mass m as $a_p^\dagger$. As annihilation and creation operators (also called Fock operators), they satisfy the commutation relations:

$$\begin{aligned}
\left[a_{p'}, a_p^\dagger\right] &= \left[b_{p'}, b_p^\dagger\right] = (2\pi)^3 2E_p\, \delta^{(3)}(\boldsymbol{p}' - \boldsymbol{p}), \\
\left[a_{p'}, a_p\right] &= \left[b_{p'}, b_p\right] = \left[a_{p'}^\dagger, a_p^\dagger\right] = \left[b_{p'}^\dagger, b_p^\dagger\right] = 0, \\
\left[a_{p'}^\dagger, b_p^\dagger\right] &= \left[a_{p'}, b_p^\dagger\right] = \left[a_{p'}^\dagger, b_p\right] = 0.
\end{aligned} \tag{5.16}$$

Note that according to the first commutation relation, the operators a_p and b_p have the dimension of $[\text{energy}]^{-1}$ (since the $\delta^{(3)}(\boldsymbol{p})$ carries the units of $[\text{energy}]^{-3}$ in natural units). Hence, the Klein–Gordon field (5.15) has the dimensions of energy. The vacuum state $|0\rangle$

[2] The reader can consult a quantum field theory textbook, such as Peskin and Schroeder (1995), for the details. In Chapter 6, however, a brief summary is provided.

is the normalised state $\langle 0|0\rangle = 1$ of zero particles. Do not confuse the vacuum $|0\rangle$ with the vector of the Hilbert space with zero norm denoted by 0. With our normalisation,[3] the action of the operators on the states is

$$
\begin{aligned}
a_p\,|0\rangle &= 0, & a_p^\dagger\,|0\rangle &= |1_a,\boldsymbol{p}\rangle,\\
a_{p'}\,|1_a,\boldsymbol{p}\rangle &= (2\pi)^3 2E_p\,\delta^{(3)}(\boldsymbol{p}'-\boldsymbol{p})\,|0\rangle, & a_p^\dagger\,|1_a,\boldsymbol{p}\rangle &= |2_a,\boldsymbol{p}\rangle.
\end{aligned}
$$

The state $|1_a,\boldsymbol{p}\rangle$ corresponds to a state with one particle of type a, with a momentum $\boldsymbol{p}$ and an energy $E_p = \sqrt{m^2+|\boldsymbol{p}|^2}$. Likewise, $|2_a,\boldsymbol{p}\rangle$ contains two particles of type a with the same momentum $\boldsymbol{p}$. Similar relations are obtained with b operators for particles of type b. Moreover, since the operators a and b commute,

$$
b_p\,|1_a,\boldsymbol{p}\rangle = b_p a_p^\dagger\,|0\rangle = a_p^\dagger b_p\,|0\rangle = 0.
$$

Supplement 5.1. Relativistic normalisation of states

In Section 5.2.1, we saw that the density of particles deduced from the Klein–Gordon equation was consistent with a relativistic normalisation that is independent of the inertial frame. Let us check here that the commutation rules (5.16) lead to an appropriate normalisation of particle states. From the commutation rules, we deduce the following normalisation:

$$
\langle 1_a,\boldsymbol{p}_1|1_a,\boldsymbol{p}_2\rangle = \langle 0|\,a_{p_1}a_{p_2}^\dagger\,|0\rangle = \left\langle 0\left|\,\left[a_{p_1},a_{p_2}^\dagger\right]\,\right|0\right\rangle = (2\pi)^3 2E_{p_1}\,\delta^{(3)}(\boldsymbol{p}_1-\boldsymbol{p}_2). \tag{5.17}
$$

Let us check that this normalisation is valid in any inertial frame. Consider, for instance, a frame with a boost γ in the x-direction that changes $(E_p,\boldsymbol{p})$ into $(E_p',\boldsymbol{p}')$. Since the y and z coordinates are not affected by the boost, it follows that

$$
\begin{aligned}
\langle 1_a,\boldsymbol{p}_1'|1_a,\boldsymbol{p}_2'\rangle &= (2\pi)^3\,2E_{p_1}'\,\delta^{(3)}(\boldsymbol{p}_1'-\boldsymbol{p}_2')\\
&= (2\pi)^3\,2E_{p_1}'\,\delta(p_1'^1-p_2'^1)\,\delta(p_1^2-p_2^2)\delta(p_1^3-p_2^3).
\end{aligned} \tag{5.18}
$$

Now, $p_1'^1 = -\beta\gamma E_{p_1} + \gamma p_1^1 = f(p_1^1)$ with

$$
f(x) = \gamma\left(-\beta\sqrt{m^2+x^2+(p_1^2)^2+(p_1^3)^2}+x\right).
$$

The last two delta functions in the norm (5.18) impose $p_2^2 = p_1^2$ and $p_2^3 = p_1^3$. Therefore, we also have $p_2'^1 = f(p_2^1)$. Consequently, according to the identity (E.7),

$$
\delta(p_1'^1-p_2'^1) = \delta(f(p_1^1)-f(p_2^1)) = \frac{\delta(p_1^1-p_2^1)}{\left|\frac{\partial f}{\partial x}(p_1^1)\right|} = \frac{\delta(p_1^1-p_2^1)}{\left|-\beta\gamma\frac{p_1^1}{E_{p_1}}+\gamma\right|} = \frac{E_{p_1}}{\gamma E_{p_1}-\beta\gamma p_1^1}\delta(p_1^1-p_2^1),
$$

where the absolute value was removed since $\gamma E_{p_1} - \beta\gamma p_1^1 = E_{p_1}' > 0$. Hence, we finally conclude

$$
\langle 1_a,\boldsymbol{p}_1'|1_a,\boldsymbol{p}_2'\rangle = (2\pi)^3\,2E_{p_1}\,\delta(p_1^1-p_2^1)\delta(p_1^2-p_2^2)\delta(p_1^3-p_2^3) = \langle 1_a,\boldsymbol{p}_1|1_a,\boldsymbol{p}_2\rangle.
$$

This shows that the normalisation of states is Lorentz invariant.

[3] In some textbooks, the operators used, a_p' and b_p', are such that with respect to our notation, $a_p' = a_p/\sqrt{2E_p}$ and $b_p' = b_p/\sqrt{2E_p}$. The field expansion is then $\phi(x) = \int \frac{d^3p}{(2\pi)^3\sqrt{2E_p}}\left[a_p'\phi_p^{(+)}(x) + b_p'^\dagger\phi_p^{(-)}(x)\right]$ and the commutators $[a_{p'}',a_p'^\dagger] = [b_{p'}',b_p'^\dagger] = (2\pi)^3\,\delta^{(3)}(\boldsymbol{p}'-\boldsymbol{p})$.

Supplement 5.1. (cont.)

Note that according to the normalisation (5.17), the norm of the state $|1_a, \boldsymbol{p}\rangle$ is

$$\langle 1_a, \boldsymbol{p}|1_a, \boldsymbol{p}\rangle = (2\pi)^3 2E_{p_1}\, \delta^{(3)}(\boldsymbol{0}),$$

which is mathematically undefined. The problem appears here because we implicitly use plane waves (with definite momentum) in an infinite volume (the whole 3D space). Indeed, according to the Formula (E.9), the delta function is

$$\delta^{(3)}(\boldsymbol{p}) = \frac{1}{(2\pi)^3}\int e^{i\boldsymbol{p}\cdot\boldsymbol{x}}\, \mathrm{d}^3\boldsymbol{x} = \lim_{L\to\infty} \Delta_L(\boldsymbol{p}),$$

with

$$\Delta_L(\boldsymbol{p}) = \frac{1}{(2\pi)^3}\int_{-L/2}^{L/2} \mathrm{d}x\, e^{ip_x x}\int_{-L/2}^{L/2} \mathrm{d}y\, e^{ip_y y}\int_{-L/2}^{L/2} \mathrm{d}z\, e^{ip_z z} = \frac{\mathcal{V}}{(2\pi)^3}\frac{\sin\left(p_x\frac{L}{2}\right)}{p_x\frac{L}{2}}\frac{\sin\left(p_y\frac{L}{2}\right)}{p_y\frac{L}{2}}\frac{\sin\left(p_z\frac{L}{2}\right)}{p_z\frac{L}{2}},$$

where $\mathcal{V} = L^3$ is the volume of the space used for the integration (a box with the sides of length L). We conclude that $\Delta_L(\boldsymbol{0}) \equiv \mathcal{V}/(2\pi)^3$ (since $\lim_{x\to 0}\sin x/x = 1$) is equivalent to $\delta^{(3)}(\boldsymbol{0})$ in the limit $\mathcal{V}\to\infty$ and hence,

$$\langle 1_a, \boldsymbol{p}|1_a, \boldsymbol{p}\rangle = 2E_p\mathcal{V}. \tag{5.19}$$

This result is a consequence of our choice of normalisation to $2E_p$ particles per unit volume. Note that this implies that any one-particle state has a non-trivial energy dimension dim $|1, \boldsymbol{p}\rangle = -1$ (whereas the vacuum $|0\rangle$ is dimensionless). What is called the one-particle state, $|1_a, \boldsymbol{p}\rangle$, does not describe a single particle with a momentum $\boldsymbol{p}$, but a uniform density of particles, $2E_P$ per unit volume, all of them carrying momentum $\boldsymbol{p}$.

By generalising Eq. (5.8), it can be easily shown that there is a density operator, $\rho = i\phi^\dagger\frac{\partial}{\partial t}\phi +$ h.c., whose expectation value in the state $|1_a, \boldsymbol{p}\rangle$ is $2E_p$ (which does not have the appropriate dimension for a density because of the normalisation of the one-particle state). A state containing one single particle in volume $\mathcal{V}$ is $|1_a, \boldsymbol{p}\rangle/\sqrt{2E_p\mathcal{V}}$.

With the creation and annihilation operators, one can express other useful operators such as the Hamiltonian, momentum and charge operators,[4]

$$\begin{aligned} H &= \int \frac{\mathrm{d}^3\boldsymbol{p}}{(2\pi)^3 2E_p}\, E_p\, (a_p^\dagger a_p + b_p^\dagger b_p), \\ \boldsymbol{P} &= \int \frac{\mathrm{d}^3\boldsymbol{p}}{(2\pi)^3 2E_p}\, \boldsymbol{p}\, (a_p^\dagger a_p + b_p^\dagger b_p), \\ Q &= \int \frac{\mathrm{d}^3\boldsymbol{p}}{(2\pi)^3 2E_p} (a_p^\dagger a_p - b_p^\dagger b_p). \end{aligned} \tag{5.20}$$

Notice that they are Hermitian (as they should be since they describe quantities that can be measured). We obviously expect that the action of these operators on the state $|1_a, \boldsymbol{p}\rangle = a_p^\dagger|0\rangle$ produces the energy, momentum and charge of the state. To check it, let us first evaluate the quantity

[4] These expressions suppose the normal ordering of operators, as it will be explained in Chapter 6.

$$(a^\dagger_{p'}a_{p'} \pm b^\dagger_{p'}b_{p'})\,|1_a,\boldsymbol{p}\rangle = a^\dagger_{p'}a_{p'}a^\dagger_p\,|0\rangle \pm a^\dagger_p b^\dagger_{p'}b_{p'}\,|0\rangle = a^\dagger_{p'}([a_{p'},a^\dagger_p] + a^\dagger_p a_{p'})\,|0\rangle$$
$$= a^\dagger_{p'}[a_{p'},a^\dagger_p]\,|0\rangle.$$

But according to the commutation rule (5.16),

$$a^\dagger_{p'}[a_{p'},a^\dagger_p]\,|0\rangle = a^\dagger_{p'}(2\pi)^3 2E_p\,\delta^{(3)}(\boldsymbol{p}'-\boldsymbol{p})\,|0\rangle = (2\pi)^3 2E_p\,\delta^{(3)}(\boldsymbol{p}'-\boldsymbol{p})\,|1_a,\boldsymbol{p}'\rangle.$$

It follows that the action of the Hamiltonian on the one-particle state gives

$$H\,|1_a,\boldsymbol{p}\rangle = \int \frac{\mathrm{d}^3\boldsymbol{p}'}{(2\pi)^3 2E_{p'}}\,E'_p\,(2\pi)^3 2E_p\,\delta^{(3)}(\boldsymbol{p}'-\boldsymbol{p})\,|1_a,\boldsymbol{p}'\rangle = E_p\,|1_a,\boldsymbol{p}\rangle.$$

The action of the momentum operator yields similarly

$$\boldsymbol{P}\,|1_a,\boldsymbol{p}\rangle = \int \frac{\mathrm{d}^3\boldsymbol{p}'}{(2\pi)^3 2E_{p'}}\,\boldsymbol{p}'\,(2\pi)^3 2E_p\,\delta^{(3)}(\boldsymbol{p}'-\boldsymbol{p})\,|1_a,\boldsymbol{p}'\rangle = \boldsymbol{p}\,|1_a,\boldsymbol{p}\rangle.$$

As for that of the charge operator,

$$Q\,|1_a,\boldsymbol{p}\rangle = \int \frac{\mathrm{d}^3\boldsymbol{p}'}{(2\pi)^3 2E_{p'}}(2\pi)^3 2E_p\,\delta^{(3)}(\boldsymbol{p}'-\boldsymbol{p})\,|1_a,\boldsymbol{p}'\rangle = |1_a,\boldsymbol{p}\rangle.$$

The same procedure can be applied to the states $|1_b,\boldsymbol{p}\rangle$. Globally, we find

$$\begin{aligned} H\,|1_a,\boldsymbol{p}\rangle &= E_p\,|1_a,\boldsymbol{p}\rangle, & H\,|1_b,\boldsymbol{p}\rangle &= E_p\,|1_b,\boldsymbol{p}\rangle,\\ \boldsymbol{P}\,|1_a,\boldsymbol{p}\rangle &= \boldsymbol{p}\,|1_a,\boldsymbol{p}\rangle, & \boldsymbol{P}\,|1_b,\boldsymbol{p}\rangle &= \boldsymbol{p}\,|1_b,\boldsymbol{p}\rangle,\\ Q\,|1_a,\boldsymbol{p}\rangle &= +\,|1_a,\boldsymbol{p}\rangle, & Q\,|1_b,\boldsymbol{p}\rangle &= -\,|1_b,\boldsymbol{p}\rangle. \end{aligned}$$

Hence, the two states $|1_a,\boldsymbol{p}\rangle$ and $|1_b,\boldsymbol{p}\rangle$ have the same energy, same momentum (and thus same mass since $m^2 = E_p^2 - |\boldsymbol{p}|^2$) but opposite charges: $|1_b,\boldsymbol{p}\rangle$ is just the antiparticle of $|1_a,\boldsymbol{p}\rangle$ (and vice versa). Moreover, since we can apply the creation operator several times, we can have states like $|n_a,\boldsymbol{p}\rangle$ with $n_a > 1$. Having several particles occupying the same quantum state is only possible for bosons, according to the Pauli exclusion principle. The bosonic nature of the particles generated by the operators $a^\dagger_p$ and $b^\dagger_p$ is implemented by their commutation relations. This makes the wave function of two identical particles automatically symmetric. Indeed, assuming that the two particles are of a-type, the wave function is the following

$$\psi_{\boldsymbol{p}_1\boldsymbol{p}_2}(\boldsymbol{x}_1,\boldsymbol{x}_2) = \langle\boldsymbol{x}_1,\boldsymbol{x}_2|1,\boldsymbol{p}_1;1,\boldsymbol{p}_2\rangle = \langle\boldsymbol{x}_1,\boldsymbol{x}_2|\,a^\dagger_{p_1}a^\dagger_{p_2}\,|0\rangle\,.$$

But the operators $a^\dagger_{p_1}$ and $a^\dagger_{p_2}$ thus commute

$$\psi_{\boldsymbol{p}_1\boldsymbol{p}_2}(\boldsymbol{x}_1,\boldsymbol{x}_2) = \langle\boldsymbol{x}_1,\boldsymbol{x}_2|\,a^\dagger_{p_2}a^\dagger_{p_1}\,|0\rangle = \langle\boldsymbol{x}_1,\boldsymbol{x}_2|1,\boldsymbol{p}_2;1,\boldsymbol{p}_1\rangle = \psi_{\boldsymbol{p}_2\boldsymbol{p}_1}(\boldsymbol{x}_1,\boldsymbol{x}_2).$$

As expected, the wave function of two identical bosons is symmetric under the interchange of the two particles. We notice that a spatial rotation or a change of reference frame does not affect the internal structure of the field since it is a scalar field. Consequently, the Klein–Gordon field describes spinless particles that can have an electric charge. The field can also describe a neutral boson: according to the expression of the charge operator Q in Eq. (5.20), the charge of a state containing a single particle is zero if the two operators a and b are equal. Then, the scalar field is Hermitian, i.e. $\phi(x) = \phi^\dagger(x)$. The particle is then its own antiparticle, as for example the neutral pion π^0.

Now that we have a better understanding of the operators, the field Formula (5.15) can be interpreted as follows: $\phi(x)$ destroys a particle a of positive charge at the spacetime point x or creates at x an antiparticle of a of negative charge. The net effect of ϕ is thus to decrease the total charge of the system by one unit. On the contrary, the effect of $\phi^\dagger$ is to increase the total charge of the system by one unit. A term involving $\phi^\dagger(x)\phi(x)$, appearing in the Lagrangians of Chapter 6, does not change the global charge of the system. Such a quadratic term can describe, at the spacetime point x, any of the following charge-conserving processes: creation and annihilation of a particle, creation and annihilation of an antiparticle, creation of a particle–antiparticle pair or annihilation of a particle–antiparticle pair.

Supplement 5.2. Deriving the wave function from the field

One may wonder what is the relationship between the basis of wave functions (5.13) $\phi_{\boldsymbol{p}}^{(\pm)}(x) = e^{-i(\pm p\cdot x)}$, eigenfunctions of the momentum and the quantum field $\phi(x)$ of the Formula (5.15)? Since a wave function deals with a fixed number of particles, taking the example of one single particle, the field must be projected onto a single-state particle. More specifically, it turns out that

$$\begin{aligned} \phi_{\boldsymbol{p}}^{+}(x) = e^{-ip\cdot x} &= \langle 0|\phi(x)|1_a, \boldsymbol{p}\rangle &= \langle 1_a, \boldsymbol{p}|\phi^\dagger(x)|0\rangle^*, \\ \phi_{\boldsymbol{p}}^{-}(x) = e^{+ip\cdot x} &= \langle 1_b, \boldsymbol{p}\,|\phi(x)|0\rangle &= \langle 0|\phi^\dagger(x)|1_b, \boldsymbol{p}\rangle^*. \end{aligned}$$

Let us check, for instance, the first equality above with $|1_a, \boldsymbol{p}\rangle = a_p^\dagger\,|0\rangle$:

$$\begin{aligned} \langle 0|\phi(x)|1_a, \boldsymbol{p}\rangle &= \int \frac{\mathrm{d}^3 p'}{(2\pi)^3 2E_{p'}} \left[\langle 0|a_{p'}a_p^\dagger|0\rangle\, \phi_{\boldsymbol{p}'}^{(+)}(x) + \langle 0|b_{p'}^\dagger a_p^\dagger|0\rangle\, \phi_{\boldsymbol{p}'}^{(-)}(x) \right] \\ &= \int \frac{\mathrm{d}^3 p'}{(2\pi)^3 2E_{p'}} \left[\langle 0|0\rangle\, (2\pi)^3 2E_{p'} \delta^{(3)}(\boldsymbol{p}' - \boldsymbol{p}) \phi_{\boldsymbol{p}'}^{(+)}(x) \right] \\ &= \phi_{\boldsymbol{p}}^{(+)}(x). \end{aligned}$$

A very similar calculation yields $\langle 1_b, \boldsymbol{p}\,|\phi(x)|0\rangle = \phi_{\boldsymbol{p}}^{-}(x)$.

5.2.4 Parity of the Scalar Field

Looking at the Klein–Gordon equation (5.7), the parity transformation $x = (t, \boldsymbol{x}) \to x' = (t, -\boldsymbol{x})$ introduced in Section 4.3.3, leaves second-order derivatives invariant and thus the d'Alembertian as well. Hence, if $\phi(x)$ is a solution of Eq. (5.7), $\phi(x')$ is a solution as well. Denoting by $\hat{\mathcal{P}}$ the unitary (and Hermitian) parity operator acting in the Hilbert space, according to the transformation (4.4) of operators under symmetries, the following transformation of a quantised scalar field is expected:

$$\hat{\mathcal{P}}\phi(x)\hat{\mathcal{P}}^{-1} = \eta\phi(x'), \tag{5.21}$$

where $\eta = \pm 1$ is the *intrinsic parity* of the scalar field. In order to determine the effect of the parity transformation on the creation and annihilation operators, let us insert the expansion (5.15) in the previous equation. The left-hand side yields

$$\hat{\mathcal{P}}\phi(x)\hat{\mathcal{P}}^{-1} = \int \frac{\mathrm{d}^3 \boldsymbol{p}}{(2\pi)^3 2E_p} \left[\hat{\mathcal{P}} a_p \hat{\mathcal{P}}^{-1} e^{-i(Et - \boldsymbol{p}\cdot\boldsymbol{x})} + \hat{\mathcal{P}} b_p^\dagger \hat{\mathcal{P}}^{-1} e^{i(Et - \boldsymbol{p}\cdot\boldsymbol{x})} \right],$$

while $\phi(x')$ is expressed as a function of $\boldsymbol{x}$ as

$$\phi(x') = \int \frac{d^3\boldsymbol{p}}{(2\pi)^3 2E_p} \left[a_p e^{-i(Et+\boldsymbol{p}\cdot\boldsymbol{x})} + b_p^\dagger e^{i(Et+\boldsymbol{p}\cdot\boldsymbol{x})}\right].$$

But nothing prevents us from changing the variables of integration $\boldsymbol{p}$ into $-\boldsymbol{p}$. Therefore, denoting by a_{-p} and b_{-p}, the annihilation operators for a state of momentum $-\boldsymbol{p}$ and energy $E_p = (m^2 + |-\boldsymbol{p}|^2)^{1/2}$, the transformed field reads

$$\phi(x') = \int \frac{d^3\boldsymbol{p}}{(2\pi)^3 2E_p} \left[a_{-p} e^{-i(Et-\boldsymbol{p}\cdot\boldsymbol{x})} + b_{-p}^\dagger e^{i(Et-\boldsymbol{p}\cdot\boldsymbol{x})}\right].$$

Comparing the expressions of $\hat{\mathcal{P}}\phi(x)\hat{\mathcal{P}}^{-1}$ and $\phi(x')$, we thus identify, with Eq. (5.21),

$$\hat{\mathcal{P}} a_p \hat{\mathcal{P}}^{-1} = \eta\, a_{-p} \quad , \quad \hat{\mathcal{P}} b_p^\dagger \hat{\mathcal{P}}^{-1} = \eta\, b_{-p}^\dagger. \tag{5.22}$$

The Hermitian conjugate of these equalities (recalling that $(\hat{\mathcal{P}}^{-1})^\dagger = \hat{\mathcal{P}}$ and $\eta = \pm 1$) yields

$$\hat{\mathcal{P}} a_p^\dagger \hat{\mathcal{P}}^{-1} = \eta\, a_{-p}^\dagger \quad , \quad \hat{\mathcal{P}} b_p \hat{\mathcal{P}}^{-1} = \eta\, b_{-p}. \tag{5.23}$$

Hence, both operators a_p and b_p, respectively destroying a boson and an anti-boson, have the same transformation rule. Assuming that the vacuum is parity-even $\hat{\mathcal{P}}^{-1}|0\rangle = \hat{\mathcal{P}}|0\rangle = |0\rangle$, we deduce the intrinsic parity of the one-particle state from the equations above:

$$\begin{aligned}
\hat{\mathcal{P}} a_p^\dagger |0\rangle &= \eta\, a_{-p}^\dagger |0\rangle \Leftrightarrow \hat{\mathcal{P}}|1a, \boldsymbol{p}\rangle = \eta\, |1a, -\boldsymbol{p}\rangle, \\
\hat{\mathcal{P}} b_p^\dagger |0\rangle &= \eta\, b_{-p}^\dagger |0\rangle \Leftrightarrow \hat{\mathcal{P}}|1b, \boldsymbol{p}\rangle = \eta\, |1b, -\boldsymbol{p}\rangle.
\end{aligned}$$

Consequently, both bosons and anti-bosons have the same intrinsic parity. We will see that this conclusion is not true for fermions.

5.2.5 Charge Conjugation of the Scalar Field

In Section 4.4.1, we explained that the charge-conjugation symmetry transforms a particle into its antiparticle. Looking at the scalar field decomposition (5.15), it means that the Fock operators a_p (operator for the particle) and b_p (operator for the antiparticle) have then to be interchanged. Therefore, the charge conjugation transforms the scalar field as follows:

$$\hat{C}\phi(x)\hat{C}^{-1} = \eta_C \phi^\dagger(x). \tag{5.24}$$

Inserting the field decomposition (5.15) in Eq. (5.24) allows us to deduce

$$\hat{C} a_p \hat{C}^{-1} = \eta_C b_p \, , \quad \hat{C} b_p \hat{C}^{-1} = \eta_C^* a_p. \tag{5.25}$$

Assuming that the vacuum is invariant under the charge-conjugation transformation, i.e. $\hat{C}|0\rangle = |0\rangle$, it follows from the Hermitian conjugate of Eq. (5.25) that

$$\hat{C} a_p^\dagger |0\rangle = \hat{C} a_p^\dagger \hat{C}^{-1} \hat{C} |0\rangle = \eta_C^* b_p^\dagger |0\rangle \; , \quad \hat{C} b_p^\dagger |0\rangle = \hat{C} b_p^\dagger \hat{C}^{-1} \hat{C} |0\rangle = \eta_C a_p^\dagger |0\rangle.$$

This confirms that the charge conjugation transforms a one-particle state into a one-antiparticle state and vice versa. Obviously, applying the charge conjugation twice recovers the same state, i.e. $\hat{C}^2 = \mathbb{1}$, $\hat{C}$ is involutive and $\hat{C} = \hat{C}^{-1}$.

The charge parity, η_C, is only meaningful for real scalar fields $\phi(x) = \phi^\dagger(x)$ that describe a particle that is its own antiparticle. In that case, $\eta_C = \pm 1$, since applying the transformation (5.24) twice yields

$$\phi(x) = \hat{C}^2\phi(x)\hat{C}^{-2} = \eta_C\hat{C}\phi(x)\hat{C}^{-1} = \eta_C^2\phi(x).$$

Otherwise, for complex scalar fields, two successive transformations give

$$\phi(x) = \hat{C}^2\phi(x)\hat{C}^{-2} = \eta_C\hat{C}\phi^\dagger(x)\hat{C}^{-1} = |\eta_C|^2\phi(x),$$

imposing $|\eta_C| = 1$. However, we saw in Section 4.4.1 that this phase can be absorbed by a proper definition of the particle and antiparticle state,[5] showing that η_C is arbitrary for complex fields.

5.3 Spin-1/2 Particles

5.3.1 The Dirac Equation

In 1928, P. A. M. Dirac proposed his famous equation in order to avoid solutions of the Klein–Gordon equation with a negative energy, which lead to negative probability densities. It was six years before the correct interpretation of the Klein–Gordon equation was known. Even if his initial goal was different, Dirac finally found that his equation was able to describe particles and antiparticles with half-integer spin.

Dirac realised that the solution of the Klein–Gordon equation with negative energy was due to the second derivative $\partial^2/\partial t^2$, leading to a probability density involving a single derivative $\partial/\partial t$, and thus allowing a negative probability. Hence, he looked for an equation having $\partial/\partial t$ dependence, as in the Schrödinger equation. The equation should be covariant under Lorentz transformations, and hence, the dependence must also be linear with the spatial derivatives. Moreover, since special relativity requires that the energy verifies the relation $E^2 = \boldsymbol{p}^2 + m^2$, the wave function sought for must also satisfy the relativistic Klein–Gordon equation,

$$(\partial^\mu\partial_\mu + m^2)\psi = 0.$$

The idea is thus to factorise the previous equation,

$$(\gamma^\kappa\partial_\kappa - im)(\beta^\lambda\partial_\lambda + im)\psi = 0,$$

where γ^μ and β^λ would be a priori two sets of four numbers. If ψ satisfies

$$(\gamma^\kappa\partial_\kappa - im)\psi = 0, \tag{5.26}$$

or

$$(\beta^\lambda\partial_\lambda + im)\psi = 0, \tag{5.27}$$

[5] If $\eta_C = e^{i\varphi}$, it suffices to redefine a_p as $e^{-i\varphi/2}a_p$ and b_p as $e^{i\varphi/2}b_p$ to eliminate η_C from Eq. (5.25). Obviously, this works only if $a_p \neq b_p$, i.e. for charged particles.

then the Klein–Gordon equation is also satisfied. Equations (5.26) and (5.27) fulfil the linearity condition with respect to the derivatives. Let us expand and identify the different terms:

$$(\partial^\mu\partial_\mu + m^2)\psi = (\gamma^\kappa\partial_\kappa - im)(\beta^\lambda\partial_\lambda + im)\psi = (\gamma^\kappa\beta^\lambda\partial_\kappa\partial_\lambda + im(\gamma^\kappa\partial_\kappa - \beta^\lambda\partial_\lambda) + m^2)\psi.$$

In order to cancel the linear term with m, we see that $\gamma^\kappa = \beta^\kappa$. The identification with the D'Alembertian term $(\partial^\mu\partial_\mu)$ then imposes $\partial^\mu\partial_\mu = \gamma^\kappa\gamma^\lambda\partial_\kappa\partial_\lambda$. The explicit expansion of the indices yields

$$\begin{aligned}\partial_0^2 - \partial_1^2 - \partial_2^2 - \partial_3^2 = {} & (\gamma^0)^2\partial_0^2 + (\gamma^1)^2\partial_1^2 + (\gamma^2)^2\partial_2^2 + (\gamma^3)^2\partial_3^2 \\ & + (\gamma^0\gamma^1 + \gamma^1\gamma^0)\partial_0\partial_1 + (\gamma^0\gamma^2 + \gamma^2\gamma^0)\partial_0\partial_2 + (\gamma^0\gamma^3 + \gamma^3\gamma^0)\partial_0\partial_3 \\ & + (\gamma^1\gamma^2 + \gamma^2\gamma^1)\partial_1\partial_2 + (\gamma^1\gamma^3 + \gamma^3\gamma^1)\partial_1\partial_3 \\ & + (\gamma^2\gamma^3 + \gamma^3\gamma^2)\partial_2\partial_3.\end{aligned}$$

If the γs were complex numbers, the first line of the equality would impose $\gamma^0 = \pm 1$ and $\gamma^{k=1,2,3} = \pm i$. However, it would be impossible to cancel the last three lines. Dirac then proposed to interpret the γs as matrices satisfying

$$\begin{aligned} &(\gamma^0)^2 = \mathbb{1}, && (\gamma^{k=1,2,3})^2 = -\mathbb{1}, \\ &\gamma^\mu\gamma^\nu + \gamma^\nu\gamma^\mu = 0, && \text{for } \mu \neq \nu. \end{aligned} \tag{5.28}$$

These equalities can be written compactly as

$$\{\gamma^\mu, \gamma^\nu\} = 2g^{\mu\nu}\mathbb{1}, \tag{5.29}$$

where the symbol $\{\}$ is the anticommutator, i.e. $\{a, b\} = ab + ba$ and the symbol $\mathbb{1}$ denotes a unit matrix. Relation (5.29) is known to define a Clifford algebra. What is the minimal rank of the γ matrices? Given that $\text{Tr}(AB) = \text{Tr}(BA)$ and since $\gamma^0\gamma^0 = \mathbb{1}$, it follows that the trace of the $\gamma^{k=1,23}$ matrices verifies

$$\text{Tr}(\gamma^k) = \text{Tr}(\gamma^k\gamma^0\gamma^0) = \text{Tr}(\gamma^0\gamma^k\gamma^0) = -\text{Tr}(\gamma^0\gamma^0\gamma^k) = -\text{Tr}(\gamma^k).$$

Notice that the negative sign is due to the anti-commutation of γ^k with γ^0. We thus conclude that $\text{Tr}(\gamma^k) = 0$. Similarly, since $\gamma^k\gamma^k = -\mathbb{1}$,

$$\text{Tr}(\gamma^0) = -\text{Tr}(\gamma^0\gamma^k\gamma^k) = -\text{Tr}(\gamma^k\gamma^0\gamma^k) = \text{Tr}(\gamma^k\gamma^k\gamma^0) = -\text{Tr}(\gamma^0).$$

Conclusion: the four Dirac matrices are traceless. In addition, since $\gamma^\mu\gamma^\mu = \pm\mathbb{1}$ and $\gamma^\mu\gamma^\nu = -\mathbb{1}\gamma^\nu\gamma^\mu$, for $\mu \neq \nu$, it implies that $\det(\gamma^\mu) \neq 0$ and $\det(\gamma^\mu)\det(\gamma^\nu) = \det(-\mathbb{1})\det(\gamma^\nu)\det(\gamma^\mu)$. Therefore, $\det(-\mathbb{1}) = (-1)^n = 1$, which shows that the rank n of the γ^μ matrices is necessarily an even number. Rank $n = 2$ is excluded since there are only three independent traceless matrices (and not four) in the vector space generated by the 2×2 matrices. Consequently, the minimal rank of γ^μ matrices is $n = 4$. Several choices of 4×4 Dirac matrices are possible, and the most common one, called the Dirac representation, is

$$\gamma^0 = \begin{pmatrix} \mathbb{1} & \mathbb{0} \\ \mathbb{0} & -\mathbb{1} \end{pmatrix}, \quad \gamma^i = \begin{pmatrix} \mathbb{0} & \sigma^i \\ -\sigma^i & \mathbb{0} \end{pmatrix}. \tag{5.30}$$

where the symbols $\mathbb{1}$ and $\mathbb{0}$ denote respectively the 2×2 identity matrix and zero matrix, and σ^i are the Pauli matrices seen in Chapter 2,

$$\sigma^1 = \begin{pmatrix} 0 & 1 \\ 1 & 0 \end{pmatrix}, \quad \sigma^2 = \begin{pmatrix} 0 & -i \\ i & 0 \end{pmatrix}, \quad \sigma^3 = \begin{pmatrix} 1 & 0 \\ 0 & -1 \end{pmatrix}.$$

Actually, any matrix $\gamma'^{\mu} = U\gamma^{\mu}U^{-1}$, where U is a 4×4 unitary matrix is a valid choice since γ'^{μ} still satisfies the Clifford algebra[6] (5.29). We can easily check that the definition of the Dirac matrices in the Dirac representation implies for their Hermitian conjugates[7]

$$\gamma^{0\dagger} = \gamma^0, \quad \gamma^{1\dagger} = -\gamma^1, \quad \gamma^{2\dagger} = -\gamma^2, \quad \gamma^{3\dagger} = -\gamma^3, \tag{5.31}$$

or in a more concise form,

$$\gamma^{\mu\dagger} = \gamma^0\gamma^{\mu}\gamma^0. \tag{5.32}$$

As soon as the matrix U is unitary, the matrices $\gamma'^{\mu} = U\gamma^{\mu}U^{-1} = U\gamma^{\mu}U^{\dagger}$ have the same hermiticity properties as γ^{μ}. Hence, Formulas (5.29) and (5.32) are the relations that define the Dirac matrices regardless of their particular representation.

We can now come back to the two Eqs. (5.26) and (5.27), 'square roots' of the Klein–Gordon equation. By convention, the Dirac equation is Eq. (5.27) (with $\beta = \gamma$) multiplied by i (it makes the momentum operator visible, $\hat{P}_{\mu} = i\partial_{\mu}$), i.e.

$$\boxed{(i\partial\!\!\!/ - m)\psi = 0, \quad \text{with} \quad \partial\!\!\!/ = \gamma^{\mu}\partial_{\mu}.} \tag{5.33}$$

The notation $\gamma^{\mu}\partial_{\mu}$ might be confusing: it seems to suggest a scalar product of 4-vectors. Although the Dirac equation is covariant, the γ^{μ} are matrices and *not* 4-vectors (but ∂_{μ} is). They remain invariant under a change of frame. To add to the confusion, we will encounter the notation

$$\gamma_{\mu} = g_{\mu\nu}\gamma^{\nu}. \tag{5.34}$$

It will be useful because bilinears of the form $\bar{\psi}\gamma^{\mu}\psi$, such as the Dirac current we shall see in the next section (see Supplement 5.3), transform like Lorentz vectors. The definition (5.34) of γ_{μ} is thus a convenient trick to keep this in mind. Notice that γ_{μ} is equivalent to the Hermitian conjugate of γ^{μ} according to the equalities (5.31).

Since the γs are 4×4 matrices, the ψ wave function is actually a wave function with four components, $\psi = (\psi_1, \psi_2, \psi_3, \psi_4)^{\mathsf{T}}$. It is called a Dirac-spinor or 4-spinor.[8] At this stage, we can already infer that the extra multiplicity is likely to have something to do with an angular momentum degree of freedom as suggested by the presence of the Pauli matrices in the definition of the γ matrices.

[6] Actually, the U matrix just needs to be invertible to guarantee that $\gamma'^{\mu} = U\gamma^{\mu}U^{-1}$ satisfies the Clifford algebra. Nevertheless, the unitarity property is required so that γ'^{μ} is Hermitian or anti-Hermitian. The hermiticity properties (5.31) are required to make sure that the Hamiltonian deduced from the Dirac equation [see definition (5.64) in Section 5.3.4] is Hermitian.

[7] The Hermitian conjugate of a matrix A is $A^{\dagger} = (A^*)^{\mathsf{T}}$, where the symbol * is the complex conjugate operation and ^{T} is the matrix transposition.

[8] It is also called a bi-spinor (with four components), which might be confusing. Actually, a bi-spinor in this context means an object made with two 2-spinors, where 2-spinors are the usual spinors (with two components) of non-relativistic quantum mechanics.

5.3.2 Covariance of the Dirac Equation

The Dirac equation has been initially built to be consistent with special relativity. Therefore, we expect it to have the same mathematical form, whatever the inertial frame used. This property is called the (Lorentz) covariance.[9] So, if $\psi(x)$ is a solution of the Dirac equation in a given frame,

$$(i\gamma^\nu \partial_\nu - m)\psi(x) = 0,$$

we expect that in another frame that transforms x into x', there exists a $\psi'(x')$ deduced from $\psi(x)$ that satisfies

$$(i\gamma^\mu \partial'_\mu - m)\psi'(x') = 0. \tag{5.35}$$

Let us make the ansatz that $\psi'(x')$ corresponds to a linear transformation of $\psi(x)$, i.e. there exists a 4×4 matrix $S(\Lambda)$ that only depends on the Lorentz transformation Λ such that

$$\psi'(x') = S(\Lambda)\psi(x).$$

Given that the product of two Lorentz transformations is also a Lorentz transformation (recall that Lorentz transformations constitute a group; see Appendix D), we expect that if

$$\psi(x) \xrightarrow{S(\Lambda_1)} \psi'(x') \xrightarrow{S(\Lambda_2)} \psi''(x''),$$

then

$$\psi''(x'') = S(\Lambda_2\Lambda_1)\psi(x) = S(\Lambda_2)S(\Lambda_1)\psi(x).$$

In the language of group theory, this means that $S(\Lambda)$ must be a representation of the Lorentz group. It implies that the inverse of $S(\Lambda)$ is necessarily defined. Then, the initial Dirac equation can be written as

$$(i\gamma^\nu \partial_\nu - m)\, S(\Lambda)^{-1}\psi'(x') = 0.$$

Since $x'^\mu = \Lambda^\mu{}_\nu x^\nu$, the derivative is $\partial_\nu = \frac{\partial}{\partial x^\nu} = \frac{\partial x'^\mu}{\partial x^\nu}\frac{\partial}{\partial x'^\mu} = \Lambda^\mu{}_\nu \partial'_\mu$, and multiplying the above equation by $S(\Lambda)$ from the left yields

$$(iS(\Lambda)\gamma^\nu S(\Lambda)^{-1}\Lambda^\mu{}_\nu \partial'_\mu - m)\psi'(x') = 0.$$

A comparison to the desired equation (5.35) requires that $S(\Lambda)\gamma^\nu S(\Lambda)^{-1}\Lambda^\mu{}_\nu = \gamma^\mu$ or equivalently

$$S(\Lambda)^{-1}\gamma^\mu S(\Lambda) = \Lambda^\mu{}_\nu \gamma^\nu. \tag{5.36}$$

Although the matrices γ^μ are not 4-vectors, they transform as such when 'sandwiched' by $S(\Lambda)^{-1}$ and $S(\Lambda)$. This is another justification for the notation of those matrices.

The next step is to find the explicit form of $S(\Lambda)$. Let us determine it for an infinitesimal Lorentz transformation

$$\Lambda^\mu{}_\nu = \delta^\mu_\nu + \omega^\mu{}_\nu.$$

[9] By an abuse of language, some authors refer to invariance instead of covariance. Normally, invariance suggests no change at all, which is not the case here since, as we shall see, spinors change under a change of frame.

Obviously, if $\omega^{\mu}_{\ \nu} = 0$, $S(\mathbb{1}) = \mathbb{1}$. In Appendix D.1, it is shown that the infinitesimal parameters are antisymmetric, parametrising the infinitesimal Lorentz transformation in Eq. (D.11) as

$$\Lambda = \mathbb{1} - \frac{i}{2}\omega_{\sigma\rho}J^{\sigma\rho}.$$

Since $S(\Lambda)$ constitutes a representation of the Lorentz group (see again Appendix D.1), let us parametrise $S(\Lambda)$ in a similar form as above, i.e.

$$S(\Lambda) = \mathbb{1} - \frac{i}{2}\omega_{\sigma\rho}S^{\sigma\rho}, \tag{5.37}$$

where $S^{\mu\nu}$ are 4×4 anti-symmetric matrices that have to be determined. Inserting the two previous equations into Eq. (5.36) and keeping up to first order in $\omega_{\sigma\rho}$ yields the relation

$$\left[\gamma^{\mu}, S^{\sigma\rho}\right] = (J^{\sigma\rho})^{\mu}_{\ \nu}\gamma^{\nu} = ig^{\mu\sigma}\gamma^{\rho} - ig^{\mu\rho}\gamma^{\sigma}, \tag{5.38}$$

where the expression of $(J^{\sigma\rho})^{\mu}_{\ \nu}$ established in Eq. (D.9) has been inserted to produce the right-hand side of Formula (5.38). In order to identify $S^{\sigma\rho}$, we need to transform the right-hand side of Eq. (5.38) into a commutator with γ^{μ}. This can be done by replacing $g^{\mu\sigma}$ and $g^{\mu\rho}$ by anticommutators of γ matrices, using the Clifford algebra (5.29),

$$\begin{aligned}
\left[\gamma^{\mu}, S^{\sigma\rho}\right] &= \frac{i}{2}(\gamma^{\mu}\gamma^{\sigma} + \gamma^{\sigma}\gamma^{\mu})\gamma^{\rho} - \frac{i}{2}(\gamma^{\mu}\gamma^{\rho} - \gamma^{\rho}\gamma^{\mu})\gamma^{\sigma} \\
&= \frac{i}{4}\left((\gamma^{\mu}\gamma^{\sigma} + \gamma^{\sigma}\gamma^{\mu})\gamma^{\rho} + \gamma^{\rho}(\gamma^{\mu}\gamma^{\sigma} + \gamma^{\sigma}\gamma^{\mu})\right. \\
&\quad \left. -(\gamma^{\mu}\gamma^{\rho} - \gamma^{\rho}\gamma^{\mu})\gamma^{\sigma} - \gamma^{\sigma}(\gamma^{\mu}\gamma^{\rho} - \gamma^{\rho}\gamma^{\mu})\right) \\
&= \frac{i}{4}\left(\gamma^{\mu}[\gamma^{\sigma}, \gamma^{\rho}] + [\gamma^{\rho}, \gamma^{\sigma}]\gamma^{\mu}\right).
\end{aligned}$$

Hence,

$$S^{\sigma\rho} = \frac{i}{4}\left[\gamma^{\sigma}, \gamma^{\rho}\right] = \begin{cases} \mathbb{0} & \text{if } \sigma = \rho, \\ \frac{i}{2}\gamma^{\sigma}\gamma^{\rho} & \text{if } \sigma \neq \rho. \end{cases} \tag{5.39}$$

(The last equality when $\sigma \neq \rho$ is deduced from the Clifford algebra.) At this stage, we have found the expression of the matrices $S^{\sigma\rho}$ that satisfies the relation (5.36). We still have to check that they form a representation of the Lorentz group. This is the case if they satisfy the commutation rule (D.18) established in Appendix D.1[10], i.e.

$$[S^{\mu\nu}, S^{\sigma\rho}] = i(g^{\nu\sigma}S^{\mu\rho} - g^{\mu\sigma}S^{\nu\rho} + g^{\mu\rho}S^{\nu\sigma} - g^{\nu\rho}S^{\mu\sigma}).$$

It is not very complicated to check it, so we leave it as an exercise (see Problem 5.4 that explains how to proceed). Thus, we have shown that for an infinitesimal Lorentz transformation, the Formula (5.37) of $S(\Lambda)$ is the matrix that has to be applied to the Dirac spinor to guarantee the covariance of the Dirac equation, where the matrices $S^{\sigma\rho}$ are given in Eq. (5.39). As recalled in Appendix D.1, a finite Lorentz transformation is obtained simply by 'exponentiating' the infinitesimal transformation, and thus $S(\Lambda)$ becomes

$$S(\Lambda) = \exp\left(-\frac{i}{2}\omega_{\sigma\rho}S^{\sigma\rho}\right). \tag{5.40}$$

[10] The readers are strongly encouraged to read Appendix D.1 if they are not familiar with the structure of the Lorentz group.

In summary, under a Lorentz transformation Λ, the Dirac spinor $\psi(x)$ transforms as $\psi'(x') = S(\Lambda)\psi(x)$, ensuring the covariance of the Dirac equation.

5.3.3 The Dirac Adjoint and the four-Current

In Sections 5.1.2 and 5.2.1, we found the expression of the 4-current from the Schrödinger and Klein–Gordon wave equations with the following procedure: we took the complex conjugate of the wave equation, multiplied it by the wave function and subtracted the obtained result from the wave equation times the complex conjugate of the wave function. This procedure was adapted to generate the continuity equation from which the 4-current is deduced.

This time, however, the complex conjugate must be replaced by the Hermitian conjugate because of the presence of matrices in the Dirac equation. The Hermitian conjugate of the Dirac equation (5.33) is

$$-i\partial_\mu \psi^\dagger \gamma^{\mu\dagger} - m\psi^\dagger = 0,$$

where the components of the Hermitian adjoint spinor $\psi^\dagger$ are

$$\psi^\dagger = \left(\psi_1^*,\ \psi_2^*,\ \psi_3^*,\ \psi_4^*\right). \tag{5.41}$$

Now, using the Hermiticity properties of the Dirac matrices (5.31), it follows that

$$-i\partial_0 \psi^\dagger \gamma^0 + i\partial_k \psi^\dagger \gamma^k - m\psi^\dagger = 0.$$

Multiplying the previous equation by γ^0 from the right-hand side and using the anti-commutation of γ^0 with γ^k yields

$$i\partial_\mu \psi^\dagger \gamma^0 \gamma^\mu + m\psi^\dagger \gamma^0 = 0.$$

Let us define a row spinor called the *Dirac adjoint* such that

$$\boxed{\bar{\psi} = \psi^\dagger \gamma^0.} \tag{5.42}$$

With the Dirac representation, the Dirac adjoint has the components

$$\bar{\psi} = \left(\psi_1^*,\ \psi_2^*,\ -\psi_3^*,\ -\psi_4^*\right).$$

The Dirac adjoint $\bar{\psi}$ then satisfies the *Dirac adjoint equation*

$$i\partial_\mu \bar{\psi}\gamma^\mu + m\bar{\psi} = 0. \tag{5.43}$$

It is often written as

$$i\bar{\psi}\overleftarrow{\not\partial} + m\bar{\psi} = 0,$$

with the left arrow on $\overleftarrow{\not\partial}$ indicating that the derivation operator is applied on the left. We can now derive the continuity equation by multiplying the Dirac adjoint equation from the right by ψ, multiplying the Dirac equation (5.33) from the left by $\bar{\psi}$ and adding both results. This yields

$$i(\partial_\mu \bar{\psi})\gamma^\mu \psi + i\bar{\psi}\gamma^\mu \partial_\mu \psi = i\partial_\mu(\bar{\psi}\gamma^\mu \psi) = 0.$$

The conserved 4-current satisfying the previous continuity equation is thus

$$j^{\mu} = \bar{\psi}\gamma^{\mu}\psi. \tag{5.44}$$

Supplement 5.3. Proof that the Dirac 4-current is a Lorentz vector

I claim that the Dirac 4-current (5.44) is a Lorentz vector. In other words, if Λ is the Lorentz transformation matrix for moving from one frame to another, i.e. $x'^{\mu} = \Lambda^{\mu}{}_{\nu}x^{\nu}$, the current in the new frame must transform as

$$j^{\mu}(x) = \overline{\psi}(x)\gamma^{\mu}\psi(x) \to j'^{\mu}(x') = \overline{\psi'}(x')\gamma^{\mu}\psi'(x') = \Lambda^{\mu}{}_{\nu}j^{\nu}(x).$$

The transformation of $\psi(x)$ is already known, i.e. $\psi'(x') = S(\Lambda)\psi(x)$, where the matrix $S(\Lambda)$ is given in Eq. (5.40). Hence, the Dirac adjoint $\bar{\psi}$ transforms as

$$\overline{\psi'}(x') = \psi^{\dagger}(x)S(\Lambda)^{\dagger}\gamma^{0},$$

with

$$S(\Lambda)^{\dagger} = \exp\left(+\frac{i}{2}\omega_{\sigma\rho}(S^{\sigma\rho})^{\dagger}\right).$$

According to the expression of $S^{\sigma\rho}$ (5.39) and using the identity (5.32) of the Hermitian conjugate of γ^{μ} matrices as well as the property $(\gamma^{0})^{2} = \mathbb{1}$, it follows that

$$(S^{\sigma\rho})^{\dagger} = -\frac{i}{4}[(\gamma^{\rho})^{\dagger},(\gamma^{\sigma})^{\dagger}] = -\frac{i}{4}[\gamma^{0}\gamma^{\rho}\gamma^{0},\gamma^{0}\gamma^{\sigma}\gamma^{0}] = -\frac{i}{4}\gamma^{0}[\gamma^{\rho},\gamma^{\sigma}]\gamma^{0} = \gamma^{0}S^{\sigma\rho}\gamma^{0}.$$

Therefore,

$$S(\Lambda)^{\dagger} = \mathbb{1} + \sum_{n=1}^{\infty}\frac{1}{n!}\left(\frac{i}{2}\omega_{\sigma\rho}\right)^{n}(\gamma^{0}S^{\sigma\rho}\gamma^{0})^{n} = \mathbb{1} + \sum_{n=1}^{\infty}\frac{1}{n!}\left(\frac{i}{2}\omega_{\sigma\rho}\right)^{n}\gamma^{0}(S^{\sigma\rho})^{n}\gamma^{0},$$

and so,

$$S(\Lambda)^{\dagger} = \gamma^{0}\exp\left(+\frac{i}{2}\omega_{\sigma\rho}S^{\sigma\rho}\right)\gamma^{0} = \gamma^{0}S(\Lambda)^{-1}\gamma^{0}.$$

Hence, the transformed Dirac adjoint is

$$\overline{\psi'}(x') = \overline{\psi}(x)S(\Lambda)^{-1}.$$

In passing, we see that the expression $\overline{\psi'}(x')\psi'(x') = \overline{\psi}(x)S(\Lambda)^{-1}S(\Lambda)\psi(x) = \overline{\psi}(x)\psi(x)$ is a Lorentz scalar. Now, coming back to the transformed Dirac 4-current, its expression is

$$j'^{\mu}(x') = \overline{\psi'}(x')\gamma^{\mu}\psi'(x') = \overline{\psi}(x)S(\Lambda)^{-1}\gamma^{\mu}S(\Lambda)\psi(x).$$

But in Section 5.3.2, we have determined the expression of $S(\Lambda)$ requiring that it satisfies the constraint (5.36) and thus,

$$j'^{\mu}(x') = \overline{\psi}(x)\Lambda^{\mu}{}_{\nu}\gamma^{\nu}\psi(x) = \Lambda^{\mu}{}_{\nu}J^{\nu}(x).$$

This confirms that the Dirac 4-current is a Lorentz vector.

The Dirac 4-current (5.44) leads to the density

$$\rho = \bar{\psi}\gamma^0\psi = \psi^\dagger\gamma^0\gamma^0\psi = \psi^\dagger\psi = \sum_{i=1}^{4}|\psi_i|^2. \tag{5.45}$$

This time, the density is always positive and can therefore be interpreted as a probability density. Note that a charge current can also be defined by simply multiplying the 4-current (5.44) by the charge q of the particle,

$$j_q^\mu = qj^\mu = q\bar{\psi}\gamma^\mu\psi. \tag{5.46}$$

5.3.4 Free-Particle Solutions

The way the (free) Dirac equation has been built shows that a solution of the equation is also a solution of the (free) Klein–Gordon equation, or more precisely, each component of the 4-spinor is a solution of the Klein–Gordon equation. It is then natural to look for solutions in which the spacetime behaviour is that of a plane wave,

$$\psi(x) = u(p)e^{-ip\cdot x}, \tag{5.47}$$

where, this time, u is a 4-spinor that depends on the 4-momentum p. A general solution of the Dirac equation can always be expressed as a linear combination of plane-wave solutions. Inserting the plane wave (5.47) in the Dirac equation (5.33) yields the *Dirac equation in momentum space*,

$$\boxed{(\not{p} - m)u(p) = 0,} \tag{5.48}$$

where $\not{p} = \gamma^\mu p_\mu$. The Dirac adjoint spinor, $\bar{u}(p)$, defined in Eq. (5.48) also satisfies an equation in momentum space. In order to confirm this, let us take the Hermitian conjugate of Eq. (5.48)

$$u^\dagger(p)(\gamma^{\mu\dagger}p_\mu - m) = u^\dagger(\gamma^0\gamma^\mu\gamma^0 p_\mu - m) = 0,$$

where the expression of $\gamma^{\mu\dagger}$ from Eq. (5.32) was used. Multiplying from the right by γ^0 and using $(\gamma^0)^2 = 1$, it follows that

$$\bar{u}(p)(\not{p} - m) = 0. \tag{5.49}$$

Solutions for a Particle at Rest

It is very instructive to look first at the solutions for a particle at rest where only $p^0 = E$ is non-zero, with Eq. (5.48) being reduced to $E\gamma^0 u - mu = 0$. Let us write the 4-spinor u in terms of two-component spinors,

$$u = \begin{pmatrix} u_a \\ u_b \end{pmatrix}, \tag{5.50}$$

where u_a and u_b are two 2-spinors. Using the Dirac representation of γ matrices, Eq. (5.48) yields the two relations: $Eu_a = mu_a$ and $Eu_b = -mu_b$. Each has two independent solutions, whose basis can be given by the eigenstates of the σ^3 Pauli matrix, i.e.

$$u_a = \begin{pmatrix} 1 \\ 0 \end{pmatrix} \text{ or } u_a = \begin{pmatrix} 0 \\ 1 \end{pmatrix} \text{ with } E = m,$$

$$u_b = \begin{pmatrix} 1 \\ 0 \end{pmatrix} \text{ or } u_b = \begin{pmatrix} 0 \\ 1 \end{pmatrix} \text{ with } E = -m.$$

Therefore, Eq. (5.48) admits four independent solutions,

$$u_1(m,\mathbf{0}) = \begin{pmatrix} 1 \\ 0 \\ 0 \\ 0 \end{pmatrix}, \; u_2(m,\mathbf{0}) = \begin{pmatrix} 0 \\ 1 \\ 0 \\ 0 \end{pmatrix}, \; u_3(-m,\mathbf{0}) = \begin{pmatrix} 0 \\ 0 \\ 1 \\ 0 \end{pmatrix}, \; u_4(-m,\mathbf{0}) = \begin{pmatrix} 0 \\ 0 \\ 0 \\ 1 \end{pmatrix}, \tag{5.51}$$

which, in terms of plane waves (5.47), are, up to a normalisation factor:

$$\psi_1 = \begin{pmatrix} 1 \\ 0 \\ 0 \\ 0 \end{pmatrix} e^{-imt}, \; \psi_2 = \begin{pmatrix} 0 \\ 1 \\ 0 \\ 0 \end{pmatrix} e^{-imt}, \; \psi_3 = \begin{pmatrix} 0 \\ 0 \\ 1 \\ 0 \end{pmatrix} e^{+imt}, \; \psi_4 = \begin{pmatrix} 0 \\ 0 \\ 0 \\ 1 \end{pmatrix} e^{+imt}. \tag{5.52}$$

The first two solutions ψ_1 and ψ_2 of the Dirac equation (5.33) have $E = m > 0$, whereas ψ_3 and ψ_4 have $E = -m < 0$. Consequently, even for the Dirac equation, we have negative energy solutions! However, unlike the Klein–Gordon equation, they do not lead to a negative probability density. These negative energy solutions will be interpreted in the next section as solutions for antiparticles.

General Solutions

Let us come back to the general case with non zero momentum. Inserting the 4-spinors (5.50) in the Dirac equation (5.48) yields

$$(\gamma^\mu p_\mu - m)u = \begin{pmatrix} (E-m) & -\boldsymbol{\sigma}\cdot\boldsymbol{p} \\ \boldsymbol{\sigma}\cdot\boldsymbol{p} & -(E+m) \end{pmatrix} \begin{pmatrix} u_a \\ u_b \end{pmatrix}. \tag{5.53}$$

Hence, u_a and u_b satisfy the two coupled equations,

$$u_a = \frac{\boldsymbol{\sigma}\cdot\boldsymbol{p}}{E-m} u_b, \tag{5.54}$$

$$u_b = \frac{\boldsymbol{\sigma}\cdot\boldsymbol{p}}{E+m} u_a, \tag{5.55}$$

that impose a constraint on E and $\boldsymbol{p}$. Indeed, inserting Eq. (5.55) in (5.54) gives

$$u_a = \frac{(\boldsymbol{\sigma}\cdot\boldsymbol{p})^2}{E^2-m^2} u_a. \tag{5.56}$$

But since

$$\boldsymbol{\sigma}\cdot\boldsymbol{p} = \begin{pmatrix} 0 & 1 \\ 1 & 0 \end{pmatrix} p_x + \begin{pmatrix} 0 & -i \\ i & 0 \end{pmatrix} p_y + \begin{pmatrix} 1 & 0 \\ 0 & -1 \end{pmatrix} p_z = \begin{pmatrix} p_z & p_x - ip_y \\ p_x + ip_y & -p_z \end{pmatrix}, \tag{5.57}$$

this implies that $(\boldsymbol{\sigma}\cdot\boldsymbol{p})^2 = |\boldsymbol{p}|^2\mathbb{1}$. Therefore, the non-trivial solution ($u_a \neq 0$) of Eq. (5.56) requires

$$\frac{|\boldsymbol{p}|^2}{E^2-m^2} = 1, \text{ i.e. } E = \pm\sqrt{|\boldsymbol{p}|^2 + m^2}.$$

Equations (5.54) and (5.55) tell us that as soon as one of the two 2-spinors (u_a or u_b) is chosen, the other is unambiguously determined. We can choose for u_a any two orthogonal 2-spinors ϕ_1 and ϕ_2, and u_b will be given by Eq. (5.55). Similarly, we can decide to choose first u_b from any two orthogonal 2-spinors χ_1 and χ_2, and then u_a will have to satisfy Eq. (5.54). In both cases, the relativistic energy constraint must be fulfilled, i.e. $E^2 = |\boldsymbol{p}|^2 + m^2$. Hence, four independent solutions of the Dirac equation are given by these 4-spinors:

$$u_1(E,\boldsymbol{p}) = N_1 \begin{pmatrix} \phi_1 \\ \frac{\sigma \cdot p}{E+m}\phi_1 \end{pmatrix}, \quad u_2(E,\boldsymbol{p}) = N_2 \begin{pmatrix} \phi_2 \\ \frac{\sigma \cdot p}{E+m}\phi_2 \end{pmatrix},$$
$$u_3(E,\boldsymbol{p}) = N_3 \begin{pmatrix} \frac{\sigma \cdot p}{E-m}\chi_1 \\ \chi_1 \end{pmatrix}, \quad u_4(E,\boldsymbol{p}) = N_4 \begin{pmatrix} \frac{\sigma \cdot p}{E-m}\chi_2 \\ \chi_2 \end{pmatrix}, \tag{5.58}$$

where $N_{i=\{1,4\}}$ are the normalisation factors of the spinors. In the limit where $\boldsymbol{p}$ approaches 0, we should recover the solutions (5.51) for a particle at rest. It is straightforward to conclude that u_1 and u_2 would match the first two solutions of Eq. (5.51), while u_3 and u_4 would match the last two.[11] We thus deduce that the solutions with positive energy $E = +\sqrt{|\boldsymbol{p}|^2 + m^2}$ are described by u_1 and u_2, while the ones with negative energy $E = -\sqrt{|\boldsymbol{p}|^2 + m^2}$ are given by u_3 and u_4.

Interpretation of Negative Energies

Historically, Dirac considered that the vacuum was full of negative energy states satisfying the Pauli exclusion principle. Electrons can then populate all negative energy states (two per level) with energy levels $-Nm_ec^2$, with N being a positive integer. This picture is usually referred to as the 'Dirac sea'. A photon having an energy $2m_ec^2$ can then excite an electron from the energy level $-m_ec^2$ to the positive energy state $+m_ec^2$. The hole created in the negative energy states was then interpreted as an antiparticle with positive energy and opposite charge. This interpretation was successful in describing the reaction $\gamma \to e^- + e^+$, leading Dirac to predict the existence of the positron in 1931, finally discovered by Anderson one year later. However, the theory had many drawbacks. For instance, the vacuum being full of fermions with negative energy states did not explain why bosons could not populate the vacuum as well (which would be striking since they are not subject to the Pauli exclusion principle). Actually, Stückelberg in 1941 and Feynman in 1948 proposed the correct interpretation that is still valid nowadays with quantum field theory and briefly mentioned in the previous case of the Klein–Gordon equation.

Consider the two solutions u_3 and u_4 with negative energy in (5.58). They can be expressed in terms of the positive quantity $E = \sqrt{|\boldsymbol{p}|^2 + m^2}$,

$$u_3(-E,\boldsymbol{p}) = N_3 \begin{pmatrix} -\frac{\sigma \cdot p}{E+m}\chi_1 \\ \chi_1 \end{pmatrix}, \quad u_4(-E,\boldsymbol{p}) = N_4 \begin{pmatrix} -\frac{\sigma \cdot p}{E+m}\chi_2 \\ \chi_2 \end{pmatrix}.$$

Both are associated with a propagation term $\exp(-i(-Et - \boldsymbol{p}\cdot\boldsymbol{x})) = \exp(-i(E(-t) - \boldsymbol{p}\cdot\boldsymbol{x}))$, such that it looks like a particle having a positive energy E but travelling backwards in time from the spacetime point (t_2, x_1) to (t_1, x_2), where $t_2 > t_1$ as shown in scheme (b) of

[11] Simply take $\phi_1 = \chi_1 = \begin{pmatrix} 1 \\ 0 \end{pmatrix}$ and $\phi_2 = \chi_2 = \begin{pmatrix} 0 \\ 1 \end{pmatrix}$.

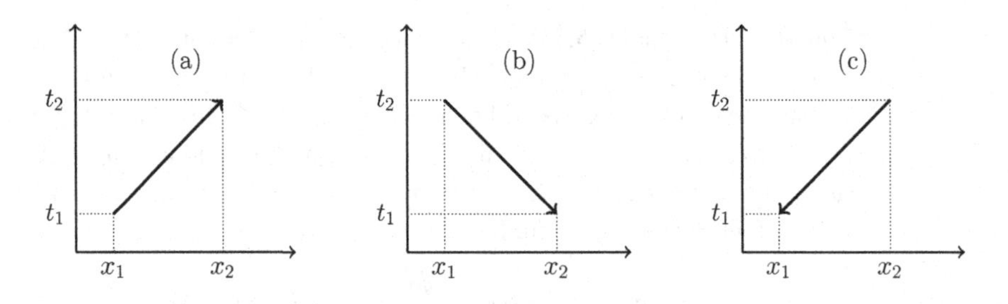

Fig. 5.1 Feynman's approach of antiparticles: (a) a particle travelling from x_1 to x_2. (b) time is reversed. (c) antiparticle as a particle with $-t$ and $-\boldsymbol{p}$.

Fig. 5.1. In contrast, a particle travelling 'normally' forward in time from (t_1, x_1) to (t_2, x_2) as described by the positive energy solutions $u_1(E,\boldsymbol{p})$ or $u_2(E,\boldsymbol{p})$ is depicted in scheme (a). Note that for a 4-momentum $p = (E,\boldsymbol{p})$, the propagation term in scheme (b) is no more Lorentz invariant since $\exp(-i(-Et - \boldsymbol{p}\cdot\boldsymbol{x})) \neq \exp(-ip\cdot x)$. In order to restore the invariance, the direction of the momentum has to be changed to its opposite $\boldsymbol{p} \to -\boldsymbol{p}$ (as we did in the context of the Klein–Gordon equation), obtaining the propagation term

$$\exp(-i(-Et - (-\boldsymbol{p})\cdot\boldsymbol{x})) = \exp(-i(E(-t) - \boldsymbol{p}\cdot(-\boldsymbol{x}))) = \exp(+ip\cdot x).$$

Hence, it looks like the particle in scheme (c), where the final point in the spacetime diagram is exchanged with the initial point. Using $-\boldsymbol{p}$ is not surprising: since $\boldsymbol{p} = \gamma m\,\mathrm{d}\boldsymbol{x}/\,\mathrm{d}t$, if t becomes $-t$, then $\boldsymbol{p}$ becomes $-\boldsymbol{p}$. We have the freedom to change $\boldsymbol{p}$ to $-\boldsymbol{p}$ because $\sqrt{|\boldsymbol{p}|^2 + m^2}$ is not affected by this change, ensuring that we still describe a negative energy solution of the Dirac equation with the same mass. Feynman's approach consists in interpreting scheme (c) as the antiparticle of scheme (a) with positive energy and travelling 'normally' from (t_1, x_1) to (t_2, x_2). The antiparticle is thus equivalent to a particle travelling backwards in time in the opposite direction. The direction of the arrow in Feynman's diagrams is representative of the direction of the *charge of the particle* (understanding the charge as the set of quantum numbers carried by the particle, the electric charge of charged particles, the weak isospin, etc.) and not that of the antiparticle. If the particle in (a) has a charge q, then this charge is brought from x_1 to x_2, increasing the net charge in x_2 from Q_{x_2} to $Q_{x_2} + q$. In scheme (c), the charge q is transferred from x_2 to x_1, with the charge in x_2 being reduced to $Q_{x_2} - q$. But this is equivalent to considering the antiparticle with charge $-q$ moving from x_1 to x_2, increasing the charge in x_2 to $Q_{x_2} + (-q)$.

The four independent solutions of the Dirac equation are then built with $u_1(E,\boldsymbol{p})e^{-ip\cdot x}$, $u_2(E,\boldsymbol{p})e^{-ip\cdot x}$ for the particle and $u_3(-E,-\boldsymbol{p})e^{+ip\cdot x}$, $u_4(-E,-\boldsymbol{p})e^{+ip\cdot x}$ for the antiparticle. Note that a plane wave described by $\exp(-ip\cdot x)$ or $\exp(+ip\cdot x)$ propagates in the same direction since only the relative sign between Et and $\boldsymbol{p}\cdot\boldsymbol{x}$ matters. It is more convenient to define two new spinors v_1 and v_2 specific to the antiparticle in which the physical quantities, i.e. a positive energy E, and the appropriate momentum direction $\boldsymbol{p}$ appear clearly:

$$v_{1,2}(E,\boldsymbol{p}) = u_{3,4}(-E,-\boldsymbol{p}).$$

It follows from Eq. (5.58) that

$$v_1(E,\boldsymbol{p}) = N_3 \begin{pmatrix} \frac{\boldsymbol{\sigma}\cdot\boldsymbol{p}}{E+m}\chi_1 \\ \chi_1 \end{pmatrix}, \quad v_2(E,\boldsymbol{p}) = N_4 \begin{pmatrix} \frac{\boldsymbol{\sigma}\cdot\boldsymbol{p}}{E+m}\chi_2 \\ \chi_2 \end{pmatrix}. \tag{5.59}$$

The insertion of $v(p)e^{+ip\cdot x}$ in the Dirac equation (5.33) leads to the Dirac equation in momentum space satisfied by v-spinors,

$$\boxed{(\not{p}+m)v(p) = 0.} \tag{5.60}$$

With respect to Eq. (5.48) satisfied by u-spinors of particles, the sign in front of the mass is thus simply inverted. Similarly, the equation satisfied by $\bar{v}(p)$ is obtained by changing $-m$ into $+m$ in Eq. (5.49), yielding

$$\bar{v}(p)(\not{p}+m) = 0. \tag{5.61}$$

Solutions with Normalisation

So far, we have left out the normalisation factors of the spinors. As for the boson case, we normalise the relativistic wave function to $2E$ particles per unit volume.[12] The density is given in Eq. (5.45), where $\rho(x) = \bar{\psi}(x)\gamma^0\psi(x) = \psi^\dagger(x)\psi(x)$. This requires that ψ has the dimensions of $[\text{energy}]^{3/2}$, such that ρ has the dimensions of $[\text{volume}]^{-1}$ in the international system of units. Let us write

$$\psi(x) = a\, w(p)e^{\pm ip\cdot x},$$

where $w(p)$ is a generic spinor standing for u_1, u_2, v_1 or v_2, and a is an irrelevant constant, which is here only for the consistency of the units. The relativistic normalisation gives

$$\rho(x) = \psi^\dagger(x)\psi(x) = w^\dagger w|a|^2 = 2E|a|^2.$$

The constant a has thus the dimensions of energy. We will simply choose $a = 1$ (in energy units) for simplicity and forget it from now on. Both ρ and E are the time component of a 4-vector. Hence, they transform the same way under a Lorentz transformation. This normalisation is thus appropriate for relativistic particles. As an example, let us do the explicit calculation for $w = v_1$ given in Eq. (5.59). It gives

$$v_1^\dagger v_1 = |N_3|^2 \left(\left(\frac{\boldsymbol{\sigma}\cdot\boldsymbol{p}}{E+m}\chi_1\right)^\dagger, \chi_1^\dagger \right) \begin{pmatrix} \frac{\boldsymbol{\sigma}\cdot\boldsymbol{p}}{E+m}\chi_1 \\ \chi_1 \end{pmatrix} = |N_3|^2 \left(\chi_1^\dagger \frac{(\boldsymbol{\sigma}\cdot\boldsymbol{p})^\dagger(\boldsymbol{\sigma}\cdot\boldsymbol{p})}{(E+m)^2}\chi_1 + \chi_1^\dagger\chi_1 \right).$$

Given that $(\boldsymbol{\sigma}\cdot\boldsymbol{p})^\dagger\boldsymbol{\sigma}\cdot\boldsymbol{p} = (\boldsymbol{\sigma}\cdot\boldsymbol{p})^2 = |\boldsymbol{p}|^2\mathbb{1}$ [see the expression (5.57) of $\boldsymbol{\sigma}\cdot\boldsymbol{p}$] and assuming that the 2-spinor χ_1 is properly normalised, i.e. $\chi_1^\dagger\chi_1 = 1$, it follows that

$$v_1^\dagger v_1 = |N_3|^2 \left(\frac{|\boldsymbol{p}|^2}{(E+m)^2} + 1 \right) = |N_3|^2 \frac{2E}{E+m}.$$

[12] Choosing $N = 1$ in Eq. (5.13) implies for the 4-current (5.12), $j^\mu = \pm 2\begin{pmatrix} E \\ \boldsymbol{p} \end{pmatrix}$ and hence $2E$ particles per unit volume.

Thus, the normalisation requires $N_3 = \pm\sqrt{E+m}$. We would have found the same values for N_1, N_2 and N_4. Hence, according to Eqs. (5.58) and (5.59), the four solutions of the Dirac equation are

$$\boxed{\begin{aligned} &E = +\sqrt{|\boldsymbol{p}|^2 + m^2} \\ &\psi_1(x) = u_1(p)e^{-ip\cdot x}, \quad \psi_2(x) = u_2(p)e^{-ip\cdot x}, \\ &\psi_{\bar{1}}(x) = v_1(p)e^{+ip\cdot x}, \quad \psi_{\bar{2}}(x) = v_2(p)e^{+ip\cdot x}, \\ &u_1(p) = \sqrt{E+m}\begin{pmatrix} \phi_1 \\ \frac{\boldsymbol{\sigma}\cdot\boldsymbol{p}}{E+m}\phi_1 \end{pmatrix}, \quad u_2(p) = \sqrt{E+m}\begin{pmatrix} \phi_2 \\ \frac{\boldsymbol{\sigma}\cdot\boldsymbol{p}}{E+m}\phi_2 \end{pmatrix}, \\ &v_1(p) = \sqrt{E+m}\begin{pmatrix} \frac{\boldsymbol{\sigma}\cdot\boldsymbol{p}}{E+m}\chi_1 \\ \chi_1 \end{pmatrix}, \quad v_2(p) = -\sqrt{E+m}\begin{pmatrix} \frac{\boldsymbol{\sigma}\cdot\boldsymbol{p}}{E+m}\chi_2 \\ \chi_2 \end{pmatrix}. \end{aligned}} \tag{5.62}$$

The two solutions ψ_1 and ψ_2 correspond to a particle, while $\psi_{\bar{1}}$ and $\psi_{\bar{2}}$ correspond to an antiparticle. You may wonder why we choose the normalisation factor $+\sqrt{E+m}$ for $v_1(p)$ while $v_2(p)$ has $-\sqrt{E+m}$. This choice will be justified in Section 5.3.5.

Example of an Explicit Formula

Let us consider, for example,

$$\phi_1 = \chi_2 = \begin{pmatrix} 1 \\ 0 \end{pmatrix}, \quad \phi_2 = \chi_1 = \begin{pmatrix} 0 \\ 1 \end{pmatrix}.$$

Those 2-spinors are eigenstates of σ_3, and thus, in non-relativistic quantum mechanics, for a measurement along the z-axis, they correspond to a spin-up state (for ϕ_1 and χ_2) or spin-down state (for ϕ_2 and χ_1).[13] Using the Formula (5.57) of $\boldsymbol{\sigma}\cdot\boldsymbol{p}$, the four solutions of our example are then

$$\begin{aligned} &\psi_1 = \sqrt{E+m}\begin{pmatrix} 1 \\ 0 \\ \frac{p_z}{E+m} \\ \frac{p_x+ip_y}{E+m} \end{pmatrix} e^{-ip\cdot x}, \quad \psi_2 = \sqrt{E+m}\begin{pmatrix} 0 \\ 1 \\ \frac{p_x-ip_y}{E+m} \\ -\frac{p_z}{E+m} \end{pmatrix} e^{-ip\cdot x}, \\ &\psi_{\bar{1}} = \sqrt{E+m}\begin{pmatrix} \frac{p_x-ip_y}{E+m} \\ -\frac{p_z}{E+m} \\ 0 \\ 1 \end{pmatrix} e^{+ip\cdot x}, \quad \psi_{\bar{2}} = -\sqrt{E+m}\begin{pmatrix} \frac{p_z}{E+m} \\ \frac{p_x+ip_y}{E+m} \\ 1 \\ 0 \end{pmatrix} e^{+ip\cdot x}. \end{aligned} \tag{5.63}$$

[13] With this choice, solutions for particle and antiparticle with the same index will have a similar interpretation in term of spin as we shall see in the next section.

Interpretation in Terms of Spin and Helicity

For a particle or an antiparticle, we always have two solutions that are degenerate in energy. There must be an operator that commutes with the energy operator $\hat{H}_D$ of the Dirac equation whose eigenvalues would distinguish between the two solutions. The application of the Ehrenfest theorem[14] would lead to the conservation of these values over time. First, let us determine the Dirac Hamiltonian, $\hat{H}_D$. The Dirac equation (5.33) reads

$$i\gamma^0 \frac{\partial}{\partial t}\psi + (i\boldsymbol{\gamma} \cdot \nabla - m)\psi = 0,$$

and thus a multiplication from the left by γ^0 and the use of the momentum operator $\hat{\boldsymbol{P}} = -i\nabla$ yield

$$i\frac{\partial}{\partial t}\psi = \hat{H}_D\,\psi,$$

where the Dirac Hamiltonian is

$$\hat{H}_D = \gamma^0\boldsymbol{\gamma} \cdot \hat{\boldsymbol{P}} + \gamma^0 m = \begin{pmatrix} \mathbb{0} & \boldsymbol{\sigma} \\ \boldsymbol{\sigma} & \mathbb{0} \end{pmatrix} \cdot \hat{\boldsymbol{P}} + \gamma^0 m. \tag{5.64}$$

Spin States

Given the Dirac Hamiltonian, let us see whether the spin operators commute with $\hat{H}_D$. A simple generalisation of the spin operators (2.61), which act on 2-spinors, to an operator acting on 4-spinors is

$$\hat{\boldsymbol{S}} = \frac{1}{2}\boldsymbol{\Sigma} = \frac{1}{2}\begin{pmatrix} \boldsymbol{\sigma} & \mathbb{0} \\ \mathbb{0} & \boldsymbol{\sigma} \end{pmatrix}. \tag{5.65}$$

However, $\hat{H}_D$ and $\hat{\boldsymbol{S}}$ do not commute because of the presence of $\hat{\boldsymbol{P}}$ in $\hat{H}_D$. Indeed, the commutator is

$$[\hat{\boldsymbol{S}}, \hat{H}_D] = \sum_{l=1}^{3}[\hat{\boldsymbol{S}}, \gamma^0\gamma^l\hat{P}^l] + m[\hat{\boldsymbol{S}}, \gamma^0],$$

where $\hat{P}^l$ is the momentum operator of the component l, i.e. $\hat{P}^l = -i\partial_l$. With the standard Dirac representation, it is easy to check that $[\hat{\boldsymbol{S}}, \gamma^0] = 0$, reducing the expression above to

$$[\hat{\boldsymbol{S}}, \hat{H}_D] = \sum_{l=1}^{3}[\hat{\boldsymbol{S}}, \gamma^0\gamma^l\hat{P}^l] = \sum_{l=1}^{3}[\hat{\boldsymbol{S}}, \gamma^0\gamma^l]\hat{P}^l.$$

In this representation, the commutator with the kth component of the spin reads

$$[\hat{S}^k, \gamma^0\gamma^l] = \frac{1}{2}\left[\begin{pmatrix} \sigma^k & 0 \\ 0 & \sigma^k \end{pmatrix}, \begin{pmatrix} 0 & -\sigma^l \\ \sigma^l & 0 \end{pmatrix}\right] = \frac{1}{2}\begin{pmatrix} 0 & -[\sigma^k, \sigma^l] \\ [\sigma^k, \sigma^l] & 0 \end{pmatrix}.$$

[14] More appropriately, the time evolution of operators in the Heisenberg picture is given by $\frac{d\hat{A}}{dt} = i[\hat{A}, \hat{H}]$. See Section 6.2.1.

Given that the commutator of Pauli matrices is $[\sigma^k, \sigma^l] = 2i\epsilon_{klm}\sigma^m$, with an implicit summation over the repeated index m [ϵ_{klm} is the spatial antisymmetric tensor (2.53)], the commutator with the Hamiltonian simply reads

$$[\hat{S}^k, \hat{H}_D] = \sum_{l=1}^{3} i\epsilon_{klm} \begin{pmatrix} 0 & -\sigma^m \\ \sigma^m & 0 \end{pmatrix} \hat{P}^l = \sum_{l=1}^{3} i\epsilon_{klm}\gamma^0\gamma^m\hat{P}^l. \tag{5.66}$$

This commutator is, in general, not equal to zero implying that, for a moving particle, the spin is not a good quantum number to distinguish amongst the solutions of the Dirac equation. Or in other words, the spin states are not eigenstates of the Dirac Hamiltonian. The same reasoning can be developed for the orbital angular momentum $\hat{\boldsymbol{L}} = \hat{\boldsymbol{r}} \times \hat{\boldsymbol{P}}$, with

$$[\hat{\boldsymbol{L}}, \hat{H}_D] = \sum_{l=1}^{3} [\hat{\boldsymbol{L}}, \gamma^0\gamma^l\hat{P}^l] = \sum_{l=1}^{3} \gamma^0\gamma^l[\hat{\boldsymbol{L}}, \hat{P}^l].$$

The orbital angular momentum does not commute with the momentum (because the derivative in $\hat{P}^l$ acts on the spatial coordinate in $\hat{\boldsymbol{L}}$). More specifically, it is not difficult to check that $[\hat{L}^k, \hat{P}^l] = i\epsilon_{klm}\hat{P}^m$, which yields

$$[\hat{L}^k, \hat{H}_D] = \sum_{l=1}^{3} i\epsilon_{klm}\gamma^0\gamma^l\hat{P}^m = \sum_{m=1}^{3} i\epsilon_{kml}\gamma^0\gamma^m\hat{P}^l = -\sum_{m=1}^{3} i\epsilon_{klm}\gamma^0\gamma^m\hat{P}^l, \tag{5.67}$$

where the dummy indices l and m were swapped in the second equality and the antisymmetry of the tensor ϵ_{kml} used. Therefore, here again, the orbital angular momentum is not a good quantum number to distinguish the solutions of Dirac equation. However, the total angular momentum $\hat{\boldsymbol{J}} = \hat{\boldsymbol{L}} + \hat{\boldsymbol{S}}$ does commute with the Hamiltonian $\hat{H}_D$. Looking at the two commutators (5.66) and (5.67), since there are implicit summations over m and l respectively in Eq. (5.66) and (5.67), it is obvious that

$$[\hat{L}^k + \hat{S}^k, \hat{H}_D] = 0.$$

In conclusion, neither the spin nor the orbital angular momenta are good quantum numbers, but the total angular momentum is a good quantum number to distinguish the solutions of the Dirac equation. It is another illustration that only the total angular momentum is a conserved quantity.

It is worth looking at the specific case of a particle at rest where the total angular momentum is reduced to the spin. Such a particle is characterised by a propagation term in the solutions of the Dirac equation reduced to $\exp(\pm imt)$. The action of the momenta operators $\hat{P}^l$ (a derivative with respect to spatial coordinates) present in the commutators (5.66) and (5.67) on such a term then gives zero, confirming that, for a particle at rest, the eigenstates of the Dirac Hamiltonian must be eigenstates of the spin operator. Take for example the explicit solutions (5.63) with $p_x = p_y = p_z = 0$. They are indeed the eigenstates of $\hat{S}_z$[15] with eigenvalues

$$\hat{S}_z\psi_1 = +\frac{1}{2}\psi_1 \ , \quad \hat{S}_z\psi_2 = -\frac{1}{2}\psi_2 \ , \quad \hat{\bar{S}}_z\psi_{\bar{1}} = +\frac{1}{2}\psi_{\bar{1}} \ , \quad \hat{\bar{S}}_z\psi_{\bar{2}} = -\frac{1}{2}\psi_{\bar{2}}. \tag{5.68}$$

[15] I use interchangeably σ_x for σ^1, σ_y for σ^2 and σ_z for σ^3 and equivalently $\hat{S}^1$ for $\hat{S}_x$,$\hat{S}^2$ for $\hat{S}_y$ and $\hat{S}^3$ for $\hat{S}_z$.

Note that the spin operator for the antiparticles is here defined as

$$\hat{\bar{S}} = -\hat{S}. \tag{5.69}$$

This is justified because, with antiparticles, the operator returning the physical momentum (positive energy, appropriate direction for $\boldsymbol{p}$) is changed from $\hat{\boldsymbol{P}} \to -\hat{\boldsymbol{P}}$ (because antiparticles have a propagation mode $e^{+ip\cdot x}$), so that $\hat{\boldsymbol{L}} = \hat{\boldsymbol{r}} \times \hat{\boldsymbol{P}} \to -\hat{\boldsymbol{L}}$. Thus, the conservation of the total angular momentum $\hat{\boldsymbol{J}} = \hat{\boldsymbol{L}} + \hat{\boldsymbol{S}}$ requires $\hat{\boldsymbol{S}} \to -\hat{\boldsymbol{S}}$. According to the equalities (5.68), we see that ψ_1 and $\psi_{\bar{1}}$ would have spin up, while ψ_2 and $\psi_{\bar{2}}$ would have spin down when $p_x = p_y = p_z = 0$. This result also holds if the (anti)particle travels in the z-direction ($p_x = p_y = 0$, $p_z = \pm|\boldsymbol{p}|$) since, according to Eq. (5.66), when the spin is measured along the z-axis, the commutator $[\hat{S}^3, \hat{H}_D]$ depends only on $\hat{P}^1$ and $\hat{P}^2$, namely on the components p_x and p_y.

Helicity States

For any momentum $\boldsymbol{p}$ of the particle, the two degenerate solutions can be classified according to their total angular momentum. The expression of the commutator (5.66) suggests however another possibility. Because of the properties of the anti-symmetric tensor ϵ_{klm}, if we are interested in the spin along the axis corresponding to the index k, there cannot be a term on the right-hand side of Eq. (5.66) containing the operator $\hat{P}^k$. Therefore, if the particle travels only along the axis corresponding to the index k, the action of the commutator on the solution of the Dirac equation will be 0 and that solution will also be an eigenstate of $\hat{S}^k$.

Helicity is precisely defined as the projection of the spin along the direction of motion (introduced in Section 4.3.3). Beware however that *helicity is not generally Lorentz invariant*. Actually, for a massive particle, one can always find a reference frame where the particle moves in the opposite direction. The helicity operator is logically given for fermion spinors by

$$\hat{h} = \hat{\boldsymbol{S}} \cdot \frac{\hat{\boldsymbol{P}}}{|\boldsymbol{p}|} = \frac{1}{2}\boldsymbol{\Sigma} \cdot \frac{\hat{\boldsymbol{P}}}{|\boldsymbol{p}|} = \frac{1}{2}\begin{pmatrix} \boldsymbol{\sigma} \cdot \frac{\hat{\boldsymbol{P}}}{|\boldsymbol{p}|} & \mathbb{0} \\ \mathbb{0} & \boldsymbol{\sigma} \cdot \frac{\hat{\boldsymbol{P}}}{|\boldsymbol{p}|} \end{pmatrix}. \tag{5.70}$$

Note that for anti-fermions, the same operator can be used since both $\hat{\boldsymbol{S}}$ and $\hat{\boldsymbol{P}}$ are inverted. Replacing the momentum operator $\hat{\boldsymbol{P}}$ by $-i\boldsymbol{\nabla}$, the helicity operator then reads

$$\hat{h} = \frac{1}{2}\boldsymbol{\Sigma} \cdot \frac{(-i\boldsymbol{\nabla})}{|\boldsymbol{p}|}, \tag{5.71}$$

which when acting on the fermion wave function $\psi(x) = u(p)e^{-ip\cdot x}$ or anti-fermion wave function $\psi(x) = v(p)e^{+ip\cdot x}$ gives

$$\hat{h}\left(u(p)e^{-ip\cdot x}\right) = \frac{1}{2}\boldsymbol{\Sigma} \cdot \frac{\boldsymbol{p}}{|\boldsymbol{p}|}u(p)e^{-ip\cdot x}, \quad \hat{h}\left(v(p)e^{+ip\cdot x}\right) = -\frac{1}{2}\boldsymbol{\Sigma} \cdot \frac{\boldsymbol{p}}{|\boldsymbol{p}|}v(p)e^{+ip\cdot x}.$$

If $\psi(x)$ is an eigenstate of the helicity operator with the eigenvalue λ, it satisfies by definition $\hat{h}\psi(x) = \lambda\psi(x)$. This imposes on the spinors u and v the relations

$$\frac{1}{2}\boldsymbol{\Sigma} \cdot \frac{\boldsymbol{p}}{|\boldsymbol{p}|}u(p) = \lambda u(p), \quad -\frac{1}{2}\boldsymbol{\Sigma} \cdot \frac{\boldsymbol{p}}{|\boldsymbol{p}|}v(p) = \lambda v(p). \tag{5.72}$$

The *helicity operator in the momentum space* then reads, for fermions,

$$\frac{1}{2}\boldsymbol{\Sigma}\cdot\frac{\boldsymbol{p}}{|\boldsymbol{p}|}=\frac{1}{2}\begin{pmatrix}\boldsymbol{\sigma}\cdot\frac{\boldsymbol{p}}{|\boldsymbol{p}|} & 0\\ 0 & \boldsymbol{\sigma}\cdot\frac{\boldsymbol{p}}{|\boldsymbol{p}|}\end{pmatrix},\tag{5.73}$$

and the opposite for anti-fermions. The helicity operator (5.70) clearly commutes with the hamiltonian $\hat{H}_D$ (5.64) since both involve the product $\boldsymbol{\sigma}\cdot\hat{\boldsymbol{P}}$. Hence, there exists a common basis of 4-spinors that are eigenstates of both operators. Since the spin projection on any axis (and thus along the direction of motion) can only be $\pm 1/2$, the helicity quantum number,[16] λ, can only be $\pm 1/2$. Depending on the sign of λ, the state is called

Left-handed helicity state if $\lambda < 0$, Right-handed helicity state if $\lambda > 0$.

After some calculations (see Problem 5.6), the general solution for helicity states (modulo one phase) can be realised:

$$\psi_{+\frac{1}{2}}=\sqrt{E+m}\begin{pmatrix}\cos\frac{\theta}{2}\\ e^{i\phi}\sin\frac{\theta}{2}\\ \frac{|\boldsymbol{p}|}{E+m}\cos\frac{\theta}{2}\\ \frac{|\boldsymbol{p}|}{E+m}e^{i\phi}\sin\frac{\theta}{2}\end{pmatrix}e^{-ip\cdot x},\quad \psi_{-\frac{1}{2}}=\sqrt{E+m}\begin{pmatrix}-\sin\frac{\theta}{2}\\ e^{i\phi}\cos\frac{\theta}{2}\\ \frac{|\boldsymbol{p}|}{E+m}\sin\frac{\theta}{2}\\ -\frac{|\boldsymbol{p}|}{E+m}e^{i\phi}\cos\frac{\theta}{2}\end{pmatrix}e^{-ip\cdot x},$$

$$\psi_{+\bar{\frac{1}{2}}}=\sqrt{E+m}\begin{pmatrix}\frac{|\boldsymbol{p}|}{E+m}\sin\frac{\theta}{2}\\ -\frac{|\boldsymbol{p}|}{E+m}e^{i\phi}\cos\frac{\theta}{2}\\ -\sin\frac{\theta}{2}\\ e^{i\phi}\cos\frac{\theta}{2}\end{pmatrix}e^{+ip\cdot x},\quad \psi_{-\bar{\frac{1}{2}}}=-\sqrt{E+m}\begin{pmatrix}\frac{|\boldsymbol{p}|}{E+m}\cos\frac{\theta}{2}\\ \frac{|\boldsymbol{p}|}{E+m}e^{i\phi}\sin\frac{\theta}{2}\\ \cos\frac{\theta}{2}\\ e^{i\phi}\sin\frac{\theta}{2}\end{pmatrix}e^{+ip\cdot x},\tag{5.74}$$

where θ and ϕ are respectively the polar and azimuthal angles of the momentum, i.e. $\boldsymbol{p}=|\boldsymbol{p}|(\sin\theta\cos\phi,\sin\theta\sin\phi,\cos\theta)^{\mathsf{T}}$.

5.3.5 Operations on Spinors

Normalisation of Spinors and Completeness Relations

Spinors depend on the 4-momentum of the particle and on a degree of freedom that can be chosen as its spin (for a particle at rest) or its helicity. We already saw in Section 5.3.4 that for relativistic particles, it is more appropriate to normalise spinors to $2E$ particles per unit volume, i.e.

$$w_i^\dagger(p)w_i(p)=2E,$$

where $w_i(p)$ denotes a generic spinor (of u- or v-type) with a polarisation (spin or helicity) index $i = 1,2$ and 4-momentum $p = (E,\boldsymbol{p})$. We are going to determine other relations among spinors, which shall be useful for calculating cross sections or particle lifetimes in

[16] Note that some authors define the helicity quantum number as twice that number in order to have an integer for both fermions and bosons.

Chapter 6. Given the normalised spinors appearing in the Formula (5.62), assuming that $\phi_i^\dagger \phi_j = \chi_i^\dagger \chi_j = \delta_{ij}$, one can easily check that

$$u_i^\dagger(p)u_j(p) = \bar{u}_i(p)\gamma^0 u_j(p) = v_i^\dagger(p)v_j(p) = \bar{v}_i(p)\gamma^0 v_j(p) = 2E\delta_{ij}, \tag{5.75}$$

$$u_i^\dagger(E,\boldsymbol{p})v_j(E,-\boldsymbol{p}) = \bar{u}_i(E,\boldsymbol{p})\gamma^0 v_j(E,-\boldsymbol{p}) = 0. \tag{5.76}$$

The proof simply relies on the relation

$$(\boldsymbol{\sigma}\cdot\boldsymbol{p})^\dagger \boldsymbol{\sigma}\cdot\boldsymbol{p} = (\boldsymbol{\sigma}\cdot\boldsymbol{p})^2 = |\boldsymbol{p}|^2 \mathbb{1}. \tag{5.77}$$

Moreover, the quantity $\sum_i w_i(p)\overline{w}_i(p)$, called the *completeness relation*, turns out to be useful. Let us calculate it, for instance, for u spinor types. Using the Dirac representation of γ matrices and Eq. (5.62), it follows that

$$\begin{aligned}\sum_{i=1}^{2} u_i(p)\bar{u}_i(p) &= (E+m)\sum_{i=1}^{2}\begin{pmatrix}\phi_i \\ \frac{\boldsymbol{\sigma}\cdot\boldsymbol{p}}{E+m}\phi_i\end{pmatrix}\begin{pmatrix}\phi_i^\dagger & \phi_i^\dagger\frac{\boldsymbol{\sigma}\cdot\boldsymbol{p}}{E+m}\end{pmatrix}\begin{pmatrix}\mathbb{1} & \mathbb{0} \\ \mathbb{0} & -\mathbb{1}\end{pmatrix} \\ &= (E+m)\begin{pmatrix}\sum_i \phi_i\phi_i^\dagger & -(\sum_i \phi_i\phi_i^\dagger)\frac{\boldsymbol{\sigma}\cdot\boldsymbol{p}}{E+m} \\ \frac{\boldsymbol{\sigma}\cdot\boldsymbol{p}}{E+m}\sum_i \phi_i\phi_i^\dagger & -\frac{\boldsymbol{\sigma}\cdot\boldsymbol{p}}{E+m}\sum_i \phi_i\phi_i^\dagger\frac{\boldsymbol{\sigma}\cdot\boldsymbol{p}}{E+m}\end{pmatrix}.\end{aligned}$$

Since $\{\phi_i\}$ constitutes a basis of 2-spinors, $\sum_{i=1}^{2}\phi_i\phi_i^\dagger = \mathbb{1}$, and thus using again Eq. (5.77), we conclude

$$\sum_{i=1}^{2} u_i(p)\bar{u}_i(p) = \begin{pmatrix}(E+m)\mathbb{1} & -\boldsymbol{\sigma}\cdot\boldsymbol{p} \\ \boldsymbol{\sigma}\cdot\boldsymbol{p} & -\frac{|\boldsymbol{p}|^2}{E+m}\mathbb{1}\end{pmatrix}.$$

Noticing that $-|\boldsymbol{p}|^2 = (-E+m)(E+m)$, it follows that

$$\sum_{i=1}^{2} u_i(p)\bar{u}_i(p) = \begin{pmatrix}E\mathbb{1} & -\boldsymbol{\sigma}\cdot\boldsymbol{p} \\ \boldsymbol{\sigma}\cdot\boldsymbol{p} & -E\mathbb{1}\end{pmatrix} + m\begin{pmatrix}\mathbb{1} & \mathbb{0} \\ \mathbb{0} & \mathbb{1}\end{pmatrix} = p^\mu\gamma_\mu + m\begin{pmatrix}\mathbb{1} & \mathbb{0} \\ \mathbb{0} & \mathbb{1}\end{pmatrix}.$$

A similar calculation allows us to determine $\sum_{i=1}^{2} v_i(p)\bar{v}_i(p)$. Dropping the 4×4 identity matrix to adopt an abbreviated (but sloppy) mathematics notation, the completeness relations finally read

$$\sum_{i=1}^{2} u_i(p)\bar{u}_i(p) = \not{p} + m, \quad \sum_{i=1}^{2} v_i(p)\bar{v}_i(p) = \not{p} - m. \tag{5.78}$$

Charge Conjugation

In classical electrodynamics, the motion of a charged particle with a charge q in an electromagnetic field $A^\mu = (V, \boldsymbol{A})$ is obtained by making the substitutions:

$$\boldsymbol{p} \to \boldsymbol{p} - q\boldsymbol{A}, \quad E \to E - qV,$$

which, with the 4-vector formalism, correspond to $p^\mu \to p^\mu - qA^\mu$. Applying the correspondence principle, $p^\mu \leftrightarrow i\partial^\mu$, the quantum version is then simply

$$\partial^\mu \to \partial^\mu + iqA^\mu. \tag{5.79}$$

Therefore, in the presence of an electromagnetic field, the Dirac equation becomes

$$\left[\gamma^\mu(i\partial_\mu - qA_\mu) - m\right]\psi = 0. \tag{5.80}$$

Let us denote the charge conjugate of the wave function by $\psi^c(x)$. It is expected to satisfy the Dirac equation with the opposite charge,

$$\left[\gamma^\mu(i\partial_\mu + qA_\mu) - m\right]\psi^c = 0. \tag{5.81}$$

Naturally, if $\psi(x)$ describes a particle, $\psi(x) = u(p)e^{-ip\cdot x}$, we expect $\psi^c(x)$ to describe its antiparticle and thus $\psi^c(x) = v(p)e^{+ip\cdot x}$, where the v spinor is related to the u spinor. So let us postulate that

$$\psi^c(x) = \hat{C}\psi^{\dagger\mathsf{T}}(x), \tag{5.82}$$

with $\hat{C}$, a 4×4 unitary matrix acting on spinors. It is recommended to use $\psi^{\dagger\mathsf{T}}$ instead of ψ^*. If $\psi(x)$ is a wave function as in this section, both are equivalent. However, when $\psi(x)$ will be a quantised field, i.e. a mixture of spinors and operators still satisfying the Dirac equation (see Section 5.3.6), the Hermitian conjugate is not a simple complex conjugation for operators. Hence, '$\dagger\mathsf{T}$' clearly shows that it is the Hermitian conjugate that matters, while for the spinors, it shows that it would still be a vector column (and not a row vector as in $\psi^\dagger$). In order to find $\hat{C}$, let us first conjugate Eq. (5.80) and then multiply by $\hat{C}$ from the left. It follows that

$$\left[-\hat{C}\gamma^{\mu *}(i\partial_\mu + qA_\mu) - m\hat{C}\right]\psi^* = 0.$$

If $\hat{C}$ satisfies

$$-\hat{C}\gamma^{\mu *} = \gamma^\mu\hat{C} \Leftrightarrow \hat{C}^\dagger\gamma^\mu\hat{C} = -\gamma^{\mu *}, \tag{5.83}$$

then ψ^c necessarily satisfies the Dirac equation (5.81). The explicit expression for the matrix $\hat{C}$ depends on the representation of the Clifford algebra used for the γ matrices. In the Dirac representation, only γ^2 is a complex matrix satisfying $\gamma^{2*} = -\gamma^2$ and thus, according to the condition (5.83), $\hat{C}$ must commute with γ^2 and anti-commute with the other γ matrices. A basic choice is $\hat{C} = \gamma^2$. If ψ represents the wave function of an electron, we would thus expect $\psi^c = \gamma^2\psi^*$ to be that of a positron. Let us check it, for instance, with the electron described by ψ_1 in Eq. (5.63):

$$\gamma^2\psi_1^* = \sqrt{E+m}\begin{pmatrix} 0 & 0 & 0 & -i \\ 0 & 0 & i & 0 \\ 0 & i & 0 & 0 \\ -i & 0 & 0 & 0 \end{pmatrix}\left[\begin{pmatrix} 1 \\ 0 \\ \frac{p_z}{E+m} \\ \frac{p_x+ip_y}{E+m} \end{pmatrix} e^{-ip\cdot x}\right]^*$$

$$= -i\sqrt{E+m}\begin{pmatrix} \frac{p_x-ip_y}{E+m} \\ -\frac{p_z}{E+m} \\ 0 \\ 1 \end{pmatrix} e^{+ip\cdot x}.$$

This is the positron solution described by $\psi_{\bar{1}}$ in Eq. (5.63), except that there is this extra factor $-i$. Hence, by using

$$\psi^c(x) = \hat{C}\psi^{\dagger\mathsf{T}}(x) = i\gamma^2\psi^{\dagger\mathsf{T}}(x), \tag{5.84}$$

we obtain the correct description. We chose $-\sqrt{E+m}$ (and not $\sqrt{E+m}$) as the normalisation of $\psi_{\bar{2}}$ in Eq. (5.63) because it precisely ensures that $\hat{C}\psi_{\bar{2}}^* = \psi_2$ (and vice versa). Transformation (5.84) implies that the spinors transform as

$$\hat{C}u_r^*(p) = v_r(p), \quad \hat{C}v_r^*(p) = u_r(p). \tag{5.85}$$

In some textbooks, the charge-conjugation matrix C is defined with respect to $\overline{\psi}(x)$, and hence, $\psi^c(x) = \hat{C}\overline{\psi}^{\mathsf{T}}(x)$ (with ψ^c being a spinor and not an adjoint spinor, one has to take the transpose of $\overline{\psi}$). Now, since $\overline{\psi}^{\mathsf{T}} = \gamma^0\psi^{\dagger\mathsf{T}}$, the equality (5.84) imposes that the new matrix $\hat{C}$ satisfies $\hat{C}\gamma^0 = i\gamma^2$. Since $(\gamma^0)^2 = 1$, it follows that

$$\psi^c(x) = \hat{C}\overline{\psi}^{\mathsf{T}}(x) = i\gamma^2\gamma^0\overline{\psi}^{\mathsf{T}}(x) = -i\gamma^0\gamma^2\overline{\psi}^{\mathsf{T}}(x). \tag{5.86}$$

When the charge conjugation is applied twice, we expect to recover the initial wave function. Let us check it, for example, with the definition (5.84), the other definition giving the same results:

$$((\psi)^c)^c = \hat{C}\left(\hat{C}\psi^{\dagger\mathsf{T}}\right)^{\dagger\mathsf{T}} = \hat{C}\hat{C}^{\dagger\mathsf{T}}\psi = \hat{C}\hat{C}^*\psi = (i\gamma^2)(-i\gamma^{2*})\psi = \psi. \tag{5.87}$$

Parity

The parity transformation (Section 4.3.3) changes the sign of the space coordinates, $x = (t, \boldsymbol{x}) \to x' = (t, -\boldsymbol{x})$. The spinors that satisfy the Dirac equation (5.33) have to transform, $\psi(x) \to \psi'(x')$, so as to still satisfy

$$(i\gamma^\mu\partial'_\mu - m)\psi'(x') = 0.$$

Let us introduce a 4×4 unitary matrix $\hat{P}$ acting on spinors such as

$$\psi'(x') = \hat{P}\psi(x).$$

The matrix $\hat{P}$ is necessarily unitary in order to conserve the normalisation of $\psi(x)$. Let us see how we can recover the Dirac equation satisfied by $\psi(x)$ starting with the Dirac equation of $\psi'(x')$:

$$\begin{aligned}(i\gamma^\mu\partial'_\mu - m)\psi'(x') &= \left(i\gamma^0\frac{\partial}{\partial t}\frac{\partial t}{\partial t'} + i\gamma^1\frac{\partial}{\partial x}\frac{\partial x}{\partial x'} + i\gamma^2\frac{\partial}{\partial y}\frac{\partial y}{\partial y'} + i\gamma^3\frac{\partial}{\partial z}\frac{\partial z}{\partial z'} - m\right)\psi'(x') \\ &= \left(i\gamma^0\frac{\partial}{\partial t} - i\gamma^1\frac{\partial}{\partial x} - i\gamma^2\frac{\partial}{\partial y} - i\gamma^3\frac{\partial}{\partial z} - m\right)\hat{P}\psi(x).\end{aligned}$$

In order to invert the sign in front of $\gamma^{i=1,2,3}$, let us multiply the above equation by γ^0 from the left and use the anti-commutation algebra. This gives

$$\left(i\gamma^0\frac{\partial}{\partial t} + i\gamma^1\frac{\partial}{\partial x} + i\gamma^2\frac{\partial}{\partial y} + i\gamma^3\frac{\partial}{\partial z} - m\right)\gamma^0\hat{P}\psi(x) = (i\gamma^\mu\partial_\mu - m)\gamma^0\hat{P}\psi(x).$$

We recover the Dirac equation if $\gamma^0\hat{P}$ is a multiple of the identity matrix, i.e. $\gamma^0\hat{P} = \eta\mathbb{1}$, and thus $\hat{P} = \eta\gamma^0$ (since $(\gamma^0)^2 = \mathbb{1}$). Given that necessarily $\hat{P}^2 = \mathbb{1}$, it follows that $\eta^2 = 1$, and we thus identify

$$\hat{P} = \eta\gamma^0, \tag{5.88}$$

where the parameter $\eta = \pm 1$ is the *intrinsic parity*. Moreover, one easily checks that the spinors (5.62) that are a solution of the Dirac equation satisfy the equalities

$$\hat{P}u(\boldsymbol{p}) = \eta u(-\boldsymbol{p}), \quad \hat{P}v(\boldsymbol{p}) = -\eta v(-\boldsymbol{p}). \tag{5.89}$$

Consequently, the intrinsic parity is $\eta_f^P = \eta$ for fermions and $\eta_{\bar{f}}^P = -\eta$ for anti-fermions. The convention is to set $\eta = 1$ in order to be consistent with the historical (and arbitrary) choice made for the proton and the neutron. Therefore,

$$\boxed{\eta_f^P = +1, \quad \eta_{\bar{f}}^P = -1.} \tag{5.90}$$

Chirality

Consider the chirality matrix defined as[17]

$$\boxed{\gamma^5 = i\gamma^0\gamma^1\gamma^2\gamma^3,} \tag{5.91}$$

In the Dirac representation, it reads

$$\gamma^5 = \begin{pmatrix} \mathbb{0} & \mathbb{1} \\ \mathbb{1} & \mathbb{0} \end{pmatrix}. \tag{5.92}$$

Given the anti-commutation algebra of the γ^μ matrices (5.29), it is easy to check that γ^5 anti-commutes with all γ^μ matrices,

$$\{\gamma^5, \gamma^\mu\} = 0. \tag{5.93}$$

This also implies that it is a unitary Hermitian matrix, i.e.

$$(\gamma^5)^2 = 1, \quad \gamma^{5\dagger} = \gamma^5. \tag{5.94}$$

Therefore, the eigenvalues of γ^5 are ± 1. The eigenstates of γ^5 are called the *chirality states*. They play an important role because they are unaffected by a Lorentz transformation in the sense that, after a Lorentz transformation, they are still chirality eigenstates, with the same eigenvalue. Indeed, if the spinor wave function, $\psi(x)$, is a solution of the Dirac equation, we saw in Section 5.3.2 that the covariance of the Dirac equation requires the following transformation of $\psi(x)$:

$$\psi(x) \to \psi'(x') = \exp\left(-\frac{i}{2}\omega_{\sigma\rho}S^{\sigma\rho}\right)\psi(x),$$

where the matrices $S^{\sigma\rho}$, given in Eq. (5.39), are the functions of the commutator $[\gamma^\sigma, \gamma^\rho]$. Let us assume that $\psi(x)$ is an eigenstate of γ^5. Since the matrix γ^5 anti-commutes with any γ^μ matrices, it commutes with $S^{\sigma\rho}$, and thus

$$\gamma^5\psi'(x') = \exp\left(-\frac{i}{2}\omega_{\sigma\rho}S^{\sigma\rho}\right)\gamma^5\psi(x),$$

confirming that $\psi'(x')$ is also an eigenstate of γ^5 with the same eigenvalue. Chirality states are denoted with the labels L or R for left or right, for a reason that will be clarified in

[17] One may encounter the notation γ_5 in the literature. Actually, $\gamma_5 = \gamma^5$.

a few lines. Spinors with a positive or negative chirality are defined respectively as right-handed or left-handed chiral states. The convention used in this textbook is the opposite for antiparticle spinors, i.e.

$$\gamma^5 u_R = +u_R,\ \ \gamma^5 u_L = -u_L,\ \ \gamma^5 v_R = -v_R,\ \ \gamma^5 v_L = +v_L. \tag{5.95}$$

The reason why chiral antiparticle spinors have opposite eigenvalues with respect to the chiral spinor comes from the requirement that the chirality states match the helicity states for a massless particle.[18] Indeed, consider the helicity states (5.74) in the ultra-relativistic limit, where $E \gg m$ or equivalently when $m = 0$. They read

$$\begin{aligned} u_{+\frac{1}{2}} &\approx \sqrt{E}\begin{pmatrix} \cos\frac{\theta}{2} \\ e^{i\phi}\sin\frac{\theta}{2} \\ \cos\frac{\theta}{2} \\ e^{i\phi}\sin\frac{\theta}{2} \end{pmatrix}, \quad & u_{-\frac{1}{2}} &\approx \sqrt{E}\begin{pmatrix} -\sin\frac{\theta}{2} \\ e^{i\phi}\cos\frac{\theta}{2} \\ \sin\frac{\theta}{2} \\ -e^{i\phi}\cos\frac{\theta}{2} \end{pmatrix}, \\ v_{+\frac{1}{2}} &\approx \sqrt{E}\begin{pmatrix} \sin\frac{\theta}{2} \\ -e^{i\phi}\cos\frac{\theta}{2} \\ -\sin\frac{\theta}{2} \\ e^{i\phi}\cos\frac{\theta}{2} \end{pmatrix}, \quad & v_{-\frac{1}{2}} &\approx -\sqrt{E}\begin{pmatrix} \cos\frac{\theta}{2} \\ e^{i\phi}\sin\frac{\theta}{2} \\ \cos\frac{\theta}{2} \\ e^{i\phi}\sin\frac{\theta}{2} \end{pmatrix}. \end{aligned} \tag{5.96}$$

Using the representation (5.92) of γ^5, it is easy to check that these helicity states are now chirality eigenstates, with

$$\gamma^5 u_{+\frac{1}{2}} = u_{+\frac{1}{2}},\ \ \gamma^5 u_{-\frac{1}{2}} = -u_{-\frac{1}{2}},\ \ \gamma^5 v_{+\frac{1}{2}} = -v_{+\frac{1}{2}},\ \ \gamma^5 v_{-\frac{1}{2}} = v_{-\frac{1}{2}}.$$

Therefore, we identify

$$u_{+\frac{1}{2}} = u_R\,,\ \ u_{-\frac{1}{2}} = u_L\,,\ \ v_{+\frac{1}{2}} = v_R\,,\ \ v_{-\frac{1}{2}} = v_L. \tag{5.97}$$

Hence, a right-handed helicity state becomes a right-handed chirality state for massless particles (and similarly for left-handed helicity becoming a left-handed chirality).

Chiral states play an important role in the Standard Model, in particular for the weak interaction. We shall see in Chapter 9 that the charged current interactions, mediated by $W^{\pm}$ bosons, act only on the left-handed chirality states of particles and the right-handed chirality states of antiparticles. Note that for massless particles, according to Eq. (5.96), left-handed chirality states of fermions are equivalent to right-handed chirality states of anti-fermions (the difference of sign has no physical effect). For a general solution of the Dirac equation, one can always decompose a Dirac spinor into its left-handed and right-handed components using the following chiral projectors:

$$\boxed{P_L = \frac{1}{2}\left(1-\gamma^5\right),\ \ P_R = \frac{1}{2}\left(1+\gamma^5\right).} \tag{5.98}$$

[18] It is clearly a matter of convention. Some authors called consistently v_R and u_R the state having the positive eigenvalue of γ^5, but in that case, the right-handed chirality becomes a left-handed helicity when the mass of the particle is neglected.

Given that $(\gamma^5)^2 = \mathbb{1}$, it is straightforward to check that P_L and P_R have the properties of projectors, i.e. $P_L^2 = P_L$, $P_R^2 = P_R$, $P_L P_R = 0$, $P_L + P_R = \mathbb{1}$. Moreover, P_R and P_L satisfy

$$\gamma^5 P_R = P_R, \;\; \gamma^5 P_L = -P_L.$$

Therefore, according to the definition (5.95), P_R projects out the right-handed states of particles (u spinors) and the left-handed states of antiparticles (v spinors). Similarly, P_L projects out the left-handed states of particles and the right-handed states of antiparticles. In summary,

$$\begin{aligned} u &= u_L + u_R, \quad u_L = P_L\, u, \quad u_R = P_R\, u, \\ v &= v_L + v_R, \quad v_L = P_R\, v, \quad v_R = P_L\, v. \end{aligned} \tag{5.99}$$

Helicity states and chirality states are often confused. The confusion is probably due to the use of the same denomination of states, i.e. left- or right-handed states, and to the fact that both concepts are identical for massless particles or equivalently when the mass of particles is neglected (which is often the case with the energetic collisions produced by modern accelerators). We saw, however, that helicity is not Lorentz invariant,[19] whereas the chirality is. For a massless particle, however, helicity is invariant since no boost can invert the momentum of the particle. There is another important difference. Consider, the Dirac Hamiltonian H_D (5.64). Using the anti-commutation property (5.93) of γ^5, the commutator of H_D with γ^5 reduces to[20]

$$\left[H_D, \gamma^5\right] = m\left[\gamma^0, \gamma^5\right].$$

It is different from zero, except when $m = 0$. Therefore, for massive particles, the chirality states are not eigenstates of the Hamiltonian. Consequently, the chirality eigenvalues are not conserved in time, and the physical states having a defined energy (so the eigenstates of the Hamiltonian) are a superposition of the two chirality states evolving in time.

5.3.6 A Few Words about the Quantised Field

As for the scalar case, I give here basic ideas without proofs. I invite the reader to consult a quantum field theory textbook (e.g., Peskin and Schroeder, 1995) for the details. I remind again that we have to give up the notion of a single relativistic particle described by the Dirac equation (see the explanations in the Klein–Gordon section).

The four wave functions in Eq. (5.62), $u_1(p)e^{-ip\cdot x}$, $u_2(p)e^{-ip\cdot x}$, $v_1(p)e^{+ip\cdot x}$ and $v_2(p)e^{+ip\cdot x}$, constitute a basis of solutions of the Dirac equation for a given 4-momentum p. Hence, a general wave function in spacetime can be written as

$$\psi(x) = \int \frac{\mathrm{d}^3 \boldsymbol{p}}{(2\pi)^3 2E_p} \sum_{r=1}^{2} \left[c_{p,r} u_r(p) e^{-ip\cdot x} + d^*_{p,r} v_r(p) e^{+ip\cdot x}\right],$$

[19] Even if helicity is generally not Lorentz invariant, since it involves a scalar product of two 3-vectors, $\boldsymbol{\Sigma}\cdot\hat{\boldsymbol{P}}$ or $\boldsymbol{\Sigma}\cdot\boldsymbol{p}$, it is invariant under spatial rotation.

[20] We saw that the Dirac Hamiltonian, H_D, commutes with the helicity operator. It is easy to check that the chirality operator, γ^5, commutes with the helicity operator. It is then tempting to conclude that H_D must commute with γ^5. However, this is wrong. The commutativity property is not transitive. Here is a simple counterexample with matrices A and B and the identity $\mathbb{1}$: we have $[A, \mathbb{1}] = [B, \mathbb{1}] = 0$, but $[A, B]$ has no reason to be zero.

where $c_{p,r}$ and $d^*_{p,r}$ are the complex Fourier coefficients and $p^0 = E_p = \sqrt{m^2 + |\boldsymbol{p}|^2}$. Since $\psi(x)$ has the dimensions of [energy]$^{3/2}$ (see discussion in Section 5.3.4), and the spinors $u_r(p)$ and $v_r(p)$ have the dimensions of [energy]$^{1/2}$, the Fourier coefficients have the dimensions [energy]$^{-1}$. The quantisation of fields transforms the coefficients into operators, i.e.

$$\psi(x) = \int \frac{\mathrm{d}^3\boldsymbol{p}}{(2\pi)^3 2E_p} \sum_{r=1}^{2} \left[c_{p,r} u_r(p) e^{-ip\cdot x} + d^\dagger_{p,r} v_r(p) e^{+ip\cdot x}\right], \tag{5.100}$$

$$\overline{\psi}(x) = \int \frac{\mathrm{d}^3\boldsymbol{p}}{(2\pi)^3 2E_p} \sum_{r=1}^{2} \left[c^\dagger_{p,r} \bar{u}_r(p) e^{+ip\cdot x} + d_{p,r} \bar{v}_r(p) e^{-ip\cdot x}\right]. \tag{5.101}$$

Since $c_{p,r}$ and $d_{p,r}$ are respectively associated with a fermion and anti-fermion wave function, we wish to interpret $c^\dagger_{p,r}$ as an operator creating a fermion with a given momentum $\boldsymbol{p}$, energy E_p, polarisation r and the corresponding operator for the anti-fermion $d^\dagger_{p,r}$. Now, consider the wave function of a state with 2-fermions

$$\psi_{\boldsymbol{p}_1\boldsymbol{p}_2}(\boldsymbol{x}_1, \boldsymbol{x}_2) = \langle \boldsymbol{x}_1, \boldsymbol{x}_2 | 1, \boldsymbol{p}_1, r_1; 1, \boldsymbol{p}_2, r_2\rangle = (\langle \boldsymbol{x}_1| \otimes \langle \boldsymbol{x}_2|) c^\dagger_{p_1,r_1} c^\dagger_{p_2,r_2} |0\rangle.$$

Clearly, if the $c^\dagger$ operators commute, we will end up with a symmetric function as for the boson case. The only way to get an anti-symmetric function (and then to be consistent with the spin-statistic theorem) is to postulate that fermion operators anti-commute. In that case,

$$\psi_{\boldsymbol{p}_1\boldsymbol{p}_2}(\boldsymbol{x}_1, \boldsymbol{x}_2) = (\langle \boldsymbol{x}_1| \otimes \langle \boldsymbol{x}_2|) c^\dagger_{p_1,r_1} c^\dagger_{p_2,r_2} |0\rangle = -(\langle \boldsymbol{x}_1| \otimes \langle \boldsymbol{x}_2|) c^\dagger_{p_2,r_2} c^\dagger_{p_1,r_1} |0\rangle = -\psi_{\boldsymbol{p}_2\boldsymbol{p}_1}(\boldsymbol{x}_1, \boldsymbol{x}_2).$$

With the normalisation used in this textbook, the complete set of anti-commutation rules then reads

$$\begin{aligned} &\{c_{p',r'}, c^\dagger_{p,r}\} = \{d_{p',r'}, d^\dagger_{p,r}\} = (2\pi)^3 2E_p\, \delta^{(3)}(\boldsymbol{p}' - \boldsymbol{p})\delta_{r'r};\\ &\{c_{p',r'}, c_{p,r}\} = \{d_{p',r'}, d_{p,r}\} = \{c^\dagger_{p',r'}, c^\dagger_{p,r}\} = \{d^\dagger_{p',r'}, d^\dagger_{p,r}\} = 0;\\ &\{c^\dagger_{p',r'}, d^\dagger_{p,r}\} = \{c_{p',r'}, d^\dagger_{p,r}\} = \{c^\dagger_{p',r'}, d_{p,r}\} = 0. \end{aligned} \tag{5.102}$$

Operators satisfying these anti-commutation rules are then interpreted as follows: $c^\dagger_{p,r}$ creates a fermion with momentum $\boldsymbol{p}$, energy $E_p = \sqrt{m^2 + |\boldsymbol{p}|^2}$ and polarisation r, while $c_{p,r}$ destroys it. Similarly, $d^\dagger_{p,r}$ creates a anti-fermion with momentum $\boldsymbol{p}$, energy $E_p = \sqrt{m^2 + |\boldsymbol{p}|^2}$ and polarisation r, while $d_{p,r}$ destroys it. Since $\{c^\dagger_{p,r}, c^\dagger_{p,r}\} = 2c^\dagger_{p,r} c^\dagger_{p,r} = 0$, necessarily, $c^\dagger_{p,r} c^\dagger_{p,r} |0\rangle = |0\rangle$. Therefore, 2-fermions or anti-fermions cannot occupy the same state, and we recover the Pauli exclusion principle.

As for the scalar field, the Hamiltonian, momentum and charge operators can be expressed in terms of creation and annihilation operators. They read (in a normal-ordered form)

$$\begin{aligned} H &= \int \frac{\mathrm{d}^3\boldsymbol{p}}{(2\pi)^3 2E_p} \textstyle\sum_{r=1,2} E_p\, (c^\dagger_{p,r} c_{p,r} + d^\dagger_{p,r} d_{p,r}),\\ \boldsymbol{P} &= \int \frac{\mathrm{d}^3\boldsymbol{p}}{(2\pi)^3 2E_p} \textstyle\sum_{r=1,2} \boldsymbol{p}\, (c^\dagger_{p,r} c_{p,r} + d^\dagger_{p,r} d_{p,r}),\\ Q &= \int \frac{\mathrm{d}^3\boldsymbol{p}}{(2\pi)^3 2E_p} \textstyle\sum_{r=1,2} q\, (c^\dagger_{p,r} c_{p,r} - d^\dagger_{p,r} d_{p,r}). \end{aligned} \tag{5.103}$$

The 1-fermion state,[21]

$$|1\boldsymbol{p},r\rangle = c^{\dagger}_{p,r}\,|0\rangle\,, \tag{5.104}$$

has then the energy E_p, momentum $\boldsymbol{p}$ and charge q, whereas the 1-anti-fermion state,

$$|\bar{1}\boldsymbol{p},r\rangle = d^{\dagger}_{p,r}\,|0\rangle\,, \tag{5.105}$$

has the same energy and momentum but an opposite charge $-q$. The Hamiltonian operator in Eq. (5.103) should not be confused with H_D in Eq. (5.64). The latter was interpreted as the one-particle Hamiltonian, where $\psi(x)$ was the wave function of a single particle. In quantum field theory, the Hamiltonian (5.103) is derived from the quantised field $\psi(x)$ through a procedure that shall be explained in Section 6.1. It is an operator acting on the Hilbert space via the creation and annihilation operators.

Finally, as for the scalar field (cf. Supplement 5.2), one may wonder what is the connection between the wave functions $u_r(p)e^{-ip\cdot x}$, $v_r(p)e^{+ip\cdot x}$ and the quantum field $\psi(x)$ of the Formula (5.100). The answer is very similar to the scalar case and can be easily checked using the anti-commutation rules of the operators. The wave functions read

$$\begin{aligned} u_r(p)e^{-ip\cdot x} &= \langle 0|\psi(x)|1\boldsymbol{p},r\rangle\,, \\ v_r(p)e^{+ip\cdot x} &= \langle \bar{1}\boldsymbol{p},r\,|\psi(x)|0\rangle\,. \end{aligned}$$

5.3.7 Considerations about the Handedness of the Dirac Field

A Dirac field, $\psi(x)$, can always be decomposed into its chiral components $\psi_L(x)$ and $\psi_R(x)$, where $\psi_L(x)$ is a chiral left-handed field and $\psi_R(x)$ is a chiral right-handed one. By definition, they satisfy

$$\gamma^5\psi_L(x) = -\psi_L(x),\ \ \gamma^5\psi_R(x) = +\psi_R(x). \tag{5.106}$$

It follows that

$$P_L\psi_L(x) = \psi_L(x),\ \ P_R\psi_L(x) = 0,\ \ P_R\psi_R(x) = \psi_R(x),\ \ P_L\psi_R(x) = 0, \tag{5.107}$$

where the projectors, P_L and P_R, are given by Eq. (5.98). Let us apply, for example, the left-handed chiral projector P_L on the field $\psi(x)$ of Eq. (5.100). For a moving particle, in particular when it is relativistic, it is more convenient to use the helicity basis (5.74) of spinors and anti-spinors rather than the total angular momentum basis. The projected field reads

$$\psi_L(x) = P_L\psi(x) = \int \frac{d^3\boldsymbol{p}}{(2\pi)^3 2E_p} \sum_{r=\pm\frac{1}{2}} \left[c_{p,r}P_L u_r(p)e^{-ip\cdot x} + d^{\dagger}_{p,r}P_L v_r(p)e^{+ip\cdot x}\right]. \tag{5.108}$$

The field $\psi_L(x)$ is obviously a chiral left-handed field since $P_L\psi_L(x) = P_LP_L\psi(x) = P_L\psi(x) = \psi_L(x)$. In Eq. (5.108), we see that by construction $\psi_L(x)$ annihilates a particle associated with the chiral spinor $P_Lu_r(p)$, an eigenstate of γ^5 with the eigenvalue

[21] As explained in Supplement 5.1, because of the relativistic normalisation, a state containing one single particle in volume $\mathcal{V}$ is actually $|1_a,\boldsymbol{p}\rangle\,/\sqrt{2E_p\mathcal{V}}$.

-1 [see the definition (5.106)] and creates an antiparticle associated with the chiral anti-spinor $P_L v_r(p)$, still with the eigenvalue -1. This result is independent of the helicity spinor chosen, $r = 1/2$ or $r = -1/2$.

It is instructive to consider the chiral field of a massless particle. In that case, we saw in Section 5.3.5 that the helicity spinors become an eigenstate of the chirality matrix, and hence,

$$P_L u_{\frac{1}{2}}(p) = 0, \;\; P_L v_{-\frac{1}{2}}(p) = 0, \;\; P_L u_{-\frac{1}{2}}(p) = u_{-\frac{1}{2}}(p), \;\; P_L v_{\frac{1}{2}}(p) = v_{\frac{1}{2}}(p).$$

Consequently, $\psi_L(x)$ is simply reduced to

$$\psi_L(x) = \int \frac{\mathrm{d}^3\boldsymbol{p}}{(2\pi)^3 2E_p}\left[c_{p,-\frac{1}{2}} u_{-\frac{1}{2}}(p) e^{-ip\cdot x} + d^\dagger_{p,\frac{1}{2}} v_{\frac{1}{2}}(p) e^{+ip\cdot x}\right]. \tag{5.109}$$

Thus, a massless chiral left-handed field destroys a particle with a left-handed helicity and creates an antiparticle with a right-handed helicity. With a massless right-handed field, we would have obtained

$$\psi_R(x) = \int \frac{\mathrm{d}^3\boldsymbol{p}}{(2\pi)^3 2E_p}\left[c_{p,\frac{1}{2}} u_{\frac{1}{2}}(p) e^{-ip\cdot x} + d^\dagger_{p,-\frac{1}{2}} v_{-\frac{1}{2}}(p) e^{+ip\cdot x}\right]. \tag{5.110}$$

In other words, a massless chiral field does no longer depend on two spinors and two anti-spinors but just on one single spinor and one anti-spinor of opposite helicity. A massless chiral field can then be written as

$$\psi(x) = \int \frac{\mathrm{d}^3\boldsymbol{p}}{(2\pi)^3 2E_p}\left[c_{p,r} u_r(p) e^{-ip\cdot x} + d^\dagger_{p,-r} v_{-r}(p) e^{+ip\cdot x}\right],$$

with $r = -1/2$ for a chiral left-handed field and $r = 1/2$ for a chiral right-handed field. This is a consequence of the fact that helicity and chirality become the same for a massless particle. Let us make clear the connection between the two. The spinors and anti-spinors of a massless particle or antiparticle satisfy the Dirac equation in momentum space, i.e. Eqs. (5.48) and (5.60), with $m = 0$ giving $\gamma^\mu p_\mu w(p) = 0$, where w stands for u_r or v_r. Let us expand the $\gamma^\mu p_\mu$ product in terms of Pauli matrices, using the Dirac representation, and keeping in mind that $p^0 = |\boldsymbol{p}|$. After a division by $|\boldsymbol{p}|$, this yields

$$\begin{pmatrix} \mathbb{1} & -\frac{\sigma\cdot p}{|p|} \\ \frac{\sigma\cdot p}{|p|} & -\mathbb{1} \end{pmatrix} w(p) = 0.$$

Consequently,

$$w(p) = \begin{pmatrix} \mathbb{0} & \boldsymbol{\sigma}\cdot\frac{\boldsymbol{p}}{|\boldsymbol{p}|} \\ \boldsymbol{\sigma}\cdot\frac{\boldsymbol{p}}{|\boldsymbol{p}|} & \mathbb{0} \end{pmatrix} w(p).$$

Let us multiply by γ^5 [given by Eq. (5.92)] the expression above,

$$\gamma^5 w(p) = \begin{pmatrix} \mathbb{0} & \mathbb{1} \\ \mathbb{1} & \mathbb{0} \end{pmatrix}\begin{pmatrix} \mathbb{0} & \boldsymbol{\sigma}\cdot\frac{\boldsymbol{p}}{|\boldsymbol{p}|} \\ \boldsymbol{\sigma}\cdot\frac{\boldsymbol{p}}{|\boldsymbol{p}|} & \mathbb{0} \end{pmatrix} w(p) = \begin{pmatrix} \boldsymbol{\sigma}\cdot\frac{\boldsymbol{p}}{|\boldsymbol{p}|} & \mathbb{0} \\ \mathbb{0} & \boldsymbol{\sigma}\cdot\frac{\boldsymbol{p}}{|\boldsymbol{p}|} \end{pmatrix} w(p).$$

We thus conclude

$$\gamma^5 w(p) = \boldsymbol{\Sigma}\cdot\frac{\boldsymbol{p}}{|\boldsymbol{p}|} w(p),$$

Table 5.1 Chirality and helicity eigenvalues for massless chiral fields.

Field	spinors	Chirality γ^5	Helicity $\hat{h}$
$\psi_L(x)$	$u(p)$	-1	$-\frac{1}{2}$
	$v(p)$	-1	$+\frac{1}{2}$
$\psi_R(x)$	$u(p)$	$+1$	$+\frac{1}{2}$
	$v(p)$	$+1$	$-\frac{1}{2}$

with $\mathbf{\Sigma}$, the spin matrix appearing in Eq. (5.65). We recognise on the right-hand side the helicity operator in momentum space (5.72), i.e. $\frac{1}{2}\mathbf{\Sigma} \cdot \frac{\boldsymbol{p}}{|\boldsymbol{p}|}$ for fermions and $-\frac{1}{2}\mathbf{\Sigma} \cdot \frac{\boldsymbol{p}}{|\boldsymbol{p}|}$ for anti-fermions. Therefore,

$$\gamma^5 u_r(p) = 2\,\hat{h} u_r(p), \quad \gamma^5 v_r(p) = -2\,\hat{h} v_r(p). \tag{5.111}$$

For a massless field, Eq. (5.111) shows that if the field is an eigenstate of the chirality operator, then the spinors and antiparticle spinors are necessarily eigenstates of the helicity operator. It confirms that helicity and chirality are equivalent concepts for massless particles. Moreover, Eq. (5.111) tells us that if the field is an eigenstate of γ^5 with the eigenvalue η, the helicity of the fermion is $\eta/2$, while that of the anti-fermion is $-\eta/2$. Thus, for a left-handed chirality field $\eta = -1$, i.e. $\gamma^5 u_r(p) = -u_r(p)$ and $\gamma^5 v_r(p) = -v_r(p)$, but the particle is left-handed in terms of helicity, whereas the antiparticle is right-handed. This is precisely the reason why we chose the opposite convention for v spinors with respect to u spinors and referred to a right-handed chirality v spinors the ones satisfying $\gamma^5 v_r(p) = -v_r(p)$. Table 5.1 summarises the chirality and helicity eigenvalues for the spinors and antiparticle spinors used in the development of a massless chiral field. Now that we have a better understanding of the handedness of a field, we can address the case of massive particles. A Dirac field can be decomposed into its chiral components,

$$\psi(x) = \psi_L(x) + \psi_R(x),$$

where $\psi_L(x) = P_L\psi(x)$ and $\psi_R(x) = P_R\psi(x)$. The expression of the left-handed field reads

$$\psi_L(x) = \int \frac{d^3\boldsymbol{p}}{(2\pi)^3 2E_p} \sum_{r=\pm\frac{1}{2}} \left[c_{p,r} u_L^r(p) e^{-ip\cdot x} + d_{p,r}^\dagger v_R^r(p) e^{+ip\cdot x}\right], \tag{5.112}$$

where the spinors $u_L^r(p) = P_L u_r(p)$ and the anti-spinors $v_R^r(p) = P_L v_r(p)$ are the eigenstates of the chirality with the eigenvalue -1. For massive fields, none of them is an eigenstate of the helicity operator. The field $\psi_L(x)$ destroys a particle with a chirality -1 and creates an antiparticle with a chirality -1. The Dirac adjoint field $\overline{\psi_L}(x)$ creates a particle with a chirality -1 and destroys an antiparticle with a chirality -1. Similarly, the right-handed field reads

$$\psi_R(x) = \int \frac{d^3\boldsymbol{p}}{(2\pi)^3 2E_p} \sum_{r=\pm\frac{1}{2}} \left[c_{p,r} u_R^r(p) e^{-ip\cdot x} + d_{p,r}^\dagger v_L^r(p) e^{+ip\cdot x}\right], \tag{5.113}$$

where the spinors $u_R^r(p) = P_R u_r(p)$ and the anti-spinors $v_L^r(p) = P_R v_r(p)$ are the eigenstates of the chirality with the eigenvalue $+1$. They do not have a definite helicity. The field $\psi_R(x)$

destroys a particle with a chirality +1 and creates an antiparticle with a chirality +1. The Dirac adjoint field $\overline{\psi_R}(x)$ creates a particle with a chirality +1 and destroys an antiparticle with a chirality +1. Obviously, summing $\psi_L(x)$ and $\psi_R(x)$ recovers $\psi(x)$ of Eq. (5.100), given that $u_r(p) = u_L^r(p) + u_R^r(p)$ and $v_r(p) = v_L^r(p) + v_R^r(p)$.

Supplement 5.4. Parity transformation of the Dirac field

We have already seen in Eq. (5.89) that spinors are transformed by the γ^0 matrix. The corresponding field transformation is given by

$$\hat{\mathcal{P}}\psi(x)\hat{\mathcal{P}}^{-1} = +\gamma^0\psi(x'), \tag{5.114}$$

with $x' = (x^0, -\boldsymbol{x})$ and $\hat{\mathcal{P}}$ being the unitary operator applied in the Hilbert space, corresponding to the parity transformation of the field. This equation means that the field ψ has an intrinsic parity +1. Inserting the expression (5.100) of the field $\psi(x)$ in Eq. (5.114) yields

$$\int \frac{d^3\boldsymbol{p}}{(2\pi)^3 2E_p} \sum_{r=1}^{2} \left[\hat{\mathcal{P}} c_{E_p,\boldsymbol{p},r} \hat{\mathcal{P}}^{-1} u_r(p) e^{-ip\cdot x} + \hat{\mathcal{P}} d^\dagger_{E_p,\boldsymbol{p},r} \hat{\mathcal{P}}^{-1} v_r(p) e^{+ip\cdot x}\right]$$

$$= \int \frac{d^3\boldsymbol{p}}{(2\pi)^3 2E_p} \sum_{r=1}^{2} \left[c_{E_p,\boldsymbol{p},r} \gamma^0 u_r(p) e^{-i(E_p t + \boldsymbol{x}\cdot\boldsymbol{p})} + d^\dagger_{E_p,\boldsymbol{p},r} \gamma^0 v_r(p) e^{+i(E_p t + \boldsymbol{x}\cdot\boldsymbol{p})}\right].$$

In order to invert the sign of $\boldsymbol{x}\cdot\boldsymbol{p}$ on the right-hand side of the equality, let us change the variables of integration from $\boldsymbol{p}$ to $-\boldsymbol{p}$. Given that $\gamma^0 u_r(E_p, -\boldsymbol{p}) = u_r(E_p, \boldsymbol{p}) = u_r(p)$ and $\gamma^0 v_r(E_p, -\boldsymbol{p}) = -v_r(E_p, \boldsymbol{p}) = -v_r(p)$ [see Eq. (5.89)], we then identify the different terms

$$\hat{\mathcal{P}} c_{E_p,\boldsymbol{p},r} \hat{\mathcal{P}}^{-1} = +c_{E_p,-\boldsymbol{p},r}, \quad \hat{\mathcal{P}} d^\dagger_{E_p,\boldsymbol{p},r} \hat{\mathcal{P}}^{-1} = -d^\dagger_{E_p,-\boldsymbol{p},r}.$$

The same expression would be obtained for the transformation of $c^\dagger_{E_p,\boldsymbol{p},r}$ and $d_{E_p,\boldsymbol{p},r}$ (just take the Hermitian conjugate of the equalities above). The vacuum is expected to be invariant under the parity transformation, with the conventional choice of an even parity, i.e. $\hat{\mathcal{P}}|0\rangle = \hat{\mathcal{P}}^{-1}|0\rangle = |0\rangle$. Hence, applying the vacuum on both sides of the transformation of the operators $c^\dagger_{E_p,\boldsymbol{p},r}$ and $d^\dagger_{E_p,\boldsymbol{p},r}$, it follows that

$$\hat{\mathcal{P}} c^\dagger_{E_p,\boldsymbol{p},r}|0\rangle = +c^\dagger_{E_p,-\boldsymbol{p},r}|0\rangle,$$
$$\hat{\mathcal{P}} d^\dagger_{E_p,\boldsymbol{p},r}|0\rangle = -d^\dagger_{E_p,-\boldsymbol{p},r}|0\rangle.$$

This is exactly what we expected: the transformation of a 1-fermion state under the parity is $c^\dagger_{E_p,-\boldsymbol{p},r}|0\rangle$, while the 1-anti-fermion state is transformed in $-d^\dagger_{E_p,-\boldsymbol{p},r}|0\rangle$. In other words, a state with $\boldsymbol{p}$ becomes a state with $-\boldsymbol{p}$. Similarily, the application of the parity transformation to a left-handed chiral field gives

$$\hat{\mathcal{P}}\psi_L(x)\hat{\mathcal{P}}^{-1} = \hat{\mathcal{P}} P_L \psi(x)\hat{\mathcal{P}}^{-1} = P_L \hat{\mathcal{P}}\psi(x)\hat{\mathcal{P}}^{-1} = P_L\gamma^0\psi(x') = \gamma^0 P_R\psi(x') = \gamma^0\psi_R(x').$$

In this sense, the parity transform of ψ_L is ψ_R, meaning that the chirality changes under a parity transformation, as we would expect from its denomination: chirality comes from the Greek, meaning 'hand', and a left hand in a mirror is seen as a right hand.

5.3.8 Charge Conjugation of the Dirac Field

Inspired by what we did in Section 5.2.5 for the scalar field, it would be tempting to propose the following transformation under charge conjugation of the Dirac field:

$$\hat{C}\psi(x)\hat{C}^{-1} = \eta_C \psi^\dagger(x).$$

However, do not forget that the Dirac field is actually made of four fields because of the spinor structure. The left-hand side of the equation would be a column vector (of fields), while the right-hand side would be a row vector. The appropriate transformation is thus[22]:

$$\hat{C}\psi(x)\hat{C}^{-1} = \eta_C \hat{\mathrm{C}}\psi^{\dagger\mathsf{T}}(x), \tag{5.115}$$

where $\hat{\mathrm{C}}$ is the charge-conjugation matrix already introduced in Eq. (5.84), which is applied on the spinors. The transposition in (5.115) being also applied on the spinors, the object $\psi^{\dagger\mathsf{T}}(x)$ is appropriately a column vector. Let us insert the Dirac field (5.100) in Eq. (5.115). On the right-hand side, we would have

$$\eta_C \hat{\mathrm{C}}\psi^{\dagger\mathsf{T}}(x) = \eta_C \int \frac{\mathrm{d}^3\boldsymbol{p}}{(2\pi)^3 2E_p} \sum_{r=1,2} \left[c^\dagger_{p,r} \hat{\mathrm{C}} u^*_r(p) e^{+ip\cdot x} + d_{p,r} \hat{\mathrm{C}} v^*_r(p) e^{-ip\cdot x} \right].$$

Given that the spinors are transformed by the charge conjugation matrix, $\hat{\mathrm{C}}$, according to Eq. (5.85), It follows that

$$\eta_C \hat{\mathrm{C}}\psi^{\dagger\mathsf{T}}(x) = \eta_C \int \frac{\mathrm{d}^3\boldsymbol{p}}{(2\pi)^3 2E_p} \sum_{r-1,2} \left[c^\dagger_{p,r} v_r(p) e^{+ip\cdot x} + d_{p,r} u_r(p) e^{-ip\cdot x} \right].$$

The left-hand side of equality (5.115) is

$$\hat{C}\psi(x)\hat{C}^{-1} = \int \frac{\mathrm{d}^3\boldsymbol{p}}{(2\pi)^3 2E_p} \sum_{r=1,2} \left[\hat{C} c_{p,r} \hat{C}^{-1} u_r(p) e^{-ip\cdot x} + \hat{C} d^\dagger_{p,r} \hat{C}^{-1} v_r(p) e^{+ip\cdot x} \right].$$

Comparing both sides, we thus conclude that

$$\hat{C} c_{p,r} \hat{C}^{-1} = \eta_C d_{p,r}\,, \quad \hat{C} d^\dagger_{p,r} \hat{C}^{-1} = \eta_C c^\dagger_{p,r}. \tag{5.116}$$

Hence, the charge conjugation interchanges, as expected, the operators of the fermion with those of the anti-fermion, leading to the property that the charge conjugation transforms a one-particle state into a one-antiparticle state and vice versa.

5.4 Spin 1 Particles: The Massless Photon

5.4.1 Solutions of Maxwell's Equations

In Chapter 2, Section 2.2.4, we saw that Maxwell's equations can be written in terms of the electromagnetic tensor $F^{\mu\nu}$, where $F^{\mu\nu} = \partial^\mu A^\nu - \partial^\nu A^\mu$ satisfies Eq. (2.42),

$$\partial_\mu F^{\mu\nu} = \mu_0 j^\nu,$$

[22] We could have used $\overline{\psi}^{\mathsf{T}}(x)$ instead of $\psi^{\dagger T}(x)$ with the other conjugation matrix given by Eq. (5.86).

with j^ν being the 4-current of charge. Inserting the expression of $F^{\mu\nu}$ in the equality above yields the wave equation satisfied by A^μ, i.e.

$$\Box A^\nu - \partial^\nu(\partial_\mu A^\mu) = \mu_0 j^\nu. \tag{5.117}$$

However, A^μ is not uniquely defined. Let us consider an arbitrary (Lorentz) scalar function $\chi(x)$ that depends only on the spacetime coordinate. The new potential, defined by

$$A^\mu \to A'^\mu = A^\mu + \partial^\mu \chi, \tag{5.118}$$

still leads to the same electromagnetic tensor. Transformation (5.118) is called a gauge transformation. With the electromagnetic tensor being invariant under this gauge transformation, Maxwell's equations are invariant as well. Now, χ can be chosen such that

$$\partial_\mu A'^\mu = 0. \tag{5.119}$$

This choice of gauge is called the Lorenz gauge. It simplifies the wave equation, yielding

$$\boxed{\Box A^\mu = \mu_0 j^\mu,} \tag{5.120}$$

where the new electromagnetic field is written without the prime symbol. For a free photon, in the absence of charges, $j^\mu = 0$, the plane-wave form

$$A^\mu = N\epsilon^\mu(\lambda, p)e^{-ip\cdot x} \tag{5.121}$$

is a solution of the wave equation, provided that $p^2 = 0$. The photons then are massless. In Eq. (5.121), N is a normalisation factor and $\epsilon^\mu(\lambda, p)$, is a 4-vector polarisation. A possible basis of 4-vectors is

$$\epsilon(\lambda = 0) = \begin{pmatrix} 1 \\ 0 \\ 0 \\ 0 \end{pmatrix}, \quad \epsilon(\lambda = 1) = \begin{pmatrix} 0 \\ 1 \\ 0 \\ 0 \end{pmatrix}, \quad \epsilon(\lambda = 2) = \begin{pmatrix} 0 \\ 0 \\ 1 \\ 0 \end{pmatrix}, \quad \epsilon(\lambda = 3) = \begin{pmatrix} 0 \\ 0 \\ 0 \\ 1 \end{pmatrix}. \tag{5.122}$$

With the solution (5.121), the Lorenz gauge condition (5.119) implies that

$$\epsilon \cdot p = \epsilon^0 p^0 - \boldsymbol{\epsilon} \cdot \boldsymbol{p} = 0. \tag{5.123}$$

Hence, it seems that only three degrees of freedom are needed to define the polarisation vector (as soon as $\boldsymbol{\epsilon} \cdot \boldsymbol{p}$ is known, $\epsilon^0 p^0$ is fixed). Actually, we can go further. Nothing prevents us to redo the same kind of gauge transformation, i.e.

$$A^\mu \to A'^\mu = A^\mu + \partial^\mu \chi.$$

But this time, χ must satisfy $\Box\chi = 0$, since the field A^μ satisfies the Lorenz gauge. A possible choice of the scalar function is

$$\chi(x) = iNae^{-ip\cdot x},$$

with a being a constant that does not depend on x. The new field definition then becomes

$$A'^\mu = N(\epsilon^\mu - ap^\mu)e^{-ip\cdot x},$$

implying that A' has the new polarisation vector $\epsilon'^{\mu} = \epsilon^{\mu} - ap^{\mu}$. We can choose a such that $\epsilon'^{0} = 0$. Consequently, Eq. (5.123) reduces to the spatial components of the polarisation vector,

$$\boldsymbol{\epsilon} \cdot \boldsymbol{p} = 0. \tag{5.124}$$

Therefore, as expected from classical electromagnetism, the polarisation of a free photon is orthogonal to its direction of motion. There are only two degrees of freedom for a real free photon. The particular choice imposing $\epsilon^0 = 0$ is clearly not covariant. It is called the Coulomb gauge. Note that the set of polarisation vectors (5.122) satisfies the four-dimensional orthogonality relation in the Minkowski space

$$\epsilon(\lambda) \cdot \epsilon(\lambda') = g_{\mu\nu}\epsilon^{\mu}(\lambda)\epsilon^{\nu}(\lambda') = g_{\lambda\lambda'}, \tag{5.125}$$

and the completeness relation

$$\sum_{\lambda=0}^{3} g_{\lambda\lambda}\epsilon^{\mu}(\lambda)\epsilon^{\nu}(\lambda) = g^{\mu\nu}. \tag{5.126}$$

In Eqs. (5.125) and (5.126), beware that $g_{\lambda\lambda'}$ or $g_{\lambda\lambda}$ should not be understood as a Lorentz tensor (unlike $g^{\mu\nu}$). They simply stand for ± 1 factors. Notice that the polarisation vectors in (5.122) are real valued. However, in Supplement 5.5, circular polarisations are introduced, which have complex components on the x- and y-axes. The relations (5.125) and (5.126) must then be generalised to

$$\epsilon(\lambda) \cdot \epsilon(\lambda')^* = g_{\mu\nu}\epsilon^{\mu}(\lambda)\epsilon^{\nu}(\lambda')^* = g_{\lambda\lambda'} \tag{5.127}$$

and

$$\sum_{\lambda=0}^{3} g_{\lambda\lambda}\epsilon^{\mu}(\lambda)\epsilon^{\nu}(\lambda)^* = g^{\mu\nu}. \tag{5.128}$$

It is conventional to choose the basis of polarisations such that the 4-vector polarisations with $\lambda = 1$ and 2 are transverse to the direction of motion. The longitudinal polarisation 4-vector, which has its spatial component in the direction of the photon momentum, is labelled with $\lambda = 3$. The set of 4-vector polarisations is completed with the time-like polarisation with $\lambda = 0$. Then, a possible basis for a photon with a 4-momentum $p = (|\boldsymbol{p}|, \boldsymbol{p})$ in an arbitrary direction is[23]

$$\epsilon(\boldsymbol{p},0) = \hat{n} = \begin{pmatrix} 1 \\ \boldsymbol{0} \end{pmatrix}, \quad \epsilon(\boldsymbol{p},1) = \begin{pmatrix} 0 \\ \boldsymbol{\epsilon}(\boldsymbol{p},1) \end{pmatrix}, \quad \epsilon(\boldsymbol{p},2) = \begin{pmatrix} 0 \\ \boldsymbol{\epsilon}(\boldsymbol{p},2) \end{pmatrix}, \quad \epsilon(\boldsymbol{p},3) = \begin{pmatrix} 0 \\ \frac{\boldsymbol{p}}{|\boldsymbol{p}|} \end{pmatrix}, \tag{5.129}$$

with the constraints from Eqs. (5.124) and (5.127) imposing

$$\boldsymbol{\epsilon}(\boldsymbol{p},i) \cdot \boldsymbol{p} = 0\,, \quad \text{and} \quad \boldsymbol{\epsilon}(\boldsymbol{p},i) \cdot \boldsymbol{\epsilon}(\boldsymbol{p},j)^* = \delta_{ij}, \tag{5.130}$$

[23] The set (5.122) is appropriate for photons propagating along the z-axis. The appropriate Lorentz transformation for a momentum $\boldsymbol{p}$ would lead to the set (5.129).

where i and j are 1 or 2. Problem 5.7 shows that the polarisations in Eq. (5.129) satisfy Eq. (5.128). For real photons, only the transverse polarisations matter. Then, the completeness relation, reduced to the physical polarisations, reads

$$\sum_{\lambda=1}^{2} \epsilon^{\mu}(p,\lambda)\epsilon^{\nu}(p,\lambda)^{*} = -g^{\mu\nu} - \frac{p^{\mu}p^{\nu}}{(p\cdot\hat{n})^2} + \frac{p^{\mu}\hat{n}^{\nu} + p^{\nu}\hat{n}^{\mu}}{p\cdot\hat{n}}, \tag{5.131}$$

where $\hat{n}$ is the unit time-like 4-vector defined in Eq. (5.129).

Supplement 5.5. Relationship between the polarisation and the spin

Polarisation and spin are often confused, in particular for photons. Let us consider a photon travelling in the z-direction. The circular polarisations based on the two linear polarisations, $\epsilon(1)$ and $\epsilon(2)$, respectively, along the x and y axes, are given by

$$\epsilon_R = \frac{1}{\sqrt{2}}(\epsilon(1) + i\epsilon(2)),$$

$$\epsilon_L = \frac{1}{\sqrt{2}}(\epsilon(1) - i\epsilon(2)),$$

where ϵ_R and ϵ_L denote respectively the right-handed and left-handed polarisations. They carry this name because a rotation of an angle θ around the z-axis transforms the basis as

$$\epsilon(1) \to \epsilon(1)\cos\theta - \epsilon(2)\sin\theta,$$

$$\epsilon(2) \to \epsilon(1)\sin\theta + \epsilon(2)\cos\theta,$$

and thus,

$$\epsilon_R \to \epsilon_R\, e^{i\theta},$$

$$\epsilon_L \to \epsilon_L\, e^{-i\theta}.$$

Therefore, under a rotation of angle $\theta > 0$, ϵ_R turns clockwise (from the point of view of the source of the photon) and ϵ_L turns anticlockwise. Now, in quantum mechanics, the rotation operator about the z-axis is given by $e^{iJ_z\theta}$, with J_z being the projection of the total angular momentum on the z-axis. A photon moving along this axis cannot have an orbital angular momentum along this direction (since $\boldsymbol{r} \times \boldsymbol{p} = r\boldsymbol{u_z} \times p\boldsymbol{u_z} = 0$), and thus, only its spin contributes to the angular momentum along the z-axis. Looking at the rules of transformation of ϵ_R and ϵ_L under a rotation, we conclude that the polarisation states ϵ_R and ϵ_L are the eigenstates of the rotation operator, with respectively $J_z = 1$ and $J_z = -1$, and thus, a spin projection $+1$ and -1 along its direction of motion. Therefore, ϵ_R and ϵ_L are the helicity eigenstates.

The circular polarisations are transverse polarisations in the sense that A^{μ} is transverse to the direction of motion (so are the electric or magnetic fields), even if the spin itself is necessarily aligned along the direction of propagation. Moreover, a rotation about the z-axis does not change $\epsilon(3)$, and thus, $\epsilon(3)$ is an eigenstate of the rotation operator with $J_z = 0$. However, since a real photon travelling along the z-axis has its polarisation vector that does not depend on $\epsilon(3)$ (remember: $\boldsymbol{\epsilon}\cdot\boldsymbol{p} = 0$), the value $J_z = 0$ is necessarily forbidden.

5.4.2 A Few Words about the Quantised Field

For real photons, the polarisation is orthogonal to the direction of motion. Any polarisation can then be expressed as the linear combination of the two polarisation vectors $\epsilon(\boldsymbol{p}, \lambda = 1)$ and $\epsilon(\boldsymbol{p}, \lambda = 2)$ of the basis. The general solution of the (classical) photon field is then

$$A^\mu(x) = \int \frac{d^3\boldsymbol{p}}{(2\pi)^3 2E_p} \sum_{\lambda=1,2} \left[\epsilon^\mu(\boldsymbol{p}, \lambda)\alpha_{\boldsymbol{p},\lambda} e^{-ip\cdot x} + \epsilon^{*\mu}(\boldsymbol{p}, \lambda)\alpha^*_{\boldsymbol{p},\lambda} e^{+ip\cdot x}\right],$$

where $E_p = |\boldsymbol{p}|$ and $\alpha_{\boldsymbol{p},\lambda}$, with $\lambda = 1, 2$, are the coefficients of the Fourier transform. In addition, the Coulomb gauge imposes $\epsilon^0(\boldsymbol{p}, 1) = \epsilon^0(\boldsymbol{p}, 2) = 0$. The norm was chosen as before with the scalar and spin-1/2 fields. Notice that the photon field, $A^\mu(x)$, is a real field, reflecting the fact that the photon is its own antiparticle.

The quantisation of the field promotes the coefficients $\alpha_{\boldsymbol{p},\lambda}$ to operators, but there are many subtleties in the quantisation procedure of the photon field, mainly due to the translation of the gauge condition in the quantum field language. In this introductory course, we will not go through this complicated process that would lead us beyond the scope of this textbook. A few indications are however given in Section 6.1.4. The interested reader is invited to consult Weinberg (1995) to know more. The field after quantisation reads

$$A^\mu(x) = \int \frac{d^3\boldsymbol{p}}{(2\pi)^3 2E_p} \sum_{\lambda=0}^{3} \left[\epsilon^\mu(\boldsymbol{p}, \lambda)\alpha_{\boldsymbol{p},\lambda} e^{-ip\cdot x} + \epsilon^{*\mu}(\boldsymbol{p}, \lambda)\alpha^\dagger_{\boldsymbol{p},\lambda} e^{+ip\cdot x}\right], \tag{5.132}$$

where the operator $\alpha^\dagger_{\boldsymbol{p},\lambda}$ creates a photon of momentum $\boldsymbol{p}$ and polarisation λ, and $\alpha_{\boldsymbol{p},\lambda}$ is the corresponding annihilator. Note that we sum over the four polarisation states: $\lambda = 0$ is a 'time-like' polarisation, $\lambda = 1, 2$ are the two transverse polarisations, while $\lambda = 3$ is the longitudinal polarisation. One of the subtleties mentioned above leads to the (strange) commutation rules

$$\begin{aligned} \left[\alpha_{\boldsymbol{p}',\lambda'}, \alpha^\dagger_{\boldsymbol{p},\lambda}\right] &= -g_{\lambda\lambda'}(2\pi)^3 2E_p \delta^{(3)}(\boldsymbol{p}' - \boldsymbol{p}), \\ \left[\alpha_{\boldsymbol{p}',\lambda'}, \alpha_{\boldsymbol{p},\lambda}\right] &= \left[\alpha^\dagger_{\boldsymbol{p}',\lambda'}, \alpha^\dagger_{\boldsymbol{p},\lambda}\right] = 0. \end{aligned} \tag{5.133}$$

This looks quite similar to the case of scalar bosons, still using commutators as expected for bosons, except that there is the extra $-g_{\lambda\lambda'}$. For the polarisations $\lambda = 1, 2, 3$, the commutation rules of operators correspond to those of the harmonic oscillators as in the scalar case (since $-g_{\lambda\lambda} = 1$). However, it is not the case for $\lambda = 0$, where $-g_{00} = -1$. In fact, after properly taking into account the gauge constraint, one can show that for physical states (on which we can do measurements), only the transverse polarisations matter.

5.4.3 Parity of the Photon

The photon field $A^\mu(x)$ behaves as a 4-vector under Lorentz transformations, with a time component A^0 resulting from the scalar potential and the space components $A^{k=1,2,3}$ from the vector potential. Following the same arguments as for the scalar field in Section 5.2.4 (i.e., the invariance of the d'Alembertian under the parity transformation, etc.), under the parity transformation, which transforms the spacetime coordinates x into $x' = (t, -\boldsymbol{x})$, we expect the following transformation of the photon field (read the Supplement 4.2):

- If the scalar potential behaves as a scalar and the vector potential as a (polar-)vector, then

$$\hat{\mathcal{P}}A^0(x)\hat{\mathcal{P}}^{-1} = A^0(x'), \ \ \hat{\mathcal{P}}A^k(x)\hat{\mathcal{P}}^{-1} = -A^k(x').$$

- If the scalar potential behaves as a pseudo-scalar and the vector potential as a pseudo-vector, then

$$\hat{\mathcal{P}}A^0(x)\hat{\mathcal{P}}^{-1} = -A^0(x'), \ \ \hat{\mathcal{P}}A^k(x)\hat{\mathcal{P}}^{-1} = A^k(x').$$

We can thus summarise this behaviour, by introducing the *intrinsic parity* of the photon, η, as such

$$\hat{\mathcal{P}}A^0(x)\hat{\mathcal{P}}^{-1} = -\eta A^0(x'), \ \ \hat{\mathcal{P}}A^k(x)\hat{\mathcal{P}}^{-1} = \eta A^k(x'), \tag{5.134}$$

where $\eta = -1$ would imply a scalar and vector behaviour, and $\eta = 1$ would entail the other possibility. Actually, the value of η has to be determined by the experimental observations. They show that the electromagnetic interaction between photons and matter respects the space inversion symmetry. Thus, the Hamiltonian (or equivalently the Lagrangian) describing this interaction must be invariant under the parity transformation. It is well known from classical electrodynamics [see Jackson (1998) for instance] that the interaction is described by a coupling between the photon potential and the electromagnetic current. We shall see in Chapter 6 that this is also valid in quantum electrodynamics with an interaction term proportional to

$$\mathcal{H}_{\text{em}} \propto j_\mu A^\mu, \tag{5.135}$$

where j^μ is the current given by Eq. (5.12) for spinless bosons and by Eq. (5.46) for spin-1/2 fermions. Let us see which condition on η imposes the invariance of the interaction term (5.135) under the parity transformation. The transformation of the term (5.135) reads

$$\hat{\mathcal{P}}j_\mu(x)A^\mu(x)\hat{\mathcal{P}}^{-1} = \hat{\mathcal{P}}j_\mu(x)\hat{\mathcal{P}}^{-1}\hat{\mathcal{P}}A^\mu(x)\hat{\mathcal{P}}^{-1} = -\eta\left(\hat{\mathcal{P}}j_0(x)\hat{\mathcal{P}}^{-1}A^0(x') - \hat{\mathcal{P}}j_k(x)\hat{\mathcal{P}}^{-1}A^k(x')\right).$$

Clearly, if j^μ behaves as a scalar for its time component $[\hat{\mathcal{P}}j_0(x)\hat{\mathcal{P}}^{-1} = j_0(x')]$ and as a vector for its space components $[\hat{\mathcal{P}}j_k(x)\hat{\mathcal{P}}^{-1} = -j_k(x')]$, the interaction term will be invariant if $-\eta = 1$. For spinless bosons, the current was given by Eq. (5.12) in the context of the Klein–Gordon wave function. The quantum field version of (5.12) requires the replacement of the complex conjugate by its Hermitian conjugate,

$$j^\mu_{\text{em}} = i(\phi^\dagger\partial^\mu\phi - \phi\partial^\mu\phi^\dagger). \tag{5.136}$$

Given that the parity operator $\mathcal{P}$ is unitary (recall Section 4.3.3), the term $\phi^\dagger\partial^\mu\phi$ transforms as

$$\hat{\mathcal{P}}\phi^\dagger(x)\partial^\mu\phi(x)\hat{\mathcal{P}}^{-1} = \hat{\mathcal{P}}\phi^\dagger(x)\hat{\mathcal{P}}^{-1}\hat{\mathcal{P}}\partial^\mu\hat{\mathcal{P}}^{-1}\hat{\mathcal{P}}\phi(x)\hat{\mathcal{P}}^{-1} = \eta_B^2\phi^\dagger(x')\hat{\mathcal{P}}\partial^\mu\hat{\mathcal{P}}^{-1}\phi(x'),$$

where $\eta_B = \pm 1$ is the intrinsic parity of the boson. Since $\partial^0 = \partial/\partial t$ is not affected by the parity transformation, whereas $\partial^{i=1,2,3} = \partial/\partial x^i$ is inverted, we conclude that for spinless bosons,

$$\hat{\mathcal{P}}j_0(x)\hat{\mathcal{P}}^{-1} = j_0(x')\,, \ \ \hat{\mathcal{P}}j_k(x)\hat{\mathcal{P}}^{-1} = -j_k(x').$$

For fermions, the current (5.46) has the functional form $\bar{\psi}\gamma^\mu\psi = \psi^\dagger\gamma^0\gamma^\mu\psi$, and given the fermion field transformation (5.114), i.e. $\hat{\mathcal{P}}\psi(x)\hat{\mathcal{P}}^{-1} = \gamma^0\psi(x')$, the parity transformed current is

$$\begin{aligned}\hat{\mathcal{P}}\psi^\dagger(x)\gamma^0\gamma^\mu\psi(x)\hat{\mathcal{P}}^{-1} &= \hat{\mathcal{P}}\psi^\dagger(x)\hat{\mathcal{P}}^{-1}\gamma^0\gamma^\mu\hat{\mathcal{P}}\psi(x)\hat{\mathcal{P}}^{-1}\\ &= \left(\hat{\mathcal{P}}\psi(x)\hat{\mathcal{P}}^{-1}\right)^\dagger\gamma^0\gamma^\mu\hat{\mathcal{P}}\psi(x)\hat{\mathcal{P}}^{-1}\\ &= \psi^\dagger(x')(\gamma^0)^\dagger\gamma^0\gamma^\mu\gamma^0\psi(x')\\ &= \psi^\dagger(x')\gamma^\mu\gamma^0\psi(x'),\end{aligned} \tag{5.137}$$

where the properties of the γ^0 matrix were used in the third line. Since γ^μ anti-commutes with γ^0 when $\mu \neq 0$, we reach the same conclusion for the current of fermions as that of bosons in terms of behaviour with respect to the parity transformation.

In conclusion, the electromagnetic current behaves as a scalar for its time component and as a vector for its space components, and hence, we conclude that since experimentally the electromagnetic interaction respects the space inversion symmetry, necessarily the intrinsic parity of the photon is

$$\eta(\gamma) = -1. \tag{5.138}$$

5.4.4 Charge-Conjugation Parity of the Photon

It is well known from classical physics that the Hamiltonian of a charged particle in an electromagnetic field is

$$H = \frac{|\boldsymbol{p}|^2}{2m} - \frac{q}{m}\boldsymbol{A}\cdot\boldsymbol{p}.$$

Since charge conjugation inverts the sign of the electric charge q, the Hamiltonian is invariant (and thus the electromagnetic interaction respects the charge-conjugation symmetry) if the sign of $\boldsymbol{A}$ is also inverted. We thus conclude that the photon must have an odd charge parity,

$$\eta_C(\gamma) = -1. \tag{5.139}$$

This conclusion is also valid in quantum field theory. Consider the Hamiltonian describing the interaction between photons and matter (5.135). Let us examine how the charge current is transformed under the charge-conjugation symmetry. We proceed as in Section 5.4.3. Given the transformation of the scalar field under charge conjugation (5.24) and the unitary operator $\hat{C}$ (see Section 4.4.1), the first term of the scalar current (5.136) transforms as

$$\hat{C}i\phi^\dagger(x)\partial^\mu\phi(x)\hat{C}^{-1} = i\hat{C}\phi^\dagger(x)\hat{C}^{-1}\partial^\mu\hat{C}\phi(x)\hat{C}^{-1} = i\eta_C^2\phi(x)\partial^\mu\phi^\dagger(x) = i\phi(x)\partial^\mu\phi^\dagger(x),$$

while the second gives $-i\phi^\dagger(x)\partial^\mu\phi(x)$. The transformed current is thus

$$\hat{C}j^\mu_{\text{em}}\hat{C}^{-1} = -j^\mu_{\text{em}}. \tag{5.140}$$

Therefore, we conclude that the interaction Hamiltonian remains invariant if the photon field transforms as

$$\hat{C}A^\mu\hat{C}^{-1} = -A^\mu. \tag{5.141}$$

This explains why the photon necessarily has odd charge parity (5.139). The same conclusion holds with the charge current of spin-1/2 fermions (5.46). It is, however, a bit more complicated to prove. Given the charge conjugation of the Dirac field (5.115) with $\eta_C^2 = 1$, the current transforms as follows:

$$\hat{C}q\overline{\psi}\gamma^\mu\psi\hat{C}^{-1} = q\hat{C}\psi^\dagger\hat{C}^{-1}\gamma^0\gamma^\mu\hat{C}\psi\hat{C}^{-1} = q\left(\eta_C\hat{\mathrm{C}}\psi^{\dagger\mathsf{T}}\right)^\dagger\gamma^0\gamma^\mu\eta_C\hat{\mathrm{C}}\psi^{\dagger\mathsf{T}} = q\psi^\mathsf{T}M^\mu\psi^{\dagger\mathsf{T}},$$

where the 4×4 matrices $M^\mu = \hat{\mathrm{C}}^\dagger\gamma^0\gamma^\mu\hat{\mathrm{C}}$ have been introduced to simplify the notations. We can rearrange this term in a more familiar form. The Dirac field is actually a vector of four components, the components of the spinors, which are multiplied by a Fock operator. Hence, with the explicit components of the Dirac field, the term $\psi^\mathsf{T}M^\mu\psi^{\dagger\mathsf{T}}$ reads

$$\psi^\mathsf{T}(x)M^\mu\psi^{\dagger\mathsf{T}}(x) = \sum_{i,j}\psi_i(x)M^\mu_{ij}\psi_j^\dagger(x) = -\sum_{i,j}\psi_j^\dagger(x)M^\mu_{ij}\psi_i(x) + \sum_{i,j}\{\psi_i(x),\psi_j^\dagger(x)\}M^\mu_{ij}.$$

We will ignore the contribution of the commutator, which turns out to be mathematically not defined with a $\delta(0)$. (Similarly, the anti-commutator of the Fock operator at equal momenta produces a $\delta(0)$.) In quantum field theory, this procedure is called the normal ordering (briefly mentioned in Chapter 6). Paying attention to the order of the indices, we thus conclude that

$$\hat{C}\overline{\psi}\gamma^\mu\psi\hat{C}^{-1} = \psi^\mathsf{T}M^\mu\psi^{\dagger\mathsf{T}} = -\psi^\dagger(M^\mu)^\mathsf{T}\psi, \quad \text{with} \quad M^\mu = \hat{\mathrm{C}}^\dagger\gamma^0\gamma^\mu\hat{\mathrm{C}}. \tag{5.142}$$

Using the explicit expression of the charge conjugation matrix (5.84), i.e. $\hat{\mathrm{C}} = i\gamma^2$, and playing with the properties of gamma matrices, we find that $(M^\mu)^\mathsf{T} = (M^\mu)^{\dagger *} = \gamma^0\gamma^\mu$. This result is true, independently of the representation of the gamma matrices. We thus conclude that

$$\psi^\mathsf{T}M^\mu\psi^{\dagger\mathsf{T}} = -\overline{\psi}\gamma^\mu\psi.$$

Consequently, as expected, it follows that the transformation of the current is given by

$$\hat{C}q\overline{\psi}\gamma^\mu\psi\hat{C}^{-1} = -q\overline{\psi}\gamma^\mu\psi. \tag{5.143}$$

Finally, the electromagnetic current for fermions behaves in the same way as that of scalars under the charge-conjugation transformation. It thus confirms that the photon has the odd charge parity (5.139).

5.5 Spin 1 Particles: Massive Bosons

The equations characterising massive vector bosons are quite similar to those for photons, with some subtleties highlighted in this section. Massive vector bosons are described by the Proca equation, a generalisation of the Maxwell equation (2.42) with a mass term, i.e.

$$\partial_\mu F^{\mu\nu} + m^2 W^\nu = \mu_0 j^\nu, \tag{5.144}$$

where W^ν is the massive vector field with mass m and $F^{\mu\nu} = \partial^\mu W^\nu - \partial^\nu W^\mu$. Inserting $F^{\mu\nu}$ in Eq. (5.144) leads to the equation of propagation in the vacuum of the field,

$$\left(\Box + m^2\right)W^\nu - \partial^\nu(\partial_\mu W^\mu) = 0. \tag{5.145}$$

Equation (5.145) is thus a generalisation of both the Klein–Gordon equation (5.7) and the photon wave equation (5.117). Taking its divergence (i.e., multiplying by ∂_ν) imposes the constraint $m^2\partial_\nu W^\nu = 0$. As $m \neq 0$, for massive fields,

$$\partial_\nu W^\nu = 0, \tag{5.146}$$

but unlike the electromagnetic field, it is not a consequence of the choice of the Lorenz gauge because the transformation (5.118) does not leave invariant Eq. (5.144). Note that Eq. (5.146) implies that the field W^μ has only three independent components. They cannot be reduced to the two transversal components, as for photons. Thanks to the constraint (5.146), Eq. (5.145) is simplified to

$$\left(\Box + m^2\right) W^\nu = 0. \tag{5.147}$$

This equation is thus the equation of propagation of the massive carriers of the weak interaction $W^\pm$ and Z^0. A solution for a free massive boson in terms of plane waves is

$$W^\mu = \epsilon^\mu(p)e^{-ip\cdot x}, \tag{5.148}$$

where the constraint (5.146) and Eq. (5.147) imply

$$\epsilon \cdot p = 0 \text{ and } p^2 = m^2. \tag{5.149}$$

A possible choice of the three physical polarisations for a massive boson with 4-momentum $p = (E, \boldsymbol{p})$ is then

$$\epsilon(\boldsymbol{p},1) = \begin{pmatrix} 0 \\ \boldsymbol{\epsilon}(\boldsymbol{p},1) \end{pmatrix}, \quad \epsilon(\boldsymbol{p},2) = \begin{pmatrix} 0 \\ \boldsymbol{\epsilon}(\boldsymbol{p},2) \end{pmatrix}, \quad \epsilon(\boldsymbol{p},3) = \begin{pmatrix} \frac{|\boldsymbol{p}|}{m} \\ \frac{E_p}{m}\frac{\boldsymbol{p}}{|\boldsymbol{p}|} \end{pmatrix}, \tag{5.150}$$

with $\boldsymbol{\epsilon}(\boldsymbol{p}, i) \cdot \boldsymbol{p} = 0$ for $i = 1$ or 2. Then, the three physical polarisation vectors satisfy $\epsilon \cdot p = 0$ in (5.149). Both $\epsilon(\boldsymbol{p}, 1)$ and $\epsilon(\boldsymbol{p}, 2)$ polarisations are transverse polarisations, while $\epsilon(\boldsymbol{p}, 3)$ is the longitudinal polarisation since its spatial part is along the momentum direction. Hence, a massive vector boson has three possible spin projections, ± 1 and 0, whereas for photons the projection 0 is forbidden. The set of polarisation vectors (5.150) can be completed by a time-like polarisation vector to form a vector basis in the Minkowski space. The appropriate choice is $\epsilon(\boldsymbol{p}, 0) = p/m$, since it is orthogonal to the three other polarisation vectors. Moreover, with this choice, the completeness relation (5.128) remains valid for massive vector bosons.

The quantisation procedure is similar to that of photons but simpler because it is sufficient to quantise the three space-like components of the vector field (i.e., those with $\lambda = 1, 2$ and 3 that have $\epsilon(\boldsymbol{p}, \lambda) \cdot \epsilon(\boldsymbol{p}, \lambda) = -1$). Indeed, the time component of the field, W^0, is not independent of the spatial components due to Proca's equation (5.144), because in the vacuum, $W^0 = -\partial_\mu F^{\mu 0}/m^2 = -\partial_i F^{i0}/m^2$. Therefore, the quantised massive vector field can be expressed as

$$W^\mu(x) = \int \frac{\mathrm{d}^3\boldsymbol{p}}{(2\pi)^3 2E_p} \sum_{\lambda=1}^{3} \left[\epsilon^\mu(\boldsymbol{p},\lambda) a_{\boldsymbol{p},\lambda} e^{-ip\cdot x} + \epsilon^{*\mu}(\boldsymbol{p},\lambda) a^\dagger_{\boldsymbol{p},\lambda} e^{+ip\cdot x}\right]. \tag{5.151}$$

For a charged particle as the $W^{\pm}$ boson, one of the carrier of the weak interaction, the field must be complex valued. Conventionally, the negative charged boson W^- is considered as the particle, and thus, the field reads

$$W^{-\mu}(x) = \int \frac{d^3\boldsymbol{p}}{(2\pi)^3 2E_p} \sum_{\lambda=1}^{3} \left[\epsilon^{\mu}(\boldsymbol{p},\lambda) a_{p,\lambda} e^{-ip\cdot x} + \epsilon^{*\mu}(\boldsymbol{p},\lambda) b^{\dagger}_{p,\lambda} e^{+ip\cdot x}\right], \quad (5.152)$$

with $a_{p,\lambda} \neq b_{p,\lambda}$. The Fock operator $a_{p,\lambda}$ destroys a W^- with 4-momentum p and polarisation vector $\epsilon^{\mu}(\boldsymbol{p},\lambda)$, whereas $b^{\dagger}_{p,\lambda}$ creates a W^+ with p and $\epsilon^{\mu}(\boldsymbol{p},\lambda)$. Similarly, $(W^{-\mu})^{\dagger}(x) = W^{+\mu}(x)$ destroys a W^+ and creates a W^-. The other carrier of the weak interaction, the Z^0 boson, being a neutral particle, has $b_{p,\lambda} = a_{p,\lambda}$ and is therefore described by the real field (5.151). The commutation relation for the operator $a_{p,\lambda}$ (and similarly for $b_{p,\lambda}$) is

$$\left[a_{p',\lambda'}, a^{\dagger}_{p,\lambda}\right] = \delta_{\lambda\lambda'}(2\pi)^3 2E_p \delta^{(3)}(\boldsymbol{p}' - \boldsymbol{p}), \quad (5.153)$$

with all the other commutators being 0.

Problems

5.1 Check that the γ matrices in the Dirac representation (5.30) satisfy the Clifford algebra (5.29).

5.2 Using the spinors $v_1(p)$ and $v_2(p)$ given by Eq. (5.59), check that $v_1(p)e^{+ip\cdot x}$ and $v_2(p)e^{+ip\cdot x}$ satisfy $(\not{p} + m)v_i(p) = 0$, with $i = 1, 2$.

5.3 Starting from Eq. (5.52), describing the spinors of a particle or antiparticle at rest $\boldsymbol{p} = \boldsymbol{0}$, show that a boost transformation of the spinors with momentum $\boldsymbol{p} \neq \boldsymbol{0}$ given by Eq. (5.40) yields the general Formula (5.63). Do not forget the normalisation of the spinors (5.52).

5.4 Covariance of the Dirac equation. The first step of the proof is shown in Section 5.3.2, where the matrices $S^{\sigma\rho}$ are determined, i.e.

$$S^{\sigma\rho} = \frac{i}{4}\left[\gamma^{\sigma}, \gamma^{\rho}\right] = \begin{cases} \mathbb{0} & \text{if } \sigma = \rho, \\ \frac{i}{2}\gamma^{\sigma}\gamma^{\rho} & \text{if } \sigma \neq \rho. \end{cases}$$

This problem aims at showing that the matrices $S^{\sigma\rho}$ form a representation of the Lorentz group by satisfying its Lie algebra, i.e.

$$[S^{\mu\nu}, S^{\sigma\rho}] = i(g^{\nu\sigma}S^{\mu\rho} - g^{\mu\sigma}S^{\nu\rho} + g^{\mu\rho}S^{\nu\sigma} - g^{\nu\rho}S^{\mu\sigma}).$$

1. Establish the following identities, valid for any matrices A, B and C (whose multiplication is possible),

$$[A, BC] = [A, B]C + B[A, C], \quad [A, BC] = \{A, B\}C - B\{A, C\}.$$

2. Check that the Lie algebra is trivially verified if $\mu = \nu$ or $\sigma = \rho$.
3. Assuming that $\sigma \neq \rho$, show that $[S^{\mu\nu}, S^{\sigma\rho}] = \frac{i}{2}\left([S^{\mu\nu}, \gamma^{\sigma}]\gamma^{\rho} + \gamma^{\sigma}[S^{\mu\nu}, \gamma^{\rho}]\right)$.
4. Finally, assuming that $\mu \neq \nu$ and using the Clifford algebra (5.29), conclude that the matrices $S^{\mu\nu}$ satisfy the Lie algebra of the Lorentz group.

5.5 Inspired by the proof given in Supplement 5.3, check that under a Lorentz transformation, the bilinear forms,
1. $\overline{\psi}(x)\gamma^5\psi(x)$ transforms as a Lorentz scalar.
2. $\overline{\psi}(x)\gamma^5\gamma^\mu\psi(x)$ transforms as a Lorentz vector.
3. $\overline{\psi}(x)\gamma^\mu\gamma^\nu\psi(x)$ transforms as a Lorentz tensor.

Then, show that under a parity transformation $[x = (t, \boldsymbol{x}) \to x' = (t, -\boldsymbol{x})]$,

4. $\overline{\psi}(x)\gamma^5\psi(x)$ transforms into $\overline{\psi'}(x')\gamma^5\psi'(x') = -\overline{\psi}(x)\gamma^5\psi(x)$.
5. $\overline{\psi}(x)\gamma^5\gamma^\mu\psi(x)$ transforms into $\overline{\psi'}(x')\gamma^5\gamma^\mu\psi'(x')$ with $\overline{\psi'}(x')\gamma^5\gamma^\mu\psi'(x') = -\overline{\psi}(x)\gamma^5 \gamma^\mu\psi(x)$ when $\mu = 0$ and $\overline{\psi}(x)\gamma^5\gamma^\mu\psi(x)$ when $\mu \neq 0$.

The combination of those properties justifies the name pseudo-scalar for $\overline{\psi}(x)\gamma^5\psi(x)$ and axial-vector (or pseudo-vector) for $\overline{\psi}(x)\gamma^5\gamma^\mu\psi(x)$.

5.6 Determination of the helicity states. In the following, u_λ denotes the spinor of a particle eigenstate of the helicity operator [Eq. (5.70)], with a helicity $\lambda = \pm 1/2$. I recall the general formula of a u-spinor,

$$u_\lambda = \sqrt{E+m}\begin{pmatrix} \phi_\lambda \\ \frac{\boldsymbol{\sigma}\cdot\boldsymbol{p}}{E+m}\phi_\lambda \end{pmatrix}.$$

1. Show that $\boldsymbol{\sigma}\cdot\boldsymbol{p}\,\phi_\lambda = 2\lambda|\boldsymbol{p}|\phi_\lambda$.
2. Using polar coordinates, show that ϕ_λ can take the form for $\lambda = +1/2$,
$\phi_{\frac{1}{2}} = \begin{pmatrix} \cos\frac{\theta}{2} \\ e^{i\phi}\sin\frac{\theta}{2} \end{pmatrix}.$
3. Finally, show that the helicity state is given by

$$\psi_{+\frac{1}{2}} = \sqrt{E+m}\begin{pmatrix} \cos\frac{\theta}{2} \\ e^{i\phi}\sin\frac{\theta}{2} \\ \frac{|\boldsymbol{p}|}{E+m}\cos\frac{\theta}{2} \\ \frac{|\boldsymbol{p}|}{E+m}e^{i\phi}\sin\frac{\theta}{2} \end{pmatrix} e^{-ip\cdot x}.$$

4. Using the same procedure, show that the three other helicity states are given by the Formula (5.74).

5.7 The basis (5.129) results from a rotation of the basis (5.122) from the z-axis to the axis of $\boldsymbol{p}$. Check first that when $\boldsymbol{p} = p\boldsymbol{e}_z$, then (5.129) gives (5.122). Since a rotation is a Lorentz transformation, show that if the polarisations in (5.122) satisfy Eq. (5.128), then it is also the case for the polarisations in (5.129). *Hint: one needs the definition (2.14) of Lorentz transformations.*

6 A Brief Overview of Quantum Electrodynamics

In the previous chapters, we saw how to calculate measurable quantities such as the cross section of a reaction or the lifetime of particles when the amplitude of the process is known. We also saw that the general solutions of the wave equations lead to quantum fields, a well-adapted framework to treat states composed of many particles that can be created or annihilated when they interact. This chapter now briefly introduces how we can use those quantum fields to access amplitudes and thus measurable quantities. More specifically, we will do an educational tour of quantum electrodynamics (QED), which describes the interaction of electrons (or any charged particles) with photons. I want to warn the reader that although this chapter uses concepts from quantum field theory, it is not a course on that topic. Rather, the aim here is to expose the concepts and prepare the reader to be able to do simple calculations of processes at lowest order.

6.1 Action and Lagrangians

6.1.1 The Least Action Principle

Classical Case

In the Hamiltonian formulation of classical mechanics, the equations of motion of a system are deduced by minimising a quantity called the action, S, which has the units of the product of an energy and a time ($J \cdot s$). Let us recall how it works using a simple example. Consider a non-relativistic particle moving in one dimension x, with a kinetic energy $T = \frac{1}{2}m\mathring{x}^2$, and a potential energy V. The symbol $\mathring{x}$ denotes the time derivative of x, i.e. the velocity $\mathrm{d}x/\mathrm{d}t$. The quantity, $L = T - V$, is called *Lagrangian* with the units of energy. It is a function that depends on x, $\mathring{x}$ and possibly on time. The conjugate momentum of x (also called the canonical momentum) and the total energy are given by the formulas

$$p = \frac{\partial L}{\partial \mathring{x}}, \quad E = p\mathring{x} - L.$$

The equation of motion between the time t_1 and t_2 is deduced by minimising the action, S, defined by the integral of the Lagrangian,

$$S = \int_{t_1}^{t_2} L(x, \mathring{x}) \, \mathrm{d}t.$$

Among all trajectories $x(t)$ that can be imagined starting from x_1 at t_1 and finishing to x_2 at t_2, the one that has the lowest value of the action is the one adopted by the particle. Let us call $x^0(t)$ this trajectory. Any trajectory can be written as $x(t) = x^0(t) + \delta x(t)$. Then, the action reads

$$S = \int_{t_1}^{t_2} L(x^0 + \delta x, \overset{\circ}{x}{}^0 + \frac{\mathrm{d}}{\mathrm{d}t}\delta x)\,\mathrm{d}t = \int_{t_1}^{t_2} \left(L(x^0, \overset{\circ}{x}{}^0) + \frac{\partial L}{\partial x}\delta x + \frac{\partial L}{\partial \overset{\circ}{x}}\frac{\mathrm{d}}{\mathrm{d}t}\delta x \right) \mathrm{d}t.$$

Given that $\delta x(t_1) = \delta x(t_2) = 0$ (all trajectories start from and finish at the same point), it follows, after an integration by parts, that

$$\int_{t_1}^{t_2} \frac{\partial L}{\partial \overset{\circ}{x}}\frac{\mathrm{d}}{\mathrm{d}t}\delta x\;\mathrm{d}t = \left[\frac{\partial L}{\partial \overset{\circ}{x}}\delta x\right]_{t_1}^{t_2} - \int_{t_1}^{t_2} \frac{\mathrm{d}}{\mathrm{d}t}\left(\frac{\partial L}{\partial \overset{\circ}{x}}\right)\delta x\;\mathrm{d}t = -\int_{t_1}^{t_2} \frac{\mathrm{d}}{\mathrm{d}t}\left(\frac{\partial L}{\partial \overset{\circ}{x}}\right)\delta x\;\mathrm{d}t.$$

Hence, the variation of the action is just given by

$$\delta S = \int_{t_1}^{t_2} L(x, \overset{\circ}{x})\,\mathrm{d}t - \int_{t_1}^{t_2} L(x^0, \overset{\circ}{x}{}^0)\,\mathrm{d}t = \int_{t_1}^{t_2} \left(\frac{\partial L}{\partial x} - \frac{\mathrm{d}}{\mathrm{d}t}\frac{\partial L}{\partial \overset{\circ}{x}}\right)\delta x\;\mathrm{d}t.$$

Therefore, its minimisation imposes $\delta S = 0$, yielding the Euler–Lagrange equation,

$$\frac{\partial L}{\partial x} - \frac{\mathrm{d}}{\mathrm{d}t}\frac{\partial L}{\partial \overset{\circ}{x}} = 0.$$

For example, it is easy to check that the Euler–Lagrange equation applied to the classical Lagrangian $L = \frac{1}{2}m\overset{\circ}{x}{}^2 - V(x)$ yields the Newton's second law, $m\;\mathrm{d}^2x/\,\mathrm{d}t^2 = -\partial V/\partial x = F$.

Simple Quantum Case

In quantum mechanics, the action is quantised, with the minimum quantum of action being the Planck constant, h. Let us call q_i, $i = 1, \ldots, n$, the generalised coordinates (there was just one single variable x before). In quantum mechanics, the observables' position and momentum are described by operators that have the additional constraint in natural units,

$$[\hat{q}_i, \hat{p}_j] = i\delta_{i,j}, \tag{6.1}$$

where $\hat{q}_i$ and $\hat{p}_j$ are the Hermitian operators acting on a Hilbert space. The Lagrangian is now an operator that depends on $\hat{q}_1, \ldots, \hat{q}_n$ and $\overset{\circ}{\hat{q}}_1, \ldots, \overset{\circ}{\hat{q}}_n$. Notice that the operators $\hat{q}_i$ and $\overset{\circ}{\hat{q}}_i$ explicitly depend on time. In other words, the usual time dependence carried by the states, such as $|\psi(t)\rangle$, has been transferred to the operators. This is the Heisenberg representation of quantum mechanics that shall be detailed in Section 6.2.1. The conjugate momenta and the total energy of the system become a generalisation of the classical case, with

$$\hat{p}_i = \frac{\partial L}{\partial \overset{\circ}{\hat{q}}_i}, \quad \hat{H} = \left(\sum_{i=1}^{n} \hat{p}_i \overset{\circ}{\hat{q}}_i\right) - L. \tag{6.2}$$

The Hamiltonian $\hat{H}$ is Hermitian since the energy must be a real number. The equation of motion is still given by the Euler–Lagrange equations, with for $\hat{q}_i$,

$$\frac{\partial L}{\partial \hat{q}_i} - \frac{\mathrm{d}}{\mathrm{d}t}\frac{\partial L}{\partial \overset{\circ}{\hat{q}}_i} = 0. \tag{6.3}$$

Quantum Field Case

The extension from discrete operators, $\hat{q}_i$, $i = 1, \ldots, n$, to continuous operators, leads to the notion of quantum fields. Instead of $\hat{q}_i$, one introduces the field $\phi(\boldsymbol{x}, t) = \phi(x)$, which depends on spacetime coordinates (that are continuous variables). Therefore, instead of considering the Lagrangian L itself, we consider the Lagrangian density $\mathcal{L}$, i.e.

$$L = \int \mathcal{L}\, \mathrm{d}^3\boldsymbol{x}. \tag{6.4}$$

Hence, the action becomes

$$S = \int \mathcal{L}\, \mathrm{d}^4 x. \tag{6.5}$$

As before, $\mathcal{L}$ depends on ϕ and its time derivative $\partial_0\phi = \mathring{\phi}$, but since ϕ is a continuous function of spacetime, $\mathcal{L}$ can also depend on its space derivatives $\partial_1\phi$, $\partial_2\phi$ and $\partial_3\phi$. Therefore, we will have to keep in mind that

$$\mathcal{L} \equiv \mathcal{L}(\phi, \partial_\mu\phi). \tag{6.6}$$

Several constraints are imposed on the Lagrangians (it is a common practice to designate the Lagrangian density simply as the Lagrangian). Physics is assumed to be Poincaré invariant, thus respecting Lorentz invariance and translation invariance (see Appendix D about the Poincaré group). Since physics is deduced from the action (6.5) and the element of integration, d^4x, is Poincaré invariant (Lorentz invariance has been proven in Problem 2.3, and translation invariance is straightforward), $\mathcal{L}$ must be Poincaré invariant. The translation invariance is satisfied if Lagrangians do not depend explicitly on x^μ. Of course, the derivatives ∂_μ being translation invariant, they can appear in $\mathcal{L}$ as specified in Eq. (6.6). Lorentz invariance imposes only Lorentz invariant combinations of fields, such as $\phi^\dagger\phi$, $A^\mu A_\mu$ or $\bar{\psi}\psi$ in $\mathcal{L}$. Another property is imposed: the Lagrangian has to be Hermitian (as the Hamiltonian). The generalisation of the momentum and Hamiltonian in Eq. (6.2) to the quantum field case is given by the 'conjugate momentum field',

$$\pi(x) = \frac{\partial\mathcal{L}}{\partial\mathring{\phi}(x)}, \tag{6.7}$$

and the Hamiltonian (density) becomes

$$\mathcal{H} = \pi(x)\mathring{\phi}(x) - \mathcal{L}. \tag{6.8}$$

6.1.2 The Quantisation Procedure

The quantisation procedure of fields generalises what we had in the simple quantum case. Indeed, Eq. (6.1) becomes[1]

$$\left[\phi(x), \pi(y)\right]_{x^0=y^0} = i\delta^{(3)}(\boldsymbol{x} - \boldsymbol{y}). \tag{6.9}$$

[1] One has to use the anti-commutator for the fermion field as explained in Chapter 5.

For example, the Lagrangian of a massive real scalar field is

$$\mathcal{L}_{\mathrm{KG}} = \frac{1}{2}\left(\partial_\mu\phi\partial^\mu\phi - m^2\phi^2\right). \tag{6.10}$$

The term involving the derivatives of the fields is the kinetic term. Since $\partial_\mu\phi\partial^\mu\phi = \mathring{\phi}^2 - (\nabla\phi)^2$, it is straightforward to check that $\pi(y) = \mathring{\phi}(y)$. Hence, the commutation rule (6.9) implies that

$$\left[\phi(x), \mathring{\phi}(y)\right]_{x^0=y^0} = i\delta^{(3)}(\boldsymbol{x}-\boldsymbol{y}), \tag{6.11}$$

that is, one cannot simultaneously know the properties of the field and its rate of change at the same place. This is a clear generalisation of Eq. (6.1). What about the commutation of two ϕs or two πs? The commutation rules must be related to the notion of causality in special relativity that demands that two measurements carried out in causally disconnected regions of spacetime cannot interfere with each other. So, for two points x and y of spacetime separated by a space-like interval $(x-y)^2 < 0$ (and obviously $(x-y)^2 < 0$ if $x^0 = y^0$), we require

$$\left[\phi(x), \phi(y)\right] = \left[\pi(x), \pi(y)\right] = 0, \quad (x-y)^2 < 0. \tag{6.12}$$

One might argue that the fields ϕ and π are not necessarily directly observables, and thus there is no need for the previous commutations. However, operators for observables are made with quadratic products of the fields and because of the two following identities:

$$[AB, CD] = A[B,C]D + AC[B,D] + [A,C]DB + C[A,D]B,$$
$$\{AB, CD\} = A\{B,C\}D - AC\{B,D\} + CA\{B,D\} - C\{A,D\}B,$$

if the fields commute (in the case of bosons) or anti-commute (in the case of fermions), the operators of observables will commute or anti-commute as well. It is this procedure, requiring conditions (6.9) and (6.12), which yields the quantised fields of Chapter 5, the commutator between the creation and annihilation operators being a simple consequence of these relations.

The last piece is the generalisation of the Euler–Lagrange equation (6.3) that for a quantum field becomes

$$\frac{\partial\mathcal{L}}{\partial\phi} - \partial_\mu\frac{\partial\mathcal{L}}{\partial(\partial_\mu\phi)} = 0. \tag{6.13}$$

For instance, its application to the Klein–Gordon Lagrangian (6.10) requires us to calculate

$$\begin{cases} \dfrac{\partial\mathcal{L}_{\mathrm{KG}}}{\partial\phi} = -m^2\phi, \\ \dfrac{\partial\mathcal{L}_{\mathrm{KG}}}{\partial(\partial_\mu\phi)} = \dfrac{1}{2}g^{\mu\nu}\dfrac{\partial}{\partial(\partial_\mu\phi)}\left(\partial_\mu\phi\partial_\nu\phi\right) = \dfrac{1}{2}g^{\mu\nu}\left(\partial_\nu\phi + \partial_\mu\phi\dfrac{\partial(\partial_\nu\phi)}{\partial(\partial_\mu\phi)}\right) \\ \qquad = \dfrac{1}{2}g^{\mu\nu}\left(\partial_\nu\phi + \partial_\mu\phi\delta^\mu_\nu\right) = g^{\mu\nu}\partial_\nu\phi, \end{cases}$$

which, after substituting into Eq. (6.13), gives $-m^2\phi - \partial^\nu\partial_\nu\phi = 0$, i.e. the Klein–Gordon equation.

6.1.3 Free-Electron Lagrangian

With the electron being an elementary particle described by the Dirac equation, the free-electron Lagrangian density is given by the following Dirac Lagrangian density:

$$\boxed{\mathcal{L}_D = \bar{\psi}(i\not{\partial} - m)\psi.} \tag{6.14}$$

At first sight, it might be surprising that the proposed Lagrangian is *not* Hermitian. However, the Hermitian version of Eq. (6.14), given the property (5.32), is

$$\mathcal{L}'_D = (\mathcal{L}_D + \mathcal{L}_D^\dagger)/2 = i(\bar{\psi}\gamma^\mu\partial_\mu\psi - (\partial_\mu\bar{\psi})\gamma^\mu\psi)/2 - m\bar{\psi}\psi = \mathcal{L}_D - i\partial_\mu(\bar{\psi}\gamma^\mu\psi)/2.$$

Hence, $\mathcal{L}'_D$ and $\mathcal{L}_D$ differ only by a total derivative, of the form $\partial_\mu f^\mu$, i.e. a total divergence. Both actually lead to the same action (6.5) and thus the same physics. Indeed, the action being the integral over the spacetime of the Lagrangian, we can use the Gauss theorem to transform the integral over d^4x into an integral over the spacetime surface of the Minkowski space. Since all fields are assumed to be zero at infinity, a total divergence thus does not contribute to the action. Consequently, we can use the simplified version of the Lagrangian given by Eq. (6.14).

Let us check that the Dirac Lagrangian (6.14) leads to the Dirac equation. A Dirac field is made of four components, thus four complex fields, ψ_α, with $\alpha = 1-4$. Since ψ_α is complex valued, we can independently vary its real and imaginary parts or equivalently ψ_α and $\bar{\psi}_\alpha$. Hence, the variables that have to be considered in the Euler–Lagrange equation are the four ψ_α, the four adjoint $\bar{\psi}_\alpha$ and their derivatives $\partial_\mu\psi_\alpha$ and $\partial_\mu\bar{\psi}_\alpha$. Let us expand the Lagrangian in terms of independent components. It reads

$$\mathcal{L}_D = \sum_{\alpha,\beta} \bar{\psi}_\alpha i\gamma^\mu_{\alpha\beta}\partial_\mu\psi_\beta - m\delta_{\alpha\beta}\bar{\psi}_\alpha\psi_\beta, \tag{6.15}$$

where $\delta_{\alpha\beta}$ is the Kronecker symbol. Applying the Euler–Lagrange equation (6.13) for $\phi = \bar{\psi}_\alpha$, it follows that

$$\left.\begin{aligned} \frac{\partial\mathcal{L}_D}{\partial\bar{\psi}_\alpha} &= \sum_\beta i\gamma^\mu_{\alpha\beta}\partial_\mu\psi_\beta - m\delta_{\alpha\beta}\psi_\beta \\ \frac{\partial\mathcal{L}_D}{\partial(\partial_\mu\bar{\psi}_\alpha)} &= 0 \end{aligned}\right\} \sum_\beta i\gamma^\mu_{\alpha\beta}\partial_\mu\psi_\beta - m\psi_\alpha = 0.$$

As this equation holds for all four α, we recover the Dirac equation, $i\gamma^\mu\partial_\mu\psi - m\psi = 0$. Problem 6.1 shows that the Dirac adjoint equation (5.43) can also be recovered.

Since the expression of the Lagrangian (6.14) is equivalent to $\mathcal{L}_D = \psi^\dagger\gamma^0(i\gamma^0\partial_0 + i\gamma^i\partial_i - m)\psi$, the conjugate momentum field of ψ is

$$\pi(x) = \frac{\partial\mathcal{L}_D}{\partial\mathring{\psi}(x)} = i\psi^\dagger(x),$$

(recalling that $(\gamma^0)^2 = \mathbb{1}$). The canonical quantisation promotes the classical field and its conjugate momentum to operators via the canonical commutation relations (6.9) and (6.12) written with anti-commutators. We saw in Section 5.3.6 that in order to satisfy the Fermi–Dirac statistics, the creation and annihilation operators must anti-commute. In addition, the Dirac field has four components. Hence, the canonical quantisation actually requires

$$\{\psi_\alpha(x), \pi_\beta(y)\}_{x^0=y^0} = i\{\psi_\alpha(x), \psi_\beta^\dagger(y)\}_{x^0=y^0} = i\delta^{(3)}(\boldsymbol{x}-\boldsymbol{y})\delta_{\alpha\beta},$$

where α and β are the spinor indices. The complete set of anti-commutation rules then reads

$$\begin{aligned} \{\psi_\alpha(x), \psi_\beta^\dagger(y)\}_{x^0=y^0} &= \delta^{(3)}(\boldsymbol{x}-\boldsymbol{y})\delta_{\alpha\beta}\,; \\ \{\psi_\alpha(x), \psi_\beta(y)\}_{x^0=y^0} &= \{\psi_\alpha^\dagger(x), \psi_\beta^\dagger(y)\}_{x^0=y^0} = 0. \end{aligned} \tag{6.16}$$

The anti-commutation rules of the creation and annihilation operators (5.102) are simply a direct consequence of the anti-commutation rules above. Let us quickly check it. Actually, it is easier to do the other way around, assuming Eq. (5.102) and verifying that it leads to Eq. (6.16). From the field expansion in Eq. (5.100), it follows that

$$\psi_\alpha(x) = \int \frac{\mathrm{d}^3\boldsymbol{p}}{(2\pi)^3 2E_p} \sum_{r=1}^{2} \left[c_{p,r}\,(u_r(p))_\alpha\, e^{-ip\cdot x} + d_{p,r}^\dagger\,(v_r(p))_\alpha\, e^{+ip\cdot x} \right],$$

$$\psi_\beta^\dagger(y) = \int \frac{\mathrm{d}^3\boldsymbol{p}'}{(2\pi)^3 2E_{p'}} \sum_{r'=1}^{2} \left[c_{p',r'}^\dagger \left(u_{r'}^\dagger(p')\right)_\beta e^{+ip'\cdot y} + d_{p',r'} \left(v_{r'}^\dagger(p')\right)_\beta e^{-ip'\cdot y} \right].$$

Given the anti-commutation rules of the Fock operators (5.102), the anticommutator of ψ_α and $\psi_\beta^\dagger$ is then

$$\begin{aligned} \{\psi_\alpha(x), \psi_\beta^\dagger(y)\} &= \sum_{r,r'} \int \frac{\mathrm{d}^3\boldsymbol{p}}{(2\pi)^3 2E_p} \frac{\mathrm{d}^3\boldsymbol{p}'}{(2\pi)^3 2E_{p'}} (2\pi)^3 2Ep\, \delta^{(3)}(\boldsymbol{p}-\boldsymbol{p}')\delta_{rr'} \\ &\quad \times \left\{ (u_r(p))_\alpha \left(u_{r'}^\dagger(p')\right)_\beta e^{-i(p\cdot x - p'\cdot y)} + (v_r(p))_\alpha \left(v_{r'}^\dagger(p')\right)_\beta e^{+i(p\cdot x - p'\cdot y)} \right\} \\ &= \sum_r \int \frac{\mathrm{d}^3\boldsymbol{p}}{(2\pi)^3 2E_p} \left\{ (u_r(p))_\alpha \left(u_r^\dagger(p)\right)_\beta e^{-ip\cdot(x-y)} + (v_r(p))_\alpha \left(v_r^\dagger(p)\right)_\beta e^{+ip\cdot(x-y)} \right\}. \end{aligned}$$

However, noticing that $u_r(p)$ is a column vector and $u_r^\dagger(p)$ is a row vector, the product $(u_r(p))_\alpha (u_r^\dagger(p))_\beta = (u_r(p))_\alpha \left(\bar{u}_r(p)\gamma^0\right)_\beta$ is just the matrix component $\left(u_r(p)\bar{u}_r(p)\gamma^0\right)_{\alpha\beta}$. A similar equality is obtained for the term with v spinors. Therefore, we can use the completeness relation (5.78) to perform the summation over r, yielding

$$\begin{aligned} \{\psi_\alpha(x), \psi_\beta^\dagger(y)\}_{x^0=y^0} &= \int \frac{\mathrm{d}^3\boldsymbol{p}}{(2\pi)^3 2E_p} \left\{(\not{p}+m)\gamma^0\right\}_{\alpha\beta} e^{-i\boldsymbol{p}\cdot(\boldsymbol{x}-\boldsymbol{y})} + \int \frac{\mathrm{d}^3\boldsymbol{p}}{(2\pi)^3 2E_p} \left\{(\not{p}-m)\gamma^0\right\}_{\alpha\beta} e^{+i\boldsymbol{p}\cdot(\boldsymbol{x}-\boldsymbol{y})} \\ &= \int \frac{\mathrm{d}^3\boldsymbol{p}}{(2\pi)^3 2E_p} \left\{(E_p\gamma^0 + p_i\gamma^i + m)\gamma^0\right\}_{\alpha\beta} e^{-i\boldsymbol{p}\cdot(\boldsymbol{x}-\boldsymbol{y})} \\ &\quad + \int \frac{\mathrm{d}^3\boldsymbol{p}}{(2\pi)^3 2E_p} \left\{(E_p\gamma^0 + p_i\gamma^i - m)\gamma^0\right\}_{\alpha\beta} e^{+i\boldsymbol{p}\cdot(\boldsymbol{x}-\boldsymbol{y})}. \end{aligned}$$

Finally, changing the variable $\boldsymbol{p}$ by $-\boldsymbol{p}$ in the second integral, and using $(\gamma^0)^2 = \mathbb{1}$, reduces the expression to

$$\{\psi_\alpha(x), \psi_\beta^\dagger(y)\}_{x^0=y^0} = \int \frac{\mathrm{d}^3\boldsymbol{p}}{(2\pi)^3} \{\mathbb{1}\}_{\alpha\beta} e^{-i\boldsymbol{p}\cdot(\boldsymbol{x}-\boldsymbol{y})} = \delta_{\alpha\beta}\delta^{(3)}(\boldsymbol{x}-\boldsymbol{y}),$$

where the equality (E.9) has been used to produce the delta function. We thus recover the first line of the anti-commutation (6.16). A very similar calculation would yield the second line.

The Dirac Lagrangian allows us to recover the Hamiltonian operator given in Eq. (5.103) (as well as the momentum and the charge operators). The Hamiltonian density of the free spinor field reads

$$\mathcal{H}_\psi = \pi\mathring{\psi} - \mathcal{L}_D = \bar{\psi}(-i\gamma^i\partial_i + m)\psi. \tag{6.17}$$

Inserting the expressions of ψ (5.100) and $\bar{\psi}$ (5.101), and taking $H_\psi = \int d^3\boldsymbol{x}\,\mathcal{H}_\psi$, leads finally to the Hamiltonian of Eq. (5.103). There are, however, several subtleties that are addressed in Supplement 6.1.

Supplement 6.1. Hamiltonian and vacuum energy

Inserting the expressions of the Dirac fields (5.100) and (5.101) into the Dirac Hamiltonian density (6.17), the Hamiltonian $H_\psi = \int d^3\boldsymbol{x}\,\mathcal{H}_\psi$ reads

$$H_\psi = \sum_{r,r'} \int d^3\boldsymbol{x} \frac{d^3\boldsymbol{p}\, d^3\boldsymbol{p}'}{(2\pi)^6 4E_p E'_p} \left[c^\dagger_{p',r'}\bar{u}_{r'}(p')e^{+ip'\cdot x} + d_{p',r'}\bar{v}_{r'}(p')e^{-ip'\cdot x}\right]$$
$$\times \left[c_{p,r}(-\gamma^i p_i + m)u_r(p)e^{-ip\cdot x} + d^\dagger_{p,r}(\gamma^i p_i + m)v_r(p)e^{+ip\cdot x}\right].$$

The spinors satisfy the Dirac equations in momentum space (5.48) and (5.60),

$$(-\gamma^i p_i + m)u_r(p) = \gamma^0 E_p u_r(p), \quad (\gamma^i p_i + m)v_r(p) = -\gamma^0 E_p v_r(p).$$

Hence, it follows that

$$H_\psi = \sum_{r,r'} \int \frac{d^3\boldsymbol{p}\, d^3\boldsymbol{p}'}{(2\pi)^3 4E'_p}\Big[c^\dagger_{p',r'}c_{p,r}\bar{u}_{r'}(p')\gamma^0 u_r(p)e^{i(E_{p'}-E_p)t}\delta^{(3)}(\boldsymbol{p}'-\boldsymbol{p})$$
$$- c^\dagger_{p',r'}d^\dagger_{p,r}\bar{u}_{r'}(p')\gamma^0 v_r(p)e^{i(E_{p'}+E_p)t}\delta^{(3)}(\boldsymbol{p}'+\boldsymbol{p})$$
$$+ d_{p',r'}c_{p,r}\bar{v}_{r'}(p')\gamma^0 u_r(p)e^{-i(E_{p'}+E_p)t}\delta^{(3)}(\boldsymbol{p}'+\boldsymbol{p})$$
$$- d_{p',r'}d^\dagger_{p,r}\bar{v}_{r'}(p')\gamma^0 v_r(p)e^{-i(E_{p'}-E_p)t}\delta^{(3)}(\boldsymbol{p}'-\boldsymbol{p})\Big],$$

where the integration over $d^3\boldsymbol{x}$ of the exponentials has yielded the delta functions via Eq. (E.9). The integration of the delta functions imposes $E_{p'} = E_p$, and the inner product Formula (5.76) eliminates the second and third lines, which leaves only

$$H_\psi = \sum_{r,r'} \int \frac{d^3\boldsymbol{p}}{(2\pi)^3 4E_p}\left[c^\dagger_{p,r'}c_{p,r}\bar{u}_{r'}(p)\gamma^0 u_r(p) - d_{p,r'}d^\dagger_{p,r}\bar{v}_{r'}(p)\gamma^0 v_r(p)\right]$$
$$= \int \frac{d^3\boldsymbol{p}}{(2\pi)^3 2E_p} E_p\left[c^\dagger_{p,r}c_{p,r} - d_{p,r}d^\dagger_{p,r}\right],$$

where the Formula (5.75) has been used to obtain the last line. Notice that the Hamiltonian does not depend on time, as expected, which was not obvious with the very first formula of this supplement.

Supplement 6.1. (cont.)

Let us change the order of the operators in the second term using the anti-commutator (5.102). This yields

$$H_\psi = \int \frac{d^3\boldsymbol{p}}{(2\pi)^3 2E_p} E_p \left[c^\dagger_{p,r} c_{p,r} + d^\dagger_{p,r} d_{p,r}\right] - \int d^3\boldsymbol{p}\, E_p\, \delta^{(3)}(\boldsymbol{0}).$$

Forgetting for the moment the second integral with the delta function, we see that the first integral applied on the vacuum $|0\rangle$ gives zero, since by definition of the vacuum, $c_{p,r}|0\rangle = d_{p,r}|0\rangle = 0$. Therefore, states containing particles have larger energy, given that $E_p = \sqrt{|\boldsymbol{p}|^2 + m^2}$ is positive. Notice that if commutators were used, we would have a disastrous minus sign, leading to an increasingly negative energy of the system for an increasing number of particle states. In other words, the Hamiltonian would be unbounded from below, rendering the theory nonsensical. Now, we must address the divergent integral

$$E_{\text{vac}} = -\int d^3\boldsymbol{p}\, E_p\, \delta^{(3)}(\boldsymbol{0}).$$

First issue, the $\delta^{(3)}(\boldsymbol{0})$ term is infinite. However, we saw in Supplement 5.1 that actually, $\delta^{(3)}(\boldsymbol{0}) = \mathcal{V}/(2\pi)^3$, where $\mathcal{V}$ is the volume of the space. The problem arises because the field spreads over the whole space, which is considered infinite. If instead of the total energy of the vacuum, E_{vac}, we consider the energy density,

$$\frac{E_{\text{vac}}}{\mathcal{V}} = -\int \frac{d^3\boldsymbol{p}}{(2\pi^3)} \sqrt{|\boldsymbol{p}|^2 + m^2},$$

then the volume is no longer problematic. However, we face a second issue: the integral still diverges. Note that implicitly we assume that the theory is valid up to arbitrarily high energies, where other interactions (including gravity!) should be taken into account. This assumption is unrealistic. Since we do not really know how to cope with this problem, we escape from this dilemma, by considering that only differences in energies can be measured.[a] Therefore, we redefine the Hamiltonian by simply forgetting the diverging integral,

$$H_\psi \equiv \int \frac{d^3\boldsymbol{p}}{(2\pi)^3 2E_p} E_p \left[c^\dagger_{p,r} c_{p,r} + d^\dagger_{p,r} d_{p,r}\right].$$

It is equivalent to reorder the operators with all annihilation operators placed to the right, ignoring the contribution of the (anti-)commutators. In quantum field theory, this procedure is called the *normal ordering*.

[a] Actually, this is not true. Gravity in general relativity couples to all kinds of energies. The very large density should largely contribute to the *cosmological constant* of the universe, whose measurement is however consistent with a flat universe. This inconsistency is still not understood.

6.1.4 Free-Photon Lagrangian

As shown in Chapter 5, the free photon obeys the wave equation (5.117), namely $\Box A^\mu - \partial^\mu(\partial_\nu A^\nu) = 0$. The corresponding Lagrangian is then

$$\boxed{\mathcal{L}_\gamma = -\frac{1}{4} F_{\mu\nu} F^{\mu\nu} = \frac{1}{2}(|\boldsymbol{E}|^2 - |\boldsymbol{B}|^2),} \tag{6.18}$$

where $F^{\mu\nu}$ is the usual electromagnetic tensor (2.37) and $\boldsymbol{E}$ and $\boldsymbol{B}$ are the electric and magnetic fields respectively (see Problem 6.2 to check the latter expression). In order to recover the free-photon wave equation, let us rewrite the Lagrangian as follows:

$$\mathcal{L}_\gamma = -\frac{1}{4}(\partial_\mu A_\nu F^{\mu\nu} + \partial_\nu A_\mu F^{\nu\mu}) = -\frac{1}{2}\partial_\mu A_\nu F^{\mu\nu} = -\frac{1}{2}g^{\mu\alpha}g^{\nu\beta}\partial_\mu A_\nu(\partial_\alpha A_\beta - \partial_\beta A_\alpha),$$

where the antisymmetry of the tensor was used, and the dummy indices μ and ν interchanged to obtain the second equality. Let us apply the Euler–Lagrange equation (6.13) with $\phi = A_\nu$. Since this Lagrangian depends only on fields derivatives, the term $\partial\mathcal{L}_\gamma/\partial A_\nu$ is necessarily zero. The other term is

$$\frac{\partial\mathcal{L}_\gamma}{\partial(\partial_\mu A_\nu)} = -\frac{1}{2}g^{\mu\alpha}g^{\nu\beta}(\partial_\alpha A_\beta - \partial_\beta A_\alpha) - \frac{1}{2}g^{\mu\alpha}g^{\nu\beta}\partial_\mu A_\nu\left\{\frac{\partial(\partial_\alpha A_\beta)}{\partial(\partial_\mu A_\nu)} - \frac{\partial(\partial_\beta A_\alpha)}{\partial(\partial_\mu A_\nu)}\right\}.$$

The quantity in the braces is simply $\delta^\mu_\alpha\delta^\nu_\beta - \delta^\mu_\beta\delta^\nu_\alpha$, and thus,

$$\frac{\partial\mathcal{L}_\gamma}{\partial(\partial_\mu A_\nu)} = -\frac{1}{2}(\partial^\mu A^\nu - \partial^\nu A^\mu) - \frac{1}{2}\partial^\alpha A^\beta\left\{\delta^\mu_\alpha\delta^\nu_\beta - \delta^\mu_\beta\delta^\nu_\alpha\right\} = -(\partial^\mu A^\nu - \partial^\nu A^\mu). \quad (6.19)$$

Finally, the Euler–Lagrange equation then reads

$$\frac{\partial\mathcal{L}_\gamma}{\partial A_\nu} - \partial_\mu\frac{\partial\mathcal{L}_\gamma}{\partial(\partial_\mu A_\nu)} = \partial_\mu\partial^\mu A^\nu - \partial_\mu(\partial^\nu A^\mu) = 0,$$

which is the wave equation of a free photon.

The quantisation of the photon field is not straightforward. In order to have a glimpse of the difficulty, let us consider the conjugate momentum field of the photon. According to Eq. (6.19),

$$\Pi^\mu = \frac{\partial\mathcal{L}_\gamma}{\partial\mathring{A}_\mu} = \frac{\partial\mathcal{L}_\gamma}{\partial(\partial_0 A_\mu)} = -F^{0\mu}.$$

We see that $\Pi^0 = 0$, and consequently, the canonical quantisation (6.9), requiring

$$\left[A^0(x), \Pi^0(y)\right]_{x^0=y^0} = i\delta^{(3)}(\boldsymbol{x} - \boldsymbol{y}),$$

cannot be satisfied for A^0. Is this really surprising? Well, we saw in Section 5.4.1 that because of the gauge freedom, a real photon represents only two independent dynamical degrees of freedom (the two transverse polarisations) and not four as we would have expected from a 4-vector field. Thus, it is not surprising to be in trouble with certain components of A^μ. We could try to quantise only the physical degrees of freedom, but we would sacrifice manifest Lorentz invariance. Actually, instead of imposing the gauge condition and trying to quantise the field, one possibility is to change the theory such that the equation of motion corresponds to the one obtained after the gauge condition. To be more specific, let us consider the modified Lagrangian

$$\mathcal{L} = -\frac{1}{4}F_{\mu\nu}F^{\mu\nu} - \frac{1}{2\zeta}(\partial_\mu A^\mu)^2, \quad (6.20)$$

where ζ is an arbitrary number called the *gauge fixing parameter*. The application of the Euler–Lagrange equation yields the equation of motion,

$$\partial_\mu\partial^\mu A^\nu - \left(1 - \frac{1}{\zeta}\right)\partial_\mu(\partial^\nu A^\mu) = 0.$$

The value $\zeta = 1$ is called the *Feynman gauge*. It leads to $\partial_\mu \partial^\mu A^\nu = 0$, i.e. the equation of motion in the Lorenz gauge. The interest of the modified Lagrangian is that now the conjugate-momentum fields can be used for the canonical quantisation since with $\zeta = 1$,

$$\Pi^0 = -\partial_\mu A^\mu \,, \quad \Pi^i = -F^{0i}. \tag{6.21}$$

We can now impose

$$\begin{aligned} \left[A_\mu(x), \Pi_\nu(y)\right]_{x^0=y^0} &= i g_{\mu\nu} \delta^{(3)}(\boldsymbol{x} - \boldsymbol{y}); \\ \left[A_\mu(x), A_\nu(y)\right]_{x^0=y^0} &= \left[\Pi_\mu(x), \Pi_\nu(y)\right]_{x^0=y^0} = 0, \end{aligned} \tag{6.22}$$

or equivalently (see Problem 6.3),

$$\begin{aligned} \left[A_\mu(x), \mathring{A}_\nu(y)\right]_{x^0=y^0} &= -i g_{\mu\nu} \delta^{(3)}(\boldsymbol{x} - \boldsymbol{y}); \\ \left[A_\mu(x), A_\nu(y)\right]_{x^0=y^0} &= \left[\mathring{A}_\mu(x), \mathring{A}_\nu(y)\right]_{x^0=y^0} = 0, \end{aligned} \tag{6.23}$$

which is very close to the commutation rules of the scalar field in Eq. (6.11), at least for $\mu > 0$ and $\nu > 0$. These conditions lead to the commutation rules (5.133) of the creation and annihilation operators of the photon and to the field expansion (5.132). With the four conjugate momenta, the Hamiltonian density reads

$$\mathcal{H}_\gamma = \Pi^\mu \partial_0 A_\mu - \mathcal{L}_\gamma. \tag{6.24}$$

Following a similar procedure to that outlined in Supplement 6.1, we arrive at the Hamiltonian

$$H_\gamma = \int \mathrm{d}^3\boldsymbol{x}\, \mathcal{H}_\gamma = \int \frac{\mathrm{d}^3\boldsymbol{p}}{(2\pi)^3 2E_p} |\boldsymbol{p}| \left(\alpha^\dagger_{\boldsymbol{p},1}\alpha_{\boldsymbol{p},1} + \alpha^\dagger_{\boldsymbol{p},2}\alpha_{\boldsymbol{p},2} + \alpha^\dagger_{\boldsymbol{p},3}\alpha_{\boldsymbol{p},3} - \alpha^\dagger_{\boldsymbol{p},0}\alpha_{\boldsymbol{p},0}\right), \tag{6.25}$$

where $E_p = |\boldsymbol{p}|$ and $\alpha_{\boldsymbol{p},\lambda}$ is the annihilation operator for the polarisation λ. The price to pay with this quantisation is that we introduce spurious degrees of freedom, and consequently, the Hamiltonian does not seem to be bounded from below because of the $-\alpha^\dagger_{\boldsymbol{p},0}\alpha_{\boldsymbol{p},0}$ contribution. We cannot eliminate them by imposing the Lorenz gauge $\partial_\mu A^\mu = 0$ on the operators because we would be back to the initial situation where $\Pi^0 = 0$ in Eq. (6.21). Actually, the idea is rather to restrict the admissible states (i.e., the physical states) to those satisfying $\partial_\mu A^\mu |\psi\rangle = 0$. The details of the procedure would lead us beyond the scope of this textbook. See, for example, Chapter 8 of Weinberg (1995). Those admissible states turn out to satisfy

$$(\alpha_{\boldsymbol{p},0} - \alpha_{\boldsymbol{p},3}) |\psi\rangle = 0.$$

According to the commutation rules (5.133), a state of photons with transverse polarisations, for instance, $|\psi\rangle = \alpha^\dagger_{\boldsymbol{p},1}\alpha^\dagger_{\boldsymbol{p}',2}|0\rangle$, obviously satisfies this condition. The contributions from the time-like and longitudinal photons cancel out in the Hamiltonian (6.25), leaving only the positive contribution to the energy from the transverse photons.

The complicated quantisation of the electromagnetic field is easier to achieve using the path integral formulation of quantum field theory. However, in this particle physics textbook, only the 'classical' quantisation with canonical commutators (historically named the *second quantisation*) is outlined, as the reader is supposed to be more or less familiar with this approach already used in non-relativistic quantum mechanics (e.g., for the quantisation of the harmonic oscillator).

6.1.5 Consequences of Gauge Invariance

In Chapter 4, we saw several applications of Emmy Noether's theorem, such as the conservation of the momentum or the angular momentum of a system. In this section, we are going to study another kind of application.

Global Phase Invariance

Let us consider the transformation of the phase of the Dirac field,

$$\psi(x) \to \psi'(x) = e^{-iq\chi}\psi(x), \tag{6.26}$$

where q and χ are two real numbers. This transformation leaves the Dirac Lagrangian (6.14) invariant,

$$\mathcal{L}_D \to \mathcal{L}' = e^{+iq\chi}\bar{\psi}(i\partial\!\!\!/ - m)e^{-iq\chi}\psi = \mathcal{L}_D.$$

The set of transformations (6.26) forms a $U(1)$ group. (Read Appendix A for more details about groups. All its elements commute with each other, i.e. the order of transformations does not matter: $U(1)$ is thus an Abelian group.) Let us consider an infinitesimal transformation of parameter $\delta\chi$. The variation of the Dirac field by this transformation is then

$$\delta\psi = \psi' - \psi = -iq\delta\chi\psi,$$

which induces the following variation of the Lagrangian:

$$\begin{aligned}\delta\mathcal{L} &= \frac{\partial\mathcal{L}}{\partial\psi}\delta\psi + \frac{\partial\mathcal{L}}{\partial(\partial_\mu\psi)}\delta(\partial_\mu\psi) + \delta\bar{\psi}\frac{\partial\mathcal{L}}{\partial\bar{\psi}} + \delta(\partial_\mu\bar{\psi})\frac{\partial\mathcal{L}}{\partial(\partial_\mu\bar{\psi})} \\ &= -iq\delta\chi\left\{\frac{\partial\mathcal{L}}{\partial\psi}\psi + \frac{\partial\mathcal{L}}{\partial(\partial_\mu\psi)}\partial_\mu\psi - \bar{\psi}\frac{\partial\mathcal{L}}{\partial\bar{\psi}} - \partial_\mu\bar{\psi}\frac{\partial\mathcal{L}}{\partial(\partial_\mu\bar{\psi})}\right\}.\end{aligned}$$

But the Euler–Lagrange equation applied to ψ and $\bar{\psi}$ implies that

$$\frac{\partial\mathcal{L}}{\partial\psi}\psi = \partial_\mu\left(\frac{\partial\mathcal{L}}{\partial(\partial_\mu\psi)}\right)\psi\,, \quad \bar{\psi}\frac{\partial\mathcal{L}}{\partial\bar{\psi}} = \bar{\psi}\partial_\mu\left(\frac{\partial\mathcal{L}}{\partial(\partial_\mu\bar{\psi})}\right).$$

It follows that

$$\frac{\delta\mathcal{L}}{\delta\chi} = -iq\partial_\mu\left\{\frac{\partial\mathcal{L}}{\partial(\partial_\mu\psi)}\psi - \bar{\psi}\frac{\partial\mathcal{L}}{\partial(\partial_\mu\bar{\psi})}\right\}.$$

According to Noether's theorem, since the transformation (6.26) is continuous, there must be a conservation law. In other words, the Lagrangian must not be sensitive to an infinitesimal transformation, implying the continuity equation $\delta\mathcal{L}/\delta\chi = \partial_\mu j^\mu = 0$, with the conserved current

$$j^\mu = -iq\left(\frac{\partial\mathcal{L}}{\partial(\partial_\mu\psi)}\psi - \bar{\psi}\frac{\partial\mathcal{L}}{\partial(\partial_\mu\bar{\psi})}\right).$$

According to the Dirac Lagrangian (6.14), $\partial\mathcal{L}/\partial(\partial_\mu\psi) = i\bar{\psi}\gamma^\mu$ and $\partial\mathcal{L}/\partial(\partial_\mu\bar{\psi}) = 0$. Therefore, we finally identify

$$j^\mu = q\bar{\psi}\gamma^\mu\psi, \tag{6.27}$$

which is precisely the charge current we introduced in Eq. (5.46). Hence, the quantity that is conserved by the phase transformation of the Dirac field is the charge $Q = \int d^3x\, j^0 = q \int d^3x\, \rho$, where ρ is the number density (5.45). If we interpret q as the electric charge of the elementary particle, then the phase transformation implies the conservation of the total electric charge Q of the system (i.e., $dQ/dt = 0$).

Local Phase Invariance

The identification of q as the electric charge of the particle is more striking if we assume that χ in the phase transformation (6.26) is not just a real number but a real scalar function, $\chi(x)$, depending on the spacetime coordinates. The phase transformation of the Dirac field is now said to be *local*,

$$\psi(x) \to \psi'(x) = e^{-iq\chi(x)}\psi(x). \tag{6.28}$$

How is the Dirac Lagrangian transformed? Now, we have to take into account the derivative of the function $\chi(x)$ itself. It yields

$$\mathcal{L}_D \to \mathcal{L}' = \mathcal{L}_D + j^\mu \partial_\mu \chi(x).$$

Thus, the Dirac Lagrangian is no more invariant, and if we wish to impose invariance under the local transformation, we need to add another term $\mathcal{L}_{\text{int}}$ to the Dirac Lagrangian. This additional term must be of the form

$$\boxed{\mathcal{L}_{\text{int}} = -j^\mu A_\mu = -q\bar{\psi}\gamma^\mu\psi A_\mu,} \tag{6.29}$$

such that its local transformation will compensate the term $j^\mu \partial_\mu \chi(x)$. The object A_μ is necessarily a Lorentz vector (given that j^μ is a vector) since the Lagrangian itself must be a Lorentz scalar. Let us see how the modified Lagrangian now transforms:

$$\mathcal{L}_D + \mathcal{L}_{\text{int}} \to \mathcal{L}' = \mathcal{L}_D + j^\mu \partial_\mu \chi(x) - j'^\mu A'_\mu.$$

But $j'^\mu = q\bar{\psi}'\gamma^\mu\psi' = q\bar{\psi}\gamma^\mu\psi = j^\mu$. It follows that

$$\mathcal{L}_D + \mathcal{L}_{\text{int}} \to \mathcal{L}' = \mathcal{L}_D + j^\mu \left(\partial_\mu \chi(x) - A'_\mu\right).$$

Hence, if the vector A_μ transforms as

$$A_\mu \to A'_\mu = A_\mu + \partial_\mu \chi(x), \tag{6.30}$$

then the Lagrangian $\mathcal{L}_D + \mathcal{L}_{\text{int}}$ is invariant. We have already met the above transformation in Eq. (5.118): it is the well-known gauge invariance arising from the Maxwell equation. Therefore, we can identify A_μ as the electromagnetic field and q as the charge of the particle.

6.1.6 QED Lagrangian

Let us recap what we have found: promoting the free Dirac Lagrangian $\mathcal{L}_D$ (6.14) to be invariant under a local phase transformation requires an additional term $\mathcal{L}_{\text{int}}$ (6.29). This term couples a vector field A^μ to the electron charge current, consistent with the photon field (i.e., having the same transformations). The Lagrangian $\mathcal{L}_D + \mathcal{L}_{\text{int}}$ is then invariant, provided the simultaneous transformations of Eqs. (6.28) and (6.30). This Lagrangian reads

$$\mathcal{L}_D + \mathcal{L}_{\text{int}} = \bar{\psi}(i\gamma^\mu \partial_\mu - q\gamma^\mu A_\mu - m)\psi = \bar{\psi}(i\gamma^\mu[\partial_\mu + iqA_\mu] - m)\psi.$$

The term in brackets, $\partial_\mu + iqA_\mu$, is precisely the term introduced in Eq. (5.79) to take into account the presence of an electromagnetic field in the Dirac equation. So, we move from a non-interacting theory, invariant under a global phase transformation of the Dirac field, to an interacting theory, invariant under a local phase transformation by doing the substitution

$$\boxed{\partial_\mu \to \mathrm{D}_\mu = \partial_\mu + iqA_\mu,} \tag{6.31}$$

with the Dirac Lagrangian then becoming

$$\mathcal{L}_D \to \bar{\psi}(i\gamma^\mu \mathrm{D}_\mu - m)\psi = \mathcal{L}_D + \mathcal{L}_{\text{int}}.$$

The quantity, D_μ, here is called the *covariant derivative*. The theory, initially invariant under global phase transformations, induced the electric charge conservation of free electrons (or of any elementary charged particle with a spin-1/2). The local phase invariance requirement has generated a new field, identified as the photon field, which couples to electrons through their charge via the Lagrangian $\mathcal{L}_{\text{int}}$. This is an example of the 'gauge principle' that lifts a global symmetry ($U(1)$ here) into a local one. It requires the replacement of the normal derivative by the covariant derivative to identify the gauge field.

We have now all the ingredients to specify the complete QED Lagrangian. It reads

$$\boxed{\mathcal{L} = \mathcal{L}_D + \mathcal{L}_\gamma + \mathcal{L}_{\text{int}} = \bar{\psi}(i\not{\partial} - m)\psi - \frac{1}{4}F_{\mu\nu}F^{\mu\nu} - q\bar{\psi}\gamma^\mu\psi A_\mu.} \tag{6.32}$$

It is the sum of the free-electron Lagrangian (i.e., the Dirac Lagrangian), the free-photon Lagrangian, and the interaction Lagrangian between electrons[2] and photons. It is invariant under a simultaneous $U(1)$ local phase transformation of the Dirac field (6.28) and a gauge transformation of the photon field, with the term containing $F^{\mu\nu}$ being invariant under the transformation (6.30).

Since the Lagrangian of the interaction, $\mathcal{L}_{\text{int}}$ in Eq. (6.29), does not involve field derivatives, the conjugate momentum fields are still those of the free fields, and the QED Hamiltonian density then reads $\mathcal{H} = \mathcal{H}_\psi + \mathcal{H}_\gamma + \mathcal{H}_{\text{int}}$, where $\mathcal{H}_\psi$ is given by Eqs. (6.17) and $\mathcal{H}_\gamma$ is the contribution from the electromagnetic field. The Hamiltonian density describing the interaction between photons and charged fermions is thus simply

$$\mathcal{H}_{\text{int}} = -\mathcal{L}_{\text{int}} = j^\mu A_\mu = q\bar{\psi}\gamma^\mu\psi A_\mu. \tag{6.33}$$

[2] For electrons, $q = -e$ has to be used.

6.2 Perturbation with Interacting Fields: The *S*-matrix

We saw in the previous section that the electron field and photon field interact. We expect processes to occur where, for example, the electron will be scattered by a photon. Hence, we need to be able to calculate measurable quantities, such as the cross section introduced in Section 3.5. This section aims to establish a link between the fields and the matrix element appearing in formulas such as the transition rate in Eq. (3.2).

6.2.1 Schrödinger, Heisenberg and Interaction Representations

In the well-known *Schrödinger representation* of non-relativistic quantum mechanics, the time evolution of a system is given by

$$i\frac{\mathrm{d}}{\mathrm{d}t}\,|\psi_{\mathrm{S}}(t)\rangle = H\,|\psi_{\mathrm{S}}(t)\rangle, \tag{6.34}$$

where the subscript s is used to emphasise that we work in the Schrödinger representation. Note that even if Eq. (6.34) is usually used in non-relativistic quantum mechanics, it also gives the evolution of a state in the relativistic case, provided that the Hamiltonian is relativistic. If H does not depend explicitly on time (the fundamental interactions are not supposed to vary with time), $|\psi_{\mathrm{S}}\rangle$ can be computed for any time t, by integrating Eq. (6.34), provided the state is known at another time, let us say $t = 0$, yielding[3]

$$|\psi_{\mathrm{S}}(t)\rangle = e^{-itH}\,|\psi_{\mathrm{S}}(0)\rangle. \tag{6.35}$$

Thus, the state $|\psi_{\mathrm{S}}\rangle$ carries the entire time dependence. The free fields used in the Hamiltonian can then be expanded in terms of creation and annihilation operators at the fixed time, $t = 0$. However, the problem with the Schrödinger representation is this particular time, $t = 0$, which is singled out and makes the theory not manifestly Lorentz invariant. It is, therefore, preferable to find a representation where the time dependence is carried out by operators, which usually also depend on spatial variables.

Now, note that the matrix element of an operator A between the states $|\psi_{\mathrm{S}}\rangle$ and $|\psi'_{\mathrm{S}}\rangle$ can be estimated at the time t with

$$\langle\psi'_{\mathrm{S}}(t)|A|\psi_{\mathrm{S}}(t)\rangle = \langle\psi'_{\mathrm{S}}(0)|\left(e^{-itH}\right)^{\dagger} A\, e^{-itH}|\psi_{\mathrm{S}}(0)\rangle = \langle\psi'_{\mathrm{S}}(0)|e^{+itH}A\, e^{-itH}|\psi_{\mathrm{S}}(0)\rangle.$$

Such an expression can be interpreted differently, with an operator A_{H} depending on time,

$$A_{\mathrm{H}}(t) = e^{+itH}A\, e^{-itH}, \tag{6.36}$$

and states $|\psi_{\mathrm{H}}\rangle = |\psi_{\mathrm{S}}(0)\rangle$, $|\psi'_{\mathrm{H}}\rangle = |\psi'_{\mathrm{S}}(0)\rangle$ without time dependence, such as

$$\langle\psi'_{\mathrm{S}}(t)|A|\psi_{\mathrm{S}}(t)\rangle = \langle\psi'_{\mathrm{H}}|A_{\mathrm{H}}(t)|\psi_{\mathrm{H}}\rangle.$$

[3] The exponential means $e^{-itH} = \sum_{k=0}^{\infty}(-itH)^k/k!$

This new representation is the *Heisenberg representation* (thus the subscript H), where operators depend on time and states do not. According to Eq. (6.35), since H is Hermitian and commutes with itself $(e^{-itH})^\dagger e^{-itH} = e^{+itH} e^{-itH} = \mathbb{1}$, the state $|\psi_{\rm H}\rangle = |\psi_{\rm S}(0)\rangle$ reads

$$|\psi_{\rm H}\rangle = e^{itH}\, |\psi_{\rm S}(t)\rangle. \tag{6.37}$$

The time dependence is entirely transferred to operators and is obtained by differentiating Eq. (6.36). Taking care of the order of the operators, it yields the Heisenberg equation,

$$i\frac{\mathrm{d}}{\mathrm{d}t} A_{\rm H}(t) = [A_{\rm H}(t), H]. \tag{6.38}$$

Note that the Hamiltonian H is identical in the two representations since it commutes with itself. A quantum field $\phi(x, t)$, which is an operator, evolves with time according to Eq. (6.36), i.e.

$$\phi(x, t) = e^{+itH} \phi(x, 0)\, e^{-itH}. \tag{6.39}$$

Finally, there is the *interaction representation* that is a hybrid of the two previous representations. It is well adapted for perturbative calculations. In this representation, both the states and the operators have time dependence. We divide the Hamiltonian into two parts,

$$H = H_0 + H_{\rm int}.$$

The Hamiltonian H_0 is supposed to be exactly solvable, while $H_{\rm int}$ is necessarily not. For conveniance, H_0 would describe a free system and $H_{\rm int}$ the perturbation due to interactions. In the interaction picture, the state $|\psi_{\rm I}\rangle$ is defined as

$$|\psi_{\rm I}(t)\rangle = e^{itH_0}\, |\psi_{\rm S}(t)\rangle. \tag{6.40}$$

Notice the use of H_0 instead of H, as in the Heisenberg representation (6.37). Obviously, in the absence of perturbation, $H = H_0$, and the interaction representation coincides with that of Heisenberg. Since $|\psi_{\rm S}(t)\rangle$ obeys Eq. (6.34), it is easy to conclude that $|\psi_{\rm I}(t)\rangle$ obeys

$$i\frac{\mathrm{d}}{\mathrm{d}t}\, |\psi_{\rm I}(t)\rangle = e^{itH_0}\, H_{\rm int}\, e^{-itH_0}\, |\psi_{\rm I}(t)\rangle.$$

Hence, defining the expression of the Hamiltonian in the interaction representation as

$$H_{\rm int\,I}(t) = e^{itH_0}\, H_{\rm int}\, e^{-itH_0}, \tag{6.41}$$

the state $|\psi_{\rm I}(t)\rangle$ satisfies the equation

$$i\frac{\mathrm{d}}{\mathrm{d}t}\, |\psi_{\rm I}(t)\rangle = H_{\rm int\,I}(t)\, |\psi_{\rm I}(t)\rangle. \tag{6.42}$$

As for $H_{\rm int}$ in Eq. (6.41), any operator in the interaction picture can be expressed from the corresponding operator in the Schrödinger picture as

$$A_{\rm I}(t) = e^{itH_0} A\, e^{-itH_0}, \tag{6.43}$$

or from the corresponding operator in the Heisenberg picture with the help of Eq. (6.36),

$$A_{\rm I}(t) = e^{itH_0} e^{-itH} A_{\rm H}\, e^{itH} e^{-itH_0}. \tag{6.44}$$

Then, differentiating Eq. (6.43), the time evolution of the operator satisfies the Heisenberg equation

$$i\frac{\mathrm{d}}{\mathrm{d}t}A_{\mathrm{I}}(t) = [A_{\mathrm{I}}(t), H_0], \tag{6.45}$$

with H being replaced by H_0. Only the knowledge of H_0 is needed to predict the time evolution of an operator. Field operators can then evolve in time as free fields, even in the presence of interactions. This constitutes the main advantage of this representation.

What about the impact on the canonical quantisation of fields? The attentive reader will have noticed that when we specified the fundamental commutators (6.9) and (6.12), we implicitly worked in the Heisenberg representation since the fields depend on time (we even have to specify the commutator at equal times). The Formula (6.44) tells us that $A_{\mathrm{I}} = UA_{\mathrm{H}}U^{-1}$, where it is easy to show that

$$U = e^{itH_0}e^{-itH}$$

is a unitary operator (the Hamiltonians being Hermitian). Denoting generically the field and its canonical conjugate respectively by $\phi_{\mathrm{H}}(x)$ and $\pi_{\mathrm{H}}(x)$ in the Heisenberg representation, and by $\phi_{\mathrm{I}}(x)$ and $\pi_{\mathrm{I}}(x)$ in the interaction representation, it follows that

$$[\phi_{\mathrm{I}}(x), \pi_{\mathrm{I}}(y)]_{x^0=y^0} = U\,[\phi_{\mathrm{H}}(x), \pi_{\mathrm{H}}(y)]_{x^0=y^0}\,U^{-1}.$$

Since $[\phi_{\mathrm{H}}(x), \pi_{\mathrm{H}}(y)]_{x^0=y^0}$ is typically of the form $i\delta^{(3)}(\boldsymbol{x}-\boldsymbol{y})$ (with possibly additional Kronecker symbols for the other degrees of freedom), it commutes with U, and hence,

$$[\phi_{\mathrm{I}}(x), \pi_{\mathrm{I}}(y)]_{x^0=y^0} = [\phi_{\mathrm{H}}(x), \pi_{\mathrm{H}}(y)]_{x^0=y^0}. \tag{6.46}$$

This shows that even when there is an interaction, the interaction picture fields obey the canonical commutation relation.

6.2.2 Dyson Expansion

Equation (6.45) gives us the time evolution of the operators in the interaction representation. Because the states have time dependence in this representation, however, we need to find the evolution operator $U(t, t_0)$ that allows calculating the state at any time t, given the state at a time t_0, i.e.

$$|\psi_{\mathrm{I}}(t)\rangle = U(t, t_0)\,|\psi_{\mathrm{I}}(t_0)\rangle. \tag{6.47}$$

Differentiating this equation and using Eq. (6.42), it follows that

$$i\frac{\mathrm{d}}{\mathrm{d}t}U(t, t_0)\,|\psi_{\mathrm{I}}(t_0)\rangle = H_{\mathrm{int\,I}}(t)\;|\psi_{\mathrm{I}}(t)\rangle = H_{\mathrm{int\,I}}(t)\,U(t, t_0)\,|\psi_{\mathrm{I}}(t_0)\rangle.$$

Since this equation holds for any $|\psi_{\mathrm{I}}(t_0)\rangle$, we conclude that

$$i\frac{\mathrm{d}}{\mathrm{d}t}U(t, t_0) = H_{\mathrm{int}}(t)\,U(t, t_0), \tag{6.48}$$

where the subscript I, referring to the interaction representation, was dropped (from now on, we will always work in this representation). Even if the interaction Hamiltonian did not depend on time in the Schrödinger representation, we have to take into account its time dependence in the interaction representation [cf. Eq. (6.41)]. If $H_{\rm int}(t)$ were just a function, the solution of Eq. (6.48) would simply be

$$\exp\left(-i\int_{t_0}^t dt' H_{\rm int}(t')\right) \equiv 1 - i\int_{t_0}^t dt' H_{\rm int}(t') + \frac{(-i)^2}{2!}\left(\int_{t_0}^t dt' H_{\rm int}(t')\right)^2 + \cdots .$$

However, $H_{\rm int}(t)$ is an operator, and if we differentiate the exponential above and multiply by i, we actually obtain

$$i\frac{d}{dt}\exp\left(-i\int_{t_0}^t dt' H_{\rm int}(t')\right) = H_{\rm int}(t) - \frac{i}{2}\left\{\left(\int_{t_0}^t dt' H_{\rm int}(t')\right)H_{\rm int}(t) + H_{\rm int}(t)\left(\int_{t_0}^t dt' H_{\rm int}(t')\right)\right\} + \cdots .$$

Now, in the expression above, $H_{\rm int}(t')$ and $H_{\rm int}(t)$ do not necessarily commute. Therefore, the right-hand side of the equality is not $H_{\rm int}(t)\exp\left(-i\int_{t_0}^t dt' H_{\rm int}(t')\right)$, and thus, Eq. (6.48) is finally not satisfied. We have to use another approach and treat more carefully the order of operators. This approach is based on an iterative procedure. The first order of the solution of Eq. (6.48) is

$$U(t,t_0) = U(t_0,t_0) - i\int_{t_0}^t dt_1 H_{\rm int}(t_1)\, U(t_1,t_0).$$

But $U(t_0,t_0) = 1$ by definition [see Eq. (6.47)], so that expressing $U(t_1,t_0)$ with the equality above yields the following second order:

$$\begin{aligned} U(t,t_0) &= 1 - i\int_{t_0}^t dt_1 H_{\rm int}(t_1)\left[1 - i\int_{t_0}^{t_1} dt_2\, H_{\rm int}(t_2)\, U(t_2,t_0)\right] \\ &= 1 - i\int_{t_0}^t dt_1 H_{\rm int}(t_1) + (-i)^2\int_{t_0}^t dt_1\int_{t_0}^{t_1} dt_2\, H_{\rm int}(t_1)\, H_{\rm int}(t_2)\, U(t_2,t_0). \end{aligned}$$

Pursuing the procedure at higher orders, it follows that

$$U(t,t_0) = \sum_{n=0}^{\infty}(-i)^n\int_{t_0}^t dt_1\int_{t_0}^{t_1} dt_2 \cdots \int_{t_0}^{t_{n-1}} dt_n\, H_{\rm int}(t_1)\, H_{\rm int}(t_2)\cdots H_{\rm int}(t_n).$$

We can rearrange the integrals to have the same range of integration $[t_0, t]$ thanks to the insertion of the step function $\theta(x)$, where $\theta(x) = 1$ for positive x and zero otherwise. Let us consider, for example, the integrals of the second-order term. It gives

$$I_2 = \int_{t_0}^t dt_1\int_{t_0}^{t_1} dt_2\, H_{\rm int}(t_1)\, H_{\rm int}(t_2) = \int_{t_0}^t dt_1\int_{t_0}^{t} dt_2\, \theta(t_1-t_2)\, H_{\rm int}(t_1)\, H_{\rm int}(t_2).$$

The presence of $\theta(t_1 - t_2)$ ensures the implicit condition $t_1 > t_2$. Now, I_2 can be split into two identical terms, where for the second, we change the order of the integrals, yielding

$$I_2 = \frac{1}{2}\int_{t_0}^t dt_1\int_{t_0}^{t} dt_2\, \theta(t_1-t_2)\, H_{\rm int}(t_1)\, H_{\rm int}(t_2) + \frac{1}{2}\int_{t_0}^t dt_2\int_{t_0}^{t} dt_1\, \theta(t_1-t_2)\, H_{\rm int}(t_1)\, H_{\rm int}(t_2).$$

Renaming the integration variables $t_1 \leftrightarrow t_2$ in the second term, it follows that

$$I_2 = \frac{1}{2}\int_{t_0}^t dt_1\int_{t_0}^{t} dt_2\, \{\theta(t_1-t_2)\, H_{\rm int}(t_1)\, H_{\rm int}(t_2) + \theta(t_2-t_1)\, H_{\rm int}(t_2)\, H_{\rm int}(t_1)\}. \quad (6.49)$$

The trick now is to define the T-product or *time-ordered product* of operators, $T\{\}$, as

$$\begin{aligned} T\{H_{\rm int}(t_1)\, H_{\rm int}(t_2)\} &= H_{\rm int}(t_1)\, H_{\rm int}(t_2)\theta(t_1-t_2) + H_{\rm int}(t_2)\, H_{\rm int}(t_1)\theta(t_2-t_1) \\ &= \begin{cases} H_{\rm int}(t_1)\, H_{\rm int}(t_2) & \text{if } t_1 > t_2, \\ H_{\rm int}(t_2)\, H_{\rm int}(t_1) & \text{if } t_2 > t_1. \end{cases} \end{aligned} \quad (6.50)$$

The T-product of n operators A simply generalises to

$$T\{A(t_1)A(t_2)\cdots A(t_n)\} = A(t_{i_1})A(t_{i_2})\cdots A(t_{i_n}), \text{ with } t_{i_1} > t_{i_2} > \cdots > t_{i_n}. \tag{6.51}$$

Hence, Eq. (6.49) becomes

$$\int_{t_0}^{t} \mathrm{d}t_1 \int_{t_0}^{t_1} \mathrm{d}t_2\, H_{\mathrm{int}}(t_1)\, H_{\mathrm{int}}(t_2) = \frac{1}{2}\int_{t_0}^{t} \mathrm{d}t_1 \int_{t_0}^{t} \mathrm{d}t_2\, T\{H_{\mathrm{int}}(t_1)\, H_{\mathrm{int}}(t_2)\}.$$

With the higher order terms, one has to take into account the number of permutations in the T-product, and the final formula is

$$U(t,t_0) = \sum_{n=0}^{\infty} \frac{(-i)^n}{n!} \int_{t_0}^{t} \mathrm{d}t_1 \int_{t_0}^{t} \mathrm{d}t_2 \cdots \int_{t_0}^{t} T\{H_{\mathrm{int}}(t_1)\, H_{\mathrm{int}}(t_2)\cdots\, H_{\mathrm{int}}(t_n)\}, \tag{6.52}$$

which can be simply written in the symbolic form

$$U(t,t_0) = T\left\{\exp\left(-i\int_{t_0}^{t} \mathrm{d}\tau\, H_{\mathrm{int}}(\tau)\right)\right\}. \tag{6.53}$$

Equipped with the general Formula (6.53), we can go a little further by assuming that long before the interaction, say at $t \to -\infty$, H_{int} is negligible. Hence, according to Eq. (6.42), the state $|\psi_{\mathrm{I}}\rangle$ is a steady state in the interaction representation (since $H_{\mathrm{int}} = 0$). The fact that when there is an interaction the kets evolve only as a function of time is one of the most useful features of this representation. Let us denote this initial state by $|i\rangle = |\psi_{\mathrm{I}}(-\infty)\rangle$. Experimentally, $|i\rangle$ is typically the state of two particles that collide. As time goes by, the particles described by $|i\rangle$ may scatter, and H_{int} is no longer negligible. The time evolution of $|\psi_{\mathrm{I}}\rangle$ is then given by

$$|\psi_{\mathrm{I}}(t)\rangle = U(t,-\infty)\,|i\rangle.$$

Long after the interaction, for the same reason, the system potentially constituted of many particles resulting from the collision, is in another steady-state $|\psi_{\mathrm{I}}(+\infty)\rangle$. The probability amplitude for finding a given final state $|f\rangle$ is then

$$S_{fi} = \langle f|\psi_{\mathrm{I}}(+\infty)\rangle = \langle f|U(+\infty,-\infty)|i\rangle.$$

The operator

$$S = U(+\infty,-\infty) = T\left\{\exp\left(-i\int_{-\infty}^{+\infty} \mathrm{d}\tau\, H_{\mathrm{int}}(\tau)\right)\right\} \tag{6.54}$$

is called the *scattering matrix* (S-matrix). Its element, $S_{fi} = \langle f|S|i\rangle$, gives the probability amplitude for having a transition from the state $|i\rangle$ to $|f\rangle$. According to Eq. (6.52), the S-matrix expression is thus

$$S = \sum_{n=0}^{\infty} S^{[n]}, \tag{6.55}$$

where the nth-order term is given by

$$S^{[n]} = \frac{(-i)^n}{n!} \int_{-\infty}^{+\infty} \cdots \int_{-\infty}^{+\infty} \mathrm{d}t_1 \cdots \mathrm{d}t_n\, T\{H_{\mathrm{int}}(t_1)\cdots\, H_{\mathrm{int}}(t_n)\}. \tag{6.56}$$

6.2.3 Connection with the Transition Rate

Let us consider an initial state $|i\rangle$ and a final state $|f\rangle$ made of several particles with definite momenta. They are the states corresponding to the asymptotic conditions at $t = -\infty$ and $t = +\infty$. The probability of a transition from $|i\rangle$ to $|f\rangle$ is then given by

$$P_{i\to f} = \frac{|\langle f|S|i\rangle|^2}{\langle i|i\rangle\,\langle f|f\rangle}. \tag{6.57}$$

The division by $\langle i|i\rangle\,\langle f|f\rangle$ is there to take into account the relativistic normalisation (the famous $2E$ particles per unit volume). Even if the theory contains the possibility of interactions, the particles have some probability of simply missing each other, leaving the initial state unchanged. Therefore, we expect S to be written as $S = \mathbb{1} + \cdots$, where the dots are an expression of the operator for a non-trivial interaction. We do not need to have an explicit formulation of S, which could be complicated since it must contain information about all possible transitions. Only the matrix element $\langle f|S|i\rangle$ for the initial and final states envisaged is required. How can we write it? It is convenient to factor out a delta function implementing 4-momentum conservation and define the Lorentz invariant scattering amplitude, $\mathcal{M}_{fi}$, such as

$$\langle f|S|i\rangle = \langle f|i\rangle + (2\pi)^4\delta^{(4)}(P_f - P_i)\, i\mathcal{M}_{fi},$$

where P_f and P_i are respectively the total 4-momentum of the final state and the initial state. The factor $(2\pi)^4$ is conveniently factorised so that $i\mathcal{M}_{fi}$ will be directly the result of a Feynman diagram calculation. Since we are interested in non-trivial interactions, where $|i\rangle$ and $|f\rangle$ differ, the matrix element is simply

$$\langle f|S|i\rangle = (2\pi)^4\delta^{(4)}(P_f - P_i)\, i\mathcal{M}_{fi},\quad i \neq f. \tag{6.58}$$

Given that $\langle f|S|i\rangle$ is defined at an order n by Eq. (6.56), it follows that

$$(2\pi)^4\delta^{(4)}(P_f - P_i)\, i\mathcal{M}_{fi}^{[n]} = \frac{(-i)^n}{n!}\int_{-\infty}^{+\infty} \mathrm{d}t_1 \cdots \mathrm{d}t_n\ \langle f|\, T\{H_{\mathrm{int}}(t_1)\cdots H_{\mathrm{int}}(t_n)\}\,|i\rangle.$$

In order to make explicit the Lorentz invariance, it is better to use the Hamiltonian density $\mathcal{H}$ (with $H = \int \mathrm{d}^3x\mathcal{H}(x)$). The final formula for the matrix elements then becomes

$$\langle f|S^{[n]}|i\rangle = (2\pi)^4\delta^{(4)}(P_f - P_i)\, i\mathcal{M}_{fi}^{[n]} = \frac{(-i)^n}{n!}\int \mathrm{d}^4x_1 \cdots \mathrm{d}^4x_n \langle f|T\{\mathcal{H}_{\mathrm{int}}(x_1)\cdots\mathcal{H}_{\mathrm{int}}(x_n)\}|i\rangle, \tag{6.59}$$

where, once again, $\mathcal{H}_{\mathrm{int}}$ is the interaction Hamiltonian given in the interaction representation (meaning that the product of fields constituting the Hamiltonian are themselves given in the interaction representation). Because of the time-ordered product in Eq. (6.59), it is not so obvious that $\mathcal{M}_{fi}^{[n]}$ is Lorentz invariant. However, notice that if two spacetime points x_1 and x_2 are such that $x_1 - x_2$ is time-like (i.e., $(x_1 - x_2)^2 > 0$), the time-ordering is Lorentz invariant, since there exists no frame inverting the time order of events. So, the question arises only for space-like intervals, where events are not causally related. But we saw in Section 6.1.2 that the fields must commute in that case [cf. Eqs. (6.9) and (6.12)]. Hence, the interaction Hamiltonians in Eq. (6.59) commute, and thus, the T-product has no incidence.

Now, let us make explicit the terms of Eq. (6.57). According to the relativistic normalisation (see Supplement 5.1 obtained for bosons, but fermions would lead to the same results), if the initial state contains n particles and the final state n' particles, the Formula (5.19) generalises to

$$\langle i|i\rangle = \prod_{k=1}^{n} 2E_k \mathcal{V}, \quad \langle f|f\rangle = \prod_{k=1}^{n'} 2E'_k \mathcal{V},$$

where the volume $\mathcal{V}$ is arbitrary. We can then do all our calculations with a finite volume $\mathcal{V}$, and for any physics calculations, the volume should cancel out so that the infinite volume limit can be taken safely. The last term in Eq. (6.57) to be determined is $|\langle f|S|i\rangle|^2$. According to Eq. (6.58), it reads

$$|\langle f|S|i\rangle|^2 = (2\pi)^4\delta^{(4)}(P_F - P_I)(2\pi)^4\delta^{(4)}(P_F - P_I)|\mathcal{M}_{fi}|^2.$$

Here, we have to evaluate the square of the delta function $\delta^{(4)}(P_F - P_I)$. The first delta-function, being applied twice, imposes $P_F = P_I$ into the second, and hence, the quantity $\delta^{(4)}(0) = \delta^{(3)}(\mathbf{0})\delta(0)$ has to be interpreted, since it is mathematically ill defined. In Supplement 5.1, we saw that $\delta^{(3)}(\mathbf{0})$ can be approximated by $\mathcal{V}/(2\pi)^3$. Following the same approach,

$$\delta(E) = \frac{1}{(2\pi)}\int_{-\infty}^{+\infty} e^{iEt}\,\mathrm{d}t = \frac{1}{(2\pi)}\lim_{T\to\infty}\int_{-T/2}^{T/2}\mathrm{d}t\, e^{iEt} = \frac{1}{(2\pi)}\lim_{T\to\infty} T\frac{\sin(ET/2)}{ET/2},$$

and hence we conclude that $\delta(0) = T/(2\pi)$, where T is the time during which there is an interaction. It follows that

$$|\langle f|S|i\rangle|^2 = (2\pi)^4\delta^{(4)}(P_F - P_I)\mathcal{V}T|\mathcal{M}_{fi}|^2.$$

Therefore, the probability (6.57) finally reads

$$P_{i\to f} = \mathcal{V}^{1-n-n'}T\,(2\pi)^4\delta^{(4)}(P_F - P_I)|\mathcal{M}_{fi}|^2\prod_{k=1}^{n}\frac{1}{2E_k}\prod_{k=1}^{n'}\frac{1}{2E'_k}.$$

The transition rate, i.e. the probability per unit of time, is thus given by

$$\Gamma_{i\to f} = \mathcal{V}^{1-n-n'}\,(2\pi)^4\delta^{(4)}(P_F - P_I)|\mathcal{M}_{fi}|^2\prod_{k=1}^{n}\frac{1}{2E_k}\prod_{k=1}^{n'}\frac{1}{2E'_k}.$$

We are almost there. We forgot that the particles do not have an infinitely accurate momentum or energy but should rather be described by a set of final states belonging to a given phase-space element (i.e., belonging to $p_x \pm \delta p_x$, $p_y \pm \delta p_y$ and $p_z \pm \delta p_z$). Let us consider an infinitesimal phase-space element such that we can assume that $|\mathcal{M}_{fi}|$ remains constant for those $\mathrm{d}N$ states. The infinitesimal transition rate for particles in this infinitesimal phase space is then simply

$$\mathrm{d}\Gamma_{i\to f} = \mathcal{V}^{1-n-n'}\,(2\pi)^4\delta^{(4)}(P_F - P_I)|\mathcal{M}_{fi}|^2\left(\prod_{k=1}^{n}\frac{1}{2E_k}\prod_{k=1}^{n'}\frac{1}{2E'_k}\right)\mathrm{d}N.$$

As is known in quantum mechanics, the components of the momentum of a particle confined in a box of side L are quantised, $p_{x,y,z} = (2\pi/L)\, n_{x,y,z}$, where n_x, n_y and n_z are integers. Therefore, the number of states whose momentum lies in the momentum space volume $\mathrm{d}^3\boldsymbol{p} = \mathrm{d}p_x\, \mathrm{d}p_y\, \mathrm{d}p_z$ around $\boldsymbol{p}$ is $\mathrm{d}N = \mathrm{d}n_x\, \mathrm{d}n_y\, \mathrm{d}n_z = \mathcal{V}\, \mathrm{d}^3\boldsymbol{p}/(2\pi)^3$. Hence, for n' particles in the final state, $\mathrm{d}N = \prod_{k=1}^{n'} \mathcal{V}\, \mathrm{d}^3\boldsymbol{p}'_k/(2\pi)^3$, yielding the final formula of the infinitesimal transition rate,

$$\mathrm{d}\Gamma_{i\to f} = \mathcal{V}^{1-n}\,(2\pi)^4\delta^{(4)}(P_F - P_I)|\mathcal{M}_{fi}|^2 \prod_{k=1}^{n} \frac{1}{2E_k} \prod_{k=1}^{n'} \frac{\mathrm{d}^3\boldsymbol{p}'_k}{(2\pi)^3 2E'_k}. \tag{6.60}$$

This formula is precisely the one used in Chapter 3, Eq. (3.2), to establish the formulas for the decay rate and cross section, where, as it should, the arbitrary volume $\mathcal{V}$ cancels out.[4]

6.3 Feynman Rules and Diagrams

The $\mathcal{S}$-matrix element at the nth-order (6.56) depends only on the Hamiltonian density of the interaction, $\mathcal{H}_{\text{int}}$ in Eq. (6.33), where all operators are given in the interaction representation. Let us exploit Formulas (6.56) and (6.59) to infer how to calculate the amplitudes of processes.

6.3.1 Electron-Photon Vertex

Consider the first order in Eq. (6.59), where by convention, q, is the charge of the particle, not that of the antiparticle,[5]

$$S^{[1]} = -i\int \mathrm{d}^4x\,\mathcal{H}_{\text{int}}(x) = -iq\int \mathrm{d}^4x\,\bar{\psi}(x)\gamma^\mu\psi(x)A_\mu(x).$$

Now, inserting the field expansion of ψ (5.100), $\bar{\psi}$ (5.101) and A_μ (5.132), it yields

$$S^{[1]} = -iq\int \mathrm{d}^4x \int \frac{\mathrm{d}^3\boldsymbol{p}}{(2\pi)^3 2E_p}\frac{\mathrm{d}^3\boldsymbol{p}'}{(2\pi)^3 2E_{p'}}\frac{\mathrm{d}^3\boldsymbol{q}}{(2\pi)^3 2E_q}\sum_{r=1,2}\sum_{r'=1,2}\sum_{\lambda=0}^{3}$$

$$\left[c^\dagger_{pr}\bar{u}_r(p)e^{+ip\cdot x} + d_{pr}\bar{v}_r(p)e^{-ip\cdot x}\right]\gamma^\mu\left[c_{p'r'}u_{r'}(p')e^{-ip'\cdot x} + d^\dagger_{p'r'}v_{r'}(p')e^{+ip'\cdot x}\right]$$

$$\times\left[\epsilon_\mu(q,\lambda)\alpha_{q\lambda}e^{-iq\cdot x} + \epsilon_\mu(q,\lambda)^*\alpha^\dagger_{q\lambda}e^{+iq\cdot x}\right].$$

[4] The arbitrary volume is an artefact arising from the use of plane waves with a definite momentum. Using instead wave packets to represent particles avoids this issue. However, the mathematical treatment is more complicated. See, for example, Weinberg (1995).

[5] For both electrons and positrons, $q = -e < 0$ is used. The distinction between particles and antiparticles is automatically taken into account by the quantised field expansion. See, for example, the charge operator in Eq. (5.103).

Integrating over x [using (E.9)], we obtain

$$S^{[1]} = (2\pi)^4 \int \frac{d^3p}{(2\pi)^3 2E_p} \frac{d^3p'}{(2\pi)^3 2E_{p'}} \frac{d^3q}{(2\pi)^3 2E_q} \sum_{r=1,2} \sum_{r'=1,2} \sum_{\lambda=0}^{3}$$

$$\begin{aligned}
&\Big\{ c^\dagger_{pr} c_{p'r'} \alpha_{q\lambda}\, (\bar{u}_r(p)\,(-iq\gamma^\mu)\, u_{r'}(p'))\epsilon_\mu(q,\lambda)\, \delta^{(4)}(p - p' - q) \\
&+ c^\dagger_{pr} d^\dagger_{p'r'} \alpha_{q\lambda}\, (\bar{u}_r(p)\,(-iq\gamma^\mu)\, v_{r'}(p'))\epsilon_\mu(q,\lambda)\, \delta^{(4)}(p + p' - q) \\
&+ d_{pr} c_{p'r'} \alpha_{q\lambda}\, (\bar{v}_r(p)\,(-iq\gamma^\mu)\, u_{r'}(p'))\epsilon_\mu(q,\lambda)\, \delta^{(4)}(p + p' + q) \\
&+ d_{pr} d^\dagger_{p'r'} \alpha_{q\lambda}\, (\bar{v}_r(p)\,(-iq\gamma^\mu)\, v_{r'}(p'))\epsilon_\mu(q,\lambda)\, \delta^{(4)}(p' - p - q) \\
&+ c^\dagger_{pr} c_{p'r'} \alpha^\dagger_{q\lambda}\, (\bar{u}_r(p)\,(-iq\gamma^\mu)\, u_{r'}(p'))\epsilon^*_\mu(q,\lambda)\, \delta^{(4)}(p - p' + q) \\
&+ c^\dagger_{pr} d^\dagger_{p'r'} \alpha^\dagger_{q\lambda}\, (\bar{u}_r(p)\,(-iq\gamma^\mu)\, v_{r'}(p'))\epsilon^*_\mu(q,\lambda)\, \delta^{(4)}(p + p' + q) \\
&+ d_{pr} c_{p'r'} \alpha^\dagger_{q\lambda}\, (\bar{v}_r(p)\,(-iq\gamma^\mu)\, u_{r'}(p'))\epsilon^*_\mu(q,\lambda)\, \delta^{(4)}(q - p - p') \\
&+ d_{pr} d^\dagger_{p'r'} \alpha^\dagger_{q\lambda}\, (\bar{v}_r(p)\,(-iq\gamma^\mu)\, v_{r'}(p'))\epsilon^*_\mu(q,\lambda)\, \delta^{(4)}(p' - p + q) \Big\}.
\end{aligned} \tag{6.61}$$

Thus, eight terms constitute $S^{[1]}$ (three fields with two operators give $2^3 = 8$ combinations), with each having a similar form: a product of three operators, a charge current (an adjoint spinor and a spinor sandwiching a γ^μ matrix), a photon polarisation vector and a delta function. Looking at the operators, we notice that the electric charge remains unchanged. For example, the first line with the product of operators $\hat{A} = c^\dagger_{pr} c_{p'r'} \alpha_{q\lambda}$ creates an electron, and annihilates an electron and a photon. It gives a non-zero contribution only if it is sandwiched between an initial state containing a photon and an electron, $|e_i, \gamma\rangle = c^\dagger_{p_i r_i} \alpha^\dagger_{q_\gamma \lambda_\gamma} |0\rangle$, that can be annihilated by $\hat{A}$, and a final state containing an electron, $|e_f\rangle = c^\dagger_{p_f r_f} |0\rangle$, that can be created by $\hat{A}$. Using the commutation rules (5.102) and (5.133) yields

$$\begin{aligned}
\langle e_f | c^\dagger_{pr} c_{p'r'} \alpha_{q\lambda} | e_i, \gamma\rangle &= \langle 0 | c_{p_f r_f} c^\dagger_{pr} c_{p'r'} c^\dagger_{p_i r_i} \alpha_{q\lambda} \alpha^\dagger_{q_\gamma \lambda_\gamma} |0\rangle \\
&= \langle 0 | \{c_{p_f r_f}, c^\dagger_{pr}\}\{c_{p'r'}, c^\dagger_{p_i r_i}\}[\alpha_{q\lambda} \alpha^\dagger_{q_\gamma \lambda_\gamma}] |0\rangle \\
&= -(2\pi)^3 2E_f \delta^{(3)}(\boldsymbol{p_f} - \boldsymbol{p})\delta_{r_f r}(2\pi)^3 2E_i \delta^{(3)}(\boldsymbol{p'_i} - \boldsymbol{p'})\delta_{r_i r'} \\
&\quad \times g_{\lambda\lambda_\gamma}(2\pi)^3 2E_\gamma \delta^{(3)}(\boldsymbol{q} - \boldsymbol{q_\gamma}).
\end{aligned} \tag{6.62}$$

Inserting this result into the $S^{[1]}$ expression, and performing the integrations over the momenta and summations over the spins and polarisations, we obtain a term[6]

$$(2\pi)^4 \delta^{(4)}(p_f - p_i - q_\gamma) \left(\bar{u}_{r_f}(p_f)\,(-iq\gamma^\mu)\, u_{r_i}(p_i)\right) \epsilon_\mu(q_\gamma, \lambda_\gamma). \tag{6.63}$$

A comparison with Eq. (6.59) leads us to conclude that the corresponding amplitude is

$$i\mathcal{M} = \left(\bar{u}_{r_f}(p_f)\,(-iq\gamma^\mu)\, u_{r_i}(p_i)\right) \epsilon_\mu(q_\gamma, \lambda_\gamma). \tag{6.64}$$

The creation of the final electron is thus associated with an adjoint spinor $\bar{u}$, while the annihilation of the initial electron is associated with a spinor u. Similarly, the annihilation of the photon is associated with a polarisation vector. It yields our first Feynman diagram shown in Fig. 6.1. The arrow of time on the horizontal axis goes from the left to the right,

[6] Real photons have only transverse polarisation, i.e. $\lambda_\gamma = 1$ or 2. Thus, $g_{\lambda_\gamma \lambda_\gamma} = -1$.

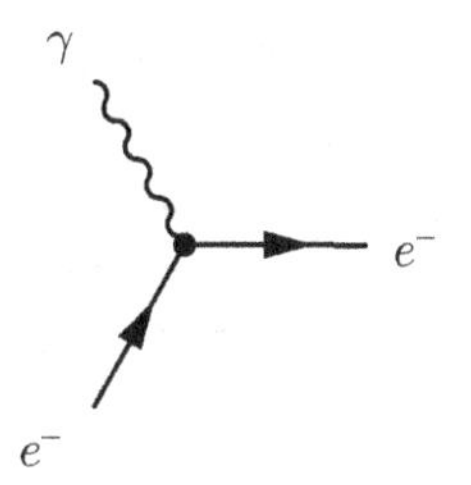

Fig. 6.1 A basic Feynman diagram in QED with one vertex.

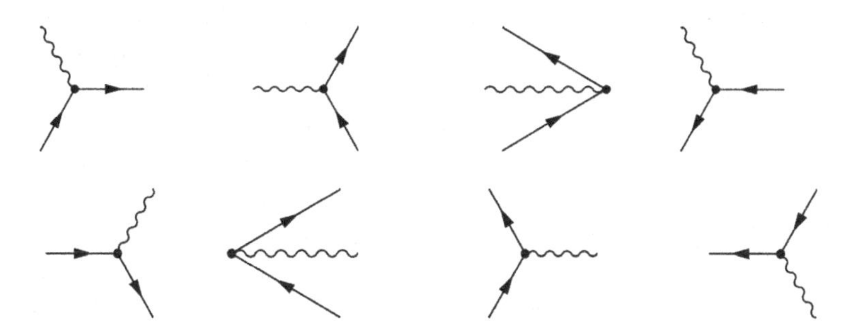

Fig. 6.2 The vertices of QED at the first order of perturbation.

while the spatial dimension is vertical. The photon is symbolised by a wavy line and the electron (actually any fermion) by a solid line. The intersection of the lines is called a vertex. According to Eq. (6.64), it is associated with the factor $-iq\gamma^{\mu}$. The direction of the arrow on the electron line has a meaning: for electrons in the initial state, it goes towards the vertex, while for electrons in the final state, it points away from the vertex. In other words, the arrow of a fermion is always oriented from the past to the future.

The expansion we established to the first order of perturbation theory leads to a diagram with only one vertex. The number of vertices represents the order of the perturbation expansion. The seven other lines of $S^{[1]}$ in Eq. (6.61) yield similar kinds of diagrams. They are all shown in Fig. 6.2. Positrons (or any antifermions) are represented by the same plain line as electrons. What differs is only the direction of the arrow. For example, the second diagram on the first row depicts a photon that annihilates into an electron (the line going up) and a positron. Note that for the outgoing positron, the arrow 'goes backwards in time'. Actually, the direction of the arrow always corresponds to the direction of the charge of the particle. For an antiparticle, since its charge is opposite to that of the particle, and since by convention only the charge of particles is used, the arrow for an antiparticle must go backwards in time. In the fourth diagram of the first row, the incoming particle is a positron, with the arrow going backwards in time. There is no arrow on the photon wavy line since it is its own antiparticle. According to Eq. (6.61) and following the same procedure that yields the amplitude (6.64), we see that anti-fermions in the initial state are associated with the adjoint antiparticle spinors $\bar{v}$, whereas anti-fermions in the final state are associated with antiparticle spinors v. For photons, a polarisation vector ϵ_{μ} is associated with an incoming photon and ϵ^*_{μ} for an outgoing one.

In the Formula (6.61) and thus in the amplitude (6.64), the factor $-iq\gamma^\mu$ is associated with the contribution of the vertex in the Feynman diagram. The charge $q = -e$ is the coupling between electrons and photons. The *same* value $q = -e$ must be used to describe the coupling between positrons and photons (to convince yourself, look at Formula (6.61), where q is the *same* at each line). Expansion at nth order would produce n vertices, leading to a factor e^n in the amplitude. The size of the expansion terms depends on $e \simeq \sqrt{4\pi/137} \simeq 0.3$ (in natural units), which finally represents the 'strength' of the perturbation.

Finally, there is the delta function that we have not yet exploited in Eq. (6.63). It imposes energy–momentum conservation at the vertex. Well, it appears that none of these eight terms is physically possible with free particles! An obvious example is the third diagram of the first row of Fig. 6.2. The energy conservation imposed by $\delta^{(4)}(p + p' + q)$ means that $E_{e^-} + E_\gamma + E_{e^+} = 0$, i.e. $E_{e^-} = E_\gamma = E_{e^+} = 0$, which is impossible ($E_{e^-} > 0$ since electrons are massive). Even more conventional diagrams like the first one are impossible. Both energy conservation and momentum conservation cannot be simultaneously satisfied (because the photon is massless). The only possibility would be to envisage that one of these particles does not have its normal mass value or, more specifically, that it does not respect the relativistic relation $E^2 = |\boldsymbol{p}|^2 + m^2$. Such a particle is called a virtual particle and is allowed by the uncertainty inherent in quantum mechanics, $\Delta E \geq \hbar/(2\Delta t)$. The relation can be interpreted as follows: the greater the energy (or mass) is shifted from its nominal value, the shorter the particle will live. In other words, virtual particles (thus not satisfying $E^2 = |\boldsymbol{p}|^2 + m^2$, with m being the usual mass) must live a short time: they have to be quickly reabsorbed (annihilated) by another process.[7] Consequently, another vertex is needed, and one has to go to the second order of the perturbation expansion. By definition, a virtual particle thus corresponds to an internal line of a Feynman diagram.[8]

6.3.2 Photon and Electron Propagators

The second order of the perturbation expansion of Eq. (6.59) is given by

$$S^{[2]} = \frac{(-i)^2}{2} q^2 \int \mathrm{d}^4x\, \mathrm{d}^4y\, T\{\bar{\psi}(x)\gamma^\mu \psi(x) A_\mu(x) \bar{\psi}(y)\gamma^\nu \psi(y) A_\nu(y)\}. \tag{6.65}$$

If we develop all quantities as in the previous section, we have $2^6 = 64$ terms and thus 64 diagrams. Two such diagrams are shown in Fig. 6.3. In the first diagram, a (virtual) photon is created at the spacetime point x (assuming $t_x < t_y$) and annihilated at y. The initial and final states are composed of an electron–positron pair. Similarly, the diagram on the right-hand side shows a virtual electron exchanged between the initial and final states made of an electron and a photon. The first order of the expansion (6.62) of the previous section shows that the only contributing terms are necessarily a function

[7] In fact, all particles can be considered as virtual: for instance, a photon emitted by a star is detected on earth by a photomultiplier that annihilates the photon. Its 'degree of virtuality' is, however, ridiculously tiny if the particle lives during a million of years!

[8] Following the previous footnote, any particle can be ultimately considered as an internal line of a big Feynman diagram. For instance, an electron produced by an accelerator will be absorbed by an atom somewhere and hence becomes an internal leg of a diagram.

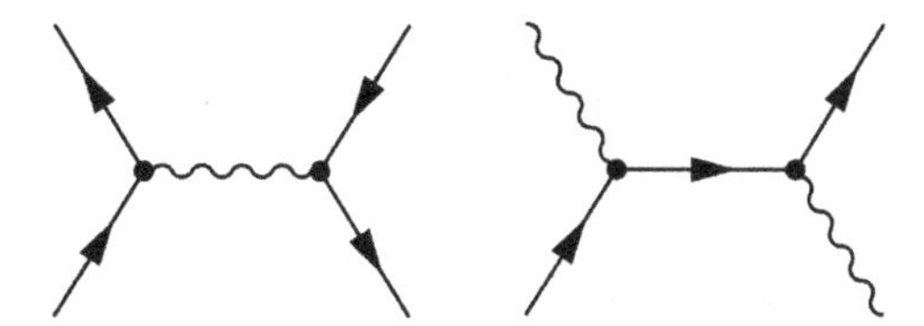

Fig. 6.3 Example of QED diagrams at the second order.

of $\langle 0|\left[\alpha,\alpha^\dagger\right]|0\rangle = \langle 0|\alpha\alpha^\dagger|0\rangle$, $\langle 0|\left\{c,c^\dagger\right\}|0\rangle = \langle 0|cc^\dagger|0\rangle$ or $\langle 0|\left\{d,d^\dagger\right\}|0\rangle = \langle 0|dd^\dagger|0\rangle$. The photon contribution, $\langle 0|\alpha\alpha^\dagger|0\rangle$, can only emerge from the contraction of the photon fields:

$$iD_{F\mu\nu}(x-y) \equiv \langle 0|T\{A_\mu(x)A_\nu(y)\}|0\rangle = \begin{cases} \langle 0|A_\mu(x)A_\nu(y)|0\rangle & \text{if } x^0 > y^0, \\ \langle 0|A_\nu(y)A_\mu(x)|0\rangle & \text{if } y^0 > x^0. \end{cases} \tag{6.66}$$

The insertion of the field expansion (5.132) easily confirms this. Similarly, for the fermions, $\langle 0|\left\{c,c^\dagger\right\}|0\rangle = \langle 0|cc^\dagger|0\rangle$ or $\langle 0|\left\{d,d^\dagger\right\}|0\rangle = \langle 0|dd^\dagger|0\rangle$ is coming from the contraction of the fermion fields:

$$i\Delta_F(x-y) \equiv \langle 0|T\{\psi(x)\bar{\psi}(y)\}|0\rangle = \begin{cases} \langle 0|\psi(x)\bar{\psi}(y)|0\rangle & \text{if } x^0 > y^0, \\ -\langle 0|\bar{\psi}(y)\psi(x)|0\rangle & \text{if } y^0 > x^0. \end{cases} \tag{6.67}$$

The minus sign when $y^0 > x^0$ arises from the interchange of the anti-commuting operators of fermions. The two quantities in Eqs. (6.66) and (6.67) are called the Feynman propagators (hence the subscript F) of the photon and the fermion respectively. They represent the amplitude corresponding to the propagation of a (virtual or real) particle from a space-time point x to a point y (i.e., its creation followed by its annihilation). Their calculation is presented in Supplements 6.2 and 6.3. They only depend on $x - y$. The expression of the photon propagator in momentum space[9] is

$$\text{Photon propagator: } iD_{F\mu\nu}(p) = \frac{-ig_{\mu\nu}}{p^2 + i\epsilon}, \tag{6.68}$$

where p is the 4-momentum of the (virtual or real) photon and ϵ is an infinitesimal positive real number to avoid singularities when $p \to 0$. (In practice, one should check that the limit $\epsilon \to 0$ can be taken at the end of the calculation.) This formula corresponds to the particular choice of the Feynman gauge (see Section 6.1.4). For the electron, the propagator is

$$\text{Spin-1/2 propagator: } i\Delta_F(p) = i\frac{\not{p} + m}{p^2 - m^2 + i\epsilon}, \tag{6.69}$$

where again p is the 4-momentum of the (virtual or real) spin-1/2 fermion and m its mass (i.e., the nominal mass used in $E^2 = |\mathbf{p}|^2 + m^2$). Since $\not{p} = \gamma^\mu p_\mu$ is a matrix, this propagator is also a matrix.

[9] Generally, the positions of particles in a reaction are not known but their momenta are.

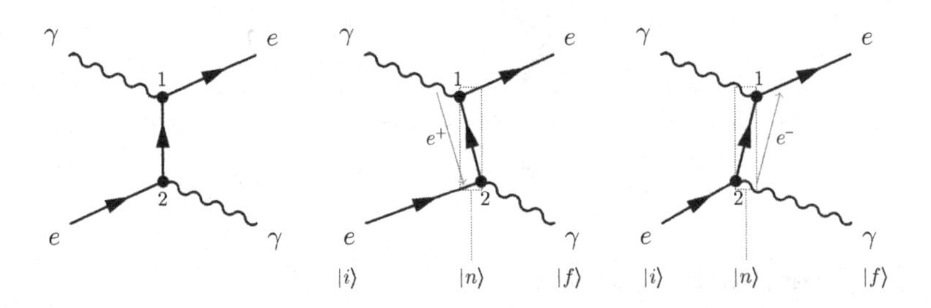

Fig. 6.4 Left: the $e^-\gamma \to e^-\gamma$ Feynman diagram. Centre and right: the corresponding time-ordered diagrams.

We have now all the ingredients to mathematically describe the diagrams in Fig. 6.3. For instance, the amplitude of the first diagram is

$$i\mathcal{M} = \left[\bar{v}_{r_{e^+}}(p_{e^+})\,(ie\gamma^{\mu})\,u_{r_{e^-}}(p_{e^-})\right] \frac{-ig_{\mu\nu}}{p^2 + i\epsilon} \left[\bar{u}_{r'_{e^-}}(p'_{e^-})\,(ie\gamma^{\nu})\,v_{r'_{e^+}}(p'_{e^+})\right]. \tag{6.70}$$

The two electromagnetic currents in square brackets are the terms already obtained with the first order of the perturbative calculation. The propagator $-ig_{\mu\nu}/(p^2 + i\epsilon)$ corresponds to the propagation of the virtual photon. A naive interpretation of this diagram is that the electron and positron of the initial state scatter via the exchange of a single photon. However, the propagator describes a virtual photon, which is not physical. It cannot be measured in a detector. Only the initial and final states can be measured. What is in-between is just an element of a calculation: here, the lowest order of the electromagnetic interaction.

Before concluding this section, let us examine the left diagram in Fig. 6.4. At first sight, it seems impossible to interpret it in terms of a particle exchange since the particle would be created and absorbed at the same time coordinate (look at the position of vertices labelled 1 and 2, given that time goes from the left to right here). But consider the two other diagrams in Fig. 6.4. When $t_1 < t_2$, in the centre diagram, a virtual positron (i.e., an electron moving backwards in time in the diagram) is created first at vertex 1 and then absorbed at vertex 2, whereas when $t_1 > t_2$ as in the right-hand side diagram, a virtual electron is created first at vertex 2 and then absorbed at vertex 1. As in the definition of the propagators (6.66) and (6.67), there is a T-product of field operators, the Feynman diagram on the left-hand side includes the two possibilities depicted by the centre and right pictures. Both time-orderings are always included in each propagator line. This makes sense since there is no way to distinguish the two modalities experimentally.

It is enlightening to see how we can recover the propagator dependence in $1/(q^2 - m^2)$ by explicitly considering time-ordered diagrams. In Appendix H, we show that, using the time-dependant perturbation theory of quantum mechanics (as opposed to the quantum field approach), in the second-order perturbation theory (namely with one intermediate state), the transition matrix element is

$$T^{(2)}_{fi} = \sum_{n\neq i} \frac{V_{fn} V_{ni}}{E_i - E_n} = \sum_{n\neq i} \frac{\langle f|V|n\rangle\,\langle n|V|i\rangle}{E_i - E_n}. \tag{6.71}$$

In Fig. 6.4, the initial state is $|i\rangle = |e_i^-, \gamma_i\rangle$, with energy and momentum $E_i = E_{e_i} + E_{\gamma_i}$ and $\boldsymbol{p}_i = \boldsymbol{p}_{e_i} + \boldsymbol{p}_{\gamma_i}$. Similarly, the final state is $|f\rangle = |e_f^-, \gamma_f\rangle$, with $E_f = E_{ef} + E_{\gamma_f}$ and $\boldsymbol{p}_f = \boldsymbol{p}_{e_f} + \boldsymbol{p}_{\gamma_f}$. Now, the intermediate state $|n\rangle$ depends on the diagram. In the centre diagram, $|n\rangle = |e_i^-, e^+, e_f^-\rangle$ (see the dotted rectangle) with $E_n = E_{e_i} + E_{e^+} + E_{e_f}$. Hence, this diagram corresponds to

$$\frac{\langle e_f^-, \gamma_f|V|e_i^-, e^+, e_f^-\rangle\,\langle e_i^-, e^+, e_f^-|V|e_i^-, \gamma_i\rangle}{(E_{e_i} + E_{\gamma_i}) - (E_{e_i} + E_{e^+} + E_{e_f})} = \frac{\langle \gamma_f|V|e_i^-, e^+\rangle\,\langle e_f^-|e_f^-\rangle\,\langle e^+, e_f^-|V|\gamma_i\rangle\,\langle e_i^-|e_i^-\rangle}{E_{\gamma_i} - E_{e^+} - E_{e_f}},$$

where $|e_i^-\rangle$ and $|e_f^-\rangle$ are the spectator states at the vertex interaction 1 and 2, respectively. In non-relativistic quantum mechanics, the states are normalised to one particle, thus $\langle e_i^-|e_i^-\rangle = \langle e_f^-|e_f^-\rangle = 1$ [see Eq. (H.2)]. In the right-hand side diagram of Fig. 6.4, $|n\rangle = |\gamma_i, e^-, \gamma_f\rangle$, with $E_n = E_{\gamma_i} + E_{e^-} + E_{\gamma_f}$. The corresponding transition is

$$\frac{\langle e_f^-, \gamma_f|V|\gamma_i, e^-, \gamma_f\rangle\,\langle \gamma_i, e^-, \gamma_f|V|e_i^-, \gamma_i\rangle}{(E_{e_i} + E_{\gamma_i}) - (E_{\gamma_i} + E_{e^-} + E_{\gamma_f})} = \frac{\langle e_f^-|V|\gamma_i, e^-\rangle\,\langle e^-, \gamma_f|V|e_i^-\rangle}{E_{e_i} - E_{e^-} - E_{\gamma_f}},$$

where, this time, the spectator states are $|\gamma_i\rangle$ and $|\gamma_f\rangle$ at the vertex interaction 1 and 2 respectively. Note that in the framework of this old-fashioned perturbation theory, the energy is not conserved by intermediate states since $E_i = E_f \neq E_n$, whereas the momentum is conserved. Besides, all particles, including the intermediate ones, are considered on the mass shell. Now, if we express the processes as a function of the Lorentz invariant amplitudes $\mathcal{M}$, we have to introduce the relativistic normalisation with the corresponding $\sqrt{2E}$ factors (see Eq. [H.26]). Since both time-ordered diagrams describe the same initial and final state, the total transition matrix element is

$$\frac{\mathcal{M}^{(2)}_{e_i^- + \gamma_i \to e_f^- + \gamma_f}}{\sqrt{(2E_{e_i})(2E_{\gamma_i})(2E_{e_f})(2E_{\gamma_f})}} = \frac{\dfrac{\mathcal{M}_{e_i^- + e^+ \to \gamma_f}}{\sqrt{(2E_{\gamma_f})(2E_{e_i})(2E_{e^+})}}\dfrac{\mathcal{M}_{\gamma_i \to e^+ + e_f^-}}{\sqrt{(2E_{e^+})(2E_{e_f})(2E_{\gamma_i})}}}{E_{\gamma_i} - E_{e^+} - E_{e_f}} + \frac{\dfrac{\mathcal{M}_{\gamma_i + e^- \to e_f^-}}{\sqrt{(2E_{e_f})(2E_{\gamma_i})(2E_{e^-})}}\dfrac{\mathcal{M}_{e_i^- \to e^- + \gamma_f}}{\sqrt{(2E_{e^-})(2E_{\gamma_f})(2E_{e_i})}}}{E_{e_i} - E_{e^-} - E_{\gamma_f}}.$$

According to the crossing reaction property (see Section 3.7), necessarily

$$\mathcal{M}_{e_i^- + e^+ \to \gamma_f} = \mathcal{M}_{e_i^- \to e^- + \gamma_f} \quad \text{and} \quad \mathcal{M}_{\gamma_i \to e^+ + e_f^-} = \mathcal{M}_{\gamma_i + e^- \to e_f^-}.$$

It follows that

$$\mathcal{M}^{(2)}_{e_i^- + \gamma_i \to e_f^- + \gamma_f} = \mathcal{M}_{e_i^- \to e^- + \gamma_f} \times \Pi \times \mathcal{M}_{\gamma_i + e^- \to e_f^-}, \tag{6.72}$$

where

$$\Pi = \frac{1}{2E_{e^+}(E_{\gamma_i} - E_{e^+} - E_{e_f})} + \frac{1}{2E_{e^-}(E_{e_i} - E_{e^-} - E_{\gamma_f})}. \tag{6.73}$$

Momentum conservation imposes $\boldsymbol{p}_{\gamma_i} = \boldsymbol{p}_{e^+} + \boldsymbol{p}_{e_f}$ and $\boldsymbol{p}_{\gamma_i} + \boldsymbol{p}_{e^-} = \boldsymbol{p}_{e_f}$. Hence, $\boldsymbol{p}_{e^+} = \boldsymbol{p}_{e^-}$, and since in this procedure, particles are always on the mass shell, $E_{e^+} = E_{e^-} \equiv E_e$. This, with the energy conservation between the initial and the final states, $E_{\gamma_i} + E_{e_i} = E_{\gamma_f} + E_{e_f}$, allows us to simplify Π in Eq. (6.73), giving

$$\Pi = \frac{1}{(E_{\gamma_i} - E_{e_f})^2 - E_e^2} = \frac{1}{(p_{\gamma_i} - p_{e_f})^2 - m_e^2},$$

where we used $E_e^2 = |\boldsymbol{p}_{e^+}|^2 + m_e^2 = |\boldsymbol{p}_{\gamma_i} - \boldsymbol{p}_{e_f}|^2 + m_e^2$ to obtain the last equality. Consequently, we recover the usual propagator dependence. In summary, either intermediate particles are assumed to be on the mass shell, and only momentum conservation is required (this is the old-fashioned approach of perturbation theory in quantum mechanics), or both energy–momentum conservations are required, $q = p_{\gamma_i} - p_{e_f}$, but the intermediate particles are virtual particles that cannot be on the mass shell, $q^2 \neq m^2$, with propagators of the form

$$\Pi(q^2) \propto \frac{1}{q^2 - m^2}.$$

Such a propagator is manifestly Lorentz invariant, whereas a single transition matrix element in Eq. (6.71) is not. It is the sum over the time-ordered diagrams that produces a Lorentz-invariant amplitude. In contrast, a time-ordered diagram is intrinsically not Lorentz invariant since the order of events in time depends on the frame considered. The modern approach, mostly due to Feynman, assuming energy–momentum conservation and virtual particles is thus more convenient than the old-fashioned approach: it avoids considering all the time-ordered diagrams for a given reaction.

Supplement 6.2. Photon propagator

The calculation is presented using the Feynman gauge (see Section 6.1.4). We start from the photon field expansion,

$$A^\mu(x) = \int \frac{d^3\boldsymbol{p}}{(2\pi)^3 2E_p} \sum_{\lambda=0}^{3} \left[\epsilon^\mu(\boldsymbol{p},\lambda)\alpha_{\boldsymbol{p},\lambda} e^{-ip\cdot x} + \epsilon^{\mu *}(\boldsymbol{p},\lambda)\alpha^\dagger_{\boldsymbol{p},\lambda} e^{+ip\cdot x}\right],$$

$$A^\nu(y) = \int \frac{d^3\boldsymbol{p}'}{(2\pi)^3 2E_{p'}} \sum_{\lambda'=0}^{3} \left[\epsilon^\nu(\boldsymbol{p}',\lambda')\alpha_{\boldsymbol{p}',\lambda'} e^{-ip'\cdot y} + \epsilon^{\nu *}(\boldsymbol{p}',\lambda')\alpha^\dagger_{\boldsymbol{p}',\lambda'} e^{+ip'\cdot y}\right],$$

where at this stage, $p^0 = E_p = |\boldsymbol{p}|$. The vacuum expectation value of $A_\mu(x)A_\nu(y)$ reads

$$\langle 0|A^\mu(x)A^\nu(y)|0\rangle = \int \frac{d^3\boldsymbol{p}}{(2\pi)^3 2E_p}\frac{d^3\boldsymbol{p}'}{(2\pi)^3 2E_{p'}} \sum_{\lambda,\lambda'} \langle 0|\alpha_{\boldsymbol{p},\lambda}\alpha^\dagger_{\boldsymbol{p}',\lambda'}|0\rangle\, \epsilon^\mu(\boldsymbol{p},\lambda)\epsilon^{\nu *}(\boldsymbol{p}',\lambda') e^{-ip\cdot x + ip'\cdot y}.$$

The term $\langle 0|\alpha_{\boldsymbol{p},\lambda}\alpha^\dagger_{\boldsymbol{p}',\lambda'}|0\rangle$ is equal to the commutator $[\alpha_{\boldsymbol{p},\lambda}, \alpha^\dagger_{\boldsymbol{p}',\lambda'}]$, and the insertion of its expression (5.133) allows us to integrate over $\boldsymbol{p}'$ and sum over λ'. It follows that

$$\langle 0|A^\mu(x)A^\nu(y)|0\rangle = \int \frac{d^3\boldsymbol{p}}{(2\pi)^3 2E_p} \sum_{\lambda} (-g_{\lambda\lambda})\epsilon^\mu(\boldsymbol{p},\lambda)\epsilon^{\nu *}(\boldsymbol{p}',\lambda) e^{-ip\cdot(x-y)}.$$

Supplement 6.2. (cont.)

In the Feynman gauge, we can choose momentum-independent polarisation vectors that obey the completeness relations (5.128). Then, it is easy to sum over λ, giving

$$\begin{aligned}\langle 0|A^\mu(x)A^\nu(y)|0\rangle &= (-g^{\mu\nu})\int \frac{\mathrm{d}^3\boldsymbol{p}}{(2\pi)^3 2E_p} e^{-ip\cdot(x-y)} \\ &= (-g^{\mu\nu})\int \frac{\mathrm{d}^3\boldsymbol{p}}{(2\pi)^3 2E_p} e^{-iE_p(x^0-y^0)} e^{+i\boldsymbol{p}\cdot(\boldsymbol{x}-\boldsymbol{y})}.\end{aligned}$$

A similar calculation yields

$$\langle 0|A^\nu(y)A^\mu(x)|0\rangle = (-g^{\mu\nu})\int \frac{\mathrm{d}^3\boldsymbol{p}}{(2\pi)^3 2E_p} e^{+iE_p(x^0-y^0)} e^{-i\boldsymbol{p}\cdot(\boldsymbol{x}-\boldsymbol{y})}.$$

In order to evaluate the propagator $i\,D_F^{\mu\nu}(x-y) = \langle 0|T\{A^\mu(x)A^\nu(y)\}|0\rangle$, let us use the step function $\theta(t)$ to express the time-ordered product,

$$\begin{aligned}iD_F^{\mu\nu}(x-y) &= \langle 0|A^\mu(x)A^\nu(y)|0\rangle\,\theta(x^0-y^0) + \langle 0|A^\nu(y)A^\mu(x)|0\rangle\,\theta(y^0-x^0) \\ &= (-g^{\mu\nu})iD_F(x-y),\end{aligned}$$

with

$$\begin{aligned}iD_F(x-y) = \int \frac{\mathrm{d}^3\boldsymbol{p}}{(2\pi)^3 2E_p}\Big[&e^{-iE_p(x^0-y^0)}\theta(x^0-y^0)e^{+i\boldsymbol{p}\cdot(\boldsymbol{x}-\boldsymbol{y})} \\ &+e^{+iE_p(x^0-y^0)}\theta(y^0-x^0)e^{-i\boldsymbol{p}\cdot(\boldsymbol{x}-\boldsymbol{y})}\Big].\end{aligned} \tag{6.74}$$

A mathematical trick is to replace the step function with its integral form (F.9), i.e. (see Appendix F)

$$\theta(t) = \lim_{\epsilon\to 0^+}\frac{i}{2\pi}\int_{-\infty}^{+\infty}\frac{e^{-irt}}{r+i\epsilon}\,\mathrm{d}r.$$

The terms depending on $\pm(x^0-y^0)$ in the square brackets of Eq. (6.74) then read (dropping the limit symbol)

$$e^{-iE_p(x^0-y^0)}\theta(x^0-y^0) = \frac{i}{2\pi}\int_{-\infty}^{+\infty}\frac{e^{-i(E_p+r)(x^0-y^0)}}{r+i\epsilon}\,\mathrm{d}r = \frac{i}{2\pi}\int_{-\infty}^{+\infty}\frac{e^{-ip^0(x^0-y^0)}}{p^0-E_p+i\epsilon}\,\mathrm{d}p^0,$$

$$e^{+iE_p(x^0-y^0)}\theta(y^0-x^0) = \frac{i}{2\pi}\int_{-\infty}^{+\infty}\frac{e^{+i(E_p+r)(x^0-y^0)}}{r+i\epsilon}\,\mathrm{d}r = \frac{i}{2\pi}\int_{-\infty}^{+\infty}\frac{e^{+ip^0(x^0-y^0)}}{p^0-E_p+i\epsilon}\,\mathrm{d}p^0,$$

where the dummy variable p^0 now stands for $r+E_p$. Beware that p^0 is no longer $E_p = |\boldsymbol{p}|$, as at the beginning of this supplement. It is now an unrestricted variable. Hence, we can form the 4-vector $p = (p^0,\boldsymbol{p})$, but p^2 is no longer 0. In Eq. (6.74), $iD_F(x-y)$ now becomes the four-dimensional integral,

$$\begin{aligned}iD_F(x-y) &= i\int\frac{\mathrm{d}^4p}{(2\pi)^4 2E_p}\left[\frac{e^{-ip\cdot(x-y)}}{p^0-E_p+i\epsilon}+\frac{e^{+ip\cdot(x-y)}}{p^0-E_p+i\epsilon}\right] \\ &= i\int\frac{\mathrm{d}^4p}{(2\pi)^4 2E_p}\left[\frac{e^{-ip\cdot(x-y)}}{p^0-E_p+i\epsilon}+\frac{e^{-ip\cdot(x-y)}}{-p^0-E_p+i\epsilon}\right],\end{aligned}$$

Supplement 6.2. (cont.)

where we changed p to $-p$ in the second term of the last equality. Keeping the first order in ϵ, the sum of fractions in the square brackets is

$$\frac{1}{p^0 - (E_p - i\epsilon)} - \frac{1}{p^0 + (E_p - i\epsilon)} = \frac{2E_p}{(p^0)^2 - E_p^2 + 2i\epsilon}. \tag{6.75}$$

As $E_p = |\boldsymbol{p}|$, $(p^0)^2 - E_p^2 = p^2$, and substituting 2ϵ for ϵ, it follows that

$$iD_F(x - y) = \int \frac{d^4p}{(2\pi)^4} \left(\frac{i}{p^2 + i\epsilon} \right) e^{-ip\cdot(x-y)}. \tag{6.76}$$

We identify the Fourier transform of the term in parenthesis. Therefore, the photon propagator in momentum space reads

$$iD_F^{\mu\nu}(p) = -g^{\mu\nu} iD_F(x - y) = \frac{-ig^{\mu\nu}}{p^2 + i\epsilon}, \tag{6.77}$$

where ϵ is an infinitesimal positive real number.

6.3.3 Summary of QED Feynman Rules

The recipe for calculating amplitudes of processes is as follows:

1. Draw, to a given order, all diagrams that describe the transition between the initial state and the final state. A QED vertex must always connect two fermions (or anti-fermions) to one photon. For each vertex, the energy–momentum is conserved, and the charge of the incoming state is always equal to the charge of the outgoing state (QED is a flow of an electric charge).
2. Calculate $i\mathcal{M}$ of each diagram using the factors given in Table 6.1. The factors of incoming and outgoing particles are listed in the 'External lines' section of the table, those of the exchanging virtual particles in the 'Propagators' section and the vertex factors are in the 'Vertex' section. For completeness, the factors for massive spinless charged bosons are also given.
3. Combine the amplitudes of all diagrams to get the total amplitude of the process. If two diagrams differ by the exchange of two external identical fermions only (two outgoings, or two incomings, or one incoming fermion and one outgoing anti-fermion, or one incoming anti-fermion and one outgoing fermion), subtract their amplitudes.[10] Otherwise, add all other amplitudes.

[10] This is due to the spin-statistics theorem: when two identical fermions are interchanged, the wave function gets a minus sign.

Supplement 6.3. Dirac propagator

We proceed similarly to the photon propagator in Supplement 6.2. The Dirac field and its adjoint are

$$\psi(x) = \int \frac{d^3\boldsymbol{p}}{(2\pi)^3 2E_p} \sum_{r=1}^{2} \left[c_{p,r} u_r(p) e^{-ip\cdot x} + d^\dagger_{p,r} v_r(p) e^{+ip\cdot x}\right],$$

$$\bar{\psi}(y) = \int \frac{d^3\boldsymbol{p}'}{(2\pi)^3 2E_{p'}} \sum_{r'=1}^{2} \left[c^\dagger_{p',r'} \bar{u}_{r'}(p') e^{+ip'\cdot y} + d_{p',r'} \bar{v}_{r'}(p') e^{-ip'\cdot y}\right],$$

giving the expectation values

$$\langle 0|\psi_\alpha(x)\bar{\psi}_\beta(y)|0\rangle = \int \frac{d^3\boldsymbol{p}}{(2\pi)^3 2E_p} \frac{d^3\boldsymbol{p}'}{(2\pi)^3 2E_{p'}} \sum_{r,r'} \langle 0|c_{p,r} c^\dagger_{p',r'}|0\rangle \left(u_r(p)\right)_\alpha \left(\bar{u}_{r'}(p')\right)_\beta e^{-ip\cdot x + ip'\cdot y},$$

$$\langle 0|\bar{\psi}_\beta(y)\psi_\alpha(x)|0\rangle = \int \frac{d^3\boldsymbol{p}}{(2\pi)^3 2E_p} \frac{d^3\boldsymbol{p}'}{(2\pi)^3 2E_{p'}} \sum_{r,r'} \langle 0|d_{p',r'} d^\dagger_{p,r}|0\rangle \left(v_r(p)\right)_\alpha \left(\bar{v}_{r'}(p')\right)_\beta e^{-ip'\cdot y + ip\cdot x}.$$

Then, we use the anti-commutation rules (5.102) and the completeness relation (5.78) to integrate over $\boldsymbol{p}'$ and sum over r', giving after a similar calculation as in Section 6.1.3,

$$\langle 0|\psi_\alpha(x)\bar{\psi}_\beta(y)|0\rangle = \int \frac{d^3\boldsymbol{p}}{(2\pi)^3 2E_p} \left(\not{p} + m\right)_{\alpha\beta} e^{-ip\cdot(x-y)} = \left(i\not{\partial} + m\right)_{\alpha\beta} \int \frac{d^3\boldsymbol{p}}{(2\pi)^3 2E_p} e^{-ip\cdot(x-y)},$$

$$\langle 0|\bar{\psi}_\beta(y)\psi_\alpha(x)|0\rangle = \int \frac{d^3\boldsymbol{p}}{(2\pi)^3 2E_p} \left(\not{p} - m\right)_{\alpha\beta} e^{+ip\cdot(x-y)} = -\left(i\not{\partial} + m\right)_{\alpha\beta} \int \frac{d^3\boldsymbol{p}}{(2\pi)^3 2E_p} e^{+ip\cdot(x-y)}.$$

Hence, the Dirac propagator $i\Delta_F(x-y) = \langle 0|T\{\psi(x)\bar{\psi}(y)\}|0\rangle$ takes the form

$$\begin{aligned} i\Delta_F(x-y) &= \langle 0|\psi(x)\bar{\psi}(y)|0\rangle\,\theta(x^0 - y^0) - \langle 0|\bar{\psi}(y)\psi(x)|0\rangle\,\theta(y^0 - x^0) \\ &= \left(i\not{\partial} + m\right) \int \frac{d^3\boldsymbol{p}}{(2\pi)^3 2E_p} \left[e^{-ip\cdot(x-y)}\theta(x^0 - y^0) + e^{+ip\cdot(x-y)}\theta(y^0 - x^0)\right]. \end{aligned}$$

The integral looks similar to $D_F(x-y)$ in Eq. (6.74), except that this time $E_p^2 = |\boldsymbol{p}|^2 + m^2$. Thus, the term $(p^0)^2 - (E_p)^2$ in Eq. (6.75) is now equal to $p^2 - m^2$. We conclude that

$$\begin{aligned} i\Delta_F(x-y) &= \left(i\not{\partial} + m\right) \int \frac{d^4p}{(2\pi)^4} \left(\frac{i}{p^2 - m^2 + i\epsilon}\right) e^{-ip\cdot(x-y)} \\ &= \int \frac{d^4p}{(2\pi)^4} \left(\frac{i(\not{p} + m)}{p^2 - m^2 + i\epsilon}\right) e^{-ip\cdot(x-y)}, \end{aligned}$$

and, therefore, the Dirac propagator in momentum space reads

$$i\Delta_F(p) = i\frac{\not{p} + m}{p^2 - m^2 + i\epsilon}.$$

Table 6.1 Feynman rules in QED.

	External lines		
Spin 0	Incoming boson		1
	Outgoing boson		1
	Incoming anti-boson		1
	Outgoing anti-boson		1
Spin 1/2	Incoming fermion		u spinor
	Outgoing fermion		$\bar{u}$ spinor
	Incoming anti-fermion		$\bar{v}$ spinor
	Outgoing anti-fermion		v spinor
Spin 1	Incoming photon		ϵ_μ
	Outgoing photon	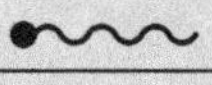	ϵ^*_μ
	Propagators		
Spin 0	Boson	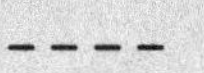	$\frac{i}{p^2-m^2+i\epsilon}$
Spin 1/2	Fermion		$i\frac{\not{p}+m}{p^2-m^2+i\epsilon}$
Spin 1	Photon		$\frac{-ig_{\mu\nu}}{p^2+i\epsilon}$
	Vertex		
	Spin 0 $-iq(p^\mu+p'^\mu)$ 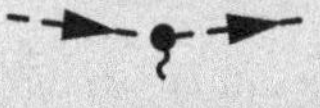	Spin $\frac{1}{2}$ $-iq\gamma^\mu$ 	

$q = eQ$, with Q being the electric charge in units of e. Warning: $q = -e$ for both e^- and e^+.

6.4 Examples of Graph Calculation

6.4.1 Starter: Squaring and Averaging Amplitudes

As a simple example, let us consider the reaction $e^- + \mu^- \to e^- + \mu^-$. Electrons and muons carry an electric charge. Therefore, they can interact via the exchange of a virtual photon, as shown in Fig. 6.5. Only the t-channel is allowed for this reaction (hence, the vertical wavy line of the virtual photon). Indeed, the s-channel would imply a vertex $e^- + \mu^- \to \gamma$, which is forbidden by the charge conservation, but also because QED vertices always connect a photon to two particles of the same flavour, the interaction term (6.33) using twice the *same* fermion field. Let us denote by p and k the respective 4-momenta of the electron and the muon in the initial state. The same labels with the prime symbol denote the corresponding

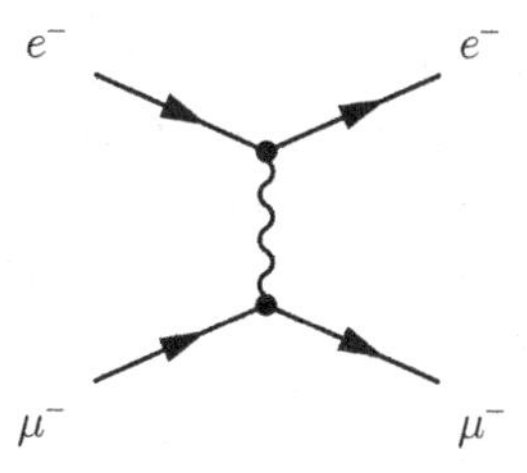

Fig. 6.5 The lowest order diagram for $e^-\mu^- \to e^-\mu^-$.

4-momenta in the final state. The photon has a momentum $q = p - p' = k' - k$, whose square is just the Mandelstam variable t (hence the t-channel). Following the Feynman rules of Table 6.1, the amplitude is

$$i\mathcal{M} = (\bar{u}_{r'}(k')(ie\gamma^\mu)u_r(k))\frac{-ig_{\mu\nu}}{q^2}(\bar{u}_{s'}(p')(ie\gamma^\nu)u_s(p)) = i\frac{e^2}{q^2}(\bar{u}_{r'}(k')\gamma^\mu u_r(k))(\bar{u}_{s'}(p')\gamma_\mu u_s(p)),$$

where r, s, r' and s' are the helicity or spin indices. The cross-section formula is proportional to the probability of the process and thus to $|\mathcal{M}|^2$. Generally, the incoming particles are not polarised. Thus, there is a probability of $1/2$ for an incoming particle to have spin-up (or down), or equivalently in terms of helicity, to be left- (or right)-handed. We should then average over the spin/helicity configurations of the initial state. Detectors usually only measure the momentum and energy of particles, not their spin or helicity. Hence, we sum over the spin/helicity configurations for the particles in the final state. Overall, the measured cross section for unpolarised particles thus corresponds to an average over the spins/helicities of the initial state and a sum over the spins/helicities of the final state, i.e.

$$\overline{|\mathcal{M}|^2} = \underbrace{\frac{1}{2}\sum_s \frac{1}{2}\sum_r}_{\text{initial state}} \underbrace{\sum_{s'}\sum_{r'}}_{\text{final state}} |\mathcal{M}|^2. \tag{6.78}$$

If the particles of the initial state were polarised (thus have a defined helicity), only the quantity

$$\overline{|\mathcal{M}|^2} = \sum_{s'}\sum_{r'} |\mathcal{M}|^2 \tag{6.79}$$

would matter. Note that we average or sum $|\mathcal{M}|^2$, not $\mathcal{M}$. The key here is that a spin-up or spin-down particle defines two *different* spin states that are in principle distinguishable (even if the spin is not measured). Therefore, one should not add or average the amplitudes $\mathcal{M}$ but only the probabilities and thus $|\mathcal{M}|^2$.

6.4.2 Spin Summations: Traces Theorems

The calculation of a cross section requires the explicit computation of the amplitudes squared (6.78) or (6.79). The brute-force approach would be to insert the spinors that are the eigenstates of the helicity operator (5.74) into the amplitude $\mathcal{M}$ and evaluate the

16 terms of the amplitudes squared (for unpolarised incoming particles). In the ultra-relativistic regime, as we shall see in Section 6.5, some terms vanish. However, when the particle masses cannot be neglected, all terms must be computed. Fortunately, it turns out that general formulas can be established once and for all. Let us see how. The unpolarised amplitude squared (6.78) can be expressed as a product of two Lorentz tensors,

$$\overline{|\mathcal{M}|^2} = \frac{e^4}{q^4} L^{\mu\nu}(\mu^-) L_{\mu\nu}(e^-), \tag{6.80}$$

where

$$\begin{aligned} L^{\mu\nu}(\mu^-) &= \frac{1}{2}\sum_{r,r'} \left[\bar{u}_{r'}(k')\gamma^\mu u_r(k)\right]^* \left[\bar{u}_{r'}(k')\gamma^\nu u_r(k)\right], \\ L_{\mu\nu}(e^-) &= \frac{1}{2}\sum_{s,s'} \left[\bar{u}_{s'}(p')\gamma_\mu u_s(p)\right]^* \left[\bar{u}_{s'}(p')\gamma_\nu u_s(p)\right]. \end{aligned} \tag{6.81}$$

Since $\bar{u}_{r'}(k')\gamma^\mu u_r(k)$ is a complex number, its complex conjugate is also its Hermitian conjugate leading to

$$(\bar{u}_{r'}(k')\gamma^\mu u_r(k))^* = u_r^\dagger(k)\gamma^{\mu\dagger}\bar{u}_{r'}^\dagger(k') = u_r^\dagger(k)\gamma^0\gamma^\mu\gamma^0(u_{r'}^\dagger(k')\gamma^0)^\dagger = \bar{u}_r(k)\gamma^\mu u_{r'}(k').$$

Therefore, $L^{\mu\nu}(\mu^-)$ reads

$$L^{\mu\nu}(\mu^-) = \frac{1}{2}\sum_r \bar{u}_r(k)\gamma^\mu \left(\sum_{r'} u_{r'}(k')\bar{u}_{r'}(k')\right)\gamma^\nu u_r(k) = \frac{1}{2}\sum_r \bar{u}_r(k)\gamma^\mu \left(\not{k}' + m'\right)\gamma^\nu u_r(k),$$

where the completeness relation (5.78) was used. The tensor, $L^{\mu\nu}$, being a complex number and not a matrix, is equal to its trace. Given the properties of traces,

$$\mathrm{Tr}(\alpha A + \beta B) = \alpha \mathrm{Tr}(A) + \beta \mathrm{Tr}(B)\ , \ \ \mathrm{Tr}(AB) = \mathrm{Tr}(BA), \tag{6.82}$$

it follows that

$$L^{\mu\nu}(\mu^-) = \frac{1}{2}\mathrm{Tr}\left[\sum_r u_r(k)\bar{u}_r(k)\gamma^\mu\left(\not{k}' + m'\right)\gamma^\nu\right] = \frac{1}{2}\mathrm{Tr}\left[(\not{k} + m)\gamma^\mu\left(\not{k}' + m'\right)\gamma^\nu\right].$$

If instead of the scattering of a particle (here a muon or an electron) we had an antiparticle, we would have to use the corresponding completeness relation for v-spinors. It would lead to a similar formula with a minus sign before the mass. We thus obtain the following four expressions of tensors:

$$\begin{aligned} L^{\mu\nu}_{uu} &= \frac{1}{2}\sum_{r,r'} \left[\bar{u}_{r'}(k')\gamma^\mu u_r(k)\right]^* \left[\bar{u}_{r'}(k')\gamma^\nu u_r(k)\right] = \frac{1}{2}\mathrm{Tr}\left[(\not{k} + m)\gamma^\mu\left(\not{k}' + m'\right)\gamma^\nu\right], \\ L^{\mu\nu}_{vv} &= \frac{1}{2}\sum_{r,r'} \left[\bar{v}_{r'}(k')\gamma^\mu v_r(k)\right]^* \left[\bar{v}_{r'}(k')\gamma^\nu v_r(k)\right] = \frac{1}{2}\mathrm{Tr}\left[(\not{k} - m)\gamma^\mu\left(\not{k}' - m'\right)\gamma^\nu\right], \\ L^{\mu\nu}_{vu} &= \frac{1}{2}\sum_{r,r'} \left[\bar{v}_{r'}(k')\gamma^\mu u_r(k)\right]^* \left[\bar{v}_{r'}(k')\gamma^\nu u_r(k)\right] = \frac{1}{2}\mathrm{Tr}\left[(\not{k} + m)\gamma^\mu\left(\not{k}' - m'\right)\gamma^\nu\right], \\ L^{\mu\nu}_{uv} &= \frac{1}{2}\sum_{r,r'} \left[\bar{u}_{r'}(k')\gamma^\mu v_r(k)\right]^* \left[\bar{u}_{r'}(k')\gamma^\nu v_r(k)\right] = \frac{1}{2}\mathrm{Tr}\left[(\not{k} - m)\gamma^\mu\left(\not{k}' + m'\right)\gamma^\nu\right]. \end{aligned} \tag{6.83}$$

We can easily generalise the above calculations to more complicated cases we will encounter in the following chapters. Denoting by w and w', two generic spinors (either a u-spinor or a v-spinor) describing particles with a respective mass m and m', and considering two 4×4 matrices, Γ_1 and Γ_2, it follows that

$$\sum_{r,r'} \left[\bar{w}_{r'}(k')\Gamma_1 w_r(k)\right]^* \left[\bar{w}_{r'}(k')\Gamma_2 w_r(k)\right] = \mathrm{Tr}\left[(\not{k} \pm m)\gamma^0\Gamma_1^\dagger\gamma^0 \left(\not{k}' \pm m'\right)\Gamma_2\right]. \quad (6.84)$$

In the formula above, a plus sign is used for u-spinors and a minus sign for v-spinors. We recover the Formula (6.83) with $\Gamma_1 = \gamma^\mu$ and $\Gamma_2 = \gamma^\nu$ since $\gamma^0\gamma^{\mu\dagger}\gamma^0 = \gamma^0\gamma^0\gamma^\mu\gamma^0\gamma^0 = \gamma^\mu$.

Formulas (6.83) and (6.84) require the evaluation of the trace of products of γ matrices. Since they satisfy the Clifford algebra (5.29), i.e. $\{\gamma^\mu, \gamma^\nu\} = \gamma^\mu\gamma^\nu + \gamma^\nu\gamma^\mu = 2g^{\mu\nu}\mathbb{1}$, it follows that

$$\mathrm{Tr}(\gamma^\mu\gamma^\nu) = g^{\mu\nu}\mathrm{Tr}(\mathbb{1}) = 4g^{\mu\nu}. \quad (6.85)$$

In passing, this implies

$$\mathrm{Tr}(\not{p}\not{k}) = 4p \cdot k. \quad (6.86)$$

The trace of an arbitrary number of γ matrices satisfies a symmetry property. Indeed, given that $(\gamma^5)^2 = 1$ and the trace property Eq. (6.82), we deduce that

$$\mathrm{Tr}(\gamma^{\mu_1}\cdots\gamma^{\mu_n}) = \mathrm{Tr}(\gamma^{\mu_1}\cdots\gamma^{\mu_n}\gamma^5\gamma^5) = \mathrm{Tr}(\gamma^5\gamma^{\mu_1}\cdots\gamma^{\mu_n}\gamma^5).$$

As $\gamma^5\gamma^\mu = -\gamma^\mu\gamma^5$, it follows that

$$\mathrm{Tr}(\gamma^{\mu_1}\cdots\gamma^{\mu_n}) = (-1)^n\mathrm{Tr}(\gamma^{\mu_1}\cdots\gamma^{\mu_n}\gamma^5\gamma^5) = (-1)^n\mathrm{Tr}(\gamma^{\mu_1}\cdots\gamma^{\mu_n}).$$

Hence, for an odd number of γ matrices (where γ is $\gamma^{\mu=0\cdots3}$ and not γ^5),

$$\mathrm{Tr}(\text{odd nb of } \gamma) = 0. \quad (6.87)$$

The case of four matrices necessitates playing again with the algebra and Eq. (6.85). One finds (see Problem 6.4)

$$\mathrm{Tr}(\gamma^\rho\gamma^\mu\gamma^\eta\gamma^\nu) = 4(g^{\rho\mu}g^{\eta\nu} - g^{\rho\eta}g^{\mu\nu} + g^{\rho\nu}g^{\mu\eta}), \quad (6.88)$$

which also leads to

$$\mathrm{Tr}(\not{p}_1\not{p}_2\not{p}_3\not{p}_4) = 4(p_1.p_2\, p_3.p_4 - p_1.p_3\, p_2.p_4 + p_1.p_4\, p_2.p_3). \quad (6.89)$$

Many other useful formulas with the γ matrices and their traces are given in Appendix G.

6.4.3 Electron–Muon Scattering Differential Cross Section

The previous section gives us the tools to calculate the unpolarised amplitudes squared. Coming back to the reaction $e^-(p) + \mu^-(k) \to e^-(p') + \mu^-(k')$, both the tensors (6.81) are of type $L_{uu}^{\mu\nu}$ in (6.83) since there is no antiparticle in this reaction. Moreover, in this simple case of elastic scattering, the particle species does not change, and thus $m = m'$. It follows that

$$\begin{aligned} L_{uu}^{\mu\nu}\Big|_{m=m'} &= \frac{1}{2}\left(k_\rho k'_\eta \mathrm{Tr}(\gamma^\rho\gamma^\mu\gamma^\eta\gamma^\nu) + mk_\rho \mathrm{Tr}(\gamma^\rho\gamma^\mu\gamma^\nu) + mk'_\eta \mathrm{Tr}(\gamma^\mu\gamma^\eta\gamma^\nu)\right. \\ &\left. + m^2 \mathrm{Tr}(\gamma^\mu\gamma^\nu)\right) \\ &= 2(k^\mu k'^\nu + k^\nu k'^\mu + (m^2 - k\cdot k')g^{\mu\nu}), \end{aligned} \tag{6.90}$$

where properties (6.85), (6.87) and (6.88) were used to obtain the last line. Notice that for antiparticles, since m is replaced by $-m$ in the tensor Formulas (6.83), we also have

$$L_{vv}^{\mu\nu}\Big|_{m=m'} = L_{uu}^{\mu\nu}\Big|_{m=m'}. \tag{6.91}$$

The unpolarised amplitude squared (6.80) is proportional to $L^{\mu\nu}(\mu^-)L_{\mu\nu}(e^-)$, which reads thanks to Eq. (6.90),

$$\begin{aligned} L^{\mu\nu}(\mu^-)L_{\mu\nu}(e^-) &= 2(k^\mu k'^\nu + k^\nu k'^\mu + (m_\mu^2 - k\cdot k')g^{\mu\nu}) \times 2(p_\mu p'_\nu + p_\nu p'_\mu \\ &+ (m_e^2 - pp')g_{\mu\nu}) \\ &= 8\left(k\cdot p\, k'\cdot p' + k\cdot p'\, k'\cdot p - m_e^2 k\cdot k' - m_\mu^2 p\cdot p' + 2m_e^2 m_\mu^2\right). \end{aligned} \tag{6.92}$$

It is generally more appropriate to express this using Mandelstam variables (cf. Section 3.6.3). Here, they verify

$$\begin{aligned} s &= (k+p)^2 = (k'+p')^2 = m_\mu^2 + m_e^2 + 2k\cdot p = m_\mu^2 + m_e^2 + 2k'\cdot p', \\ t &= q^2 = (p-p')^2 = (k'-k)^2 = 2m_e^2 - 2p\cdot p' = 2m_\mu^2 - 2k\cdot k', \\ u &= (p-k')^2 = (p'-k)^2 = m_\mu^2 + m_e^2 - 2p\cdot k' = m_\mu^2 + m_e^2 - 2p'\cdot k. \end{aligned}$$

Hence, using Eq. (6.92), the unpolarised amplitudes squared (6.80) becomes

$$\overline{|\mathcal{M}|^2} = \frac{2e^4}{t^2}\left[(s - m_e^2 - m_\mu^2)^2 + (u - m_e^2 - m_\mu^2)^2 + 2t(m_e^2 + m_\mu^2)\right]. \tag{6.93}$$

An alternative expression may be more convenient depending on the frame in which $\overline{|\mathcal{M}|^2}$ is evaluated. Problem 6.7 proposes to show that

$$\overline{|\mathcal{M}|^2} = \frac{4e^4}{t^2}\left[(s - m_e^2 - m_\mu^2)(m_e^2 + m_\mu^2 - u) + t(m_e^2 + m_\mu^2) + \frac{t^2}{2}\right]. \tag{6.94}$$

It is important to realise that both expressions (6.93) and (6.94) are valid in any frame since $\mathcal{M}$ is Lorentz invariant by construction.

Calculating the cross section becomes easier on making some assumptions. Let us envisage two extreme cases: when both electrons and muons are ultra-relativistic and when the incoming muon is at rest, the electrons still being ultra-relativistic (the muon is 200 times heavier than the electron).

Ultra relativistic e^-, μ^-: in this case, all masses can be neglected. In the CM frame, the Mandelstam variables are

$$s = (E_e + E_\mu)^2 = 4|\boldsymbol{p}^*|^2, \;\; t = -2|\boldsymbol{p}^*|^2(1 - \cos\theta^*), \;\; u = -2|\boldsymbol{p}^*|^2(1 + \cos\theta^*),$$

where θ^* is the angle in the CM frame between the incident e^- and the outgoing e^-, and $\boldsymbol{p}^*$ is the momentum in that frame [the four particles have the same momentum norm since the masses are neglected; see Eq. (3.34)]. The unpolarised amplitudes squared (6.93) takes the form

$$\overline{|\mathcal{M}|^2} = \frac{2e^4}{t^2}\left[s^2+u^2\right] = \frac{2e^4}{(1-\cos\theta^*)^2}\left[4+(1+\cos\theta^*)^2\right] = \frac{2e^4}{\sin^4\frac{\theta^*}{2}}\left[1+\cos^4\frac{\theta^*}{2}\right]. \tag{6.95}$$

The differential cross section is obtained using the Formula (3.42),

$$\frac{d\sigma}{d\Omega^*} = \frac{1}{64\pi^2 s}\frac{2e^4}{\sin^4\frac{\theta^*}{2}}\left[1+\cos^4\frac{\theta^*}{2}\right] = \frac{\alpha^2}{2s}\left[\frac{1+\cos^4\frac{\theta^*}{2}}{\sin^4\frac{\theta^*}{2}}\right], \tag{6.96}$$

where the fine structure constant (2.80) was inserted. In the expression of the differential cross section (6.96), s is simply $(E_e+E_\mu)^2$, where the energies of the incident particles are calculated in the CM frame.

Muon at rest: if we still neglect the electron mass, the Mandelstam variables in the lab frame, where the muon is at rest, are

$$s = m_\mu^2 + 2E_e m_\mu,\quad t = q^2 = -2E_e E_e'(1-\cos\theta) = -4E_e E_e' \sin^2\frac{\theta}{2},\quad u = m_\mu^2 - 2E_e' m_\mu,$$

with θ, the scattering angle between the two electrons. Inserting these variables into the absolute amplitude squared (6.94) yields, with $m_e = 0$,

$$\begin{aligned}\overline{|\mathcal{M}|^2} &= \frac{16e^4 E_e E_e' m_\mu^2}{q^4}\left[1-\sin^2\frac{\theta}{2}+\frac{2E_e E_e'}{m_\mu^2}\sin^4\frac{\theta}{2}\right]\\ &= \frac{16e^4 E_e E_e' m_\mu^2}{q^4}\cos^2\frac{\theta}{2}\left[1-\frac{q^2}{2m_\mu^2}\tan^2\frac{\theta}{2}\right].\end{aligned} \tag{6.97}$$

In order to calculate the differential cross section, it is convenient to use Eq. (3.43) since

$$\frac{d\sigma}{d\Omega} = \frac{d\sigma}{dt}\frac{dt}{d\Omega} = \frac{1}{128\pi^2 s}\frac{\overline{|\mathcal{M}|^2}}{|\boldsymbol{p}^*|^2}\frac{dt}{d\cos\theta}.$$

As we neglect the electron mass, Eq. (3.34) leads to $|\boldsymbol{p}^*|^2 = (s-m_\mu^2)^2/(4s) = E_e^2 m_\mu^2/s$. The variable t given above is also

$$t = (k'-k)^2 = 2m_\mu^2 - 2E_\mu' m_\mu = 2m_\mu^2 - 2(E_e - E_e' + m_\mu)m_\mu,$$

where energy conservation was used in the last equality. The only variable that depends on $\cos\theta$ is E_e', which is easy to check by exploring the energy–momentum conservation (see Problem 6.8), i.e.

$$E_e' = \frac{E_e m_\mu}{m_\mu + E_e(1-\cos\theta)}. \tag{6.98}$$

Hence, $dt/d\cos\theta = 2m_\mu\, dE_e'/d\cos\theta$, and using the expression of E_e' from the problem, it follows that $dE_e'/d\cos\theta = E_e'^2/m_\mu$. Putting everything together, we obtain

$$\frac{d\sigma}{d\Omega} = \frac{1}{64\pi^2}\frac{\overline{|\mathcal{M}|^2}}{m_\mu^2}\left(\frac{E_e'}{E_e}\right)^2. \tag{6.99}$$

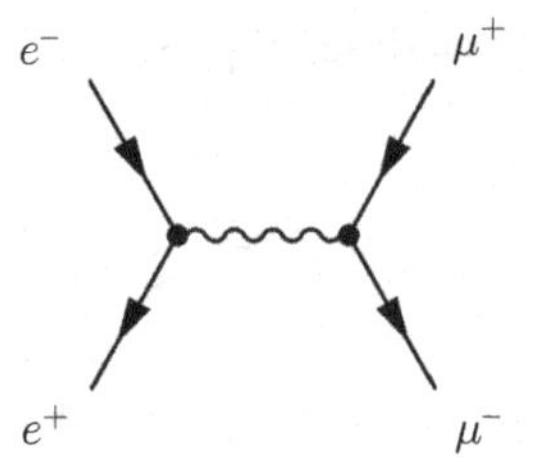

Fig. 6.6 The lowest order diagram of the reaction $e^- + e^+ \to \mu^+ + \mu^-$.

Finally, inserting the expression of $\overline{|\mathcal{M}|^2}$ above, with $q^4 = 16E_e^2 E_e'^2 \sin^4 \frac{\theta}{2}$, and using the fine structure constant (2.80) yield

$$\frac{d\sigma}{d\Omega} = \frac{\alpha^2}{4E_e^2 \sin^4 \frac{\theta}{2}} \frac{E_e'}{E_e} \cos^2 \frac{\theta}{2} \left[1 - \frac{q^2}{2m_\mu^2} \tan^2 \frac{\theta}{2} \right]. \tag{6.100}$$

This formula turns out to be useful in another context, namely when we will explore the proton substructure in Chapter 7.

6.4.4 Electron–Positron Annihilation: $e^- + e^+ \to \mu^+ + \mu^-$

In this example, we are confronted for the first time with an amplitude involving antiparticles. The Feynman diagram for $e^-(p) + e^+(k) \to \mu^+(k') + \mu^-(p')$ is shown in Fig. 6.6. We can follow the same kind of calculation as in the previous example, but instead of u-spinors, v-spinors must be used for e^+ and μ^+. Therefore, the lepton tensor $L_{\mu\nu}^{\mu\nu}$ in (6.83) should be used. However, instead of following all the previous calculations (do it as an exercise!), we can do better by profiting from the power of the crossed reactions presented in Section 3.7. Let us compare the diagram of the reaction $e^- + e^+ \to \mu^- + \mu^+$ in Fig. 6.6 with that of $e^- + \mu^- \to e^- + \mu^-$ in Fig. 6.5. The t-channel of the reaction $e^- + e^+ \to \mu^- + \mu^+$ is just the s-channel of $e^- + \mu^- \to e^- + \mu^-$, as illustrated in Fig. 6.7. Hence, if we still neglect the masses of the electron and positron, we just have to interchange in Eq. (6.95) the variables s and t. It yields the differential cross section [still using Formula (3.42)],

$$\frac{d\sigma}{d\Omega^*} = \frac{1}{64\pi^2 s} 2e^4 \frac{(t^2 + u^2)}{s^2} \frac{|\boldsymbol{p}'^*|}{|\boldsymbol{p}^*|}.$$

As the masses are neglected, the momenta in the CM frame are $|\boldsymbol{p}'^*| = |\boldsymbol{p}^*| = \sqrt{s}/2$, according to Eq. (3.34). The three Mandelstam variables are

$$s = (k+p)^2 = 4|\boldsymbol{p}^*|^2,\ t = (p-k')^2 = -2|\boldsymbol{p}^*|^2(1-\cos\theta),\ u = (p-p')^2 = -2|\boldsymbol{p}^*|^2(1+\cos\theta),$$

where θ is the angle (in the CM frame) between the incident e^- and the outgoing μ^+. Using $\alpha = e^2/(4\pi)$, we obtain for the differential cross section,

$$\frac{d\sigma}{d\Omega^*} = \frac{\alpha^2}{4s}(1 + \cos^2\theta), \tag{6.101}$$

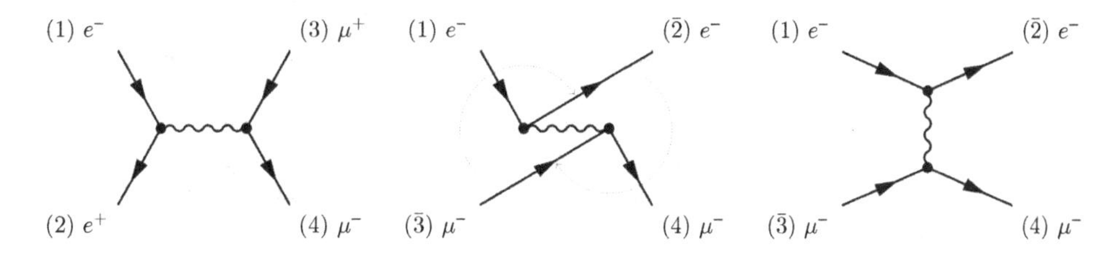

Fig. 6.7 Left: the Feynman diagram of $e^- + e^+ \to \mu^- + \mu^+$ considered as the s-channel, i.e. $s = (p_1 + p_2)^2$, $t = (p_1 - p_3)^2, u = (p_1 - p_4)^2$. Centre: the particles labelled (2) and (3) are exchanged, and their corresponding momenta become $p_2 \leftrightarrow -p_3$. The affected particles are then interpreted as the antiparticles of those of the previous diagram. The variable s becomes t (and vice versa): the diagram is thus the t-channel of the one on the left-hand side. Right: the same diagram as in the centre but redrawn more conventionally (after 'vertical stretching').

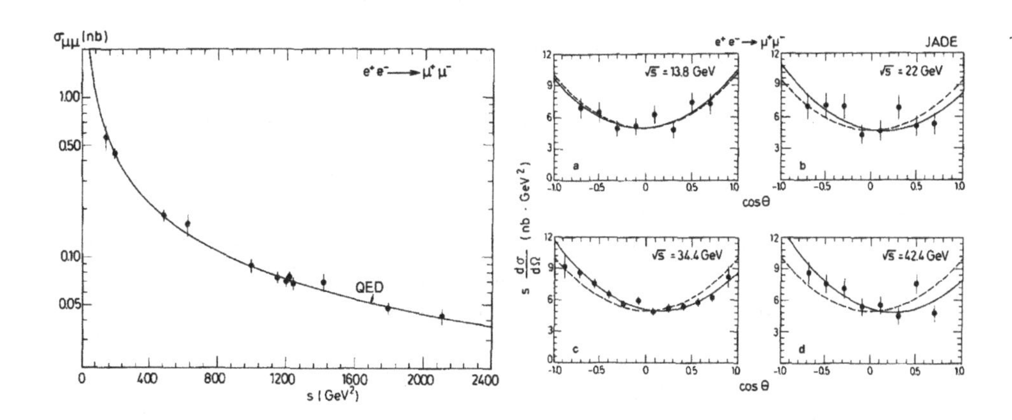

Fig. 6.8 Left: total cross section for $e^+e^- \to \mu^+\mu^-$. Right: angular distribution, i.e. $s\, \mathrm{d}\sigma / \mathrm{d}\Omega$ as a function of $\cos\theta$. See text for the description of solid and dashed lines. Reprinted by permission from Springer Nature, (1985).

while the total cross section is

$$\sigma = \int_0^{2\pi} \mathrm{d}\phi \int_{-1}^{1} \mathrm{d}(\cos\theta)\, \frac{\alpha^2}{4s}(1 + \cos^2\theta) = \frac{4\pi\alpha^2}{3s}. \tag{6.102}$$

Figure 6.8 presents the cross section measured by the JADE experiment. This experiment operated in the 1980s at the particle accelerator PETRA at the Deutsches Elektronen-Synchrotron (DESY) in Hamburg. It collected data from e^-e^+ collisions at various CM energy, $\sqrt{s}$. The left-hand side figure shows the total cross section as a function of $\sqrt{s}$. The solid line results from a calculation in QED to order α^3. It is, however, very close (within 1 percent) to Formula (6.102) obtained at order α (to compare the values obtained with the formula, do not forget the conversion factor from natural units to nanobarns, about 0.389×10^6). The right-hand side figure shows the differential cross section as a function of $\cos\theta$, where θ is the angle between the incoming e^+ and the outgoing μ^+ (which is $\theta - \pi$ in Eq. (6.101) and thus, leaves the formula unchanged). The dashed line corresponds to QED prediction to order α^3, which is very close to Eq. (6.101). The solid line

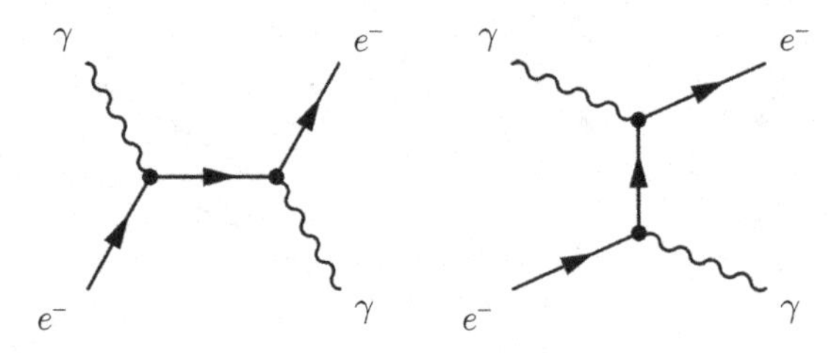

Fig. 6.9 The two lowest order Compton diagrams.

shows an asymmetry at high energy in the $\cos\theta$ distribution due to electroweak corrections, where $e^+ + e^- \to \mu^+ + \mu^-$ can occur via a Z boson exchange. (We shall see this in Chapter 10.)

6.4.5 Compton Scattering: $e^- + \gamma \to e^- + \gamma$

Compton scattering is the scattering of a photon by an electron,

$$e^-(p, s) + \gamma(k, \lambda) \to e^-(p', s') + \gamma(k', \lambda').$$

The labels s and λ denote the spin/helicity of the electron and the polarisation of the photon respectively, while p and k are their momenta. The labels with the prime symbols are reserved for the particles in the final states. There are two diagrams, shown in Fig. 6.9, contributing to this process at lowest order (i.e., at α order): the s-channel (left-hand side of 6.9) and u-channel[11] (right-hand side). Since photons are bosons, the total amplitude is the sum of the two. Following the Feynman rules of Table 6.1, the amplitude of the first diagram is

$$i\mathcal{M}_1 = \bar{u}_{s'}(p')(-iq\gamma^\mu)\epsilon^*_\mu(k', \lambda')\, i\frac{\not p + \not k + m}{(p+k)^2 - m^2}(-iq\gamma^\nu)\epsilon_\nu(k, \lambda)u_s(p).$$

There is no ambiguity in the position of the different terms in the equation: as the amplitude must be a complex number, the adjoint spinor has to be on the left of the electron propagator. Similarly, the second amplitude is

$$i\mathcal{M}_2 = \bar{u}_{s'}(p')(-iq\gamma^\nu)\epsilon_\nu(k, \lambda)\, i\frac{\not p - \not k' + m}{(p-k')^2 - m^2}(-iq\gamma^\mu)\epsilon^*_\mu(k', \lambda')u_s(p).$$

As the incoming and outgoing particles are on the mass shell ($p^2 = p'^2 = m^2$ and $k^2 = k'^2 = 0$), the total amplitude becomes, using $q = -e$,

$$\mathcal{M} = -e^2\epsilon^*_\mu(k', \lambda')\epsilon_\nu(k, \lambda)\, \bar{u}_{s'}(p')\left(\gamma^\mu \frac{\not p + \not k + m}{2p\cdot k}\gamma^\nu - \gamma^\nu\frac{\not p - \not k' + m}{2p\cdot k'}\gamma^\mu\right)u_s(p).$$

[11] It may not be obvious that the two contributions are s- and u-channels. Let us assign to the particles in $e^-(p) + \gamma(k) \to e^-(p') + \gamma(k')$ the labels $(1) + (2) \to (3) + (4)$. Then the u-channel is obtained by swapping (2) and (4) (namely the two photons) in the left-hand side diagram of Fig. 6.9. It gives the diagram on the right after a vertical 'stretching'.

Now, notice that $\not{p}\gamma^\nu = \gamma^\rho p_\rho \gamma^\nu = p_\rho(-\gamma^\nu\gamma^\rho + 2g^{\rho\nu}) = -\gamma^\nu \not{p} + 2p^\nu$. Hence, due to the Dirac equation in momentum space, $(\not{p} - m)u_s(p) = 0$,

$$\not{p}\gamma^\nu u_s(p) = -m\gamma^\nu u_s(p) + 2p^\nu u_s(p).$$

This equality cancels the mass term in the numerator of both propagators. The scalar product in the denominators can be easily expressed as a function of the Mandelstam variables, giving $2p \cdot k = (s - m^2)$ and $2p \cdot k' = -(u - m^2)$. Overall, this yields the total amplitude

$$\mathcal{M} = -e^2 \epsilon^*_\mu(k', \lambda')\epsilon_\nu(k, \lambda)\, \bar{u}_{s'}(p') \left(\gamma^\mu \frac{2p^\nu + \not{k}\gamma^\nu}{s - m^2} + \gamma^\nu \frac{2p^\mu - \not{k}'\gamma^\mu}{u - m^2} \right) u_s(p).$$

For the calculation of the unpolarised cross section, we have to average over the two transverse polarisations of the incoming photon, since the photon here is a real photon (and not virtual), and average over the two spins/helicities of the electron. We also have to sum over the final polarisations and spin/helicity states. Hence, the unpolarised amplitude squared reads

$$\overline{|\mathcal{M}|^2} = \frac{1}{2}\sum_{\lambda=1,2} \frac{1}{2}\sum_{s} \sum_{\lambda'=1,2} \sum_{s'} |\mathcal{M}|^2 = \overline{|\mathcal{M}_1|^2} + \overline{|\mathcal{M}_2|^2} + \overline{\mathcal{M}_1\mathcal{M}_2^*} + \left(\overline{\mathcal{M}_1\mathcal{M}_2^*}\right)^*,$$

with

$$\overline{|\mathcal{M}_1|^2} = \frac{e^4}{4(s - m^2)^2} \sum_{\lambda,\lambda',s,s'} \left[\epsilon^*_\mu(k', \lambda')\epsilon_\nu(k, \lambda)\, \bar{u}_{s'}(p')\gamma^\mu(2p^\nu + \not{k}\gamma^\nu)u_s(p)\right]$$
$$\times \left[\epsilon^*_\rho(k', \lambda')\epsilon_\eta(k, \lambda)\, \bar{u}_{s'}(p')\gamma^\rho(2p^\eta + \not{k}\gamma^\eta)u_s(p)\right]^*,$$

$$\overline{|\mathcal{M}_2|^2} = \frac{e^4}{4(u - m^2)^2} \sum_{\lambda,\lambda',s,s'} \left[\epsilon^*_\mu(k', \lambda')\epsilon_\nu(k, \lambda)\, \bar{u}_{s'}(p')\gamma^\nu(2p^\mu - \not{k}'\gamma^\mu)u_s(p)\right]$$
$$\times \left[\epsilon^*_\rho(k', \lambda')\epsilon_\eta(k, \lambda)\, \bar{u}_{s'}(p')\gamma^\eta(2p^\rho - \not{k}'\gamma^\rho)u_s(p)\right]^*,$$

$$\overline{\mathcal{M}_1\mathcal{M}_2^*} = \frac{e^4}{4(s - m^2)^2(u - m^2)^2} \sum_{\lambda,\lambda',s,s'} \left[\epsilon^*_\mu(k', \lambda')\epsilon_\nu(k, \lambda)\, \bar{u}_{s'}(p')\gamma^\mu(2p^\nu + \not{k}\gamma^\nu)u_s(p)\right]$$
$$\times \left[\epsilon^*_\rho(k', \lambda')\epsilon_\eta(k, \lambda)\, \bar{u}_{s'}(p')\gamma^\eta(2p^\rho - \not{k}'\gamma^\rho)u_s(p)\right]^*.$$

Just the first calculation will be detailed. No need to calculate $\overline{|\mathcal{M}_2|^2}$ since the second graph is just the u-channel of the first: interchanging s and u variables will give the result. For the interference term, there is no other solution than doing the calculation (which is very similar to the first term). The difficulty here is the summation or average over the physical polarisations. In Eq. (5.131), we saw the corresponding completeness relation. It is rather complicated but fortunately, in all QED processes, the two terms with p^μ or p^ν in Eq. (5.131) necessarily vanish when combined with the other elements of a QED amplitude. In other words, one can show that the following simplified completeness relation can always be used [see, for instance, Peskin and Schroeder (1995, p. 159)]

$$\sum_{\lambda=1}^{2} \epsilon^*_\mu(\lambda)\epsilon_\nu(\lambda) \rightsquigarrow -g_{\mu\nu}. \tag{6.103}$$

The symbol $\rightsquigarrow$ emphasises the fact that it is not an actual equality but merely a simplification when combined with the rest of the amplitude.[12] Therefore, according to Eq. (6.103), the unpolarised amplitudes squared $\overline{|\mathcal{M}_1|^2}$ now reads

$$\overline{|\mathcal{M}_1|^2} = \frac{e^4}{4(s-m^2)^2} g_{\nu\eta} g_{\mu\rho} \sum_{s,s'} \left[\bar{u}_{s'}(p')\gamma^\mu(2p^\nu + \not{k}\gamma^\nu)u_s(p)\right]\left[\bar{u}_{s'}(p')\gamma^\rho(2p^\eta + \not{k}\gamma^\eta)u_s(p)\right]^*.$$

The square brackets are just complex numbers. Thus, taking the complex conjugate is equivalent to taking the Hermitian conjugate. Hence, the second square brackets simplify to

$$u_s^\dagger(p)(2p^\eta + \gamma^{\eta\dagger}k_\alpha\gamma^{\alpha\dagger})\gamma^{\rho\dagger}\gamma^{0\dagger}u_{s'}(p') = \bar{u}_s(p)(2p^\eta + \gamma^\eta\not{k})\gamma^\rho u_{s'}(p'),$$

where the relation (5.32) of the Hermitian conjugate of γ matrices has been used as well as the property $(\gamma^0)^2 = \mathbb{1}$ to obtain the last equality. Now, we can use the completeness relation for spinors to sum over the spins of $u_s(p)\bar{u}_s(p)$ and $u_{s'}(p')\bar{u}_{s'}(p')$. It yields

$$\overline{|\mathcal{M}_1|^2} = \frac{e^4}{4(s-m^2)^2}\mathrm{Tr}\left((\not{p}' + m)\gamma^\mu(2p^\nu + \not{k}\gamma^\nu)(\not{p} + m)(2p_\nu + \gamma_\nu\not{k})\gamma_\mu\right).$$

According to Eq. (6.87), the trace of an odd number of γ matrices is 0. This reduces $\overline{|\mathcal{M}_1|^2}$ to

$$\overline{|\mathcal{M}_1|^2} = \frac{e^4}{4(s-m^2)^2}\left(\mathrm{Tr}(A) + m^2\mathrm{Tr}(B)\right),$$

with $A = \not{p}'\gamma^\mu(2p^\nu + \not{k}\gamma^\nu)\not{p}(2p_\nu + \gamma_\nu\not{k})\gamma_\mu$ and $B = \gamma^\mu(2p^\nu + \not{k}\gamma^\nu)(2p_\nu + \gamma_\nu\not{k})\gamma_\mu$. Using standard properties of γ matrices, one can show in Problem 6.9 that

$$\mathrm{Tr}(A) = 32\left[-m^2 p'\cdot p + (p'\cdot k)(p\cdot k - m^2)\right] \quad \text{and} \quad \mathrm{Tr}(B) = 64(m^2 + p\cdot k).$$

It follows that

$$\overline{|\mathcal{M}_1|^2} = \frac{8e^4}{(s-m^2)^2}\left(-m^2 p'\cdot p + (p'\cdot k)(p\cdot k) - m^2 p'\cdot k + 2m^4 + 2m^2 p\cdot k\right).$$

Now, let us replace the scalar products with expressions of the Mandelstam variables s and u. Since the variables are not independent, $s + t + u = \sum_{i=1}^4 m_i^2$, the variable t is $t = 2m^2 - s - u$. As $s = (p+k)^2, t = (p'-p)^2$ and $u = (k-p')^2$, we obtain

$$p\cdot k = \frac{s-m^2}{2}, \quad p'\cdot p = m^2 - \frac{t}{2} = \frac{s+u}{2}, \quad p'\cdot k = \frac{m^2-u}{2}.$$

The insertion of the above expressions finally yields

$$\overline{|\mathcal{M}_1|^2} = 4e^4\left[\frac{2m^4}{(s-m^2)^2} + \frac{m^2}{s-m^2} - \frac{1}{2}\frac{u-m^2}{s-m^2}\right]. \tag{6.104}$$

The expression for $\overline{|\mathcal{M}_2|^2}$ is obtained by swapping s and u,

$$\overline{|\mathcal{M}_2|^2} = 4e^4\left[\frac{2m^4}{(u-m^2)^2} + \frac{m^2}{u-m^2} - \frac{1}{2}\frac{s-m^2}{u-m^2}\right], \tag{6.105}$$

[12] To understand quickly where this simplification is coming from, the gauge freedom of the electromagnetic field in Eq (5.118) leads in momentum space (via the Fourier transform) to the freedom to redefine $\epsilon^\mu \to \epsilon^\mu + p^\mu$. Since an amplitude is of the form $\mathcal{M} = \epsilon^\mu T_\mu$, it remains invariant only if $p^\mu T_\mu = 0$, hence the conclusion of Eq. (6.103).

while for the interference, a similar calculation yields

$$\overline{\mathcal{M}_1\mathcal{M}_2^*} = 2e^4\left[\frac{m^2}{s-m^2} + \frac{m^2}{u-m^2} + \frac{4m^4}{(s-m^2)(u-m^2)}\right]. \tag{6.106}$$

Therefore, the full unpolarised amplitudes squared $\overline{|\mathcal{M}|^2} = \overline{|\mathcal{M}_1|^2} + \overline{|\mathcal{M}_2|^2} + \overline{\mathcal{M}_1\mathcal{M}_2^*} + (\overline{\mathcal{M}_1\mathcal{M}_2^*})^*$ finally reads

$$\overline{|\mathcal{M}|^2} = -2e^4\left[\frac{u-m^2}{s-m^2} + \frac{s-m^2}{u-m^2} - 4\left(\frac{m^2}{s-m^2} + \frac{m^2}{u-m^2} + \left(\frac{m^2}{s-m^2} + \frac{m^2}{u-m^2}\right)^2\right)\right]. \tag{6.107}$$

Note that, as expected, the expression is symmetric between s and u. In order to obtain the cross section, let us express the variables using the measured quantities in the lab frame, where the electron is conventionally considered at rest. The energies of the photons are denoted $\omega = |\boldsymbol{k}|$ and $\omega' = |\boldsymbol{k}'|$. Hence, $p = (m, \boldsymbol{0})$, $k = (\omega, \boldsymbol{k})$, $p' = (E', \boldsymbol{p}')$, $k' = (\omega', \boldsymbol{k}')$. From $s = (p+k)^2$ and $u = (p-k')^2$, we obtain $s - m^2 = 2m\omega$ and $u - m^2 = -2m\omega'$, respectively. Moreover, $m^2 = p'^2 = (p + k - k')^2$, so that

$$m^2 = m^2 - 2k\cdot k' + 2p\cdot k - 2p\cdot k' = m^2 - 2\omega\omega'(1-\cos\theta) + 2m\omega - 2m\omega',$$

with θ being the scattering angle of the photon (between $\boldsymbol{k}$ and $\boldsymbol{k}'$). Therefore,

$$\frac{1}{\omega'} - \frac{1}{\omega} = \frac{1}{m}(1-\cos\theta). \tag{6.108}$$

Consequently, the unpolarised amplitude squared reads

$$\overline{|\mathcal{M}|^2} = 2e^4\left[\frac{\omega'}{\omega} + \frac{\omega}{\omega'} + 4\left(\frac{m}{2\omega} - \frac{m}{2\omega'} + (\frac{m}{2\omega} - \frac{m}{2\omega'})^2\right)\right] = 2e^4\left[\frac{\omega'}{\omega} + \frac{\omega}{\omega'} - \sin^2\theta\right].$$

The differential cross section in the lab frame is rather easy to calculate from Eq. (3.44), giving

$$\frac{\mathrm{d}\sigma}{\mathrm{d}t} = \frac{e^4}{32\pi s|\boldsymbol{p}^*|^2}\left[\frac{\omega'}{\omega} + \frac{\omega}{\omega'} - \sin^2\theta\right].$$

As $|\boldsymbol{p}^*|^2 = (s-m^2)/4s$ [see Eq. (3.34)], and $s - m^2 = 2m\omega$, this yields

$$\frac{\mathrm{d}\sigma}{\mathrm{d}t} = \frac{e^4}{32\pi\, m^2\omega^2}\left[\frac{\omega'}{\omega} + \frac{\omega}{\omega'} - \sin^2\theta\right] = \frac{\pi\alpha^2}{2m^2\omega^2}\left[\frac{\omega'}{\omega} + \frac{\omega}{\omega'} - \sin^2\theta\right]. \tag{6.109}$$

The last step consists in expressing $\mathrm{d}t$ as a function of ω, ω' and θ. The variable t is $t = (k-k')^2 = -2\omega\omega'(1-\cos\theta)$. Since ω' depends on θ via Eq. (6.108),

$$\mathrm{d}t = 2\omega\left[-\mathrm{d}\omega'(1-\cos\theta) + 2\omega\omega'\,\mathrm{d}(\cos\theta)\right],$$

where Eq. (6.108) tells us that

$$\mathrm{d}\left(\frac{1}{\omega'}\right) = -\frac{1}{m}\,\mathrm{d}(\cos\theta) = -\frac{\mathrm{d}\omega'}{\omega'^2}.$$

Hence,

$$\mathrm{d}t = 2\omega\left[-\frac{\omega'^2}{m}(1-\cos\theta) + \omega'\right]\mathrm{d}(\cos\theta) = 2\omega\left[-\omega'^2\left(\frac{1}{\omega'} - \frac{1}{\omega}\right) + \omega'\right]\mathrm{d}(\cos\theta) = 2\omega'^2\,\mathrm{d}(\cos\theta),$$

where Eq. (6.108) has been used again to replace $(1 - \cos\theta)/m$ in the first equality. Inserting this result into Eq. (6.109) yields the Compton differential cross section

$$\frac{d\sigma}{d(\cos\theta)} = \frac{\pi\alpha^2}{m^2}\left(\frac{\omega'}{\omega}\right)^2\left[\frac{\omega'}{\omega} + \frac{\omega}{\omega'} - \sin^2\theta\right]. \tag{6.110}$$

This formula is known as the Klein–Nishina formula.

At low energy, when ω is close to 0, multiplying Eq. (6.108) by ω shows that ω/ω' becomes close to 1, i.e. the scattering becomes elastic. Therefore, the differential cross section converges to

$$\frac{d\sigma}{d(\cos\theta)} \simeq \frac{\pi\alpha^2}{m^2}\left[2 - \sin^2\theta\right] = \frac{\pi\alpha^2}{m^2}\left[1 + \cos^2\theta\right],$$

and the cross section tends to

$$\sigma = \int_{-1}^{1} \frac{\pi\alpha^2}{m^2}\left[1 + \cos^2\theta\right] d(\cos\theta) = \frac{8\pi\alpha^2}{3m^2} = \frac{8\pi}{3}r_e^2,$$

where $r_e = \alpha/m$ is the classical radius of the electron.[13] This formula is the Thomson cross-section (its value is about 0.6 barns), which can be obtained by classical electromagnetism for an elastic scattering between a photon and an electron (see, e.g., the textbook Jackson, 1998). Note that the Thomson cross section is finally the 'simplest' cross-section with an electron since, from purely dimensional considerations, the cross section (an effective area) is just proportional to the square of the classical radius of the electron.

6.5 QED and Helicity/Chirality

An interaction in QED couples a photon to charged fermions via the electromagnetic current, $q\bar{w}'\gamma^\mu w$, where w and w' denote generic spinors (u-type for fermions or v-type for anti-fermions). We saw in Section 5.3.5 that any spinor can be decomposed into left- and right-handed components of chirality through the projectors (5.98). It follows that

$$\bar{w}'\gamma^\mu w = (\bar{w}'_L + \bar{w}'_R)\gamma^\mu(w_L + w_R) = \bar{w}'_L\gamma^\mu w_L + \bar{w}'_R\gamma^\mu w_R + \bar{w}'_L\gamma^\mu w_R + \bar{w}'_R\gamma^\mu w_L.$$

Given that the projectors P_L and P_R are Hermitian, and $P_{L,R}\gamma^\mu = \gamma^\mu P_{R,L}$ (since γ^5 anti-commutes with γ^μ), it turns out that two of those four contributions necessarily vanish. Let us take the example of the current with two fermions for which the term $\bar{u}'_L\gamma^\mu u_R$ reads

$$\bar{u}'_L\gamma^\mu u_R = u'^\dagger P_L\gamma^0\gamma^\mu P_R u = \bar{u}' P_R P_L \gamma^\mu u.$$

This term vanishes since $P_R P_L = 0$. The other vanishing contribution is $\bar{u}'_R\gamma^\mu u_L = 0$. Overall, the following eight terms vanish (remember that for anti-fermions, $v_L = P_R v$ and $v_R = P_L v$):

$$\bar{u}'_L\gamma^\mu u_R = \bar{u}'_R\gamma^\mu u_L = \bar{v}'_L\gamma^\mu v_R = \bar{v}'_R\gamma^\mu v_L = 0;$$
$$\bar{u}'_L\gamma^\mu v_L = \bar{u}'_R\gamma^\mu v_R = \bar{v}'_L\gamma^\mu u_L = \bar{v}'_R\gamma^\mu u_R = 0.$$

[13] The classical electron radius corresponds to the radius of a distribution of charge totalling the electron charge if its electrostatic self-energy is equal to the electron mass, i.e. in natural units: $E = e^2/(4\pi r_e) = m$.

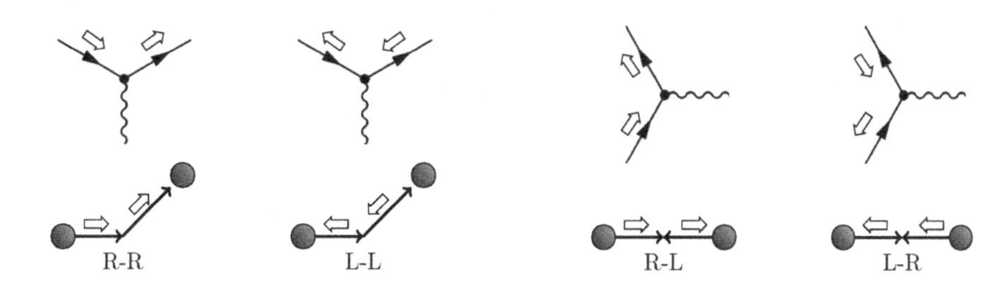

Fig. 6.10 The Helicity configuration in QED when masses of particles are neglected. The thick arrows represent the spins of the particle. The t- and s-channel processes are respectively depicted on the left- and right-hand sides. For t-channels, two other diagrams exist with particles replaced by antiparticles. For s-channels, the two other diagrams correspond to the final state configuration. The graphics on the lower part represent the direction of motion of the particles (in the CM frame on the right-hand side), and the labels denote the helicity configuration.

We then conclude that for diagrams involving t-channels or s-channels (see Section 3.7), the only non-zero contributions of the currents are the chirality configurations

$$\begin{aligned} t\text{-channel:}\ & \bar{u}'_L\gamma^\mu u_L,\ \bar{u}'_R\gamma^\mu u_R,\ \bar{v}'_L\gamma^\mu v_L,\ \bar{v}'_R\gamma^\mu v_R; \\ s\text{-channel:}\ & \bar{u}'_L\gamma^\mu v_R,\ \bar{u}'_R\gamma^\mu v_L,\ \bar{v}'_L\gamma^\mu u_R,\ \bar{v}'_R\gamma^\mu u_L. \end{aligned} \tag{6.111}$$

At high energy, in the ultra-relativistic regime where $E \gg m$, the helicity states coincide with chirality states (see Section 5.3.5). We thus conclude from Eq. (6.111) that only specific helicity configurations are possible. They are depicted in Fig. 6.10.

For t-channels (mainly involved in scattering processes), the helicity of the outgoing particle (or antiparticle) is the same as that of the incoming particle. QED then conserves helicity in the ultra-relativistic limit. For s-channels (mainly involved in pair creation or annihilation), the particle helicity is the opposite to that of the antiparticle. Hence, in the CM frame of the pair, the spins of the particle and antiparticle must be aligned, leading necessarily to $J_z = \pm 1$ and excluding the value $J_z = 0$. For instance, when incoming e^+ and e^- beams collide, the mediating virtual photon (assuming only the QED process) is necessarily transverse.

6.6 A Few Words about Renormalisation

Let us again consider the electron–muon scattering, $e^-(p_1)+\mu^-(p_2) \to e^-(p'_1)+\mu^-(p'_2)$, for which the amplitude is (see Fig. 6.5),

$$i\mathcal{M}_1 = 4\pi\alpha\ \bar{u}(p'_1)\gamma^\mu u(p_1)\frac{ig_{\mu\nu}}{q^2}\bar{u}(p'_2)\gamma^\nu u(p_2),$$

with $\alpha = e^2/4\pi$. Suppose you want to measure α with this process. Let us imagine, we have a beam of e^- and μ^-, and we detect the scattered e^- and μ^-. By playing with the beam energy and the scattering angle, we can count the number of recorded events and thus measure the cross section at a given momentum transfer q^2. From the measurement, we would then obtain the value of $\alpha_{\text{meas}} = \alpha$. Now, there are higher order corrections to this

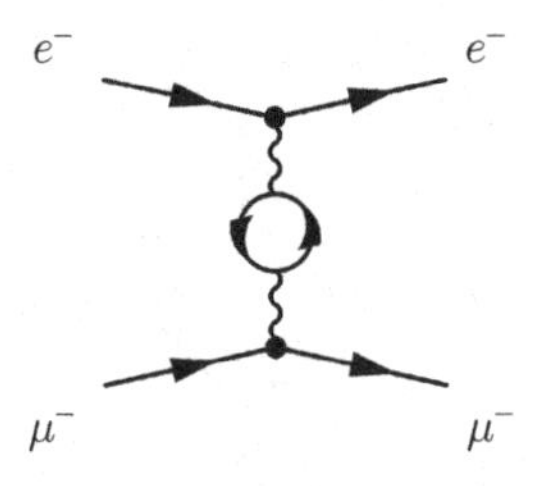

Fig. 6.11 A second-order correction to the electron–muon scattering.

process. One is shown in Fig. 6.11. The virtual photon splits into a virtual electron–positron pair, forming a fermion loop. If the photon has a momentum q, then one member of the pair carries a momentum p and the other $q-p$, such that energy–momentum is conserved. However, there is no constraint on the value of p. Therefore, all possible values have to be taken into account. The amplitude of this diagram can be calculated, but it requires additional rules with respect to those given in Section 6.1. First, there is a multiplicative factor -1 for a closed fermion loop related to the anti-commutation of fermionic operators. More specifically, if, as in Section 6.3, we insert in the scattering matrix the Dirac fields corresponding to the interaction terms due to the loop and try to move all annihilation operators to the right, there would be an odd number of permutations, which is the source of the -1 factor. Second, there is a trace encompassing the two fermionic propagators corresponding to the loop. The loop is sandwiched between two photons. A process such as $\gamma \to$ loop $\to \gamma$ cannot depend on spinor indices since the photon does not carry a spinor index, hence the trace. More detailed explanations can be found in quantum field theory books (for instance, Peskin and Schroeder, 1995; Weinberg, 1995). The amplitude is then given by

$$i\mathcal{M}_2 = \bar{u}(p_1')ie\gamma^\mu u(p_1)\left\{\frac{-ig_{\mu\mu'}}{q^2}\int\frac{\mathrm{d}^4p}{(2\pi)^4}(-1)\mathrm{Tr}\left[ie\gamma^{\mu'}i\frac{\not{p}+m_e}{p^2-m_e}ie\gamma^{\nu'}i\frac{\not{p}-\not{q}+m_e}{(p-q)^2-m_e}\right]\frac{-ig_{\nu'\nu}}{q^2}\right\}$$
$$\times\,\bar{u}(p_2')ie\gamma^\nu u(p_2).$$

The full amplitude up to the second order in α is the sum of $\mathcal{M}_1$ and $\mathcal{M}_2$, yielding

$$i\mathcal{M} = -\bar{u}(p_1')e\gamma^\mu u(p_1)\left\{\frac{-ig_{\mu\nu}}{q^2}+\frac{-i\Pi^{[2]}_{\mu\nu}(q^2)}{q^4}\right\}\bar{u}(p_2')e\gamma^\nu u(p_2), \tag{6.112}$$

where

$$i\Pi^{[2]}_{\mu\nu}(q^2) = \int\frac{\mathrm{d}^4p}{(2\pi)^4}(-1)(ie)^2\mathrm{Tr}\left[\gamma_\mu i\frac{\not{p}+m_e}{p^2-m_e}\gamma_\nu i\frac{\not{p}-\not{q}+m_e}{(p-q)^2-m_e}\right].$$

Equation (6.112) tells us that the second-order contribution modifies the initial photon propagator as

$$-ig_{\mu\nu}/q^2 \to -ig_{\mu\nu}/q^2 - i\Pi^{[2]}_{\mu\nu}(q^2)/q^4.$$

However, a simple dimensional analysis shows that the integral in $\Pi_{\mu\nu}$ diverges! More specifically, one can show that $\Pi_{\mu\nu}$ can be reduced to the only contributing term (after a procedure called dimensional regularisation)

$$\Pi^{[2]}_{\mu\nu}(q^2) = -q^2 g_{\mu\nu}\frac{\alpha}{3\pi}\left(\left[\int_{m_e^2}^{\infty}\frac{\mathrm{d}p^2}{p^2}\right] - f(q^2)\right) + \cdots + O(\alpha),$$

where m_e is the electron mass, $\alpha = e^2/(4\pi)$, and $f(q^2)$ is a finite function given by

$$f(q^2) = 6\int_0^1 \mathrm{d}z\, z(1-z)\ln\left(1 - \frac{q^2}{m_e^2}z(1-z)\right). \tag{6.113}$$

Clearly, $\int_{m_e^2}^{\infty}\mathrm{d}p^2/p^2$ diverges logarithmically. This divergence with increasing momentum is called an *ultraviolet divergence*. Let us introduce a cut-off Λ^2, namely a maximum value for the upper bound of the integral. Is it problematic to truncate the integral this way? After all, any theory must have its domain of validity. Finally, Λ is a parametrisation of our ignorance, above which new physics must appear. Then, with this prescription, the amplitude becomes

$$\mathcal{M} = \bar{u}(p_1')\gamma^{\mu}u(p_1)\frac{g_{\mu\nu}}{q^2}4\pi\alpha\left\{1 - \frac{\alpha}{3\pi}\ln\left(\frac{\Lambda^2}{m_e^2}\right) + \frac{\alpha}{3\pi}f(q^2)\right\}\bar{u}(p_2')\gamma^{\nu}u(p_2).$$

At this stage, let us rename α above as α_0. The index 0 refers to the fact that this is the parameter used in the calculation in the absence of loops, as at the beginning of this section. The parameter α_0 is called the *bare* constant. How can we interpret the measurement of the cross section we make? The actual scattering amplitude $\mathcal{M}$ is not supposed to depend on an arbitrary parameter Λ. If Λ is changed, we do not want the predicted cross section to change. In other words, we would have to change α_0 itself to compensate for the change of Λ. Consequently, α_0, the constant of the QED Lagrangian, is *not* a physical parameter! The renormalisation procedure consists in expressing the calculations with physical (i.e., measured) parameters. Here, with our second-order expansion, if our measurement is performed at $q^2 = q_0^2$, we would identify

$$\alpha_{\text{meas}}(q_0^2) = \alpha_0\left(1 - \frac{\alpha_0}{3\pi}\ln\left(\frac{\Lambda^2}{m_e^2}\right) + \frac{\alpha_0}{3\pi}f(q_0^2)\right) = \alpha_0 - \frac{\alpha_0^2}{3\pi}\left(\ln\left(\frac{\Lambda^2}{m_e^2}\right) - f(q_0^2)\right) + O(\alpha_0^3). \tag{6.114}$$

Let us invert the previous equation,

$$\begin{aligned}\alpha_0 &= \alpha_{\text{meas}}(q_0^2) + \frac{\alpha_0^2}{3\pi}\left(\ln\left(\frac{\Lambda^2}{m_e^2}\right) - f(q_0^2)\right) + O(\alpha_0^3)\\ &= \alpha_{\text{meas}}(q_0^2) + \frac{\alpha_{\text{meas}}^2(q_0^2)}{3\pi}\left(\ln\left(\frac{\Lambda^2}{m_e^2}\right) - f(q_0^2)\right) + O(\alpha_{\text{meas}}^3(q_0^2)).\end{aligned}$$

The second equality is equivalent to the first one, to the second-order used here. But what if, instead of q_0^2, we would have chosen an arbitrary value q^2? Starting from Eq. (6.114), we would have, up to the second order in α (dropping the O symbols for clarity),

$$\begin{aligned}
\alpha_{\text{meas}}(q^2) &= \alpha_0 - \frac{\alpha_0^2}{3\pi}\left(\ln\left(\frac{\Lambda^2}{m_e^2}\right) - f(q^2)\right) \\
&= \alpha_{\text{meas}}(q_0^2) + \frac{\alpha_{\text{meas}}^2(q_0^2)}{3\pi}\left(\ln\left(\frac{\Lambda^2}{m_e^2}\right) - f(q_0^2)\right) - \frac{\alpha_{\text{meas}}^2(q_0^2)}{3\pi}\left(\ln\left(\frac{\Lambda^2}{m_e^2}\right) - f(q^2)\right) \\
&= \alpha_{\text{meas}}(q_0^2) + \frac{\alpha_{\text{meas}}^2(q_0^2)}{3\pi}\left(f(q^2) - f(q_0^2)\right).
\end{aligned} \tag{6.115}$$

The good news now is that the arbitrary Λ no longer appears! We can now calculate a process at any energy (here $\sqrt{|q^2|}$) if we have measured α_{meas} at a given reference point (here q_0^2). But there is a price to pay: what we call a coupling constant is *not* a constant! The strength of electromagnetism, α, actually depends on the energy scale. Note that $q^2 < 0$ for scattering processes. We can define the positive quantity $Q^2 = -q^2$. In the approximation where $Q^2 \gg m_e^2$ (valid for high energy physics experiments), Problem 6.10 shows that $\alpha_{\text{meas}}(Q^2)$ that we now simply denote by $\alpha(Q^2)$ becomes

$$\alpha(Q^2) = \alpha(Q_0^2)\left[1 + \frac{\alpha(Q_0^2)}{3\pi}\ln\left(\frac{Q^2}{Q_0^2}\right)\right] + O(\alpha^2). \tag{6.116}$$

In Eqs. (6.115) and (6.116), we have managed to eliminate Λ up to the second order (α^2) of perturbation theory. The renormalisation works if we can eliminate it to any order. In QED, this turns out to be the case. When higher orders are taken into account (adding more loop contributions), the dependence of α on Q^2 when $Q^2 \gg m_e$ then becomes

$$\alpha(Q^2) = \frac{\alpha(Q_0^2)}{1 - \frac{\alpha(Q_0^2)}{3\pi}\ln\left(\frac{Q^2}{Q_0^2}\right)}. \tag{6.117}$$

Let us interpret this formula, only valid in the regime $Q^2 \gg m_e^2$. It gives the value of the fine structure constant at a given energy $\sqrt{Q^2}$, provided it has been measured at another energy scale. Its variation is only logarithmic with the scale. It is the reference at $Q^2 = m_e^2$, which is used in books to quote the value of $\alpha = 1/137.03599\ldots$, or equivalently, the value of the elementary electric charge e. For higher Q^2, α varies, and, for instance, at $\sqrt{Q^2}$ close to the Z^0 boson mass (about 91 GeV/c^2), α is about 1/128. The variation of the charge (or α) as a function of the energy scale is an observed physical effect. It is illustrated in Fig. 6.12a. This variation of the charge is due to the vacuum polarisation. Imagine you want to test the charge of an electron by approaching a probe charge. In the vicinity of the electron, virtual charged pairs can be created living only a time $\Delta t \sim \hbar/m_e c^2$. They spread apart at a maximum distance $c\Delta t$, creating a dipole of that length. Attracted by the electron bare charge, the virtual positrons spend more time near the negative bare charge before annihilation, whereas the virtual electrons are repelled. This is illustrated in Fig. 6.12b. The cloud of virtual positrons tends to produce a *charge screening* of the bare charge. If the probe charge does not approach the bare charge enough, for instance, at a distance illustrated by the largest dashed circle in Fig. 6.12b, it will 'feel' mainly the surrounding clouds of positrons, yielding to a measurement underestimating the real charge of the electron. In other words, at a small Q (i.e., at a large distance), the charge seems

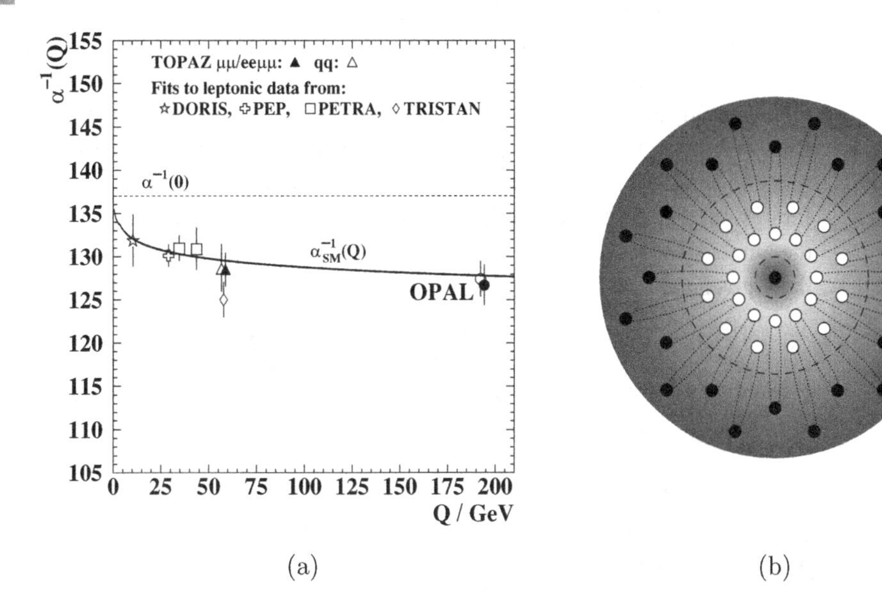

Fig. 6.12 (a) Variation of the inverse of α as a function of Q, with the associated measurements. Figure from OPAL Collaboration (2004). (b) The charge screening effect. Virtual electrons are represented by the black circles and positrons by the open circles. They form transient dipoles (the dotted ellipses). The radii of the dashed circles centred on the bare electron charge at the centre represent two distances of interaction (see text). (a) Reprinted by permission from Springer Nature (2004).

smaller: $|e_{\text{meas}}| \simeq |-e + \epsilon \sum e|$, where ϵ represents the fraction due to the screening. Conversely, at high Q (i.e., close to the bare charge, as illustrated by the small dashed circle in Fig. 6.12b), the probe charge penetrates the virtual cloud, which minimises its effect ($\epsilon \simeq 0$).

In summary, renormalisation is a procedure that absorbs all infinite or arbitrary quantities by a redefinition of a finite number of parameters. In QED, they are the electron charge (or equivalently the fine structure constant), the electron mass (yes, the mass!) and the normalisation of the fields (thus, the name renormalisation).

6.7 A Major Test of QED: $g-2$

There are many tests that validate QED, both at high energy (in accelerators) and low energy. One of the most impressive tests is probably the prediction of the gyromagnetic ratio, which is related to the magnetic moment of the electron.

6.7.1 Prediction with Dirac Equation

In classical mechanics, an electron placed in a magnetic field $\boldsymbol{B}$ has additional potential energy due to the field given by (see Problem 6.12)

$$E_B = -\boldsymbol{\mu} \cdot \boldsymbol{B} = -\frac{q}{2m} \boldsymbol{L} \cdot \boldsymbol{B},$$

where $\boldsymbol{\mu}$ is the magnetic moment, $\boldsymbol{L}$, the orbital angular momentum, m, the electron mass, and $q = -e$, its charge. The gyromagnetic ratio, i.e. the ratio between the magnetic moment and the orbital angular momentum is thus $-g\mu_B$, where $\mu_B = e/(2m)$ is the Bohr magneton and $g = 1$, the g-factor. Since electrons carry an intrinsic angular momentum, the spin $\boldsymbol{S}$, we would also expect additional energy due to the spin $E_B = -g\frac{q}{2m}\boldsymbol{S}\cdot\boldsymbol{B}$. It turns out that the Dirac equation predicts the value $g = 2$. To check this, let us start from the Dirac equation in momentum space, Eq. (5.48). In the presence of an electromagnetic field, given the usual substitution $p_\mu \to p_\mu - qA_\mu$ with $A^\mu = (V, \boldsymbol{A})$, it reads

$$\left(\gamma^\mu p_\mu - q\gamma^\mu A_\mu - m\right) u(p) = \left(\gamma^0 E - q\gamma^0 V - \boldsymbol{\gamma}\cdot\boldsymbol{p} + q\boldsymbol{\gamma}\cdot\boldsymbol{A} - m\right) u(p) = 0.$$

Expanding the γ matrices in terms of Pauli matrices yields the two coupled equations

$$\left.\begin{aligned}(\boldsymbol{\sigma}\cdot\boldsymbol{p} - q\boldsymbol{\sigma}\cdot\boldsymbol{A})u_b(p) &= (E - qV - m)u_a(p)\\ (\boldsymbol{\sigma}\cdot\boldsymbol{p} - q\boldsymbol{\sigma}\cdot\boldsymbol{A})u_a(p) &= (E - qV + m)u_b(p)\end{aligned}\right\} \quad \text{with } u(p) = \begin{pmatrix} u_a(p) \\ u_b(p)\end{pmatrix}.$$

In the non-relativistic limit, $E \simeq m$, and the energy qV of usual electromagnetic fields is negligible with respect to m. It follows that

$$\begin{aligned}(\boldsymbol{\sigma}\cdot\boldsymbol{p} - q\boldsymbol{\sigma}\cdot\boldsymbol{A})u_b &= (E_{\text{n.r.}} - qV)u_a,\\ (\boldsymbol{\sigma}\cdot\boldsymbol{p} - q\boldsymbol{\sigma}\cdot\boldsymbol{A})u_a &\simeq 2m\, u_b,\end{aligned}$$

where $E_{\text{n.r.}} = E - m$ is the non-relativistic energy ($E_{\text{n.r.}} \ll m$). Inserting the second equation into the first leads to

$$\left[(\boldsymbol{\sigma}\cdot\boldsymbol{p})^2 + q^2(\boldsymbol{\sigma}\cdot\boldsymbol{A})^2 - q\left((\boldsymbol{\sigma}\cdot\boldsymbol{p})(\boldsymbol{\sigma}\cdot\boldsymbol{A}) + (\boldsymbol{\sigma}\cdot\boldsymbol{A})(\boldsymbol{\sigma}\cdot\boldsymbol{p})\right)\right] u_a = 2m(E_{\text{n.r.}} - qV)u_a.$$

According to Eq. (5.77), $(\boldsymbol{\sigma}\cdot\boldsymbol{p})^2 = |\boldsymbol{p}|^2\mathbb{1}$ and similarly, $(\boldsymbol{\sigma}\cdot\boldsymbol{A})^2 = |\boldsymbol{A}|^2\mathbb{1}$. Moreover, using the identity $(\boldsymbol{\sigma}\cdot\boldsymbol{x})(\boldsymbol{\sigma}\cdot\boldsymbol{y}) = \boldsymbol{x}\cdot\boldsymbol{y} + i\boldsymbol{\sigma}\cdot(\boldsymbol{x}\times\boldsymbol{y})$, it follows that

$$\left[(\boldsymbol{p} - q\boldsymbol{A})^2 - iq\boldsymbol{\sigma}\cdot(\boldsymbol{p}\times\boldsymbol{A} + \boldsymbol{A}\times\boldsymbol{p})\right] u_a = 2m(E_{\text{n.r.}} - qV)u_a.$$

This equation is in momentum space. We can interpret it in the position space by simply substituting $u_a(p)$ with $\psi(x) = u_a(p)e^{-ip\cdot x}$ and $\boldsymbol{p}$ with $\hat{\boldsymbol{p}} = -i\boldsymbol{\nabla}$. This gives

$$(-i\boldsymbol{\nabla} - q\boldsymbol{A}(x))^2\psi(x) - q\boldsymbol{\sigma}\cdot[\boldsymbol{\nabla}\times(\boldsymbol{A}(x)\psi(x)) + \boldsymbol{A}(x)\times\boldsymbol{\nabla}\psi(x)] = 2m(E_{\text{n.r.}} - qV(x))\psi(x).$$

The term in square brackets is just $(\boldsymbol{\nabla}\times\boldsymbol{A}(x))\,\psi(x) = \boldsymbol{B}(x)\psi(x)$. Hence, it follows that

$$\left[\frac{(-i\boldsymbol{\nabla} - q\boldsymbol{A}(x))^2}{2m} - \frac{q}{2m}\boldsymbol{\sigma}\cdot\boldsymbol{B}(x) + qV(x)\right]\psi(x) = E_{\text{n.r.}}\,\psi(x)$$

The quantity in square brackets,

$$H_{E\simeq m} = \frac{(\hat{\boldsymbol{p}} - q\boldsymbol{A})^2}{2m} - \frac{q}{2m}\boldsymbol{\sigma}\cdot\boldsymbol{B} + qV,$$

corresponds to the non-relativistic Hamiltonian of the Dirac equation in the presence of an electromagnetic field. Recalling that the spin operator for 2-components spinors is $\boldsymbol{S} = \frac{1}{2}\boldsymbol{\sigma}$, we identify the contribution to the energy due to the magnetic field,

$$E_B = -2\frac{q}{2m}\boldsymbol{S}\cdot\boldsymbol{B},$$

which means that the non-relativistic limit of the Dirac equation implies a g-factor of 2.

6.7.2 Higher-Order Corrections

The value $g = 2$ of the g-factor predicted by the Dirac equation only takes into account the first-order calculation in e of the perturbative series expansion in QED. To convince ourselves, let us consider the electromagnetic current connected to the basic vertex in QED. The diagram is shown on the left-hand side of Fig. 6.13.

If $\psi_i(x) = u_i(p_i)e^{-ip_i\cdot x}$ and $\psi_f(x) = u_f(p_f)e^{-ip_f\cdot x}$ are the wave functions of the initial and final electron, respectively, the current in the three-dimensional space reads

$$j^\mu(x) = -e\overline{u_f}(p_f)\gamma^\mu u_i(p_i)e^{-i(p_i-p_f)\cdot x} = i\left[\overline{u_f}(p_f)(ie\gamma^\mu)u_i(p_i)\right]e^{-i(p_i-p_f)\cdot x}. \tag{6.118}$$

The term in the square brackets would result directly from the Feynman rules (see Table 6.1). The spinors satisfy the Dirac equation in momentum space, and thus, $u_i(p_i) = \not{p}_i u_i(p_i)/m$ and $\overline{u_f}(p_f) = \overline{u_f}(p_f)\not{p}_f/m$. Therefore, we can decompose the current into two parts,

$$j^\mu(x) = -\frac{e}{2m}\left[\overline{u_f}(p_f)\not{p}_f\gamma^\mu u_i(p_i) + \overline{u_f}(p_f)\gamma^\mu\not{p}_i u_i(p_i)\right]e^{-i(p_i-p_f)\cdot x}.$$

Writing $\not{p}_f = p_{f\nu}\gamma^\nu$ and $\not{p}_i = p_{i\nu}\gamma^\nu$, and using the Clifford algebra, $\gamma^\mu\gamma^\nu + \gamma^\nu\gamma^\mu = 2g^{\mu\nu}$, yields

$$j^\mu(x) = -\frac{e}{2m}\overline{u_f}(p_f)\left[2p_f^\mu + \gamma^\mu\gamma^\nu(p_{i\nu} - p_{f\nu})\right]u_i(p_i)\,e^{-i(p_i-p_f)\cdot x}.$$

Now, let us introduce the matrix

$$\sigma^{\mu\nu} = \frac{i}{2}\left(\gamma^\mu\gamma^\nu - \gamma^\nu\gamma^\mu\right), \tag{6.119}$$

such that, combined with the Clifford algebra, $\gamma^\mu\gamma^\nu = g^{\mu\nu} - i\sigma^{\mu\nu}$, we obtain the current

$$j^\mu(x) = i\left[\frac{1}{2m}\overline{u_f}(p_f)(ie)(p_f^\mu + p_i^\mu)u_i(p_i) + \overline{u_f}(p_f)\frac{e}{2m}\sigma^{\mu\nu}(p_{i\nu} - p_{f\nu})u_i(p_i)\right]e^{-i(p_i-p_f)\cdot x}. \tag{6.120}$$

This expression of the current (known as Gordon decomposition) is interesting because according to the Feynman rules in Table 6.1, p. 198, we see that the first term in brackets would correspond to the vertex factor of a spin 0 particle with the electric charge $-e$. We then expect the second term,

$$j_s^\mu(x) = i\overline{u_f}(p_f)\frac{e}{2m}\sigma^{\mu\nu}(p_{i\nu} - p_{f\nu})u_i(p_i)\,e^{-i(p_i-p_f)\cdot x}, \tag{6.121}$$

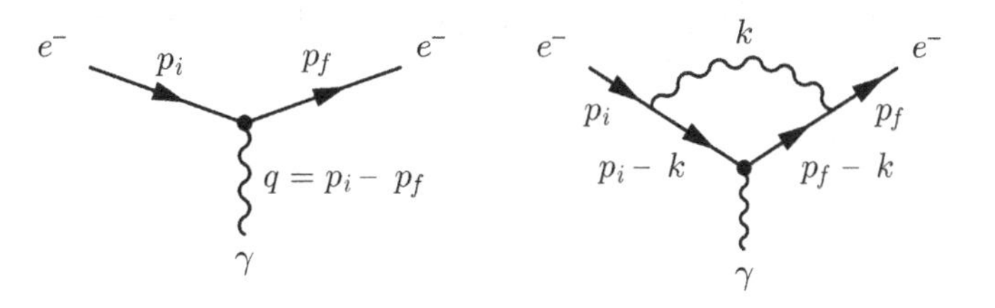

Fig. 6.13 Left: The simplest QED vertex, at order e. Right: a possible correction to that vertex at order e^3.

to be related to the interaction involving the electron spin. To check this assumption, consider, to the first order of perturbation, the amplitude given by the interaction of such a current with an electromagnetic field A_μ. It reads

$$S^{[1]} = -i \int \mathrm{d}^4x\, j_s^\mu(x) A_\mu(x) = \int \mathrm{d}^4x\, \overline{u_f} \frac{e}{2m} \sigma^{\mu\nu} (p_{i_\nu} - p_{f_\nu}) u_i\, e^{-i(p_i - p_f)\cdot x} A_\mu(x).$$

Let us consider a static field (Coulomb scattering) for simplification. Then A_μ does not depend on time, and we can integrate over time, using Eq. (E.9). This gives

$$S^{[1]} = 2\pi\, \delta(E_i - E_f) \int \mathrm{d}^3x\, \overline{u_f} \frac{e}{2m} \sigma^{\mu\nu} (p_{i_\nu} - p_{f_\nu}) u_i\, e^{i(\boldsymbol{p}_i - \boldsymbol{p}_f)\cdot \boldsymbol{x}} A_\mu(\boldsymbol{x}).$$

Because of the Dirac-delta function, the energy component of $p_{i_\nu} - p_{f_\nu}$ in the integral is zero, and hence, only the three momentum components matter. After re-arrangement, $S^{[1]}$ thus reads

$$S^{[1]} = -2\pi i\, \delta(E_i - E_f) \int \mathrm{d}^3x\, \overline{u_f} \frac{e}{2m} \sigma^{\mu j} u_i\, \partial_j \left(e^{i(\boldsymbol{p}_i - \boldsymbol{p}_f)\cdot \boldsymbol{x}} \right) A_\mu(\boldsymbol{x}),$$

which, after an integration by parts (the field A_μ vanishing at infinity), yields

$$S^{[1]} = 2\pi i\, \delta(E_i - E_f) \int \mathrm{d}^3x\, \overline{u_f} \frac{e}{2m} \sigma^{\mu j} u_i\, e^{i(\boldsymbol{p}_i - \boldsymbol{p}_f)\cdot \boldsymbol{x}}\, \partial_j A_\mu(\boldsymbol{x}). \tag{6.122}$$

As in the previous section, in order to see clearly the spin contribution, we are going to consider the low energy limit, where the mass dominates over the 3-momentum. Then, only the upper component of the spinors matters since, according to the Formula (5.62),

$$u = \begin{pmatrix} u_a \\ \frac{\sigma\cdot p}{E+m} u_a \end{pmatrix} \xrightarrow{m \gg |p|} \begin{pmatrix} u_a \\ 0 \end{pmatrix}.$$

Now, one can check in Problem 6.13 that

$$\sigma^{0j} = i \begin{pmatrix} \mathbb{0} & \sigma^j \\ \sigma_j & \mathbb{0} \end{pmatrix}, \quad \sigma^{kj} = \sum_l \epsilon_{kjl} \begin{pmatrix} \sigma^l & \mathbb{0} \\ \mathbb{0} & \sigma^l \end{pmatrix}, \tag{6.123}$$

where ϵ_{kjl} is the antisymmetric tensor ($\epsilon_{kjl} = 1$ for cyclic permutation of 123, -1 for anticyclic permutation, 0 otherwise). Therefore, the low energy limit implies that the following term in Eq. (6.122) reduces to

$$\overline{u_f} \sigma^{\mu j} u_i \partial_j A_\mu = \overline{u_f} \sigma^{0j} u_i \partial_j A_0 + \overline{u_f} \sigma^{kj} u_i \partial_j A_k \approx \overline{u_f} \sigma^{kj} u_i \partial_j A_k \approx u_{a_f}^\dagger \left[\sum_l \epsilon_{kjl}\, \sigma^l \partial_j A_k \right] u_{a_i}. \tag{6.124}$$

It turns out that the term in square brackets is $-[\boldsymbol{\sigma} \times \boldsymbol{\partial}]^k A_k$ (Problem 6.14). Thus, we conclude that

$$\overline{u_f} \sigma^{\mu j} u_i \partial_j A_\mu = -u_{a_f}^\dagger [\boldsymbol{\sigma} \times \boldsymbol{\partial}]^k A_k\, u_{a_i} = -u_{a_f}^\dagger [\boldsymbol{\sigma} \times \boldsymbol{\partial}] \cdot \boldsymbol{A}\, u_{a_i} = -u_{a_f}^\dagger [\boldsymbol{\partial} \times \boldsymbol{A}] \cdot \boldsymbol{\sigma}\, u_{a_i}.$$

As the magnetic field is $\boldsymbol{B} = \boldsymbol{\partial} \times \boldsymbol{A}$, we finally obtain from Eq. (6.122) the first-order calculation,

$$S^{[1]} = 2\pi i\, \delta(E_i - E_f)\, u_{a_f}^\dagger(p_f) \left(\int \mathrm{d}^3x \left(-\frac{e}{2m} \right) \boldsymbol{\sigma} \cdot \boldsymbol{B}\, e^{i(\boldsymbol{p}_i - \boldsymbol{p}_f)\cdot \boldsymbol{x}} \right) u_{a_i}(p_i).$$

We identify in the integral the magnetic moment of the electron

$$\boldsymbol{\mu} = -\frac{e}{2m}\boldsymbol{\sigma} = -2\frac{e}{2m}\mathbf{S} \tag{6.125}$$

coupled to the magnetic field. As announced at the beginning of this section, the current (6.121) naturally yields the gyromagnetic ratio with the g-factor 2.

Now, consider the second diagram on the right-hand side in Fig. 6.13. The new virtual photon introduces a correction to the initial current (6.118). According to Feynman rules, the square brackets in Eq. (6.118) are now $\overline{u_f}(p_f)(ie\Gamma^\mu)u_i(p_i)$ with

$$\Gamma^\mu = \int \frac{d^4k}{(2\pi)^4}(ie\gamma^\nu)i\frac{\not{p}_i - \not{k} + m}{(p_i-k)^2 - m^2 + i\epsilon}\gamma^\mu i\frac{\not{p}_f - \not{k} + m}{(p_f-k)^2 - m^2 + i\epsilon}(ie\gamma^\eta)\frac{-ig_{\nu\eta}}{k^2 + i\epsilon}.$$

The quantity Γ^μ corresponds to the second-order correction to the standard vertex of QED. The full contribution, up to second-order, is thus the sum of both diagrams in Fig. 6.13, leading to a vertex factor $ie(\gamma^\mu + \Gamma^\mu)$. The calculation of Γ^μ is complicated for two reasons. First, for the high 4-momenta of the photon loop, the integral diverges since it behaves as $\int d^4k/k^4$. This is an ultraviolet divergence similar to what we encountered in the previous section. Second, because the photon is massless, when the loop momentum approaches zero, there is another divergence, called *infrared.* The actual details of the regularisation of Γ^μ are beyond the scope of this book but can be found, for example, in Peskin and Schroeder (1995, Chapter 6) or Weinberg (1995, Chapter 11). The important conclusion is that, for a small momentum transfer q, the term proportional to $\sigma^{\mu\nu}$ in the current (6.120) is modified to

$$i\frac{e}{2m}\sigma^{\mu\nu}q_\nu \to i\frac{e}{2m}\sigma^{\mu\nu}q_\nu + \frac{\alpha}{2\pi}i\frac{e}{2m}\sigma^{\mu\nu}q_\nu,$$

Therefore, we can immediately deduce that the gyromagnetic ratio in Eq. (6.125) must be modified to

$$\boldsymbol{\mu} = -2\left(1 + \frac{\alpha}{2\pi}\right)\frac{e}{2m}\mathbf{S},$$

or equivalently, a_e, the anomalous magnetic moment corresponding to the relative deviation with respect to the value $g = 2$, is

$$a_e = \frac{g-2}{2} = \frac{\alpha}{2\pi}.$$

Numerically, we obtain $a_e \simeq 0.00116$, to be compared with the most precise experimental value obtained so far (see Odom et al., 2006; Hanneke et al., 2008, 2011):

$$a_e^{\text{exp}} = 1\,159\,652\,180.73\,(0.28) \times 10^{-12} \tag{6.126}$$

the numbers in parentheses denoting measurement uncertainty in the last two digits at one standard deviation. Pretty close! Such experimental accuracy pushes the theory very far and requires a calculation at least to the 8th order (i.e., α^4). The calculation, which represents 12672 vertex-type Feynman diagrams up to 10th order (!!), has been recently evaluated (Aoyama et al., 2012, 2014). Hadronic (vacuum polarisation), electroweak effects and small QED contributions from virtual muon and tau–lepton loop contributions have also to be taken into account. Fortunately, the diagrams are now evaluated numerically, and the computed value of a_e from Aoyama et al. (2014) is

$$a_e^{\text{th}} = 1\,159\,652\,181.643\,(25)(23)(16)(763) \times 10^{-12},$$

where the first three uncertainties are from the 8th-order term, 10th-order term, and hadronic and electroweak terms, respectively. It is worth noticing that the largest uncertainty, the fourth, originates from the measurement of the fine structure constant. Both theory and experiment are in very good agreement since

$$a_e^{\text{exp}} - a_e^{\text{th}} = -0.91\,(0.82) \times 10^{-12}.$$

As the largest source of uncertainty of a_e^{th} is the experimental value used for the fine structure constant (obtained from Cesium or Rubidium atom experiments), and no longer an uncertainty coming from the calculation itself, it now makes sense to determine α from the theory and the measured value of a_e instead of the other way around. Such accuracy in the computation of the g-factor is one of the most impressive triumphs of quantum electrodynamics theory. However, discrepancies between the measured value and the theoretical prediction of the anomalous muon magnetic moment have recently been reported. They might be the first indications of physics beyond the Standard Model, as we shall see in Chapter 12.

6.7.3 Measurement of the Electron *g*-Factor

We have already seen that a system with an electron in a magnetic field gets an additional contribution to the energy with a term $-\boldsymbol{\mu}\cdot\boldsymbol{B} = g\mu_B\,\boldsymbol{S}\cdot\boldsymbol{B}$ with $\mu_B = e/(2m)$. Since the electron spin projection is $\hbar m_s$ with $m_s = \pm 1/2$, the energy contribution is $h\nu_s m_s$, where ν_s is the spin frequency related to the cyclotron frequency $\nu_c = e|\boldsymbol{B}|/(2\pi m)$ by $\nu_s = g\nu_c/2$. If an experiment manages to measure both ν_c and ν_s, or more specifically ν_c and $\nu_a = \nu_s - \nu_c$, the 'anomaly' frequency, then the g-factor would be accessible, since

$$\frac{g}{2} = \frac{\nu_s}{\nu_c} = 1 + \frac{\nu_s - \nu_c}{\nu_c} = 1 + \frac{\nu_a}{\nu_c}.$$

The experimental set-up [described in Hanneke et al. (2011)] is based on a Penning trap (see Fig. 6.14) which suspends a single electron using a strong magnetic field along the z-axis and an electrostatic quadrupole potential. The electric field confines the electron in a potential well, in which it can make small vertical oscillations. The Penning trap is used to artificially bind the electron in an orbital state as if it was an electron of an atom. Actually, the electrostatic potential shifts the cyclotron frequency from ν_c to $\bar{\nu}_c$ and thus ν_a to $\bar{\nu}_a = \nu_s - \bar{\nu}_c$. The lowest energy level, including the leading relativistic correction, can be approximated by the formula (Hanneke et al., 2011)

$$E_{n,m_s} = h\nu_s m_s + \left(n + \frac{1}{2}\right) h\bar{\nu}_c - \frac{1}{2}\left(n + \frac{1}{2} + m_s\right)^2 h\delta,$$

where the quantum number n is an integer ($n = 0, 1$, etc.), and δ is a frequency of the order $\delta \approx 10^{-9}\nu_c$, taking into account relativistic corrections. The determination of the transitions between the energy levels

$$E_{1,\frac{1}{2}} - E_{0,\frac{1}{2}} = h\left(\bar{\nu}_c - \frac{3}{2}\delta\right), \quad E_{0,\frac{1}{2}} - E_{1,-\frac{1}{2}} = h\bar{\nu}_a, \quad E_{1,-\frac{1}{2}} - E_{0,-\frac{1}{2}} = h\left(\bar{\nu}_c - \frac{1}{2}\delta\right),$$

allows measuring $\bar{\nu}_c$, δ and ν_s and hence $g/2$. It turns out that $\bar{\nu}_c \approx 150$ GHz while $\nu_s \approx 150.2$ GHz. Since $\bar{\nu}_c \simeq \nu_c$ is proportional to B, a high magnetic field is necessary to increase the spacing between the cyclotron energy levels. The cavity of the Penning trap

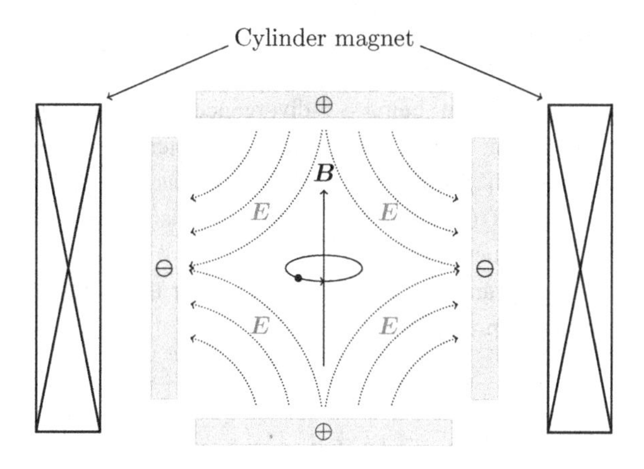

Fig. 6.14 Sketch of a Penning trap. A constant electric field (dotted grey lines) generated by the quadrupole in grey traps the electron (black circle) inside a cavity (a vacuum chamber). The superposed magnetic field is generated by the surrounding cylinder magnet.

is maintained at low temperature (100 mK) to avoid transitions from the ground state to other states due to blackbody photons. The electron is initially prepared in the state $|0, \frac{1}{2}\rangle$. The spin-up state is obtained by playing with the electrostatic potential (which induces an additional magnetic field). The higher cyclotron states are artificially excited by inserting microwaves in the trap cavity. Then, the measurement detects the anomaly transition $|0, \frac{1}{2}\rangle \to |1, -\frac{1}{2}\rangle$ followed by the decay to the ground state $|1, -\frac{1}{2}\rangle \to |0, -\frac{1}{2}\rangle$. Measurements are also performed at slightly different cyclotron frequencies (by varying the B field) to estimate the frequency shift due to the trap cavity (the shape of the trap cavity modifies the density of states of the radiation modes of free space). This overly simplistic description of the experimental setup does not do justice to the complexity of an experiment aiming to obtain a measurement accuracy of the order of one part in 10^{13} as in Eq. (6.126). The reader is thus invited to read the description given in Hanneke et al. (2011).

Problems

6.1 Apply the Euler–Lagrange equation to ψ_β in Eq. (6.15) and show that it gives the Dirac adjoint equation (5.43).

6.2 The electromagnetic Lagrangian is given by $\mathcal{L}_\gamma = -\frac{1}{4}F_{\mu\nu}F^{\mu\nu}$. Show that $\mathcal{L}_\gamma = -\frac{1}{2}F_{0i}F^{0i} - \frac{1}{4}F_{ji}F^{ji}$. Using the components of the electric and magnetic field given by Eq. (2.40), deduce that $\mathcal{L}_\gamma = \frac{1}{2}(|\boldsymbol{E}|^2 - |\boldsymbol{B}|^2)$.

6.3 In this problem, we prove the implication

$$\left.\begin{aligned} \left[A_\mu(x), \Pi_\nu(y)\right] &= ig_{\mu\nu}\delta^{(3)}(\boldsymbol{x}-\boldsymbol{y}) \\ \left[A_\mu(x), A_\nu(y)\right] &= \left[\Pi_\mu(x), \Pi_\nu(y)\right] = 0 \end{aligned}\right\} \Rightarrow \left\{\begin{aligned} \left[A_\mu(x), \mathring{A}_\nu(y)\right] &= -ig_{\mu\nu}\delta^{(3)}(\boldsymbol{x}-\boldsymbol{y}) \\ \left[A_\mu(x), A_\nu(y)\right] &= \left[\mathring{A}_\mu(x), \mathring{A}_\nu(y)\right] = 0 \end{aligned}\right.$$

where all commutators are taking at equal time, i.e. $x^0 = y^0$.

1. In the Feynman gauge ($\zeta = 1$), show that the Lagrangian (6.20) is $\mathcal{L} = -\frac{1}{2}\partial_\mu A_\nu \partial^\mu A^\nu + \frac{1}{2}\partial_\mu [A_\nu(\partial^\nu A^\mu) - A^\mu(\partial_\nu A^\nu)]$.
2. The second term, being a 4-divergence, can be ignored since it would not contribute to the action. Hence, show that the first term leads to the conjugate momentum field $\Pi_\nu = -\partial^0 A_\nu$. Conclude that the commutators with $\mathring{A}_\nu$ are those given above.

6.4 Starting from $\mathrm{Tr}(\gamma^\rho\gamma^\mu\gamma^\eta\gamma^\nu)$, show that it is $-\mathrm{Tr}(\gamma^\rho\gamma^\mu\gamma^\eta\gamma^\nu) + 8g^{\rho\mu}g^{\eta\nu} - 8g^{\eta\rho}g^{\mu\nu} + 8g^{\rho\nu}g^{\mu\eta}$ to deduce Eq. (6.88).

6.5 Draw the diagrams at the minimal order for the following reactions and write down the corresponding probability amplitudes.

$$1.\ e^+ + \mu^- \to e^+ + \mu^-, \quad 2.\ e^+ + \mu^+ \to e^+ + \mu^+, \quad 3.\ e^+ + e^- \to \mu^+ + \mu^-.$$

What about $e^+ + \mu^- \to e^- + \mu^+$?

6.6 Draw all diagrams to the first-order in α_e for the reactions below. If there are several diagrams per reaction, precise whether you have to add or subtract the diagrams.

1. $e^- + e^- \to e^- + e^-$ (Møller scattering),
2. $e^+ + e^- \to e^+ + e^-$ (Bhabba scattering),
3. $e^- + e^+ \to \gamma + \gamma$ (pair annihilation).

6.7 Show that the absolute squared amplitude (6.93) of the scattering $e^- + \mu^- \to e^- + \mu^-$,

$$\overline{|\mathcal{M}|^2} = \frac{2e^4}{t^2}\left[(s - m_e^2 - m_\mu^2)^2 + (u - m_e^2 - m_\mu^2)^2 + 2t(m_e^2 + m_\mu^2)\right],$$

may be written

$$\overline{|\mathcal{M}|^2} = \frac{4e^4}{t^2}\left[(s - m_e^2 - m_\mu^2)(m_e^2 + m_\mu^2 - u) + t(m_e^2 + m_\mu^2) + \frac{t^2}{2}\right].$$

6.8 Show that when the electron mass is neglected and the initial muon is at rest, the energy of the scattered electron in the process $e^- + \mu^- \to e^- + \mu^-$ is given by Eq. (6.98).

6.9 In Compton scattering (see Section 6.4.5), show that

$$\mathrm{Tr}(\not{p}'\gamma^\mu(2p^\nu + \not{k}\gamma^\nu)\not{p}(2p_\nu + \gamma_\nu\not{k})\gamma_\mu) = 32\left[-m^2 p'\cdot p + (p'\cdot k)(p\cdot k - m^2)\right],$$

where p and p' are the 4-momenta of the electrons, and k and k', those of the photons. Hint: use the properties $\not{p}^2 = p^2 = \not{p}'^2 = p'^2 = m^2$ and $k^2 = 0$. Also, use the γ properties (G.6), (6.85), (6.89). Similarly, show that

$$\mathrm{Tr}(\gamma^\mu(2p^\nu + \not{k}\gamma^\nu)(2p_\nu + \gamma_\nu\not{k})\gamma_\mu) = 64m^2 + 64p\cdot k.$$

6.10 For $Q^2 \gg m^2$, show that $f(Q^2)$ in Eq. (6.113) converges to $-\ln(m^2/Q^2) - 5/3$. Conclude that Eq. (6.115) becomes Eq. (6.116).

6.11 Consider the ultra-relativistic regime of the reaction $e^-(p) + e^+(k) \to \mu^-(p') + \mu^+(k')$.

1. List the helicity combinations that contribute to the process.
2. For each, compute the amplitude $\mathcal{M}$. (Use the Formulas (5.96) for the helicity states spinors).
3. In order to determine the spin-averaged amplitude squared $\overline{|\mathcal{M}_{\text{tot}}|^2}$ of the process, do you have to add first the different amplitudes and square the result or do you have to add the individual squared amplitudes? Justify your answer.

4. Using the Mandelstam variables, conclude that $\overline{|\mathcal{M}_{\rm tot}|^2}$ is $2e^4(t^2+u^2)/s^2$.
5. Using the spin summations and trace theorems, redo the calculation of $\overline{|\mathcal{M}_{\rm tot}|^2}$. Check the consistency of your result.

6.12 Starting from the non-relativistic Hamiltonian in the presence of a magnetic field, $\hat{H} = (\hat{p} - q\boldsymbol{A})^2/(2m)$, with $\boldsymbol{A}$, the vector potential and q, the particle charge, show that the additional potential energy due to the field is $E_B = -q/(2m)\hat{\boldsymbol{L}}$, where $\hat{\boldsymbol{L}}$ is the orbital angular momentum operator. The equalities $\boldsymbol{B} = \nabla \times \boldsymbol{A}$ and $\nabla \cdot \boldsymbol{A} = 0$ may be useful.

6.13 Check Eq. (6.123) using the Dirac representation of γ matrices.

6.14 Check that in Eq. (6.124), $\sum_l \epsilon_{kjl}\, \sigma^l \partial_j A_k = -\,[\boldsymbol{\sigma} \times \boldsymbol{\partial}]^k A_k$. Hint: proceed by components.

7 From Hadrons to Partons

In this chapter, the notion of partons is introduced. Evidence of the substructure of the nucleon is given, and the formalism of the deep inelastic scattering is presented.

7.1 Electron–Proton Scattering

Nowadays, we know that nucleons are not elementary particles, but composed of quarks, to the first approximation. Already in the early 1930s, there were indications that the proton was not an elementary particle. The first measurements of its magnetic moment by Stern and Rabi revealed an unexpectedly large value. A value of the order of $|\boldsymbol{\mu_p}| = 2.79\,\mu_N$ was measured, where

$$\mu_N = \frac{e}{2M} \tag{7.1}$$

is the nuclear magneton and M is the proton mass. Given that the proton is a spin-1/2 particle, this implies a g-factor, $g_p = 5.59$ since in natural units,

$$|\boldsymbol{\mu_p}| = g_p \frac{e}{2M}|\boldsymbol{S}| = g_p|\boldsymbol{S}|\,\mu_N = \frac{g_p}{2}\mu_N. \tag{7.2}$$

Therefore, g_p differs significantly from the value 2 expected for elementary fermions described by the Dirac equation (see Chapter 6). A possible explanation for such a large value was to assume that the proton is made of elementary fermions, with its magnetic moment resulting from the vector sums of the moments of its constituents. (We shall see in Chapter 8, an example of the calculation of hadron magnetic moments assuming non-relativistic sub-constituents.) Therefore, probing the internal structure of the proton became rapidly interesting. Experiments probing the internal structure of particles are conceptually similar to the famous Rutherford experiment. One observes the scattering of a probe particle, whose properties are well known, on the particle under study. In the proton case, one has to use a probe that simplifies the interpretation of the scattering experiment. Such a probe is, for example, an electron that interacts mostly via the electromagnetic interaction with the proton.[1] By the mid-1950s, accelerators were powerful enough to envisage these studies. Electron beams with energy between 100 and 200 MeV were achieved, producing elastic scatterings. The reaction we consider to probe the proton substructure is thus

$$e^-(k) + \mathrm{p}(p) \to e^-(k') + X(p'), \tag{7.3}$$

[1] We consider the electron energy small enough, say below 20 GeV, so that the weak interaction can be neglected.

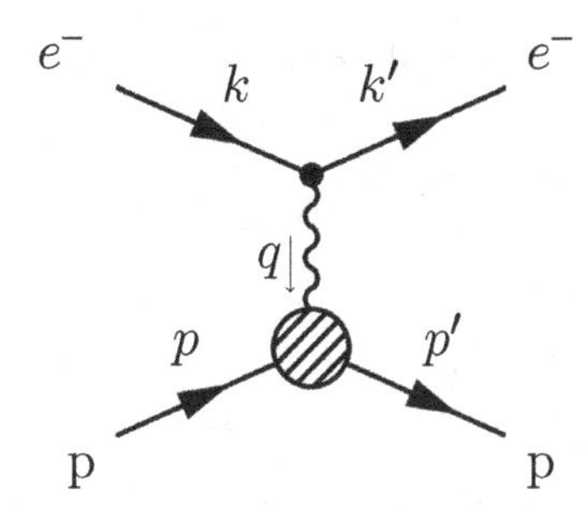

Fig. 7.1 Elastic scattering of an electron on a proton.

where X is a proton if the scattering remains elastic or a set of multi-hadrons in the case of a deep inelastic scattering (see Section 7.1.3). The 4-momentum transfer is $q = k - k' = p' - p$. For a scattering process, with q being space-like ($q^2 < 0$), there exists a frame where the energy transfer is zero, and only momentum is transferred. Hence, q is usually simply called the momentum transfer. It is convenient to introduce the positive variable,

$$Q^2 = -q^2 > 0. \tag{7.4}$$

Large values of Q^2 thus indicate a large momentum transfer from the incoming electron to the proton.

7.1.1 Elastic Scattering with a Point-Like Proton

As a starting point, let us consider small momentum transfers. When the momentum transfer between the electron and the proton is small enough, the wavelength of the virtual photon (see Fig. 7.1) is large since $\lambda \propto 1/\sqrt{Q^2}$, and the proton size cannot be resolved. The proton is then seen as an elementary point-like particle. In that case, the blob representing the proton-virtual photon vertex in Fig. 7.1 corresponds to the usual $ie\gamma^\mu$ factor. The proton in the initial state is considered at rest, i.e. $p = (M, \mathbf{0})$, and the electron mass is neglected. In such a case, the calculation of the cross section is the same as that of the process $e^- + \mu^- \to e^- + \mu^-$ in Sections 6.4.1–6.4.3. Let us recall the calculation we did but replace the muon with the proton. The energies of the incident and outgoing electron are now denoted E and E', respectively. The absolute amplitude square is [see Eq. (6.80)]

$$\overline{|\mathcal{M}|^2} = \frac{e^4}{q^4} L^{\mu\nu}(\mathrm{p}) L_{\mu\nu}(e^-) = \frac{16\pi^2\alpha^2}{q^4} L^{\mu\nu}(\mathrm{p}) L_{\mu\nu}(e^-),$$

where

$$q^2 = (k - k')^2 = -2EE' \sin^2\frac{\theta}{2}. \tag{7.5}$$

The tensor product $L^{\mu\nu}(\mathrm{p})L_{\mu\nu}(e^-)$ is deduced from Eq. (6.97) yielding

$$L^{\mu\nu}(\mathrm{p}) L_{\mu\nu}(e^-) = 16EE'M \cos^2\frac{\theta}{2}\left[1 - \frac{q^2}{2M^2}\tan^2\frac{\theta}{2}\right]. \tag{7.6}$$

According to Eq. (6.99), the differential cross section now reads

$$\left(\frac{d\sigma}{d\Omega}\right)_{\text{point-like}} = \frac{\alpha^2}{4M^2q^4}L^{\mu\nu}(\text{p})L_{\mu\nu}(e^-)\left(\frac{E'}{E}\right)^2, \tag{7.7}$$

yielding [see Eq. (6.100)]

$$\left(\frac{d\sigma}{d\Omega}\right)_{\text{point-like}} = \frac{\alpha^2}{4E^2\sin^4\frac{\theta}{2}}\frac{E'}{E}\cos^2\frac{\theta}{2}\left[1 - \frac{q^2}{2M^2}\tan^2\frac{\theta}{2}\right], \tag{7.8}$$

where the energy E' is constrained by Eq. (6.98) to be

$$E' = \frac{E}{1+\frac{2E}{M}\sin^2\frac{\theta}{2}}. \tag{7.9}$$

The term in Eq. (7.8),

$$\left(\frac{d\sigma}{d\Omega}\right)_{\text{Mott}} = \frac{\alpha^2}{4E^2\sin^4\frac{\theta}{2}}\frac{E'}{E}\cos^2\frac{\theta}{2}, \tag{7.10}$$

is the Mott cross section. It describes the scattering of a charged spin-1/2 particle by a charged spin-0 particle (the charge being $\pm e$). This is proven in Problem 7.2, assuming that the proton is spinless. Therefore, in Mott scattering, only the charge of the proton matters. (When the electron spin is also ignored, the Mott cross section becomes the classical Rutherford cross section, as shown in Problem 7.1.) The factor E'/E just reflects the proton recoil (in the absence of recoil, the energy conservation imposes $E' = E$). The additional term in the square bracket of Eq. (7.8) is thus due to the interaction of the electron with the spin of the proton and hence with its magnetic moment. Now, recall that the tensor $L^{\mu\nu}$ is defined in Eq. (6.83) as

$$L^{\mu\nu}(\text{p}) = \frac{1}{2}\sum_{r,r'}\left[\bar{u}_{p',r'}\gamma^\mu u_{p,r}\right]^*\left[\bar{u}_{p',r'}\gamma^\nu u_{p,r}\right]. \tag{7.11}$$

The term in the brackets is the electromagnetic current of the proton (the charge e is dropped) that we can express with the Gordon decomposition in Eq. (6.120) as

$$j^\mu_{\text{p.l.}} = \bar{u}(p')\gamma^\mu u(p) = \bar{u}(p')\left(\frac{1}{2M}(p'^\mu + p^\mu) + i\frac{1}{2M}\sigma^{\mu\nu}(p'_\nu - p_\nu)\right)u(p), \tag{7.12}$$

where the subscript p.l. stands for point-like. We showed in Chapter 6 that the last term with $\sigma^{\mu\nu}$ gives rise to the spin–spin interaction. It is thus this term that is responsible for the factor $-\frac{q^2}{2M^2}\tan^2\frac{\theta}{2}$ in Eqs. (7.7) and (7.8).

In the following sections, we will compare the double differential cross sections $\frac{d\sigma}{dE'\,d\Omega}$ of the scattering $e^- + \text{p} \rightarrow e^- + X$ obtained under various assumptions. Since in Eqs. (7.7) or (7.8) E' is fixed to the value calculated by Eq. (7.9), we have

$$\frac{d\sigma}{d\Omega} = \int dE'\frac{d\sigma}{dE'\,d\Omega} = \int dE'\frac{d\sigma}{d\Omega}\delta\left(\frac{E}{1+\frac{2E}{M}\sin^2\frac{\theta}{2}} - E'\right).$$

Introducing the variable q^2 from Eq. (7.5), the delta function reads

$$\delta\left(\frac{E}{1+\frac{2E}{M}\sin^2\frac{\theta}{2}} - E'\right) = \delta\left(\frac{E - E' + \frac{q^2}{2M}}{1+\frac{2E}{M}\sin^2\frac{\theta}{2}}\right) = \left(1+\frac{2E}{M}\sin^2\frac{\theta}{2}\right)\delta\left(\frac{q^2}{2M} + E - E'\right),$$

where for the last equality, the property (E.3) of the delta function has been used. However, according to Eq. (7.9), $1 + 2E/M \times \sin^2(\theta/2)$ in the first bracket is simply E/E'. Hence, inserting Eqs. (7.7) or (7.8), we identify the double differential cross section for a point-like proton to be

$$\begin{aligned}\left(\frac{d\sigma}{dE'\,d\Omega}\right)_{\text{point-like}} &= \frac{\alpha^2}{4M^2q^4}\delta\left(\frac{q^2}{2M}+E-E'\right)L^{\mu\nu}(\mathrm{p})L_{\mu\nu}(e^-)\frac{E'}{E} \\ &= \frac{\alpha^2}{4E^2\sin^4\frac{\theta}{2}}\delta\left(\frac{q^2}{2M}+E-E'\right)\cos^2\frac{\theta}{2}\left(1-\frac{q^2}{2M^2}\tan^2\frac{\theta}{2}\right).\end{aligned} \tag{7.13}$$

In Problem 7.3, the reader directly derives this formula, starting from the general elementary cross section (3.39).

7.1.2 Elastic Scattering with a Finite-Size Proton: Nucleon Form Factor

How is the scattering cross section modified to take into account a finite extent of the charge distribution of the proton? The electron leg in Fig. 7.1 has no reason to be affected by the finite size of the proton. Therefore, if we still assume a process with one virtual photon exchange, the previous question is equivalent to 'how must the electromagnetic current of the proton be modified?' We know that the amplitude of the process must be Lorentz invariant. Therefore, the electron tensor $L^{\mu\nu}(e^-)$ must still be contracted by the Lorentz tensor $K^{\mu\nu}(\mathrm{p})$ of the proton, where $K^{\mu\nu}(\mathrm{p})$ is the result of the product of the modified electromagnetic current of the proton $j^{\mu}{}_{\text{f.s.}}$ (here f.s. stands for finite size), i.e.

$$K^{\mu\nu}(\mathrm{p}) = \frac{1}{2}\sum_{r,r'}\left[j^{\mu}_{\text{f.s.}}\right]^* j^{\nu}_{\text{f.s.}}, \tag{7.14}$$

with a sum over the proton spin/helicity degrees of freedom. Since $j^{\mu}_{\text{f.s.}}$ must still be a 4-vector that possibly depends on the spin state of the proton and its momentum, it takes the form

$$j^{\mu}_{\text{f.s.}} = \bar{u}(p')\Gamma^{\mu}u(p), \tag{7.15}$$

where Γ^{μ} can only depend on the two 4-vectors p^{μ} and p'^{μ}, or equivalently, $p^{\mu}+p'^{\mu}$ and $p^{\mu}-p'^{\mu}$ and on the γ^{μ} matrix. The blob in Fig. 7.1 representing the proton-virtual photon vertex corresponds to the $ie\Gamma^{\mu}$ factor. Other contributions to Γ^{μ}, such as $p^{\mu}\not{p}$, $\sigma^{\mu\nu}p_{\nu}$ and $p^{\mu}i\sigma^{\rho\eta}p_{\rho}p'_{\eta}$, may be envisaged, but they can always be reduced to linear combinations of $p+p', p-p'$ and γ^{μ}. For instance, the Dirac equation allows us to replace $\not{p}u(p)$ with $Mu(p)$. Note that Γ^{μ} can also depend on scalar functions built with scalar products of p and p'. Those scalar functions are thus necessarily a function of $p^2 = M^2$, $p'^2 = M^2$ and $p\cdot p' = M^2 - q^2/2$ (since $q = p' - p$). As M is a constant, they finally only depend on the variable q^2. Taking into account all these considerations, the generalised form of Γ^{μ} is thus

$$\Gamma^{\mu} = f_1(q^2)\gamma^{\mu} + f_2(q^2)(p'+p)^{\mu} + f_3(q^2)(p'-p)^{\mu}, \tag{7.16}$$

where the f_is are scalar functions depending on q^2. One could think of other terms involving the γ^5 matrix, such as $\gamma^{\mu}\gamma^5$, but this would lead to parity violation in QED. These terms will be, however, present for the weak-interaction current where parity is not conserved (see Chapter 9). Finally, exploiting the conservation of the current $\partial_\mu j'^\mu = 0$, Problem 7.4 shows that the current can be expressed as

$$j^\mu_{\text{f.s.}} = \bar{u}(p')\left[\mathcal{F}_1(q^2)\gamma^\mu + i\frac{\kappa}{2M}\mathcal{F}_2(q^2)\sigma^{\mu\nu}\left(p'_\nu - p_\nu\right)\right]u(p), \tag{7.17}$$

or equivalently

$$\Gamma^\mu = \mathcal{F}_1(q^2)\gamma^\mu + i\frac{\kappa}{2M}\mathcal{F}_2(q^2)\sigma^{\mu\nu}\left(p'_\nu - p_\nu\right), \tag{7.18}$$

where $\mathcal{F}_i(q^2)$ are two dimensionless scalar functions called *form factors*,[2] while κ is a parameter that takes into account the proton magnetic moment. Indeed, replacing $\bar{u}(p')\gamma^\mu u(p)$ by the Gordon decomposition (7.12) leads to the expression of the current

$$j^\mu_{\text{f.s.}} = \bar{u}(p')\left[\frac{\mathcal{F}_1(q^2)}{2M}\left(p'^\mu + p^\mu\right) + i\frac{\mathcal{F}_1(q^2) + \kappa\mathcal{F}_2(q^2)}{2M}\sigma^{\mu\nu}\left(p'_\nu - p_\nu\right)\right]u(p). \tag{7.19}$$

Since the term proportional to $\sigma^{\mu\nu}$ in the current (7.12) gave rise to the g-factor, $g = 2$, the modified current above leads to [see Eq. (7.2)] $g = 2\mu_p/\mu_N = 2[\mathcal{F}_1(0) + \kappa\mathcal{F}_2(0)]$. In the limit $q^2 \to 0$, the proton must be seen as point-like, and hence $\mathcal{F}_1(0) = \mathcal{F}_2(0) = 1$ to recover the point-like current. The value of $\kappa \simeq 1.79$ is, however, chosen to reproduce the measured value of $g_p = 2(1 + \kappa \times 1) \simeq 5.59$ and $\mu_p/\mu_N = 1 + \kappa = 2.79$. In the case of the neutron, since it is neutral, its charge cannot contribute to the scattering of an electron, and, therefore, one has $\mathcal{F}_1(0) = 0$ and $\mathcal{F}_2(0) = 1$. The value $\kappa = -1.91$ must be chosen to match the measured value of the neutron g-factor, $g_n = -3.82 \simeq 2(0 + \kappa \times 1)$ and $\mu_n/\mu_N = \kappa = -1.91$.

The rest of the cross section calculation follows the same steps as for the point-like proton. First, it requires evaluating the tensor $K^{\mu\nu}$ defined in Eq. (7.14), which, according to the expression of the current $j^\mu_{\text{f.s.}}$ in Eq. (7.19) and Casimir's trick (6.84), reads

$$K^{\mu\nu} = \text{Tr}\left[(\not{p} + M)\gamma^0\Gamma^{\mu\dagger}\gamma^0\left(\not{p}' + M\right)\Gamma^\nu\right]. \tag{7.20}$$

The double differential cross section is then obtained by

$$\left(\frac{d\sigma}{dE'\,d\Omega}\right)_{\text{finite size}} = \frac{\alpha^2}{4M^2q^4}\delta\left(\frac{q^2}{2M} + E - E'\right)K^{\mu\nu}(\text{p})L_{\mu\nu}(e^-)\frac{E'}{E}.$$

The calculation of $K^{\mu\nu}(\text{p})L_{\mu\nu}(e^-)$ is rather lengthy but does not present difficulties. One finally obtains the following result:

$$\left(\frac{d\sigma}{dE'\,d\Omega}\right)_{\text{finite size}} = \frac{\alpha^2}{4E^2\sin^4\frac{\theta}{2}}\delta\left(\frac{q^2}{2M} + E - E'\right)\cos^2\frac{\theta}{2}\left(\mathcal{F}_1^2 - \frac{\kappa^2q^2}{4M^2}\mathcal{F}_2^2 - \frac{q^2}{2M^2}(\mathcal{F}_1 + \kappa\mathcal{F}_2)^2\tan^2\frac{\theta}{2}\right), \tag{7.21}$$

which after integration over E' yields

$$\left(\frac{d\sigma}{d\Omega}\right)_{\text{finite size}} = \left(\frac{d\sigma}{d\Omega}\right)_{\text{Mott}}\left(\mathcal{F}_1^2 - \frac{\kappa^2q^2}{4M^2}\mathcal{F}_2^2 - \frac{q^2}{2M^2}(\mathcal{F}_1 + \kappa\mathcal{F}_2)^2\tan^2\frac{\theta}{2}\right). \tag{7.22}$$

This formula is known as the Rosenbluth cross section. It describes the electron–nucleon elastic scattering, assuming pure electromagnetic interaction in the one-photon exchange

[2] $\mathcal{F}_1$ and $\mathcal{F}_2$ are called the Dirac and Pauli form factors respectively.

approximation. Notice that we recover the point-like cross section in Eq. (7.8) when $\mathcal{F}_1 = 1$ and $\kappa = 0$.

Supplement 7.1. Physical significance of the form factors

Let us consider the simple case of an electron scattered by a Coulomb potential created by a static source having a spatial extension (i.e., not point-like). Let us denote by $\rho(\boldsymbol{x})$ the source charge density ($\int \rho(\boldsymbol{x})\, \mathrm{d}^3\boldsymbol{x} = 1$) and by Q its total charge. The Coulomb potential is the solution of $\nabla^2 A^\mu(x) = -j^\mu(x)$ (in natural units), with $j^0(x) = Q\rho(x)$ and $\boldsymbol{j}(x) = 0$ giving

$$A^0(\boldsymbol{x}) = \int \mathrm{d}^3\boldsymbol{x}' \, \frac{Q\rho(\boldsymbol{x}')}{4\pi |\boldsymbol{x} - \boldsymbol{x}'|}, \qquad \boldsymbol{A}(\boldsymbol{x}) = 0.$$

Denoting by $k = (E, \boldsymbol{k})$ the 4-momentum of the incoming electron and by $k' = (E, \boldsymbol{k}')$ that of the scattered electron, the first order of perturbation reads

$$S^{[1]} = -i \int \mathrm{d}^4x\, j_e^\mu(x) A_\mu(x) = -i \int \mathrm{d}^4x\, j_e^0(x) A_0(\boldsymbol{x}) = -i \int \mathrm{d}^4x\, j_e^0(x) A^0(\boldsymbol{x}).$$

Let us consider the classical expressions of j_e and A (i.e., they are not quantised here). Then, inserting $j_e^0(x) = -e\overline{\psi'}(x)\gamma^0\psi(x)$, with $\psi(x) = u(k)\exp(-ik\cdot x)$ and $\psi'(x) = u(k)\exp(-ik'\cdot x)$, leads to

$$S^{[1]} = ie\bar{u}(k')\gamma^0 u(k) \int \mathrm{d}^4x\, e^{-iq\cdot x} A^0(\boldsymbol{x}) = 2\pi i \delta(E - E')\, e\bar{u}(k')\gamma^0 u(k) \int \mathrm{d}^3\boldsymbol{x}\, e^{i\boldsymbol{q}\cdot\boldsymbol{x}} A^0(\boldsymbol{x}),$$

where $q = k - k'$ and the integration over time yields the delta function since $A^0(\boldsymbol{x})$ does not depend on time. The term in the last integral is $A^0(\boldsymbol{q})$, i.e. the three-dimensional Fourier transform of the potential $A^0(\boldsymbol{x})$. It reads

$$A^0(\boldsymbol{q}) = \iint \mathrm{d}^3\boldsymbol{x}\, \mathrm{d}^3\boldsymbol{x}'\, e^{i\boldsymbol{q}\cdot\boldsymbol{x}} \frac{Q\rho(\boldsymbol{x}')}{4\pi|\boldsymbol{x} - \boldsymbol{x}'|} = \int \mathrm{d}^3\boldsymbol{r}\, e^{i\boldsymbol{q}\cdot\boldsymbol{r}} \frac{Q}{4\pi|\boldsymbol{r}|} \times \int \mathrm{d}^3\boldsymbol{x}'\, e^{i\boldsymbol{q}\cdot\boldsymbol{x}'} \rho(\boldsymbol{x}'),$$

where we have changed the variable of integration from $\boldsymbol{x}$ to $\boldsymbol{r} = \boldsymbol{x} - \boldsymbol{x}'$. The first integral corresponds to the Fourier transform of a potential of a point-like charge (a charge Q placed in $\boldsymbol{r} = \boldsymbol{0}$), while the second is the Fourier transform of the charge density. Hence, the scattering by a static charge with a finite-size distribution is related to that of a point-like distribution by

$$S^{[1]} = S^{[1]}_{\text{point-like}} F(\boldsymbol{q}), \quad \text{with } F(\boldsymbol{q}) = \int \mathrm{d}^3\boldsymbol{x}'\, e^{i\boldsymbol{q}\cdot\boldsymbol{x}'} \rho(\boldsymbol{x}').$$

In this simple example, the form factor F is thus interpreted as the Fourier transform of the charge distribution. Notice that since the charge density is normalised, $F(0) = 1$. Similarly, when $\rho(\boldsymbol{x}') = \delta(\boldsymbol{x}')$, the charge is point-like distributed and $F(\boldsymbol{q}) = 1$. From the expression of $S^{[1]}$, one would thus obtain the cross section

$$\left(\frac{\mathrm{d}\sigma}{\mathrm{d}\Omega}\right)_{\text{finite size}} = \left(\frac{\mathrm{d}\sigma}{\mathrm{d}\Omega}\right)_{\text{point-like}} |F(\boldsymbol{q})|^2.$$

What is the physical significance of the form factors? In Supplement 7.1, we show that in the simple case of a static Coulomb potential, the form factor is simply the three-dimensional Fourier transform of the charge distribution. For more complicated cases, such as the potential due to a nucleon, the physicist Robert G. Sachs introduced the following linear combinations:

$$G_E(q^2) = \mathcal{F}_1(q^2) + \frac{\kappa q^2}{4M^2}\mathcal{F}_2(q^2), \quad G_M(q^2) = \mathcal{F}_1(q^2) + \kappa\mathcal{F}_2(q^2),$$

that are now called the *Sachs form factors*. Notice that $G_E(0) = 1$ and $G_M(0) = 1 + \kappa = \mu_p/\mu_N = 2.79$ for the proton, while for the neutron $G_E(0) = 0$ and $G_M(0) = \kappa = \mu_n/\mu_N = -1.91$. He showed in Sachs (1962) that in the *Breit frame*, also called the brick-wall frame, where the momentum of the scattered electron is $\boldsymbol{k}' = -\boldsymbol{k}$ as if it had bounced off, the Fourier transforms of the charge distribution $\rho(\boldsymbol{x})$ and the magnetic moment distribution $\mu(\boldsymbol{x})$ correspond to $G_E(|\boldsymbol{q}|^2)$ and $G_M(|\boldsymbol{q}|^2)$ respectively. Note that in that frame, since $E' = E$, $q^0 = E - E' = 0$ and thus $q^2 = -|\boldsymbol{q}|^2$. Therefore, in principle, once $G_E(|\boldsymbol{q}|^2)$ and $G_M(|\boldsymbol{q}|^2)$ have been measured, one can perform the three-dimensional inverse Fourier transform to access the spatial distribution of the charge and the magnetic moment in the nucleon. This interpretation is valid at low momentum transfer, but when q^2 becomes large, the recoiling nucleon becomes relativistic, and its speed is of the same order of magnitude as that of the probe electron. Hence, the nucleon motion cannot be neglected and tends to blur its physical spatial structure (or equivalently, its structure looks different in different reference frames). In such a case, the form factors do not necessarily represent the charge and magnetic moment distributions.

Using the Sachs form factors, the expression of the Rosenbluth cross section (7.22) is

$$\left(\frac{d\sigma}{d\Omega}\right)_{\text{finite size}} = \left(\frac{d\sigma}{d\Omega}\right)_{\text{Mott}} \left(\frac{G_E^2 - \frac{q^2}{4M^2}G_M^2}{1 - \frac{q^2}{4M^2}} - \frac{q^2}{2M^2}G_M^2 \tan^2\frac{\theta}{2}\right). \tag{7.23}$$

Its measurement at various q^2 and scattering angles θ gives access to the Sachs form factors. For instance, at low $Q^2 = -q^2$, the cross section is dominated by G_E contribution, whereas G_M dominates the high Q^2 regime. Figure 7.2 shows the measurements of the form factors compiled in Hughes et al. (1965) as a function of Q^2. By varying the scattering angle θ and the incoming electron energy E in such a way that Q^2 remains constant, the slope of the Rosenbluth cross section normalised to the Mott cross section, as a function of $\tan^2(\theta/2)$ gives $G_M^2Q^2/(2M^2)$ and the ordinate intercept, $[G_E^2 + G_M^2Q^2/(4M^2)]/[1 + Q^2/(4M^2)]$. The shape of G_E and G_M are reasonably well described by the dipole formula

$$G(Q^2) = \frac{G(0)}{\left(1 + \frac{Q^2}{0.71\ \text{GeV}^2}\right)^2}, \tag{7.24}$$

at least for Q^2 below a few $(\text{GeV})^2$, where $G_M/G_E \simeq 2.79$ for protons. Note that at high Q^2, the measurements, consistent with the Rosenbluth Formula (7.23), differ radically from the scattering on a point-like proton [Mott Formula (7.10)], since the form factor (7.24) imposes an additional Q^{-6} dependence due to the $q^2G_M^2$ terms in Eq. (7.23) beyond that of σ_{Mott}. This is a clear evidence of the finite size of the proton.

From the empirical shape (7.24), one can take the inverse Fourier transform to access the charge and magnetic moment spatial distributions when $Q^2 \simeq |\boldsymbol{q}^2|$ and finally deduce a proton radius of about 0.8 fm (see Problem 7.5). Surprisingly, nowadays, the chapter about the proton size is still not closed. While electron–proton elastic scattering is still measured to infer the proton size, another approach, followed in atomic physics, is to measure the

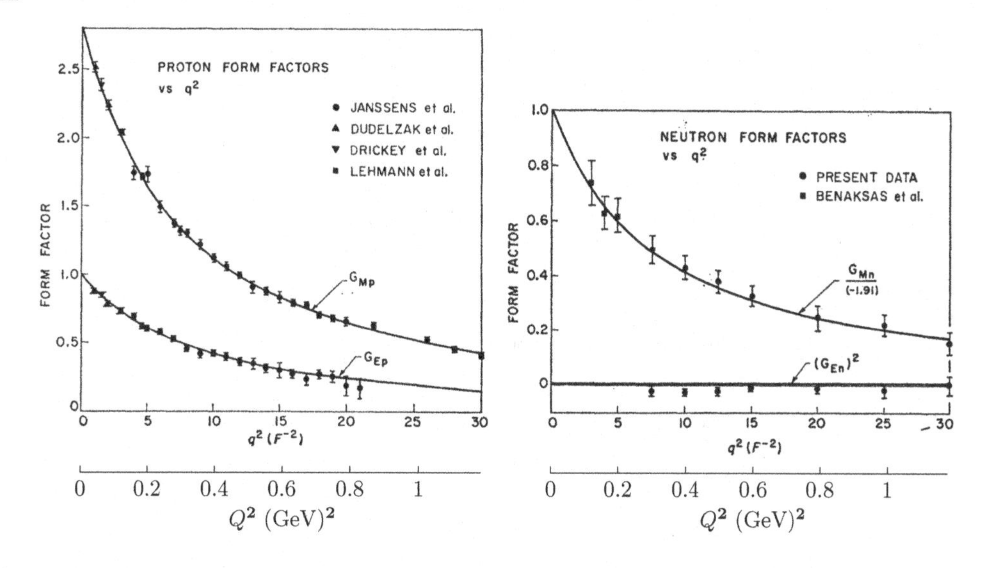

Fig. 7.2 Electromagnetic form factors G_E and G_M as a function of Q^2 for the proton (left) and the neutron (right). The fit corresponds to a generalisation of the simple dipole function (7.24). Figures from Hughes et al. (1965). Horizontal axes in $(\text{GeV})^2$ are superimposed to the original figures where Q^2 (denoted q^2) is in $(\text{fermi})^{-2}$. Reprinted figures with permission from Hughes E. B., Griffy T. A., Yearian M. R., and Hofstadter R., Phys. Rev. **139**, B458, (1965). Copyright 1965 by the American Physical Society.

hyperfine structure of the hydrogen atom by replacing the orbiting electron with a muon. It is worth mentioning that despite two decades of increasing accuracy of both kinds of experiments, there is a significant discrepancy between the radii obtained by these two approaches (atomic physics measures a smaller proton, with a much higher accuracy than elastic scattering).

7.1.3 Electron–Proton Deep Inelastic Scattering

When the momentum transfer $Q^2 = -q^2 > 0$ between the electron and the proton increases, the proton can be excited into higher mass baryons during a very short time, producing an enhancement of the $e^- + p$ cross section at the baryon mass (i.e., a resonance). For instance, $e^- + \text{p} \to e^- + \Delta^+$, where the Δ^+ baryon (a bounded state of *uud* quarks but in the isospin configuration $I = 3/2$) decays promptly, by the strong interaction, into $\Delta^+ \to \text{p} + \pi^0$. With the notations of Fig. 7.3, in this example, the recoiling hadronic system is $X = \text{p} + \pi^0$, and its invariant mass is $W = m_{\Delta^+}$. Other resonances exist, such as the series of N baryons (with isospin $I = 1/2$). In this region of Q^2, where resonances are produced, the scattering is inelastic, i.e. the kinetic energy between the initial state and the final state is not conserved ($T_e + T_\text{p} \neq T'_e + T_X$).

When Q^2 increases further, the proton breaks up in many hadrons. Then, the 4-momentum of the recoiling hadronic system p' results from many contributions, smoothing the W distribution. Therefore, the presence of resonances is no longer visible for the large

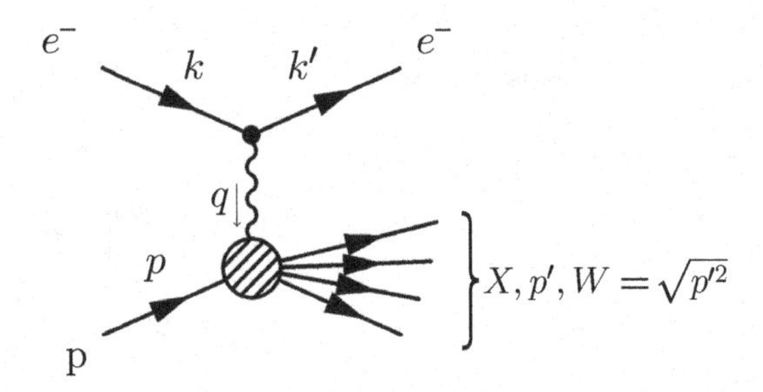

Fig. 7.3 Deep inelastic scattering of an electron on a proton.

W mass. This is the deep inelastic scattering regime. Experimentally, only the final electron is detected. No attempt is made to select a specific hadronic channel (hence, the X symbol in Fig. 7.3). The reaction is inclusive (in contrast to the previous case, where the reaction was exclusive since the hadronic state was perfectly known, as for $X = \mathrm{p} + \pi^0$). The hadronic mass is, however, inferred by the conservation of 4-momentum since (assuming the proton is at rest)

$$W^2 = p'^2 = (p + k - k')^2 = M^2 - Q^2 + 2M(E - E'). \tag{7.25}$$

Once the energies E and E' of the incoming and outgoing electrons and the scattering angle θ are measured, Q^2 is fixed, as well as W. It is important to realise that E, E' and Q^2 (or θ) are three independent variables. Unlike the elastic scattering, the relation (7.9) giving E' does no longer hold since the hadronic mass is not fixed at the proton mass M.

How can we write the inclusive cross section? As in the previous section, for a given hadronic state, X_i, we expect the unpolarised absolute amplitude square to be given by the product of two tensors, i.e.

$$\overline{|M|^2}_{e^-+\mathrm{p}\to e^-+X_i} = \frac{16\pi^2\alpha^2}{q^4} L_{\mu\nu}(e^-)K^{\mu\nu}(X_i).$$

If X_i contains n_i particles with 4-momenta $p_1^i = (E_1^i, \boldsymbol{p}_1^i), \ldots, p_{n_i}^i = (E_{n_i}^i, \boldsymbol{p}_{n_i}^i)$, the cross section should be [Eq. (3.29)]

$$\mathrm{d}\sigma_i = \frac{4\pi^2\alpha^2}{\sqrt{(p\cdot k)^2 - (mM)^2}\,q^4}(2\pi)^4\delta^{(4)}(p' + k' - p - k)L_{\mu\nu}(e^-)K^{\mu\nu}(X_i)\frac{\mathrm{d}^3\boldsymbol{k}'}{(2\pi)^3 2E'}\prod_{n=1}^{n_i}\frac{\mathrm{d}^3\boldsymbol{p}_n^i}{(2\pi)^3 2E_n^i},$$

where $p' = \sum_{n=1}^{n_i} p_n^i$. Since we are interested in the inclusive cross section, we have to sum over all possible X_is and integrate over the corresponding 4-momenta. Let us denote by $W^{\mu\nu}$ the result of these summations and integrations

$$4\pi M\, W^{\mu\nu} = \sum_i \int \cdots \int (2\pi)^4\delta^{(4)}(p' + k' - p - k)K^{\mu\nu}(X_i)\prod_{n=1}^{n_i}\frac{\mathrm{d}^3\boldsymbol{p}_n^i}{(2\pi)^3 2E_n^i},$$

where the factor $4\pi M$ is there to stick to the usual convention. Therefore, the elementary inclusive cross section reads

$$\mathrm{d}\sigma = \frac{4\pi^2\alpha^2}{\sqrt{(p\cdot k)^2 - (mM)^2}\,q^4} L_{\mu\nu}(e^-)\, 4\pi M\, W^{\mu\nu}\frac{\mathrm{d}^3\boldsymbol{k}'}{(2\pi)^3 2E'}.$$

Neglecting the electron mass and calculating the cross section in the proton rest frame, where $p \cdot k = ME$, it follows that

$$d\sigma = \frac{\alpha^2}{q^4} L_{\mu\nu}(e^-) W^{\mu\nu} \frac{d^3\boldsymbol{k}'}{EE'} = \frac{\alpha^2}{q^4} L_{\mu\nu}(e^-) W^{\mu\nu} \frac{E'}{E} \, dE' \, d\Omega. \tag{7.26}$$

The tensor $W^{\mu\nu}$ seems complicated. However, it results from the integration and average over the degrees of freedom of the hadronic system, so it can only depend on the two independent 4-momenta: p and q (since $p' = p+q$). Moreover, since $L^{\mu\nu}(e^-)$ is symmetric [see Eq. (6.90)], antisymmetric contributions in $W^{\mu\nu}$ would cancel the amplitude, and thus, only symmetric ones need to be kept. Therefore, only linear combinations of $g^{\mu\nu}$, $p^\mu p^\nu$, $q^\mu q^\nu$ and $(p^\mu q^\nu + p^\nu q^\mu)$ weighted by scalar functions can be envisaged. Those scalar functions can only depend on 4-scalars, i.e. $p^2 = M^2, q^2$ and $p \cdot q$. It is convenient to introduce the variable

$$\nu = \frac{p \cdot q}{M}. \tag{7.27}$$

In the proton rest frame (thus in the lab frame if the proton is a fixed target), ν is just the energy transfer $\nu = E - E'$. As the proton mass M is a constant, the scalar functions finally only depend on the variables q^2 and ν. The conservation of the hadron electromagnetic current further constrains the form of $W^{\mu\nu}$. Indeed, in the previous section, we saw that $K^{\mu\nu}$ results from the product of two 4-currents that are conserved, $q_\mu J^\mu = 0$. This implied $q_\mu K^{\mu\nu} = q_\nu K^{\mu\nu} = 0$. Similarly, we require

$$q_\mu W^{\mu\nu} = q_\nu W^{\mu\nu} = 0. \tag{7.28}$$

Problem 7.6 shows that the constraint (7.28) leads to the expression of the tensor:

$$W^{\mu\nu}(p,q) = W_1(q^2,\nu)\left(-g^{\mu\nu} + \frac{q^\mu q^\nu}{q^2}\right) + \frac{W_2(q^2,\nu)}{M^2}\left(p^\mu - \frac{p \cdot q}{q^2} q^\mu\right)\left(p^\nu - \frac{p \cdot q}{q^2} q^\nu\right). \tag{7.29}$$

The scalar functions $W_{1,2}$ are the form factors called the *structure functions*. They are called like this because they will tell us something about the internal structure of the proton. Unlike the elastic case where the functions equivalent to $W_{1,2}$ depend only on q^2 via the form factors $\mathcal{F}_{1,2}(q^2)$ (see the next section), the structure functions W_1 and W_2 depend on two Lorentz scalars q^2 and ν [because E' is no longer constrained by Eq. (7.9)]. The rest of the calculation of the cross section is very similar to the elastic case. We insert Eq. (7.29) in Eq. (7.26) and, given Eq. (6.90), finally obtain

$$\frac{d\sigma}{dE' \, d\Omega} = \frac{\alpha^2}{4E^2 \sin^4 \frac{\theta}{2}} \cos^2 \frac{\theta}{2} \left[W_2(q^2,\nu) + 2W_1(q^2,\nu) \tan^2 \frac{\theta}{2} \right]. \tag{7.30}$$

7.1.4 Summary

Assuming that the proton is an elementary fermion with no spatial extension, we found the elastic cross section (7.13) (replacing $E - E'$ by ν)

$$\frac{d\sigma}{dE' \, d\Omega} = \frac{\alpha^2}{4E^2 \sin^4 \frac{\theta}{2}} \cos^2 \frac{\theta}{2} \delta\left(\frac{q^2}{2M} + \nu\right) \left[1 - 2\frac{q^2}{4M^2} \tan^2 \frac{\theta}{2}\right].$$

For a real proton with a finite size and an anomalous magnetic moment, the elastic cross section is given by Eq. (7.21), i.e.

$$\frac{d\sigma}{dE'\,d\Omega} = \frac{\alpha^2}{4E^2\sin^4\frac{\theta}{2}}\cos^2\frac{\theta}{2}\delta\left(\frac{q^2}{2M}+\nu\right)$$
$$\left[\mathcal{F}_1^2(q^2) - \frac{\kappa^2 q^2}{4M^2}\mathcal{F}_2^2(q^2) - 2\frac{q^2}{4M^2}\left(\mathcal{F}_1(q^2)+\kappa\mathcal{F}_2(q^2)\right)^2\tan^2\frac{\theta}{2}\right].$$

Finally, for higher momentum transfer, the deep inelastic scattering cross section is [Eq. (7.30)]

$$\frac{d\sigma}{dE'\,d\Omega} = \frac{\alpha^2}{4E^2\sin^4\frac{\theta}{2}}\cos^2\frac{\theta}{2}\left[W_2(q^2,\nu) + 2W_1(q^2,\nu)\tan^2\frac{\theta}{2}\right].$$

We see a manifestly similar structure, where moving from the deep inelastic scattering cross section to the elastic scattering cross section with an elementary proton, the structure functions would have to be successively replaced by

$$\begin{aligned}
W_1(q^2,\nu) &\xrightarrow{\text{elastic}} -\frac{q^2}{4M^2}\left(\mathcal{F}_1(q^2)+\kappa\mathcal{F}_2(q^2)\right)^2\delta\left(\frac{q^2}{2M}+\nu\right) \xrightarrow{\text{elementary}} -\frac{q^2}{4M^2}\delta\left(\frac{q^2}{2M}+\nu\right),\\
W_2(q^2,\nu) &\xrightarrow{\text{elastic}} \left(\mathcal{F}_1^2(q^2) - \frac{\kappa^2 q^2}{4M^2}\mathcal{F}_2^2(q^2)\right)\delta\left(\frac{q^2}{2M}+\nu\right) \xrightarrow{\text{elementary}} \delta\left(\frac{q^2}{2M}+\nu\right).
\end{aligned} \tag{7.31}$$

7.2 The Parton Model

7.2.1 Bjorken Scaling

After the measurements of the elastic scattering cross section revealing the proton finite size in the 1960s, there were theoretical conjectures that suggested that the inelastic spectra might decrease rapidly with increasing 4-momentum transfer squared, Q^2. People more or less expected a behaviour similar to the elastic scattering, where G_E and G_M behave as Q^{-8}. Therefore, in the late 1960s and early 1970s, at the Stanford Linear Accelerator Center (SLAC), several experiments led by J. Friedman, H. Kendall (MIT) and R. Taylor (SLAC) were performed to measure the deep inelastic cross section and the structure functions $W_{1,2}$. What they found led them to be awarded the Nobel prize in 1990 since they revealed the substructure of the nucleons. They measured the scattering cross section of an electron beam (energy from 4 to 21 GeV) on a fixed target consisting either of liquid hydrogen or deuterium. Both the scattering angle θ and the energy of the scattered electrons E' were measured. Hence, given E, E' and θ, the values of $Q^2 = -q^2$ [Eq. (7.5)], ν [Eq. (7.27)], and the mass of the hadronic system W [Eq. (7.25)] were known.

Figure 7.4 shows their measurements of the differential cross section $d\sigma/(d\Omega\,dE')$ normalised to the Mott cross section $d\sigma_{\text{Mott}}/d\Omega$. First, we see that the measured cross section largely dominates the elastic cross section for $Q^2 \gtrsim 1$ GeV. Hence, the structure functions $W_{1,2}$ extracted from data behave very differently from the elastic form factors. That was a real surprise. Second, the Q^2 dependence of the inelastic scattering at a constant hadronic

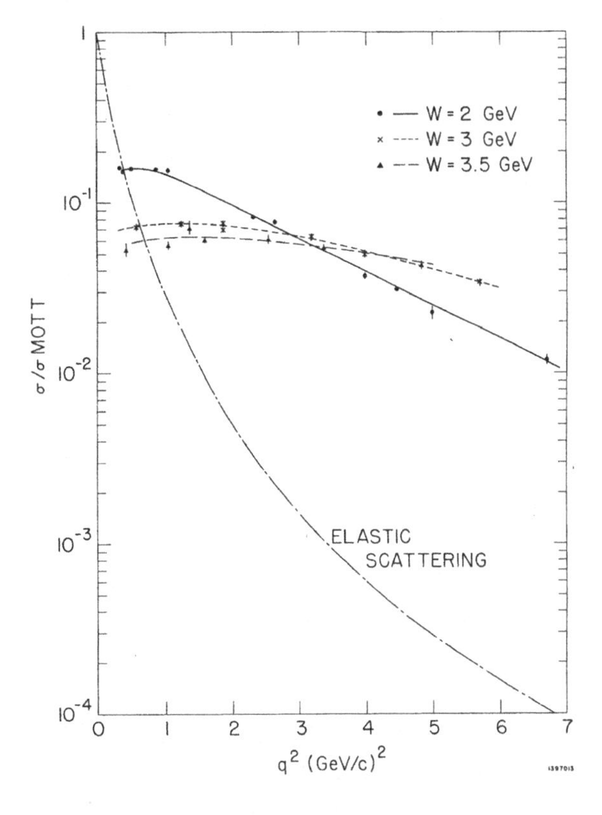

Fig. 7.4 Differential cross section $d\sigma/(d\Omega\, dE')$ of $e^- + p$ scattering normalised to the Mott cross section $d\sigma_{\mathrm{Mott}}/d\Omega$ as a function of the momentum transfer Q^2 for various masses of the hadronic system. Reprinted figure with permission from Breidenbach M. et al. (1969). Copyright 1969 by the American Physical Society.

mass W is rather weak. It even becomes almost independent of Q^2 for high W. High masses of the hadronic system mean necessarily a high energy transfer ν [see Eq. (7.25), where $E' - E = \nu$].

Those experimental facts corroborated a prediction made by Bjorken (based on theoretical considerations concerning the current algebra), who claimed that at high Q^2 and high ν, the structure functions are only a function of their ratio via the dimensionless Bjorken-x variable[3]

$$x = \frac{Q^2}{2M\nu}. \tag{7.32}$$

Note that x is a positive quantity since $Q^2 > 0$. Moreover, the conservation of the baryon number implies that the final state contains at least one baryon, and thus, necessarily, $W \geq M$. Therefore, $W^2 = M^2 - Q^2 + 2M\nu \geq M^2$, which leads to $x \leq 1$. The case $x = 1$ corresponds to elastic scattering ($W = M$), whereas for inelastic scattering necessarily,

[3] This phenomenon is known as Bjorken scaling. A function is said to *scale* when it depends on dimensionless kinematic quantities and not on the absolute energy or momentum.

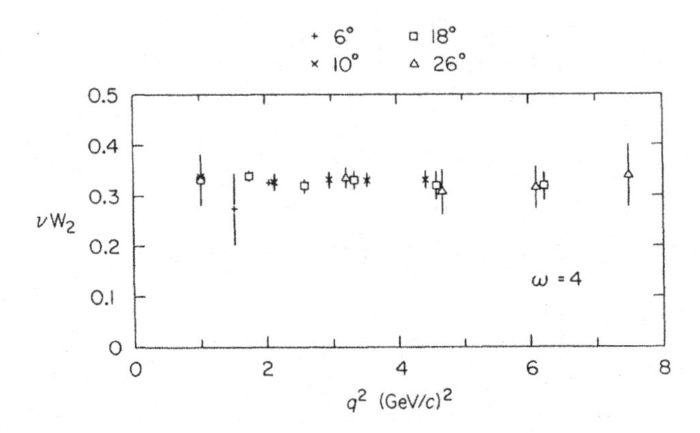

Fig. 7.5 The νW_2 structure function measured as a function of Q^2 for a fixed value of Bjorken-x, $x = 0.25$. (The variable ω is $1/x$.) Reprinted figure with permission from Kendall (1991). Copyright 1991 by the American Physical Society.

$0 < x < 1$. Bjorken specifically predicted in Bjorken (1969) that for a constant ratio Q^2/ν, the structure functions should behave as

$$MW_1(Q^2,\nu) \xrightarrow[x \text{ fixed}]{(Q^2,\nu)\to\infty} F_1(x) \quad , \quad \nu W_2(Q^2,\nu) \xrightarrow[x \text{ fixed}]{(Q^2,\nu)\to\infty} F_2(x), \tag{7.33}$$

where F_1 and F_2 are two finite functions. In April 1968, at the suggestion of Bjorken, Kendall and his team experimentally checked the behaviour of νW_2. Figure 7.5 presents νW_2 as a function of Q^2 for a fixed value of Bjorken-x ($x = 0.25$ in this figure). The behaviour is compatible with no dependence on Q^2 as predicted by Bjorken.

7.2.2 The Partons

How is Bjorken scaling interpreted? According to Eqs. (7.31), we have shown that the structure functions for an elementary fermion would be

$$MW_1(Q^2,\nu) \xrightarrow{\text{elementary}} \frac{-q^2}{4M}\delta\left(\frac{q^2}{2M}+\nu\right) = \frac{Q^2}{4M}\delta\left(\nu-\frac{Q^2}{2M}\right) = \frac{Q^2}{4M\nu}\delta\left(1-\frac{Q^2}{2M\nu}\right) = \frac{x}{2}\delta(1-x),$$

$$\nu W_2(Q^2,\nu) \xrightarrow{\text{elementary}} \nu\delta\left(\frac{q^2}{2M}+\nu\right) = \nu\delta\left(\nu-\frac{Q^2}{2M}\right) = \delta\left(1-\frac{Q^2}{2M\nu}\right) = \delta(1-x).$$

Structure functions of elementary fermions, thus of point-like fermions, correspond to functions depending only on x. Moreover, when Q^2 and ν are large enough, we observe that the proton structure functions depend only on x. The proton, however, has a spatial extension revealed by the elastic cross section measurements. The logical interpretation proposed in 1969 by Feynman, Bjorken and Paschos after the first results of the deep inelastic cross section (Bjorken-Paschos, 1969; Feynman, 1969) is that the probe electron, or rather the virtual photon, interacts with elementary point-like fermions[4] that would be inside the proton. They called those elementary objects the *partons*. The scaling behaviour predicted by Bjorken is then easy to understand qualitatively. Indeed, the wavelength of the

[4] The cross section formulas of point-like objects are based on Feynman's rules with spinors that are used to describe spin-1/2 fermions.

virtual photon is $\hbar/\sqrt{Q^2}$. However, the resolving power becomes irrelevant when photons scatter on point-like objects. Therefore, the structure function should become independent of Q^2. Let us examine this assumption more closely. Let $q_i e$ be the charge of parton i. This parton should carry only a fraction z_i of the proton 4-momentum. Implicitly, we neglect the internal motions of partons inside the proton, considering large proton 4-momentum (in Feynman's paper, the infinite momentum frame of reference is considered, in which the relativistic time dilation slows down the motion of the constituents nearly to rest). Thus, we assume that the parton momentum is aligned with that of the proton, neglecting any transverse motion, and therefore, any interaction among the partons. In other words, we assume the partons inside the proton to be free particles.[5] In this context, the parton 4-momentum is $p_i = z_i p$, and thus, $m_i = z_i M$. Of course, a parton with a variable mass is meaningless. In fact, we have to consider the masses as negligible since we work in a frame where the particle energies are infinite. The scattering cross section of a parton of charge q_i is then the elastic one (7.13), replacing M with $z_i M$ and one α with $q_i^2\alpha$, yielding

$$\frac{d\sigma}{dE'\,d\Omega} = \frac{\alpha^2 q_i^2}{4E^2 \sin^4\frac{\theta}{2}} \cos^2\frac{\theta}{2} \delta\left(\nu - \frac{Q^2}{2z_i M}\right)\left[1 + \frac{Q^2}{2z_i^2 M^2}\tan^2\frac{\theta}{2}\right].$$

Comparing this formula with the deep inelastic cross section (7.30), we would have for this particular parton

$$MW_1(Q^2,\nu) \xrightarrow[x\ \text{fixed}]{(Q^2,\nu)\to\infty} q_i^2 \tfrac{Q^2}{4z_i^2 M}\delta\left(\nu - \tfrac{Q^2}{2z_i M}\right) = \tfrac{q_i^2 x}{2z_i^2}\delta\left(1 - \tfrac{x}{z_i}\right) = \tfrac{q_i^2 x}{2z_i}\delta(z_i - x),$$
$$\nu W_2(Q^2,\nu) \xrightarrow[x\ \text{fixed}]{(Q^2,\nu)\to\infty} \nu q_i^2 \delta\left(\nu - \tfrac{Q^2}{2z_i M}\right) = q_i^2\delta\left(1 - \tfrac{x}{z_i}\right) = q_i^2 z_i \delta(z_i - x).$$

Now, there must be a probability density $f_i(z_i)$ that the parton carries a fraction z_i of the proton momentum. Therefore, we have to weight z_i by this probability. Moreover, assuming that there are several types of partons in the proton, the virtual photon can interact with any of these charged partons. It follows that the structure functions must converge to

$$MW_1(Q^2,\nu) \xrightarrow[x\ \text{fixed}]{(Q^2,\nu)\to\infty} F_1(x) = \textstyle\sum_i \int_0^1 dz_i\, f_i(z_i) \frac{q_i^2 x}{2z_i}\delta(z_i - x) = \frac{1}{2}\sum_i q_i^2 f_i(x),$$
$$\nu W_2(Q^2,\nu) \xrightarrow[x\ \text{fixed}]{(Q^2,\nu)\to\infty} F_2(x) = \textstyle\sum_i \int_0^1 dz_i\, f_i(z_i) q_i^2 z_i \delta(z_i - x) = x\sum_i q_i^2 f_i(x). \tag{7.34}$$

This scenario is consistent with Bjorken's prediction (7.33), where F_1 and F_2 depend only on x. In addition, Eq. (7.34) predicts the Callan–Gross formula of the parton model, i.e.

$$2xF_1(x) = F_2(x). \tag{7.35}$$

Is this verified experimentally? Figure 7.6 presents the ratio $2xF_1(x)/F_2(x)$ measured in electron–proton inelastic scattering at SLAC (Bodek et al., 1979). For elementary fermions, Eq. (7.35) predicts this ratio to be equal to 1. For an elastic scattering on spin-0 particles, since the term proportional to $\tan^2(\theta/2)$, i.e. W_1 and thus F_1, would be absent, this ratio should be 0. The data displayed in Fig 7.6 are compatible with the Callan–Gross formula and therefore validate the interpretation of an elastic scattering on elementary fermions.

Let us see now the physical meaning of the Bjorken-x variable. The parton i, carrying a fraction z_i of the proton 4-momentum, has a 4-momentum $z_i p$. After the elastic scattering

[5] This strong assumption will be justified in Chapter 8.

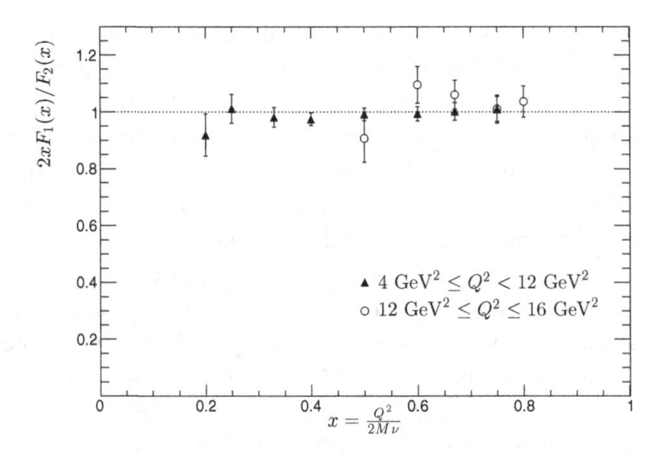

Fig. 7.6 Measurement of the ratio $2xF_1/F_2$ versus x. Data from Bodek et al. (1979).

with the virtual photon, its 4-momentum becomes $z_i p + q$. Since in the infinite momentum frame, all masses are neglected as well as transverse motions, necessarily $(z_i p + q)^2 \approx 0$ and $p^2 \approx 0$. It follows that $q^2 + 2z_i p \cdot q \approx 0$, i.e. $z_i = Q^2/(2p \cdot q) = x$ [see Eqs. (7.32) and (7.27)]. Hence, in the infinite momentum frame, Bjorken-x is just the fraction of longitudinal momentum carried by the parton that has scattered.

Let us recap: the scattering experiments have shown that the proton contains point-like fermions called partons. They necessarily carry an electric charge since the scattering involves the electromagnetic interaction. Since the proton is a fermion, it must contain an odd number of partons. A single parton is excluded; otherwise, the proton would be a point-like particle. Hence, the smallest possible number is 3. Anticipating Chapter 8, we will assume these partons to be quarks and antiquarks (at the first approximation).

7.3 Parton Distribution Functions

The functions $f_i(x)$ appearing in Eq. (7.34) are called the parton distribution functions or parton density functions (PDFs). They represent the probability densities (strictly speaking, they rather represent the number densities as they are normalised to the number of partons) to find a parton of type i carrying a momentum fraction x of the proton. In other words, $f_i(x)\,\mathrm{d}x$ is the number of partons of type i within the proton carrying a momentum fraction between x and $x + \mathrm{d}x$. There is not yet a theory that can predict the values of the PDFs, even if we will see that quantum chromodynamics can predict (partially) the evolution of the PDFs as a function of Q^2 or x. Hence, our knowledge of PDFs relies on experimental measurements based on the deep inelastic scattering cross sections that give access to $F_{1,2}(x)$, and thus $f_i(x)$.

Let us imagine that the proton consists only of the three valence quarks *uud* responsible for its quantum number (see Table 1.1). They have similar masses, and assuming that they

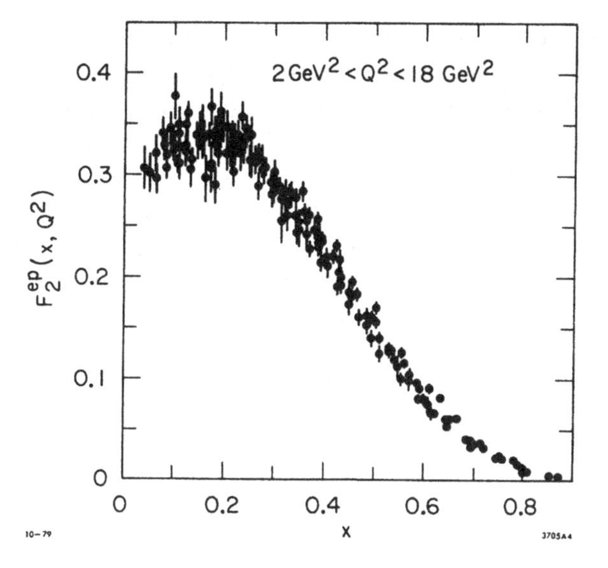

Fig. 7.7 Measurement of F_2 as a function of x. Figure from Atwood et al. (1982). Reprinted by permission from Springer Nature (1982)

do not interact with each other, we would expect them all to share the same fraction of the proton momentum. Denoting by $u(x)$ and $d(x)$ the PDF of the u and d quarks respectively, we would thus expect

$$u(x) = 2\delta(x - 1/3), \quad d(x) = \delta(x - 1/3),$$

with the normalisation $\int_0^1 u(x)\, dx = 2$ and $\int_0^1 d(x)\, dx = 1$ in agreement with the quark content of the proton. The mean fraction of proton momentum carried by the two u quarks and the d quark would be $\int_0^1 xu(x)\, dx = 2/3$ and $\int_0^1 xd(x)\, dx = 1/3$, respectively. Because of the δ function, the functions $F_1(x)$ and $F_2(x)$ in Eq. (7.34) should present a huge peak at $x = 1/3$. Well, Fig 7.7 shows the measurement of $F_2(x)$. Clearly, our naive expectation does not match the data. We have forgotten that because of quantum fluctuations, deep inelastic scattering can result from the scattering of an electron on any type of quark or antiquark. More specifically, we will see in Chapter 8 that a quark can radiate a gluon which can then split into a quark–antiquark pair (as an electron can radiate a photon producing an electron–positron pair), i.e.

$$g \to q + \bar{q}.$$

Hence, from time to time, the scattering can occur on a member of those pairs. Since s, c, b and t quarks are much heavier than the other quarks (even heavier than the proton for the last three), it is reasonable to neglect the contribution of such quantum fluctuations in the proton (the lifetime of those quantum fluctuations would be of the order of $\hbar/mc^2$).[6] We shall see in the next section that this is fully justified. Therefore, we expect $u(x), d(x), \bar{u}(x)$

[6] Another way of expressing that the probability decreases with the mass of the virtual quark is to recall that the denominator of the quark propagator is a function of m_q^2. The heavier the quark is, the lower the probability.

and $\bar{d}(x)$ to be the dominant contributions to $F_1(x)$ and $F_2(x)$. To distinguish valence quarks from the others, let us use the subscript V (for valence) for the former and S (for sea) for the latter. Sea quarks and antiquarks denote all (virtual) quarks and antiquarks produced by quantum fluctuations. The PDF of the u quark thus results from $u(x) = u_V(x) + u_S(x)$, whereas antiquark $\bar{u}$ can only originate from $\bar{u}_S(x)$. Note that a quark can emit an unlimited number of gluons (as an electron can radiate an unlimited number of photons) that can potentially convert into a pair of quark–antiquark q_S–$\bar{q}_S$. Therefore, the number of sea quarks is not well defined, meaning that potentially

$$\int_0^1 [q_S(x) + \bar{q}_S(x)]\, \mathrm{d}x = \infty.$$

However, we shall see that those sea quarks carry only a small fraction of the proton momentum (low x), meaning that

$$\int_0^1 x[q_S(x) + \bar{q}_S(x)]\, \mathrm{d}x \ll 1.$$

Let us be more quantitative. Assuming that only u and d quarks significantly contribute to the proton structure functions, according to Eq. (7.34), the function $F_2(x)$ reads

$$F_2^p(x) = x\left[\frac{4}{9}\left(u^p(x) + \bar{u}_S^p(x)\right) + \frac{1}{9}\left(d^p(x) + \bar{d}_S^p(x)\right)\right], \tag{7.36}$$

where the superscript p is added to notify that we are dealing with the PDFs in the proton. The area defined by F_2^p is then

$$A^p = \int_0^1 \mathrm{d}x\, F_2^p(x) = \frac{4}{9}\int_0^1 \mathrm{d}x\, x(u^p(x) + \bar{u}_S^p(x)) + \frac{1}{9}\int_0^1 \mathrm{d}x\, x(d^p(x) + \bar{d}_S^p(x)) = \frac{4}{9}f_u + \frac{1}{9}f_d,$$

where the momentum fractions of quarks and antiquarks in the proton are denoted f_u and f_d for u and d respectively. This can be measured directly in Fig. 7.7. It is about 0.18. We need, however, a complementary measurement to determine the individual contributions. For instance, the deep inelastic scattering of an electron on a neutron (udd). Since both protons and neutrons contain u and d quarks, Eq. (7.34) tells us that $F_2^n(x)$ should have the same form as Eq. (7.36), except that the PDFs in the proton must be replaced with PDFs in the neutron, i.e.

$$F_2^n(x) = x\left[\frac{4}{9}\left(u^n(x) + \bar{u}_S^n(x)\right) + \frac{1}{9}\left(d^n(x) + \bar{d}_S^n(x)\right)\right].$$

Moreover, assuming the isospin symmetry, the proton and the neutron can be considered different states of the *same* particle (the nucleon) with respect to the strong interaction (see Chapter 8). They behave in the same way when they undergo this interaction. Therefore, the u-quark PDF in the proton should be the same as the d-quark PDF in the neutron (and vice versa). Then we can replace u^n with d^p and d^n with u^p in $F_2^n(x)$, yielding

$$F_2^n(x) = x\left[\frac{4}{9}\left(d^p(x) + \bar{d}_S^p(x)\right) + \frac{1}{9}\left(u^p(x) + \bar{u}_S^p(x)\right)\right]. \tag{7.37}$$

The corresponding area can then be determined and compared with the data. The result is

$$A^n = \frac{4}{9}f_d + \frac{1}{9}f_u \simeq 0.12.$$

From the values of A^p and A^n, we then deduce that $f_u \simeq 0.36$ and $f_d \simeq 0.18$. Hence, we find that, on average, u quarks carry twice as much momentum as the d quarks in the proton. This is expected for a proton made of uud quarks. However, we learn something new: $f_u + f_d \simeq 0.54$. There are 46% missing! One could argue that this is due to the heavier quarks we neglected. But this is not true, their masses being too large to compensate for the 46% missing, as will be confirmed in the next section. Since the scattering with electrons probes only the charged component of the nucleon, we must conclude that the missing fraction is due to neutral partons in the proton: the most natural candidates in the quantum chromodynamics framework (see Chapter 8) are the gluons.

7.4 What Is a Proton?

7.4.1 PDFs of the Proton

The proton (or any hadron) has a more complicated structure than naively expected: it is made of its valence quarks responsible for its quantum number, the sea quarks and antiquarks, and the gluons, with the last three emerging from quantum fluctuations. Ultimately, the PDFs of each kind of partons (including the gluons) are disentangled from fits to several types of data: deep inelastic scattering of electrons or neutrinos on several types of hadrons, for instance. However, the PDFs are constrained by the hadron quantum numbers. The proton or neutron have no strangeness, i.e. they do not contain a net excess of strange quarks or antiquarks. Similarly, there is no net excess of heavier quarks (c, b, t). Therefore, necessarily,

$$\int_0^1 \mathrm{d}x \left[q_S(x) - \bar{q}_S(x)\right] = 0, \quad q = s, c, b, t. \tag{7.38}$$

Moreover, assuming the isospin symmetry, i.e. $u(x) \equiv u^p(x) = d^n(x)$ and $d(x) \equiv d^p(x) = u^n(x)$, the electric charge of the proton and neutron imposes

$$\begin{aligned} \textstyle\int_0^1 \mathrm{d}x \left(\frac{2}{3}\left[u(x) - \bar{u}_S(x)\right] - \frac{1}{3}\left[d(x) - \bar{d}_S(x)\right]\right) &= 1, \\ \textstyle\int_0^1 \mathrm{d}x \left(\frac{2}{3}\left[d(x) - \bar{d}_S(x)\right] - \frac{1}{3}\left[u(x) - \bar{u}_S(x)\right]\right) &= 0. \end{aligned}$$

The combinations of these two constraints are equivalent to

$$\begin{aligned} \textstyle\int_0^1 \mathrm{d}x \left[u(x) - \bar{u}_S(x)\right] &= 2, \\ \textstyle\int_0^1 \mathrm{d}x \left[d(x) - \bar{d}_S(x)\right] &= 1. \end{aligned} \tag{7.39}$$

Note that if we assume $u_S(x) = \bar{u}_S(x)$ and $d_S(x) = \bar{d}_S(x)$ (justified by the fact that quarks and antiquarks from the sea are produced by pairs), given that $u(x) = u_V(x) + u_S(x)$ and $d(x) = d_V(x) + d_S(x)$, it follows that

$$\begin{aligned} \textstyle\int_0^1 \mathrm{d}x\, u_V(x) &= 2, \\ \textstyle\int_0^1 \mathrm{d}x\, d_V(x) &= 1. \end{aligned} \tag{7.40}$$

The constraints (7.38), (7.39) or (7.40) are called *sum rules*. They simplify the expressions of the structure functions and thus ease their interpretation. For instance, if we further assume that $u_S(x) = \bar{u}_S(x) = d_S(x) = \bar{d}_S(x)$, which is reasonable because of the isospin symmetry, the functions $F_2^p(x)$ and $F_2^n(x)$ in Eqs. (7.36) and (7.37) then read

$$\begin{aligned} F_2^p(x) &= x\left[\tfrac{4}{9}u_V(x) + \tfrac{1}{9}d_V(x) + \tfrac{10}{9}u_S(x)\right], \\ F_2^n(x) &= x\left[\tfrac{4}{9}d_V(x) + \tfrac{1}{9}u_V(x) + \tfrac{10}{9}u_S(x)\right]. \end{aligned}$$

Therefore, their ratio is

$$\frac{F_2^n(x)}{F_2^p(x)} = \frac{4d_V(x) + u_V(x) + 10u_S(x)}{4u_V(x) + d_V(x) + 10u_S(x)}. \tag{7.41}$$

Experimentally, at low x, this ratio tends to 1. This could be achieved if the numerator and the denominator of Eq. (7.41) are equal, and thus $u_V(x) = d_V(x)$, which would contradict the sum rule (7.40). The other possibility is that $u_S(x)$ dominates at low x in this ratio. This is a reasonable assumption since, qualitatively, the lifetime of a quark–antiquark pair from the sea is about $\hbar/E$, where E is the energy of the pair. Since the probability to scatter with one member of the pair must be proportional to its lifetime, this limits the energy and thus the momentum, which biases x towards low values. Therefore, we conclude that the sea contribution dominates in the low x region. Conversely, the quark or antiquark contribution from the sea must be marginal in high x regions, and thus valence quarks should dominate in those regions.

Several groups in the world provide the parton distribution functions from global fits of data. They differ by the data and by the parametrisation of the PDF used. However, they give consistent results, and the general trend is similar. An example is shown in Fig. 7.8 for $Q^2 = 10$ GeV. The reason why the value of Q^2 must be specified will be explained in Section 7.5 and Chapter 8. According to the figure, at low x, the proton is mostly dominated by partons from the sea, with a large contribution from gluons. A qualitative explanation is the following: being a massless particle, a gluon produced by a quantum fluctuation from a valence quark $q_V \to q_V + g$ can have very low energy during a significant time ($\tau \simeq \hbar/E$), and therefore it can carry a very small fraction of the proton momentum. Similarly, since the sea antiquarks originate from virtual gluons, their density functions necessarily peak towards low values of x. For $x \simeq 1/3$, Fig. 7.8 shows that the valence quarks play the dominant role, more or less corresponding to the naive expectation, where $u_V(x) \simeq 2d_V(x)$ (twice as much as more u quarks than d quarks in a proton). The region where x is very close to 1 is only populated by valence quarks, with a dominance of the u_V quark. In terms of momentum carried by the different kinds of partons, the integral of the $xf(x)$ displayed in Fig. 7.8 gives approximatively (see, e.g., Salam, 2015):

$$\begin{aligned} \textstyle\int_0^1 x u_V(x)\,\mathrm{d}x &= 0.267 \\ \textstyle\int_0^1 x d_V(x)\,\mathrm{d}x &= 0.111 \\ \textstyle\int_0^1 x u_S(x)\,\mathrm{d}x &= 0.053 \\ \textstyle\int_0^1 x d_S(x)\,\mathrm{d}x &= 0.066 \\ \textstyle\int_0^1 x s_S(x)\,\mathrm{d}x &= 0.033 \\ \textstyle\int_0^1 x c_S(x)\,\mathrm{d}x &= 0.016 \\ \text{total} &= 0.546 \end{aligned}$$

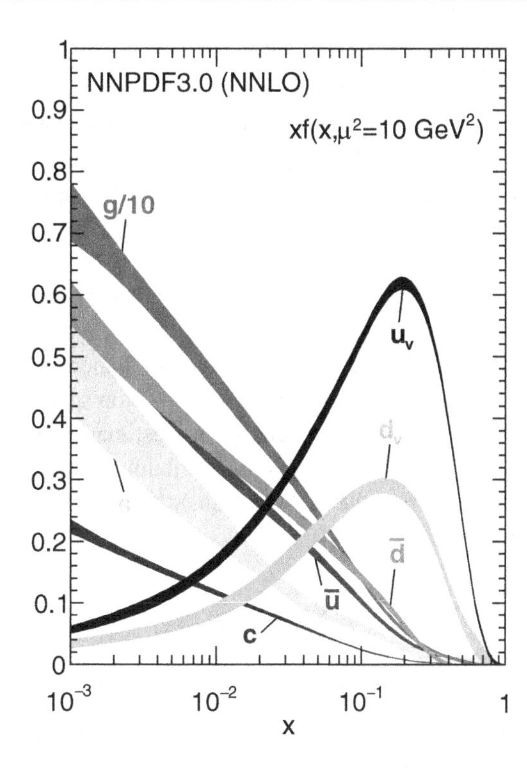

Fig. 7.8 Distributions of $xf(x)$ for the valence quarks, $f = u_V, d_V$, the sea (anti-)quarks, $f = \bar{u}, \bar{d}, s, c$, and the sea gluons, $f = g$. The contribution of gluons is divided by 10. Figure from Zyla et al. (Particle Data Group) (2020).

Since the total momentum of the proton is necessarily carried by partons, the momentum sum rule requires

$$\sum_i \int_0^1 xf_i(x)\,\mathrm{d}x = 1,$$

where the sum is over all possible partons. Consequently, $1 - 0.546 = 0.454$ is mostly carried by gluons. One may wonder how the contribution of the antiquarks from the sea can be evaluated (see Fig. 7.8) since the deep inelastic scattering of an electron is only sensitive to the square of parton charges [see Eq. (7.34)], simply because the virtual photon interacts with particles and antiparticles in exactly the same way. We will see, however, in Chapter 9 that the weak interaction makes a difference between qs and $\bar{q}$s. The solution is, therefore, to analyse the deep inelastic scattering of a neutrino on the proton to disentangle quark and antiquark contributions. Problem 7.7 shows important neutrino deep inelastic scattering results.

7.4.2 Proton Spin

If the proton were a non-relativistic bound system, its spin would be due to the addition of the spins of its three valence quarks (in the ground state, there is no orbital angular momentum between its constituents). This is the naive static model where we assume that two of the quarks have opposite spins, and the spin of the third is parallel to the proton spin. However, the proton is clearly a relativistic system: the u and d quark masses (below 10 MeV) are negligible compared to the proton mass.[7] Moreover, the proton size is about 1 fm, and the Heisenberg uncertainty principle implies that the valence quarks must have typical momenta of the order of 200 MeV. Given the small mass of the u and d quarks, they are therefore relativistic.

Since the late 1980s, several experiments have measured the contribution of quarks to the proton spin using deep inelastic scattering with polarised beams. The very first measurements made by the European Muon Collaboration (EMC) at CERN in 1987 gave the surprising result that almost none of the proton spin was due to its quark constituents (see EMC Collaboration, 1988)! Today, it is estimated that about 30% of the proton spin (De Florian et al., 2014) is due to the contribution of quark spins (valence quarks and quark–antiquark pairs from the sea). This quantity is usually referred to as Σ in the literature and corresponds to

$$\Sigma = \sum_{q,\bar{q}} \int_0^1 \mathrm{d}x\, [q_\uparrow(x) - q_\downarrow(x)] + [\bar{q}_\uparrow(x) - \bar{q}_\downarrow(x)],$$

where $q_\uparrow$ denotes the number density of valence quarks and quarks from the sea with spin parallel to the proton spin, and $q_\downarrow$ denotes the density of valence quarks and quarks from the sea with spin antiparallel to the proton spin. A similar notation is used for antiquarks. Where are the other 70% coming from? Well, there are the quark orbital angular momentum L_q, the gluon spin[8] Δ_G,

$$\Delta_G = \int_0^1 \mathrm{d}x\, [g_\uparrow(x) - g_\downarrow(x)],$$

and the gluon orbital angular momentum L_G. The sum of all those contributions must match the spin of the proton, yielding the angular momentum sum rule,

$$\frac{1}{2} = \frac{1}{2}\Sigma + L_q + \Delta_G + L_G. \tag{7.42}$$

The contribution Δ_G is poorly known (most of the constraints are coming from polarised proton–proton collisions at the Relativistic Heavy Ion Collider, RHIC), but it is estimated to be about 0.2. The other contributions L_q and L_G have never been measured. The decomposition in Eq. (7.42) is, however, not unambiguous (related to gauge invariance and the repartition between Δ_G and L_G) and is still a subject of debate [see, for instance, the review Leader and Lorce (2014)].

[7] For hadrons made of heavy quarks as b or c quarks, the static approximation remains however valid since the mass of those hadrons is dominated by the mass of the heavy quarks.

[8] In Chapter 8, we shall see that gluons are spin-1 bosons.

7.5 Going beyond the Simple Parton Model

With the constant rise of the energy of the accelerators, one can access larger momentum transfers and hence probe smaller scales. In the 1990s and for almost the last two decades, the HERA collider at the Deutsches Elektronen-SYnchrotron (DESY) laboratory in Germany strongly contributed to the measurements of structure functions at large Q^2. A wide range of Bjorken-x values, especially small values, were probed. HERA produced collisions of electrons or positrons of 27.5 GeV on protons of 820 GeV. Two experiments operated at HERA: H1 and ZEUS. A summary of the measurements of F_2^p is provided in Fig. 7.9, where HERA's measurements are presented in addition to the previous measurements at lower Q^2. Almost five orders of magnitude in Q^2 are scanned.

First, we observe in Fig. 7.9 that at high Q^2, there is no evidence of a rapid decrease of the cross section. If the quarks were eventually composite particles, we would expect their elastic cross section to fall rapidly at a large Q^2 because quarks would have form factors following a mechanism similar to the elastic cross section of the proton. Since regions up

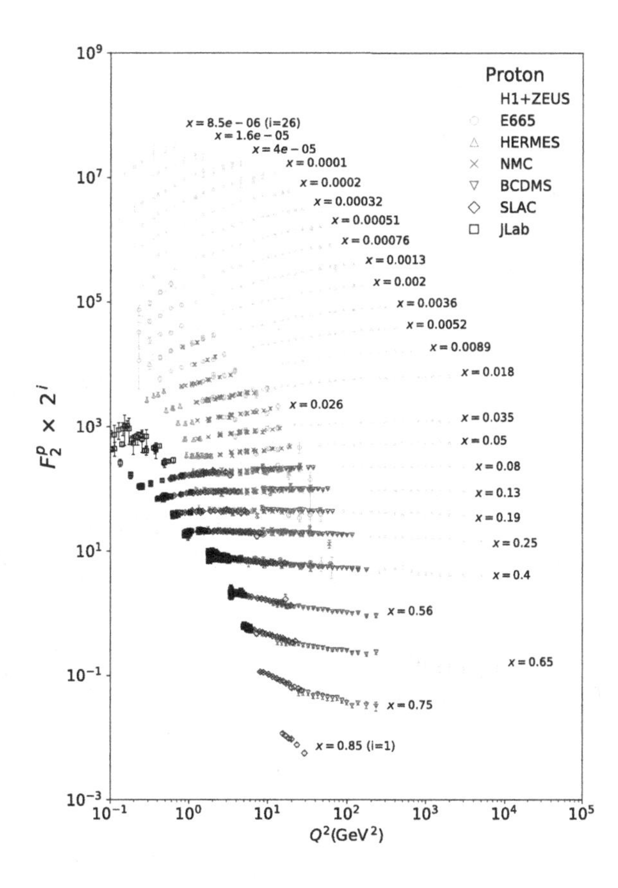

Fig. 7.9 The proton function structure F_2 for different x as a function of Q^2. F_2 has been multiplied by a number 2^i depending on x to ease the reading of the plot (i ranges from 1 to 26). Figure from Workman et al. (Particle Data Group) (2022).

to $Q^2 \simeq 10^4$ GeV^2 are probed, we can guess that the quark radius (if any) must be below the wavelength of the virtual boson[9] $\lambda = 1/\sqrt{Q^2}$, giving about 10^{-18} m [see Eq. (2.79)]. More sophisticated analyses assume that the quark size does not change its coupling, such that

$$\frac{d\sigma}{dQ^2} = \left(\frac{d\sigma}{dQ^2}\right)_{\text{point-like}} \left|F(Q^2)\right|^2,$$

where $F(Q^2)$ is the effective quark form factor (recall Supplement 7.1). Assuming $Q^2 = -|\boldsymbol{q}|^2$ (since q is space-like, we can always find a reference frame where $q^0 = 0$) , the effective form factor is the three-dimensional Fourier transform of the charge density $\rho(\boldsymbol{x})$, whose expansion is

$$F(\boldsymbol{q}) = \int d^3\boldsymbol{x} \exp(i\boldsymbol{q}\cdot\boldsymbol{x})\rho(\boldsymbol{x}) = \int d^3\boldsymbol{x} \left(1 + i\boldsymbol{q}\cdot\boldsymbol{x} - (\boldsymbol{q}\cdot\boldsymbol{x})^2/2 + \cdots\right)\rho(\boldsymbol{x}).$$

If we suppose that the density is spherically symmetric, $\rho(\boldsymbol{x}) = \rho(|\boldsymbol{x}|)$, and defining the charge RMS radius as

$$R^2 = \langle|\boldsymbol{x}|^2\rangle = \int |\boldsymbol{x}|^2\rho(|\boldsymbol{x}|)\,|\boldsymbol{x}|^2 d|\boldsymbol{x}| d(\cos\theta)\, d\phi = 4\pi \int |\boldsymbol{x}|^4\rho(|\boldsymbol{x}|)d|\boldsymbol{x}|,$$

the form factor expansion reads

$$F(\boldsymbol{q}) = 1 - R^2|\boldsymbol{q}|^2/6 + \cdots,$$

where the linear term cancels by symmetry, and factor 1/6 stems from the integration over $\cos\theta$ of $(\boldsymbol{q}\cdot\boldsymbol{x})^2/2 = |\boldsymbol{q}|^2|\boldsymbol{x}|^2\cos^2\theta/2$. Therefore, assuming that, at the Q^2 scale probed by HERA, the expected deviations with respect to the point-like cross section are small, we conclude that

$$\frac{d\sigma}{dQ^2} = \left(\frac{d\sigma}{dQ^2}\right)_{\text{point-like}} \left(1 - \frac{R^2}{6}Q^2\right)^2. \tag{7.43}$$

The ZEUS experiment at HERA has recently used the measured inclusive deep inelastic cross section to extract R^2 from the Formula (7.43), using a χ^2-minimisation procedure. The parameter R^2 was treated as a test statistic to obtain its probability distribution (assuming true values of R^2 and generating the corresponding cross section). Then, the R^2 distribution can be used for limit setting (see Chapter 1). The resulting 95% confidence level upper limit on the effective quark radius determined in ZEUS Collaboration (2016) is

$$R_{\text{quark}} < 0.43 \times 10^{-18}\ \text{m}. \tag{7.44}$$

Figure 7.9 leads us to a second observation. Outside the mid-range of Bjorken-x values, there is a clear dependence on Q^2, where F_2 increases at small x (typically for $x < 0.02$) and decreases at high x ($x > 0.65$). Hence, Bjorken scaling is no longer true in those regions, and F_2 is in general a function of the two variables Q^2 and x, i.e.

$$F_2(x, Q^2) \xrightarrow{x \simeq [0.02, 0.65]} F_2(x).$$

[9] For high Q^2, the virtual photon interaction must be complemented with the virtual Z^0 boson interaction allowed by the weak interaction.

The violation of Bjorken scaling can be explained qualitatively. A quark is never alone: it emits gluons that are reabsorbed. Moreover, a quark within the proton interacts with the other valence quarks by exchanging gluons. We will see in Chapter 8 that gluons can split into a quark–antiquark pair or even split into other gluons. In short, a quark is always surrounded by a cloud of partons. When Q^2 increases, the cloud gets bigger, or more appropriately, the ability to probe the details of the cloud improves. At high x, the fraction of momentum carried by the emission of gluons reduces x for the interacting quark: the structure function decreases for increasing Q^2. At low x, those gluons having small x may split into pairs of quark–antiquark, and one of the pair members may interact with the virtual photon. In other words, at low x for increasing Q^2, there are more and more quarks that may be probed by the photon: the structure function thus increases. This is the behaviour observed in Fig. 7.9. The scaling violation is, therefore, a consequence of the interaction of quarks within the proton, the initial assumption that they are free particles being too simplistic. We will come back on the scaling violation in Chapter 8.

Problems

7.1 Spinless scattering: the Rutherford cross section.
We consider the elastic electron–proton scattering $e^-(k)+\mathrm{p}(p) \to e^-(k')+\mathrm{p}(p')$ assuming that both the electron and the proton are spinless. We assume that the proton is at rest and neglect its recoil. The electron mass is also neglected.

1. Show that the amplitude is

$$\mathcal{M} = \frac{4\pi\alpha M}{E\sin^2\frac{\theta}{2}},$$

where M is the proton mass, θ is the scattering angle of the electron and E is the energy of the incoming electron.

2. Conclude that the differential cross section [hint: use the Formula (3.44)] can be expressed when the proton recoil is neglected as

$$\frac{\mathrm{d}\sigma}{\mathrm{d}\Omega} = \frac{\alpha^2}{4E^2\sin^4\frac{\theta}{2}},$$

which is the well-known Rutherford formula that is usually derived using a spinless particle (historically alpha) in a static Coulomb field.

7.2 Spinless scattering: the Mott cross section.
This problem is the continuation of Problem 7.1. We still assume the proton to be spinless, but the electron is now a spin-1/2 particle. The proton recoil is no longer neglected.

1. Show that the spin-averaged amplitude squared is

$$\overline{|\mathcal{M}|} = 16\frac{e^4}{q^4}EE'M^2\cos^2\frac{\theta}{2},$$

with $q^2 = -4EE'\sin^2(\theta/2)$. *Hint: use Eq. (7.9) to replace $E' - E$ with an expression depending on* $\sin^2(\theta/2)$.

2. Deduce the Mott differential cross section given in Eq. (7.10).

7.3 In this problem, the scattering $e^-(k) + \mathrm{p}(p) \to e^-(k') + \mathrm{p}(p')$ is considered, assuming that the proton is a point-like elementary particle at rest and neglecting the electron mass. Start from the general elementary cross section (3.39) and integrate over d^4p' instead of $\mathrm{d}^3\boldsymbol{p}'$ [using Eq. (E.6)] to show that

$$\frac{\mathrm{d}\sigma}{\mathrm{d}E'\,\mathrm{d}\Omega} = \frac{1}{8M(2\pi)^2}\delta\left(q^2 + 2M(E-E')\right)\overline{|\mathcal{M}|^2}\frac{E'}{E} = \frac{1}{16M^2(2\pi)^2}\delta\left(\frac{q^2}{2M} + E - E'\right)\overline{|\mathcal{M}|^2}\frac{E'}{E},$$

where M is the proton mass, $q^2 = -4EE'\sin^2(\theta/2)$ (with θ being the scattering angle of the electron), and E and E' the energy of the incoming and outgoing electron respectively.

7.4 Writing the current in spacetime coordinates $j^\mu_{\text{f.s.}} = \overline{\psi_{p'}(x)\Gamma^\mu\psi_p(x)}$, with $\psi_{p'}(x) = u(p')e^{-ip'\cdot x}$, $\psi_p(x) = u(p)e^{-ip\cdot x}$ and Γ^μ defined in Eq. (7.16), show that the conservation of the current $\partial_\mu j^\mu_{\text{f.s.}} = 0$ leads to Eq. (7.17).

7.5 Estimation of the proton radius from the form factor.

1. Check that if the charge distribution $\rho(\boldsymbol{r})$ has a spherical symmetry [$\rho(\boldsymbol{r}) = \rho(r)$], the form factor $G(\boldsymbol{q}) = \int \rho(\boldsymbol{r})e^{i\boldsymbol{q}.\boldsymbol{r}}\,\mathrm{d}^3\boldsymbol{r}$ can be written as $G(|\boldsymbol{q}|) = \frac{4\pi}{|\boldsymbol{q}|}\int r\rho(r)\sin(|\boldsymbol{q}|r)\,\mathrm{d}r$.
2. Assume that $\rho(r) = Ae^{-\alpha r}$ with $\alpha \geq 0$. Now, determine A as a function of α.
3. The measurement of the proton form factor is compatible with the dipole Formula (7.24). Show that for $Q^2 \simeq |\boldsymbol{q}^2|$, it is compatible with $\alpha = \sqrt{0.71}$ and deduce the mean value of the proton radius defined as $r_p = \sqrt{\langle r^2\rangle}$.

7.6 Write the hadronic tensor as

$$W^{\mu\nu} = -W_1 g^{\mu\nu} + \frac{W_2}{M^2}p^\mu p^\nu + \frac{W_3}{M^2}q^\mu q^\nu + \frac{W_4}{M^2}(p^\mu q^\nu + p^\nu q^\mu),$$

(we divide by M^2 such that all scalar functions have the same units) and show that the Eq. (7.28) implies Eq. (7.29). Hint: express W_4 and W_3 as a function of W_1 and W_2. Remark: another term proportional to $i\epsilon_{\mu\nu\rho\eta}p^\rho q^\eta/M^2$ is possible, but the Levi–Civita tensor usually emerges from a trace involving a γ^5 matrix, which would violate parity.

7.7 The deep inelastic scattering cross section of neutrinos on protons depends on two structure functions $F_2^{\nu p}(x)$ and $F_3^{\nu p}(x)$, where $F_3^{\nu p}$ is sensitive to the difference between quark and antiquark distributions (it originates from a parity-violating term). In the parton model, neglecting heavy quarks contribution (s, c, etc.), $F_2^{\nu p}(x)$ is found to be

$$F_2^{\nu p}(x) = 2x[d(x) + \bar{u}(x)].$$

The nucleon structure functions F_2^{eN} and $F_2^{\nu N}$ are defined by averaging the proton and neutron structure functions. Assuming the isospin symmetry, show that

$$\frac{F_2^{eN}}{F_2^{\nu N}} = \frac{1}{2}\left[\left(\frac{2}{3}\right)^2 + \left(\frac{1}{3}\right)^2\right] = \frac{5}{18}.$$

The ratio of the structure functions has been measured with deuterium, an isoscalar target (total isospin 0, so the same number of neutrons as protons). The data are consistent with the previous equation. Show that this ratio would at least be 1/2 if the quark charges were not fractional. Conclude that necessarily the quark electric charge is fractional. Since the ratio is constant, note that it implies that electron and neutrino deep inelastic scattering probe the same parton distribution.

8 Quantum Chromodynamics

This chapter is divided into two parts. The first introduces the quark model, following more or less the historical developments. It led to the symmetry group SU(3) flavour, where the u, d and s quarks are the three degrees of freedom. This symmetry is, however, only approximate, with the three quarks being not fully equivalent since they have different masses. The second part introduces the quantum chromodynamics (QCD), i.e. the formal gauge theory of the strong interaction. Here again, the symmetry group is SU(3), but the degrees of freedom are the three quark colours. This symmetry is assumed to be exact which has consequences on the existence of gluons and their properties, the carriers of the strong interaction at the elementary particle level, briefly mentioned in Chapter 7.

8.1 Quark Model

8.1.1 Isospin Number

In the early 1930s, Heisenberg noticed troubling similitudes between nuclei of equal mass numbers: their energy levels have a similarities with equal spacing between excited states, same spins and same parities. This is, for instance, the case of ${}^{14}_{6}\mathrm{C}$, ${}^{14}_{7}\mathrm{N}$ and ${}^{14}_{8}\mathrm{O}$, where the energy levels for states having the same spin and parity are very close. It looked as if the nuclei properties only depend on the mass number and not on the individual number of protons or neutrons separately. Since the interaction between neutrons and protons is dominated by the strong interaction (even if the electromagnetic interaction plays a role between protons to take into account the Coulomb repulsion), this suggested that the strong interaction cannot distinguish a neutron from a proton. The masses of the proton and neutron are so close (938.3 MeV and 939.5 MeV respectively) that Heisenberg, in 1932, proposed to interpret them, as far as nuclear interactions are concerned, as two states of the same object, the nucleon. Since the masses are (almost) equal, the two states would be degenerate. The situation is then similar to that of the spin $-1/2$ system, for which the two states $s_z = \pm 1/2$ are degenerate in absence of a magnetic field. The nucleon state $|N\rangle$ is then described as an object having an *isotopic spin*[1] (*isospin* for short) $I = 1/2$. It is a superposition of two degenerate states, $|p\rangle$ with the third component of the isospin $I_3 = +1/2$ (isospin up) and $|n\rangle$ with $I_3 = -1/2$ (isospin down), with a corresponding isospinor two-component notation

[1] Unlike the spin, the isospin is a simple number. It has no units (i.e., there is no $\hbar$).

$$|N\rangle = \begin{pmatrix} \alpha \\ \beta \end{pmatrix} = \alpha \begin{pmatrix} 1 \\ 0 \end{pmatrix} + \beta \begin{pmatrix} 0 \\ 1 \end{pmatrix} = \alpha\,|p\rangle + \beta\,|n\rangle,$$

with α and β being respectively the amplitude for the nucleon to have isospin up (i.e., be a proton) and down (i.e., be a neutron). Since $|p\rangle$ and $|n\rangle$ are degenerate, any linear transformation $|N'\rangle = U\,|N\rangle$, where U is a unitary matrix (recall Section 4.2, p. 101), should give the same physical results. The unitary conditions constrain the possible value of the determinant of U, since

$$\det(U^\dagger U) = \left[\det(U^\mathsf{T})\right]^* \det(U) = |\det(U)|^2 = 1.$$

Therefore, $\det(U) = e^{i\varphi}$, where φ is a real number. In terms of a symmetry group, the U matrices form a group called U(2), the 2×2 unitary matrices (see Appendix A). The U matrices can then result from the product of two matrices belonging to two different categories: those having $\det(U) = 1$ and those equal to $e^{i\varphi/2}\,\mathbb{1}$ whose determinant is $\left(e^{i\varphi/2}\right)^2 = e^{i\varphi}$. The matrices of the former comprise the SU(2) group, and the letter S stands for *special* because a special value of the determinant is chosen (i.e., 1), while the matrices of the latter are just determined by a single complex number of modulus 1 (such numbers, which can be seen as 1×1 unitary matrices, correspond to the U(1) group). Therefore, U(2) can be decomposed as

$$\mathrm{U(2)} = \mathrm{SU(2)} \times \mathrm{U(1)}. \tag{8.1}$$

The SU(2) transformation in the decomposition (8.1) corresponds to the isospin transformation. This group must already be well known to the reader (from quantum mechanics courses). It is, in particular, a *Lie group* (see Appendices A and B); the natural representation of its elements is

$$U = e^{-i\boldsymbol{\alpha}\cdot\boldsymbol{I}}, \tag{8.2}$$

with α_i, $i = 1, 2, 3$, being three real numbers and I_i the generators satisfying

$$[I_i, I_j] = i\epsilon_{ijk} I_k,$$

where ϵ_{ijk} is the spatial antisymmetric tensor. The natural representation of the generators as matrices of rank 2 is given by the three Pauli matrices [Eq. (2.62)]:

$$I_i = \frac{1}{2}\sigma_i \quad \text{with} \quad \sigma_1 = \begin{pmatrix} 0 & 1 \\ 1 & 0 \end{pmatrix}, \quad \sigma_2 = \begin{pmatrix} 0 & -i \\ i & 0 \end{pmatrix}, \quad \sigma_3 = \begin{pmatrix} 1 & 0 \\ 0 & -1 \end{pmatrix}. \tag{8.3}$$

A rotation of an isospin multiplet [via Formula (8.2)] does not affect how it couples to the strong force. Therefore, there are necessarily conserved quantities (Noether's theorem): I_3 and I (analogous to conservations of J_3 and J for the angular momentum), where I_3 can take $2I + 1$ values ranging in $-I, -(I-1), \ldots, I-1, I$. An isospin state is thus labelled by the values of I and I_3, with the proton and neutron being $|1/2, 1/2\rangle$ and $|1/2, -1/2\rangle$ respectively. Conventionally, the most positively charged particle has the largest value of I_3. Similar to the nucleon doublet, the masses of the three pions π^-, π^0 and π^+ are so close that they correspond to an isotriplet with $I = 1$. The four $\Delta^-, \Delta^0, \Delta^+, \Delta^{++}$ particles (and the four anti-Δ)[2] belong to the isomultiplet $I = 3/2$. It is worth mentioning that no particles form an isomultiplet with $I > 3/2$ (this will contribute to the edification of

[2] Warning: $\overline{\Delta^+} \equiv |\bar{u}\bar{u}\bar{d}\rangle$ is not the $\Delta^- \equiv |ddd\rangle$!

the quark model, as we shall see in Section 8.1.4). Now, if we believe in the tetra- and pentaquark nature of some newly discovered states at LHC (briefly mentioned in Section 8.2.3), then $I > 3/2$ is technically possible.

Since I and I_3 are conserved numbers by the strong interaction, the initial state and final state of a reaction induced by this interaction must have the same I_3 and I. In addition to this selection rule, it allows us to predict the relative strength of strong processes. As an example, let us consider the decay of the Δ^0 baryon. How can we compare the two channels $\Delta^0 \to \pi^0 + n$ and $\Delta^0 \to \pi^- + p$? Given that

$$\Delta^0 \equiv |3/2, -1/2\rangle, \pi^0 \equiv |1, 0\rangle, \pi^- \equiv |1, -1\rangle, p \equiv |1/2, 1/2\rangle, n \equiv |1/2, -1/2\rangle,$$

we must combine an isospin doublet (the nucleons) with a triplet (the pions) to recover the Δ multiplet. The rules for combining isospin are identical to those of the angular momentum (both use the $\mathfrak{su}(2)$ algebra). Hence, we can use the Clebsch–Gordan coefficients in the table labelled $1 \times 1/2$ of Fig. 2.1, yielding

$$\left|\frac{3}{2}, -\frac{1}{2}\right\rangle = \sqrt{\frac{2}{3}}\,|1, 0\rangle \otimes \left|\frac{1}{2}, -\frac{1}{2}\right\rangle + \sqrt{\frac{1}{3}}\,|1, -1\rangle \otimes \left|\frac{1}{2}, \frac{1}{2}\right\rangle.$$

Therefore, $|\Delta^0\rangle = \sqrt{2/3}\,|\pi^0 n\rangle + \sqrt{1/3}\,|\pi^- p\rangle$, and we conclude that

$$\frac{\mathrm{BR}(\Delta^0 \to \pi^0 + n)}{\mathrm{BR}(\Delta^0 \to \pi^- + p)} = \left|\frac{\langle \pi^0 n|\Delta^0\rangle}{\langle \pi^- p|\Delta^0\rangle}\right|^2 = \left|\frac{\sqrt{2/3}}{\sqrt{1/3}}\right|^2 = 2.$$

Another example of application is given in Problem 8.1.

8.1.2 Baryon Number

In the decomposition (8.1), the SU(2) part has been attributed to the isospin symmetry. But what about the U(1) part? Notice that the U(1) quantum number is shared by both members of the isospin doublet (in terms of matrix multiplication, the $e^{i\varphi/2}$ factor is a global factor). Hence, the neutron and proton have the same quantum number under this U(1) symmetry. This quantum number is the baryon number that we have already introduced in Chapter 4 (Section 4.4.2). The scale of this number being arbitrary (because one can write $e^{i\varphi/2} = e^{iq\varphi'/2}$, with $\varphi' = \varphi/q$, where q is the conserved number), the convention is to assign the value 1 to the proton and the neutron, or equivalently 1/3 to the u and d quarks and the opposite values for their antiparticles. The baryon number is conserved since the strong interaction is invariant under the global transformation (8.1). We saw, however, in Section 4.4.2 that experimentally, all interactions conserve this number. So this global U(1) symmetry is extended beyond the isospin doublet, and all quark flavours carry this quantum number.

8.1.3 Strangeness and Hypercharge

The discovery of the kaons ($K^{\pm}$ and K^0), the Λ and Σ baryons and the possibility of studying them at accelerator facilities in the 1950s and 1960s led to the observation that they were copiously produced, but their lifetime was much longer than that of other particles, such as the Δ baryon. Moreover, it was noted that those long-living particles were

Supplement 8.1. Anti-nucleon iso-doublet

There is a subtle point concerning the definition of the anti-nucleon state from the nucleon one. Consider an infinitesimal isospin transformation of the nucleon isodoublet. According to the relation (8.2), it reads

$$\begin{pmatrix} p' \\ n' \end{pmatrix} = e^{-i\epsilon\cdot\frac{\sigma}{2}} \begin{pmatrix} p \\ n \end{pmatrix} \simeq \left(1 - i\left[\epsilon_1\frac{\sigma_1}{2} + \epsilon_2\frac{\sigma_2}{2} + \epsilon_3\frac{\sigma_3}{2}\right]\right)\begin{pmatrix} p \\ n \end{pmatrix} = \begin{pmatrix} 1 - i\frac{\epsilon_3}{2} & -i\frac{\epsilon_1 - i\epsilon_2}{2} \\ -i\frac{\epsilon_1 + i\epsilon_2}{2} & 1 + i\frac{\epsilon_3}{2} \end{pmatrix}\begin{pmatrix} p \\ n \end{pmatrix},$$

giving

$$\begin{aligned} p' &= (1 - i\frac{\epsilon_3}{2})p - i\frac{\epsilon_1 - i\epsilon_2}{2}n, \\ n' &= (1 + i\frac{\epsilon_3}{2})n - i\frac{\epsilon_1 + i\epsilon_2}{2}p. \end{aligned} \tag{8.4}$$

In group theory, the antiparticles are assigned to the complex conjugate of the representation to which the particles belong. In other words, when particle doublets, such as the nucleon, belong to the fundamental representation **2**, anti-doublets belong to $\bar{\mathbf{2}}$ and transform as

$$\begin{pmatrix} \bar{p}' \\ \bar{n}' \end{pmatrix} = \left(e^{-i\epsilon\cdot\frac{\sigma}{2}}\right)^* \begin{pmatrix} \bar{p} \\ \bar{n} \end{pmatrix} \simeq \begin{pmatrix} 1 + i\frac{\epsilon_3}{2} & i\frac{\epsilon_1 + i\epsilon_2}{2} \\ i\frac{\epsilon_1 - i\epsilon_2}{2} & 1 - i\frac{\epsilon_3}{2} \end{pmatrix}\begin{pmatrix} \bar{p} \\ \bar{n} \end{pmatrix},$$

and therefore

$$\begin{aligned} \bar{p}' &= (1 + i\frac{\epsilon_3}{2})\bar{p} + i\frac{\epsilon_1 + i\epsilon_2}{2}\bar{n}, \\ \bar{n}' &= (1 - i\frac{\epsilon_3}{2})\bar{n} + i\frac{\epsilon_1 - i\epsilon_2}{2}\bar{p}. \end{aligned} \tag{8.5}$$

Note that it is consistent having $\bar{p}$ in the upper position (even if its charge is lower than $\bar{n}$ charge) because, as

$$\left(e^{-i\epsilon\cdot\frac{\sigma}{2}}\right)^* = e^{-i\epsilon\cdot\left(-\frac{\sigma^*}{2}\right)},$$

the generator of the third component of the isospin in the $\bar{\mathbf{2}}$ representation is

$$I_3 = -\frac{\sigma_3^*}{2} = -\frac{\sigma_3}{2} = \frac{1}{2}\begin{pmatrix} -1 & 0 \\ 0 & 1 \end{pmatrix}.$$

Hence, it gives correctly the isospin $-1/2$ and $+1/2$ for the anti-proton and anti-neutron, respectively, i.e. the opposite sign of that of particles. Now, SU(2) is specific in the sense that both **2** and $\bar{\mathbf{2}}$ can be seen as the same representation. Indeed, in both cases, only the quantum numbers $I = 1/2$ and $I_3 = \pm 1/2$ are used for doublets. Therefore, we can define doublets for antiparticles in such a way that they transform according to **2** (and not $\bar{\mathbf{2}}$), and thus follow Eq. (8.4) instead of Eq. (8.5). In the **2** representation, the particle having $I_3 = +1/2$ (or the largest charge) is in the upper position of the doublet. Thus, a doublet with the anti-neutron in the upper position must be looked for. We notice that Eq. (8.5) can be rewritten as follows:

$$\begin{aligned} (-\bar{n}') &= (1 - i\frac{\epsilon_3}{2})(-\bar{n}) - i\frac{\epsilon_1 - i\epsilon_2}{2}\bar{p}, \\ \bar{p}' &= (1 + i\frac{\epsilon_3}{2})\bar{p} - i\frac{\epsilon_1 + i\epsilon_2}{2}(-\bar{n}). \end{aligned}$$

It has exactly the required form of transformation as in Eq. (8.4). Therefore, by defining the anti-nucleon doublet as

Supplement 8.1. (cont.)

$$|\bar{N}\rangle = \begin{pmatrix} -\bar{n} \\ \bar{p} \end{pmatrix}, \tag{8.6}$$

we can use the **2** representation and, thus, the same transformations as for the nucleon. This trick is useful because the same Clebsch–Gordan coefficients are then used to combine particles and antiparticles (and, e.g., quarks and antiquarks forming mesons).

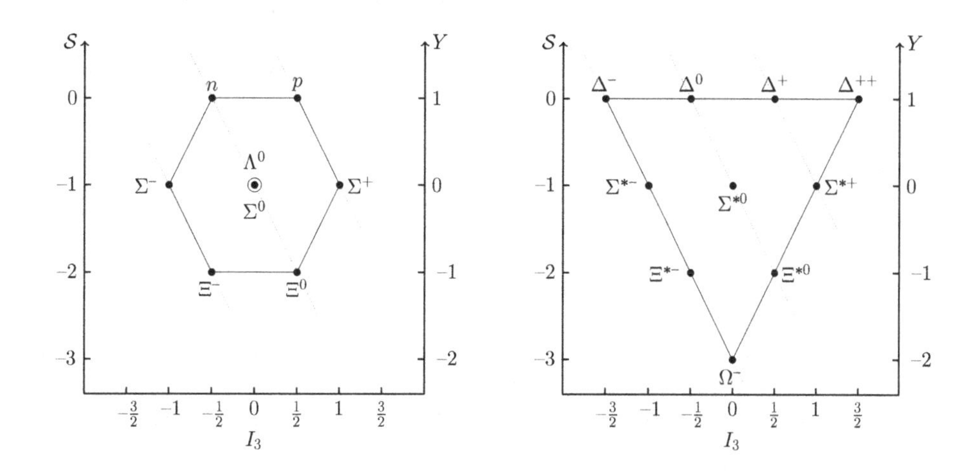

Fig. 8.1 Classification of baryons into an octet ($J^P = 1/2^+$) in the left diagram and into a decuplet ($J^P = 3/2^+$) in the right. The isospin projection number is on the x-axis, while the strangeness or the hypercharge is on the y-axis. Hadrons having the same electric charge are on the downward sloping diagonal lines of the multiplets.

always produced in pairs. The hypothesis of a production mechanism different from that of the decay was, then, suggested: those hadrons were produced by strong interaction but decayed by weak interaction. The question was then to understand what prevented them from decaying via strong interaction. In 1953, Gell-Mann and Nishijima introduced a new additive quantum number, the *strangeness*. It was postulated that it is conserved by strong and electromagnetic interactions but *not* by weak interaction. Hence, only weak decays were allowed for particles carrying a non-zero strangeness. The pair production by strong interaction was explained by pair members carrying opposite strangeness and, thus, a pair with zero overall strangeness. Eight years later (1961), Gell-Mann realised that particles having similar properties, i.e. same spins and same parities, can be arranged into geometrical patterns in the isospin-strangeness (I_3, S) plane. Figures 8.1 and 8.2 show these patterns for the baryons and mesons respectively.[3] For baryons, similar patterns exist populated with anti-baryons. Particles of a pattern having the same strangeness and belonging to the same isospin multiplet have nearly the same masses (as the pions, or the Δs). However, for different strangeness, the spread of masses is very large. For

[3] In 1961, when Gell-Mann arranged the particles in patterns, the Ω^- and Ξ^* baryons as well as the η, η' and ϕ mesons had not yet been discovered

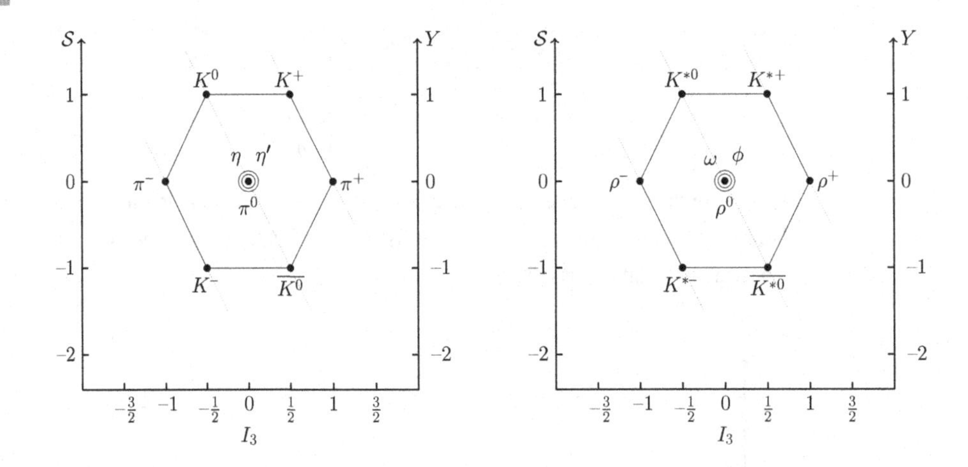

Fig. 8.2 Classification of mesons into a pseudoscalar nonet ($J^P = 0^-, s = I = 0$) in the left diagram and into a vector nonet ($J^P = 1^-, s = 1, I = 0$) in the right diagram. The isospin projection number is on the x-axis, while the strangeness or the hypercharge is on the y-axis. Hadrons having the same electric charge are on the downward sloping diagonal lines of the multiplets.

instance, $m_\pi \simeq 140$ MeV$/c^2$, whereas $m_K \simeq 495$ MeV$/c^2$, or $m_p \simeq 0.94$ GeV$/c^2$ versus $m_\Xi \simeq 1.32$ GeV$/c^2$. Instead of using the strangeness, Nishijima and Gell-Mann proposed an alternative quantum number, called the *hypercharge*. It is essentially the same as strangeness but shifted in the case of baryons, i.e.

$$Y = B + S, \tag{8.7}$$

where B is the baryon number. It has the advantage that both baryons and mesons can be treated in a unified way when multiplets are presented in the isospin–hypercharge plane (I_3, Y), as shown in Figs. 8.1 and 8.2, where, in that plane, there is always a state populating the centre at $(0, 0)$. This detail will be important when the multiplets will be recovered from fundamental group representations. With the definition (8.7) of the hypercharge, Gell-Mann and Nishijima observed that

$$Q = I_3 + \frac{Y}{2}. \tag{8.8}$$

This formula is now known as the *Gell-Mann–Nishijima* formula. It is reflected in Figs. 8.1 and 8.2 by the fact that hadrons with the same electric charge lay on the downward sloping diagonal lines of the multiplets.

8.1.4 The SU(3) Flavour Symmetry

To interpret the baryon and meson multiplet, Gell-Mann and Zweig, following the precursor work of others,[4] postulated independently in 1964 (before the advent of deep inelastic

[4] Fermi and Yang, in 1949, proposed a model based on SU(2) with the neutron and proton as the building blocks. In 1956, Sakata added to the Fermi–Yang model the Λ carrying strangeness -1. All hadrons were bound states of p, n and Λ. Neglecting their difference in masses, Sakata thus produced the first SU(3) model based on the triplet p, n, Λ. Finally, Ne'eman and Gell-Mann, in 1961, introduced as a basic unit an octet of SU(3) in the famous eightfold-way model that led to the quark model.

scattering experiments revealing the substructure of nucleons) that hadrons are made of quarks.[5] Two quarks, u and d, are the members of an isospin doublet with $\mathcal{S} = 0$, and one quark, s, is a $\mathcal{S} = -1$ isosinglet (the unfortunate sign is historical, the first meson discovered with $\mathcal{S} \neq 0$ actually contained an anti-s-quark, the $K^+ \equiv u\bar{s}$), i.e.

$$\begin{pmatrix} u \\ d \end{pmatrix} : \begin{matrix} I_3 = +\frac{1}{2}, & \mathcal{S} = 0 \\ I_3 = -\frac{1}{2}, & \mathcal{S} = 0 \end{matrix} \qquad (s) : I_3 = 0, \quad \mathcal{S} = -1. \tag{8.9}$$

The three quark flavours, u, d and s, were thought of as three degrees of freedom of the symmetry group $\mathrm{SU}(3)_f$, with f standing for flavour. The symmetry is approximate since members of a multiplet can have a significant dispersion of masses, reflecting the non-equality of the three quark masses. Baryons are composed of three quarks, while mesons are made of a quark–antiquark pair. Quarks are assumed to democratically share the baryon number, meaning that they have $B = 1/3$, and consequently, their hypercharge is from Eq. (8.7) $Y_u = Y_d = 1/3$ and $Y_s = -2/3$. Therefore, according to the formula (8.8), their electric charges are $Q_u = 2/3$ and $Q_d = Q_s = -1/3$. In this model, the three quark flavours belong to an SU(3) triplet, such as

$$|u\rangle = \begin{pmatrix} 1 \\ 0 \\ 0 \end{pmatrix}, \quad |d\rangle = \begin{pmatrix} 0 \\ 1 \\ 0 \end{pmatrix}, \quad |s\rangle = \begin{pmatrix} 0 \\ 0 \\ 1 \end{pmatrix}. \tag{8.10}$$

A general quark state is a linear combination of these elements

$$|q\rangle = q_1 |u\rangle + q_2 |d\rangle + q_3 |s\rangle = \begin{pmatrix} q_1 \\ q_2 \\ q_3 \end{pmatrix}.$$

SU(3) is a Lie group that has $3^2 - 1 = 8$ generators T_i (see Appendix C). A quark state is rotated under an SU(3) transformation with eight parameters $\boldsymbol{\alpha} = (\alpha_1, \dots, \alpha_8)$ as

$$|q'\rangle = U\,|q\rangle \quad \text{with } U = e^{-i\boldsymbol{\alpha}\cdot\boldsymbol{T}}.$$

The generators T_i take the matrix representation

$$T_i = \frac{\lambda_i}{2},$$

where λs are the 3×3 Gell-Mann matrices. They can be viewed as a generalisation of the Pauli matrices related to the SU(2) generators by $I_i = \sigma_i/2$. To identify the matrix representation, we can consider just the first two flavours u and d, for which we should recover the SU(2) generators, i.e.

$$\lambda_{i=1,2,3} = \begin{pmatrix} \multicolumn{2}{c}{\sigma_i} & 0 \\ & & 0 \\ 0 & 0 & 0 \end{pmatrix} \Rightarrow \lambda_1 = \begin{pmatrix} 0 & 1 & 0 \\ 1 & 0 & 0 \\ 0 & 0 & 0 \end{pmatrix},$$

$$\lambda_2 = \begin{pmatrix} 0 & -i & 0 \\ i & 0 & 0 \\ 0 & 0 & 0 \end{pmatrix}, \quad \lambda_3 = \begin{pmatrix} 1 & 0 & 0 \\ 0 & -1 & 0 \\ 0 & 0 & 0 \end{pmatrix}. \tag{8.11}$$

[5] The name quark was introduced by Gell-Mann and is a reference to the novel 'Finnegans Wake', by James Joyce. Zweig called these elementary objects aces.

The diagonal matrix is associated with a conserved quantum number. Defining the isospin generator as $I_3 = \lambda_3/2$ gives the isospins of the states $|u\rangle, |d\rangle, |s\rangle$:

$$I_3 = \frac{\lambda_3}{2}: \quad I_3 |u\rangle = +\frac{1}{2} |u\rangle, \quad I_3 |d\rangle = -\frac{1}{2} |d\rangle, \quad I_3 |s\rangle = 0\, |s\rangle.$$

For the other flavour pairs (u, s) and (d, s), by symmetry, we should recover the SU(2) generators for the considered couple. Hence, it suffices to insert a row and a column of zeros corresponding to the position of the third flavour (d and u, respectively), yielding

$$\begin{aligned}
\lambda_4 &= \begin{pmatrix} 0 & 0 & 1 \\ 0 & 0 & 0 \\ 1 & 0 & 0 \end{pmatrix}, \quad \lambda_5 = \begin{pmatrix} 0 & 0 & -i \\ 0 & 0 & 0 \\ i & 0 & 0 \end{pmatrix}, \quad \lambda_3' = \begin{pmatrix} 1 & 0 & 0 \\ 0 & 0 & 0 \\ 0 & 0 & -1 \end{pmatrix}, \\
\lambda_6 &= \begin{pmatrix} 0 & 0 & 0 \\ 0 & 0 & 1 \\ 0 & 1 & 0 \end{pmatrix}, \quad \lambda_7 = \begin{pmatrix} 0 & 0 & 0 \\ 0 & 0 & -i \\ 0 & i & 0 \end{pmatrix}, \quad \lambda_3'' = \begin{pmatrix} 0 & 0 & 0 \\ 0 & 1 & 0 \\ 0 & 0 & -1 \end{pmatrix}.
\end{aligned} \tag{8.12}$$

We notice that λ_3' and λ_3'' are the diagonal matrices. We could imagine that they would define the generators of two conserved quantum numbers in addition to the isospin, but it turns out that they are not linearly independent of λ_3 since $\lambda_3 = \lambda_3' - \lambda_3''$. Therefore, we can define a linear combination of λ_3', λ_3'' independent of λ_3, such that its eigenvalues are proportional to the hypercharges Y of the $|u\rangle, |d\rangle, |s\rangle$ states. The appropriate choice is

$$\lambda_8 = \frac{1}{\sqrt{3}} \left[\lambda_3' + \lambda_3''\right] = \frac{1}{\sqrt{3}} \begin{pmatrix} 1 & 0 & 0 \\ 0 & 1 & 0 \\ 0 & 0 & -2 \end{pmatrix}, \tag{8.13}$$

where the factor $1/\sqrt{3}$ is such that any eight matrices $\lambda_1, \ldots, \lambda_8$ satisfy $\mathrm{Tr}(\lambda_a \lambda_b) = 2\delta_{ab}$ (this property will be used later). The hypercharge generator is then

$$Y = \frac{1}{\sqrt{3}}\lambda_8: \quad Y|u\rangle = \frac{1}{3}|u\rangle, \quad Y|d\rangle = \frac{1}{3}|d\rangle, \quad Y|s\rangle = -\frac{2}{3}|s\rangle,$$

whose eigenvalues $(1/3, 1/3, -2/3)$ are the hypercharges of u, d and s, respectively. With the two generators I_3 and Y that commute, we can label the basis states $|u\rangle, |d\rangle, |s\rangle$ with their eigenvalues, $|i_3, y\rangle$, i.e.

$$|u\rangle = \left|\frac{1}{2}, \frac{1}{3}\right\rangle, \quad |d\rangle = \left|-\frac{1}{2}, \frac{1}{3}\right\rangle, \quad |s\rangle = \left|0, -\frac{2}{3}\right\rangle.$$

Similarly, the antiquarks have opposite quantum numbers. Indeed, the antiquark triplet

$$|\bar{u}\rangle = \begin{pmatrix} 1 \\ 0 \\ 0 \end{pmatrix}, \quad |\bar{d}\rangle = \begin{pmatrix} 0 \\ 1 \\ 0 \end{pmatrix}, \quad |\bar{s}\rangle = \begin{pmatrix} 0 \\ 0 \\ 1 \end{pmatrix} \tag{8.14}$$

transforms under a rotation in the flavour space of real parameters $\boldsymbol{\alpha} = (\alpha_1, \ldots, \alpha_8)$ according to conjugate representation, i.e. $[\exp(-i\boldsymbol{\alpha}\cdot\boldsymbol{\lambda}/2)]^* = \exp(-i\boldsymbol{\alpha}\cdot(-\boldsymbol{\lambda}^*)/2)$. It shows that the generators of the representation $\bar{\mathbf{3}}$ are given by the matrices $-\lambda_i^*/2$. Since the two diagonal matrices λ_3 and λ_8 are real matrices, we thus have $-\lambda_3^* = -\lambda_3$ and $-\lambda_8^* = -\lambda_8$, implying that $\bar{u}, \bar{d}$ and $\bar{s}$ have opposite isospin and hypercharge to those of u, d and s, respectively. The fundamental representations $\mathbf{3}$ and $\bar{\mathbf{3}}$ are represented geometrically in plane (I_3, Y) in Fig. 8.3.

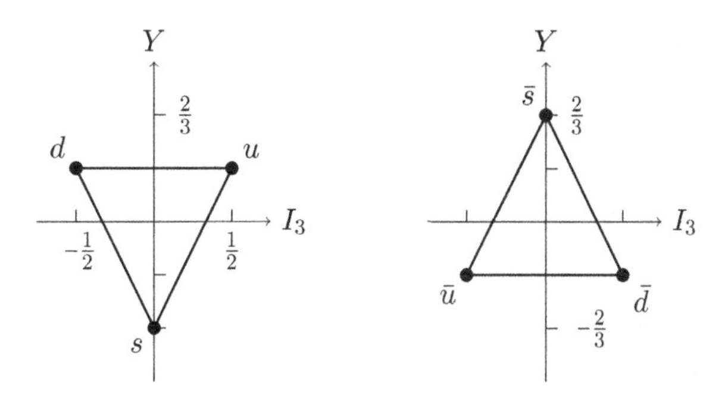

Fig. 8.3 SU(3) quark (**3**-representation, left-hand side) and antiquark ($\bar{\mathbf{3}}$-representation, right-hand side) multiplets in the (I_3, Y) plane.

Meson Multiplets

By construction, the flavour states of the mesons $q\bar{q}$ are generated by the representation $\mathbf{3} \otimes \bar{\mathbf{3}}$. The result in the plane (I_3, Y) is shown in Fig. 8.4. It suffices to superimpose the centre of gravity of the antiquark multiplet on top of every site of the quark multiplet since I_3 and Y are additive quantum numbers. Figure 8.4 suggests that we obtain the decomposition into an octet of meson and a singlet, i.e.

$$\mathbf{3} \otimes \bar{\mathbf{3}} = \mathbf{8} \oplus \mathbf{1}.$$

Formally, to determine the content of the multiplets, we introduce, as in the quantum mechanics course dealing with the angular momentum, the ladder operators

$$I_\pm = \frac{1}{2}(\lambda_1 \pm i\lambda_2), \quad U_\pm = \frac{1}{2}(\lambda_6 \pm i\lambda_7), \quad V_\pm = \frac{1}{2}(\lambda_4 \pm i\lambda_5). \tag{8.15}$$

They are the raising and lowering operators along the direction $d \leftrightarrow u$, $s \leftrightarrow d$ and $s \leftrightarrow u$, respectively. Thus, their actions on a state with isospin i_3 and hypercharge y are

$$I_+ |i_3, y\rangle = |i_3 + 1, y\rangle, \quad U_+ |i_3, y\rangle = |i_3 - 1/2, y + 1\rangle, \quad V_+ |i_3, y\rangle = |i_3 + 1/2, y + 1\rangle,$$

whereas I_-, U_- and V_- perform the reciprocal action. Using the matrix representations (8.11), (8.12) and (8.13), this is straightforward to check since the application of the operators on the states (8.10) gives $I_+ |d\rangle = |u\rangle$, $U_+ |s\rangle = |d\rangle$, $V_+ |s\rangle = |u\rangle$ and the reciprocal action for I_-, U_-, V_- operators. Since the fundamental representations **3** and $\bar{\mathbf{3}}$ lead to different quantum numbers (see Fig. 8.3, where geometrically there is no overlap of the two patterns), it is not possible to transform the antiquark triplets with **3**, using a similar trick as we did in Supplement 8.1 where the anti-nucleon can transform under **2** of SU(2). As for quarks, the ladder operators for the antiquarks are the linear combinations of the generators, which for antiquarks are based on the matrices $-\lambda_i^*$. The antiquark ladder operators are then

$$I_\pm = -\frac{1}{2}(\lambda_1^* \pm i\lambda_2^*), \quad U_\pm = -\frac{1}{2}(\lambda_6^* \pm i\lambda_7^*), \quad V_\pm = -\frac{1}{2}(\lambda_4^* \pm i\lambda_5^*). \tag{8.16}$$

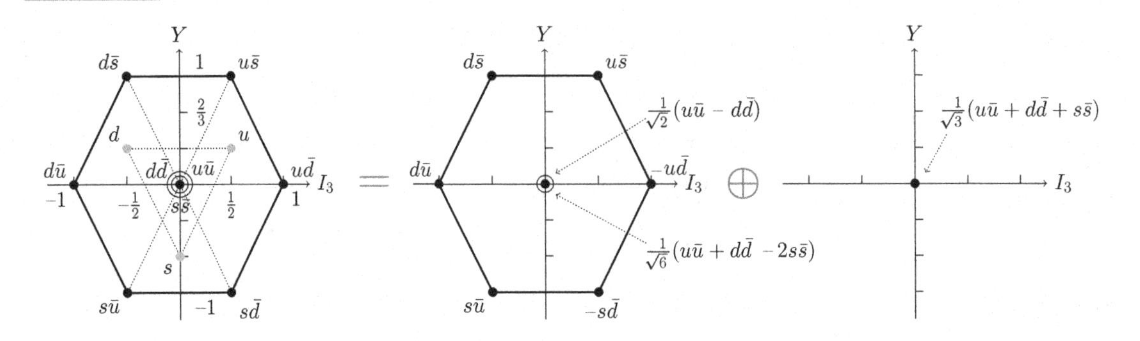

Fig. 8.4 The quark content of the mesons nonet resulting from $\mathbf{3} \otimes \bar{\mathbf{3}} = \mathbf{8} \oplus \mathbf{1}$. The relative phases ensure that the ladder operators $I_\pm$ correctly step between the states.

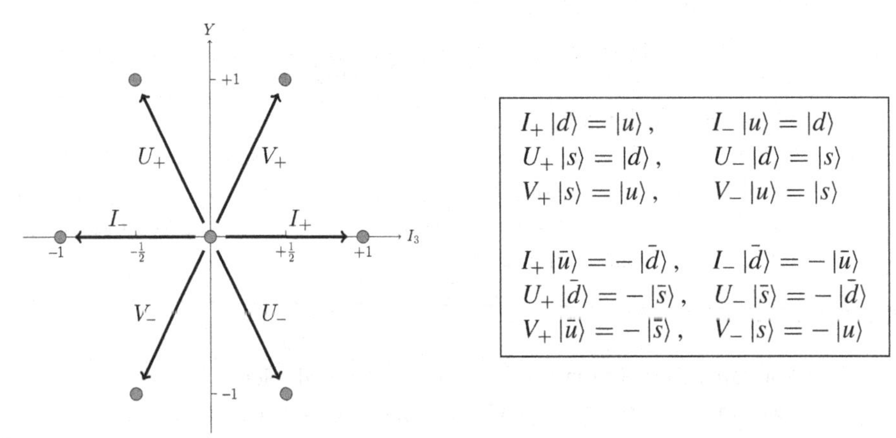

Fig. 8.5 Left: the action of the ladder operators on a state located at the centre. Right: the corresponding effect on the states of the fundamental representations $\mathbf{3}$ and $\bar{\mathbf{3}}$. All other combinations of ladder operators give zero.

It leads to $I_+ |\bar{u}\rangle = - |\bar{d}\rangle$, $U_+ |\bar{d}\rangle = - |\bar{s}\rangle$ and $V_+ |\bar{u}\rangle = - |\bar{s}\rangle$. Figrue 8.5 summarises the action of the SU(3) ladder operators.

The procedure to identify the multiplets and their content is then to start with a state having the highest weight, i.e. the state for which

$$I_+ |i_3, y\rangle = U_+ |i_3, y\rangle = V_+ |i_3, y\rangle = 0.$$

Then the successive application of I_-, U_- or V_- generates all the multiplet states to which the highest weight state belongs. For $\mathbf{3} \otimes \bar{\mathbf{3}}$, the highest weight state is $|u\bar{s}\rangle$. (See the left diagram of Fig. 8.4, where $|u\bar{s}\rangle$ is located in the upper-right corner and the direction of the increasing ladder operators in Fig. 8.5.) The action of the three lowering ladder operators on $|u\bar{s}\rangle$ yields the states $|d\bar{s}\rangle$, $- |u\bar{d}\rangle$ and $|s\bar{s}\rangle - |u\bar{u}\rangle$. As physical states are defined up to a phase, the negative sign in $- |u\bar{d}\rangle$ may be ignored. To represent a flavour state of a physical particle, the state $|s\bar{s}\rangle - |u\bar{u}\rangle$ must be normalised to $(|s\bar{s}\rangle - |u\bar{u}\rangle)/\sqrt{2}$. Let us detail, once, how the operators are applied to tensorial states. Recall that $|u\bar{s}\rangle$ actually means $|u\rangle \otimes |\bar{s}\rangle$. The operators are extended to the tensorial space thanks to the identity operators of each space. For instance, V_- when applied to $|u\bar{s}\rangle$ means $V_- \otimes \mathbb{1} + \mathbb{1} \otimes V_-$, where V_- in $V_- \otimes \mathbb{1}$ is the ladder operator of the quark space and V_- in $\mathbb{1} \otimes V_-$ is the ladder operator of the antiquark space. Therefore, its action yields

$$\begin{aligned} V_- |u\bar{s}\rangle &\equiv (V_- \otimes \mathbb{1} + \mathbb{1} \otimes V_-)(|u\rangle \otimes |\bar{s}\rangle) = V_- |u\rangle \otimes |\bar{s}\rangle + |u\rangle \otimes V_- |\bar{s}\rangle \\ &= |s\rangle \otimes |\bar{s}\rangle - |u\rangle \otimes |\bar{u}\rangle \equiv |s\bar{s}\rangle - |u\bar{u}\rangle. \end{aligned}$$

Starting with the highest weight state $|u\bar{s}\rangle$, the successive applications of I_-, U_- and V_- generate the following eight other states (after normalisation):

$$|d\bar{s}\rangle, \ -|u\bar{d}\rangle, \ \frac{1}{\sqrt{2}}(|s\bar{s}\rangle - |u\bar{u}\rangle), \ \frac{1}{\sqrt{2}}(|s\bar{s}\rangle - |d\bar{d}\rangle), \ |d\bar{u}\rangle, \ \frac{1}{\sqrt{2}}(|u\bar{u}\rangle - |d\bar{d}\rangle), \ -|s\bar{d}\rangle, \ |s\bar{u}\rangle.$$

Three have the same quantum numbers, $i_3 = y = 0$. They are

$$|\psi_1\rangle = \frac{1}{\sqrt{2}}(|s\bar{s}\rangle - |u\bar{u}\rangle), \ |\psi_2\rangle = \frac{1}{\sqrt{2}}(|s\bar{s}\rangle - |d\bar{d}\rangle), \ |\psi_3\rangle = \frac{1}{\sqrt{2}}(|u\bar{u}\rangle - |d\bar{d}\rangle). \quad (8.17)$$

Note that they are not independent since $|\psi_3\rangle = |\psi_2\rangle - |\psi_1\rangle$. Therefore, they can be reduced to two independent combinations, for instance,[6] $|\psi_3\rangle$ and $|\psi_2\rangle + |\psi_1\rangle$. The state $|\psi_3\rangle$ is the π^0 meson state, while $|\psi_2\rangle + |\psi_1\rangle$ reads after normalisation

$$|\psi_8\rangle = \frac{1}{\sqrt{6}}(|u\bar{u}\rangle + |d\bar{d}\rangle - 2\,|s\bar{s}\rangle), \quad (8.18)$$

which is the η meson state. It is easy to check that $|\pi^0\rangle$ is a member of the isospin triplet with $|\pi^\pm\rangle$ since $I_\pm\,|\pi^0\rangle$ gives, up to a normalisation factor, $|\pi^\pm\rangle$ (i.e., $-|u\bar{d}\rangle$ and $|d\bar{u}\rangle$). Similarly, $I_\pm\,|\eta\rangle = 0$, which shows that $|\eta\rangle$ is an isosinglet.

Starting with $|u\bar{s}\rangle$, we have found the eight states populating the octet in Fig. 8.4. However, $\mathbf{3} \otimes \bar{\mathbf{3}}$ corresponds to nine states. According to the first diagram of Fig. 8.4, the missing state has also $i_3 = y = 0$, and thus, it is a combination of $|u\bar{u}\rangle$, $|d\bar{d}\rangle$ and $|s\bar{s}\rangle$, i.e. $|\psi_S\rangle \propto |u\bar{u}\rangle + \alpha\,|d\bar{d}\rangle + \beta\,|s\bar{s}\rangle$ (up to a normalisation factor). This state must be orthogonal to the two other states at $i_3 = y = 0$. The orthogonality condition $\langle\psi|\pi^0\rangle = 0$ imposes $1 - \alpha = 0$, while with $\langle\psi|\eta\rangle = 0$, the constraint is $2 - 2\beta = 0$. Hence, after normalisation

$$|\psi_S\rangle = \frac{1}{\sqrt{3}}(|u\bar{u}\rangle + |d\bar{d}\rangle + |s\bar{s}\rangle). \quad (8.19)$$

This combination is an SU(3) singlet. The action of any ladder operator cannot produce another existing state if it is a singlet. Let us check, for instance, $I_+\,|\psi_S\rangle = (0 - |u\bar{d}\rangle + |u\bar{d}\rangle + 0 + 0 + 0)/\sqrt{3} = 0$. The reader can check that it is the case for the five other operators. As a singlet, it is invariant (modulo a possible phase) under a rotation in the flavour space (and in particular for any interchange $u \leftrightarrow d \leftrightarrow s$).

The pseudoscalar meson nonet in Fig. 8.2 is reasonably well described by the eight states of the octet in Fig. 8.4 and the $|\psi_S\rangle$ state of the singlet. For the vector nonet, the situation is

[6] We could have chosen $|\psi_1\rangle = |\psi_2\rangle - |\psi_3\rangle \propto |s\bar{s}\rangle - |u\bar{u}\rangle$ and $|\psi_2\rangle + |\psi_3\rangle \propto |u\bar{u}\rangle + |s\bar{s}\rangle - 2\,|d\bar{d}\rangle$ or any two orthogonal combinations of $|\psi_1\rangle, |\psi_2\rangle$ and $|\psi_3\rangle$. If the SU(3) flavour symmetry were an exact symmetry, all those combinations would be equivalent. However, the SU(3) flavour symmetry is only approximate because the u, d and s quarks have different masses. Therefore, we privilege a combination with an isospin eigenstate because the isospin symmetry is less broken than the SU(3) flavour symmetry since the u and d quark masses are almost identical. That is why $|\psi_3\rangle$ and $|\psi_8\rangle$ are chosen.

more complicated. Mesons not at the centre of the nonet match the states found in the octet. In the centre, the ρ^0 particle belongs to an isotriplet with $\rho^\pm$. It is described by $|\psi_3\rangle$ as the π^0. However, as $|\psi_8\rangle$ and $|\psi_S\rangle$ share the same quantum numbers (both are an isosinglet $I = 0$ with a hypercharge $y = 0$), they can mix to produce the physical states ϕ and ω. Thus, introducing the mixing angle θ, the ϕ and ω states read

$$|\phi\rangle = \cos\theta\,|\psi_8\rangle - \sin\theta\,|\psi_S\rangle = \left(\tfrac{1}{\sqrt{3}}\cos\theta - \sqrt{\tfrac{2}{3}}\sin\theta\right)\tfrac{|u\bar{u}\rangle + |d\bar{d}\rangle}{\sqrt{2}} - \left(\sqrt{\tfrac{2}{3}}\cos\theta + \tfrac{1}{\sqrt{3}}\sin\theta\right)|s\bar{s}\rangle,$$
$$|\omega\rangle = \sin\theta\,|\psi_8\rangle + \cos\theta\,|\psi_S\rangle = \left(\sqrt{\tfrac{2}{3}}\cos\theta + \tfrac{1}{\sqrt{3}}\sin\theta\right)\tfrac{|u\bar{u}\rangle + |d\bar{d}\rangle}{\sqrt{2}} + \left(\tfrac{1}{\sqrt{3}}\cos\theta - \sqrt{\tfrac{2}{3}}\sin\theta\right)|s\bar{s}\rangle.$$

Experimentally, it is found that $\theta = 36.5^\circ$ (Particle Data Group, 2022), corresponding to $\sin\theta \simeq 1/\sqrt{3}$. Therefore, the ϕ meson is almost a pure $|s\bar{s}\rangle$ state, while ω is $(|u\bar{u}\rangle + |d\bar{d}\rangle)/\sqrt{2}$. For pseudoscalar mesons, this mixing angle is close to zero, justifying why η is $|\psi_8\rangle$ and η', $|\psi_S\rangle$.

Baryon Multiplets

Baryons are made of three quarks (or antiquarks for anti-baryons). The SU(3) flavour symmetry should be able to recover the observed baryons containing the u, d or s quarks starting from the elementary representation **3** (see Problem 8.2). Figure 8.6 illustrates the decomposition $\mathbf{3} \otimes \mathbf{3} \otimes \mathbf{3}$. First, the product $\mathbf{3} \otimes \mathbf{3}$ generates a sextet **6** and a triplet, which turns out to have exactly the same quantum numbers as the antiquarks. This triplet is thus mathematically $\bar{\mathbf{3}}$ already encountered. Hence, the decomposition $\mathbf{3} \otimes \mathbf{3} = \mathbf{6} \oplus \bar{\mathbf{3}}$, as shown in the first row of Fig. 8.6. To obtain the content of the sextet, we start with the state having the highest weight and apply the lowering ladder operators. The content of $\bar{\mathbf{3}}$ is obtained by requiring the orthogonality of its states with those of the sextet. The next step is to add the third quark. The second and third rows of Fig. 8.6 show the results of $\mathbf{6} \otimes \mathbf{3}$ and $\bar{\mathbf{3}} \otimes \mathbf{3}$ respectively. The former generates a decuplet and an octet, while the latter generates an octet and a singlet. Their contents are determined with successive applications of ladder operators and the orthogonality condition of the states. The two states at the centre of each octet are chosen such that one forms an isosinglet and the other an isotriplet to match the observed baryons.

In terms of symmetry under particle interchanges, we observe that the decuplet is fully symmetric under the interchange of any quarks, while the first octet resulting from $\mathbf{6} \otimes \mathbf{3}$ is only symmetric under the interchange of the first two quarks. In contrast, the second octet resulting from $\bar{\mathbf{3}} \otimes \mathbf{3}$ is antisymmetric under the interchange of the first two quarks. Finally, the singlet is fully antisymmetric. Therefore, the product $\mathbf{3} \otimes \mathbf{3} \otimes \mathbf{3}$ generates

$$\mathbf{3} \otimes \mathbf{3} \otimes \mathbf{3} = (\mathbf{6} \oplus \bar{\mathbf{3}}) \otimes \mathbf{3} = (\mathbf{6} \otimes \mathbf{3}) \oplus (\bar{\mathbf{3}} \otimes \mathbf{3}) = \mathbf{10}_S \oplus \mathbf{8}_{MS} \oplus \mathbf{8}_{MA} \oplus \mathbf{1}_A, \tag{8.20}$$

where the subscripts S and A stand for symmetric and antisymmetric respectively, while M stands for mixed. The states populating the two octets carry the same quantum numbers. Therefore, linear combinations can be performed to define a new octet with different symmetry. For instance, we can write the state $(2uud - udu - duu)/\sqrt{6}$ of $\mathbf{8}_{MS}$ as

$$\frac{2uud - udu - duu}{\sqrt{6}} = \frac{1}{\sqrt{3}}\frac{udu - duu}{\sqrt{2}} + \frac{2}{\sqrt{3}}\frac{uud - udu}{\sqrt{2}}.$$

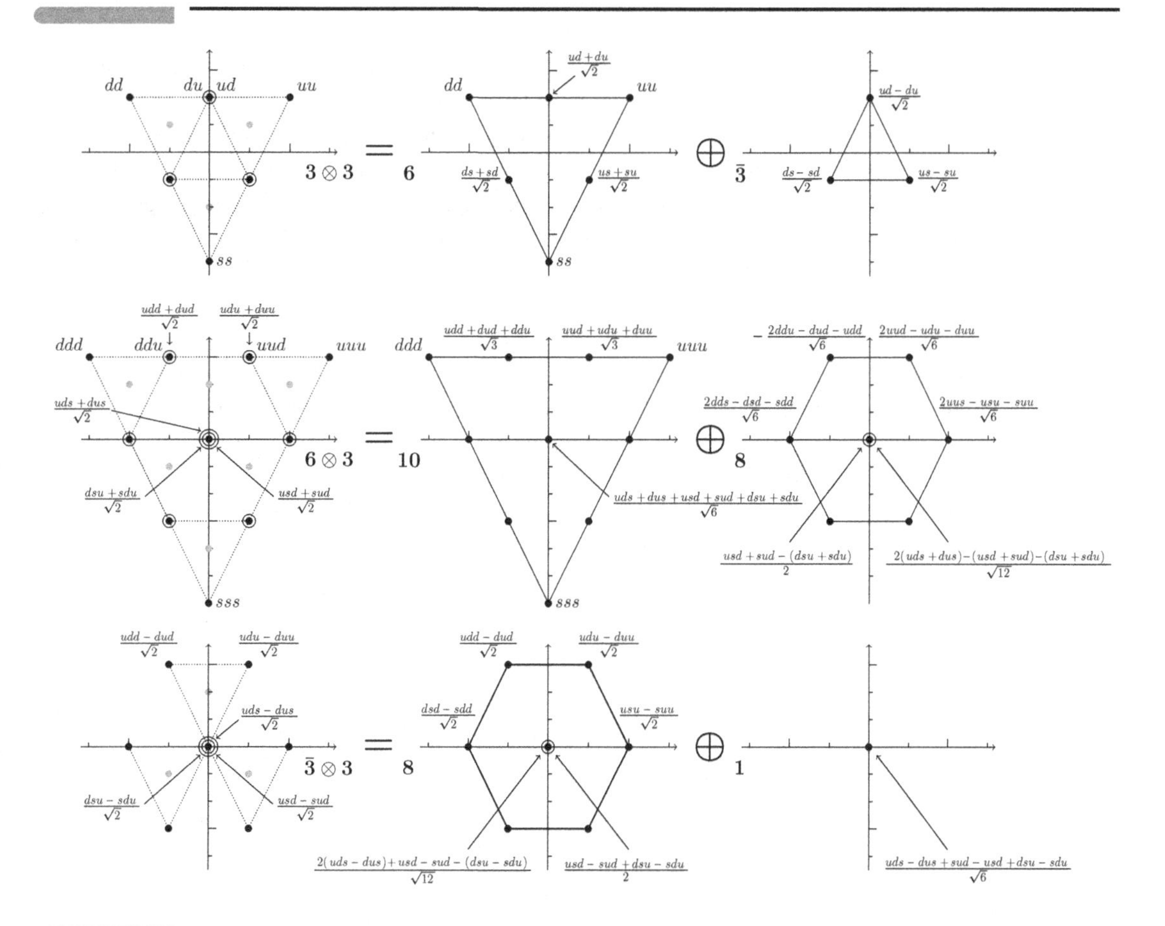

Fig. 8.6 Decomposition of $\mathbf{3} \otimes \mathbf{3} \otimes \mathbf{3}$. First row: $\mathbf{3} \otimes \mathbf{3} = \mathbf{6} \oplus \bar{\mathbf{3}}$. Second row: $\mathbf{6} \otimes \mathbf{3} = \mathbf{10} \oplus \mathbf{8}$. Third row: $\bar{\mathbf{3}} \otimes \mathbf{3} = \mathbf{8} \oplus \mathbf{1}$. The relative phases ensure that the ladder operators $I_\pm$ correctly step between the states.

The first term is antisymmetric under the interchange of the first two quarks, whereas the second term is antisymmetric under the interchange of the last two. Therefore, we could have chosen to have one octet with the symmetry of the first term and the other octet with the symmetry of the second term (all states of $\mathbf{8}_{MS}$ can be broken down as above).

Now, let us identify the observed baryons with the multiplets of Eq. (8.20). The decuplet is clearly identified as the decuplet in Fig. 8.1. The two octets are more tricky. We could imagine two sets of baryons populating the two octets, but it turns out that they are populated by the *same* set of particles, with those of the octet displayed in Fig. 8.1. The reason is the following: baryons are fermions, and therefore their wave function must be antisymmetric under the interchange of *any* two identical quarks, i.e.

$$\psi_{q_1q_2q_3} = -\psi_{q_2q_1q_3} = -\psi_{q_3q_2q_1} = -\psi_{q_1q_3q_2}.$$

In the SU(3) flavour symmetry, the three quarks, u, d and s, are considered as three states of the same particle described by the triplet. It is equivalent to what we encountered when we considered the u and d quarks as two states of the isospin doublet. Therefore, the flavour becomes a degree of freedom that has to be taken into account in the particle interchange, as well as the interchange of spins and of space positions. This is often called the *generalised Pauli principle*. We will see very soon that there is another degree of freedom, called the *colour*, which turns out to be always antisymmetric when colours are interchanged among quarks. Hence, for baryons, the state built from the tensorial product of the degrees of freedom

$$|\psi\rangle = |\psi_{\text{flavour}}\rangle \otimes |\psi_{\text{spin}}\rangle \otimes |\psi_{\text{spatial}}\rangle$$

must lead to a symmetric wave function. Baryons in Fig. 8.1 are the ground states of baryons with the u, d or s quarks. Therefore, their three quarks have zero orbital angular momenta, $(-1)^{l=0}(-1)^{l'=0} = 1$, and the position space wave function is thus symmetric under the interchange of the quark positions. What about the spin? The composition of three spins results from the SU(2) composition $\mathbf{2} \otimes \mathbf{2} \otimes \mathbf{2}$. The reader is already familiar with the composition of two spin-1/2 in quantum mechanics giving a triplet of spin-1 and a singlet of spin 0, i.e. $\mathbf{2} \otimes \mathbf{2} = \mathbf{3} \oplus \mathbf{1}$. Adding the third quark gives a total spin of 3/2 and twice a total spin of 1/2. More specifically, Problem 8.3 shows that the third spin leads to

$$\mathbf{2} \otimes \mathbf{2} \otimes \mathbf{2} = \mathbf{3} \otimes \mathbf{2} \oplus \mathbf{1} \otimes \mathbf{2} = \mathbf{4}_S \oplus \mathbf{2}_{\text{MS}} \oplus \mathbf{2}_{\text{MA}},$$

with the content

$$\mathbf{4} \equiv \begin{pmatrix} \uparrow\uparrow\uparrow \\ \frac{\downarrow\uparrow\uparrow+\uparrow\downarrow\uparrow+\uparrow\uparrow\downarrow}{\sqrt{3}} \\ \frac{\downarrow\downarrow\uparrow+\downarrow\uparrow\downarrow+\uparrow\downarrow\downarrow}{\sqrt{3}} \\ \downarrow\downarrow\downarrow \end{pmatrix}, \quad \mathbf{2}_{\text{MS}} \equiv \begin{pmatrix} \frac{2\uparrow\uparrow\downarrow-\uparrow\downarrow\uparrow-\downarrow\uparrow\uparrow}{\sqrt{6}} \\ -\frac{2\downarrow\downarrow\uparrow-\downarrow\uparrow\downarrow-\uparrow\downarrow\downarrow}{\sqrt{6}} \end{pmatrix}, \quad \mathbf{2}_{\text{MA}} \equiv \begin{pmatrix} \frac{\uparrow\downarrow\uparrow-\downarrow\uparrow\uparrow}{\sqrt{2}} \\ \frac{\uparrow\downarrow\downarrow-\downarrow\uparrow\downarrow}{\sqrt{2}} \end{pmatrix}. \tag{8.21}$$

Note that this result is obvious since SU(2) is also the symmetry group of the isospin, and therefore, it suffices to replace u with $\uparrow$ and d with $\downarrow$ in the multiplets of Fig. 8.6 and read the states obtained at $Y = 1$ (i.e., where there is no strangeness). The doublets (spin-1/2) $\mathbf{2}_{\text{MS}}$ and $\mathbf{2}_{\text{MA}}$ are symmetric and antisymmetric under the interchange of the first two spins respectively. Therefore, the generalised Pauli principle requires combining the flavours in $\mathbf{8}_{\text{MS}}$ with spins in $\mathbf{2}_{\text{MS}}$ and the flavours in $\mathbf{8}_{\text{MA}}$ with spins in $\mathbf{2}_{\text{MA}}$ to obtain fully symmetric states, i.e.[7]

$$|\psi\rangle = \frac{1}{\sqrt{2}}\left(|\psi^{\text{MS}}_{\text{flavour}}\rangle \otimes |\psi^{\text{MS}}_{\text{spin}}\rangle + |\psi^{\text{MA}}_{\text{flavour}}\rangle \otimes |\psi^{\text{MA}}_{\text{spin}}\rangle\right).$$

Let us take the example of a spin-up proton. Its state is then

$$\begin{aligned} |p,\uparrow\rangle &= \frac{1}{\sqrt{2}} \left(\frac{2uud-udu-duu}{\sqrt{6}} \otimes \frac{2\uparrow\uparrow\downarrow-\uparrow\downarrow\uparrow-\downarrow\uparrow\uparrow}{\sqrt{6}} + \frac{udu-duu}{\sqrt{2}} \otimes \frac{\uparrow\downarrow\uparrow-\downarrow\uparrow\uparrow}{\sqrt{2}}\right) \\ &= \frac{1}{3\sqrt{2}} \; [\, 2u\uparrow u\uparrow d\downarrow +2u\uparrow d\downarrow u\uparrow +2d\downarrow u\uparrow u\uparrow \\ &\quad -u\uparrow u\downarrow d\uparrow -u\downarrow d\uparrow u\uparrow -d\uparrow u\uparrow u\downarrow \\ &\quad -u\downarrow u\uparrow d\uparrow -u\uparrow d\uparrow u\downarrow -d\uparrow u\downarrow u\uparrow \,]. \end{aligned} \tag{8.22}$$

[7] The combinations $|\psi^{\text{MS}}_{\text{flavour}}\rangle \otimes |\psi^{\text{MS}}_{\text{spin}}\rangle$ and $|\psi^{\text{MA}}_{\text{flavour}}\rangle \otimes |\psi^{\text{MA}}_{\text{spin}}\rangle$ are only symmetric under the interchange of the first two quarks. However, their sum is symmetric under the interchange of any quarks. It is because $\mathbf{8}_{\text{MS}}$ can be broken down into two octets, with each being antisymmetric under the interchange of two quarks (1-3 and 2-3) complementing $\mathbf{8}_{\text{MA}}$ (antisymmetric under 1-2). The same is true for $\mathbf{2}_{\text{MS}}$.

The other baryons of the octet can be built similarly. We obtain the particles of Fig. 8.1, with, for the particle at the centre, an isosinglet Λ and a member of the isotriplet Σ^0. The flavour decuplet must be combined with the fully symmetric spin-3/2 states. Problem 8.4 proposes to determine the states of a spin-up Δ^+ and a spin-up Λ. Note that we assume here that the spin of baryons is fully carried by their valence quarks (which is not true, as we saw in Section 7.4.2).

The original quark model proposed by Gell-Mann and Zweig was remarkably successful in explaining the observed hadrons in terms of elementary constituents. Gell-Mann even predicted the existence of a particle with quantum numbers $S = -3$ and $Q = -1$ to complete the decuplet three years before it was actually observed: the Ω^-, finally discovered in 1964 by Nicholas Samios and his group at Brookhaven National Laboratory.

Baryon Magnetic Moments

The quark model allows us to determine the wave function of hadrons. An interesting application is to estimate their magnetic moments and compare the prediction of the quark model to the measurements. Mesons belonging to the pseudoscalar multiplet cannot have a magnetic moment because they are spinless. In contrast, baryons can. For baryons in the ground state and assuming non-relativistic quarks, the magnetic moment results from the sum of those of the constituent quarks,

$$\boldsymbol{\mu} = \boldsymbol{\mu}_q \otimes \mathbb{1} \otimes \mathbb{1} + \mathbb{1} \otimes \boldsymbol{\mu}_q \otimes \mathbb{1} + \mathbb{1} \otimes \mathbb{1} \otimes \boldsymbol{\mu}_q,$$

where

$$\boldsymbol{\mu}_q = \frac{Q_q\, e}{m_q} \mathbf{S},$$

the quantities Q_q and m_q being the quark charge in elementary charge units and the quark mass (see Section 6.7.1). Introducing the nuclear magneton μ_N [Eq. (7.1)], the quark magnetic moment for a quark in states $|q \uparrow\rangle$ is then $\mu_q = \langle q \uparrow | \boldsymbol{\mu}_q | q \uparrow\rangle$, yielding

$$\mu_u = \frac{2}{3}\frac{M}{m_u}\mu_N, \quad \mu_d = -\frac{1}{3}\frac{M}{m_d}\mu_N, \quad \mu_s = -\frac{1}{3}\frac{M}{m_s}\mu_N, \tag{8.23}$$

where M is the proton mass. Using, as an example, the spin-up proton state (8.22), the proton magnetic moment is given by $\langle p \uparrow | \boldsymbol{\mu} | p \uparrow\rangle$. Terms such as

$$\begin{aligned}\langle u \uparrow u \downarrow d \uparrow | \boldsymbol{\mu} | u \uparrow u \downarrow d \uparrow\rangle &= \langle u \uparrow | \boldsymbol{\mu} | u \uparrow\rangle + \langle u \downarrow | \boldsymbol{\mu} | u \downarrow\rangle + \langle d \uparrow | \boldsymbol{\mu} | d \uparrow\rangle \\ &= \mu_u - \mu_u + \mu_d = \mu_d\end{aligned}$$

must be evaluated. All cross terms (those having an initial state different from the final state) vanish. Overall, it is easy to check that

$$\mu_p \equiv \langle p \uparrow | \boldsymbol{\mu} | p \uparrow\rangle = \frac{1}{18}\left[4(2\mu_u - \mu_d) \times 3 + \mu_d \times 3 + \mu_d \times 3\right] = \frac{4}{3}\mu_u - \frac{1}{3}\mu_d.$$

If we further assume that $m_u \simeq m_d = m_0$, then, according to Eq. (8.23), $\mu_u \simeq -2\mu_d$ and $\mu_p = -3\mu_d = \mu_N M/m_0$. Similarly, by symmetry, the neutron magnetic moment would be

$$\mu_n = \frac{4}{3}\mu_d - \frac{1}{3}\mu_u \simeq -\frac{2}{3}\frac{M}{m_0}\mu_N.$$

Therefore, the quark model predicts $\mu_p/\mu_n = -3/2$. The experimental value of this ratio is 1.45, an agreement better than 10%. Moreover, given that $\mu_p = 2.79\mu_N$, we deduce $m_0 = M/2.79 = 336\ \text{MeV}/c^2$. This mass is the *constituent mass* of the proton (or neutron). Problem 8.5 proposes to calculate the Λ magnetic moment. Since this baryon contains an s quark, one can deduce an estimate of the s quark mass of the order of 513 MeV/c^2 (see Problem 8.5). It should be understood that the quark masses obtained from this approach are not the true masses (those listed in Table 1.1). They are an effective mass, mostly resulting from the QCD potential due to the colour field. All masses of light hadrons (those not containing a c or b quark) are largely dominated by this QCD potential contribution. A striking illustration of this is the proton (or neutron), for which the sum of its (true) quark masses accounts for at most 1% of its mass.[8] This means that if the up and down quarks were massless, then the mass of the proton and neutron would remain almost unchanged.

Hadron Masses

The calculation of hadron masses from first principles is complicated and obtained, so far, in the context of lattice QCD, which will be briefly outlined in Sections 8.5.1 and 8.5.2. However, at this stage, we can try to push a bit the phenomenological model previously seen based on effective constituent masses and the SU(3) flavour symmetry. One of the questions we may answer is why hadrons that have the same quantum numbers (i_3, y and so the same quark content) but belong to two different multiplets have different masses? For instance, compare the two mesons π^+, member of the pseudoscalar octet, and ρ^+, member of the vector octet (Fig. 8.2). Both are made of the $u\bar{d}$ quarks but the former have $m_\pi = 140\ \text{MeV}/c^2$ while $m_\rho = 774\ \text{MeV}/c^2$. The pseudoscalar particles differ obviously from the vector particles by their spin ($J_\pi = 0$ and $J_\rho = 1$, with zero orbital angular momentum). In quantum mechanics courses, one usually determines the energy levels of the hydrogen, and it is well known that because of the spin–spin interaction between the electron and the proton, an additional potential energy term splits the energy levels (the hyperfine structure). This term is proportional to $\langle \boldsymbol{S_e} \cdot \boldsymbol{S_p} \rangle / (m_e m_p)$. We could imagine a similar phenomenon within the hadrons between the quarks. This assumes that the QCD potential due to colour interaction is not that much different from the Coulomb potential, at least at a small distance (within the hadrons).[9] We will confirm this assumption later in this chapter. In this context, we thus expect that the mass of a meson can be phenomenologically described from the constituent masses and the spin state of the system by

$$M_{q\bar{q}} = m_q + m_{\bar{q}} + A\frac{\langle \boldsymbol{S_q} \cdot \boldsymbol{S_{\bar{q}}} \rangle}{m_q m_{\bar{q}}}, \tag{8.24}$$

where A is a constant of our model (with units of mass cubed, $\hbar = 1$) that would be common to all mesons. The spin of a meson without the internal orbital angular momentum

[8] In reality, there are additional quark–antiquark pairs which contribute to the mass. Current phenomenology estimates the total quark contribution to the nucleon mass to be about 8–10%.

[9] Classical hyperfine splitting due to QED exists in hadrons too. But its magnitude is much lower than the QCD contribution. For instance, QED hyperfine splitting must be of the order of 1 MeV to explain the mass difference between a proton and a neutron, whereas its QCD counterpart must account for the mass difference between a proton and a Δ^+ (about 300 MeV).

Table 8.1 Meson masses MeV$/c^2$. Mesons to the left of the vertical bar have $J = 0$, the others $J = 1$. Fit parameters: $m_0 = 309, m_s = 480, A = 6.1 \times 10^7\,(\text{MeV}/c^2)^3$.

Mass	π	K	η	ρ	ω	K^*	ϕ
Observed	139	496	548	775	783	894	1 019
Calculated	139	481	555	777	777	892	1 026

satisfies $\boldsymbol{J}^2 = \boldsymbol{S}_q{}^2 + \boldsymbol{S}_{\bar{q}}{}^2 + 2\boldsymbol{S}_q \cdot \boldsymbol{S}_{\bar{q}}$. Therefore, given that $\langle \boldsymbol{J}^2 \rangle = J(J+1)$ and $\langle \boldsymbol{S}^2 \rangle = 3/4$, it follows that

$$\langle \boldsymbol{S}_q \cdot \boldsymbol{S}_{\bar{q}} \rangle = \frac{1}{2}\left[J(J+1) - \frac{3}{2}\right] = \begin{cases} \frac{1}{4}, & \text{for } J = 1 \\ -\frac{3}{4}, & \text{for } J = 0. \end{cases}$$

leading to

$$M_{q\bar{q}}^{J=1} = m_q + m_{\bar{q}} + \frac{A}{4m_q m_{\bar{q}}}, \quad M_{q\bar{q}}^{J=0} = m_q + m_{\bar{q}} - \frac{3A}{4m_q m_{\bar{q}}}. \tag{8.25}$$

We observe that since a vector meson is systematically heavier than a pseudoscalar meson made of the same quarks (see Table 8.1), the constant A in Eq. (8.25) is necessarily positive. Assuming that $m_0 \equiv m_u \simeq m_d$, we can adjust the values of the parameters m_0, m_s and A to reproduce the observed mass of the light mesons. The best fit[10] is obtained for $m_0 = 309, m_s = 480$ and $A = 6.1 \times 10^7 (\text{MeV}/c^2)^3$. The comparisons of the prediction of this simple model to the measurements are presented in Table 8.1. Surprisingly, it works pretty well, within a 1% error. The only exception, not shown in the table, is for the η' meson whose mass is larger than that predicted by the quark model by 300%! This meson is specific since it is a flavour singlet. As such, it can interact with pure particular gluonic systems that are also flavour singlets. This contributes to its mass but cannot be described by the simple quark model presented here. However, the η' large mass is understood in the framework of lattice QCD.

The same task can be performed for baryons. In this simple model, their masses would be given by

$$M_{q_1q_2q_3} = m_{q_1} + m_{q_2} + m_{q_3} + A'\left(\frac{\langle \boldsymbol{S}_{q_1} \cdot \boldsymbol{S}_{q_2} \rangle}{m_{q_1} m_{q_2}} + \frac{\langle \boldsymbol{S}_{q_1} \cdot \boldsymbol{S}_{q_3} \rangle}{m_{q_1} m_{q_3}} + \frac{\langle \boldsymbol{S}_{q_2} \cdot \boldsymbol{S}_{q_3} \rangle}{m_{q_2} m_{q_3}}\right), \tag{8.26}$$

with A' being the baryon constant. As $\boldsymbol{J}^2 = \sum_i \boldsymbol{S}_{q_i}{}^2 + 2\sum_{i,j} \boldsymbol{S}_{q_i} \cdot \boldsymbol{S}_{q_j}$, it implies

$$\langle \boldsymbol{S}_{q_1} \cdot \boldsymbol{S}_{q_2} \rangle + \langle \boldsymbol{S}_{q_1} \cdot \boldsymbol{S}_{q_3} \rangle + \langle \boldsymbol{S}_{q_2} \cdot \boldsymbol{S}_{q_3} \rangle = \frac{1}{2}\left[J(J+1) - \frac{9}{4}\right] = \begin{cases} \frac{3}{4}, & \text{for } J = \frac{3}{2} \\ -\frac{3}{4}, & \text{for } J = \frac{1}{2} \end{cases}. \tag{8.27}$$

For baryons of the decuplet ($J = 3/2$), their spin state is fully symmetric under the interchange of any quarks. Thus, necessarily $\langle \boldsymbol{S}_{q_i} \cdot \boldsymbol{S}_{q_j} \rangle$ is the same for all couples $i \neq j$, leading

[10] A χ^2 minimisation has been performed using the masses and the errors quoted in Particle Data Group (2022). For states depending on the u/d and s flavours, the flavour state has to be taken. For instance, the mass of a meson described by $\alpha u\bar{u} + \beta s\bar{s}$, with $\alpha^2 + \beta^2 = 1$, is $\alpha^2 M_{u\bar{u}} + \beta^2 M_{s\bar{s}}$, where $M_{u\bar{u}}$ and $M_{s\bar{s}}$ are obtained from Eq. (8.25).

from Eq. (8.27) to $\langle \boldsymbol{S}_{q_i} \cdot \boldsymbol{S}_{q_j} \rangle = 1/4$. According to Eq. (8.26), the masses of the baryons populating the decuplet are then

$$M_{\Delta,\Omega} = 3m_{0,s} + \frac{3A'}{4m_{0,s}^2}, \quad M_{\Sigma^*} = 2m_0 + m_s + A'\left(\frac{1}{4m_0^2} + \frac{1}{2m_0 m_s}\right),$$
$$M_{\Xi^*} = 2m_s + m_0 + A'\left(\frac{1}{4m_s^2} + \frac{1}{2m_0 m_s}\right).$$

For the baryons of the octet ($J = 1/2$), the situation is slightly more complicated. The proton and neutron have three quarks of the same mass (m_0). Equations (8.26) and (8.27) imply the nucleon mass $m_N = 3m_0 - 3A'/(4m_0^2)$. Other baryons of the octet are made either of two identical quarks (*uus*, *dds*, *ssu* or *ssd*) or of *uds* quarks. In the ground state, the wave function must be symmetric under the interchange of any two quarks, i.e. both flavour and spin interchange. The interchange of two identical flavours is obviously symmetric, imposing the spin state of these two quarks to be also symmetric. Therefore, the two identical quarks system must have a total spin $S = 1$ (remember, the states resulting from two spin-1/2 with $S_z = 1, 0, -1$ are $\uparrow\uparrow$, $(\uparrow\downarrow + \downarrow\uparrow)/\sqrt{2}$ and $\downarrow\downarrow$, respectively), and thus $\langle \boldsymbol{S}_q \cdot \boldsymbol{S}_q \rangle = \frac{1}{2}\left[1(1+1) - \frac{3}{2}\right] = 1/4$. According to Eq. (8.27), the pair of two different quarks then satisfies $2\langle \boldsymbol{S}_q \cdot \boldsymbol{S}_{q'} \rangle + \langle \boldsymbol{S}_q \cdot \boldsymbol{S}_q \rangle = -3/4$, and thus $\langle \boldsymbol{S}_q \cdot \boldsymbol{S}_{q'} \rangle = -1/2$. It follows from Eq. (8.26) the following mass formulas for the Σ^+ (*uus*), Σ^- (*dds*), Ξ^0 (*ssu*) and Ξ^- (*ssd*) baryons:

$$M_{\Sigma^\pm} = 2m_0 + m_s + A'\left(\frac{1}{4m_0^2} - \frac{1}{m_0 m_s}\right), \quad M_{\Xi^0} = M_{\Xi^-} = 2m_s + m_0 + A'\left(\frac{1}{4m_s^2} - \frac{1}{m_0 m_s}\right).$$

Finally, for the states at the centre of the octet, the Σ^0, a member of the isospin triplet, is associated with a flavour state symmetric under the interchange $u \leftrightarrow d$ (see Fig. 8.6) and, therefore, must be in a spin state symmetric under the interchange of their spins. Hence, $S_{ud} = 1$, leading to $\langle \boldsymbol{S}_u \cdot \boldsymbol{S}_d \rangle = 1/4$ and $\langle \boldsymbol{S}_u \cdot \boldsymbol{S}_s \rangle = \langle \boldsymbol{S}_d \cdot \boldsymbol{S}_s \rangle = -1/2$. On the other hand, the Λ is an isosinglet, antisymmetric under the interchange $u \leftrightarrow d$, and thus necessarily $S_{ud} = 0$ (which is antisymmetric i.e., $[\uparrow\downarrow - \downarrow\uparrow]/\sqrt{2}$) and $\langle \boldsymbol{S}_u \cdot \boldsymbol{S}_d \rangle = \frac{1}{2}\left[0(0+1) - \frac{3}{2}\right] = -3/4$. Thus, Eq. (8.27) imposes $\langle \boldsymbol{S}_u \cdot \boldsymbol{S}_s \rangle + \langle \boldsymbol{S}_d \cdot \boldsymbol{S}_s \rangle = -3/4 - (-3/4) = 0$. It follows that

$$M_{\Sigma^0} = 2m_0 + m_s + A'\left(\frac{1}{4m_0^2} - \frac{1}{m_0 m_s}\right), \quad M_\Lambda = 2m_0 + m_s - \frac{3}{4m_0^2}A'.$$

Equipped with all those mass formulas, we can adjust the values of A', m_0 and m_s using the measurements of baryon masses. The result is presented in Table 8.2. The accuracy of the fit is again better than 1%. Note that the values of m_0 and m_s are significantly different than those found for mesons. Since they are effective masses related to the QCD potential, it is not surprising to have a different potential for a bound system of three quarks than for two quarks.

8.1.5 Beyond SU(3) Flavour

Between 1974 and 1995, three other quark flavours were discovered: the *charm c*, *beauty b* (often called bottom) and *top t* (sometimes called truth). The corresponding additive quantum numbers were introduced with the convention that the flavour quantum number has the same sign as the quark charge. Hence, as I_3 is positive for the u quark and negative

Table 8.2 Baryon masses MeV$/c^2$. Baryons to the left of the vertical bar have $J = 1/2$, the others $J = 3/2$. Fit parameters: $m_0 = 363, m_s = 538, A = 2.6 \times 10^7 (\text{MeV}/c^2)^3$.

Mass	p, n	Λ	Σ	Ξ	Δ	Σ^*	Ξ^*	Ω
Observed	939	1 116	1 194	1 321	1 232	1 384	1 533	1 672
Calculated (fit)	939	1 114	1 179	1 326	1 237	1 379	1 527	1 681

Table 8.3 Quark quantum numbers and current quark masses. These quantum numbers are conserved by the strong interaction.

Quarks	u	d	s	c	b	t
Mass (GeV$/c^2$)	1.7–3.1×10^{-3}	$4.1-5.7 \times 10^{-3}$	0.09	1.27	4.18	172.8
Charge	$2/3$	$-1/3$	$-1/3$	$2/3$	$-1/3$	$2/3$
Spin	$1/2$	$1/2$	$1/2$	$1/2$	$1/2$	$1/2$
Strong isospin $\|I, I_3\rangle$	$\|1/2, 1/2\rangle$	$\|1/2, -1/2\rangle$	$\|0, 0\rangle$	$\|0, 0\rangle$	$\|0, 0\rangle$	$\|0, 0\rangle$
Baryon # B	$1/3$	$1/3$	$1/3$	$1/3$	$1/3$	$1/3$
Strangeness $\mathcal{S}$	0	0	-1	0	0	0
Charm C	0	0	0	1	0	0
Beauty $\mathcal{B}$	0	0	0	0	-1	0
Truth $\mathcal{T}$	0	0	0	0	0	1
Strong hypercharge Y	$1/3$	$1/3$	$-2/3$	0	0	0
Parity η_P	1	1	1	1	1	1

for the d quark, the strangeness $\mathcal{S}$ is negative for s quark, the charm C positive for c, etc. The Gell-Mann–Nishijima formula is now generalised to the other quark flavours,

$$Q = I_3 + \frac{B + \mathcal{S} + C + \mathcal{B} + \mathcal{T}}{2}. \tag{8.28}$$

The four quarks, u, d, s and c or b, can form a fundamental representation of a symmetry group SU(4) flavour. It produces three-dimensional multiplets where the three axes are the isospin, the strangeness and the charm or the beauty quantum numbers respectively [Particle Data Group (2022) provides the diagrams of those multiplets]. For instance, the ground state of pseudoscalar mesons $J^P = 0^-$ made of the u, d, s or c quarks decomposes into (see Appendix C)

$$\mathbf{4} \otimes \bar{\mathbf{4}} = \mathbf{15} \oplus \mathbf{1}.$$

All members of those multiplets have already been found. However, they are not so useful except for bookkeeping and classification purposes since the symmetry between them is strongly broken by the large difference in quark masses. Table 8.3 summarises the quark quantum numbers relevant for the strong interaction and the quark masses [from Particle Data Group (2022)]. Since, the u and d quarks have close masses, the isospin is an excellent symmetry, and the members of an isospin multiplet differ in mass by at most 3%, well within the effects induced by electromagnetic interaction (since they do not have the same charge). We have seen that the members of multiplets made with a strange quark can have a significant difference in masses (up to 40%). Therefore, the flavour symmetry SU(3) is approximate. With quarks of larger masses, the differences are much more pronounced, frankly breaking the flavour symmetry. Almost all hadrons made the of u, d, s, c or b quarks have been discovered (one of the last hadrons found is the Ξ_{cc}^{++} made of the ccu quarks in 2017). No hadrons containing a top quark have ever been observed (so far). Their existence is very unlikely (but not forbidden by a known principle) because the decay width of the top quark is very large (about 1.4 GeV), and hence its lifetime is very short. Therefore, it decays before having time to form a hadron with other quarks or antiquarks.

8.2 Colour as a Gauge Theory

8.2.1 Need for Colours

Consider the particles in the three corners of the decuplet in Fig. 8.1. They are the Δ^-, Δ^{++} and Ω^- baryons, and according to the quark model, they are composed of three identical quark flavours (see Fig. 8.6). Since they are the ground state of spin-3/2 baryons, their three quarks have zero orbital angular momentum, and their spins must be aligned. Furthermore, because of the Pauli principle, their wave function must be antisymmetric with the interchange of any two quarks. However, $|\psi_{\text{flavour}}\rangle = |qqq\rangle$, $\psi_{\text{spin}} = |\uparrow\uparrow\uparrow\rangle$ or $|\downarrow\downarrow\downarrow\rangle$ and $|\psi_{\text{spatial}}\rangle = |l = l' = 0\rangle$ are all symmetric. We then have a global wave function that would be symmetric, in contradiction with Pauli's principle! This observation led Greenberg to propose, in 1964, a new quantum number carried by quarks, named *colour*. Needless to say, suggesting that quarks, for which there was not yet experimental evidence, carry a new hidden charge generated scepticism. When Greenberg presented his idea to Oppenheimer, his reaction was simply 'It's beautiful, but I don't believe a word of it!'.[11] Anyway, we will see that Greenberg was right. A quark is postulated to carry a new three-valued charge, commonly denoted red (r), green (g) and blue (b). Therefore, if the baryons is made of three identical quark flavours, with each quark carrying a different colour, then we can save the Pauli exclusion principle since the quarks are not, strictly speaking, in the same state, i.e. $|q_r q_g q_b\rangle$. Furthermore, if the colour is assumed to be a perfect symmetry, the exchange of quark colours should have no observable effects. Baryons being fermions imply that the interchange of the quark colours must be antisymmetric. With three colours and three quarks, the only fully antisymmetric state is

$$|\psi_{\text{colour}}^{\text{baryon}}\rangle = \frac{1}{\sqrt{6}}|rgb - rbg + gbr - grb + brg - bgr\rangle. \tag{8.29}$$

Then the spin-3/2 state of a baryon, such as the Ω^-, is

$$\begin{aligned}|\Omega^-, S_z = \tfrac{3}{2}\rangle &= \tfrac{1}{\sqrt{6}}\,|sss\rangle \otimes |rgb - rbg + gbr - grb + brg - bgr\rangle \otimes |\uparrow\uparrow\uparrow\rangle \otimes |l = l' = 0\rangle \\ &= \tfrac{1}{\sqrt{6}}\Big(|s_r \uparrow s_g \uparrow s_b \uparrow\rangle - |s_r \uparrow s_b \uparrow s_g \uparrow\rangle + |s_g \uparrow s_b \uparrow s_r \uparrow\rangle - |s_g \uparrow s_r \uparrow s_b \uparrow\rangle \\ &\quad + |s_b \uparrow s_r \uparrow s_g \uparrow\rangle - |s_b \uparrow s_g \uparrow s_r \uparrow\rangle\Big) \otimes |l = l' = 0\rangle.\end{aligned}$$

[11] Greenberg reported Oppenheimer's reaction in Greenberg (2015).

The colour state (8.29) is said to be colourless since the three colours contribute equally. One can also say that the baryon is white in colour by analogy with the visual arts. We will see very soon that in the language of group theory, Eq. (8.29) defines a colour singlet.[12] An important postulate of QCD assumes the following.

Postulate 1: all hadrons are in the colour-singlet state.

Consequently, since the colour singlet is uniquely defined, *all baryons* have the *same* colour state (8.29). Similarly, all anti-baryons are in a colour state equivalent to Eq. (8.29), where all colours are replaced by anti-colours, $\bar{r}\bar{g}\bar{b}$ etc. Since, by construction, the singlet (8.29) is antisymmetric, all baryons are in a state symmetric under the interchange of the other degrees of freedom, i.e.

$$|\psi^{\text{baryon}}\rangle = |\psi_{\text{flavour}}\rangle \otimes |\psi_{\text{spin}}\rangle \otimes |\psi_{\text{spatial}}\rangle \Rightarrow \text{symmetric}. \tag{8.30}$$

Mesons are also hadrons but made of a quark–antiquark pair. Colour states are then built with a colour and an anti-colour. The only colour-singlet state is

$$|\psi^{\text{mesons}}_{\text{colour}}\rangle = \frac{1}{\sqrt{3}}\,|r\bar{r} + g\bar{g} + b\bar{b}\rangle. \tag{8.31}$$

This time, the colour singlet is symmetric. It is important to realise that a state $r\bar{r}$ is *not* colourless, in the sense that it is not a pure colour singlet. It results from a linear combination of the singlet and some members of a colour octet, as we shall see soon.

In order to accommodate the experimental fact that no free quark has ever been observed, QCD postulates a principle even stronger than the postulate 1. It is the *colour confinement* hypothesis that states that

Postulate 2: only particles in a colour-singlet state can propagate as free particles.

This postulate has interesting consequences on possible exotic states (i.e., not baryons or mesons) that might be observed. We shall come back to this point when the colour symmetry group is introduced in Section 8.2.3.

8.2.2 Experimental Evidences of the Colour

One may wonder whether the colour of the quarks is a pure mathematical artifice to 'antisymmetrise' the baryon wave function or whether it has a real physical basis with experimental observations. Several observations validate the three-colour model. Let us consider the production of hadrons at e^+e^- colliders. Quarks, carrying an electric charge, can be produced via the s-channel of the e^+e^- annihilation as shown in Fig. 8.7. As soon as $\sqrt{s} \ll m_Z$, we can neglect the electroweak interaction (see Chapter 9) and consider only the virtual photon exchange. What is the cross section of this process? This process is similar to the process $e^- + e^+ \to \mu^- + \mu^+$ shown in Section 6.4.4, for which we found in

[12] You might argue that the other state $|rgb + rbg + gbr + grb + brg + bgr\rangle/\sqrt{6}$ is also a colour singlet since an interchange seems to leave the state untouched. In fact, no. This state belongs to a colour decuplet. By analogy with the spin, such a mistake would be equivalent to confuse the state $|1,0\rangle = |\uparrow\downarrow + \downarrow\uparrow\rangle/\sqrt{2}$ that has a spin 1 but a projection 0, with the state $|0,0\rangle = |\uparrow\downarrow - \downarrow\uparrow\rangle/\sqrt{2}$ that is really spinless.

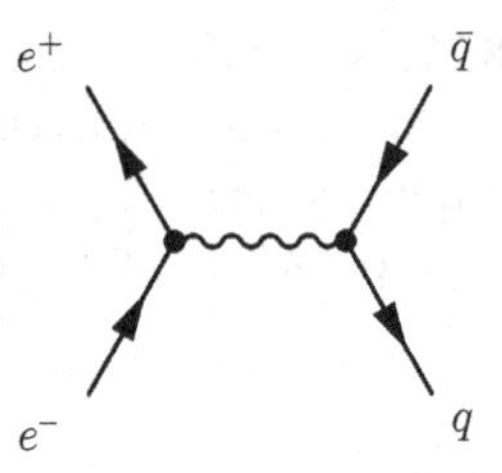

Fig. 8.7 The lowest order diagram $e^- + e^+ \to q + \bar{q}$.

Eq. (6.102), $\sigma = 4\pi\alpha^2/(3s)$. Denoting by Q_i the charge of the quark flavour i in units of the elementary charge e, we would expect

$$\sigma_{e^-e^+\to q_i\bar{q}_i} = \frac{4\pi\alpha^2 Q_i^2}{3s}.$$

However, a quark can have three different colours. We should then have three more contributions to the cross section,[13] i.e.

$$\sigma_{e^-e^+\to q_i\bar{q}_i} = 3\frac{4\pi\alpha^2 Q_i^2}{3s}.$$

Now, if $\sqrt{s}$ is large enough, several quark flavours q_i can be produced, yielding

$$\sigma_{e^-e^+\to q\bar{q}} = 3\frac{4\pi\alpha^2}{3s}\sum_{i,\, m_{q_i}\leq\sqrt{s}/2} Q_i^2.$$

Hence, the ratio of quark production to muon production is

$$R_{\text{had}} = \frac{\sigma_{e^-e^+\to q\bar{q}}}{\sigma_{e^-e^+\to\mu^-\mu^+}} = 3\sum_{i,\, m_{q_i}\leq\sqrt{s}/2} Q_i^2. \tag{8.32}$$

In detectors, we observe hadrons, not directly quarks. Let us assume that to first order, the Formula (8.32) remains valid for the hadron final states. Our prediction of R_{had} can then be compared with the experimental measurements shown in Fig. 8.8. Before the charm meson threshold ($\sqrt{s} \simeq 3$ GeV) and before and after the beauty meson threshold ($\sqrt{s} \simeq 9$ GeV), Eq. (8.32) gives

$$R_{\text{had}}^{uds} = 3\left[\left(\tfrac{2}{3}\right)^2 + 2\left(\tfrac{1}{3}\right)^2\right] = 2,\ R_{\text{had}}^{udsc} = 3\left[2\left(\tfrac{2}{3}\right)^2 + 2\left(\tfrac{1}{3}\right)^2\right] = \tfrac{10}{3},$$
$$R_{\text{had}}^{udscb} = 3\left[2\left(\tfrac{2}{3}\right)^2 + 3\left(\tfrac{1}{3}\right)^2\right] = \tfrac{11}{3},$$

while without the colour factor 3, we would have found $R_{\text{had}}^{uds} = 2/3$, $R_{\text{had}}^{udsc} = 10/9$, and $R_{\text{had}}^{udscb} = 11/9$. Data rule out the no-colour scenario. The ratio at energies below about 30 GeV (above this energy, the electroweak theory contributions cannot be neglected) appears close to the values calculated in our model with three colours. This ratio must be compared in the continuum region, namely, outside of resonance peaks, in particular at low energy (where we should have added form factors to describe the modified cross section). The

[13] Have a look at Problem 8.6 to better appreciate this factor 3.

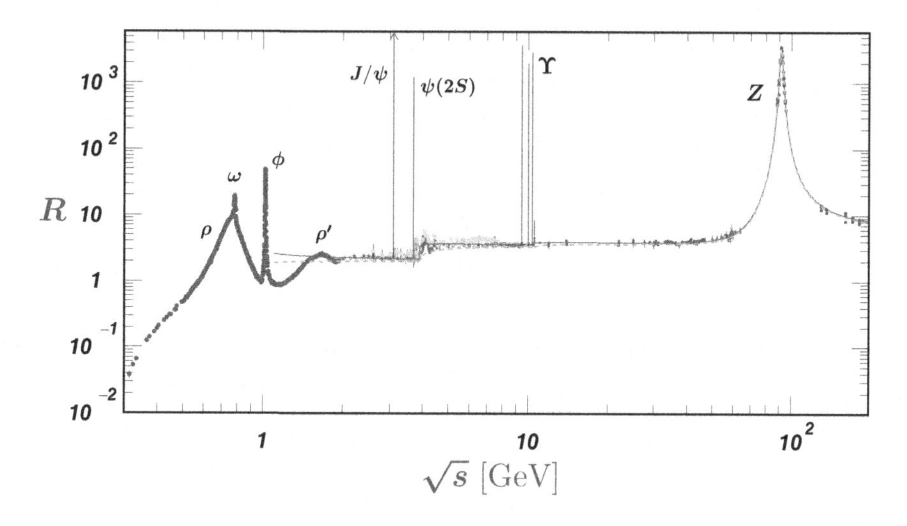

Fig. 8.8 R_{had} (see text) as a function of $\sqrt{s}$ in various experiments. The broken curve, outside of resonance peaks, corresponds to the predictions described in the text. Figure from Workman et al. (Particle Data Group)(2022).

slight discrepancy in the continuum (about 10%) is due to higher order effects not taken into account in our calculation that supposes the exchange of a single photon. For instance, the emission of gluons by quarks slightly increases the value of R_{had}.

There are many other pieces of evidence of the quark colours, such as the π^0 lifetime, τ branching ratios in hadronic decays etc. All data are consistent with a three-colour model.

8.2.3 The SU(3)$_c$ Symmetry Group

In 1965, Han and Nambu proposed to base the symmetry group of the strong interaction on the colour symmetry group SU(3)$_c$, where the subscript c is there to avoid the confusion with the SU(3) flavour group already encountered. Three colours per quark (and not two or four) are consistent with the experimental observations, hence the choice of the special unitary group SU(3).[14] The colour symmetry is assumed to be exact: the transformation corresponding to that symmetry must leave the physical observations unchanged. The hadrons found in nature are all assumed to be colourless, i.e. in the colour-singlet state. This is the hypothesis of *colour confinement*. Moreover, bound states of two quarks have never been found, whereas quark–antiquark bound states exist. Therefore, the anti-colour is necessarily different from the colour. In SU(3), the **3** and $\overline{\mathbf{3}}$ representations are indeed not equivalent. Mathematically, the three colours belong to an SU(3)$_c$ triplet

[14] Among the unitary group (a needed property to describe symmetries in quantum mechanics), only SU(2) and SU(3) can form an irreducible triplet representation. For SU(2) and SU(3), they are the spin-1 representation and fundamental representation respectively. See Appendix C.

$$|r\rangle = \begin{pmatrix}1\\0\\0\end{pmatrix}, \quad |g\rangle = \begin{pmatrix}0\\1\\0\end{pmatrix}, \quad |b\rangle = \begin{pmatrix}0\\0\\1\end{pmatrix}. \tag{8.33}$$

It constitutes the basis states of the fundamental representation **3**. The general colour state of a quark is then

$$|q\rangle = \begin{pmatrix}c_r\\c_g\\c_b\end{pmatrix}, \text{ with } \langle q|q\rangle = |c_r|^2 + |c_g|^2 + |c_b|^2 = 1.$$

The Lie group SU(3) has $3^2 - 1 = 8$ generators, T_i. The formalism is strictly the same as the SU(3) flavour group. The colour state is transformed under the $\text{SU}(3)_c$ transformation of parameter $\boldsymbol{\alpha} = (\alpha_1, \ldots, \alpha_8)$ as

$$|q'\rangle = U\,|q\rangle\,, \text{ with } \; U = e^{-i\boldsymbol{\alpha}\cdot\boldsymbol{T}}.$$

The generators T_a take the matrix representation

$$T_a = \frac{\lambda_a}{2},$$

where λ_a are the eight Gell-Mann matrices already introduced and reproduced again here

$$\lambda_1 = \begin{pmatrix}0&1&0\\1&0&0\\0&0&0\end{pmatrix}, \quad \lambda_2 = \begin{pmatrix}0&-i&0\\i&0&0\\0&0&0\end{pmatrix}, \quad \lambda_3 = \begin{pmatrix}1&0&0\\0&-1&0\\0&0&0\end{pmatrix}, \quad \lambda_4 = \begin{pmatrix}0&0&1\\0&0&0\\1&0&0\end{pmatrix},$$
$$\lambda_5 = \begin{pmatrix}0&0&-i\\0&0&0\\i&0&0\end{pmatrix}, \quad \lambda_6 = \begin{pmatrix}0&0&0\\0&0&1\\0&1&0\end{pmatrix}, \quad \lambda_7 = \begin{pmatrix}0&0&0\\0&0&-i\\0&i&0\end{pmatrix}, \quad \lambda_8 = \begin{pmatrix}\frac{1}{\sqrt{3}}&0&0\\0&\frac{1}{\sqrt{3}}&0\\0&0&\frac{-2}{\sqrt{3}}\end{pmatrix}. \tag{8.34}$$

The norm of these matrices has been chosen so that they satisfy

$$\text{Tr}(\lambda_a\lambda_b) = 2\delta_{ab}. \tag{8.35}$$

The matrices λ_3 and λ_8 are diagonal, meaning that, like the SU(3) flavour group, there are two conserved quantum numbers. Let us call them the colour-isopsin I_{3c} and the colour-hypercharge Y_c by analogy with the previous case. A colour eigenstate can then be represented by these two numbers $|i_{3c}, y_c\rangle$, and a graphical representation of the colour and anti-colour states is shown in Fig. 8.9. The generators of $\text{SU}(3)_c$ obey the commutation rules

$$[T_a, T_b] = if_{abc}T_c \Rightarrow \left[\frac{\lambda_a}{2}, \frac{\lambda_b}{2}\right] = if_{abc}\frac{\lambda_c}{2}, \tag{8.36}$$

where if_{abc} are the structure constants of $\text{SU}(3)_c$ (see Appendix C). Their values are

$$f_{abc} = -\frac{i}{4}\text{Tr}([\lambda_a, \lambda_b]\lambda_c) \begin{cases} f_{123} = 1, \\ f_{147} = f_{246} = f_{257} = f_{345} = f_{516} = f_{637} = \frac{1}{2}, \\ f_{458} = f_{678} = \frac{\sqrt{3}}{2}, \\ f_{ijk} = f_{jki} = f_{kij} = -f_{jik} = -f_{ikj}, \\ f_{iij} = f_{ijj} = 0, \\ f_{ijk} = 0, \text{ for other permutations.} \end{cases} \tag{8.37}$$

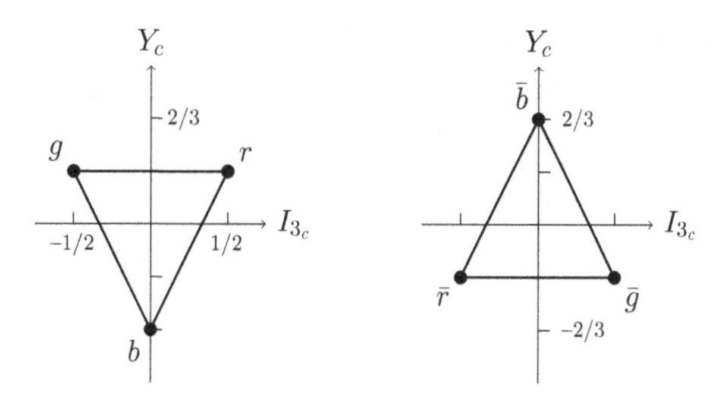

Fig. 8.9 $SU(3)_c$ representation for colour (**3** representation) and anti-colour ($\bar{\mathbf{3}}$ representation).

As colours, anti-colours $|\bar{r}\rangle\,, |\bar{g}\rangle$ and $|\bar{b}\rangle$ constitute the basis states of the fundamental conjugate representation $\bar{\mathbf{3}}$,

$$|\bar{r}\rangle = \begin{pmatrix} 1 \\ 0 \\ 0 \end{pmatrix}, \quad |\bar{g}\rangle = \begin{pmatrix} 0 \\ 1 \\ 0 \end{pmatrix}, \quad |\bar{b}\rangle = \begin{pmatrix} 0 \\ 0 \\ 1 \end{pmatrix}. \tag{8.38}$$

An antiquark state then transforms under an anti-colour rotation with real parameters $\boldsymbol{\alpha} = (\alpha_1, \ldots, \alpha_8)$ as

$$|\bar{q'}\rangle = \left(e^{-i\boldsymbol{\alpha}\cdot\frac{\boldsymbol{\lambda}}{2}}\right)^* |\bar{q}\rangle = e^{-i\boldsymbol{\alpha}\cdot\frac{-\boldsymbol{\lambda}^*}{2}} |\bar{q}\rangle.$$

The generators of $\bar{\mathbf{3}}$ are then $-\lambda_a^*/2$, giving the quantum numbers of the anti-colour states shown in Fig. 8.9.

QCD postulates that baryons and mesons must belong to the colour singlet. It is possible since we already know from the SU(3) flavour symmetry that the singlet is present in the two decompositions

$$\mathbf{3} \otimes \mathbf{3} \otimes \mathbf{3} = \mathbf{10} \oplus \mathbf{8} \oplus \mathbf{8} \oplus \mathbf{1}, \quad \mathbf{3} \otimes \bar{\mathbf{3}} = \mathbf{8} \oplus \mathbf{1}.$$

The singlet in $\mathbf{3} \otimes \bar{\mathbf{3}}$ has already been determined in the context of the SU(3) flavour symmetry in Eq. (8.19). The comparison of the representations of the two groups in Figs. 8.9 and 8.3 shows that we just have to substitute $u \to r, d \to g$ and $s \to b$. It yields the colour singlet given in Eq. (8.31). For the singlet in $\mathbf{3} \otimes \mathbf{3} \otimes \mathbf{3}$, it is easy to check that it is given by Eq. (8.29). We introduce the ladder operators whose action on colour states is deduced from those of the SU(3) flavour group in Fig. 8.5, i.e.

$$\begin{aligned}
&I_+ |g\rangle = |r\rangle\,, \quad I_- |r\rangle = |g\rangle\,, \quad && I_+ |\bar{r}\rangle = -|\bar{g}\rangle\,, \quad I_- |\bar{g}\rangle = -|\bar{r}\rangle\,, \\
&U_+ |b\rangle = |g\rangle\,, \quad U_- |g\rangle = |b\rangle\,, \quad && U_+ |\bar{g}\rangle = -|\bar{b}\rangle\,, \quad U_- |\bar{b}\rangle = -|\bar{g}\rangle\,, \\
&V_+ |b\rangle = |r\rangle\,, \quad V_- |r\rangle = |b\rangle\,, \quad && V_+ |\bar{r}\rangle = -|\bar{b}\rangle\,, \quad V_- |\bar{b}\rangle = -|\bar{r}\rangle\,.
\end{aligned}$$

The action of any ladder operator on the state (8.31) gives 0, for instance, $I_+(rgb - grb) = rrb - rrb = 0$, and similarly for the other combinations. The colour confinement postulate tells us that only particles in a singlet state can be observed. One may wonder if there are other possibilities than mesons or baryons. The answer is positive. The reader is invited to read Section C.4 in Appendix C to learn how to easily deduce the following decompositions. For instance, if two quarks or two antiquarks cannot form a bound state because no singlet is present in $\mathbf{3} \otimes \mathbf{3} = \mathbf{6} \oplus \bar{\mathbf{3}}$, bound states of two quarks and two antiquarks, commonly called *tetraquarks* are, in principle, possible:

$$\mathbf{3}\otimes\bar{\mathbf{3}}\otimes\mathbf{3}\otimes\bar{\mathbf{3}} = (\mathbf{8}\oplus\mathbf{1})\otimes(\mathbf{8}\oplus\mathbf{1}) = \mathbf{8}\otimes\mathbf{8}\oplus\mathbf{8}\oplus\mathbf{8}\oplus\mathbf{1} = \mathbf{27}\oplus\mathbf{10}\oplus\overline{\mathbf{10}}\oplus\mathbf{8}\oplus\mathbf{8}\oplus\mathbf{1}\oplus\mathbf{8}\oplus\mathbf{8}\oplus\mathbf{1}.$$

The last singlet results from the product of the two singlets in $(\mathbf{8}\oplus\mathbf{1})\otimes(\mathbf{8}\oplus\mathbf{1})$. Therefore, it can be interpreted as a bound state of two mesons or, equivalently, a molecule of mesons. The other singlet is the result of the product of the two octets. Molecules of mesons are expected to be loosely bound since the colour force must be weaker between colourless objects (this residual strong force is similar to that bounding nucleons together in nuclei). In contrast, 'genuine' tetraquarks from $\mathbf{8}\otimes\mathbf{8}$ would have colour objects as building blocks and, therefore, would be tightly bound. This difference should be reflected in their energy levels (i.e., their masses). Notice that if we had instead written $\mathbf{3}\otimes\mathbf{3}\otimes\bar{\mathbf{3}}\otimes\bar{\mathbf{3}} = (\mathbf{6}\otimes\bar{\mathbf{3}})\otimes(\bar{\mathbf{6}}\otimes\mathbf{3})$, one singlet would result from the product of the two sextets, while the other would be from the two triplets. Since the beginning of the twenty-first century, a dozen tetraquark candidates have been found in experiments (BESII, BaBar, Belle, CDF and, more recently, CMS and LHCb). They exhibit properties incompatible with the usual $q\bar{q}$ states, but their actual nature is complicated to establish. They are usually named X, Y or Z, followed by a number in parenthesis corresponding to the observed mass. For instance, the $Z(4430)^{\pm}$ particle is compatible with the quark content $c\bar{c}d\bar{u}$. When the tetraquark candidate mass is very close to the sum of the masses of two mesons, the molecule interpretation is usually favoured. Pentaquarks (bound states of $qqqq\bar{q}$), dibaryons ($qqqqqq$), baryonium ($qqq\bar{q}\bar{q}\bar{q}$) and *glueballs* (several gluons together can form a colour singlet, as we shall see soon) may also exist. But as for tetraquarks, their observation is delicate and their interpretation even more so.

8.2.4 Gauge Invariance Consequences

In 1954, Yang and Mills generalised the idea of a local phase symmetry to a symmetry group larger than U(1) (the symmetry group of QED), implying unitary rotations of several fields. Such a theory is called a *non-Abelian gauge theory* because the transformations of the fields do not commute.

Quarks being spin-1/2 elementary fermions satisfy the Dirac equation, and thus, the free field Lagrangian is given by the Dirac Lagrangian. However, as quarks carry three colours, instead of having just a single Dirac field for a given quark flavour q, there are now three fields, q_i, $i = 1, 2, 3$, one per colour of quarks (i.e., r for $i = 1$, g for $i = 2$ and b for $i = 3$). The free-field Lagrangian is assumed to be independent of the colour degrees of freedom. It is just the sum of the three Lagrangians

$$\mathcal{L} = \sum_{i=1}^{3} \bar{q}_i(i\gamma^\mu\partial_\mu - m)q_i = \bar{q}(i\gamma^\mu\partial_\mu - m)q, \text{ with } q = \begin{pmatrix} q_1 \\ q_2 \\ q_3 \end{pmatrix}. \tag{8.39}$$

Each q_i is associated with the same quark mass m. What are the symmetries of this Lagrangian? First, we notice that we have the freedom to change the phase of the quark fields, $q \to e^{i\varphi}q$, without affecting the Lagrangian. This freedom[15] is associated with the conservation of the baryon quantum number, as we saw in Section 8.1.1. Now consider a global transformation in the colour space. It yields

$$q' = Uq = e^{-ig_s\boldsymbol{\alpha}\cdot\frac{\boldsymbol{\lambda}}{2}}q = e^{-ig_s\alpha^a\frac{\lambda_a}{2}}q, \tag{8.40}$$

where it is convenient to factorise a common real constant g_s, and where α_a is a set of eight real numbers corresponding to the parameters of the $SU(3)_c$ transformation. Note that repeated indices are, as usual, summed over. The matrix U belongs to the $SU(3)_c$ group, and hence, it satisfies $UU^\dagger = U^\dagger U = 1$ and $\det U = 1$. Since U acts on the colour space, while γ^μ acts on the spinor space, they both commute. It follows that

$$\mathcal{L} \xrightarrow{\text{global}} \mathcal{L}' = q^\dagger U^\dagger \gamma^0(i\gamma^\mu\partial_\mu - m)Uq = \mathcal{L}.$$

The Lagrangian is thus invariant under a global transformation of colour. According to Noether's theorem, the colour is thus conserved (both I_{3_c} and Y_c). Following what we did for QED, if we want to have a gauge theory of the strong interaction, the Lagrangian of the theory must be invariant under a *local* transformation, i.e. a transformation where U depends on spacetime, the parameters becoming scalar functions $\alpha_a(x)$. In QED, this procedure gives rise to the photon field. The local transformation of the Lagrangian yields

$$\begin{aligned}\mathcal{L} \xrightarrow{\text{local}} \mathcal{L}' &= \bar{q}U^\dagger(i\gamma^\mu\partial_\mu(Uq) - mUq)\\ &= \bar{q}U^\dagger(i\gamma^\mu(\partial_\mu U)q + i\gamma^\mu U\partial_\mu q - mUq)\\ &= \mathcal{L} + \bar{q}i\gamma^\mu(U^\dagger\partial_\mu U)q.\end{aligned} \tag{8.41}$$

The Lagrangian is manifestly no longer invariant because of the eight additional terms

$$U^\dagger\partial_\mu U = U^\dagger(-ig_s\partial_\mu\alpha^a(x)\frac{\lambda_a}{2})U. \tag{8.42}$$

Let us restore the invariance and proceed as in QED by introducing eight new vector fields G_a^μ, the *gluons* fields via the covariant derivative

$$\boxed{\mathrm{D}_\mu = \partial_\mu + ig_s\frac{\lambda_a}{2}G_\mu^a.} \tag{8.43}$$

The equation above is just analogous to Eq. (6.31) in QED. The $SU(3)_c$ transformation of the gluon fields must compensate for the additional terms (8.42) to restore the local invariance. Hence, the new Lagrangian we now consider is

$$\mathcal{L} = \bar{q}(i\gamma^\mu\,\mathrm{D}_\mu - m)q = \bar{q}(i\gamma^\mu\partial_\mu - m)q - j_a^\mu\,G_\mu^a, \tag{8.44}$$

with eight (conserved) currents given by

$$j_a^\mu = g_s\,\bar{q}\gamma^\mu\frac{\lambda_a}{2}q. \tag{8.45}$$

[15] When all quark flavours are considered, the baryon number is a consequence of the *same* change of phase of all quark fields.

The transformation of the Lagrangian (8.44) yields

$$\mathcal{L} \xrightarrow{\text{local}} \mathcal{L}' = \bar{q}(i\gamma^\mu \partial_\mu - m)q + \bar{q}i\gamma^\mu(U^\dagger \partial_\mu U)q - j'^\mu_a G'^a_\mu$$
$$= \mathcal{L} + j^\mu_a G^a_\mu + \bar{q}i\gamma^\mu(U^\dagger \partial_\mu U)q - j'^\mu_a G'^a_\mu.$$

The invariance is thus restored if

$$j^\mu_a G^a_\mu + \bar{q}i\gamma^\mu(U^\dagger \partial_\mu U)q - j'^\mu_a G'^a_\mu = 0. \tag{8.46}$$

This constraint imposes the transformation of gluon fields. Supplement 8.2 shows that necessarily,[16]

$$G^a_\mu \xrightarrow{\text{local}} G'^a_\mu = G^a_\mu + g_s\alpha^b(x) f_{abc}\, G^c_\mu + \partial_\mu \alpha^a(x). \tag{8.47}$$

Let us recap: when the quark fields are transformed under a local $SU(3)_c$ transformation, the invariance of the Lagrangian (8.44) leads us to introduce eight gluon fields that transform as in (8.47). Can we go further to interpret the eight gluon fields? The fact that there are eight such objects (all equivalent regarding the strong interaction because of the colour symmetry) suggests that they belong to an octet of colour. If it is the case, they must transform as a member of an octet. Let us consider the octet coming from the decomposition $\mathbf{3} \otimes \bar{\mathbf{3}} = \mathbf{8} \oplus \mathbf{1}$. Introducing the colour state vector $\boldsymbol{c} = (r, g, b)^\mathsf{T}$, the eight linearly independent states of the octet can be generated by

$$\mathbf{8} \equiv \boldsymbol{w} = \begin{pmatrix} w_1 \\ w_2 \\ \vdots \\ w_8 \end{pmatrix} \quad \text{with } w_a = \boldsymbol{c}^\dagger \frac{\lambda_a}{2} \boldsymbol{c} = (\bar{r}, \bar{g}, \bar{b}) \frac{\lambda_a}{2} \begin{pmatrix} r \\ g \\ b \end{pmatrix} \quad (a = 1, 8). \tag{8.48}$$

(Here r^* and $\bar{r}$ are equivalent notation.) Also, notice that the singlet state is (modulo a factor) $\boldsymbol{c}^\dagger\, \mathbb{1}\, \boldsymbol{c}$. If these eight states correspond to an eight-dimensional representation of $SU(3)_c$ (as I claim), their transformation under $SU(3)_c$ must give specific linear combinations of themselves so that the transformed fields are still members of the same multiplet (by definition of a group). Using an infinitesimal transformation where $U \simeq 1 - ig_s\alpha^b(x)\frac{\lambda_b}{2}$, and given that the Gell-Mann matrices are Hermitian, w_a is transformed into

$$w'_a = \boldsymbol{c}^\dagger \left(1 + ig_s\alpha^b(x)\frac{\lambda_b}{2}\right) \frac{\lambda_a}{2} \left(1 - ig_s\alpha^b(x)\frac{\lambda_b}{2}\right) \boldsymbol{c} = w_a - ig_s\alpha^b(x)\boldsymbol{c}^\dagger \left(\frac{\lambda_a}{2}\frac{\lambda_b}{2} - \frac{\lambda_b}{2}\frac{\lambda_a}{2}\right) \boldsymbol{c},$$

where only the first order in $\alpha^b(x)$ is kept. The commutator of Gell-Mann matrices is given by Eq. (8.36). Given the properties of the structure constants (8.37), it follows that

$$w'_a = w_a - ig_s\alpha^b(x)(if_{abc})\, w_c = w_a + g_s\alpha^b(x) f_{abc}\, w_c = w_a - ig_s\alpha^b(x)(-if_{bac})\, w_c. \tag{8.49}$$

As announced, the transformed w'_a results from a linear combination of themselves. In vectorial notation, we can introduce $\boldsymbol{G}^{(8)}$, a vector of eight 8×8 matrices $G^{(8)}_1 \cdots G^{(8)}_8$, having, for components,

$$\left(G^{(8)}_b\right)_{ac} = -if_{bac}, \tag{8.50}$$

[16] In some textbooks, the sign is the opposite because the initial transformation is $U = e^{+ig_s\alpha^a\frac{\lambda_a}{2}}$. In others, g_s is not included in the exponent but only in the covariant derivative, leading to the final term $\alpha^b(x) f_{abc}\, G^c_\mu + \frac{1}{g_s}\partial_\mu \alpha^a(x)$. It is just a matter of convention.

such that

$$w' = w - ig_s\alpha \cdot G^{(8)} w.$$

A finite transformation would have given $w' = e^{-ig_s\alpha(x).G^{(8)}} w$. One can show that the $G^{(8)}$ matrices verify the commutation rule[17]

$$\left[G_a^{(8)}, G_b^{(8)}\right] = if_{abc} G_c^{(8)}. \qquad (8.51)$$

Hence, these matrices share the same algebra as the $SU(3)_c$ generators. Therefore, they constitute a representation of $SU(3)_c$. This is a general property of the Lie group, for which such a representation, called the *adjoint representation*, always exists and has the matrix elements given by minus the structure constants (see Appendices A and C for more details). Comparing the transformation of the gluon fields (8.47) and that of w_a in Eq. (8.49), we conclude that the gluon fields belong to the adjoint representation[18] of $SU(3)_c$. Hence, they are the members of the colour octet and carry both colour and anti-colour charges. Explicitly the expansion of the octet states (8.48) leads to the gluon states after normalisation,

$$G_1 = \tfrac{1}{\sqrt{2}}(\bar{g}r + \bar{r}g),\ G_2 = \tfrac{i}{\sqrt{2}}(\bar{g}r - \bar{r}g),\ G_3 = \tfrac{1}{\sqrt{2}}(\bar{r}r - \bar{g}g),\ G_4 = \tfrac{1}{\sqrt{2}}(\bar{b}r + \bar{r}b),$$

$$G_5 = \tfrac{i}{\sqrt{2}}(\bar{b}r - \bar{r}b),\ G_6 = \tfrac{1}{\sqrt{2}}(\bar{b}g + \bar{g}b),\ G_7 = \tfrac{i}{\sqrt{2}}(\bar{b}g - \bar{g}b),\ G_8 = \tfrac{1}{\sqrt{6}}(\bar{r}r + \bar{g}g - 2\bar{b}b).$$

Beware, however, that the physical gluons must have definite numbers for their colour isospin and colour hypercharge. They must be an eigenstate of the corresponding generators T_3 and T_8 (T_i for colour, it is $\lambda_i/2$, while for anti-colours, it is $-\lambda_i^*/2$) or equivalently, they must be represented by linear combinations $\tilde{\lambda}_a$ of Gell-Mann matrices commuting with λ_3 and λ_8. The matrices $\tilde{\lambda}_a$ are just the ladder operators defined in Eq. (8.15) up to a normalisation factor (a multiplication by $\sqrt{2}$) to ensure that $\mathrm{Tr}(\tilde{\lambda}_a\tilde{\lambda}_b) = 2\delta_{ab}$. The physical gluon states $c^\dagger \frac{\tilde{\lambda}_a}{2} c$ are then given in terms of colour and anti-colour content by

$$\begin{aligned} G_1 = \bar{g}r,\ G_2 = \bar{r}g,\ G_3 = \tfrac{1}{\sqrt{2}}(\bar{r}r - \bar{g}g),\ G_4 = \bar{b}r, \\ G_5 = \bar{r}b,\ G_6 = \bar{b}g,\ G_7 = \bar{g}b,\ G_8 = \tfrac{1}{\sqrt{6}}(\bar{r}r + \bar{g}g - 2\bar{b}b). \end{aligned} \qquad (8.52)$$

These states are strictly equivalent to those of the octet in the SU(3) flavour symmetry shown in Fig. 8.4 with the usual substitution $u \leftrightarrow r$, $d \leftrightarrow g$ and $s \leftrightarrow b$. According to the Lagrangian (8.44), the gluon field is coupled to a quark field via the current (8.45). The gluons $G_1 - G_7$ in Eq. (8.52) change the quark (antiquark) colour (anti-colour). For instance, when a red quark emits the gluon $\bar{g}r$ or absorbs the gluon $\bar{r}g$, it is changed into its green version, as shown in Fig 8.10. The two gluons G_3 and G_8 do not change the quark colour, but the different colours are treated differently (they are not in a singlet of colours). As with any gluon, they carry the strong interaction.

In conclusion, promoting the $SU(3)_c$ colour symmetry to a local gauge symmetry gives rise to eight fields, the gluons, that are members of an $SU(3)_c$ octet. They carry a colour

[17] This follows from the Jacobi identity of Gell-Mann matrices $[\lambda_a, [\lambda_b, \lambda_c]] + [\lambda_b, [\lambda_c, \lambda_a]] + [\lambda_c, [\lambda_a, \lambda_b]] = 0$.

[18] If the transformation was global, the identification of G_a^μ with w_a would be perfect since the term $\partial_\mu \alpha^a$ in (8.47) would be zero (α does not depend on x for a global transformation).

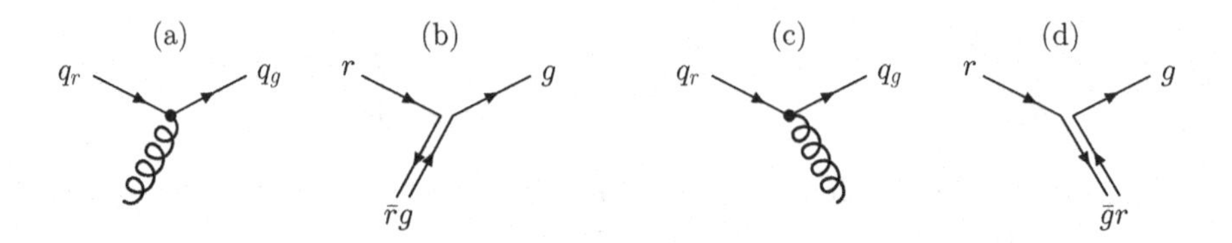

Fig. 8.10 The colour exchange of a quark from red to green. In (a) and (b), a gluon is absorbed, while in (c) and (d), it is emitted. Diagrams (a) and (c) are the classical Feynman diagrams, whereas (b) and (d) are the corresponding colour flow.

and an anti-colour charge. This is a fundamental difference with respect to the QED theory, where the photon does not carry an electric charge. Note that if a gluon has a colour and an anti-colour charge, a pair of gluons can be colourless since the decomposition of $\mathbf{8} \otimes \mathbf{8}$ yields, among other multiplets, a singlet (the calculation is presented in Appendix C, Section C.4). Bound states of gluons, the glueballs, are thus in principle possible. Another difference between photons and gluons is that photons are their own antiparticles, whereas a given gluon is not. However, as QCD respects the charge conjugation, the whole gluon octet is its own antiparticle in the sense that both gluon and the corresponding anti-gluon are the members of the octet. Finally, like photons, the gluons are spin-1 particles (their fields are vector fields) and massless since a term $m^2 G_a^\mu G_\mu^a$ would violate the gauge invariance.

Supplement 8.2. Effect of colour transformation on gluon fields

Gluons, carrying colour and anti-colour, are sensitive to a colour transformation. However, if the gluon field is denoted by G_μ^a, the new gluon field after the colour transformation, G'^a_μ, cannot be arbitrary because the QCD lagrangian must remain invariant. Equation (8.46) is the condition that the transformed gluon fields must fulfil. Let us decompose the transformed gluon fields as

$$G'^a_\mu = G^a_\mu + \delta G^a_\mu$$

and assume that $|\delta G^a_\mu| \ll |G^a_\mu|$. The condition (8.46) then becomes

$$j'^\mu_a \, \delta G^a_\mu = (j^\mu_a - j'^\mu_a)\, G^a_\mu + \bar{q} i \gamma^\mu (U^\dagger \partial_\mu U) q.$$

Given the expression (8.45) of the QCD current,

$$j'^\mu_a = g_s \bar{q} U^\dagger \gamma^\mu \frac{\lambda_a}{2} U q,$$

it follows that

$$g_s \bar{q} \gamma^\mu U^\dagger \frac{\lambda_a}{2} U \, \delta G^a_\mu \, q = g_s \bar{q} \gamma^\mu \left(\frac{\lambda_a}{2} - U^\dagger \frac{\lambda_a}{2} U \right) q \, G^a_\mu + \bar{q} i \gamma^\mu (U^\dagger \partial_\mu U) q.$$

Since the equation above must hold for all quark fields, it cannot depend on them, and it simplifies after multiplying by $(\gamma^\mu)^{-1}$ to

$$g_s U^\dagger \frac{\lambda_a}{2} U \, \delta G^a_\mu = g_s \left(\frac{\lambda_a}{2} - U^\dagger \frac{\lambda_a}{2} U \right) G^a_\mu + i (U^\dagger \partial_\mu U).$$

Supplement 8.2. (cont.)

Using the unitarity property of the U matrix, the previous equation reads

$$\frac{\lambda_a}{2} U\,\delta G^a_\mu = \left(U\frac{\lambda_a}{2} - \frac{\lambda_a}{2}U\right) G^a_\mu + \frac{i}{g_s}\partial_\mu U.$$

For simplicity, we consider an infinitesimal colour transformation, keeping only the first order in $\alpha^b(x)$, $U = 1 - ig_s\alpha^b(x)\lambda_b/2$. Neglecting the second-order terms as $[\alpha^b(x)]^2$ or $\alpha^b(x)\,\delta G^a_\mu$ (remember $|\delta G^a_\mu| \ll |G^a_\mu|$), we deduce

$$\begin{aligned}\tfrac{\lambda_a}{2}\delta G^a_\mu &= \left((1 - ig_s\alpha^b(x)\tfrac{\lambda_b}{2})\tfrac{\lambda_a}{2} - \tfrac{\lambda_a}{2}(1 - ig_s\alpha^b(x)\tfrac{\lambda_b}{2})\right) G^a_\mu + \partial_\mu\alpha^b(x)\tfrac{\lambda_b}{2} \\ &= ig_s\alpha^b(x)\left[\tfrac{\lambda_a}{2}, \tfrac{\lambda_b}{2}\right] G^a_\mu + \partial_\mu\alpha^b(x)\tfrac{\lambda_b}{2} \\ &= -g_s\alpha^b(x)f_{abc}\tfrac{\lambda_c}{2} G^a_\mu + \partial_\mu\alpha^b(x)\tfrac{\lambda_b}{2},\end{aligned}$$

where the commutation algebra (8.36) of the Gell-Mann matrices was used. Note that in this expression, there are implicit summations over colour indices a, b and c. We can simplify the expression by changing the labels of those indices: in the first term of the right-hand side, we swap a and c and in the second term, we change b for a. It follows that

$$\frac{\lambda_a}{2}\delta G^a_\mu = -g_s\alpha^b(x)f_{cba}\frac{\lambda_a}{2} G^c_\mu + \partial_\mu\alpha^a(x)\frac{\lambda_a}{2}.$$

This expression is equivalent to $\lambda_a/2 \times g^a(x) = 0$, where $g^a(x) = \delta G^a_\mu(x) + g_s\alpha^b(x)f_{cba}G^c_\mu(x) - \partial_\mu\alpha^a(x)$. Since it must hold for all x, necessarily $g^a(x) = 0$. Moreover, using the property (8.37) of the structure constants, $f_{cba} = f_{bac} = -f_{abc}$, we finally obtain the transformation rule for the gluon fields

$$G^a_\mu \xrightarrow{\text{local}} G'^a_\mu = G^a_\mu + g_s\alpha^b(x)f_{abc}G^c_\mu + \partial_\mu\alpha^a(x),$$

where the summation over b and c is implicit. This expression is the one given in Eq. (8.47).

8.2.5 QCD Lagrangian

The generalisation of the Lagrangian (8.44) to the six flavours of quarks is straightforward,

$$\mathcal{L} = \sum_{q=u,d,s,c,b,t} \bar{q}(i\gamma^\mu\partial_\mu - m_q)q - g_s\,\bar{q}\gamma^\mu\frac{\lambda_a}{2}q\,G^a_\mu\,, \quad \text{with } q = \begin{pmatrix} q_1 \\ q_2 \\ q_3 \end{pmatrix}. \tag{8.53}$$

The term

$$\mathcal{L}_{\text{int}} = \sum_q -g_s\,\bar{q}\gamma^\mu\frac{\lambda_a}{2}q\,G^a_\mu \tag{8.54}$$

describes the interaction between the gluons and the (anti-)quarks. Keep in mind that the Lagrangian (8.53) is a compact way of writing

$$\mathcal{L} = \sum_q \left(\sum_i \bar{q}_i(i\gamma^\mu\partial_\mu - m_q)\,q_i - \sum_{i,j} g_s\,\bar{q}_i\gamma^\mu\frac{1}{2}\,(\lambda_a)_{ij}\,q_j\,G^a_\mu\right),$$

where i and j are the colour indices (from 1 to 3), and each q_i is a Dirac field (of four components) associated with the flavour q. As in QED, a term describing the propagation of the gluons, namely the kinetic (but accompanied with a potential energy term), must be added. In QED, this term is

$$\mathcal{L} = -\frac{1}{4}F^{\mu\nu}F_{\mu\nu}, \quad \text{with} \quad F^{\mu\nu} = \partial^\mu A^\nu - \partial^\nu A^\mu.$$

It would be tempting to propose

$$\mathcal{L} = -\frac{1}{4}G_a^{\mu\nu}G^a_{\mu\nu}, \quad \text{with} \quad G^a_{\mu\nu} = \partial_\mu G^a_\nu - \partial_\nu G^a_\mu.$$

However, this term is not gauge invariant. Indeed, the derivative of the fields transforms as

$$\begin{aligned}
\partial_\mu G^a_\nu \to \partial_\mu G'^a_\nu &= \partial_\mu G^a_\nu + g_s(\partial_\mu\alpha^b)f_{abc}\, G^c_\nu + g_s\alpha^b f_{abc}\,\partial_\mu G^c_\nu + \partial_\mu\partial_\nu\alpha^a; \\
\partial_\nu G^a_\mu \to \partial_\nu G'^a_\mu &= \partial_\nu G^a_\mu + g_s(\partial_\nu\alpha^b)f_{abc}\, G^c_\mu + g_s\alpha^b f_{abc}\,\partial_\nu G^c_\mu + \partial_\nu\partial_\mu\alpha^a.
\end{aligned}$$

Consequently,

$$\begin{aligned}
\delta(\partial_\mu G^a_\nu - \partial_\nu G^a_\mu) &= \left(\partial_\mu G'^a_\nu - \partial_\nu G'^a_\mu\right) - \left(\partial_\mu G^a_\nu - \partial_\nu G^a_\mu\right) \\
&= g_s f_{abc}\left[(\partial_\mu\alpha^b)G^c_\nu - (\partial_\nu\alpha^b)G^c_\mu\right] + g_s f_{abc}\alpha^b\left[\partial_\mu G^c_\nu - \partial_\nu G^c_\mu\right].
\end{aligned} \tag{8.55}$$

Then, a Lagrangian based on $G_a^{\mu\nu}G^a_{\mu\nu}$, with a simple kinetic term $G^a_{\mu\nu} = \partial_\mu G^a_\nu - \partial_\nu G^a_\mu$, would not be invariant. Actually, to compensate for these additional terms, one should use instead

$$G^a_{\mu\nu} = \partial_\mu G^a_\nu - \partial_\nu G^a_\mu - g_s f_{abc} G^b_\mu G^c_\nu. \tag{8.56}$$

The interested reader can consult Supplement 8.3 that shows the invariance of the Lagrangian based on the gluon tensor (8.56). It is important to realise that in addition to the purely kinetic terms $\partial_\mu G^a_\nu - \partial_\nu G^a_\mu$, the term $-g_s f_{abc} G^b_\mu G^c_\nu$ adds an interaction between two gluon fields. We will stress this point in the next section.

In conclusion, the whole QCD Lagrangian is

$$\boxed{\mathcal{L}^{\text{QCD}} = \sum_{q=u,d,s,c,b,t} \bar{q}(i\gamma^\mu\partial_\mu - m_q)q - g_s\,\bar{q}\gamma^\mu\frac{\lambda_a}{2}q\,G^a_\mu - \frac{1}{4}G_a^{\mu\nu}G^a_{\mu\nu},} \tag{8.57}$$

where the tensor $G_a^{\mu\nu}$ is given by Eq. (8.56).

Supplement 8.3. Proof of the gauge invariance of the gluon field Lagrangian

Let us check that the gluon field-strength tensor given by

$$G^a_{\mu\nu} = \partial_\mu G^a_\nu - \partial_\nu G^a_\mu - g_s f_{abc} G^b_\mu G^c_\nu$$

leads to an invariant Lagrangian $G_a^{\mu\nu}G^a_{\mu\nu}$ under the colour transformation (8.47). The variation of $G^a_{\mu\nu}$ due to the colour transformation is

$$\delta G^a_{\mu\nu} = \delta(\partial_\mu G^a_\nu - \partial_\nu G^a_\mu) - g_s f_{abc}\,\delta(G^b_\mu G^c_\nu).$$

Supplement 8.3. (cont.)

The first term on the right-hand side has already been calculated in Eq. (8.55), and the second term requires the calculation of

$$G'^b_\mu G'^c_\nu = (G^b_\mu + g_s\alpha^d f_{bde} G^e_\mu + \partial_\mu \alpha^b)(G^c_\nu + g_s\alpha^d f_{cde} G^e_\nu + \partial_\nu \alpha^c)$$
$$\simeq G^b_\mu G^c_\nu + g_s\alpha^d (f_{cde} G^e_\nu G^b_\mu + f_{bde} G^e_\mu G^c_\nu) + G^b_\mu \partial_\nu \alpha^c + G^c_\nu \partial_\mu \alpha^b,$$

where only terms up to first order in α have been kept [simply because the transformation (8.47) of the gluon field has been obtained with an infinitesimal colour transformation]. Consequently, with the insertion of Eq. (8.55), the variation of the gluon field-strength tensor is

$$\delta G^a_{\mu\nu} = g_s f_{abc}\left[(\partial_\mu \alpha^b)G^c_\nu - (\partial_\nu \alpha^b)G^c_\mu\right] + g_s f_{abc}\alpha^b\left[\partial_\mu G^c_\nu - \partial_\nu G^c_\mu\right]$$
$$- g_s^2 \alpha^d (f_{abc} f_{cde} G^e_\nu G^b_\mu + f_{abc} f_{bde} G^e_\mu G^c_\nu) - g_s f_{abc}(G^b_\mu \partial_\nu \alpha^c + G^c_\nu \partial_\mu \alpha^b).$$

Playing with the dummy indices $b \leftrightarrow c$ in the term $-g_s f_{abc} G^b_\mu \partial_\nu \alpha^c$ and $e \to c, b \to e$ in $f_{abc} f_{cde} G^e_\nu G^b_\mu$ yields, given the structure constant properties (8.37),

$$\delta G^a_{\mu\nu} = g_s f_{abc}\alpha^b\left[\partial_\mu G^c_\nu - \partial_\nu G^c_\mu\right] - g_s^2 \alpha^d (-f_{aeb} f_{dbc} + f_{abc} f_{bde}) G^e_\mu G^c_\nu.$$

We can rearrange this expression a bit by noticing $f_{bde} = f_{deb}$, such that

$$\delta G^a_{\mu\nu} = g_s f_{abc}\alpha^b\left[\partial_\mu G^c_\nu - \partial_\nu G^c_\mu\right] - g_s^2 \alpha^d (f_{deb} f_{abc} - f_{aeb} f_{dbc}) G^e_\mu G^c_\nu. \tag{8.58}$$

But using the definition (8.50),

$$f_{deb} f_{abc} - f_{aeb} f_{dbc} = \left(i\left(G^{(8)}_d\right)_{eb} i\left(G^{(8)}_a\right)_{bc} - i\left(G^{(8)}_a\right)_{eb} i\left(G^{(8)}_d\right)_{bc}\right)$$
$$= -\left(G^{(8)}_d G^{(8)}_a\right)_{ec} + \left(G^{(8)}_a G^{(8)}_d\right)_{ec}$$
$$= -i f_{daf}\left(G^{(8)}_f\right)_{ec},$$

by virtue of the implicit summation over repeated colour indices, and where, in the last line, the commutation rule (8.51) has been used. The second term of Eq. (8.58) is thus

$$i g_s^2 \alpha^d f_{daf}\left(G^{(8)}_f\right)_{ec} G^e_\mu G^c_\nu = i g_s^2 \alpha^b f_{baf}\left(G^{(8)}_f\right)_{ek} G^e_\mu G^k_\nu = i g_s^2 \alpha^b f_{bac}\left(G^{(8)}_c\right)_{ek} G^e_\mu G^k_\nu,$$

where I have played again with the dummy indices, successively changing d for b, c for k and f for c. According to the definition (8.50), $\left(G^{(8)}_c\right)_{ek} = -i f_{cek}$, and hence, Eq. (8.58) finally reads

$$\delta G^a_{\mu\nu} = g_s f_{abc}\alpha^b\left[\partial_\mu G^c_\nu - \partial_\nu G^c_\mu - g_s f_{cek} G^e_\mu G^k_\nu\right] = g_s f_{abc}\alpha^b G^c_{\mu\nu}.$$

Consequently, the variation of the Lagrangian $\frac{1}{4}G^{\mu\nu}_a G^a_{\mu\nu}$ after a colour transformation is

$$\frac{1}{4}\delta(G^{\mu\nu}_a G^a_{\mu\nu}) = \frac{1}{2}G^{\mu\nu}_a \delta G^a_{\mu\nu}$$
$$= \frac{1}{4}\left(g_s f_{abc}\alpha^b G^{\mu\nu}_a G^c_{\mu\nu} + g_s f_{abc}\alpha^b G^{\mu\nu}_a G^c_{\mu\nu}\right)$$
$$= \frac{1}{4}\left(g_s f_{abc}\alpha^b G^{\mu\nu}_a G^c_{\mu\nu} - g_s f_{abc}\alpha^b G^{\mu\nu}_c G^a_{\mu\nu}\right)$$
$$= 0,$$

where again the dummy indices a and c have been interchanged in the second term of the third line. This proves that the Lagrangian of the gluon fields is gauge invariant.

8.2.6 Discovery of Gluons

According to the interaction Lagrangian (8.54), gluons, the gauge bosons of QCD, couple to the quark fields. Therefore, highly energetic quarks may radiate energetic gluons (gluon bremsstrahlung) via, for example, the process shown below:

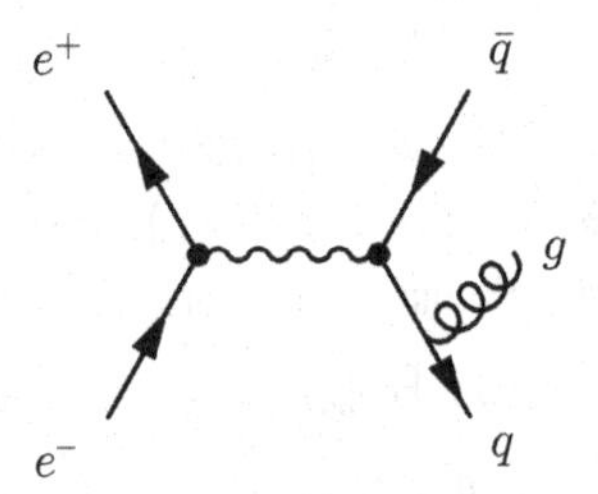

If the gluon is energetic enough and it is non-collinear to the quark direction, it should produce a jet of hadrons well separated from the quark jet. The notion of jets is explained in more detail in Section 8.5.4. At e^+e^- colliders, the topology of such an event would be characterised by the presence of three distinct jets of hadrons. The conservation of the energy-momentum implies that the three jets must be coplanar. Moreover, one of them must have a total zero electric charge if originated from a gluon (which is neutral). In 1979, at the Positron–Electron Tandem Ring Accelerator (PETRA) e^+e^- storage ring (DESY, Hambourg), the first three-jet events with such properties were observed for the first time by the Two-Arm Spectrometer SOlenoid (TASSO) experiment. A display of one of these events is shown in Fig. 8.11. The energy in the CM ranged from 13 to 31.6 GeV. The three-jet events were mostly observed at the higher energy, consistent with the scenario of energetic gluon emission.

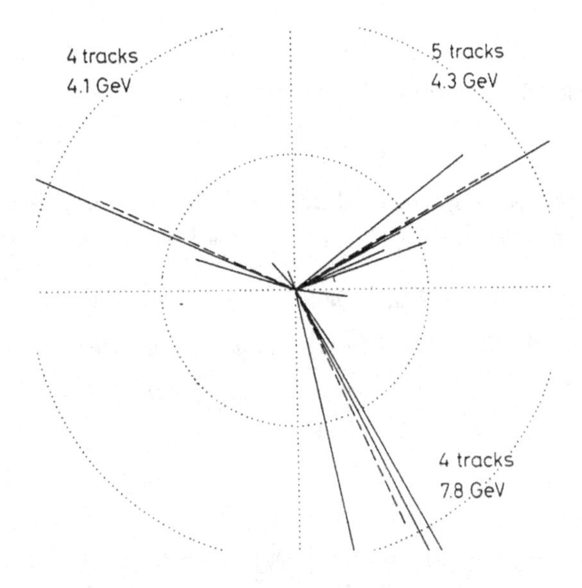

Fig. 8.11 One of the first three-jets events in the TASSO experiment demonstrating gluon emission. Solid lines: charged tracks. Dashed lines: direction of the fitted jets. Source: Bethke (2007). Reprinted with permission from Elsevier (2007).

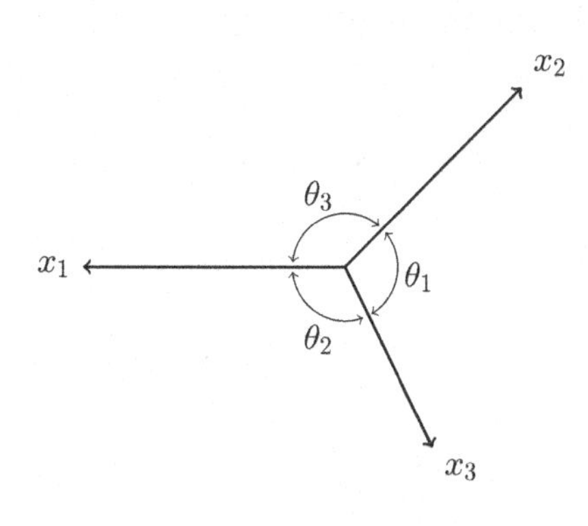

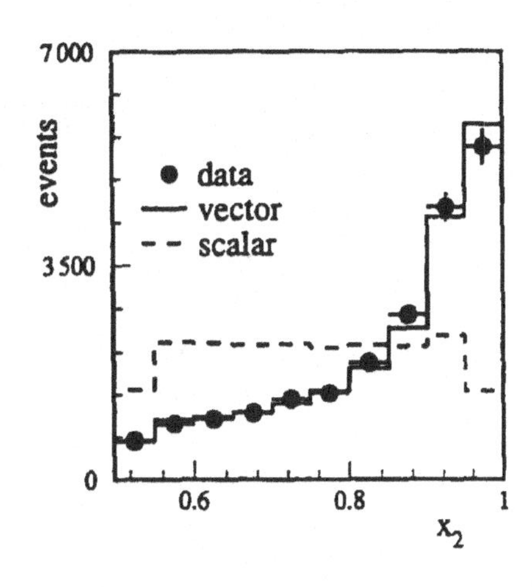

Fig. 8.12 Left: momenta and angles of the $q\bar{q}g$ final state. Right: measured and predicted distribution of the scaled energy of the second jet (x_2). The solid and dashed lines show the predictions for the vector and scalar gluons, respectively. Source: Adeva et al. (1991). The right figure reprinted with permission from Elsevier (1991).

In the process $e^+ + e^- \to q + \bar{q} + g$, given that electrons and quarks are fermions, the conservation of the angular momentum implies that the gluon could be a scalar (spin 0) or a vector particle (spin 1). The analysis of the jet angular distribution allows us to discriminate between the two scenarios as follows. Denoting by x_1, x_2 and x_3, the jet energies normalised to the beam energy, i.e. $x_1 + x_2 + x_3 = 2$, with $x_1 \geq x_2 \geq x_3$, and assuming massless quarks (and gluons), the variables x_i $(i = 1, 2, 3)$ are given by

$$x_i = \frac{E_i}{E_{\text{beam}}} = \frac{2\sin\theta_i}{\sin\theta_1 + \sin\theta_2 + \sin\theta_3},$$

where θ_i is the angle between the two jets other than jet i (see the left-hand side scheme in Fig. 8.12). Since gluons are radiated by quarks or antiquarks, we expect the less energetic jet (jet 3) to often be the gluon jet. The distributions of x_1 and x_2 are expected to follow (Ellis Gaillard and Ross, 1976)

$$\frac{d\sigma_V}{dx_1\, dx_2} \propto \frac{x_1^2 + x_2^2}{(1 - x_1)(1 - x_2)}, \quad \frac{d\sigma_S}{dx_1\, dx_2} \propto \frac{x_3^2}{(1 - x_1)(1 - x_2)},$$

where the subscripts V and S stand for the vector and scalar, respectively. Therefore, for the vector gluon model, the distribution should present a maximum at $x_{1,2} \to 1$. However, for the scalar gluon model, when $x_{1,2} \to 1$, $x_3 = 2 - x_1 - x_2$ becomes close to zero, and thus the cross section should not present such a maximum. Figure 8.12 shows the x_2 distribution obtained by the L3 collaboration at the LEP e^+e^- collider $\left(\sqrt{s} = 91.2\ \text{GeV}\right)$. It is compared with the vector (solid line) and scalar gluon (dashed line) predictions. The scalar model fails to describe the data, whereas the agreement with the vector model, having a maximum towards 1, is excellent. It thus confirms[19] that the gluon is a spin-1 boson.

[19] The first evaluation of the gluon spin was performed by the TASSO experiment in 1980, using the same approach (see TASSO Collaboration, 1980).

8.3 Perturbative QCD

8.3.1 QCD Feynman Rules

The QCD Lagrangian (8.57) expanded in terms of colour degrees of freedom is

$$\begin{aligned}\mathcal{L} &= \sum_{q=u,d,s,c,b,t}\left(\sum_{\alpha}\bar{q}_\alpha(i\gamma^\mu\partial_\mu - m_q)\,q_\alpha - \sum_{\alpha,\beta} g_s\,\bar{q}_\alpha\gamma^\mu\frac{1}{2}\,(\lambda_a)_{\alpha\beta}\,q_\beta\,G^a_\mu\right) - \frac{1}{4}G_a^{\mu\nu}G^a_{\mu\nu}\\ &= \sum_{q=u,d,s,c,b,t}\sum_{\alpha,\beta}\left(\bar{q}_\alpha(i\gamma^\mu\partial_\mu - m_q)\delta_{\alpha\beta}\,q_\beta - g_s\,\bar{q}_\alpha\gamma^\mu\frac{1}{2}\,(\lambda_a)_{\alpha\beta}\,q_\beta\,G^a_\mu\right) - \frac{1}{4}G_a^{\mu\nu}G^a_{\mu\nu},\end{aligned} \tag{8.59}$$

where α and β are the colour indices of the quarks (from 1 to 3) and a is, the colour index of the gluon (from 1 to 8). The summations over the quark colour indices are explicit, but to lighten the writing, there is also an implicit summation over the gluon index colour a. The comparison of the QCD Lagrangian (8.59) with the QED Lagrangian allows us to easily derive the QCD Feynman rules. Let us recall here the QED Lagrangian

$$\mathcal{L}^{\text{QED}} = \bar{\psi}(i\gamma^\mu\partial_\mu - m)\psi - eQ\bar{\psi}\gamma^\mu\psi A_\mu - \frac{1}{4}F^{\mu\nu}F_{\mu\nu},$$

where ψ is the Dirac field of spin-1/2 fermions. In QED, the external lines of the Feynman diagrams associated with a fermion or anti-fermion and corresponding to the field ψ are described by spinors depending on the particle momentum and its spin/helicity state, i.e. $u_s(p)$ for a particle and $v_s(p)$ for the antiparticle. In QCD, one has to take into account the additional colour index of the quark field. It is then convenient to introduce the three following colour vectors inspired by the basis of colour and anti-colour states (8.33) and (8.38):

$$c_1 = \begin{pmatrix}1\\0\\0\end{pmatrix} \equiv r \text{ or } \bar{r}, \quad c_2 = \begin{pmatrix}0\\1\\0\end{pmatrix} \equiv g \text{ or } \bar{g}, \quad c_3 = \begin{pmatrix}0\\0\\1\end{pmatrix} \equiv b \text{ or } \bar{b}. \tag{8.60}$$

An incoming or outgoing quark of colour α is thus described by $c_\alpha u_s(p)$ or $[c_\alpha u_s(p)]^\dagger\gamma^0 = \bar{u}_s(p)c_\alpha^\dagger$, respectively. Similarly, for an incoming or outgoing antiquark, we obtain $\bar{v}_s(p)c_\alpha^\dagger$ or $c_\alpha v_s(p)$, respectively.

$$\begin{aligned} &\xrightarrow{\alpha}\bullet = c_\alpha u_s(p) \qquad &\bullet\xrightarrow{\alpha} = c_\alpha^\dagger\bar{u}_s(p)\\ &\xleftarrow{\alpha}\bullet = c_\alpha^\dagger\bar{v}_s(p) \qquad &\bullet\xleftarrow{\alpha} = c_\alpha v_s(p).\end{aligned} \tag{8.61}$$

In the Lagrangian $\mathcal{L}^{\text{QED}}$, the first term $\bar{\psi}(i\gamma^\mu\partial_\mu - m)\psi$ describes a free fermion. In QCD, we have the equivalent term $\sum_{\alpha,\beta}\bar{q}_\alpha(i\gamma^\mu\partial_\mu - m_q)\delta_{\alpha\beta}\,q_\beta$. It tells us that during the propagation of a free quark, its colour does not change. The propagator associated with the quark field is the usual Dirac propagator of massive fermions with the additional requirement of the colour conservation, i.e.

$$\overset{\alpha}{\longrightarrow}\overset{\beta}{} = i\frac{\not{p}+m}{p^2-m^2+i\epsilon}\delta_{\alpha\beta}. \tag{8.62}$$

The colour vectors of the intermediate quark must be applied to the two vertices connected by the propagator.

What about the interaction term between quarks and gluons? Let us compare the interaction terms in QED and QCD given by

$$\mathcal{L}_{\text{int}}^{\text{QED}} = -e\bar{\psi}Q\gamma^{\mu}\psi A_{\mu}, \quad \mathcal{L}_{\text{int}}^{\text{QCD}} = \sum_{\alpha,\beta} -g_s\, \bar{q}_{\alpha}\gamma^{\mu}\frac{1}{2}(\lambda_a)_{\alpha\beta}\, q_{\beta}\, G_{\mu}^{a}. \tag{8.63}$$

In QED, it describes the interaction between a photon and a fermion with an electric charge eQ, giving rise to the vertex factor $-ieQ\gamma^{\mu}$. The identification of the vertex factor corresponding to the interaction of a quark with a gluon is then straightforward, yielding $-i\alpha_s\frac{(\lambda_a)_{\alpha\beta}}{2}\gamma^{\mu}$. The QCD Feynman rules learned so far are summarised in the two diagrams below:

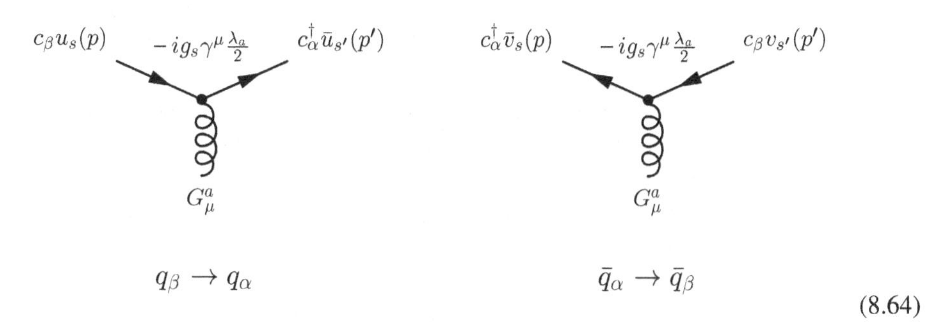

(8.64)

Note that since the vertex factor is sandwiched between the spinors and the colour vectors, the quark or antiquark leg of the diagrams contributes to the amplitudes

$$\mathcal{M}_a^{\mu}(q_{\beta} \to q_{\alpha}) = c_{\alpha}^{\dagger}\bar{u}_{s'}(p')\left(-ig_s\gamma^{\mu}\frac{\lambda_a}{2}\right)c_{\beta}u_s(p) = -ig_s\bar{u}_{s'}(p')\gamma^{\mu}\frac{1}{2}(\lambda_a)_{\alpha\beta}\, u_s(p),$$
$$\mathcal{M}_a^{\mu}(\bar{q}_{\alpha} \to \bar{q}_{\beta}) = c_{\alpha}^{\dagger}\bar{v}_s(p)\left(-ig_s\gamma^{\mu}\frac{\lambda_a}{2}\right)c_{\beta}v_{s'}(p') = -ig_s\bar{v}_s(p)\gamma^{\mu}\frac{1}{2}(\lambda_a)_{\alpha\beta}\, u_{s'}(p'),$$

consistent with the QCD interaction term (8.63). The use of colour vectors in the Feynman rules (8.61), (8.62) and (8.64) is optional. In most textbooks, the Feynman rules are given without colour vectors. However, they are a convenient way to keep track of quark colours and avoid applying the colour indices in the wrong order. They can be ignored if one directly writes the Gell-Mann matrices with the appropriate indices.

The last term in the QCD Lagrangian (8.59) is the Lagrangian for the gauge fields, which, after the insertion of Eq. (8.56), reads

$$-\frac{1}{4}G_a^{\mu\nu}G_{\mu\nu}^{a} = -\frac{1}{4}(\partial^{\mu}G_a^{\nu} - \partial^{\nu}G_a^{\mu})(\partial_{\mu}G_{\nu}^{a} - \partial_{\nu}G_{\mu}^{a}) + \mathcal{L}_{3-g} + \mathcal{L}_{4-g}, \tag{8.65}$$

where

$$\mathcal{L}_{3-g} = \frac{1}{2}g_s f_{abc}(\partial^{\mu}G_a^{\nu} - \partial^{\nu}G_a^{\mu})G_{\mu}^{b}G_{\nu}^{c}, \tag{8.66}$$
$$\mathcal{L}_{4-g} = -\frac{1}{4}g_s^2 f_{abc}f_{ade}G_b^{\mu}G_c^{\nu}G_{\mu}^{d}G_{\nu}^{e}. \tag{8.67}$$

The first term on the right-hand side of Eq. (8.65) is the pure kinetic-potential energy term, similar to $-\frac{1}{4}F^{\mu\nu}F_{\mu\nu}$ in QED. In terms of the Feynman diagrams, it corresponds to the gluon propagator, similar to the photon propagator since both are massless vector bosons,

$$a \text{ (gluon line) } b \quad = \quad -\, i\frac{g_{\mu\nu}}{p^2+i\epsilon}\delta^{ab}. \tag{8.68}$$

The only difference is the presence of the colour indices a and b (from 1 to 8) that tells us that during the gluon propagation, its colour state does not change. This propagator formula is valid in the Feynman gauge.

A gluon in the initial or final state of a process is described (as for photons) by a polarisation vector ϵ^{μ} for an incoming gluon and $\epsilon^{\mu *}$ for an outgoing gluon. Its massless nature also imposes the transversality condition $k^{\mu}\epsilon_{\mu} = 0$, where k^{μ} is the gluon 4-momentum. The colour/anti-colour state of the gluon, defined by the index $a = 1, \ldots, 8$, determines the λ_a matrix that is used at the qqg vertex.

Finally, the Lagrangians (8.66) and (8.67) are specific to the non-Abelian nature of QCD (hence, the presence of the structure constants). They describe the 3-gluon and 4-gluon interactions, respectively. Notice that the *same coupling constant*, g_s, is involved for all QCD interaction types, qqg, ggg or $gggg$. Unlike photons, which are electrically neutral and therefore cannot self-interact, gluons carry a colour and an anti-colour that allow their self-interaction. The vertex factors of the gluon self-interactions are given in Fig. 8.13. All gluon momenta point into the vertex, i.e. $\sum_i q_i = 0$. If not, change q_i for $-q_i$ in the vertex factor. For the 4-gluon vertex factor, the repeated index e runs from 1 to 8. Let us see how we obtain these factors. The 3-gluon Lagrangian (8.66) reads

$$\begin{aligned}\mathcal{L}_{3-g} &= \frac{1}{2} g_s f_{abc} [g_{\nu\rho}(\partial_\mu G_a^\nu) G_b^\mu G_c^\rho - g_{\mu\nu}(\partial_\rho G_a^\mu) G_b^\nu G_c^\rho] \\ &= \frac{1}{2} g_s f_{a'b'c'} [(g_{\mu'\rho'}\partial_{\nu'} - g_{\mu'\nu'}\partial_{\rho'}) G_{a'}^{\mu'}] G_{b'}^{\nu'} G_{c'}^{\rho'}.\end{aligned}$$

In the last equality, the dummy indices have been changed to the corresponding primed symbols to avoid confusion in what follows. This Lagrangian leads to several terms in the vertex factor because with the three identical gluons shown in the cubic interaction of Fig. 8.13, there are $3! = 6$ possibilities to associate the field operators with the gluons. Hence, let us rewrite

$$\mathcal{L}_{3-g} = \frac{1}{6}\left(6 \times \frac{1}{2} g_s f_{a'b'c'} [(g_{\mu'\rho'}\partial_{\nu'} - g_{\mu'\nu'}\partial_{\rho'}) G_{a'}^{\mu'}] G_{b'}^{\nu'} G_{c'}^{\rho'}\right)$$

and give some examples. For example, the gluons labelled $G_a^\mu, G_b^\nu, G_c^\rho$ in Fig. 8.13 can be produced by the operators of the gluon fields present in the Lagrangian when

$$\frac{1}{2} g_s f_{a'b'c'} (g_{\mu'\rho'}\partial_{\nu'} - g_{\mu'\nu'}\partial_{\rho'}) \equiv \begin{pmatrix} \mu' & \nu' & \rho' \\ a' & b' & c' \end{pmatrix} = \begin{pmatrix} \mu & \nu & \rho \\ a & b & c \end{pmatrix} \equiv \frac{1}{2} g_s f_{abc} (g_{\mu\rho}\partial_\nu - g_{\mu\nu}\partial_\rho).$$

Two other possibilities are $\begin{pmatrix} \mu' & \nu' & \rho' \\ a' & b' & c' \end{pmatrix} = \begin{pmatrix} \nu & \mu & \rho \\ b & a & c \end{pmatrix}$ and $\begin{pmatrix} \mu' & \nu' & \rho' \\ a' & b' & c' \end{pmatrix} = \begin{pmatrix} \rho & \nu & \mu \\ c & b & a \end{pmatrix}$. Feynman rules are usually given in the momentum space, where, by the Fourier transform, the spacetime derivative is replaced by the 4-momentum,[20] i.e. $\partial_\mu \to -iq_\mu$. Labelling q_1, q_2 and q_3 the

[20] More specifically, in the Fourier expansion of the field, $\int \ldots (a(q)e^{-iq\cdot x} + a^\dagger(q)e^{+iq\cdot x})$, the derivative ∂_μ yields a factor $-iq_\mu$ for a particle annihilated. This is the case in the vertex considered here since gluon momenta point into the vertex.

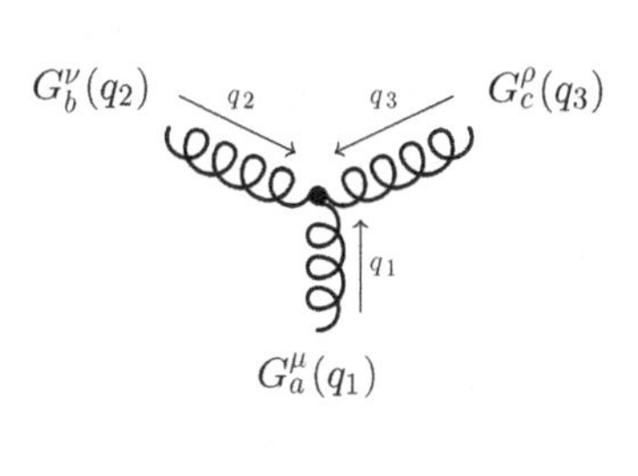

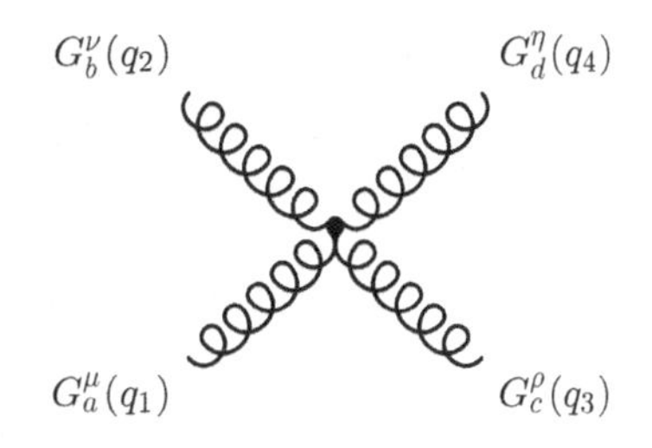

$$-g_s f_{abc}[\; g_{\mu\nu}(q_1-q_2)_\rho + g_{\nu\rho}(q_2-q_3)_\mu + g_{\rho\mu}(q_3-q_1)_\nu \;]$$

$$-ig_s^2[\; f_{abe}f_{cde}(g_{\mu\rho}g_{\nu\eta}-g_{\mu\eta}g_{\nu\rho}) + f_{ade}f_{bce}(g_{\mu\nu}g_{\rho\eta}-g_{\mu\rho}g_{\nu\eta}) + f_{ace}f_{bde}(g_{\mu\nu}g_{\rho\eta}-g_{\mu\eta}g_{\nu\rho}) \;]$$

Fig. 8.13 Gluon self-interaction vertex factors.

4-momentum of the gluon with the colour indices a, b and c, respectively, the three previous configurations lead to a vertex factor term[21]

$$3!(i)\times\frac{1}{6}(-i)\left(\frac{1}{2}g_s f_{abc}(g_{\mu\rho}q_{1_\nu}-g_{\mu\nu}q_{1_\rho})+\frac{1}{2}g_s f_{bac}(g_{\nu\rho}q_{2_\mu}-g_{\nu\mu}q_{2_\rho}) + \frac{1}{2}g_s f_{cba}(g_{\rho\mu}q_{3_\nu}-g_{\rho\nu}q_{3_\mu})\right).$$

It turns out that the three other configurations are identical to these three. For instance, $\begin{pmatrix}\mu' & \nu' & \rho' \\ a' & b' & c'\end{pmatrix}=\begin{pmatrix}\mu & \rho & \nu \\ a & c & b\end{pmatrix}$ gives

$$3!\times i\times\frac{1}{6}(-i)\left(\frac{1}{2}g_s f_{acb}(g_{\mu\nu}q_{1_\rho}-g_{\mu\rho}q_{1_\nu})\right)=3!\times i\times\frac{1}{6}(-i)\left(\frac{1}{2}g_s f_{abc}(g_{\mu\rho}q_{1_\nu}-g_{\mu\nu}q_{1_\rho})\right),$$

because of the symmetry properties of the structure constants [see Eq. (8.37)]. Therefore, overall, the vertex factor is

$$\begin{aligned}F_{3g} &= g_s f_{abc}(g_{\mu\rho}q_{1_\nu}-g_{\mu\nu}q_{1_\rho}) - g_s f_{abc}(g_{\nu\rho}q_{2_\mu}-g_{\nu\mu}q_{2_\rho}) - g_s f_{abc}(g_{\rho\mu}q_{3_\nu}-g_{\rho\nu}q_{3_\mu}) \\ &= -g_s f_{abc}[g_{\mu\nu}(q_1-q_2)_\rho + g_{\nu\rho}(q_2-q_3)_\mu + g_{\rho\mu}(q_3-q_1)_\nu],\end{aligned}$$

which is the factor given in Fig. 8.13 for the cubic self-interactions of gluons.

For the quartic self-interaction, we start from the Lagrangian (8.67) and change the dummy colour indices to have the 4-gluon fields with the colour order a, b, c and d (one has to interchange successively $c \leftrightarrow b$, $d \leftrightarrow c$ and $d \leftrightarrow e$). Given the cyclic property of the structure constants, this yields

$$\mathcal{L}_{4-g} = -\frac{1}{4}g_s^2 f_{abe}f_{cde}g_{\mu\rho}g_{\nu\eta}G_a^\mu G_b^\nu G_c^\rho G_d^\eta = -\frac{1}{4}g_s^2 f_{a'b'e'}f_{c'd'e'}g_{\mu'\rho'}g_{\nu'\eta'}G_{a'}^{\mu'}G_{b'}^{\nu'}G_{c'}^{\rho'}G_{d'}^{\eta'}.$$

Here, there are $4! = 24$ permutations. Because of the symmetry of structure constants, the four following configurations are equivalent:

$$\begin{pmatrix}\mu' & \nu' & \rho' & \eta' \\ a' & b' & c' & d'\end{pmatrix}=\begin{pmatrix}\mu & \nu & \rho & \eta \\ a & b & c & d\end{pmatrix}, \begin{pmatrix}\nu & \mu & \eta & \rho \\ b & a & d & c\end{pmatrix}, \begin{pmatrix}\rho & \eta & \mu & \nu \\ c & d & a & b\end{pmatrix}, \begin{pmatrix}\eta & \rho & \nu & \mu \\ d & c & b & a\end{pmatrix},$$

[21] Moving from the Lagrangian to the vertex factor, one has to multiply by i and by the number of permutations.

where $\begin{pmatrix} \mu' & \nu' & \rho' & \eta' \\ a' & b' & c' & d' \end{pmatrix} \equiv -\frac{1}{4} g_s^2 f_{a'b'e'} f_{c'd'e'} g_{\mu'\rho'} g_{\nu'\eta'}$. Therefore, among the 24 permutations, only 6 are really independent (but they count 4 times). They are

$$\begin{pmatrix} \mu & \nu & \rho & \eta \\ a & b & c & d \end{pmatrix}, \begin{pmatrix} \nu & \mu & \rho & \eta \\ b & a & c & d \end{pmatrix}, \begin{pmatrix} \rho & \nu & \mu & \eta \\ c & b & a & d \end{pmatrix}, \begin{pmatrix} \eta & \nu & \rho & \mu \\ d & b & c & a \end{pmatrix}, \begin{pmatrix} \mu & \rho & \eta & \nu \\ a & c & d & b \end{pmatrix}, \begin{pmatrix} \mu & \eta & \nu & \rho \\ a & d & b & c \end{pmatrix}.$$

Overall, the vertex factor is then

$$\begin{aligned} F_{4g} &= 4! \times i \times \frac{1}{24}\left(-\frac{1}{4}\right) g_s^2 [4 f_{abe} f_{cde} g_{\mu\rho} g_{\eta\nu} + 4 f_{bae} f_{cde} g_{\nu\rho} g_{\eta\mu} + 4 f_{cbe} f_{ade} g_{\rho\mu} g_{\eta\nu} \\ &\quad + 4 f_{dbe} f_{cae} g_{\eta\rho} g_{\mu\nu} + 4 f_{ace} f_{dbe} g_{\mu\eta} g_{\rho\nu} + 4 f_{ade} f_{bce} g_{\mu\nu} g_{\eta\rho}] \\ &= -i g_s^2 [f_{abe} f_{cde} (g_{\mu\rho} g_{\nu\eta} - g_{\mu\eta} g_{\nu\rho}) + f_{ade} f_{bce} (g_{\mu\nu} g_{\rho\eta} - g_{\mu\rho} g_{\nu\eta}) + f_{ace} f_{bde} (g_{\mu\eta} g_{\nu\rho} - g_{\mu\nu} g_{\rho\eta})], \end{aligned}$$

which is the factor indicated in Fig. 8.13.

The QCD Feynman rules derived in this section are valid for the tree diagrams, i.e. when there is no loop. In non-Abelian gauge theories, there are subtleties in the construction of the perturbation theory in covariant gauges. In particular, additional non-physical fields, called *Faddeev–Popov ghosts*, have to be introduced for the quantisation of massless gluon fields that suppress the non-transverse components of real gluons while preserving gauge invariance (a gauge fixing procedure). Those ghosts come with their Feynman rules, but they are only relevant when loops are considered (when a virtual gluon appears in a loop, its four polarisations are taken into account, and the ghosts compensate for the spurious modes). In this introductory book, we will ignore them, but the interested reader can consult a quantum field theory textbook, such as Peskin and Schroeder (1995) and Weinberg (1995).

8.3.2 Nucleon–Nucleon Strong Interaction

In the introductory chapter, we mentioned that at the scale of nucleons, the strong force is mediated by pion (or more generally by meson) exchange. However, we know that gluons

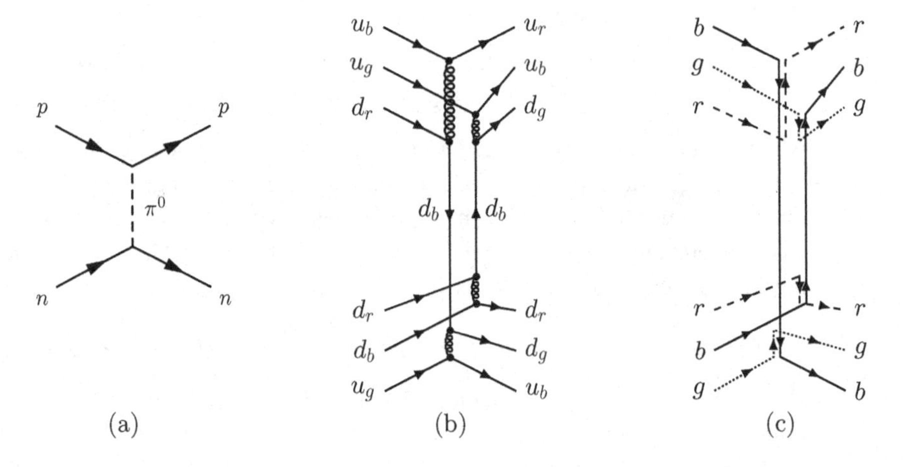

Fig. 8.14 The pion exchange between a proton and a neutron. (a) The effective pion exchange Feynman diagram. (b) A Feynman diagram with gluon exchanges contributing to the exchange of a pion (the two d_b vertical quark lines) between the nucleons. (c) The corresponding colour flow diagram with blue, green and red colours represented by the solid, dotted and dashed lines, respectively.

are the gauge bosons of QCD. How can we reconcile these two descriptions of the interaction? As an example, let us consider the proton–neutron scattering. Figure 8.14a shows the effective Feynman diagram with the neutral pion exchange. Since hadrons are colour neutral, they cannot exchange a single gluon. Moreover, colour confinement requires that only colour neutral objects can be exchanged at a long distance and thus between the two nucleons. Figure 8.14b shows a possible Feynman diagram at a more fundamental scale (there are many others), where a neutral pion (in the diagram, the pion is simply $d\bar{d}$ and not $(u\bar{u} - d\bar{d})/\sqrt{2}$) is exchanged between a proton and a neutron. Gluons are exchanged between quarks to ensure that the nucleons remain colour neutral, as can be checked with the help of the colour flow diagram in Fig. 8.14c.

8.3.3 Quark–Quark Interaction and Colour Factors

Let us consider the scattering between two quarks of different flavours, for instance, $u + d \to u + d$, as often encountered at protons colliders. The corresponding graph is

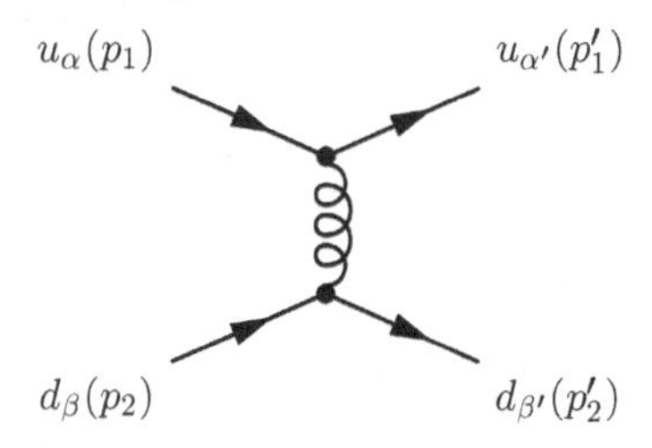

where α, β, α' and β' are the colour indices corresponding to the colour vectors c_α, c_β, $c_{\alpha'}$ and $c_{\beta'}$ defined in Eq. (8.60). The corresponding amplitude of the diagram is

$$i\mathcal{M} = \left[\bar{u}_u(p'_1)c^\dagger_{\alpha'}\left(-ig_s\gamma^\mu\tfrac{\lambda_a}{2}\right)c_\alpha u_u(p_1)\right]\left(-i\frac{g_{\mu\nu}\delta^{ab}}{(p_1-p'_1)^2}\right)$$
$$\left[\bar{u}_d(p'_2)c^\dagger_{\beta'}\left(-ig_s\gamma^\nu\tfrac{\lambda_b}{2}\right)c_\beta u_d(p_2)\right],$$

where there are implicit summations over the repeated colour indices a and b of the gluons, and the spinors of the u and d quarks are denoted u_u and u_d, respectively. Note that the eight gluons may contribute to the amplitude of this process. The amplitude simplifies to

$$\mathcal{M} = \frac{g_s^2}{(p_1 - p'_1)^2}C^{\alpha'\beta'}_{\alpha\,\beta}\left[\bar{u}_u(p'_1)\gamma^\mu u_u(p_1)\right]\left[\bar{u}_d(p'_2)\gamma_\mu u_d(p_2)\right], \tag{8.69}$$

where[22]

$$C^{\alpha'\beta'}_{\alpha\,\beta} = \frac{1}{4}\sum_{a=1}^{8} c^\dagger_{\alpha'}\lambda_a c_\alpha\, c^\dagger_{\beta'}\lambda_a c_\beta = \frac{1}{4}\sum_{a=1}^{8}(\lambda_a)_{\alpha'\alpha}\,(\lambda_a)_{\beta'\beta}. \tag{8.70}$$

Let us compare the amplitude to that of an equivalent scattering in QED, for instance, $e^- + \mu^- \to e^- + \mu^-$,

$$\mathcal{M} = \frac{e^2}{q^2}\left[\bar{u}_e(p'_1)\gamma^\mu u_e(p_1)\right]\left[\bar{u}_\mu(p'_2)\gamma_\mu u_\mu(p_2)\right]. \tag{8.71}$$

[22] The position of indices in $C^{\alpha'\beta'}_{\alpha\,\beta}$ is as follows: upper indices correspond to the first indices of Gell-Mann matrices, while the lower indices correspond to the second. In terms of group theory, upper indices transform as the **3** representation and lower as the $\bar{\mathbf{3}}$ representation (see Section C.3 in Appendix C).

We obviously have the change of the coupling constant $e \to g_s$ or equivalently the change of the fine structure constant α to

$$\alpha_s = \frac{g_s^2}{4\pi}, \tag{8.72}$$

and the change of the electric charges of the interacting particles (in elementary charge units, $Q_e = Q_\mu = -1$ in this example), for the colour factor

$$f^{(t)}_{c_\alpha c_\beta \to c_{\alpha'} c_{\beta'}} = C^{\alpha' \beta'}_{\alpha \beta}, \tag{8.73}$$

the label (t), referring to the colour factor in the t-channel. Hence, the difference between the fermion–fermion QED process and that of QCD can be summarised with the prescription

$$\boxed{\text{QED: } Q_1 Q_2 \alpha \quad \longleftrightarrow \quad \text{QCD: } f_c\, \alpha_s,} \tag{8.74}$$

with f_c being a generic colour factor [equal to Eq. (8.73) in our example]. The question is thus how to calculate the colour factor. As an example, imagine that the two quarks are red and conserve the same colours after the interaction, i.e. $u_r d_r \to u_r d_r$. According to the colour vectors (8.60), $\alpha = \beta = 1$ and $\alpha' = \beta' = 1$. Therefore, in the sum of Eq. (8.70), only $(\lambda_3)_{11}$ and $(\lambda_8)_{11}$ have a non-zero contribution [see the matrices in Eq. (8.34)]. In other words, only G_3 and G_8 in Eq. (8.52) can annihilate and create the red colour. The result of the sum (8.70) is

$$f^{(t)}_{rr \to rr} = C^{1\,1}_{1\,1} = \frac{1}{4}[(\lambda_1)_{11}(\lambda_1)_{11} + (\lambda_2)_{11}(\lambda_2)_{11}] = \frac{1}{3}.$$

The reader can easily check that we obtain the same result if we consider the two other colours, i.e. $u_g d_g \to u_g d_g$ or $u_b d_b \to u_b d_b$,

$$f^{(t)}_{rr \to rr} = C^{1\,1}_{1\,1} = f_{gg \to gg} = C^{2\,2}_{2\,2} = f_{bb \to bb} = C^{3\,3}_{3\,3} = 1/3. \tag{8.75}$$

This is expected by the colour symmetry. Now, envisage the case where *both* red quarks have their colour changed. For instance, $u_r d_r \to u_g d_g$. This time, $\alpha = \beta = 1$ and $\alpha' = \beta' = 2$. According to the colour factor (8.70), what matters are the Gell-Mann matrices satisfying $(\lambda_a)_{21} \neq 0$, i.e. $(\lambda_1)_{21} = 1$ and $(\lambda_2)_{21} = i$. Therefore, $C^{2\,2}_{1\,1} = 1^2 + i^2 = 0$ and the amplitude is 0. Again, this is expected by the colour conservation: starting from a red initial state, the final state cannot be green and must stay red. However, if we start with two quarks having a different colour, they may swap their colours, as in $u_r d_g \to u_g d_r$. In this example, the colour factor is

$$f^{(t)}_{rg \to gr} = C^{2\,1}_{1\,2} = \frac{1}{4}[(\lambda_1)_{21}(\lambda_1)_{12} + (\lambda_2)_{21}(\lambda_2)_{12}] = \frac{1}{2},$$

and the exchanged gluons associated with λ_1 and λ_2 are $\bar{g}r$ and $\bar{r}g$. By colour symmetry, this colour factor is also valid for

$$C^{2\,1}_{1\,2} = C^{1\,2}_{2\,1} = C^{3\,1}_{1\,3} = C^{1\,3}_{3\,1} = C^{3\,2}_{2\,3} = C^{2\,3}_{3\,2} = 1/2. \tag{8.76}$$

What if the two quarks have different colours and conserve the same colour? For instance, $u_r d_g \to u_r d_g$. This time, the colour factor is

$$f^{(t)}_{rg \to rg} = C^{1\,2}_{1\,2} = \frac{1}{4}[(\lambda_3)_{11}(\lambda_3)_{22} + (\lambda_8)_{11}(\lambda_8)_{22}] = -\frac{1}{6}.$$

which, by symmetry, is also

$$C_{1\,2}^{1\,2} = C_{2\,1}^{2\,1} = C_{1\,3}^{1\,3} = C_{3\,1}^{3\,1} = C_{2\,3}^{2\,3} = C_{3\,2}^{3\,2} = -1/6. \tag{8.77}$$

The comparisons of Eqs. (8.75), (8.76) and (8.77) are interesting: it shows that the strength of the colour force between two quarks depends on their colour. It may be counterintuitive for colour symmetry. However, when we combine the colour of two quarks, we have seen that we can obtain a sextet of colours whose states are

$$rr,\ (rg+gr)/\sqrt{2},\ gg,\ (gb+bg)/\sqrt{2},\ bb,\ (rb+br)/\sqrt{2},$$

or an anti-triplet, with the states

$$(rg-gr)/\sqrt{2},\ (gb-bg)/\sqrt{2},\ (rb-br)/\sqrt{2}$$

(see Fig. 8.6 and replace the flavours with colours). In the sextet, the colour conservation by QCD imposes that $rr \to rr$ and $(rg+gr)/\sqrt{2} \to (rg+gr)/\sqrt{2}$, etc. The colour factor $f_{rr\to rr}$ has already been calculated in Eq. (8.75), and it is $1/3$. For the other factor, we can simply write

$$f^{(t)}_{\frac{rg+gr}{\sqrt{2}} \to \frac{rg+gr}{\sqrt{2}}} = \frac{1}{2}\left(f^{(t)}_{rg\to rg} + f^{(t)}_{rg\to gr} + f^{(t)}_{gr\to rg} + f^{(t)}_{gr\to gr}\right) = \frac{1}{2}\left(-\frac{1}{6}+\frac{1}{2}+\frac{1}{2}-\frac{1}{6}\right) = \frac{1}{3},$$

where the intermediate colour factors have already been calculated in Eqs. (8.76) and (8.77). Hence, we find the same colour factors for all states of the sextet, which is consistent with the colour symmetry: all states of a multiplet must have the same colour factor. As an exercise, the reader can calculate the colour factor of the anti-triplet (Problem 8.7).

It is worth mentioning that the colour factor (8.70) can be calculated once and for all (see Problem 8.8), yielding the colour *Fierz identity*

$$C_{\alpha\,\beta}^{\alpha'\beta'} = \frac{1}{4}\sum_{a=1}^{8}(\lambda_a)_{\alpha'\alpha}\,(\lambda_a)_{\beta'\beta} = \frac{1}{2}\left(\delta_{\alpha'\beta}\delta_{\alpha\beta'} - \frac{1}{3}\delta_{\alpha'\alpha}\delta_{\beta'\beta}\right). \tag{8.78}$$

It shows that necessarily, the colour factors are real numbers (it is a consequence of the Hermiticity of the Gell-Mann matrices). The reader can check the previous calculations with this formula.

Once, the amplitude of a process is determined with its specific colour factor, the cross section is calculated by averaging and summing the amplitude squared over the degrees of freedom. In QCD, besides the spin, the colour degree of freedom must be taken into account. Since the two quarks can have three different colours, the colour summing/averaging implies

$$\overline{|\mathcal{M}|^2} = \frac{1}{3}\frac{1}{3}\sum_{\alpha,\alpha',\beta,\beta'}|\mathcal{M}|^2 = \left(\frac{1}{9}\sum_{\alpha,\alpha',\beta,\beta'}|C_{\alpha\,\beta}^{\alpha'\beta'}|^2\right)\overline{|\mathcal{M}_0|^2},$$

where $\mathcal{M}_0$ is the part of the amplitude that does not depend on the colour degrees of freedom (the QED-like amplitude). Equation (8.78) implies that

$$\begin{aligned}\sum_{\alpha,\alpha',\beta,\beta'}|C_{\alpha\,\beta}^{\alpha'\beta'}|^2 &= \frac{1}{4}\left(\sum_{\alpha\beta}\delta_{\beta\beta}\delta_{\alpha\alpha} + \frac{1}{9}\sum_{\alpha\beta}\delta_{\alpha\alpha}\delta_{\beta\beta} - \frac{2}{3}\sum_{\alpha}\delta_{\alpha\alpha}\right)\\ &= \frac{1}{4}\left(3\times 3 + \frac{1}{9}\times 3\times 3 - \frac{2}{3}\times 3\right) = 2.\end{aligned} \tag{8.79}$$

(In a term such as $\sum_{\alpha,\alpha',\beta,\beta'} \delta_{\alpha'\beta}\delta_{\alpha\beta'}\delta_{\alpha'\beta}\delta_{\alpha\beta'}$, the Kronecker symbols impose $\alpha' = \beta$ and $\beta' = \alpha$, reducing the summation to $\sum_{\alpha,\beta} \delta_{\beta\beta}\delta_{\alpha\alpha} = \sum_{\alpha} \delta_{\alpha\alpha} \sum_{\beta} \delta_{\beta\beta} = 3\times 3$.) An alternative proof is proposed in Problem 8.9. Therefore, once the QED-like average amplitude squared is known (i.e., replacing the gluon with the photon), one multiplies it by $(\alpha_s/\alpha)^2 \times f_c$, with $f_c = 2/9$ to obtain the QCD cross section. In Section 6.4.3, the spin-averaged amplitude squared has been calculated for the process $e^- + \mu^- \to e^- + \mu^-$. Neglecting the particle masses, Eq. (6.95) gives

$$\overline{|\mathcal{M}|^2_{e\mu}} = 2e^4 \frac{s^2 + u^2}{t^2} = 32\pi^2\alpha^2 \frac{s^2 + u^2}{t^2}.$$

The QED spin-averaged amplitude squared of $u+d \to u+d$ is thus $Q_u^2 Q_d^2 \overline{|\mathcal{M}|^2_{e\mu}}$, where Q_u and Q_d are the quark electric charges. Since, for QCD, one has to change $Q_u^2 Q_d^2 \alpha \to f_c \alpha_s^2$, the QCD amplitude squared is

$$\overline{|\mathcal{M}|^2_{ud}} = \frac{2}{9}\, 32\pi^2\alpha_s^2 \frac{s^2 + u^2}{t^2}, \tag{8.80}$$

yielding, using Eqs. (3.44) and (3.34), the differential cross section

$$\frac{d\sigma_{ud}}{dt} = \frac{1}{16\pi s^2} \overline{|\mathcal{M}|^2_{ud}} = \frac{2}{9} \frac{2\pi\alpha_s^2}{s^2} \frac{s^2 + u^2}{t^2}. \tag{8.81}$$

8.3.4 Quark–Quark Interaction of the Same Flavour

When the two quarks are of the same flavour, say $u + u \to u + u$, there are two QCD diagrams to take into account

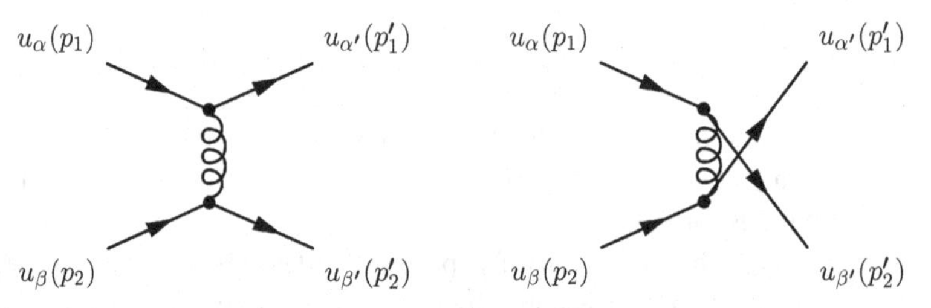

Let us call $\mathcal{M}_t$ and $\mathcal{M}_u$ the amplitudes of the first and second diagrams, respectively. The letters u and t refer to the Mandelstam variables used in the gluon propagator. Because of the crossing of the two outgoing quarks, the two amplitudes must be subtracted since quarks are fermions. Thus, to obtain the QCD cross section, we must now calculate

$$\overline{|\mathcal{M}|^2} = \overline{|\mathcal{M}_t - \mathcal{M}_u|^2} = \overline{|\mathcal{M}_t|^2} + \overline{|\mathcal{M}_u|^2} - 2\Re(\overline{\mathcal{M}_t\mathcal{M}_u^*}).$$

The term $\overline{|\mathcal{M}_t|^2}$ has already been calculated in the previous section in Eq. (8.80) with the colour-averaged colour factor $\frac{1}{9}\sum |C_{\alpha\beta}^{\alpha'\beta'}|^2 = 2/9$. The second term, $\overline{|\mathcal{M}_u|^2}$, can be deduced from the first by simply swapping the variables t and u. The colour factor is the same since, for the second diagram, what matters is

$$f^{(u)}_{c_\alpha c_\beta \to c_{\alpha'} c_{\beta'}} = \frac{1}{4} \sum_{a=1}^{8} (\lambda_a)_{\beta'\alpha} (\lambda_a)_{\alpha'\beta} = C_{\alpha\beta}^{\beta'\alpha'},$$

and according to the Formula (8.79), $\sum |C^{\alpha'\beta'}_{\alpha\beta}|^2 = \sum |C^{\beta'\alpha'}_{\alpha\beta}|^2 = 2$. The last term must be, however, calculated. It does not present any particular difficulties. As an exercise, the reader can check that in QED, this interference term (as an example take the reaction $e^- + e^- \to e^- + e^-$) is

$$-2\Re(\overline{\mathcal{M}_t \mathcal{M}_u^*}) = 64\pi^2\alpha^2 \frac{s^2}{tu}.$$

The calculation of the corresponding colour factor in QCD deserves some caution. It reads

$$C_{tu} = \frac{1}{9} \sum_{\alpha,\alpha',\beta,\beta'} C^{\alpha'\beta'}_{\alpha\beta} \left(C^{\beta'\alpha'}_{\alpha\beta}\right)^* = \frac{1}{9} \sum_{\alpha,\alpha',\beta,\beta'} C^{\alpha'\beta'}_{\alpha\beta} C^{\beta'\alpha'}_{\alpha\beta}.$$

Fortunately, the Formula (8.78) helps us in this task. The coefficient C_{tu} reads

$$\begin{aligned}
C_{tu} &= \frac{1}{36} \sum_{\alpha,\alpha',\beta,\beta'} \left(\delta_{\alpha'\beta}\delta_{\alpha\beta'} - \frac{1}{3}\delta_{\alpha'\alpha}\delta_{\beta'\beta}\right)\left(\delta_{\beta'\beta}\delta_{\alpha\alpha'} - \frac{1}{3}\delta_{\beta'\alpha}\delta_{\alpha'\beta}\right) \\
&= \frac{1}{36}\left(\sum_{\alpha}\delta_{\alpha\alpha} - \frac{1}{3}\sum_{\alpha,\beta}\delta_{\alpha\alpha}\delta_{\beta\beta} - \frac{1}{3}\sum_{\alpha,\beta}\delta_{\alpha\alpha}\delta_{\beta\beta} + \frac{1}{9}\sum_{\alpha}\delta_{\alpha\alpha}\right) \\
&= \frac{1}{36}\left(3 - \frac{2}{3} \times 3 \times 3 + \frac{1}{9} \times 3\right) \\
&= -2/27.
\end{aligned}$$

Therefore, putting everything together, the QCD cross section is [see Eq. (8.81)]

$$\frac{d\sigma}{dt} = \frac{2\pi\alpha_s^2}{s^2}\left(\frac{2}{9}\frac{s^2+u^2}{t^2} + \frac{2}{9}\frac{s^2+t^2}{u^2} - \frac{4}{27}\frac{s^2}{tu}\right).$$

8.3.5 Quark–Antiquark Interaction

Instead of a quark pair, let us consider the scattering of a quark and an antiquark of different flavours, for instance, $u + \bar{d} \to u + \bar{d}$. The corresponding graph is

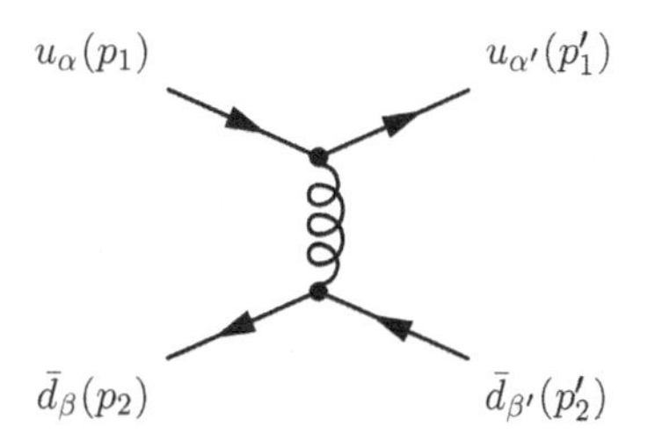

Note that only the t-channel is possible. An s-channel can only couple a quark and an antiquark of the same flavour since the gluon field in $\mathcal{L}^{\text{QCD}}_{\text{int}}$ in Eq. (8.63) couples to the same quark field q (with possibly different colours). The corresponding amplitude of the diagram is

$$\begin{aligned}
i\mathcal{M} = &\left[\bar{u}_u(p_1')c^\dagger_{\alpha'}\left(-ig_s\gamma^\mu\frac{\lambda_a}{2}\right)c_\alpha u_u(p_1)\right]\left(-i\frac{g_{\mu\nu}\delta^{ab}}{(p_1-p_1')^2}\right) \\
&\left[\bar{v}_d(p_2)c^\dagger_\beta\left(-ig_s\gamma^\nu\frac{\lambda_b}{2}\right)c_{\beta'}v_d(p_2')\right],
\end{aligned}$$

reducing to

$$\mathcal{M} = \frac{g_s^2}{(p_1 - p_1')^2}\frac{1}{4}(\lambda_a)_{\alpha'\alpha}(\lambda_a)_{\beta\beta'}\left[\bar{u}_u(p_1')\gamma^\mu u_u(p_1)\right]\left[\bar{v}_d(p_2)\gamma_\mu v_d(p_2')\right]. \tag{8.82}$$

Labelling the anti-colour vectors with a bar below (to avoid the confusion with the colour of quarks), this time, the colour factor is

$$f^{(t)}_{c_\alpha \underline{c}_\beta \to c_{\alpha'} \underline{c}_{\beta'}} = \frac{1}{4}\sum_{a=1}^{8}(\lambda_a)_{\alpha'\alpha}\,(\lambda_a)_{\beta\beta'} = C^{\alpha'\,\beta}_{\alpha\,\beta'}. \tag{8.83}$$

Note the inversion of the β and β' index positions with respect to the particle case in Eq. (8.73). Whereas for quarks the index of the incoming colour is the lower index of the C tensor, for antiquarks it is the upper index. Hence, the tensor $C^{\alpha'\,\beta}_{\alpha\,\beta'}$ also corresponds to the colour factor for particles

$$C^{\alpha'\,\beta}_{\alpha\,\beta'} = f^{(t)}_{c_\alpha \underline{c}_\beta \to c_{\alpha'} \underline{c}_{\beta'}} = f^{(t)}_{c_\alpha c_{\beta'} \to c_{\alpha'} c_\beta}.$$

More generally, for processes with a gluon exchange in the t-channel, the same C-tensor and thus the same colour factor corresponds to four situations:

$$f^{(t)}_{c_\alpha c_\beta \to c_{\alpha'} c_{\beta'}} = f^{(t)}_{\underline{c}_{\alpha'} c_\beta \to \underline{c}_\alpha c_{\beta'}} = f^{(t)}_{c_\alpha \underline{c}_{\beta'} \to c_{\alpha'} \underline{c}_\beta} = f^{(t)}_{\underline{c}_{\alpha'} \underline{c}_{\beta'} \to \underline{c}_\alpha \underline{c}_{\beta'}} = C^{\alpha'\,\beta'}_{\alpha\,\beta}. \tag{8.84}$$

Hence, the same gluons are exchanged for all these process. As an example, let us consider the case where the exchanged gluons are $\bar{g}r$ and $\bar{r}g$, i.e. $f^{(t)}_{rg\to gr} = f^{(t)}_{\bar{g}g\to\bar{r}r} = f^{(t)}_{r\bar{r}\to g\bar{g}} = f^{(t)}_{\bar{g}\bar{r}\to\bar{r}\bar{g}}$. The corresponding colour flow diagrams are shown below.

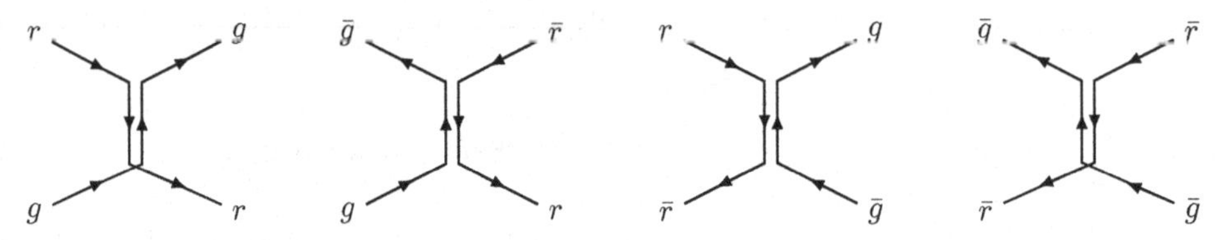

When the quark and the antiquark are of the same flavour, say u and $\bar{u}$, then the s-channel is allowed. The Feynman diagram is

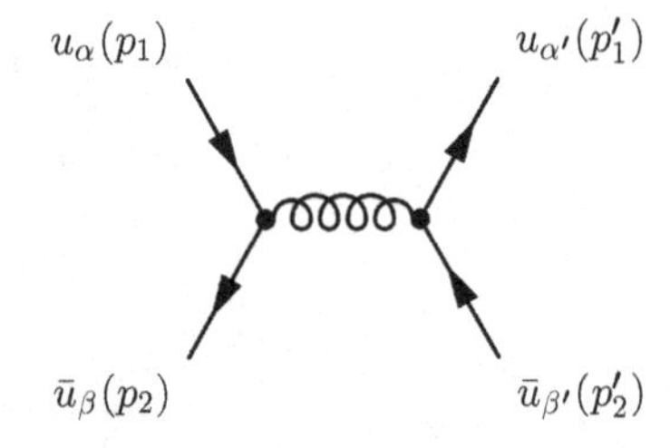

from which we deduce the amplitude:

$$i\mathcal{M} = \left[\bar{v}_u(p_2)c^\dagger_\beta\left(-ig_s\gamma^\mu\frac{\lambda_a}{2}\right)c_\alpha u_u(p_1)\right]\left(-i\frac{g_{\mu\nu}\delta^{ab}}{(p_1+p_2)^2}\right)$$
$$\left[\bar{u}_d(p_1')c^\dagger_{\alpha'}\left(-ig_s\gamma^\nu\frac{\lambda_b}{2}\right)c_{\beta'}v_d(p_2')\right].$$

Therefore, the colour factor in the s-channel is

$$f^{(s)}_{c_\alpha \underline{c}_\beta \to c_{\alpha'} \underline{c}_{\beta'}} = \frac{1}{4}\sum_a(\lambda_a)_{\beta\alpha}(\lambda_a)_{\alpha'\beta'} = C^{\beta\,\alpha'}_{\alpha\,\beta'}, \tag{8.85}$$

which is not equal to the colour factor of the t-channel (8.83), except when all colour indices are equal. However, the colour-averaged colour factor $\frac{1}{9}\sum |C^{\alpha'\beta'}_{\alpha\,\beta}|^2 = 2/9$. More generally, for reactions involving quarks or antiquarks in the initial and final states, whatever the channel t, u or s, the colour-averaged colour factor is always

$$\frac{1}{9}\sum_c |f_c^{(t)}|^2 = \frac{1}{9}\sum_c |f_c^{(u)}|^2 = \frac{1}{9}\sum_c |f_c^{(s)}|^2 = \frac{2}{9}, \tag{8.86}$$

where c is a generic colour configuration.

8.3.6 Interactions with Gluons in the Initial or Final States

As an example of such interactions, we consider now the quark–gluon scattering, i.e. the reaction $q + g \to q + g$. This process is very similar to the Compton scattering shown in Section 6.4.5, with its two Feynman diagrams, but in QCD, there is a third possible diagram due to the 3-gluon interaction, as shown in Fig. 8.15.
Applying the QCD Feynman rules, the amplitude of the first diagram is

$$i\mathcal{M}_1 = \bar{u}_{s'}(p')c^{\dagger}_{\alpha'}\left(-ig_s\gamma^{\mu}\tfrac{\lambda_{a'}}{2}\right)\epsilon^*_{\mu}(k',\lambda')\; c_{\beta'} i\tfrac{\not p+\not k+m}{(p+k)^2-m^2}$$
$$\delta_{\beta\beta'}c^{\dagger}_{\beta}\left(-ig_s\gamma^{\nu}\tfrac{\lambda_a}{2}\right)c_{\alpha}u_s(p)\epsilon_{\nu}(k,\lambda),$$

reducing to

$$\mathcal{M}_1 = g_s^2\frac{1}{4}\,(\lambda_{a'})_{\alpha'\beta'}\,(\lambda_a)_{\beta\alpha}\,\delta_{\beta\beta'}\left\{-\epsilon^*_{\mu}(k',\lambda')\epsilon_{\nu}(k,\lambda)\,\bar{u}_{s'}(p')\gamma^{\mu}\,\frac{\not p+\not k+m}{s-m^2}\gamma^{\nu}u_s(p)\right\}, \tag{8.87}$$

where s is the Mandelstam variable $s = (p+k)^2$ in the propagator, and, as usual, there is an implicit summation over the repeated colour indices β and β' of the intermediate quark. The quantity in the curly brace is the QED amplitude (up to the QED coupling constant and the electric charge of the quark). Therefore, squaring and averaging/summing this amplitude over the spin degrees of freedom gives the result already established in Eq. (6.104).

To obtain the colour-averaged colour factor, we need to average and sum over the eight gluon colours and the three quark colours, i.e.

$$f^c_{1,1} = \frac{1}{8}\frac{1}{3}\sum_{a,a'}\sum_{\alpha\alpha'}\frac{1}{4}\,(\lambda_{a'})_{\alpha'\beta'}\,(\lambda_a)_{\beta\alpha}\,\delta_{\beta\beta'}\frac{1}{4}\,(\lambda_{a'})^*_{\alpha'\delta'}\,(\lambda_a)^*_{\delta\alpha}\,\delta_{\delta\delta'}$$
$$= \frac{1}{8}\frac{1}{3}\delta_{\beta\beta'}\delta_{\delta\delta'}\sum_{a,\alpha}\frac{1}{4}\,(\lambda_a)_{\beta\alpha}\,(\lambda_a)_{\alpha\delta}\sum_{a',\alpha'}\frac{1}{4}\,(\lambda_{a'})_{\delta'\alpha'}\,(\lambda_{a'})_{\alpha'\beta'},$$

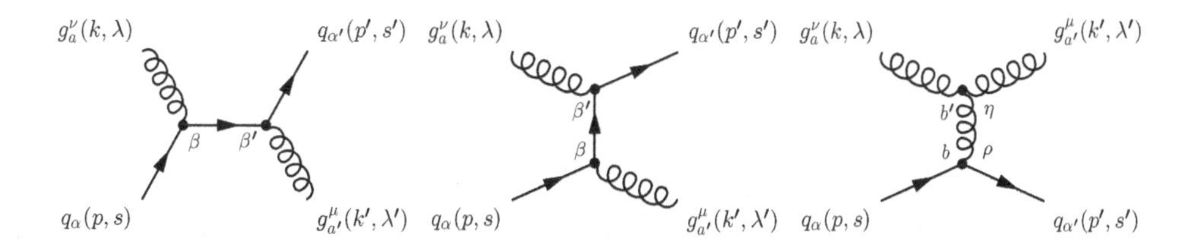

Fig. 8.15 The lowest order diagram of the gluon scattering. The third diagram is allowed by the trilinear gluon coupling.

where the last line follows from the Gell-Mann matrices being Hermitian. Fierz's identity (8.78) allows us to evaluate

$$\frac{1}{4}\sum_{\alpha}\sum_{a}(\lambda_a)_{\beta\alpha}(\lambda_a)_{\alpha\delta}=\sum_{\alpha}\frac{1}{2}\left(\delta_{\beta\delta}\delta_{\alpha\alpha}-\frac{1}{3}\delta_{\beta\alpha}\delta_{\alpha\delta}\right)=\frac{1}{2}\left(3-\frac{1}{3}\right)\delta_{\beta\delta}=C_F\delta_{\beta\delta}, \tag{8.88}$$

with

$$C_F = 4/3. \tag{8.89}$$

Therefore, the colour-averaged colour factor is (making explicit the summation)

$$f^c_{1,1}=\frac{1}{8}\frac{1}{3}C_F^2\sum_{\beta,\beta',\delta,\delta'}\delta_{\beta\beta'}\delta_{\delta\delta'}\delta_{\beta\delta}\delta_{\beta'\delta'}=\frac{1}{8}C_F^2=\frac{2}{9}. \tag{8.90}$$

The amplitude of the second diagram is

$$\mathcal{M}_2=g_s^2\frac{1}{4}(\lambda_{a'})_{\beta\alpha}(\lambda_a)_{\alpha'\beta'}\delta_{\beta\beta'}\left\{-\epsilon^*_\mu(k',\lambda')\epsilon_\nu(k,\lambda)\,\bar{u}_{s'}(p')\gamma^\nu\frac{\not p-\not k'+m}{u-m^2}\gamma^\mu u_s(p)\right\}, \tag{8.91}$$

where $u=(p-k')^2$. Note that in Eq. (8.91), the colour indices are interchanged between the two Gell-Mann matrices with respect to the first amplitude in Eq. (8.87). The colour-averaged colour factor gives the same result as in Eq. (8.90), i.e. $f^c_{2,2}=f^c_{1,1}=2/9$.

Finally, there is the scattering amplitude for the triple gluon interaction, i.e. the third diagram in Fig. 8.15. With the labels indicated in the figure, the propagator momentum is $q=p-p'=k'-k$, and thus q^2 is the Mandelstam variable t. To simplify the notation, let us denote $\epsilon(k,\lambda)\equiv\epsilon$ and $\epsilon(k',\lambda')\equiv\epsilon'$. Then, the amplitude reads

$$\begin{aligned}i\mathcal{M}_3=\bar{u}_{s'}(p')c^\dagger_{\alpha'}\left(-ig_s\gamma^\rho\frac{\lambda_b}{2}\right)c_\alpha u_s(p)\left(-i\frac{g_{\rho\eta}}{t}\delta^{bb'}\right)\epsilon'^*_\mu\epsilon_\nu\\ \times\left(-g_sf_{b'aa'}[g^{\eta\nu}(q-k)^\mu+g^{\nu\mu}(k+k')^\eta+g^{\mu\eta}(-k'-q)^\nu]\right).\end{aligned}$$

Keep in mind that the 3-gluon vertex factor given in Fig. 8.13 assumes that the 4-momenta point into the vertex, hence the inversion of the sign in front of k' in the formula above. Since physical gluons have transverse polarisations, $\epsilon'_\mu k'^\mu=\epsilon_\nu k^\nu=0$, the expression of the amplitude simplifies to

$$\mathcal{M}_3=-if_{baa'}\frac{1}{2}(\lambda_b)_{\alpha'\alpha}\left\{g_s^2\bar{u}_{s'}(p')\gamma^\rho u_s(p)\frac{1}{t}[\epsilon'^*\cdot\epsilon\,(k+k')_\rho-2(k\cdot\epsilon'^*)\epsilon_\rho-2(k'\cdot\epsilon)\epsilon'_\rho]\right\}. \tag{8.92}$$

The colour-averaged colour factor associated with $\overline{|\mathcal{M}_3|^2}$ then reads (from now on, there is no implicit summation on repeated indices)

$$f^c_{3,3}=\frac{1}{8}\frac{1}{3}\sum_{b,a,a'}f_{baa'}f_{baa'}\sum_{\alpha\alpha'}\frac{1}{4}(\lambda_b)_{\alpha',\alpha}\left(\lambda_b^*\right)_{\alpha'\alpha}.$$

The sum over the quark colour gives

$$\frac{1}{4}\sum_{\alpha,\alpha'}(\lambda_b)_{\alpha'\alpha}\left(\lambda^*_b\right)_{\alpha'\alpha}=\frac{1}{4}\sum_{\alpha\alpha'}(\lambda_b)_{\alpha'\alpha}(\lambda_b)_{\alpha\alpha'}=\frac{1}{4}\sum_{\alpha'}(\lambda_b\lambda_b)_{\alpha'\alpha'}=\frac{1}{4}\mathrm{Tr}(\lambda_b\lambda_b)=\frac{1}{2}\delta_{bb}.$$

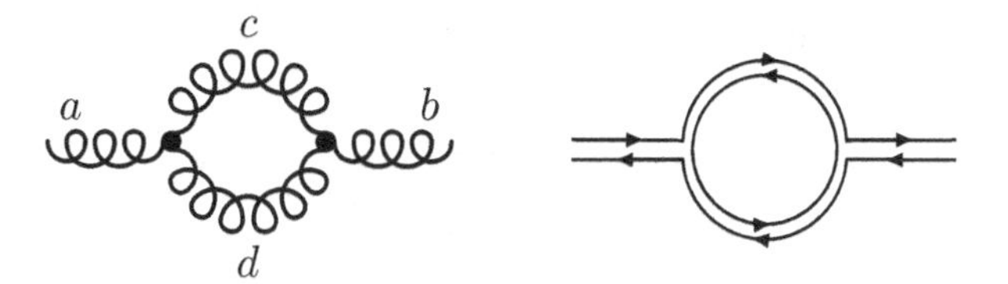

Fig. 8.16 Feynman diagram with the gluon loop (left) and the corresponding colour flow (right).

It follows that

$$f^c_{3,3} = \frac{1}{8}\frac{1}{3}\frac{1}{2}\sum_b \sum_{a,a'} f_{baa'} f_{baa'}.$$

To evaluate $f^c_{3,3}$, the following formula is useful:

$$\sum_{c,d} f_{acd} f_{bcd} = C_A \delta_{ab}, \quad \text{with} \quad C_A = 3. \tag{8.93}$$

This relation results directly from the $\mathfrak{su}(3)$ algebra and can be understood as follows. The gluon loop in the Feynman diagram of Fig. 8.16 has the colour factor $\sum_{c,d} f_{acd} f_{bcd}$. It is necessarily proportional to δ_{ab}, the colour factor of a free gluon without a loop (as in the propagator). To determine the proportionality constant, consider the colour flow in Fig. 8.16. Since the colour must be conserved between the initial gluon and the final one, there can be only three different colours for the most internal colour loop. Therefore, $C_A = 3$. Equipped with the Formula (8.93), we thus conclude

$$f^c_{3,3} = \frac{1}{8}\frac{1}{3}\frac{1}{2} C_A \sum_b \delta_{bb} = \frac{1}{3}\frac{1}{2} C_A = \frac{1}{2}. \tag{8.94}$$

The calculation of the cross section requires evaluating

$$\overline{|\mathcal{M}_1 + \mathcal{M}_2 + \mathcal{M}_3|^2} = \overline{|\mathcal{M}_1|^2} + \overline{|\mathcal{M}_2|^2} + \overline{|\mathcal{M}_3|^2} + 2\Re\left(\overline{\mathcal{M}_1\mathcal{M}_2^*} + \overline{\mathcal{M}_1\mathcal{M}_3^*} + \overline{\mathcal{M}_2\mathcal{M}_3^*}\right).$$

Denoting by $\tilde{\mathcal{M}}$ the part of the amplitude that does not depend on the colours, Eqs. (6.104)–(6.106) allow us to conclude (just change the coupling constant e for g_s)

$$\overline{|\tilde{\mathcal{M}}_1|^2} = -2g_s^4 \frac{u}{s}, \quad \overline{|\tilde{\mathcal{M}}_2|^2} = -2g_s^4 \frac{s}{u}, \quad 2\Re(\overline{\tilde{\mathcal{M}}_1\tilde{\mathcal{M}}_2^*}) = 0. \tag{8.95}$$

The calculations of $\overline{|\tilde{\mathcal{M}}_3|^2}$ and the interference terms $\overline{\tilde{\mathcal{M}}_1\tilde{\mathcal{M}}_3^*}$ and $\overline{\tilde{\mathcal{M}}_2\tilde{\mathcal{M}}_3^*}$ are lengthy. I simply give the result,

$$\overline{|\tilde{\mathcal{M}}_3|^2} = -4g_s^4 \frac{su}{t^2}, \quad 2\Re(\overline{\tilde{\mathcal{M}}_1\tilde{\mathcal{M}}_3^*}) = 4g_s^4 \frac{t+s}{t}, \quad 2\Re(\overline{\tilde{\mathcal{M}}_2\tilde{\mathcal{M}}_3^*}) = -4g_s^4 \frac{t+u}{t}. \tag{8.96}$$

The final step is to multiply the terms in Eqs. (8.95) and (8.96) by their specific average colour factors. We have already calculated those associated with $\overline{|\tilde{\mathcal{M}}_1|^2}$, $\overline{|\tilde{\mathcal{M}}_2|^2}$ and $\overline{|\tilde{\mathcal{M}}_3|^2}$, i.e. $f^c_{1,1}$, $f^c_{2,2}$ and $f^c_{3,3}$, respectively. What is missing are the colour factors associated with the interferences, $f^c_{1,3}$ and $f^c_{2,3}$. Looking at the expression of $\mathcal{M}_1$ and $\mathcal{M}_3$ in Eqs. (8.87) and (8.92), the colour factor $f^c_{1,3}$ reads

$$f^c_{1,3} = \frac{1}{8}\frac{1}{3}\sum_{a,a',b}\sum_{\alpha,\alpha',\beta} \frac{1}{4} (\lambda_{a'})_{\alpha'\beta} (\lambda_a)_{\beta\alpha} \left(-if_{baa'}\frac{1}{2}(\lambda_b)_{\alpha'\alpha}\right)^* = \frac{1}{8}\frac{1}{3}\sum_{a,a',b} if_{baa'} \frac{1}{2^3} \mathrm{Tr}\,(\lambda_{a'}\lambda_a\lambda_b).$$

It can be calculated by considering the following relation proved in Problem 8.10:

$$\sum_{a,b} f_{abd} \frac{1}{2^3} \mathrm{Tr}\,(\lambda_a \lambda_b \lambda_c) = \frac{3i}{4} \delta_{cd}. \tag{8.97}$$

Since $f_{baa'} = -f_{a'ab}$, it follows that

$$f^c_{1,3} = \frac{1}{8}\frac{1}{3}\frac{3}{4} \sum_b \delta_{bb} = \frac{1}{4}. \tag{8.98}$$

We find, similarly, $f^c_{2,3} = -f^c_{1,3} = -1/4$. Gathering all the pieces, i.e. Eqs. (8.90), (8.94)–(8.96) and (8.98), we finally obtain the gluon-scattering cross section (using $g_s^4 = 16\pi^2\alpha_s^2$):

$$\frac{d\sigma}{dt} = \frac{1}{16\pi s^2} \overline{|\mathcal{M}|^2} = \frac{\pi\alpha_s^2}{s^2}\left(\frac{2}{9}\left[-2\frac{u}{s} - 2\frac{s}{u}\right] + \frac{1}{2}\left[-4\frac{su}{t^2}\right] + \frac{1}{4}\left[4\frac{t+s}{t}\right] - \frac{1}{4}\left[-4\frac{t+u}{t}\right]\right).$$

Given that we neglect the quark mass, $s + t + u = 0$, the cross section simplifies to

$$\frac{d\sigma}{dt} = \frac{\pi\alpha_s^2}{s^2}\left(-\frac{4}{9}\frac{u^2 + s^2}{su} + \frac{u^2 + s^2}{t^2}\right). \tag{8.99}$$

The cross section for the processes $q + \bar{q} \to g + g$ and $g + g \to q + \bar{q}$ can be deduced from the cross section (8.99) by crossing $s \leftrightarrow t$, adding an overall minus sign to take into account the change of side of one fermion, and by correcting for the different colour averaging of the initial state $\left(\frac{1}{3}\frac{1}{3}\right.$ and $\frac{1}{8}\frac{1}{8}$, respectively, instead of $\frac{1}{3}\frac{1}{8}$ in $\left. q + g \to q + g\right)$.

8.3.7 Measurement of Colour Factors

The colour factors calculated in the previous sections result directly from the SU(3) colour gauge group. It is then fundamental to measure and compare them with the theory. Most QCD processes can be decomposed as a succession of basic processes, such as a quark radiating a gluon, $q \to qg$, a gluon splitting into two gluons, $g \to gg$, or a gluon splitting into a quark–antiquark pair $g \to q\bar{q}$. The colour factors associated with these three basic processes are denoted as C_F, C_A and T_F,[23] respectively. Due to confinement, we never observe individual colours. Therefore, we always have to average the squared amplitude $\mathcal{M}\mathcal{M}^*$ over incoming colours and sum over outgoing colours. Moreover, the colours of intermediate particles are always summed over. The complex conjugate of the amplitude $\mathcal{M}^* = \mathcal{M}^\dagger$ may be interpreted as a process where incoming particles become outgoing ones and vice versa.[24] Consequently, for the process, $q \to qg$, the colour-averaged factor C_F resulting from the squared amplitude is the colour factor of the diagram on the right-hand side in Fig. 8.17. Similarly, the colour factors C_A and T_F are those in the diagrams in Fig. 8.18. We have already calculated C_F and C_A in Eqs. (8.88) and (8.93), respectively,

[23] In group theory, C_F is the Casimir invariant of the fundamental representation, i.e. the eigenvalue of a Casimir operator that commutes with all generators. Similarly, C_A is the Casimir invariant of the adjoint representation. In both cases, the Casimir operator is $T_a T_a$ (implicit sum), where T_a ($a = 1, \ldots, 8$) are the generators of the representation (equal to $\lambda_a/2$ in the fundamental representation). The other parameter, T_F, is the index of the fundamental representation, corresponding to the normalisation of T_a in the fundamental representation.

[24] Simply because $\mathcal{M} \propto \langle f|i\rangle$ and $\mathcal{M}^* \propto \langle i|f\rangle$. For instance, the basic vertex factor in a QED process, $\bar{u}(p')\gamma^\mu u(p)$, becomes $[\bar{u}(p')\gamma^\mu u(p)]^* = \bar{u}(p)\gamma^\mu u(p')$, and thus, the initial incoming fermion described by the spinor $u(p)$ becomes $\bar{u}(p)$, which can be interpreted as an outgoing fermion.

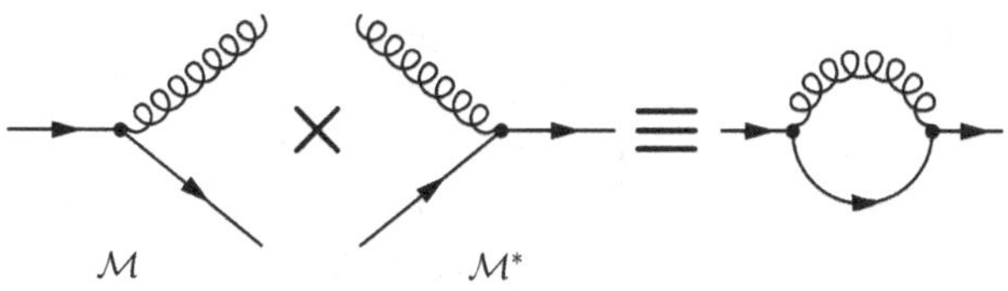

Fig. 8.17 Illustration of the equivalence between the amplitude squared and the diagram on the right.

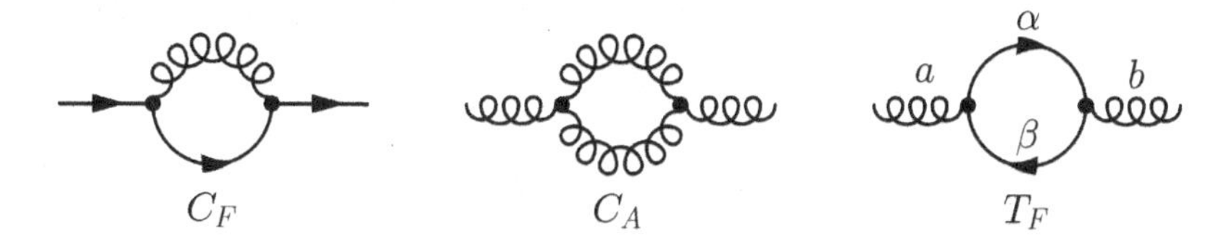

Fig. 8.18 The basic Feynman diagrams with average colour factors C_F, C_A and T_F.

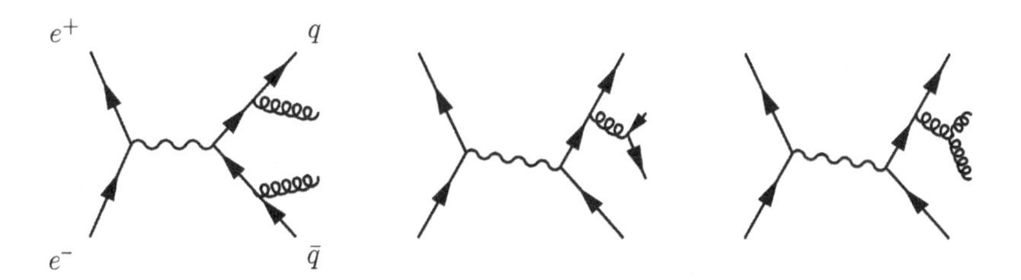

Fig. 8.19 Production of events containing four jets in the final state at e^+e^- colliders.

giving $C_F = 4/3$ and $C_A = 3$. For T_F, with the colour indices indicated in the diagram on the right-hand side of Fig. 8.18, the average colour factor is

$$T_F = \frac{1}{8} \sum_{a,b} \sum_{\alpha,\beta=1}^{3} \frac{1}{4} (\lambda_a)_{\alpha\beta} (\lambda_b)_{\beta\alpha} = \frac{1}{8} \sum_{a,b} \frac{1}{4} \mathrm{Tr}\,(\lambda_a \lambda_b) = \frac{1}{8} \sum_{a,b} \frac{1}{2} \delta_{ab} = \frac{1}{8} \frac{1}{2} \sum_{a} \delta_{aa} = \frac{1}{2}, \tag{8.100}$$

where the relation (8.35) has been used.

Experimentally, several approaches exist to measure these colour factors. For example, consider the production of four jets (of hadrons) at the e^+e^- colliders. The relevant graphs are shown in Fig. 8.19. The first graph involves C_F (twice), the second C_F and T_F and the last C_F and C_A. Moreover, the angular correlations between the four jets differ in these three graphs: in terms of spins, $q \to q+g$ results in a transition $1/2 \to 1/2+1$, $g \to g+g$ results in a transition $1 \to 1+1$ and $g \to q+\bar{q}$ results in a transition $1 \to 1/2+1/2$. Therefore, the measurement of the jet angular distribution gives a handle to access C_F, C_A and T_F. Figure 8.20 shows the values estimated for C_F and C_A from a compilation of data collected at the LEP collider e^+e^- at CERN. The combined result is shown with the solid ellipse. Different models of the gauge group are compared, and only SU(3), for which $C_F = 4/3$ and $C_A = 3$ (depicted by the solid star in the figure), is compatible with the data.

The fact that the values of C_F and C_A differ significantly has interesting experimental consequences. Since $C_A > C_F$, there should be more radiation of soft gluons in a gluon jet than in a quark jet. Consequently, gluon jets have a larger multiplicity than quark jets, and thus, hadrons in gluon jets tend to have a softer energy spectrum. We also expect gluon jets to be less collimated than quark jets. This is indeed verified experimentally, and this property is used to distinguish quark and gluon jets on a statistical basis.

8.4 Asymptotic Freedom and the Running of α_S

There is an apparent paradox in QCD: on the one hand, we have learned how to calculate QCD amplitudes, based on perturbation theory, and thus under the implicit assumption that the strong coupling constant must be weak enough to provide valid calculations. Furthermore, in Chapter 7, where the parton model was presented, we always assumed for the calculations of the deep inelastic scattering cross sections that partons (namely quarks and gluons) inside hadrons do not interact together as if they were free. This assumption was valid since the parton model, in general, compares well with experimental data in

Supplement 8.4. Diagrammatic calculation of average colour factors

Instead of calculating colour factors with index notation, it is convenient to calculate them diagrammatically. Our previous calculations showed some equivalences:

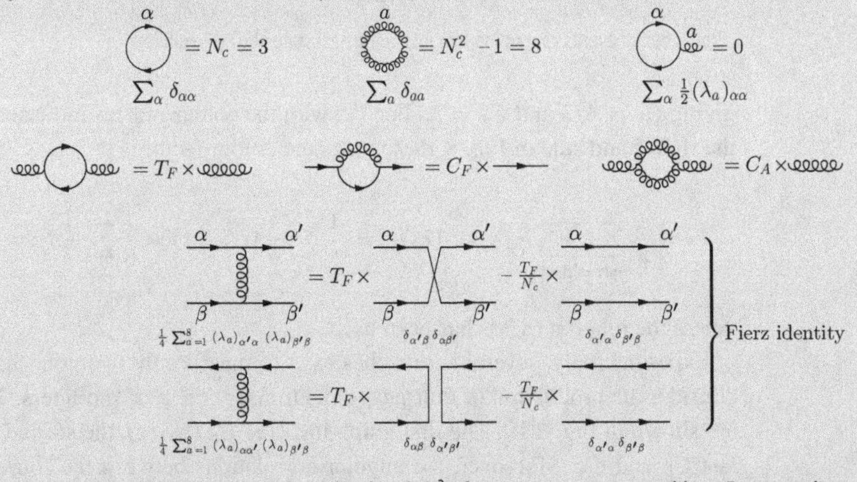

α $= N_c = 3$ $\quad$ $\sum_\alpha \delta_{\alpha\alpha}$

a $= N_c^2 - 1 = 8$ $\quad$ $\sum_a \delta_{aa}$

α a $= 0$ $\quad$ $\sum_\alpha \frac{1}{2}(\lambda_a)_{\alpha\alpha}$

$= T_F \times$ $\qquad$ $= C_F \times$ $\qquad$ $= C_A \times$

α α' β β' $= T_F \times$ α α' β β' $- \frac{T_F}{N_c} \times$ α α' β β'

$\frac{1}{4}\sum_{a=1}^{8} (\lambda_a)_{\alpha'\alpha} (\lambda_a)_{\beta'\beta}$ $\qquad$ $\delta_{\alpha'\beta}\delta_{\alpha\beta'}$ $\qquad$ $\delta_{\alpha'\alpha}\delta_{\beta'\beta}$ $\qquad$ Fierz identity

$= T_F \times$ $\qquad$ $- \frac{T_F}{N_c} \times$

$\frac{1}{4}\sum_{a=1}^{8} (\lambda_a)_{\alpha\alpha'} (\lambda_a)_{\beta'\beta}$ $\qquad$ $\delta_{\alpha\beta}\delta_{\alpha'\beta'}$ $\qquad$ $\delta_{\alpha'\alpha}\delta_{\beta'\beta}$

In Fig. 8.17, we saw that the squared amplitude $|\mathcal{M}|^2$ of a process is represented by a Feynman diagram since the Hermitian conjugation corresponds to mirroring the diagram across a vertical line and reversing all arrows. By connecting the initial state to the final state, the sum over the colour indices becomes implicit since the colours of intermediate particles are always summed over. The idea is then to simplify the obtained diagram thanks to the equivalences above. For example, the calculation of the colour factor of the quark–antiquark scattering of Section 8.3.5 gives, thanks to the Fierz identity equivalence,

Supplement 8.4. (cont.)

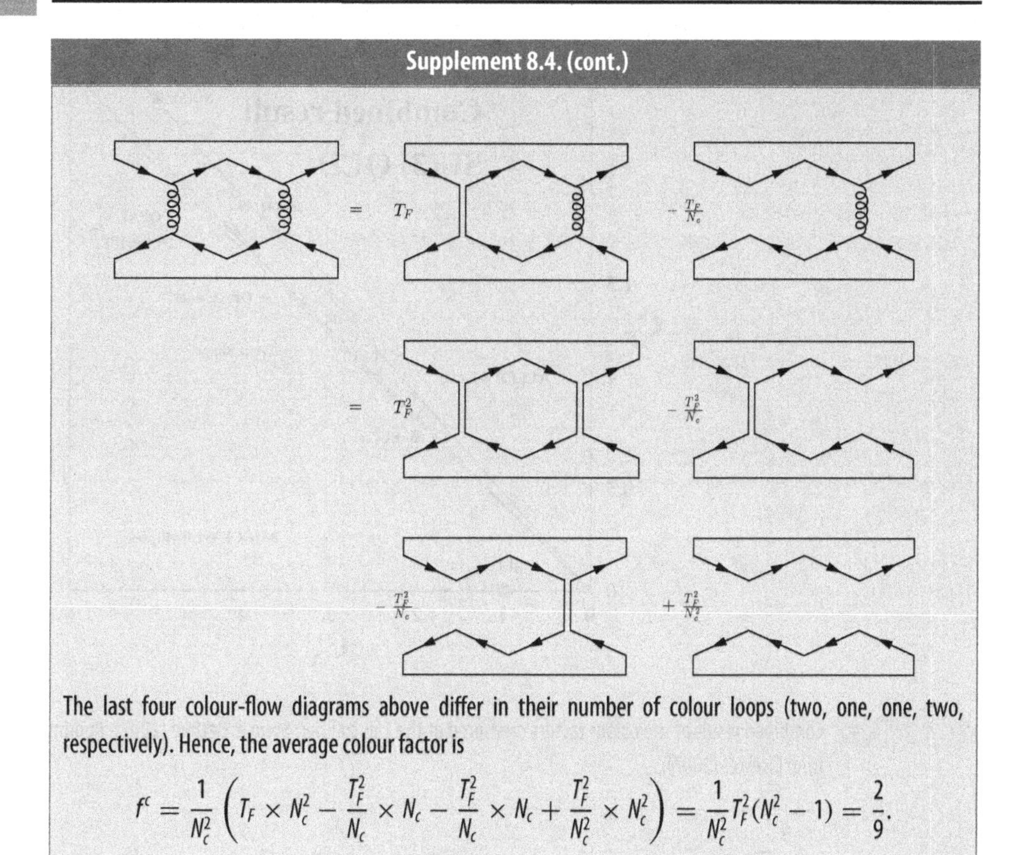

The last four colour-flow diagrams above differ in their number of colour loops (two, one, one, two, respectively). Hence, the average colour factor is

$$f^c = \frac{1}{N_c^2}\left(T_F \times N_c^2 - \frac{T_F^2}{N_c} \times N_c - \frac{T_F^2}{N_c} \times N_c + \frac{T_F^2}{N_c^2} \times N_c^2\right) = \frac{1}{N_c^2} T_F^2 (N_c^2 - 1) = \frac{2}{9}.$$

high-energy scatterings. On the other hand, we have admitted, and it has never been contradicted by the experimental facts, that QCD should be very strong to bind (anti-)quarks together such that no coloured objects can be observed as free particles. Therefore, the QCD coupling constant must be very large. The resolution of this paradox relies on a QCD property related to its non-Abelian nature, called *asymptotic freedom*. This was found by Gross, Politzer and Wilczek in 1973, who were awarded the Nobel prize in 2004 for their discovery.

In QED, we saw in Section 6.6, that the renormalisation procedure leads to the concept of a running coupling constant (and mass). In QCD, we have the same situation. The QCD vacuum polarisation (the gluon propagator) receives contributions from diagrams such as those in Fig. 8.21.

The first diagram with the fermion loop is analogous to that of QED in Fig. 6.11. It contributes, as in QED, to a screening effect. However, the last two diagrams with gluon self-interactions are specific to QCD and have the opposite effect. More specifically, the evolution equation for the QCD coupling, $\alpha_s(Q^2)$, is governed by the perturbative series[25]

$$\frac{\partial \alpha_s}{\partial \ln(Q^2)} = \beta(\alpha_s), \quad \text{with} \quad \beta(\alpha_s) = -\alpha_s^2(b_0 + \alpha_s b_1 + \alpha_s^2 b_2 + \cdots). \tag{8.101}$$

[25] The derivation of this formula is beyond the scope of this textbook. The interested reader can consult, for example, Peskin and Schroeder (1995, chapter 16).

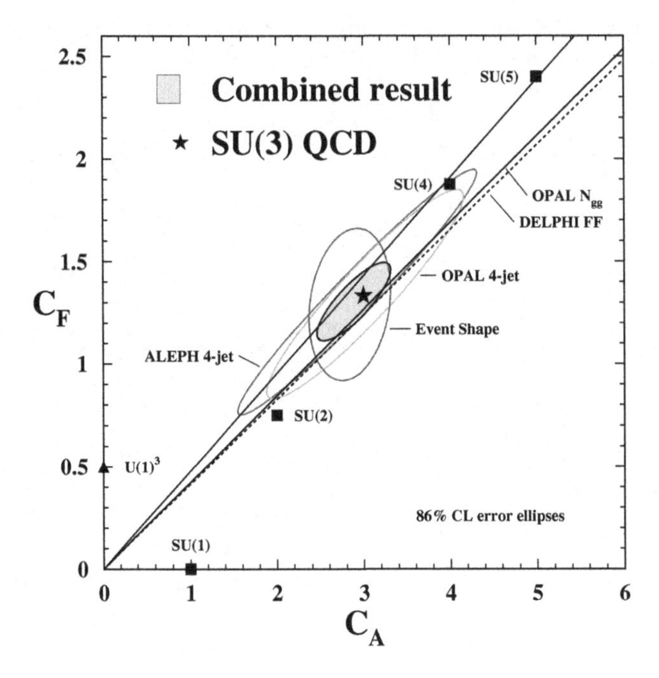

Fig. 8.20 Combined result of the colour factors measured at the LEP collider. Source: Bethke (2007). Reprinted with permission from Elsevier (2007).

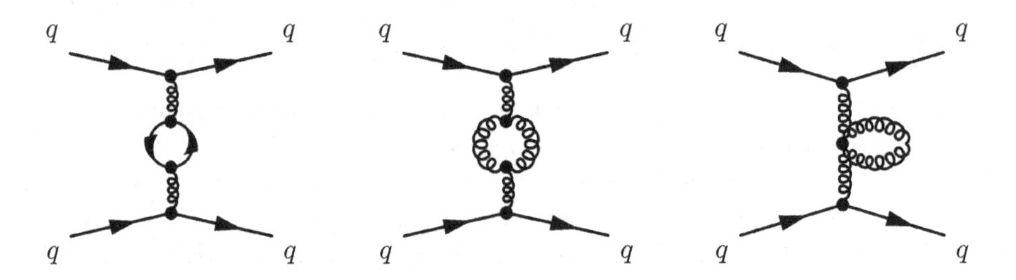

Fig. 8.21 One-loop contribution to QCD vacuum polarisation.

If α_s is small enough, one can keep only the leading-order term corresponding to one-loop contributions in diagrams such as Fig. 8.21. The leading-order coefficient, b_0, is

$$b_0 = \frac{11C_A - 4N_f T_F}{12\pi}, \tag{8.102}$$

where $C_A = 3$ and $T_F = 1/2$ are, as usual, the colour factors in Eqs. (8.93) and (8.100), while N_f is the number of quark flavours with a mass below a certain scale Q_0^2. Qualitatively, T_F originates from the first diagram in Fig. 8.21, while C_A is coming from the second (see Fig. 8.18). The last diagram turns out to have zero contribution to the coefficient b_0 of the β-function in the massless approximation. We can easily integrate Eq. (8.101) and using Eq. (8.102), thus obtaining the one-loop calculation of the coupling constant (often referred to as the *leading log approximation*)

$$\alpha_s(Q^2) = \frac{\alpha_s(Q_0^2)}{1 + \alpha_s(Q_0^2)\frac{(33-2N_f)}{12\pi}\ln\left(\frac{Q^2}{Q_0^2}\right)}. \tag{8.103}$$

If the value of α_s is known at the scale Q_0^2, Eq. (8.103) gives us its value at another scale, Q^2. In QED, we found in Eq. (6.117), a similar form of the fine structure constant. Notice that from α_s, we obtain α simply by setting $C_A = 0$ and $T_F = N_f = 1$, which is consistent with the fact that photons do not have self-coupling ($C_A = 0$).[26] There is, however, a major difference: since, at most, $N_f = 6$, the term in front of the logarithm in Eq. (8.103) is positive, whereas it is negative in QED (for all N_f). Therefore, when Q^2 becomes larger, α_s decreases, whereas α increases. For very large values of Q^2, α_s tends to zero, leading to a quasi-free behaviour of partons. This is *asymptotic freedom*, allowed because C_A in the β-function [see Eq. (8.102)] is positive in QCD, a consequence of the colour and anti-colour carried by the gluon gauge bosons.

It is instructive to reformulate Eq. (8.103) in terms of an energy scale, $Q^2 = \Lambda^2_{\text{QCD}}$, for which α_s becomes infinite, yielding

$$\alpha_s(Q^2) = \frac{12\pi}{(33-2N_f)\ln\left(\frac{Q^2}{\Lambda^2_{QCD}}\right)}, \quad \text{with} \quad \Lambda_{\text{QCD}} = Q_0 \exp\left(-\frac{6\pi}{(33-2N_f)\alpha_s(Q_0^2)}\right). \tag{8.104}$$

This expression is valid only for $Q^2 > \Lambda^2_{\text{QCD}}$ at a one-loop calculation. Of course, an infinite coupling constant is not physically possible. One should take into account the higher order terms in the β-function (8.101). However, the scale Λ_{QCD} gives an idea where the strong interaction becomes really strong and where the perturbative approach breaks down. It is a universal constant of QCD in the sense that starting from Eq. (8.103), we obtain

$$\frac{1}{\alpha_s(Q^2)} - \frac{33-2N_f}{12\pi}\ln Q^2 = \frac{1}{\alpha_s(Q_0^2)} - \frac{33-2N_f}{12\pi}\ln Q_0^2 = -\frac{33-2N_f}{12\pi}\ln \Lambda^2_{\text{QCD}},$$

implying that Λ_{QCD} is necessarily a constant that does not depend on Q^2 or Q_0^2. Starting from a measurement of α_s typically around the τ lepton mass, $m_\tau \simeq 1.78$ GeV$/c^2$, where $\alpha_s(m_\tau^2) \simeq 0.33$, one can extract a value of Λ_{QCD}. Using Eq. (8.104), it gives a one-loop result of about $\Lambda_{\text{QCD}} \simeq 200$ MeV (using $N_f = 4$ for the four quarks lighter than the τ lepton). In terms of the distance, it is equivalent to 1 fm, which is the typical size of hadrons.

How can we evaluate the running of α_s? Well, as for α in QED, we can measure α_s with processes involving various energy scales. For instance, by measuring the cross section of jet production as a function of the jet momentum transfer. The cross section of four-jets events in Fig. 8.19 is directly proportional to α_s^2 and thus provides sensitive measurements of the coupling constant at high energy. At a low energy scale, the hadronic decays of the τ lepton and the deep inelastic processes provide other measurements.

[26] In QED, one can take the reference scale Q_0 as low as possible since α does not diverge at 0. Therefore, the usual reference is $Q_0 = m_e$, and thus $N_f = 1$. Moreover, $T_F = 1$, the factor 1/2 in QCD coming from a different normalisation of the coupling constant, i.e. $\text{Tr}(T_a^2) = 1/2$, while $\text{Tr}(Q^2) = 1$, where T_a and Q are the generators of QCD and QED, respectively.

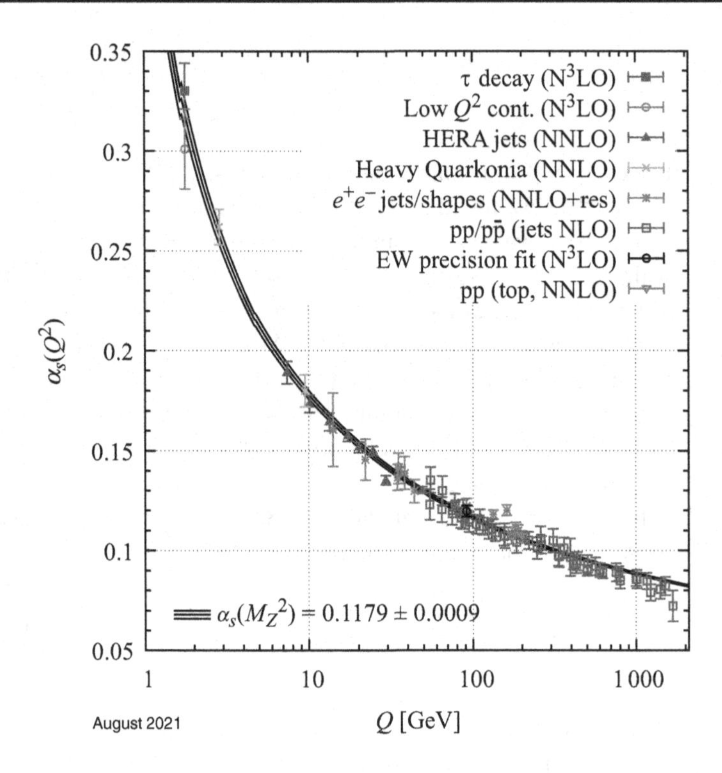

Fig. 8.22 Summary of α_s measurements as a function of the energy scale. Source: Workman et al. (Particle Data Group) (2022).

Figure 8.22 presents a compilation of many measurements at various scales. The data are compared with the running of α_s (the curve with its error band), taking as a reference the world average value of α_s at $Q_0 = m_Z \simeq 91$ GeV. Using a version of the β-function involving more loops (at least four) than in Eq. (8.103), the agreement is remarkable. The figure shows two small discontinuities in the curve at about $Q = 1.5$ and 4.5 GeV. This is due to the change in the number of flavours, N_f, when the charm and beauty quark mass thresholds are crossed. Notice that $\alpha_s(m_Z) \simeq 0.12$ is still 15 times larger than α at that mass ($\alpha(m_Z) \simeq 1/128$).

8.5 Non-perturbative QCD

If QCD is the correct theory of the strong interaction, i.e. if the Lagrangian (8.57) is correct, one should be able to calculate quantities such as the hadron masses, the effective potential felt by quarks, describe how quarks and gluon are confined in hadrons, etc. Those quantities often involve an energy scale below 1 GeV (for instance, meson masses made of light quarks). We have seen in the previous section that when the energy scale approaches $\Lambda_{\text{QCD}} \sim 200$ MeV, the coupling constant α_s becomes large (at about 1 GeV, it is already of the order of 0.5). Therefore, in this regime, the perturbative approach becomes useless since neglected terms can be of the same order of magnitude as the terms used in the calculation, and non-perturbative methods must be employed.

8.5.1 A Rough Idea of Lattice Gauge Theory

In 1974, Kenneth G. Wilson invented a new approach to quantise a gauge theory on a discrete lattice while preserving the gauge invariance on the lattice (Wilson, 1974). Using this approximation, Wilson showed that when α_s is strong enough, QCD exhibits confinement of colour, i.e. the only finite-energy states are those that are singlets of colour. Lattice QCD is an example of a gauge theory treated on a lattice, i.e. treated in a spacetime that is taken not to be a continuum, but rather a collection of lattice points. The lattice spacing, a, then introduces a minimum distance (in spacetime) between two neighbouring points. We restrict our discussion to (hyper-)cubic lattice in four dimensions, with a being the side of the cube. The idea is then to perform the calculations for various values of a and extrapolate the results to the continuum limit, where $a \to 0$. If a is small enough, i.e. smaller than the inverse of all relevant energy and momentum scales, say a fraction of fm, the results should not depend too much on a. This approach has the advantage of naturally introducing a maximum value of the momenta, π/a (see later), providing a regularisation for divergences at high momenta (ultraviolet divergences).

QCD in Euclidean Space

As we shall see, numerical calculations on the lattice are much easier in the Euclidean space, where the four dimensions are all equivalent, than in the Minkowski space. Coordinates in the Euclidean space (denoted with a tilde symbol below) are obtained by introducing a Wick rotation with $t = -i\tilde{t}$, i.e.

$$\tilde{x} = (\tilde{t}, x, y, z) = (it, x, y, z) = (\tilde{x}^0, \tilde{x}^1, \tilde{x}^2, \tilde{x}^3) = (\tilde{x}_0, \tilde{x}_1, \tilde{x}_2, \tilde{x}_3). \tag{8.105}$$

By construction, the scalar product in the Euclidean space is

$$\tilde{x} \cdot \tilde{x} = \tilde{x}^\mu \tilde{x}_\mu = \tilde{x}^\mu \tilde{x}^\mu = \tilde{x}_\mu \tilde{x}_\mu = \tilde{t}^2 + |\boldsymbol{x}|^2 = -x^\mu x_\mu. \tag{8.106}$$

Since in the Euclidean space, $\tilde{x}^\mu$ and $\tilde{x}_\mu$ denote the same quantity, there is no reason to distinguish covariant and contravariant coordinates. From now on, we will only use lower indices, $\tilde{x}_\mu$, to avoid possible confusion with the Minkowski space. Repeated indices in Euclidean space are, as usual, summed over. Obviously, for any two 4-vectors x^μ and y^μ, we have

$$\tilde{x}_\mu \tilde{y}_\mu = \tilde{x}_0 \tilde{y}_0 + \boldsymbol{x} \cdot \boldsymbol{y} = -x^\mu y_\mu. \tag{8.107}$$

Equation (8.105) implies for the action

$$S = \int \mathrm{d}^4 x\, \mathcal{L} = -i \int \mathrm{d}^4 \tilde{x}\, \mathcal{L}. \tag{8.108}$$

Similarly, Eq. (8.105) implies for the derivatives

$$\partial_0 = \frac{\partial}{\partial t} = i\frac{\partial}{\partial \tilde{t}} = i\tilde{\partial}_0,$$
$$\partial_i = \frac{\partial}{\partial x^i} = \tilde{\partial}_i. \tag{8.109}$$

The Dirac Lagrangian now reads for quark fields

$$\mathcal{L}_{\rm D} = \bar{q}\left(i\gamma^0\partial_0 + i\gamma^i\partial_i - m_q\right)q = \bar{q}\left(-\gamma^0\tilde{\partial}_0 + i\gamma^i\tilde{\partial}_i - m_q\right)q = -\bar{q}\left(\tilde{\gamma}_\mu\tilde{\partial}_\mu + m_q\right)q, \tag{8.110}$$

where the Dirac matrices in the Euclidean space are

$$\tilde{\gamma}_0 = \gamma^0, \quad \tilde{\gamma}_i = -i\gamma^i, \quad \{\tilde{\gamma}_\mu, \tilde{\gamma}_\nu\} = 2\delta_{\mu\nu}. \tag{8.111}$$

The QCD interaction term (8.54) and the gauge Lagrangian depend on G^a_μ. In QED, the field A_μ can be made a pure gauge field, $A_\mu(x) = \partial_\mu\alpha(x)$. Therefore, in the Euclidean space, it is conventional to have the gauge fields following the same transformation as the derivatives in Eq. (8.109), i.e.

$$G^a_0 = i\tilde{G}^a_0, \quad G^a_i = \tilde{G}^a_i = -G^{ai}. \tag{8.112}$$

This choice also has the advantage of preserving the structure of the covariant derivative. It implies for the tensor $G^a_{\mu\nu}$, defined in Eq. (8.56), that

$$G^a_{00} = -\tilde{G}^a_{00} = G^{a00}, \quad G^a_{0i} = i\tilde{G}^a_{0i} = -G^{a0i}, \quad G^a_{ij} = \tilde{G}^a_{ij} = G^{aij}. \tag{8.113}$$

To simplify the notations, let us define the 3×3 matrices

$$G_\mu(x) = \frac{\lambda_a}{2}G^a_\mu(x). \tag{8.114}$$

Then, with Eqs. (8.111) and (8.112), the interaction term (8.54) in the Euclidean space reads

$$\mathcal{L}_{\rm int} = -g_s\bar{q}(\gamma^0 G_0 + \gamma^i G_i)q = -ig_s\bar{q}\tilde{\gamma}_\mu\tilde{G}_\mu q. \tag{8.115}$$

For the gauge Lagrangian, according to Eq. (8.113),

$$\mathcal{L}_{\rm gauge} = -\frac{1}{4}G^{\mu\nu}_a G^a_{\mu\nu} = -\frac{1}{4}\tilde{G}^a_{\mu\nu}\tilde{G}^a_{\mu\nu}. \tag{8.116}$$

Finally, the three Lagrangians (8.110), (8.115) and (8.116), when combined with Eq. (8.108), yield the QCD action expressed with Euclidean quantities

$$S_{\rm QCD} = i\tilde{S}_{\rm QCD} = i\int {\rm d}^4\tilde{x}\left(\bar{q}(\tilde{\gamma}^\mu\tilde{\rm D}_\mu + m_q)q + \frac{1}{4}\tilde{G}^a_{\mu\nu}\tilde{G}^a_{\mu\nu}\right), \tag{8.117}$$

with the covariant derivative in the Euclidean space given by

$$\tilde{\rm D}_\mu = \tilde{\partial}_\mu + ig_s\tilde{G}_\mu. \tag{8.118}$$

Since $\tilde{\rm D}_\mu$ is expressed with the matrix $\tilde{G}_\mu$, it is convenient to express $\tilde{S}_{\rm QCD}$ only with matrices. Let us define

$$\tilde{G}_{\mu\nu} = \frac{\lambda_a}{2}\tilde{G}^a_{\mu\nu}. \tag{8.119}$$

As $\mathrm{Tr}(\lambda_a\lambda_b) = \delta_{ab}/2$, it follows that $\mathrm{Tr}(\tilde{G}_{\mu\nu}\tilde{G}_{\mu\nu}) = \tilde{G}^a_{\mu\nu}\tilde{G}^a_{\mu\nu}/2$. Hence,

$$\tilde{S}_{\mathrm{QCD}} = \int \mathrm{d}^4\tilde{x}\left(\bar{q}(\tilde{\gamma}^{\mu}\tilde{\mathrm{D}}_{\mu} + m_q)q + \frac{1}{2}\mathrm{Tr}(\tilde{G}_{\mu\nu}\tilde{G}_{\mu\nu})\right). \tag{8.120}$$

QCD on the Lattice

One of the main difficulties of the lattice approach is to discretise the QCD action (to allow numerical calculations) while still preserving the gauge invariance, whatever the lattice spacing. A position on a hypercubic lattice is represented by

$$\tilde{x} \xrightarrow{\text{lattice}} \tilde{x} = a(n_0^x, n_1^x, n_2^x, n_3^x), \tag{8.121}$$

where the n_i^xs are four integers, ranging from 0 to $N-1$ (assuming for simplicity that the four dimensions have the same lattice spacing a). The integration on the lattice reads

$$\int \mathrm{d}^4\tilde{x} \xrightarrow{\text{lattice}} a^4 \sum_{\tilde{x}}. \tag{8.122}$$

The differentiation of a generic field $\varphi(\tilde{x})$ is calculated on the lattice with the symmetric difference

$$\tilde{\partial}_{\mu}\varphi(\tilde{x}) \xrightarrow{\text{lattice}} \frac{1}{2a}\left[\varphi(\tilde{x}+a\hat{\mu}) - \varphi(\tilde{x}-a\hat{\mu})\right], \tag{8.123}$$

where $\hat{\mu}$ is the unit vector in the μ direction,

$$\hat{\mu}_i = \delta_{\mu i}. \tag{8.124}$$

The fields can be expressed in the momentum space via the Fourier transform. In such a case, imposing the boundary condition $\varphi(\tilde{x}+L) = \varphi(\tilde{x})$, with $L = aN$, the momentum is quantised with values $\tilde{p}_i = 2\pi k_i/L$, where k_i $(i = 0, 1, 2, 3)$ is an integer ranging within $(-N/2, \ldots, 0, \ldots N/2)$ (assuming N is even). Therefore, the maximum value of momenta on the lattice is $|\tilde{p}| = \pi N/L = \pi/a$, eliminating ultraviolet divergences.

Now, let us address the delicate question of the gauge invariance on the lattice. We have learned that replacing the usual derivative with the covariant derivative ensures the gauge invariance of the theory in the continuum. How can we translate this recipe into the lattice language? For this purpose, consider the operator in the Minkowski space

$$\begin{aligned} U_{\mathcal{C}}(x,y) &= \mathcal{P}\left\{\exp\left(-ig_s \oint_y^x \mathrm{d}z^{\mu}\, G_{\mu}(z)\right)\right\} \\ &= \sum_{n=0}^{\infty}\frac{(-ig_s)^n}{n!}\mathcal{P}\left\{\oint_y^x \mathrm{d}z_n^{\mu_n}\ldots\mathrm{d}z_1^{\mu_1}\, G_{\mu_n}(z_n)\ldots G_{\mu_1}(z_1)\right\}. \end{aligned} \tag{8.125}$$

This object is called a *Wilson line* or *parallel transporter*. As explained in Supplement 8.5, it transports the colour transformation law of the quark field at the point y to that of the field at the point x. The integration is performed over a curve $\mathcal{C}$ (or path), starting at y and ending at x. The symbol $\mathcal{P}$ indicates a path-ordered quantity (see Supplement 8.5 for the details), ensuring the transitivity property to the Wilson line,[27] i.e.

$$U_{C_2+C_1}(x,y) = U_{C_2}(x,z)U_{C_1}(z,y), \tag{8.126}$$

where the notation $C_2 + C_1$ means a path passing first through C_1 and then through C_2 (the operator reads from right to left). The term $\mathrm{d}z^\mu G_\mu = \mathrm{d}z^0 G_0 + \mathrm{d}z^i G_i$ in the Wilson line (8.125) has the same expression as in the Euclidean space, $\mathrm{d}\tilde{z}_\mu \tilde{G}_\mu$, because of Eq. (8.112). It follows that

$$\tilde{U}_C(\tilde{x},\tilde{y}) = \mathcal{P}\left\{\exp\left(-ig_s \oint_{\tilde{y}}^{\tilde{x}} \mathrm{d}\tilde{z}_\mu\, \tilde{G}_\mu(\tilde{z})\right)\right\}. \tag{8.127}$$

On the lattice, the nearest neighbour of $\tilde{x}$ in the μ direction is $\tilde{y} = \tilde{x} + a\hat{\mu}$, with $\hat{\mu}$ defined in Eq. (8.124). Since a is small, the Wilson line becomes

$$\tilde{U}_\mu(\tilde{x}) = \tilde{U}_C(\tilde{x},\tilde{x}+a\hat{\mu}) = \exp\left(ig_s a\tilde{G}_\mu(\tilde{x})\right) \simeq \mathbb{1} + ig_s a\tilde{G}_\mu(\tilde{x}) + \cdots, \tag{8.128}$$

where the dots hide higher order terms in the lattice spacing, a. This function is called a *link* (for the gauge link or link variable). On the lattice, it connects the site $\tilde{x}$ to the site $\tilde{x} + a\hat{\mu}$. Now, according to Supplement 8.5, the covariant derivative in the Euclidean space takes the form [see Eq. (8.134)]

$$\hat{n}^\mu \tilde{\mathrm{D}}_\mu q(\tilde{x}) = \lim_{\epsilon\to 0} \frac{\tilde{U}_C(\tilde{x},\tilde{x}+\epsilon\hat{n})q(\tilde{x}+\epsilon\hat{n}) - \tilde{U}_C(\tilde{x},\tilde{x}-\epsilon\hat{n})q(\tilde{x}-\epsilon\hat{n})}{2\epsilon}.$$

Note that on the lattice, $\tilde{U}^\dagger_\mu(\tilde{x}-a\hat{\mu}) = \tilde{U}^\dagger_C(\tilde{x}-a\hat{\mu},\tilde{x}) = \tilde{U}_C(\tilde{x},\tilde{x}-a\hat{\mu})$. Therefore,

$$\bar{q}(\tilde{x})\tilde{\gamma}_\mu \tilde{\mathrm{D}}_\mu q(\tilde{x}) \xrightarrow{\text{lattice}} \bar{q}(\tilde{x})\tilde{\gamma}_\mu \frac{\tilde{U}_\mu(\tilde{x})q(\tilde{x}+a\hat{\mu}) - \tilde{U}^\dagger_\mu(\tilde{x}-a\hat{\mu})q(\tilde{x}-a\hat{\mu})}{2a}, \tag{8.129}$$

and the first part of the QCD action (8.120) involving fermions reads

$$\int \mathrm{d}^4\tilde{x}\; \left(\bar{q}(\tilde{\gamma}^\mu \tilde{\mathrm{D}}_\mu + m_q)q\right) \xrightarrow{\text{lattice}} \tag{8.130}$$

$$\tilde{S}_F^{\text{lat.}} = a^4 \sum_{\tilde{x}} \sum_{\mu} \frac{1}{2a} \bar{q}(\tilde{x})\tilde{\gamma}_\mu \left(\tilde{U}_\mu(\tilde{x})q(\tilde{x}+a\hat{\mu}) - \tilde{U}^\dagger_\mu(\tilde{x}-a\hat{\mu})q(\tilde{x}-a\hat{\mu})\right) + a^4 \sum_{\tilde{x}} m_q \bar{q}(\tilde{x})q(\tilde{x}).$$

For the pure gauge part of the QCD action, let us consider the *Wilson loop*

$$\tilde{U}_{\mu\nu}(\tilde{x}) = \tilde{U}_\mu(\tilde{x})\tilde{U}_\nu(\tilde{x}+a\hat{\mu})\tilde{U}^\dagger_\mu(\tilde{x}+a\hat{\nu})\tilde{U}^\dagger_\nu(\tilde{x}), \tag{8.131}$$

where $\hat{\nu}$ is the unit vector in the ν direction ($\nu \neq \mu$). In the $(\hat{\mu},\hat{\nu})$ plane, $\tilde{U}_{\mu\nu}(\tilde{x})$ defines a closed square path starting and ending at $\tilde{x}$ (see Fig. 8.23). On the lattice, it is called a *plaquette*. Expanding $\tilde{U}_{\mu\nu}(\tilde{x})$ as a function of a using Eq. (8.128), it turns out that

$$\tilde{U}_{\mu\nu}(\tilde{x}) = \exp\left(ig_s a^2 \tilde{G}_{\mu\nu}(\tilde{x}) + \cdots\right) = \mathbb{1} + ig_s a^2 \tilde{G}_{\mu\nu}(\tilde{x}) - \frac{1}{2} g_s^2 a^4 \tilde{G}_{\mu\nu}(\tilde{x})\tilde{G}_{\mu\nu}(\tilde{x}) + \cdots,$$

[27] Remember, $G_\mu(x)$ is a matrix whose entries depend on the spacetime position, and thus the product of several G_μ is generally not commutative. Therefore, the order of those matrices in Eq. (8.125) matters.

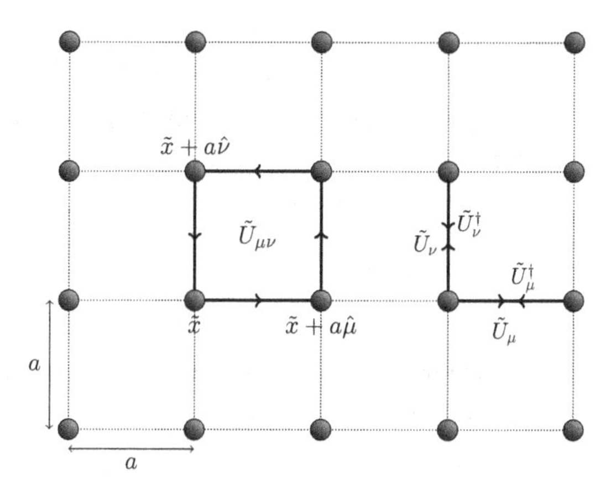

Fig. 8.23 Sketch of links and plaquettes on the lattice (two-dimensional slice in the $\mu - \nu$ plane). The arrows are representative of the order in which operators are written.

where $\tilde{G}_{\mu\nu}(\tilde{x})$ has been defined in Eq. (8.119). (The expansion above is easy to check in the Abelian case.) Since $\mathrm{Tr}(\tilde{G}_{\mu\nu}) = 0$ (the Gell-Mann matrices being traceless), the term $\frac{1}{2}\mathrm{Tr}(\tilde{G}_{\mu\nu}\tilde{G}_{\mu\nu})$ in Eq. (8.120) is

$$\frac{1}{a^4 g_s^2}\Re\mathrm{Tr}\left(1 - \tilde{U}_{\mu\nu}(\tilde{x})\right) = \frac{1}{2}\mathrm{Tr}(\tilde{G}_{\mu\nu}\tilde{G}_{\mu\nu}) + \cdots$$

Taking the real part eliminates half of the terms in the expansion and gives a real-valued lattice action. The action on the lattice is thus

$$\int \mathrm{d}^4\tilde{x}\frac{1}{2}\mathrm{Tr}(\tilde{G}_{\mu\nu}\tilde{G}_{\mu\nu}) \xrightarrow{\text{lattice}} \tag{8.132}$$

$$\tilde{S}_G^{\text{lat.}} = \sum_{\tilde{x}}\sum_{\mu<\nu}\frac{2}{g_s^2}\Re\mathrm{Tr}\left(\mathbb{1} - \tilde{U}_{\mu\nu}(\tilde{x})\right) = \sum_{\tilde{x}}\sum_{\mu<\nu}\frac{2}{g_s^2}\mathrm{Tr}\left(\mathbb{1} - \frac{\tilde{U}_{\mu\nu}(\tilde{x}) + \tilde{U}^\dagger{}_{\mu\nu}(\tilde{x})}{2}\right),$$

the factor 2 coming from the requirement $\mu < \nu$ on the lattice (limiting the computing time). Notice that $\mathrm{Tr}(\mathbb{1}) = 3$ for the fundamental representation of SU(3).

Supplement 8.5. Geometry and parallel transport of fields

Under an SU(N) gauge transformation of real parameters $\boldsymbol{\alpha}(x) = (\alpha_1(x), \ldots, \alpha_{N^2-1}(x))$, a field $\psi(x)$ is transformed into

$$\psi(x) \xrightarrow{\text{gauge}} \Omega(x)\psi(x) = e^{-ig\boldsymbol{\alpha}(x)\cdot\boldsymbol{T}}\psi(x), \tag{8.133}$$

where $\boldsymbol{T}$ are the generators and g is the associated coupling constant [$T_a = \lambda_a/2$ and $g = g_s$ for SU(3)]. Derivatives of ψ are not gauge invariant, and one has to introduce the covariant derivative to restore the gauge invariance. The reason can be traced back to the notion of derivative in the direction of the unit vector $\hat{n}^\mu$,

$$\hat{n}^\mu\partial_\mu\psi(x) = \lim_{\epsilon\to 0}\frac{\psi(x+\epsilon\hat{n}) - \psi(x-\epsilon\hat{n})}{2\epsilon},$$

Supplement 8.5. (cont.)

which depends on the field at two different spacetime points. Since the field transformation law depends on the spacetime point, the object $\partial_\mu \psi(x)$ does not have a well-defined transformation law. Therefore, to have a sensible derivative, we first have to transport the gauge transformation properties of the field ψ to the same spacetime point. It introduces the notion of covariant derivative,

$$\hat{n}^\mu \mathrm{D}_\mu \psi(x) = \lim_{\epsilon \to 0} \frac{U_C(x, x+\epsilon\hat{n})\psi(x+\epsilon\hat{n}) - U_C(x, x-\epsilon\hat{n})\psi(x-\epsilon\hat{n})}{2\epsilon}, \tag{8.134}$$

where U_C, assumed to be a Lorentz scalar quantity, is the field *parallel transporter* we are looking for. We wish $U_C(x,y)$ to transport the gauge transformation law from y to x, and thus, $U_C(x,y)\psi(y)$ must transform as $\psi(x)$, i.e.

$$U_C(x,y)\psi(y) \xrightarrow{\text{gauge}} \Omega(x)U_C(x,y)\psi(y).$$

Therefore, multiplying by the inverse transformation of $\psi(y)$, $U_C(x,y)$ transforms into

$$U_C(x,y) \xrightarrow{\text{gauge}} \Omega(x)U_C(x,y)\Omega^\dagger(y). \tag{8.135}$$

Since the SU(N) generators are Hermitian, $U_C^\dagger(x,y)$ must transform into

$$U_C^\dagger(x,y) \xrightarrow{\text{gauge}} \Omega(y)U_C^\dagger(x,y)\Omega(x),$$

i.e. the same transformation law as $U_C(y,x)$. Therefore, we identify

$$U_C^\dagger(x,y) = U_C(y,x). \tag{8.136}$$

An important requirement of U_C is the transitivity property. In other words, we wish

$$U_{C_2+C_1}(x,z) = U_{C_2}(x,y)U_{C_1}(y,z), \tag{8.137}$$

such that $U_{C_2}(x,y)U_{C_1}(y,z)\psi(z)$ transforms under a gauge transformation as $\psi(x)$. Since moving the field from one point to the other and then going back to the initial point following the *same path* should have no final effect,

$$U_C(x,y)U_C(y,x) = 1,$$

it implies with Eq. (8.136) that U_C is unitary. Consequently, as it is a Lorentz scalar and a unitary matrix, it can be represented as

$$U_C(x,y) = e^{-ig\phi_C(x,y)}, \tag{8.138}$$

with $\phi_C(x,y)$ being a Hermitian matrix, and where the coupling constant g has been factorised, and the sign is, for convenience, the same as that in Eq. (8.133). Assuming $U_C(x,y)$ is a continuous function of x and y, an infinitesimal transport is thus proportional to the difference between the two points. Since a finite path C from y to x can be divided into n infinitesimal segments, $[z_0, z_1], \ldots, [z_{n-1}, z_n]$, with $z_0 = y$ and $z_n = x$, we must have

$$\phi_C(x,y) = \lim_{n\to\infty} \sum_{i=0}^{n} (z_{i+1} - z_i)^\mu X_\mu(z_i) = \oint_y^x \mathrm{d}z^\mu X_\mu(z),$$

Supplement 8.5. (cont.)

where X_μ is a vector contracted with the separation between the elementary points. To identify X_μ, consider $U_C(x, x + \epsilon\hat{n})$, with $\epsilon \to 0$. Then, $\phi_C(x, x + \epsilon\hat{n}) = -\epsilon\hat{n}^\mu X_\mu(x)$, and $U_C(x, x + \epsilon\hat{n}) = 1 + ig\epsilon\hat{n}^\mu X_\mu(x)$. Inserting this expression into Eq. (8.135) shows that X_μ satisfies the same gauge transformation as G_μ in Eq. (8.114) [it is easy to check this when $T = 1$ as in QED, where $A_\mu \to A_\mu + \partial_\mu\alpha(x)$, but it is more tricky when T is a matrix, where we have in the example of SU(3) the transformation (8.47) of G^a_μ]. Therefore, X_μ is identified as G_μ. Moreover, the covariant derivative in Eq. (8.134) coincides with $\partial_\mu + igG_\mu$, i.e. Eq. (8.43). Therefore, it is tempting to suppose that Eq. (8.138) is the correct formula for the parallel transporter. However, the transitivity property would require that $e^{-ig\phi(x,y)} = e^{-ig\phi(x,z)}e^{-ig\phi(z,y)}$, which is not true in general when ϕ is a matrix. The introduction of the path-ordering $\mathcal{P}$ fixes this issue. If the path is parametrised by a continuous parameter $s \in [0, 1]$, i.e. $z = z(s)$, with $z(0) = y$ and $z(1) = x$, then in the power-series expansion of the exponential, the matrices in each term are ordered so that higher values of s stand to the left, i.e. if $z(s_i) = z_i$, then $s_n \geq s_{n-1} \cdots \geq s_1$. Thanks to this trick,

$$\mathcal{P}e^{-ig\phi(x,y)} = \mathcal{P}e^{-ig\phi(x,z)}\mathcal{P}e^{-ig\phi(z,y)}.$$

It yields the formula of the parallel transporter given in Eq. (8.125).

Calculating Expectation Values on the Lattice

In perturbative theories, amplitudes are calculated order by order with Feynman diagrams. In non-perturbative theories, the *path integral* approach of quantum field theory is followed. It is only flashed here. The reader should consult a quantum field theory textbook (e.g., Gelis, 2019, Chapter 5; Peskin and Schroeder, 1995) to fill the gaps of the presentation below. The generating functional of path integral is similar to the partition function in statistical physics. Restricting the discussion to QCD, it is defined as

$$Z = \int \mathcal{D}G_a^\mu \int \mathcal{D}q \int \mathcal{D}\bar{q}\, e^{iS_{\text{QCD}}} = \int \mathcal{D}G_\mu^a \int \mathcal{D}q \int \mathcal{D}\bar{q}\, e^{-\tilde{S}_{\text{QCD}}},$$

where the measure of this functional integral is denoted by the symbol $\mathcal{D}$. The symbol $\int \mathcal{D}q$ means the sum over all possible configurations of the quark field q (and a similar meaning for the adjoint field $\bar{q}$ and the gluon field G^a_μ). The measure

$$\mathcal{D}q \xrightarrow{\text{lattice}} \prod_{\tilde{x}} \prod_{\alpha,c} \mathrm{d}q^c_\alpha(\tilde{x}) \tag{8.139}$$

for one flavour of quark is thus the product of integration measures for the classical field variables at all points of the lattice for the four spinor indices α and the three colour indices c. A similar measure is understood for $\bar{q}$, considered as an independent variable. The actions S_{QCD} and $\tilde{S}_{\text{QCD}}$ have been given in Eq. (8.117). Note one of the interests in working in the Euclidean space: instead of having rapid oscillations for large actions (which are delicate to calculate with computers), we have a smooth integral. On the lattice, the action is given by the sum $\tilde{S}_F^{\text{lat.}} + \tilde{S}_G^{\text{lat.}}$ defined in Eqs. (8.130)[28] and (8.132). Note that the gauge fields do not appear explicitly: their action is encoded in the link variables that transport the properties of lattice site to its neighbour. Therefore, the dynamical variable on the lattice is $\tilde{U}_\mu$ and not G^a_μ. The path integral is thus

$$Z = \int \mathcal{D}\tilde{U} \int \mathcal{D}q \int \mathcal{D}\bar{q}\, e^{-\tilde{S}_F^{\text{lat.}} - \tilde{S}_G^{\text{lat.}}}. \tag{8.140}$$

Since $\tilde{U}_\mu(\tilde{x})$ in Eq. (8.128) is an SU(3) matrix, $\int \mathcal{D}\tilde{U}$ implies an integration over the whole group[29] SU(3),

$$\mathcal{D}\tilde{U} = \prod_{\tilde{x}} \prod_{\mu} \mathrm{d}\tilde{U}_\mu(\tilde{x}). \tag{8.141}$$

With the path integral approach, the expectation value of a gauge-invariant operator $O(q, \bar{q}, \tilde{U})$ on the lattice is then given by

$$\langle O \rangle = \frac{1}{Z} \int \mathcal{D}\tilde{U} \int \mathcal{D}q \int \mathcal{D}\bar{q}\, O(q, \bar{q}, \tilde{U})\, e^{-\tilde{S}_F^{\text{lat.}} - \tilde{S}_G^{\text{lat.}}} = \frac{1}{Z} \int \mathcal{D}\tilde{U} e^{-\tilde{S}_G^{\text{lat.}}} Z_F \langle O \rangle_F, \tag{8.142}$$

with

$$Z_F = \int \mathcal{D}q \int \mathcal{D}\bar{q}\, e^{-\tilde{S}_F^{\text{lat.}}}, \quad \langle O \rangle_F = \frac{1}{Z_F} \int \mathcal{D}q \int \mathcal{D}\bar{q}\, O(q, \bar{q}, \tilde{U})\, e^{-\tilde{S}_F^{\text{lat.}}}. \tag{8.143}$$

Therefore, the calculation can be divided into two parts: first, the calculation of $\langle O \rangle_F$, which depends on the gauge configuration $\tilde{U}$, and then, the integration over the gauge configurations. A point left out so far is the fermionic nature of q and $\bar{q}$, and hence, the necessity to incorporate Fermi statistics in the path integral formulation. The (classical) quark fields q and $\bar{q}$ are actually represented by anti-commuting numbers, the so-called *Grassmann numbers*. In other words, $q^c_\alpha(\tilde{x}) q^{c'}_{\alpha'}(\tilde{x}') = -q^{c'}_{\alpha'}(\tilde{x}') q^c_\alpha(\tilde{x})$ (and similarly for $\bar{q}$ and between q and $\bar{q}$). Notice that necessarily $\left(q^c_\alpha(\tilde{x})\right)^2 = 0$, implying that the power series

[28] The fermionic action on the lattice is actually more complicated than Eq. (8.130). One can show that the quark propagators in the Euclidean space and on the lattice are

$$\tilde{K}(p) = \frac{-i\sum_\mu \tilde{\gamma}_\mu \tilde{p}_\mu + m_q}{\sum_\mu \tilde{p}_\mu^2 + m_q^2} \quad \text{and} \quad \tilde{K}^{\text{lat.}}(p) = \frac{-i\sum_\mu \tilde{\gamma}_\mu \frac{\sin(a\tilde{p}_\mu)}{a} + m_q}{\sum_\mu \frac{\sin^2(a\tilde{p}_\mu)}{a^2} + m_q^2},$$

respectively. Taking the continuum limit $a \to 0$, $\tilde{K}^{\text{lat.}}(p) \to \tilde{K}(p)$, as expected since $\sin(a\tilde{p}_\mu)/a \to p_\mu$ for all values, except for the value $\tilde{p}_\mu = \pi/a$, which is allowed by the boundary condition imposed on the lattice $q(\tilde{x} + L) = q(\tilde{x})$. This unphysical result is called the fermion doubling problem [in four dimensions, there are actually 15 unphysical values, $p = (\pi/a, 0, 0, 0), (0, \pi/a, 0, 0), \ldots, (\pi/a, \pi/a, \pi/a, \pi/a)$]. To avoid this problem, an additional term is added to $\tilde{S}_F^{\text{lat.}}$ in Eq. (8.130) that reads [justifications for this Wilson term can be found in Gattringer and Lang (2010, chapter 5)]

$$a^4 \sum_{\tilde{x}} \sum_{\mu} \frac{1}{2a} \left(2\bar{q}(\tilde{x}) q(\tilde{x}) - \bar{q}(\tilde{x}) \tilde{U}_\mu(\tilde{x}) q(\tilde{x} + a\hat{\mu}) - \bar{q}(\tilde{x}) \tilde{U}^\dagger_\mu(\tilde{x} - a\hat{\mu}) q(\tilde{x} - a\hat{\mu})\right).$$

[29] Many mathematical aspects are not covered here, such as how can we integrate over a group? Roughly speaking, since SU(3) matrices depend on eight parameters α^a via Eq. (8.40), we expect $\mathrm{d}\tilde{U}_\mu(\tilde{x})$ to be something like $\int \mathrm{d}\alpha^1 \ldots, \mathrm{d}\alpha^8$. However, there is a non-trivial factor depending on the α^as that reflects the metric in the parameter space. An important point is that these α^a parameters are 'angular variables' that vary in a compact range, making at each $\tilde{x}$, the group integration a compact integral. This complicated measure is the *Haar measure*. Some details can be found in Gattringer and Lang (2010, chapter 4).

for a function of Grassmann numbers have a finite number of terms. Now, the fermionic action on the lattice (8.130) is quadratic in q and $\bar{q}$ (it is also true for the Wilson term mentioned in footnote 28). It reads

$$\tilde{S}_F^{\text{lat.}} = a^4 \sum_{\tilde{x},\tilde{x}'} \sum_{\alpha,\alpha',c,c'} \bar{q}_\alpha^c(\tilde{x}) D_{\alpha\,\alpha'}^{c\,c'}(\tilde{x},\tilde{x}') q_{\alpha'}^{c'}(\tilde{x}'),$$

where $D_{\alpha\,\alpha'}^{c\,c'}(\tilde{x},\tilde{x}')$ is the component of a matrix D.[30] The fermionic path integral can be explicitly calculated for operators being a product of qs and $\bar{q}$s (a result known as Wick's theorem), leading to a result that is a function of $\det(D)$ for Z_F and of D^{-1} for O_F in Eq. (8.143) (Gattringer and Lang, 2010). The challenge is then to calculate the determinant and the inverse of a huge matrix (for a lattice with N sites in each dimension, there are 3 colours $\times$ 4 spinor indices, hence $12N^4$ degrees of freedom and thus a matrix $12N^4 \times 12N^4$). To give an order of magnitude, a lattice with a size $L = 5$ fm with $N = 48$ points in each dimension (such as $a = L/N = 0.1$ fm, 10 times smaller than the proton radius), $12N^4 \simeq 6 \times 10^7$. Moreover, the path integration of the gauge part is transformed with the measure (8.141) into multiple ordinary integrals, i.e. the numerical evaluation of $4 \times 8 \times N^4 \simeq 17 \times 10^7$ integrals (four Lorentz indices and eight parameters for all SU(3) matrices)! The integrals are numerically evaluated by a Monte-Carlo sampling technique benefiting from the fact that $\tilde{S}_G \geq 0$ in the Euclidean space and interpreting $e^{-\tilde{S}_G}$ as a probability distribution of the link configuration (in a $32N^4$-dimensional space).

8.5.2 Hadron Masses

The previous section briefly explained how to estimate quantities with a non-perturbative approach based on the lattice QCD. In this section, we will see how hadron masses can be evaluated with this method. Correctly reproducing the variety of the hadron spectroscopy constitutes a powerful test for the correctness of QCD.

The problem here is to identify an operator whose expectation value calculated with the Formula (8.142) gives access to the mass of the hadron studied. Let us suppose that we find an operator O carrying the same quantum numbers as the hadron h, and consider the vacuum expectation value of the product of O at two different (Euclidean) times 0 and $\tilde{t}$, $\langle O^\dagger(0) O(\tilde{t})\rangle$. Inserting a complete basis of the QCD Hamiltonian eigenstates yields

$$\langle O^\dagger(0) O(\tilde{t})\rangle = \langle 0|O^\dagger(0) O(\tilde{t})|0\rangle = \sum_n \langle 0|O^\dagger(0)|\psi_n\rangle\,\langle\psi_n|O(\tilde{t})|0\rangle.$$

Since in the Heisenberg picture, $O(t) = e^{itH} O e^{-itH}$ [Eq. (6.36)], it follows with $t = -i\tilde{t}$,

$$\langle O^\dagger(0) O(\tilde{t})\rangle = \sum_n \langle 0|O^\dagger(0)|\psi_n\rangle\,\langle\psi_n|O(0)|0\rangle\, e^{-E_n\tilde{t}} = \sum_n |\,\langle\psi_n|O(0)|0\rangle\,|^2 e^{-E_n\tilde{t}}.$$

[30] For instance, with Eq. (8.130),

$$D_{\alpha\,\alpha'}^{c\,c'}(\tilde{x},\tilde{x}') = \sum_\mu \left(\tilde{\gamma}_\mu\right)_{\alpha\alpha'} \frac{\left(\tilde{U}_\mu(\tilde{x})\right)_{cc'} \delta(\tilde{x} + a\hat{\mu}, \tilde{x}') - \left(\tilde{U}_\mu^\dagger(\tilde{x} - a\hat{\mu})\right)_{cc'} \delta(\tilde{x} - a\hat{\mu}, \tilde{x}')}{2a} + m_q \delta_{\tilde{x}\tilde{x}'} \delta_{cc'} \delta_{\alpha\alpha'}.$$

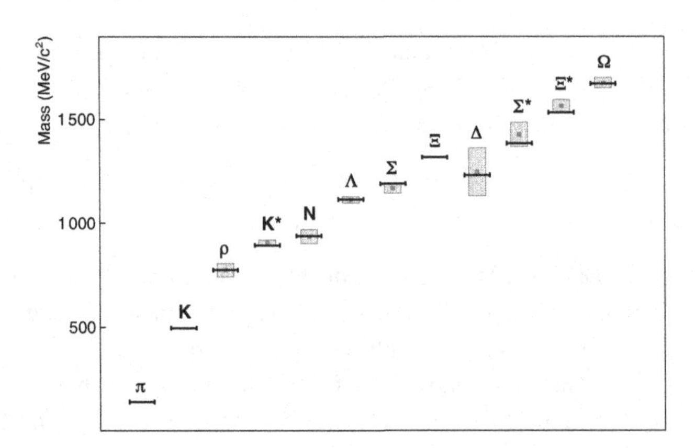

Fig. 8.24 Hadron masses prediction from the lattice QCD. Black bars are the experimental values from Particle Data Group (2022). Grey squares are the predictions from Dürr S. et al. (2008) with their uncertainties pictured with the grey boxes. The π, K and Ξ were used as references for calibrating the predictions.

Non-zero contributions of $\langle \psi_n | O(0) | 0 \rangle$ are coming from the states having the same quantum numbers as O, and hence, as the hadron h. For a large time $\tilde{t}$, the state with the lowest energy dominates in this sum. Thus, by evaluating $C(\tilde{t}) = \langle O^\dagger(0) O(\tilde{t}) \rangle$ at various $\tilde{t}$, the leading exponential decay provides the minimum energy (and hence the mass) of the hadron h sharing the same quantum number as O, i.e. $\ln \left[C(\tilde{t})/C(\tilde{t}+a) \right] \simeq a m_h + \cdots$. This method works well for non-excited states, and if all the basic parameters in the QCD Hamiltonian (including quark masses and the coupling constant) are known. Moreover, only dimensionless combinations, such as mass ratios, can be calculated. In practice, one adjusts these parameters with few observables as the pion or kaon mass and then provides predictions for other masses. An example of an operator O for pseudoscalar mesons, for instance the charged pion π^+, is $O_{\pi^+} = \bar{d}^c_\alpha(\tilde{x}) \left(\tilde{\gamma}^5\right)_{\alpha\alpha'} u^{c'}_{\alpha'}(\tilde{x})$ ($\tilde{\gamma}^5$ ensures that O transforms as a pseudoscalar under parity transformation). One usually works with the operator projected at a given momentum with $O(\tilde{t}, \tilde{\boldsymbol{p}}) = \sum_{\tilde{x}} O(\tilde{t}, \tilde{\boldsymbol{x}})$ and takes $\tilde{\boldsymbol{p}} = \mathbf{0}$. Several choices of operators are possible, but some give numerical results with better accuracy. In Fig. 8.24, the hadron masses calculated with the lattice QCD are compared with the experimental values. They are in agreement, within the limit of the accuracy of the calculation.

8.5.3 Confinement

The confinement postulate states that only colour-singlet particles can be observed. The question is then: What forces hadrons to be in a colour-singlet state? Before shedding light on this topic with results from the lattice QCD, it is instructive to have a phenomenological approach by considering a model of the QCD colour potential. Let us restrict the discussion to a system made of a quark and an antiquark. We know that for QED processes, the static potential energy has a Coulombic behaviour, i.e.

$$V(r) = Q_1 Q_2 \frac{\alpha}{r},$$

where Q_1 and Q_2 are the electric charges (in units of e) of the two particles and r, the distance separating them. For $Q_1Q_2 < 0$, the potential is attractive, the potential energy decreasing when particles get closer. Hence, the Colomb potential between a quark with an electric charge Q and its antiquark of charge $-Q$ is $V(r) = -Q^2\alpha/r$. Because of the negative sign, it is attractive. In QCD, at high energy and thus a small distance between the quark and the antiquark, the coupling constant g_s tends to zero (cf., asymptotic freedom). Hence, the non-Abelian term specific to QCD in the field-strength tensor $G^a_{\mu\nu}$ of Eq. (8.56) vanishes, leaving only $G^a_{\mu\nu} \simeq \partial_\mu G^a_\nu - \partial_\nu G^a_\mu$, i.e. the form known from QED. At small distances, we thus expect a Coulombic behaviour of the QCD static potential energy for the pair $q\bar{q}$, replacing the QED charge factors with the QCD colour factor and the QED coupling constant with α_s, i.e.

$$V_{q\bar{q}}(r) = -C\frac{\alpha_s}{r}. \tag{8.144}$$

If the colour factor C is positive, the potential would be attractive and repulsive otherwise. In Section 8.3.5, we have calculated these colour factors for various colour configurations of $q\bar{q}$. We can then compare the case where $q\bar{q}$ is in a colour-singlet state (the only colour state allowed to form a meson) to the case of the colour octet. The colour singlet is $(r\bar{r} + g\bar{g} + b\bar{b})/\sqrt{3}$. The associated colour factor is thus

$$C_{\text{singlet}} \equiv f_{\frac{r\bar{r}+g\bar{g}+b\bar{b}}{\sqrt{3}} \to \frac{r\bar{r}+g\bar{g}+b\bar{b}}{\sqrt{3}}} = \frac{1}{3}\sum_{c=r,g,b}\sum_{c'=r,g,b} f_{c\bar{c}\to c'\bar{c}'} = \frac{1}{3}\left\{\sum_{c=r,g,b} f_{c\bar{c}\to c\bar{c}} + \sum_{c,c'=r,g,b}^{c\neq c'} f_{c\bar{c}\to c'\bar{c}'}\right\}.$$

By colour symmetry, there is no need to calculate these nine terms. The three terms in the first sum are necessarily all equal, as are the six terms in the second sum. Calculating, for instance, $f_{r\bar{r}\to r\bar{r}}$ for the first sum, the colour factor formula for the t-channel exchange (8.83) tells us that $f_{r\bar{r}\to r\bar{r}} = C^{11}_{22} = 1/3$, using the Fierz identity (8.78). Similarly, with $f_{r\bar{r}\to g\bar{g}}$ for the second sum, we find $f_{r\bar{r}\to g\bar{g}} = C^{21}_{12} = 1/2$. Therefore,

$$C_{\text{singlet}} = \frac{1}{3}\left\{3 \times \frac{1}{3} + 6 \times \frac{1}{2}\right\} = \frac{4}{3},$$

which is a positive number. The potential felt by the $q\bar{q}$ pair in the colour-singlet state is thus attractive, as it should be to form a bound state. What if the $q\bar{q}$ system is in the colour octet state? Taking, for instance, the member $r\bar{g}$ of the octet [the $q\bar{q}$ colour octet has the same members as the gluon octet (8.52)], $C_{\text{octet}} \equiv f_{r\bar{g}\to r\bar{g}} = C^{12}_{12} = -1/6$. The potential felt by the colour octet state is, this time, repulsive, preventing the formation of a bound state.

Of course, these considerations are not proof of the colour confinement: we should have found an infinitely repulsive potential for any coloured state. It just gives strong indications why bound states are always found in singlet configuration. Bound states are characterised by gluon exchange at low momentum transfer, for which the coupling constant α_s becomes large. Only non-perturbative methods can be conclusive. Using the lattice QCD approach, Wilson showed that QCD exhibits the confinement of colour, with the only finite-energy states being those that are singlets of the colour $SU(3)_c$ group. In other words, the free energy of an isolated colour charge is infinite. Another way of showing this is to observe that it requires infinite energy to separate two colour charges by an infinite distance. Supplement 8.6 indicates how lattice QCD can predict the behaviour of the colour potential

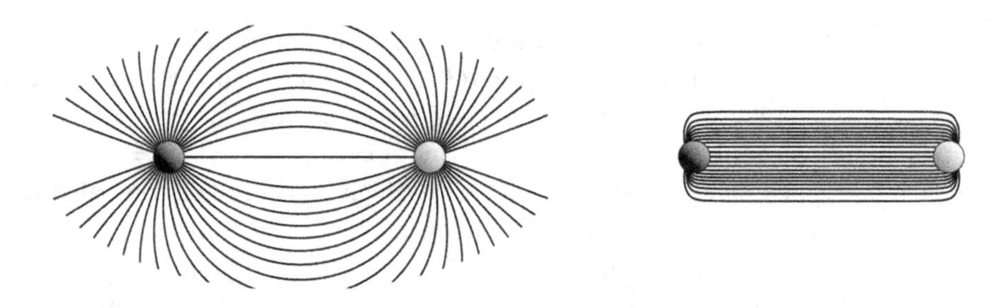

Fig. 8.25 Field lines due to a static potential. Left: Colomb field lines between two opposite electric charges. Right: Sketch of colour field lines between a colour and an anti-colour charge.

between a quark and an antiquark in the non-perturbative regime, where the coupling constant is large. It is found in Eq. (8.149) that the potential exhibits a linear behaviour with the distance r separating the $q\bar{q}$ pair. Therefore, at an infinite distance, the system energy would be infinite. Combined with the prediction in the perturbative regime, Eq. (8.144), we conclude that the potential reads over the whole range of distances

$$V_{q\bar{q}}(r) = V_0 - \frac{A}{r} + \sigma r, \tag{8.145}$$

where V_0 is a constant to set the normalisation of the energy (it has no physical effects since the force between the quark and the antiquark is derivable from the potential). The constant A is $\frac{4}{3}\alpha_s$ in the 1-gluon exchange approximation (this is what we implicitly supposed when we calculated the colour factor above) and the positive constant, σ, called the string tension, which on the lattice is proportional to $\ln(3g_s^2)/a^2$ [Eq. (8.149)], shows that the potential grows linearly with distance. Equation (8.145) matches the phenomenological potential given in the introductory chapter [cf. Eq. (1.5)]. As we explained there, since the potential is larger as one tries to pull the two constituents of the pair apart, the quarks are necessarily confined. If, however, the quark and antiquark are pulled sufficiently far apart (because produced at high energy), the energy may exceed the threshold for creating a new $q\bar{q}$ pair, which eventually recombines with the two initial constituents to form two mesons. It costs less energy to create new particles and reduce the colour potential between the members of the new pairs than to further extend the distance between the members of the original pair. This process is called the *hadronisation*, detailed in the next section. As a magnet bar cannot be broken to form monopoles (each broken piece forms a magnet with a north pole and a south pole), a meson (or hadron in general) cannot be broken into a quark and an antiquark but can only form other mesons. Qualitatively, this property arises from the formation of a colour flux with a tube shape between coloured particles. As illustrated in Fig 8.25, unlike QED where the field lines between two electric charges spread out in space, the colour field lines are squeezed in a narrow tube because the gluons exchanged between the coloured particles also interact with each other (due to the gluon self-coupling). Hence, the field lines cannot extend and are confined in a tube, such as the energy cost of separating colour sources grows proportionally with the separation distance (the energy density, $\mathrm{d}E/\mathrm{d}r$, being constant), the flux tube conserving approximately the same diameter. Notice that unlike what is often said, the flux tube is not like a spring where the attractive force increases with its elongation. In QCD, the force, deriving from the linear part of the potential when r is large enough, is constant with r. Those tubes are described by lattice QCD calculations [see, for instance, Cardoso et al. (2010) and the animations provided on the website referred in Leinweber (2003–2004)]. Typical values of the parameter σ in the potential is $(0.450\ \mathrm{GeV})^2$ or equivalently about 1 GeV/fm, i.e. a force of the order of 10^5 N (!). Such values reproduce well the spectroscopy of heavy quark–antiquark bound states (in practice, the $b\bar{b}$ and $c\bar{c}$ mesons), where the infinite mass approximation and thus the non-relativistic hypothesis are reasonably valid.

Even if we mainly discussed the quark–antiquark system, the same conclusions apply for baryons: a linear confinement potential is felt between quarks in baryons, too, the colour flux tubes being also present, however, with a more complicated pattern.

Supplement 8.6. Static quark–antiquark potential from lattice QCD

On the lattice, one has to identify an observable for the static quark potential $V(r)$. This role is played by the Wilson loop. In Eq. (8.131), the smallest Wilson loop (the plaquette) has been introduced. Now, we consider a loop of length R along a space axis, let us say, the $\tilde{x}$-axis ($\mu = 1$), and a length T along the Euclidean time axis ($\mu = 0$). It thus corresponds to a rectangle of size $R \times T$ in the $(\tilde{x}, \tilde{t})$ plane, with each side of the rectangle being a Wilson line in the $\tilde{x}$ or $\tilde{t}$ dimension,

$$W_{\text{loop}} = W^{\tilde{t}=0}(R,0)W^{\tilde{x}=R}(T,0)W^{\tilde{t}=T}(0,R)W^{\tilde{x}=0}(0,T).$$

The quantity $W^{\tilde{t}=0}(R,0)$ denotes a Wilson line from 0 to R in the spatial dimension, keeping the time fixed at 0 [and a similar meaning for $W^{\tilde{x}=R}(T,0)$ in the temporal dimension, keeping the position fixed at R]. Like plaquettes, the trace of every loop of links on the lattice is gauge invariant and can be used as observable. However, $\mathrm{Tr}(W_{\text{loop}})$ is easier to interpret in the gauge where the gluon field has no components on the time axis, i.e. $\tilde{G}_0^a = 0$, implying $\tilde{G}_0 = \lambda_a \tilde{G}_0^a/2 = 0$. In such a case, the temporal Wilson line is the identity because it is a product of several links (8.128) in the $\mu = 0$ direction, for which $\tilde{U}_0(\tilde{x}) = \exp\left(ig_s a\tilde{G}_0(\tilde{x})\right) = \mathbb{1}$. It follows that

$$\mathrm{Tr}(W_{\text{loop}}) = \mathrm{Tr}\left[W^{\tilde{t}=0}(R,0)W^{\tilde{t}=T}(0,R)\right] = \mathrm{Tr}\left[W^{0\dagger}(0,R)W^{T}(0,R)\right].$$

The expectation value of $\mathrm{Tr}(W_{\text{loop}})$ can then be calculated using Eq.(8.142) with $\mathcal{O} = \mathrm{Tr}(W_{\text{loop}})$. Since $\mathcal{O}$ does not depend on the fermionic variables, we can forget them and just integrate the term $\mathcal{O}\exp(-\tilde{S}_G^{\text{lat.}})$ over $\mathcal{D}\tilde{U}$. The important observation is that $\tilde{S}_G^{\text{lat}}$ depends on $1/g_s^2$ in Eq. (8.132). Since we work in a non-perturbative regime, where g_s is large, we can expand the exponential in powers of $1/g_s^2$. The result is, to first order [a proof can be found in Gattringer and Lang (2010, chapter 3)],

$$\langle \mathrm{Tr}(W_{\text{loop}})\rangle \propto \left(\frac{1}{3g_s^2}\right)^{RT/a^2} + \mathcal{O}(1/g_s^2). \tag{8.146}$$

The last step is to make the connection between $\langle \mathrm{Tr}(W_{\text{loop}})\rangle$ and the $q\bar{q}$ potential. Let us set a quark at position $x = R$ and an antiquark at $x = 0$ and link them with the Wilson line $W^{\tilde{t}}(0,R)$,

$$\mathcal{O}(\tilde{t}) = \bar{q}(\tilde{t}, \tilde{x} = 0)W^{\tilde{t}}(0,R)q(\tilde{t}, \tilde{x} = R). \tag{8.147}$$

Supplement 8.6. (cont.)

As in Section 8.5.2, we focus the attention on the correlation at two different times, yielding

$$\langle \mathcal{O}^\dagger(0)\mathcal{O}(T)\rangle = \sum_n |\langle \psi_n|\mathcal{O}(0)|0\rangle|^2 e^{-E_n T},$$

with $\{\psi_n\}$ being a complete basis of the QCD Hamilonian. When T is large enough, only the $q\bar{q}$ state with the lowest energy survives in the sum. Furthermore, in the approximation where the quark mass is large to consider that q and $\bar{q}$ stay still (static potential), the energy is dominated by the potential energy. It follows that

$$\lim_{T,m_q\to\infty} \langle \mathcal{O}^\dagger(0)\mathcal{O}(T)\rangle \propto e^{-V(R)T}. \tag{8.148}$$

On the other hand, $\langle \mathcal{O}^\dagger(0)\mathcal{O}(T)\rangle$ can be calculated on the lattice using Eq. (8.142). Without giving the details of the calculation, one finds that in the infinite mass approximation, it is simply proportional to $\langle \mathrm{Tr}(W_{\mathrm{loop}})\rangle$. Therefore, Eqs. (8.146) and (8.148) lead us to conclude

$$e^{-V(R)T} \propto \left(\frac{1}{3g_s^2}\right)^{RT/a^2} \Rightarrow V(R) \propto \frac{R}{a^2}\ln\left(3g_s^2\right). \tag{8.149}$$

8.5.4 From Hadronisation to Jets

Hadronisation (also frequently called *fragmentation*) is the mechanism by which quarks and gluons produced in hard processes form the colourless hadrons that are observed in the final state of a reaction. It is a non-perturbative process not well understood from first principles, which is, so far, only parametrised by phenomenological models used in simulations. For instance, the *string model* is directly inspired by the linear behaviour of the colour potential (8.145). To illustrate the hadronisation, let us consider the production in an e^+e^- collider of a quark–antiquark pair at high energy. In the CM frame, the pair is produced back to back and quickly separates because of the energetic boost. For a separation distance of the order of 1 fm or above, the colour potential energy stored in the colour tube is large, with the parameter σ in Eq. (8.145) being about 1 GeV/fm. This energy can then be converted into the production of a new $q'\bar{q}'$ pair through radiation of gluons (the coupling constant α_s increasing at large distances), breaking the initial colour tube into two colour tubes produced by $q\bar{q}'$ and $q'\bar{q}$ forming colour-singlet states. This situation is energetically more favourable than further stretching the initial pair. Since energy is conserved, the kinetic energy of the quark and antiquark produced directly from e^+e^- decreases. The production of new pairs continues until there is not enough kinetic energy to further separate the members of pairs. The hadronisation process is depicted in Fig. 8.26. All quarks and antiquarks are then bound in a collection of hadrons (the last line of Fig. 8.26) whose direction of motion approximately follows the direction of the initial $q\bar{q}$ pair. This collection is called a *jet*. At lowest order, two back-to-back jets are then expected in the CM frame of the collision. Hadronisation models are more sophisticated than the simple model presented here. They must be able to account for the production of baryons,

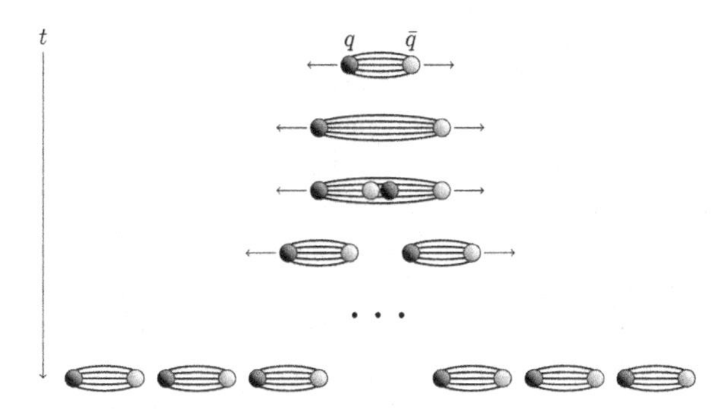

Fig. 8.26 Sketch of the hadronisation process.

hadrons of heavy flavours, and hard emission of gluons (producing extra jets) that matches the observations.

In the hadronisation process, the probability for the production of a specific hadron in the final state from a parton (quark, antiquark or gluon) is driven by the *fragmentation functions*. They are the counterpart of the PDFs for the non-perturbative parton-to-hadron transition, whereas PDFs describe the hadron-to-parton transition (see Section 7.3). They depend on the fraction, z, of the momentum of the parton that the hadron carries. More specifically, $D_p^h(z)$ describes the probability density that the parton $p = q, \bar{q}, g$ and the debris resulting from it end up producing the hadron h. Since the momentum is conserved, necessarily, the momentum carried by the parton must finally be shared by the hadrons. Therefore, we have the sum rule, separately for each parton p (and each quark or antiquark flavour),

$$\sum_h \int_0^1 \mathrm{d}z\, z D_p^h(z) = 1. \tag{8.150}$$

At high energy, in most cases, one can separate the perturbative part of the cross section involving the partons from the non-perturbative one. The exclusive cross section to produce a specific hadron is thus the cross section to produce a parton weighted by the fragmentation functions. For instance, at e^+e^- colliders, the cross section of the reaction $e^+ + e^- \to h + X$, where X stands for any states allowed with the hadron h production, reads

$$\sigma(e^+ + e^- \to h + X) = \sum_{q,\bar{q}} \int \mathrm{d}z\, \sigma(e^+ + e^- \to q\bar{q}) \left[D_q^h(z) + D_{\bar{q}}^h(z) \right].$$

Since the hadron may be produced from the quark q, the antiquark $\bar{q}$, and their debris, the two fragmentation functions D_q^h and $D_{\bar{q}}^h$ are present in the expression above. In contrast, the collision e^+e^- cannot directly generate a gluon; hence, D_g^h is absent in first order. At hadron colliders, the fragmentation function D_g^h should also be taken into account, as well as PDFs for the initial state (see Section 8.6.2). As PDFs, the fragmentation functions must be measured in experiments by performing global fits to data. Experimentally, it is found that the number of hadrons produced from the initial parton roughly follows a logarithmic dependence on the energy. Typically, a quark of 10 GeV produces about 7 hadrons, while at 100 GeV, it yields about 15 hadrons.

In real experiments, one cannot observe partons but she or he sees energy deposits and tracks left by hadrons in calorimeters and trackers (see Section 1.4.4). From these signals, hadron momenta are evaluated, and jets are reconstructed. The jet kinematics are then used to assess the initial parton (quark or gluon) kinematics. However, in order to relate the jets of hadrons, which are registered by detectors, to the jets of partons, which can be computed within perturbative QCD, one needs a robust jet algorithm. Ideally, jet algorithms should be resilient to QCD soft effects. More specifically, the set of hard jets that are found in the event should remain unchanged if a hadron momentum is split into two collinear momenta or if soft gluons are radiated (the so-called infrared radiation). These requirements are justified because the initial parton undergoes many collinear splittings as part of the fragmentation process, and there are always some emissions of soft particles in QCD events.[31] Two broad classes of jet algorithms exist: the cone algorithms and sequential recombination algorithms based on the notion of distance between pairs of particles. Historically, cone algorithms were invented first but are still used. Their concept is simple, and a modern example of the algorithm is described here. The hadron (or energy deposit) having the largest momentum (transverse momentum at hadron colliders) defines the initial cone axis. The cone radius is fixed. One computes the momentum of hadrons contained in the cone, takes the resulting momentum as a new direction and iterates until the cone is found stable. The hadrons contained in the cone are then assigned to the same jet. They are removed from the list of hadrons to be considered, and the procedure is reiterated until all hadrons are assigned to a jet. At the LHC, most of the jet algorithms are based on the second kind of algorithms that use a sequential combination of the closest particles, according to a distance measure,

$$d_{ib} = k_{Ti}^{2p}, \quad d_{ij} = \min(k_{Ti}^{2p}, k_{Tj}^{2p}) \frac{\Delta R_{ij}^2}{R^2},$$

where k_{Ti} is the transverse momentum of the particle i (with respect to the beam axis), p is a parameter (the power), R is another parameter usually called the jet radius and $\Delta R_{ij}^2 = (\eta_i - \eta_j)^2 + (\phi_i - \phi_j)^2$ is a geometrical distance, with η being the pseudorapidity (see Supplement 2.1) and ϕ the azimuthal angle. The quantity d_{ib} is the 'distance' to the beam, while d_{ij} is the relative 'distance' between the particles i and j. The algorithm computes all possibilities of d_{ij} and d_{ib} and starts with the smallest distance d. If $d = d_{ij}$, then particles i and j are recombined to form a particle k (by adding their 4-momenta), whereas if $d = d_{ib}$, the particle i is classified as a jet and removed from the list of particles. The procedure is repeated until there is no particle left. The algorithms mostly vary by the value of the power p. At LHC, $p = -1$ is often used, the algorithm being called anti-k_T, followed by a number corresponding to the value of R, usually between 0.5 and 0.8 (the k_T algorithm

[31] The finite energy and angle resolution of real experiments make the events with ultra-soft or collinear emissions indistinguishable from those with no emissions. However, in fixed-order perturbative QCD calculations used by simulations, divergent contributions, i.e. real emission of soft gluons and virtual loop diagrams at the same order of perturbation theory cancel. If the jet algorithm treats both kinds of divergent contributions differently, the comparison with real data becomes problematic.

has $p > 0$). The anti-k_T algorithm naturally favours the clustering starting from pairs of energetic particles and leads to jets with circular shapes in the (η, ϕ) plane.

8.6 Back to the Parton Model

8.6.1 QCD Corrections

At the end of Chapter 7, we noted that for small or large values of the Bjorken-x variable (i.e., the fraction of nucleon momentum carried by a parton), the Bjorken scaling law is violated (see Fig. 7.9), leading to a dependence on Q^2, the negative 4-momentum squared of the virtual photon. We can now be more quantitative on this question. The basic diagram for the deep inelastic scattering off a proton by an electron is shown below

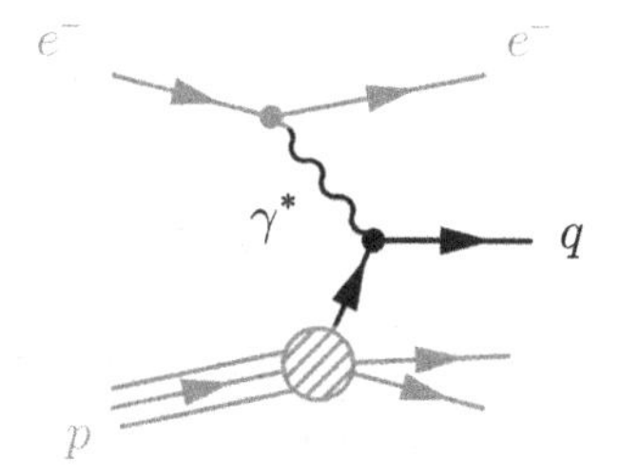

where the elementary process is the scattering of a quark by a virtual photon (represented by the unattenuated part of the diagram), $\gamma^* + q \to q$. However, as quarks can radiate gluons, there are corrections to first order of α_s, given by the diagrams of Fig. 8.27. In the diagram on the left-hand side, the initial quark carries a fraction y of the proton momentum, i.e. $p_i = yp$, radiates a gluon, which reduces the quark momentum by a fraction z, giving a final momentum $zp_i = zyp = xp$. Assuming, as in Section 7.2.2, that the quark stays on-shell, the different variables are related to the 4-momentum q of the virtual photon by

$$Q^2 = -q^2, \quad x = \frac{Q^2}{2p \cdot q}, \quad x = zy, \quad z = \frac{Q^2}{2p_i \cdot q},$$

where p is the proton 4-momentum. Since the quark radiates a gluon, its transverse momentum p_T with respect to the proton direction is no longer necessarily zero (the momentum

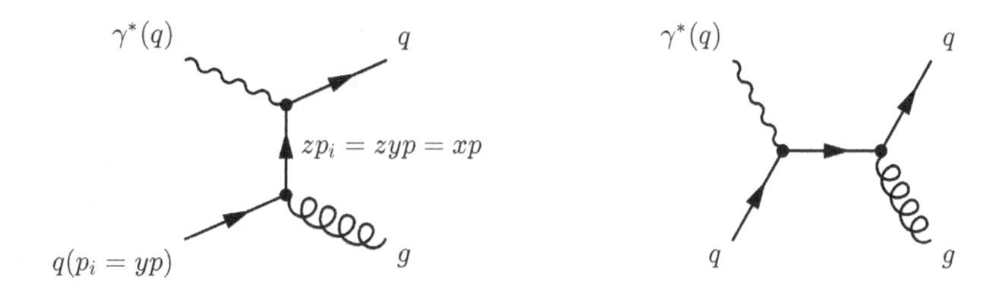

Fig. 8.27 Diagrams describing the correction to the deep inelastic scattering due to the emission of a real gluon.

is not aligned with the proton direction). The structure functions in Eq. (7.34) came from the elementary interaction $\gamma^* + q \to q$, where the initial quark was assumed to have only a longitudinal momentum. Now, we have the additional contribution $\gamma^* + q \to q + g$. By calculating the cross section of this reaction [the calculation is similar to what we did in Section 8.3.6 and can be found, for instance, in Halzen and Martin (1984, chapter 10)], one can show that these diagrams induce a modification of the structure functions F_1 and F_2, which now become dependent on both variables x and Q^2 via the PDFs, $f_i(x, Q^2)$, for parton i, i.e.

$$F_1(x, Q^2) = \sum_i e_i^2 f_i(x, Q^2), \quad F_2(x, Q^2) = xF_1(x, Q^2) = x \sum_i e_i^2 f_i(x, Q^2). \tag{8.151}$$

In Eq. (8.151), the quark charge is denoted e_i to avoid confusion with the quark distribution function that will be denoted from now on $q_i(x, Q^2)$ for a quark flavour i. To first order in α_s, the evolution of $q_i(x, Q^2)$ due to the diagrams of Fig. 8.27 is given by

$$\frac{\partial q_i(x, Q^2)}{\partial \ln Q^2} = \frac{\alpha_s}{2\pi} \int_0^1 \mathrm{d}y \int_0^1 \mathrm{d}z\, q_i(y, Q^2) P_{q\to qg}(z)\, \delta(x - zy) = \frac{\alpha_s}{2\pi} \int_x^1 \frac{\mathrm{d}y}{y} q_i(y, Q^2) P_{q\to qg}\left(\frac{x}{y}\right), \tag{8.152}$$

where the function $P_{q\to qg}(z)$ is called a *splitting function*. It represents the probability that a quark, having emitted a gluon, has its original momentum reduced by a factor z. Equation (8.152) describes the variation of the quark distribution function due to a change of scale Q^2 for a quark with momentum xp. When the scale is larger, one can resolve the emission of a soft gluon by a quark[32]. A quark with momentum xp may then originates from a parent quark with a larger momentum yp that has radiated a gluon with a probability $\frac{\alpha_s}{2\pi} P_{q\to qg}(z)$, reducing the momentum to $zyp = xp$. The splitting function is given by

$$P_{q\to qg}(z) = C_F \left(\frac{1+z^2}{1-z}\right), \qquad (z < 1). \tag{8.153}$$

As expected, the probability of emitting a quark is proportional to the colour factor relevant for the transition $q \to qg$. The fraction of momentum of the parent quark carried by the final quark is $z = x/y$. According to the Formula (8.153), the probability is larger when z is close to 1. For x close to 1, since $1 \geq y \geq x$, the ratio x/y is necessarily close to unity. Hence, the probability of radiating a gluon when $x \simeq 1$ is necessarily high. As $x = Q^2/(2p\cdot q)$, when Q^2 increases and $x \simeq 1$, x gets closer to one. Thus, as Q^2 increases, we expect the quark to radiate more and more gluons in the large x region. Therefore, more gluons are present in the nucleon, and the quark density function (and thus F_2) should then decrease in this regime. This is indeed observed in Fig. 7.9 that shows the structure function F_2 as a function of Q^2 over a wide range of x. On the other hand, we expect the opposite behaviour at small x, i.e. a larger quark density when Q^2 increases, which is confirmed by the data in Fig. 7.9. Note that the splitting function (8.153) does not depend on the quark flavour. It is only correct in the approximation where the quark masses are neglected. The splitting function (8.153) has a singularity in $z = 1$, which belongs to the integration domain of

[32] Alternatively, one can say that at a low scale, even if gluons are emitted, $1/Q$ represents a time scale (in natural units) long enough such that they are re-absorbed and hence have no impact on the quark distribution function.

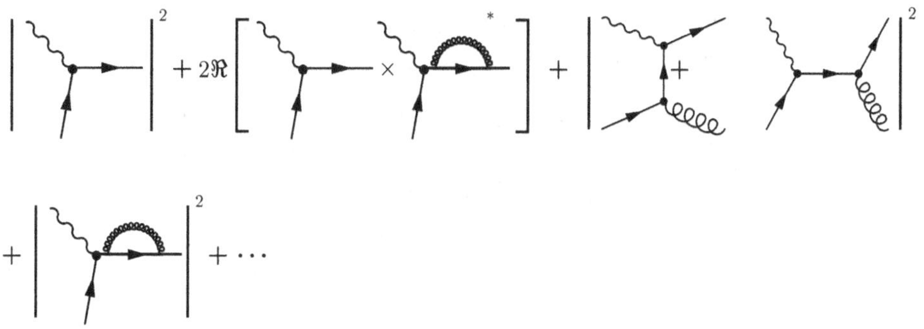

Fig. 8.28 The born diagram contribution $O(1)$ in α_s (first diagram of the first row) receives correction at order $O(\alpha_s)$ from virtual gluon emission diagrams (product in the first row) and from real gluon emission diagrams (last term of the first row). The second row is at order larger than $O(\alpha_s)$. Only one virtual gluon emission is shown. There are two others with the gluon loop on the initial state quark and with a vertex correction.

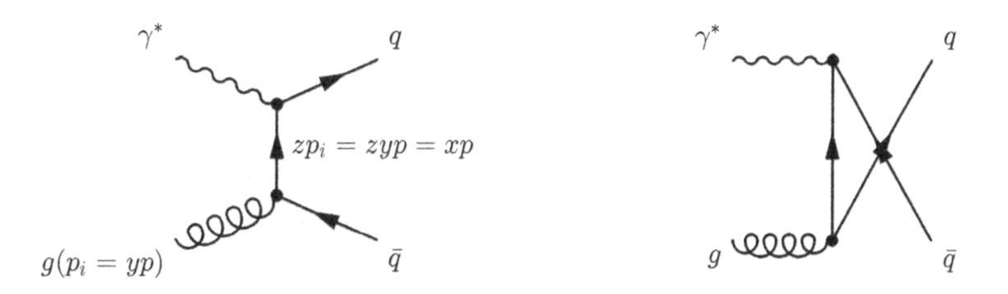

Fig. 8.29 Diagrams describing the correction to the deep inelastic scattering due to the splitting of a gluon.

Eq. (8.152). However, one can show that, at the order α_s, the interference terms between the born process (without gluons) and the virtual gluon processes cancel the divergences (see Fig. 8.28). The splitting function is then modified to the smooth function (actually a distribution),

$$P_{q\to qg}(z) = C_F \left[\frac{1+z^2}{1-z}\right]_+ = C_F\left(\frac{1+z^2}{[1-z]_+} + 2\delta(1-z)\right), \tag{8.154}$$

where $[F(z)]_+$ defines a distribution such as for any regular test function in $f(z)$,

$$\int_0^1 \mathrm{d}z\,[F(z)]_+ = \int_0^1 \mathrm{d}z\,(f(z) - f(1))F(z), \quad \text{and} \quad [F(z)]_+ = F(z) \quad \text{for} \quad z < 1.$$

At the order α_s, another process should also be taken into account. Instead of having a quark as the initial parton, a gluon (from the sea) can play this role. As shown in the diagrams of Fig. 8.29, a gluon carrying a fraction y of the proton momentum can split into a quark–antiquark pair, producing a quark carrying a fraction $z = x/y$ of the gluon momentum (and thus a fraction x of the proton momentum). The splitting function $P_{g\to q\bar{q}}$ describing this probability reads

$$P_{g\to q\bar{q}}(z) = T_F(z^2 + (1-z)^2). \tag{8.155}$$

Denoting by $g(x, Q^2)$ and $\bar{q}_i(x, Q^2)$, the gluon and the antiquark distribution functions, respectively, the evolution of the PDFs at leading order α_s now becomes

$$\frac{\partial q_i(x,Q^2)}{\partial \log Q^2} = \frac{\alpha_s}{2\pi}\int_x^1 \frac{dy}{y}\left[q_i(y,Q^2)P_{q\to qg}\left(\frac{x}{y}\right) + g(y,Q^2)P_{g\to q\bar{q}}\left(\frac{x}{y}\right)\right], \tag{8.156a}$$

$$\frac{\partial \bar{q}_i(x,Q^2)}{\partial \log Q^2} = \frac{\alpha_s}{2\pi}\int_x^1 \frac{dy}{y}\left[\bar{q}_i(y,Q^2)P_{q\to qg}\left(\frac{x}{y}\right) + g(y,Q^2)P_{g\to q\bar{q}}\left(\frac{x}{y}\right)\right], \tag{8.156b}$$

$$\frac{\partial g(x,Q^2)}{\partial \log Q^2} = \frac{\alpha_s}{2\pi}\int_x^1 \frac{dy}{y}\left[g(y,Q^2)P_{g\to gg}\left(\frac{x}{y}\right) + \sum_i \left(q_i(y,Q^2) + \bar{q}_i(y,Q^2)\right)P_{q\to qg}\left(\frac{x}{y}\right)\right], \tag{8.156c}$$

where $P_{g\to gg}$ is the splitting function[33] for a gluon splitting into two gluons. It reads

$$P_{g\to gg}(z) = 2C_A\left(\frac{z}{[1-z]_+} + \frac{1-z}{z} + z(1-z)\right) + \frac{11C_A - 4T_F N_f}{6}\delta(1-z), \tag{8.157}$$

where N_f is the number of quark flavours. In Eqs. (8.156b) and (8.156c), the same splitting function $P_{q\to qg}$ is used for quarks and antiquarks because of the invariance of QCD under charge conjugation. Equations (8.156a–8.156c) are called the DGLAP evolution equations. The acronym DGLAP stands for the five physicists who established them: Dokshitzer, Gribov, Lipatov, Altarelli and Parisi. They give the evolution of the PDFs, and once the PDFs are known at a given scale Q_0^2, their values at another scale Q^2 can be predicted. Note that if α_s vanished, there would have been no QCD interaction and, therefore, no evolution of the PDFs with Q^2. The Bjorken scaling would be restored. Since $\alpha_s(Q^2)$ tends to 0 for large Q^2, Bjorken scaling is thus a good approximation at high energy.

In summary, when the nucleon structure is probed at larger Q^2, i.e. finer scales, more partons (namely gluons and $q\bar{q}$ pairs) appear in such a way that it can be predicted by the perturbative QCD theory.

8.6.2 Application to Hadron Colliders

At hadron colliders, hard scattering occurs between partons from the two incident hadrons and produces many particles, including potentially other hadrons. For instance, let us consider the case of a proton collider such as the LHC, where the collision process can be summarised by the following diagram:

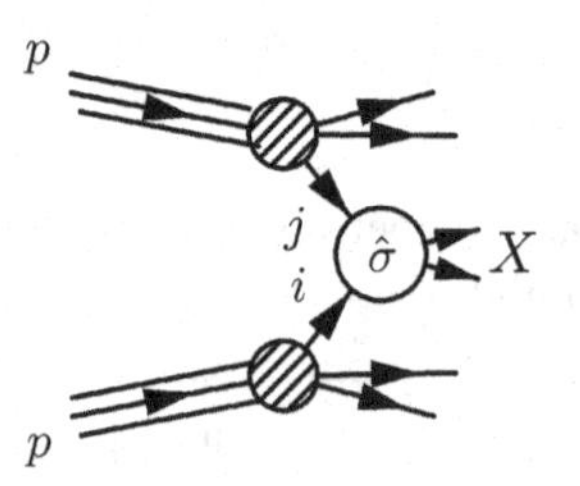

[33] Splitting functions are often denoted in the literature as P_{ab}, describing the parton splitting $b \to a$, where the parton a carries a fraction z of the parton b momentum. Hence, $P_{q\to qg} \equiv P_{qq}$, $P_{g\to q\bar{q}} \equiv P_{qg}$ and $P_{g\to gg} \equiv P_{gg}$.

Particles i and j denote the two partons, and $\hat{\sigma}$ is the cross section of the hard process corresponding to the reaction considered $i+j \to X$, with X a final state potentially containing hadrons. The cross section measured in experiments is not $\hat{\sigma}$ but $\sigma_{pp\to X}$. It is related to the cross section of the hard process by

$$\begin{aligned}\sigma_{pp\to X} &= \int_0^1 \mathrm{d}x_1\,\mathrm{d}x_2 \sum_{ij} f_i(x_1,Q^2) f_j(x_2,Q^2)\hat{\sigma}_{ij\to X} \\ &= \int_0^1 \mathrm{d}x_1\,\mathrm{d}x_2 \sum_{ij} f_i(x_1,Q^2) f_j(x_2,Q^2) \sum_k \int_0^1 \mathrm{d}z\hat{\sigma}_{ij\to k} D_k^X(z,Q^2),\end{aligned} \tag{8.158}$$

where $f_{i,j}$ are the PDFs of the partons i,j, and Q^2 is the momentum scale at which they are evaluated (via the appropriate DGLAP equation). We sum over all types of partons that could contribute to the final state X. All fractions of momentum $x_{1,2} \in [0,1]$ are also considered. In the second line of Eq. (8.158), the cross section $\hat{\sigma}_{ij\to X}$ is decomposed in cross sections, $\hat{\sigma}_{ij\to k}$, only involving partons. The fragmentation function D_k^X is thus used. The cross section $\hat{\sigma}_{ij\to k}$ is often referred to as the short distance (high momentum) cross section. It can be calculated in QCD with the perturbative series of α_s. All singularities corresponding to a long distance (low momentum) physics have been factored out and absorbed in the PDFs and fragmentation functions. The scale Q^2 thus separates the high- and low-momentum physics. It has to be representative of the hard process: the squared mass of a resonance (if X is produced via the resonance), p_T^2, etc.

As a concrete example, let us take the Drell–Yan production, where $i = q$, $j = \bar{q}$ and $X = \mu^-\mu^+$. The final state X does not contain hadrons; hence, only the first line of Eq. (8.158) is relevant. Let $\sqrt{s}$ be the energy in the CM frame of the two protons and p_1 and p_2 their 4-momenta. Neglecting their mass, $s = (p_1+p_2)^2 \simeq 2p_1\cdot p_2$. Now, denoting by p_q and $p_{\bar{q}}$, the parton 4-momenta, the actual energy $\sqrt{\hat{s}}$ available for the partons interaction is (still neglecting the masses) $\hat{s} \simeq 2p_q\cdot p_{\bar{q}} = 2x_1x_2\,p_1\cdot p_2$, and thus, $\hat{s} \simeq x_1x_2 s$. The cross section $\hat{\sigma}_{q\bar{q}\to\mu^-\mu^+}$ is easy to calculate from that of the QED process, $e^-e^+ \to \mu^-\mu^+$, already calculated in Eq. (6.102). Denoting by e_q, the quark charge and taking into account the colour factors, it yields

$$\hat{\sigma}_{q\bar{q}\to\mu^-\mu^+} = \frac{1}{3}\times\frac{1}{3}\times 3\times\frac{4\pi\alpha^2 e_q^2}{3\hat{s}} = \frac{4\pi\alpha^2}{9\hat{s}}e_q^2.$$

Then, according to Eq. (8.158), the double differential cross section for the proton collisions is

$$\frac{\mathrm{d}^2\sigma_{pp\to\mu^-\mu^+}}{\mathrm{d}x_1\,\mathrm{d}x_2} = \frac{4\pi\alpha^2}{9\hat{s}}\sum_q e_q^2\left[q(x_1,Q^2)\bar{q}(x_2,Q^2) + q(x_2,Q^2)\bar{q}(x_1,Q^2)\right], \tag{8.159}$$

where the quark and antiquark PDFs are denoted $q()$ and $\bar{q}()$, respectively. The natural scale for Q is the mass of the $\mu^+\mu^-$ pair, $M_{\mu\mu}$. Instead of using x_1 and x_2, it is convenient to express the differential cross section with variables that can be directly measured in experiments. They are, for instance, $M_{\mu\mu}$ and the rapidity y [see Eq. (2.49)] of the muon pair (or equivalently of the parton pair). In the CM of the two protons, the components of parton 4-momenta are

$$p_q = \frac{\sqrt{s}}{2}(x_1,0,0,x_1), \quad p_{\bar{q}} = \frac{\sqrt{s}}{2}(x_2,0,0,-x_2),$$

where, as usual, the two proton beams are moving along the z-axis. Hence, the energy and the longitudinal momentum read

$$E = \frac{\sqrt{s}}{2}(x_1 + x_2), \quad p_L = \frac{\sqrt{s}}{2}(x_1 - x_2),$$

implying the rapidity

$$y = \frac{1}{2}\ln\left(\frac{E + p_L}{E - p_L}\right) = \frac{1}{2}\ln\frac{x_1}{x_2}.$$

Given that $\hat{s} = x_1 x_2 s = M_{\mu\mu}^2$, we deduce the following expressions for x_1 and x_2:

$$x_1 = \sqrt{\frac{\hat{s}}{s}}e^{y}, \quad x_2 = \sqrt{\frac{\hat{s}}{s}}e^{-y}, \quad \text{or } x_{1,2} = \frac{M_{\mu\mu}}{\sqrt{s}}e^{\pm y}. \tag{8.160}$$

Therefore,

$$\mathrm{d}x_1\,\mathrm{d}x_2 = \begin{vmatrix} \frac{\partial x_1}{\partial y} & \frac{\partial x_1}{\partial M_{\mu\mu}} \\ \frac{\partial x_2}{\partial y} & \frac{\partial x_2}{\partial M_{\mu\mu}} \end{vmatrix} \mathrm{d}y\,\mathrm{d}M_{\mu\mu} = \begin{vmatrix} \frac{M_{\mu\mu}}{\sqrt{s}}e^{y} & \frac{1}{\sqrt{s}}e^{y} \\ -\frac{M_{\mu\mu}}{\sqrt{s}}e^{-y} & \frac{1}{\sqrt{s}}e^{-y} \end{vmatrix} \mathrm{d}y\,\mathrm{d}M_{\mu\mu} = \frac{2M_{\mu\mu}}{s}\mathrm{d}y\,\mathrm{d}M_{\mu\mu},$$

and the Drell–Yan cross section at LHC (8.159) finally reads

$$\frac{\mathrm{d}^2\sigma_{pp\to\mu^-\mu^+}}{\mathrm{d}y\,\mathrm{d}M_{\mu\mu}} = \frac{8\pi\alpha^2}{9M_{\mu\mu}s}\sum_q e_q^2\left[q(x_1, M_{\mu\mu}^2)\bar{q}(x_2, M_{\mu\mu}^2) + q(x_2, M_{\mu\mu}^2)\bar{q}(x_1, M_{\mu\mu}^2)\right]. \tag{8.161}$$

Different values of $M_{\mu\mu}$ and y allow probing the PDFs at different values of Bjorken-x via the correspondence given in Eq. (8.160). In particular, the antiquark PDFs can be studied, whereas the sensitivity of deep inelastic scattering experiments to these PDFs is rather weak. Note that for the cross-section measurement, as x_1 and x_2 are positive, $x_1 \leq x_1 + x_2 = 2E/\sqrt{s}$. But, according to Eq. (8.160),

$$y = \frac{1}{2}\ln\left(\frac{x_1^2 s}{M_{\mu\mu}^2}\right) \leq \ln\left(\frac{2E}{M_{\mu\mu}}\right).$$

Therefore, the larger the mass $M_{\mu\mu}$ of the produced system, the more centrally (i.e., $y \to 0$) it is produced in the detector.

Problems

8.1 Applications of the isospin symmetry.

The deuteron d is a bound state of one proton and one neutron. It is an isospin singlet.

1. The reaction $p + p \to \pi^+ + d$ is easily seen. Explain why it excludes that the pion is an isospin singlet or a member of a doublet.
2. We now assume that pions are members of an isotriplet. Using isospin invariance, show that

$$\sigma(p + p \to \pi^+ + d) = \sigma(n + n \to \pi^- + d).$$

3. What do you expect for $\sigma(n+p \to \pi^0 + d)$ with respect to the two previous cross sections?

8.2 Prove using the weight diagram that in SU(3), $\mathbf{3} \otimes \mathbf{3} \otimes \mathbf{3} = \mathbf{10} + \mathbf{8} + \mathbf{8} + \mathbf{1}$ and complete the missing states in Fig. 8.6.

8.3 Prove using the weight diagram that in SU(2), $\mathbf{2} \otimes \mathbf{2} \otimes \mathbf{2} = \mathbf{4} + \mathbf{2} + \mathbf{2}$ and use the SU(2) ladder operator J_- of Eq. (B.10) to obtain the states listed in Eq. (8.21).

8.4 Expand the states $|\Delta^+, \uparrow\rangle$ and $|\Lambda, \uparrow\rangle$ (see Figs. 8.1 and 8.6) in terms of flavour and spin products.

8.5 The Λ baryon is the isospin singlet located at the centre of the baryon octet. Using the result of the previous problem, show that its magnetic moment is $\mu_\Lambda = \mu_s$. Given that the measured value is $\mu_\Lambda = -0.61\mu_N$, estimate the s quark mass (constituent mass).

8.6 The reaction $e^+e^- \to q\bar{q}$
I recall some tensorial products relevant for QCD: $\mathbf{3} \otimes \bar{\mathbf{3}} = \mathbf{8} \oplus \mathbf{1}, \mathbf{8} \otimes \mathbf{8} = \mathbf{27} \oplus \overline{\mathbf{10}} \oplus \mathbf{8} \oplus \mathbf{1} \oplus \mathbf{10} \oplus \mathbf{8}$.

1. What are the 'colour hypercharge' and the 'colour isospin' quantum numbers of the $q\bar{q}$ pair? Justify the answer.
2. Deduce which states among the octet and the singlet fulfil the condition of the first question.
3. Show that with respect to the cross section σ_0 of the reaction $e^+e^- \to \mu^+\mu^-$, the cross section of $e^+e^- \to q\bar{q}$ can be written as $\sigma = \sigma_0 Q^2 (N_{\text{octet}} + N_{\text{singlet}})$, with $N_{\text{octet}} = 2$ and $N_{\text{singlet}} = 1$, and Q is the electric charge of the quark.
4. We assume that the $q\bar{q}$ pair was produced in the octet states. Explain why additional gluons are expected in the event? Imagine that the $q\bar{q}$ pair finally forms a meson, then what is the minimal number of gluons expected?

8.7 Following the colour factor calculations in Section 8.3.3, show that a system of two quarks in a anti-triplet colour state has a colour factor $-2/3$. Similarly, check that the colour factor corresponding to a transition between a colour sextet state to a colour anti-triplet state is zero.

8.8 To prove the colour Fierz identity (8.78),

$$\frac{1}{4}\sum_{a=1}^{8} (\lambda_a)_{\alpha'\alpha} (\lambda_a)_{\beta'\beta} = \frac{1}{2}\left(\delta_{\alpha'\beta}\delta_{\alpha\beta'} - \frac{1}{3}\delta_{\alpha'\alpha}\delta_{\beta'\beta}\right),$$

start by considering that any Hermitian matrix M can be decomposed on the basis of $\{\mathbb{1}, \lambda_1, \ldots, \lambda_8\}$, i.e. $M = c_0\mathbb{1} + c_a\lambda_a$ (implicit summation on a). Show that $c_0 = \text{Tr}(M)/3$ and $c_a = \text{Tr}(\lambda_a M)/2$ [keep in mind the normalisation (8.35)]. Then, express the component $M_{\alpha'\alpha}$ as $M_{\beta\beta'}\delta_{\alpha'\beta}\delta_{\alpha\beta'}$ and show that

$$M_{\beta\beta'}\delta_{\alpha'\beta}\delta_{\alpha\beta'} = \frac{1}{3}M_{\beta\beta'}\delta_{\beta'\beta}\delta_{\alpha\alpha'} + \frac{1}{2}(\lambda_a)_{\beta'\beta}M_{\beta\beta'}(\lambda_a)_{\alpha'\alpha}.$$

Deduce the Fierz identity.

8.9 Alternative proof of Eq. (8.79). Start from Eq. (8.70) and check that $\sum_{\alpha,\alpha',\beta,\beta'} |C^{\alpha\alpha'}_{\beta\beta'}|^2 = 2$. Hint: do not forget that the Gell-Mann matrices are Hermitian.

8.10 To prove Eq. (8.97), use first Eq. (8.36) to show (no implicit summation)

$$\sum_a f_{abd}\frac{1}{2^3}\text{Tr}\,(\lambda_a\lambda_b\lambda_c) = \frac{i}{2^4}\left[-\text{Tr}(\lambda_b\lambda_d\lambda_b\lambda_c) + \text{Tr}(\lambda_d\lambda_b\lambda_b\lambda_c)\right].$$

Then, using the Fierz identity (8.78), show that

$$\sum_b \frac{1}{2^3}(\lambda_b\lambda_d\lambda_b)_{\alpha'\beta} = \sum_{\alpha\beta'}\sum_b \frac{1}{2}(\lambda_d)_{\alpha\beta'}\frac{1}{2}(\lambda_b)_{\alpha'\alpha}\frac{1}{2}(\lambda_b)_{\beta'\beta} = -\frac{1}{6}\frac{1}{2}(\lambda_d)_{\alpha'\beta}.$$

Now, multiply by $\frac{1}{2}(\lambda_c)_{\beta\alpha'}$ the both sides and sum over $\beta\alpha'$ to take the trace and deduce

$$\sum_b \frac{1}{2^4}\mathrm{Tr}(\lambda_b\lambda_d\lambda_b\lambda_c) = -\frac{1}{12}\delta_{dc}.$$

Similarly, using Eq. (8.88), check that

$$\sum_b \frac{1}{2^4}\mathrm{Tr}(\lambda_d\lambda_b\lambda_b\lambda_c) = \frac{2}{3}\delta_{dc}.$$

Finally, deduce the result (8.97).

8.11 Colour factors for baryons. The baryon colour state is based on the singlet resulting from $\mathbf{3}\otimes\mathbf{3}\otimes\mathbf{3} = (\mathbf{6}\oplus\bar{\mathbf{3}})\otimes\mathbf{3}$. Calculate the colour factor for two quarks in $\mathbf{6}$ and two quarks in $\bar{\mathbf{3}}$. Which configuration is repulsive or attractive? What can you conclude for the singlet configuration?

8.12 Quarks annihilation into gluons.

1. Draw all diagrams at the lowest order for the process $q\bar{q} \to gg$.
2. How must these diagrams be combined?
3. Write down the amplitudes and express them when appropriate as a function of the corresponding amplitude in QED for $q\bar{q} \to \gamma\gamma$. Use the following notations $q(p_1, \text{colour } \alpha)\bar{q}(p_2, \text{colour } \beta) \to g(p_3, \text{colour } a)g(p_4, \text{colour } b)$.

9 Weak Interaction

So far, we have studied the electromagnetic and the strong interaction as two examples of gauge theories. This chapter presents the last interaction described by the Standard Model of particle physics, the weak interaction. We will follow a historical approach, trying to explain the evolution of its theoretical description. The gauge theory of the weak interaction will be described in Chapter 10, within the context of the unification of electromagnetism and the weak interaction.

9.1 Fermi Theory

9.1.1 Birth of a New Interaction

In 1934, Fermi proposed a new theory to describe the decay of a neutron into a proton, the so-called β-decay, $n \to p + e^- + \bar{\nu}$. The neutrino had not yet been discovered, but had been postulated by Pauli to explain the observed continuous energy spectrum of the electron and to satisfy the conservation of the angular momentum. By analogy with the nuclear γ-decay where a photon is emitted, $N \to N\gamma$, Fermi assumed that the electron and the neutrino were created in the same process leading to the Feynman graph:

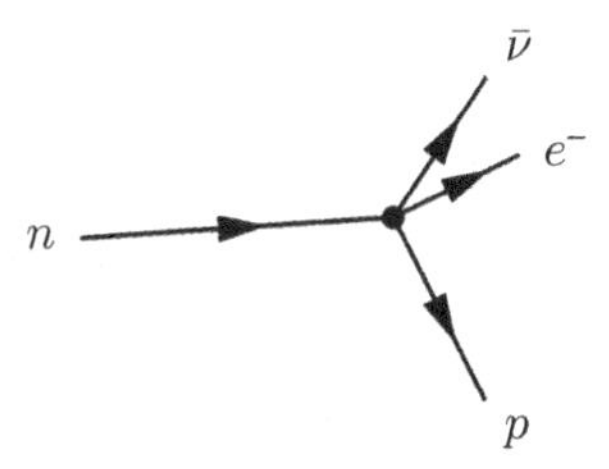

This graph is an example of a 4-fermion contact interaction (there is no propagation of gauge bosons). While in QED, the nuclear γ-decay can be described by the elementary transition $p \to p\gamma$ (the proton being bound in a nucleus) with the Lagrangian

$$\mathcal{L} = e(\bar{u}_p \gamma^\mu u_p) A_\mu,$$

Fermi replaced the proton current by a current with the transition $n \to p$ and the photon A_μ by a current expressing the creation of the electron and neutrino, i.e.

$$\mathcal{L} = G_F(\bar{u}_p \gamma^\mu u_n)(\bar{u}_e \gamma_\mu u_\nu).$$

The two currents imply a change of the electric charge in the particles, whereas, in QED, the charge remains constant. The elementary charge e is replaced by a new coupling constant, the Fermi constant, whose value $G_F \simeq 1.2\ 10^{-5}\ \mathrm{GeV}^{-2}$ was obtained to match the measurement of the β-decay rate. A new coupling constant and a new courant meant that a new force was born. The Fermi coupling constant is so small that the probability of interaction is necessarily small, leading to naming this new interaction: the *weak force*.

9.1.2 The θ^+ and τ^+ Mystery

Many experiments were made to better understand the properties of the weak interaction. In the 1950s, two particles called θ^+ and τ^+ discovered in cosmic rays (the latter was *not* the now well-known lepton of the third family) had very similar properties: the same mass and the same lifetime (within the experimental errors). However, they decayed via the weak interaction differently

$$\theta^+ \to \pi^+ + \pi^0, \qquad \tau^+ \to \pi^+ + \pi^+ + \pi^-.$$

Therefore, they were considered as two different particles because, according to the decays above, they have opposite parities (see Section 5.3.5), which were assumed to be conserved by all interactions. Indeed, given that the intrinsic parity of pions is -1, their spin is 0, as well as those of θ^+ and τ^+, the parity of θ^+ is

$$\eta(\theta^+) = \eta(\pi^+)\,\eta(\pi^0)\,\eta(\text{spatial}) = (-1)(-1)(-1)^{l=0} = +1,$$

where the orbital quantum number $l = 0$ due to the conservation of the angular momentum ($\boldsymbol{J} = \boldsymbol{L} + \boldsymbol{S}$). For the τ^+ parity, one has to consider the orbital momentum $\boldsymbol{l_1}$ between the first two πs and the orbital momentum $\boldsymbol{l_2}$ between the third π and the barycentre of the first two πs. As τ^+ is spinless, necessarily $\boldsymbol{0} = \boldsymbol{l_1} + \boldsymbol{l_2}$, implying that $l_1 = l_2$, and thus, $\eta(\text{spatial}) = (-1)^{2l_1} = +1$. Therefore, the τ^+ parity is

$$\eta(\tau^+) = \eta(\pi^+)\,\eta(\pi^+)\,\eta(\pi^-)\,\eta(\text{spatial}) = (-1)(-1)(-1)(+1) = -1.$$

In 1956, two physicists, Yang and Lee, suggested that θ^+ and τ^+ were actually the same particles K^+ in our modern nomenclature, implying that parity is not conserved by the weak interaction. A real revolution in physics! Nature distinguishes left and right[1]!

9.1.3 Wu Experiment: When Left Is Not Right

After such a revolutionary hypothesis, the challenge was to prove it experimentally. Yang and Lee published a paper (Lee and Yang, 1956) suggesting several experiments. The first one was led by Wu using a source of cobalt polarised, thanks to an external magnetic field. The cobalt has the following β-decay:

$${}^{60}\mathrm{Co} \to {}^{60}\mathrm{Ni}^* + e^- + \bar{\nu}.$$

The spins of the nuclei are $J(\mathrm{Co}) = 5$ and $J(\mathrm{Ni}^*) = 4$. The magnetic field was large enough and the temperature was low enough (about 0.01 K!) to ensure that the spin of the

[1] Indeed, your left hand appears as a right hand in a mirror. The non-conservation of parity induces the two situations to lead to different physics measurements (with particles, not your hands!).

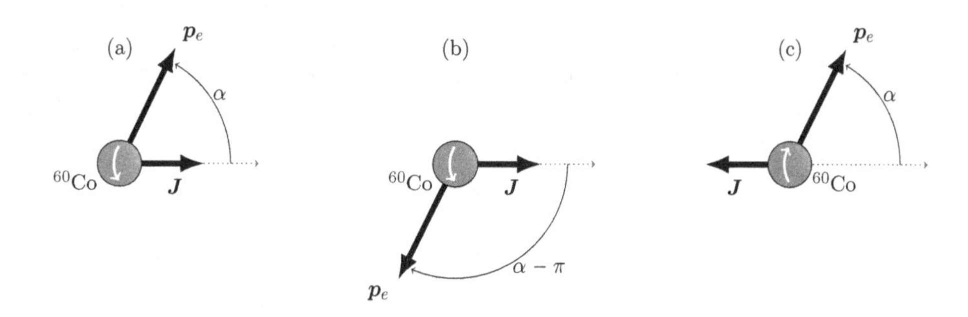

Fig. 9.1 Sketch of Wu's experiment. (a) The electron is emitted with an angle α with respect to the horizontal axis. (b) The parity transformed of (a). (c) The parity transformed of (a) with an additional rotation equivalent to the inversion of the direction of the cobalt spin.

cobalt was properly aligned. Since angular momenta (both the spin and the orbital angular momenta) are pseudo-vectors, they remain unchanged under the parity transformation [cf. Eq. (4.10)]. Figure 9.1 depicts the schematic diagram of the experiment. If parity were conserved, the rate of electrons emitted with an angle α (Fig. 9.1a) should be equal to that of the parity transformed situation, i.e. with an angle $\alpha - \pi$ (Fig. 9.1b), with the angle being measured with respect to the nuclei spin or equivalently the magnetic field axis. Experimentally, it was easier to measure the rates at the same angle α but with the direction of the magnetic field inverted (Fig. 9.1c). Wu and her team found in Wu et al. (1957) different rates and thus an asymmetry between the original setup and the parity transformed setup, indicating a clear sign of parity violation! Electrons were more likely to be emitted in a direction opposite to that of the cobalt spin (an angle $\alpha = \pi$ in Fig 9.1a). As $J(\text{Co}) = 5$ and $J(\text{Ni}^*) = 4$, the angular momentum conservation imposes that the $e^-\bar{\nu}$ system has $J(e^-\bar{\nu}) = 1$. Since $J_z(\text{Co}) = 5$ (the z-axis is chosen as the horizontal axis in Fig. 9.1), $J_z(e^-\bar{\nu}) = 1$, implying that both the electron and the neutrino have $S_z = +1/2$. The nickel nucleus is much heavier than the $e^-\bar{\nu}$ pair, and thus, its recoil can be neglected. Since the cobalt nucleus was at rest, the electron and the anti-neutrino were emitted back to back and therefore had an opposite helicity. The electron being more likely emitted in the opposite direction of its spin, the left-handed helicity is favoured, whereas the anti-neutrino should be right-handed. If their masses are neglected (a reasonable assumption here), both helicity and chirality are equivalent (see chirality section in Section 5.3.5). Consequently, Wu's experiment suggested that the weak interaction (at least a β-decay) produces a left-handed chiral particle and a right-handed chiral anti-particle.

9.2 Modification of Fermi Theory

9.2.1 The *V*–*A* Current

Parity violation signalled the death of Fermi theory. Indeed, the current involved in the Fermi Lagrangian had the same structure as that of QED (by construction) or QCD (for the spinor part), i.e. $j^\mu = u_1^\dagger \gamma^0 \gamma^\mu u_2$, where u_1 and u_2 are generic 4-spinors (u-spinors or v-spinors). We have already established in Eq. (5.137) that the QED current behaves as a scalar for its time component and as a vector for its space components, i.e. (given that the parity operator for spinors is γ^0)

$$j^0 = u_1^\dagger \gamma^0 \gamma^0 u_2 \xrightarrow{\text{parity}} (\gamma^0 u_1)^\dagger \gamma^0 \gamma^0 (\gamma^0 u_2) = j^0,$$

$$j^i = u_1^\dagger \gamma^0 \gamma^i u_2 \xrightarrow{\text{parity}} (\gamma^0 u_1)^\dagger \gamma^0 \gamma^i (\gamma^0 u_2) = u_1^\dagger \gamma^i \gamma^0 u_2 = -j^i.$$

Therefore, the amplitude, being made with the product of two such currents, j_1^μ and j_2^μ, transforms as

$$\mathcal{M} = j_1^\mu j_{2\mu} = j_1^0 j_{20} + j_1^i j_{2i} \xrightarrow{\text{parity}} j_1^0 j_{20} + (-j_1^i)(-j_{2i}) = j_1^\mu j_{2\mu}.$$

It is invariant under the parity transformation. The same conclusion is valid for QCD amplitudes since in the QCD current there is just an additional Gell-Mann matrix, which remains unchanged under the parity transformation.

How should we modify the current to include parity violation? We have just seen that electrons with left-handed helicity were favoured in the Wu experiment. We could think that inserting the helicity projector would be enough. However, in Section 5.3.4, we saw that helicity is not Lorentz invariant. Obviously, weak interaction cannot depend on the frame! Relativistic fermions are described by spinors, so a covariant current is necessarily a bilinear form of fermion fields:

$$j = \bar{\psi} \Gamma \psi = \psi^\dagger \gamma^0 \Gamma \psi,$$

where Γ is a 4×4 matrix (since ψ has four components). The constraints of Lorentz invariance and hermiticity of the current reduce the possibilities of Γ to these possibilities:

$$\Gamma = 1,\ i\gamma^5,\ \gamma^\mu,\ \gamma^\mu \gamma^5,\ i(\gamma^\mu \gamma^\nu - \gamma^\nu \gamma^\mu),$$

which correspond to 16 linearly independent 4×4 matrices. Among them, only $\Gamma = \gamma^\mu$ and $\Gamma = \gamma^\mu \gamma^5$ lead to a current that can be contracted by a vector field A_μ, with the other possibilities being Lorentz scalars (1, $i\gamma^5$) or a rank-2 tensor [$i(\gamma^\mu \gamma^\nu - \gamma^\nu \gamma^\mu)$, which has six components, and thus would need to be contracted with a spin-2 boson field, $A_{\mu\nu}$]. The case $\Gamma = \gamma^\mu$ is the usual QED case. For $\Gamma = \gamma^\mu \gamma^5$, using the anti-commutation property (5.93), it is easy to check that under the parity transformation,

$$j^\mu = u_1^\dagger \gamma^0 \gamma^\mu \gamma^5 u_2 \xrightarrow{\text{parity}} (\gamma^0 u_1)^\dagger \gamma^0 \gamma^\mu \gamma^5 \gamma^0 u_2 = -u_1^\dagger \gamma^\mu \gamma^0 \gamma^5 u_2,$$

i.e.

$$j^0 \xrightarrow{\text{parity}} -\bar{u}_1 \gamma^0 \gamma^5 u_2 = -j^0, \qquad j^i \xrightarrow{\text{parity}} +\bar{u}_1 \gamma^\mu \gamma^5 u_2 = j^i.$$

The spatial components are not affected, which implies that j^μ is an axial (pseudo) vector. Clearly, an amplitude based on the products of two axial vectors would remain invariant under the parity transformation. However, parity violation is obtained by a linear combination of a vector current $j_V^\mu = \bar{u}_1 \gamma^\mu u_2$ and an axial vector $j_A^\mu = \bar{u}_1 \gamma^\mu \gamma^5 u_2$, i.e.

$$j^\mu = c_V j_V^\mu - c_A j_A^\mu = \bar{u}_1 \gamma^\mu (c_V - c_A \gamma^5) u_2, \tag{9.1}$$

where the minus sign is purely conventional and c_V and c_A are simple coefficients. Indeed, an amplitude based on the product of two such currents,

$$\begin{aligned}\mathcal{M} &= \left(c_V j_V^\mu - c_A j_A^\mu\right)\left(c'_V j'_{V\mu} - c'_A j'_{A\mu}\right)\\ &= c_V c'_V j_V^\mu j'_{V\mu} + c_A c'_A j_A^\mu j'_{A\mu} - c_V c'_A j_V^\mu j'_{A\mu} - c_A c'_V j_A^\mu j'_{V\mu},\end{aligned}$$

transforms as (check it)

$$\mathcal{M} \xrightarrow{\text{parity}} \mathcal{M}' = \left(c_V j_V^\mu + c_A j_A^\mu\right)\left(c'_V j'_{V\mu} + c'_A j'_{A\mu}\right),$$

which is different than $\mathcal{M}$ if $c_V c'_A \neq 0$ or $c'_V c_A \neq 0$. The weak current (9.1) based on the difference between a vector current and an axial vector current implies parity violation. It is often called a $V - A$ *current*. To find the values of the coefficients c_V and c_A in Eq. (9.1), let us come back to the β-decay $n \to p\, e^-\, \bar{\nu}$, where the current between the outgoing electron and the outgoing anti-neutrino would be

$$j_{e\bar{\nu}}^\mu = \bar{u}_e \gamma^\mu (c_V - c_A \gamma^5) v_\nu. \tag{9.2}$$

Wu's experiment showed that the electron was mainly produced with a left-handed helicity, which, in the low mass limit, is equivalent to a left-handed chirality state. In Section 5.3.5, we saw that the left-handed chirality projector for fermions [Eq. (5.98)] is $P_L = (1-\gamma^5)/2$. If $c_V = c_A = 1/2$ in Eq. (9.2), given the properties (5.93) and (5.94) of γ^5,

$$j_{e\bar{\nu}}^\mu = \bar{u}_e \gamma^\mu P_L v_\nu = u_e^\dagger \gamma^0 \gamma^\mu P_L v_\nu = u_e^\dagger P_L \gamma^0 \gamma^\mu v_\nu = (P_L u_e)^\dagger \gamma^0 \gamma^\mu v_\nu = \overline{u_{e_L}} \gamma^\mu v_\nu,$$

where u_{e_L} is the (chiral) left-handed spinor of the electron. With such a current, the electron would be left-handed, as observed in Wu's experiment. Moreover, since $P_L = P_L^2$,

$$j_{e\bar{\nu}}^\mu = \bar{u}_e \gamma^\mu P_L P_L v_\nu = \overline{u_{e_L}} \gamma^\mu v_{\nu R},$$

because P_L is the right-handed chirality projector for antiparticles. Hence, the β-decay would produce a left-handed electron and a right-handed anti-neutrino. The current (9.1), with $c_V = c_A = 1/2$, was proposed in late 1957[2] in Sudarshan and Marshak (1957) and Feynman and Gell-Mann (1958). The amplitude of the β-decay then reads[3]

$$\mathcal{M} = \frac{4}{\sqrt{2}} G_F \left[\bar{u}_p \gamma^\mu \frac{1}{2}(1-\gamma^5) u_n\right] \left[\bar{u}_e \gamma_\mu \frac{1}{2}(1-\gamma^5) v_\nu\right], \tag{9.3}$$

where the factors 4 and $\sqrt{2}$ are purely conventional to keep the original value of the Fermi constant.

[2] An interesting discussion about the paternity of the $V-A$ theory can be found in Marshak (1992). It is reported that Feynman stated in 1974 (at the Neutrino 1974 conference in Pennsylvania), 'We have a conventional theory of weak interactions invented by Marshak and Sudarshan, published by Feynman and Gell-Mann, and completed by Cabbibo'.

[3] Historically, a factor 1.26 was put in front of the γ^5 of the $n \to p$ current to match the experimental rate. We know now that it comes from the composite nature of the neutron. There is no such factor when the β-decay is transposed to the quark level: $d \to u e^- \bar{\nu}$. Hence, it is omitted here.

9.2.2 Helicity of the Neutrino

In the β-decay, $n \to p + e^- + \bar{\nu}_e$, the anti-neutrino is right-handed. For the electron capture, $p + e^- \to n + \nu_e$, we expect the neutrino to be left-handed. Indeed, the amplitude from the $V - A$ theory is now [compare it with Eq. (9.3)]

$$\mathcal{M} = \frac{4}{\sqrt{2}} G_F \left[\overline{u_n} \gamma^\mu \frac{1}{2}(1 - \gamma^5) u_p \right] \left[\overline{u_\nu} \gamma_\mu \frac{1}{2}(1 - \gamma^5) u_e \right],$$

which shows that the leptonic current is simply $\overline{u_{\nu_L}} \gamma_\mu u_{e_L}$. In 1958, M. Goldhaber and his collaborators measured the neutrino helicity produced by an electron capture reaction (Goldhaber et al., 1958). The reaction used was

$$e^-(1/2) + {}^{152}\text{Eu}(0) \to {}^{152}\text{Sm}^*(1) + \nu(1/2) \quad \text{followed by} \quad {}^{152}\text{Sm}^*(1) \to {}^{152}\text{Sm}(0) + \gamma(1). \tag{9.4}$$

The numbers in the parentheses are the particle spins. In the first reaction, the europium is at rest and absorbs an electron in an orbital angular momentum S-state ($l = 0$). Therefore, the total angular momentum of the initial state is that of the electron ($\pm 1/2$), as shown in Fig. 9.2-a. As the initial state is at rest, the momentum of the excited samarium is opposite to the neutrino momentum. By conservation of the angular momentum, the spin of Sm* is aligned (i.e., the same sign of its projection) with that of the electron and opposite to the neutrino spin. Therefore, the neutrino helicity is the same (same sign) as the Sm* helicity. The Sm* lifetime is very short, about 0.035 ps. Therefore, it decays via the second reaction before its recoil energy is lost (the stopping time is of the order of 1 ps) and thus has conserved its initial direction of motion at the decay time. Now, by angular momentum conservation, the

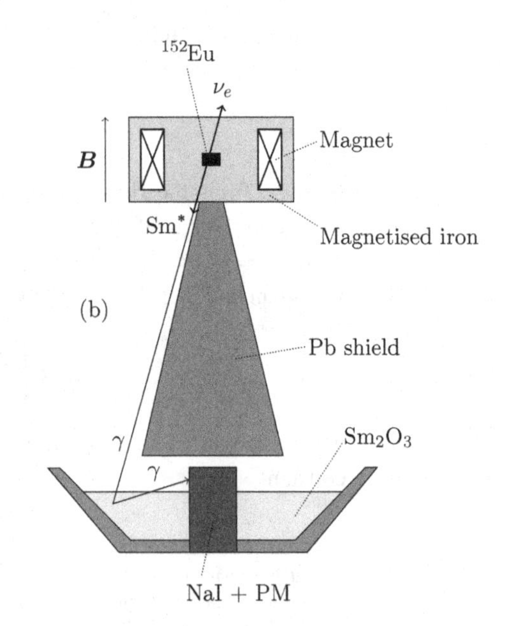

Fig. 9.2 (a) Spin (thick white arrows) and momentum (black arrows) of particles involved in Goldhaber's experiment. (b) Schematic of the experimental setup of Goldhaber's experiment.

photon spin is aligned to that of Sm*. If one finds a way to select the photons emitted along the direction of the Sm* motion (the so-called forward photon), their helicities will be that of Sm* and thus that of the neutrino. The ingenious idea of Goldhaber is to select those photons exploiting the resonant scattering method. The difference of energy level between Sm* and Sm is 963 keV. However, the photon emitted has a lower energy since Sm must recoil in order to conserve momentum. About 3.3 eV is used by the recoil (see Problem 9.1), which is much larger than the natural width of the energy level (about 0.01 eV). However, because of the initial Sm* momentum, photons emitted in the forward direction will benefit from a Doppler effect raising their energy above 963 keV. Therefore, they can excite a Sm atom, which further decays,

$$\gamma + \text{Sm} \to \text{Sm}^* \to \gamma + \text{Sm}.$$

This is the resonant scattering. The principle of the experiment is then the following: the ^{152}Eu source is surrounded by magnetised iron (see Fig. 9.2b). When the magnetic field points upward, the electron spins of the iron are then aligned downward (due to ferromagnetism). With the z-axis in the upward direction, if the initial photon produced by the second reaction of Eq. (9.4) has its spin $s_z = +1$ since the electrons of the iron have $s_z = -1/2$, the photon is likely to flip the spin of the electron to $s_z = +1/2$, reducing the photon energy. On the other hand, those having $s_z = -1$ cannot flip the electron spin and interact less. Therefore, for photons going downward, photons with negative helicity have lower energy than that of photons with positive helicity. Conversely, when the magnetic field points downward, photons going downward with positive helicity have lower energy than that of photons with negative helicity. Photons exiting the iron that reach the hollow, a truncated cone filled with Sm_2O_3 located downstream (see Fig. 9.2b), may then scatter resonantly. Only those emitted in the forward direction with respect to the Sm* direction and still having enough energy (thus, mainly right-handed when the magnetic field points upward and left-handed when it is downward) will actually scatter. Those are precisely the ones conserving the neutrino helicities. The scattered photons are then detected by a NaI scintillator equipped with a photomultiplier. The experiment counts the number of photons with the two magnetic field configurations. The helicity of the photons is determined from their count rate as a function of the magnetic field direction. Goldhaber and collaborators (1958) estimated $h_\gamma \simeq (-67 \pm 15)\%$. This is in good agreement with the theoretically expected value of -0.84, which already accounts for the angle dependence of the photon polarisation but does not include thermal effects and capture of non S-state electrons. This led Goldhaber to conclude that the neutrino is left-handed.

9.2.3 Evidence for *V–A* Current

There is ample evidence for the $V - A$ current of the weak interaction. One of the most striking of them is the branching ratio of the charged pion. Since it is the lightest meson, it cannot decay via the strong interaction. The π^- is made of $\bar{u}d$, two quarks of different electric charges. Hence, they cannot annihilate into photons as the π^0 does. Therefore, the charged pions decay via the weak interaction into lighter particles, i.e.

$$\pi^- \to e^- + \bar{\nu}_e, \qquad \pi^- \to \mu^- + \bar{\nu}_\mu.$$

As the phase space available in the electron channel is much larger than in the muon channel (given that $m_\mu \simeq 200 m_e$ and $m_\nu \simeq 0$), we would naively expect a branching ratio larger for the former than the latter. However, experimentally, the measurement gives the opposite, i.e.

$$\frac{\Gamma(\pi^- \to e^-\bar{\nu}_e)}{\Gamma(\pi^- \to \mu^-\bar{\nu}_\mu)} = 1.23 \times 10^{-4}. \tag{9.5}$$

It turns out that this is a consequence of the $V-A$ structure. Let us see why. Pions are spinless particles, so the charged lepton and the anti-neutrino have opposite spin projections. In the pion rest frame, they fly back to back, and hence, they have necessarily the same helicity as shown below:

$$\overset{\bar{\nu}_\ell}{\underset{\Leftarrow}{\longleftarrow}} \; \overset{\pi^-}{\bullet} \; \overset{\ell^-}{\underset{\Rightarrow}{\longrightarrow}}.$$

Given that neutrinos can be considered massless, their helicity state corresponds to their chirality state, which, with the $V-A$ current, is right-handed for anti-neutrinos. Consequently, the charged lepton is also in a right-handed helicity state. However, with the $V-A$ current, fermions and thus the charged leptons are in the left-handed chiral state (whereas anti-fermions are in the right-handed chiral state). Therefore, in the present case, the left-handed chiral projection of a right-handed helicity state must be taken into account. Consider the right-handed helicity state $\psi_{+\frac{1}{2}} \equiv \ell_{\frac{1}{2}}$ in Eq. (5.74) and project it on the left-handed chirality state. Choosing the axis such as $\theta = \phi = 0$, the projection of the left-handed spinor of the charged leptons then reads, with $P_L = (1-\gamma^5)/2$,

$$P_L \ell_{\frac{1}{2}} = \frac{1}{2}\begin{pmatrix} 1 & 0 & -1 & 0 \\ 0 & 1 & 0 & -1 \\ -1 & 0 & 1 & 0 \\ 0 & -1 & 0 & 1 \end{pmatrix} \sqrt{E_\ell + m_\ell} \begin{pmatrix} 1 \\ 0 \\ \frac{|\boldsymbol{p}_\ell|}{E_\ell + m_\ell} \\ 0 \end{pmatrix}$$

$$= \frac{1}{2}\sqrt{E_\ell + m_\ell}\left(1 - \frac{|\boldsymbol{p}_\ell|}{E_\ell + m_\ell}\right)\begin{pmatrix} 1 \\ 0 \\ -1 \\ 0 \end{pmatrix},$$

where $|\boldsymbol{p}_\ell|$ is the momentum (of ℓ or $\bar{\nu}_\ell$) in the CM frame. We see that if the mass of the charged lepton were 0 (and thus $|\boldsymbol{p}_\ell| = E_\ell$), the projection would give 0, and hence the decay of the pion would not occur. Since m_e or m_μ are very small, the decay is naturally suppressed (explaining the long lifetime of the charged pion, 2.6×10^{-8} s) by a factor we can evaluate. The spinor of the right-handed anti-neutrino is given by $\psi_{\bar{\frac{1}{2}}} \equiv v_{\bar{\frac{1}{2}}}$ in Eq. (5.74) and reads, with $m = 0$ and $\theta = \phi = \pi$,

$$v_{\bar{\frac{1}{2}}} = \sqrt{|\boldsymbol{p}_\ell|}\begin{pmatrix} 1 \\ 0 \\ -1 \\ 0 \end{pmatrix}.$$

The amplitude of the pion decay is necessarily proportional to the leptonic current

$$\mathcal{M} \propto j^\pi_\mu \bar{\ell}_{\frac{1}{2}} \gamma^\mu P_L v_{\bar{\frac{1}{2}}} = j^\pi_\mu \overline{P_L \ell_{\frac{1}{2}}} \gamma^\mu v_{\bar{\frac{1}{2}}} = \sqrt{E_\ell + m_\ell}\left(1 - \frac{|\boldsymbol{p}_\ell|}{E_\ell + m_\ell}\right)\sqrt{|\boldsymbol{p}_\ell|}\mathcal{M}',$$

where j^π_μ is the current associated with the pion and $\mathcal{M}'$ is

$$\mathcal{M}' = \frac{1}{2} j^\pi_\mu \begin{pmatrix} 1 & 0 & -1 & 0 \end{pmatrix} \gamma^0 \gamma^\mu \begin{pmatrix} 1 \\ 0 \\ -1 \\ 0 \end{pmatrix}.$$

The current j^π_μ is a function of the pion form factor but no need to be specific to evaluate the ratio of the decay widths in each channel. Moreover, $\mathcal{M}'$ is independent of the leptonic decay channel. The two-body decay width is given by Eq. (3.36). Therefore, it is proportional to

$$\Gamma(\pi^- \to \ell^- \bar{\nu}_\ell) \propto \frac{|\boldsymbol{p}_\ell|}{32\pi^2 m_\pi^2}\left(\sqrt{E_\ell + m_\ell}\left(1 - \frac{|\boldsymbol{p}_\ell|}{E_\ell + m_\ell}\right)\sqrt{|\boldsymbol{p}_\ell|}\right)^2 \int d\Omega |\mathcal{M}'|^2.$$

It follows that the ratio of decay widths reads

$$\frac{\Gamma(\pi^- \to e^- \bar{\nu}_e)}{\Gamma(\pi^- \to \mu^- \bar{\nu}_\mu)} = \frac{|\boldsymbol{p}_e|^2 (E_e + m_e)\left(1 - \frac{|\boldsymbol{p}_e|}{E_e + m_e}\right)^2}{|\boldsymbol{p}_\mu|^2 (E_\mu + m_\mu)\left(1 - \frac{|\boldsymbol{p}_\mu|}{E_\mu + m_\mu}\right)^2}.$$

The energy and momentum formulas in the CM frame (i.e., the pion rest frame) have been established in Section 3.6.1, yielding

$$E_\ell = \frac{m_\pi^2 + m_\ell^2}{2m_\pi}, \qquad |\boldsymbol{p}_\ell| = \frac{m_\pi^2 - m_\ell^2}{2m_\pi}.$$

This leads to

$$\frac{\Gamma(\pi^- \to e^- \bar{\nu}_e)}{\Gamma(\pi^- \to \mu^- \bar{\nu}_\mu)} = \left(\frac{m_e}{m_\mu}\right)^2 \left(\frac{m_\pi^2 - m_e^2}{m_\pi^2 - m_\mu^2}\right)^2 = 1.28 \times 10^{-4},$$

which is very close to the experimental value in Eq. (9.5). Considering that only the leading order was taken into account, it validates the $V - A$ structure of the leptonic current.

9.2.4 The Four-Fermion Contact Interaction Failure

The 4-fermion contact interaction with $V - A$ currents described well the weak interaction processes known in the 1950s–1960s, namely at low energy. Rapidly, it was realised that this theory was only an approximation, i.e. an effective theory in the low energy range. For example, consider the reaction $\nu_\mu + e^- \to \nu_e + \mu^-$, described by the 4-fermion interaction:

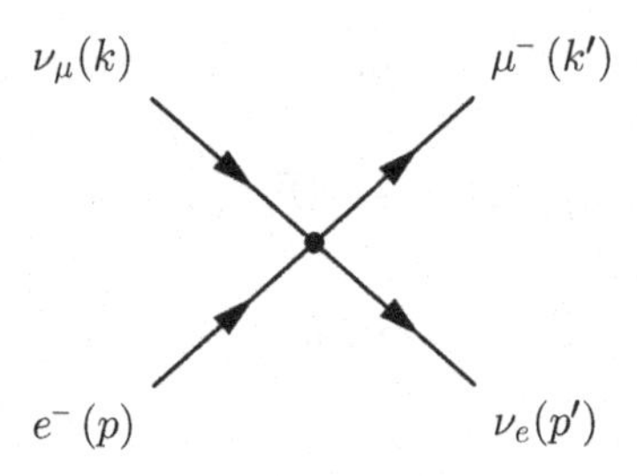

The corresponding amplitude is

$$\mathcal{M} = \frac{4}{\sqrt{2}} G_F \left[\bar{u}_{\nu_e}\gamma^\mu P_L u_e\right]\left[\bar{u}_\mu \gamma_\mu P_L u_{\nu_\mu}\right] = \frac{G_F}{\sqrt{2}}\left[\bar{u}_{\nu_e}\gamma^\mu(1-\gamma^5)u_e\right]\left[\bar{u}_\mu\gamma_\mu(1-\gamma^5)u_{\nu_\mu}\right]. \tag{9.6}$$

Let us calculate the cross section of this process. The calculation is very similar to $e^- + \mu^- \to e^- + \mu^-$ in QED, except that γ^μ is replaced by $\gamma^\mu(1-\gamma^5)$. However, spin averaging requires attention. The neutrino mass is so tiny that only the left-handed helicity state matters. Therefore, only the spin of the incoming electron has to be averaged. Applying Eq. (6.84) with $\Gamma_1 = \gamma^\mu(1-\gamma^5)$ and $\Gamma_2 = \gamma^\nu(1-\gamma^5)$, it follows given that $\gamma^0\Gamma_1^\dagger\gamma^0 = \gamma^0(1-\gamma^5)\gamma^0\gamma^\mu\gamma^0\gamma^0 = \gamma^\mu(1-\gamma^5)$,

$$\overline{|\mathcal{M}|^2} = \left(\frac{G_F^2}{2}\right)\frac{1}{2}\mathrm{Tr}\left(\not{p}\gamma^\mu(1-\gamma^5)\not{p}'\gamma^\nu(1-\gamma^5)\right)\mathrm{Tr}\left(\not{k}\gamma_\mu(1-\gamma^5)\not{k}'\gamma_\nu(1-\gamma^5)\right), \tag{9.7}$$

where the masses of the charged leptons have been neglected. Using the anti-commutation property of γ^5 and $(\gamma^5)^2 = 1$, the first trace simplifies to

$$\begin{aligned}\mathrm{Tr}\left(\not{p}\gamma^\mu(1-\gamma^5)\not{p}'\gamma^\nu(1-\gamma^5)\right) &= 2\left(\mathrm{Tr}\left(\not{p}\gamma^\mu\not{p}'\gamma^\nu\right) - \mathrm{Tr}\left(\not{p}\gamma^\mu\not{p}'\gamma^\nu\gamma^5\right)\right)\\ &= 2p_\rho p'_\eta\left[4(g^{\rho\mu}g^{\eta\nu} - g^{\rho\eta}g^{\mu\nu} + g^{\rho\nu}g^{\mu\eta}) - 4i\epsilon^{\rho\mu\eta\nu}\right],\end{aligned}$$

where the last equality has been obtained using Eqs. (6.88) and (G.22). Therefore, the traces in Eq. (9.7) read

$$\begin{aligned}\mathrm{Tr}_1 &\equiv \mathrm{Tr}\left(\not{p}\gamma^\mu(1-\gamma^5)\not{p}'\gamma^\nu(1-\gamma^5)\right) = 8(p^\mu p'^\nu - p\cdot p' g^{\mu\nu} + p^\nu p'^\mu - ip_\rho p'_\eta\epsilon^{\rho\mu\eta\nu}),\\ \mathrm{Tr}_2 &\equiv \mathrm{Tr}\left(\not{k}\gamma_\mu(1-\gamma^5)\not{k}'\gamma_\nu(1-\gamma^5)\right) = 8(k_\mu k'_\nu - k\cdot k' g_{\mu\nu} + k_\nu k'_\mu - ik^\rho k'^\eta\epsilon_{\rho\mu\eta\nu}).\end{aligned} \tag{9.8}$$

Given that $\epsilon^{\rho\mu\eta\nu}g_{\mu\nu} = 0$ (since $g_{\mu\nu} \neq 0$ only when $\mu = \nu$ for which $\epsilon^{\rho\mu\eta\nu} = 0$), the product of traces reads

$$\begin{aligned}\mathrm{Tr}_1 \times \mathrm{Tr}_2 = 64(&2\,p\cdot k\,p'\cdot k' + 2\,p\cdot k'\,p'\cdot k\\ &- ip_\rho p'_\eta k_\mu k'_\nu\epsilon^{\rho\mu\eta\nu} - ip_\rho p'_\eta k_\nu k'_\mu\epsilon^{\rho\mu\eta\nu} - ip^\mu p'^\nu k^\rho k'^\eta\epsilon_{\rho\mu\eta\nu}\\ &- ip^\nu p'^\mu k^\rho k'^\eta\epsilon_{\rho\mu\eta\nu} - p_\rho p'_\eta k^\alpha k'^\beta\epsilon^{\rho\mu\eta\nu}\epsilon_{\alpha\mu\beta\nu}).\end{aligned}$$

In the first term of the second line, we can change the dummy indices $\mu \leftrightarrow \nu$. Doing so, we have $\epsilon^{\rho\nu\eta\mu} = -\epsilon^{\rho\nu\mu\eta} = \epsilon^{\rho\mu\nu\eta} = -\epsilon^{\rho\mu\eta\nu}$. Hence, the first two terms of the second line cancel. Similarly, the last two terms of the second line cancel. Now, using properties (G.26)

$$\epsilon^{\rho\mu\eta\nu}\epsilon_{\alpha\mu\beta\nu} = \epsilon^{\mu\nu\rho\eta}\epsilon_{\mu\nu\alpha\beta} = -2(\delta^\rho_\alpha\delta^\eta_\beta - \delta^\rho_\beta\delta^\eta_\alpha),$$

it follows that

$$\begin{aligned}\mathrm{Tr}_1 \times \mathrm{Tr}_2 &= 64\left(2p\cdot k\, p'\cdot k' + 2\, p\cdot k'\, p'\cdot k + 2p_\rho p'_\eta k^\alpha k'^\beta (\delta^\rho_\alpha \delta^\eta_\beta - \delta^\rho_\beta \delta^\eta_\alpha)\right)\\ &= 256\, p\cdot k\, p'\cdot k'.\end{aligned} \tag{9.9}$$

4-momentum conservation $p+k = p'+k'$ implies when masses are neglected, $p\cdot k = p'\cdot k'$. The square of the CM energy is $s = (p+k)^2 \simeq 2p\cdot k$. Therefore, the squared amplitude (9.7) reduces to

$$\overline{|\mathcal{M}|^2} = \left(\frac{G_F^2}{2}\right)\frac{1}{2}256\frac{s^2}{4} = 16\, G_F^2 s^2. \tag{9.10}$$

Finally, the differential cross section is obtained using Eq. (3.42), which reads, given that the momenta in the centre-of-mass frame are $|\boldsymbol{p}^*|^2 = |\boldsymbol{p}'^*|^2 = s/2$ [cf. Eq. (3.34)],

$$\frac{\mathrm{d}\sigma}{\mathrm{d}\Omega^*} = \frac{1}{64\pi^2 s}|\mathcal{M}|^2\frac{|\boldsymbol{p}'^*|}{|\boldsymbol{p}^*|} = \frac{G_F^2 s}{4\pi^2} \quad \text{i.e.} \quad \sigma = \frac{G_F^2 s}{\pi}. \tag{9.11}$$

The cross section is thus proportional to the square of the CM energy and is then divergent for $s \to \infty$. In physicist's jargon, it is said that the *unitarity* is violated, meaning that the probability of observing this reaction is greater than 1. It is a sign that this theory cannot be used at high energy. It is necessarily an effective theory, i.e. an approximation of a more general theory, only valid in the low energy regime. Moreover, it can be shown that the 4-fermion contact interaction is not a renormalisable theory: all calculations to higher order diverge.

9.3 A More Modern Weak Interaction

9.3.1 The *W* Boson

In the modern interpretation of interactions, the forces are mediated by vector bosons. Since in the β-decay, we have currents that are a Lorentz vector, the W boson responsible for the weak interaction must be a spin-1 boson as the photon. However, there is an obvious difference: the weak current encountered so far is electrically charged. One unit of charge is transferred between the neutrino and the electron, for instance. This weak current is called the (weak) charged current. Therefore, the W bosons (W^- and its charge conjugate W^+) carry one unit of electric charge and are the intermediate vector bosons of the weak interaction coupled with the charged currents. The β-decay now corresponds to the graph below:

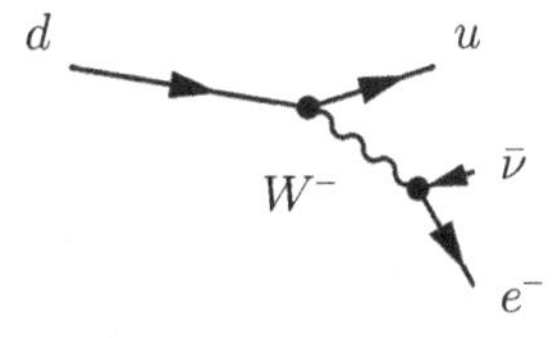

At each vertex, only left-handed chiral fermions or right-handed chiral anti-fermions are involved. It would be tempting to imagine that the Ws are just charged photons. If it were the case, we should be able to observe a similar reaction as the Compton process, replacing the outgoing photon with a W boson, i.e. $\gamma + e^- \rightarrow W^- + \nu_e$. However, it is not the case, even if nothing forbids this process. The reason is that Ws have a large mass suppressing their production at low energy. A well-known consequence of their mass is the short range of the weak interaction (see Section 1.2.3, where the connection between short-range interactions and masses of the force carrier is established in the context of Yukawa's theory).

The $W^\pm$ was discovered at CERN in 1983 in the UA1 experiment and a few days later by the UA2 experiment, both operating at the Super Proton Synchrotron – a proton–antiproton collider then operating at $\sqrt{s} = 540$ GeV. The decay channel producing an electron, $p+\bar{p} \rightarrow W^- + X \rightarrow e^- + \bar{\nu}_e + X$ (and the charge-conjugated reaction), was used to ease the identification of the final state in the context of a hadron collider producing mostly jets. One of the difficulties is evaluating the W mass, $m_W^2 = (p_e + p_{\nu_e})^2$, given that the neutrino is not detected. However, in Section 1.4.4, we explained that the neutrino momentum is evaluated thanks to the missing momentum, i.e. the opposite of the sum of all the visible momenta, including charged as well as neutral particles. As all particles cannot be identified, one assumes that all hadrons are pions, and the energy vectors measured in calorimeters are used in the reconstruction instead of the momenta measured in the tracker. Since particles emitted in the beam direction cannot be detected (the UA1 detector was hermetic down to angles 0.2°), the momentum or energy balance is only measured in the transverse plane (to the beam). At $\sqrt{s} = 540$ GeV, the Ws are mostly produced at rest. Hence, the neutrino and the electron must be emitted back to back. This gives a handle to control the accuracy of the reconstructed neutrino momentum. Moreover, for Ws at rest, $|\boldsymbol{p}_e| = m_W/2$ (neglecting the electron mass), and the electron transverse momentum is thus $p_{eT} = m_W \sin\theta/2$. The differential cross section of W production,

$$\frac{\mathrm{d}\sigma}{\mathrm{d}p_{eT}} = \frac{\mathrm{d}\sigma}{\mathrm{d}\cos\theta}\frac{\mathrm{d}\cos\theta}{\mathrm{d}p_{eT}} = \frac{\mathrm{d}\sigma}{\mathrm{d}\cos\theta}\frac{4p_{eT}/m_W^2}{\sqrt{1-4p_{eT}^2/m_W^2}},$$

thus presents a peak towards $p_{eT} = m_W/2$, about 40 GeV/c^2. Analysing about 18 nb^{-1} corresponding to 10^9 collisions, the UA1 experiment found six events compatible with W production and estimated its mass to be $m_W = 81 \pm 5$ GeV/c^2 (UA1 Collaboration, 1983). Nowadays, the mass of the W is well known and has been measured with good accuracy at the Tevatron collider (near Chicago) and, more recently, at the LHC at CERN. Its mass is (Particle Data Group, 2022)

$$m_W = 80.379 \pm 0.012 \text{ GeV}/c^2. \tag{9.12}$$

The propagation of a massive boson is then described by the propagator

$$W_{\text{propagator}} = i\frac{-g^{\mu\nu} + q^\mu q^\nu/m_W^2}{q^2 - m_W^2 + i\epsilon}. \tag{9.13}$$

This expression can be deduced following the same procedure as in Supplements 6.2 and 6.3 for the photon and Dirac propagators, given that the sum over the massive boson polarisations reads

$$\sum_\lambda \epsilon_\mu(q,\lambda)\epsilon_\nu^*(q,\lambda) = -g_{\mu\nu} + q_\mu q_\nu/m_W^2. \tag{9.14}$$

Note that setting $m_W \to 0$ does not recover the photon propagator. The reason is related to the fact that with a massive spin-1 boson, the three polarisations are allowed, unlike the photon, which has only two polarisations. It turns out that the longitudinal polarisation is actually responsible for the term $q^\mu q^\nu / m_W^2$.

9.3.2 Charged Current of Leptons

The cross section (9.11) has dimensions of area, i.e. GeV^{-2} in natural units, which shows that G_F is not dimensionless but is also in GeV^{-2}. In complete analogy with QED, let us define g_w, the analogous of e, the dimensionless weak coupling constant[4]. The charged current for leptons is then

$$j^\mu_{e^- \to \nu_e} = \frac{g_w}{\sqrt{2}} \bar{u}_{\nu_e} \gamma^\mu \frac{1}{2}(1 - \gamma^5) u_e, \tag{9.15}$$

where $\sqrt{2}$ is factored out to match the electroweak definition that we shall see later. The reaction $\nu_\mu + e^- \to \nu_e + \mu^-$ leading to the unitarity violation in Fermi theory is now described by the graph with the $W^\pm$ exchange:

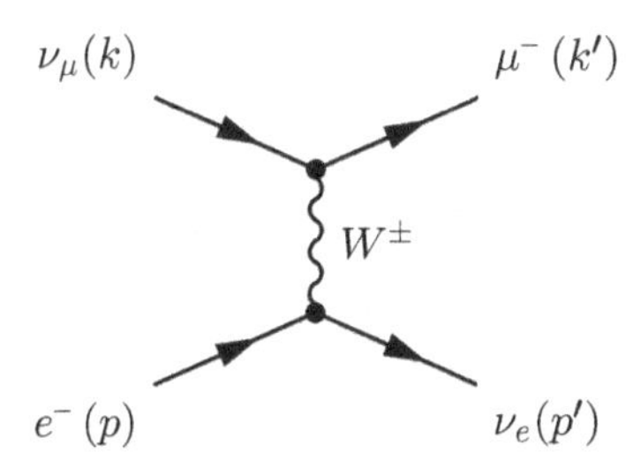

for which the amplitude reads

$$\begin{aligned} i\mathcal{M} &= j^\mu_{e^- \to \nu_e} \times i \frac{-g_{\mu\nu} + q_\mu q_\nu / m_W^2}{q^2 - m_W^2} \times j^\nu_{\nu_\mu \to \mu^-} \\ &= \left[\frac{g_w}{\sqrt{2}} \bar{u}_{\nu_e} \gamma^\mu \frac{1}{2}(1 - \gamma^5) u_e \right] i \frac{-g_{\mu\nu} + q_\mu q_\nu / m_W^2}{q^2 - m_W^2} \left[\frac{g_w}{\sqrt{2}} \bar{u}_\mu \gamma^\nu \frac{1}{2}(1 - \gamma^5) u_{\nu_\mu} \right]. \end{aligned} \tag{9.16}$$

At low energy, when $q^2 \ll m_W^2$, we should recover the effective Fermi theory. In this approximation, the amplitude becomes

$$\mathcal{M} = \frac{g_w^2}{8 m_W^2} \left[\bar{u}_{\nu_e} \gamma^\mu (1 - \gamma^5) u_e \right] \left[\bar{u}_\mu \gamma_\mu (1 - \gamma^5) u_{\nu_\mu} \right].$$

Comparing this expression to Eq. (9.6), we identify

$$\boxed{\frac{G_F}{\sqrt{2}} = \frac{g_w^2}{8 m_W^2}.} \tag{9.17}$$

At high energy, q^2 cannot be neglected with respect to m_W^2. In this regime, the fermion masses are neglected, and due to the Dirac equation in momentum space,

$$(\not{p} - m) u_e \simeq \not{p} u_e = 0, \quad \bar{u}_{\nu_e} (\not{p}' - m) \simeq \bar{u}_{\nu_e} \not{p}' = 0.$$

[4] The w in g_w stands for weak, not for the W boson.

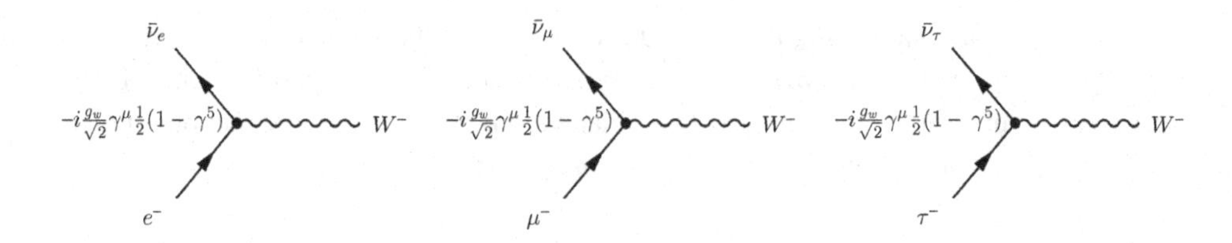

Fig. 9.3 Vertex factors of the weak interaction (charged current) in the leptonic sector.

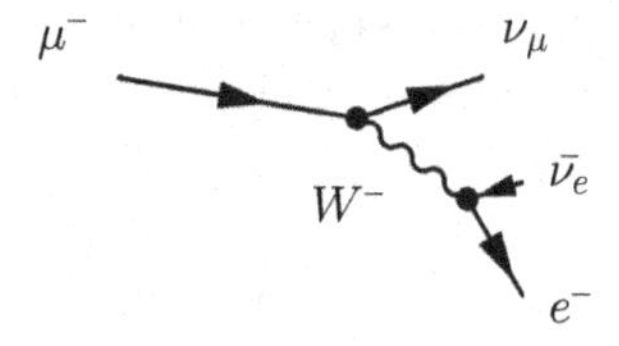

Fig. 9.4 Feynman diagram (at lowest order) of the muon decay.

As $q = p - p'$, the term $q_\mu q_\nu$ due to the W propagator in Eq. (9.16) vanishes because

$$\bar{u}_{\nu_e}\gamma^\mu\frac{1}{2}(1-\gamma^5)u_e\, q_\mu q_\nu = \bar{u}_{\nu_e}(\not{p}-\not{p}')\frac{1}{2}(1-\gamma^5)u_e q_\nu = \bar{u}_{\nu_e}\frac{1}{2}(1+\gamma^5)\not{p}u_e q_\nu = 0.$$

Hence, the amplitude reduces to

$$\mathcal{M} = \frac{g_W^2}{8(m_W^2 - q^2)}\left[\bar{u}_{\nu_e}\gamma^\mu(1-\gamma^5)u_e\right]\left[\bar{u}_\mu\gamma_\mu(1-\gamma^5)u_{\nu_\mu}\right].$$

Compared to Eq. (9.6) and the squared amplitude (9.10), we thus have now

$$\overline{|\mathcal{M}|^2} = \frac{g_W^4}{2(m_W^2 - q^2)^2}s^2.$$

The squared amplitude no longer diverges since at very high energy, for $|q^2| \gg m_W^2$, in the CM frame, $q^4 = s^2\sin^2(\theta/2)$ and so $\overline{|\mathcal{M}|^2} = g_w^4/[2\sin^2(\theta/2)]$ is finite.

The calculation we have followed assumes the same coupling constant g_w for the current $j^\mu_{e^-\to\nu_e}$ and $j^\nu_{\nu_\mu\to\mu^-}$. One may wonder if g_w is universal for the three families of leptons, as shown in Fig. 9.3 The vertex factors above are deduced by analogy with that of QED, where for the current $j^\mu = e\bar{u}\gamma^\mu u$, the factor is $-ie\gamma^\mu$. For weak interactions, as $j^\mu = \frac{g_W}{\sqrt{2}}\bar{u}\gamma^\mu\frac{1}{2}(1-\gamma^5)u$, the factor is thus $-i\frac{g_W}{\sqrt{2}}\gamma^\mu\frac{1}{2}(1-\gamma^5)$. The question is thus to test whether $g_W = g_w^e = g_w^\mu = g_w^\tau$ or equivalently $G_F = \sqrt{2}g_W^2/(8m_W^2) = G_F^e = G_F^\mu = G_F^\tau$. Let us consider, for example, the muon decay $\mu^- \to e^- + \bar{\nu}_e + \nu_\mu$ (see Fig. 9.4). Neglecting the mass of particles in the final state, one can show that (see Problem 9.2)

$$\frac{1}{\tau_\mu} = \Gamma_{\mu^-\to e^-\bar{\nu}_e\nu_\mu} = \frac{G_F^2 m_\mu^5}{192\pi^3}.$$

Assuming non-universal constants, one would have found

$$\frac{1}{\tau_\mu} = \Gamma_{\mu^-\to e^-\bar{\nu}_e\nu_\mu} = \frac{G_F^e G_F^\mu m_\mu^5}{192\pi^3}.$$

Similarly, the τ lepton can decay into the electronic or muonic channel. The corresponding decay widths are

$$\Gamma_{\tau^- \to e^- \bar{\nu}_e \nu_\tau} = \frac{G_F^e G_F^\tau m_\tau^5}{192\pi^3}, \qquad \Gamma_{\tau^- \to \mu^- \bar{\nu}_\mu \nu_\tau} = \frac{G_F^\mu G_F^\tau m_\tau^5}{192\pi^3}.$$

As the partial decay width satisfies

$$\Gamma_{\tau^- \to e^- \bar{\nu}_e \nu_\tau} = \mathrm{Br}(\tau^- \to e^- \bar{\nu}_e \nu_\tau) \times \Gamma = \frac{\mathrm{Br}(\tau^- \to e^- \bar{\nu}_e \nu_\tau)}{\tau_\tau},$$

where Γ is the total decay width and τ_τ, the τ lifetime, it follows that

$$\frac{\tau_\mu}{\tau_\tau} = \frac{192\pi^3}{G_F^e G_F^\mu m_\mu^5} \frac{\frac{G_F^e G_F^\tau m_\tau^5}{192\pi^3}}{\mathrm{Br}(\tau^- \to e^- \bar{\nu}_e \nu_\tau)} = \frac{G_F^\tau}{G_F^\mu} \left(\frac{m_\tau}{m_\mu}\right)^5 \frac{1}{\mathrm{Br}(\tau^- \to e^- \bar{\nu}_e \nu_\tau)}.$$

The lifetimes, masses and branching ratio above have been measured with an excellent accuracy (Particle Data Group, 2022),

$$\begin{aligned} &\tau_\mu = 2.196981(22) \times 10^{-6}\ \mathrm{s}, && m_\mu = 105.6583745(24)\ \mathrm{GeV}, \\ &\tau_\tau = (290.3 \pm 0.5) \times 10^{-15}\ \mathrm{s}, && m_\tau = 1776.86 \pm 0.12\ \mathrm{GeV}, \\ &\mathrm{Br}(\tau^- \to e^- \bar{\nu}_e \nu_\tau) = (17.82 \pm 0.04)\%, && \end{aligned}$$

yielding

$$G_F^\tau / G_F^\mu = 1.0026 \pm 0.0029.$$

Therefore, the ratio of coupling constants is compatible with 1. A similar combination using $\mathrm{Br}(\tau^- \to \mu^- \bar{\nu}_\mu \nu_\tau)$ also shows that G_F^τ / G_F^e is compatible with 1. We conclude that the weak coupling constant is universal for the three generations of leptons.

9.3.3 Strength of the Weak Interaction

The very small value of $G_F \simeq 1.166 \times 10^{-5}\ \mathrm{GeV}^{-2}$ compared with the fine structure constant $\alpha = 1/137$ explains why the weak interaction is qualified as 'weak' at low energy, where G_F is valid. However, inserting the value of the W mass (9.12) in Eq. (9.17) gives an estimate of g_w, or equivalently of $\alpha_w = g_w^2/(4\pi)$,

$$g_w = \sqrt{\frac{8 m_W^2 G_F}{\sqrt{2}}} = 0.65, \qquad \alpha_w = \frac{g_w^2}{4\pi} \simeq \frac{1}{29}. \tag{9.18}$$

Therefore, the coupling constant α_w turns out to be larger than the fine structure constant α! Intrinsically, the weak force is thus stronger than the electromagnetism. Its weakness, observed at low energy, is simply due to the large mass of the W boson. At high energy, where the mass of the boson becomes negligible, the weak interaction is stronger than electromagnetism. At the LHC, for instance, weak interactions are more likely than electromagnetic ones.[5]

[5] At very large energy, the distinction between the two interactions becomes irrelevant. See the electroweak model in Chapter 10.

9.3.4 Charged Current of Quarks

The Cabibbo Angle

One may wonder whether the weak coupling constant is the same for quarks and leptons. In the 1960s, when people compared the Fermi constant measured in β-decay [using the amplitude (9.3)] with the Fermi constant obtained from the muon lifetime measurement, it introduced a doubt since there was a slight but statistically significant discrepancy between the two, as shown in the following using today's numbers:

$$G_F^{\beta} = 1.1357(1) \times 10^{-5}\ \text{GeV}^{-2}, \qquad G_F^{\mu} = 1.1663787(6) \times 10^{-5}\ \text{GeV}^{-2}.$$

Moreover, there was another indication of differences between the quark and the lepton sector. The kaon decay $K^+ \to \mu^+ + \nu_\mu$ was observed, which implies a variation of the strangeness quantum number $\Delta S = 1$ (or in modern language, a coupling between u and s quarks was possible, as $K^+ \equiv u\bar{s}$). Compared with the corresponding pion decay ($\pi^+ \equiv u\bar{d} \to \mu^+ + \nu_\mu$) with no variation of the strangeness $\Delta S = 0$, the kaon decay was suppressed by a factor of about 20. In 1963, the physicist Nicola Cabibbo postulated that the strength G_F of the weak interaction was universal, but he introduced an angle θ_c, now known as the *Cabibbo angle*, such that the decay widths were a function of this angle via

$$\Gamma_{\Delta S=0} \propto G_F^2 \cos^2\theta_c \ \text{ and } \ \Gamma_{\Delta S=1} \propto G_F^2 \sin^2\theta_c.$$

Empirically, the factor 20 required an angle $\theta_c \simeq 13^\circ$ or $\sin\theta_c \simeq 0.23$. After the discovery of quarks, the reinterpretation of the Cabibbo angle was as follows: the quark that couples to the u-quark in the weak interaction is a linear combination of the d and s quarks,

$$|d'\rangle = \cos\theta_c\,|d\rangle + \sin\theta_c\,|s\rangle.$$

The quark state d' is the weak eigenstate, whereas d and s are the mass eigenstates (the ones that propagate[6]). In terms of vertex factors, we thus have

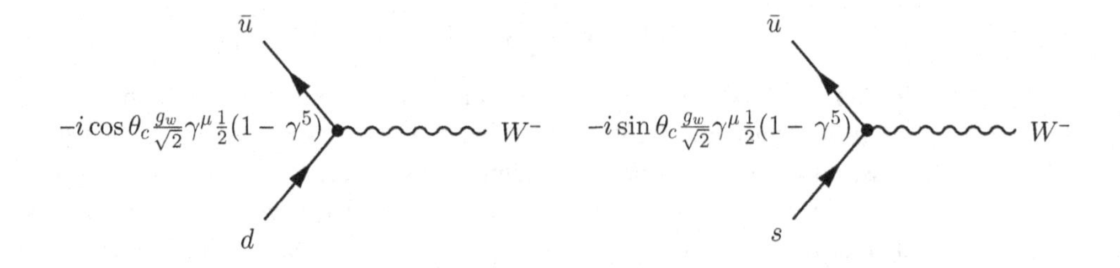

[6] Quarks cannot be seen as free particles; they always feel the strong interactions. Thus, the propagation of free quarks (pure mass eigenstates) does not really make sense. Therefore, what is called the mass eigenstate is, by definition, the eigenstate of the kinetic and strong interaction parts of the Hamiltonian.

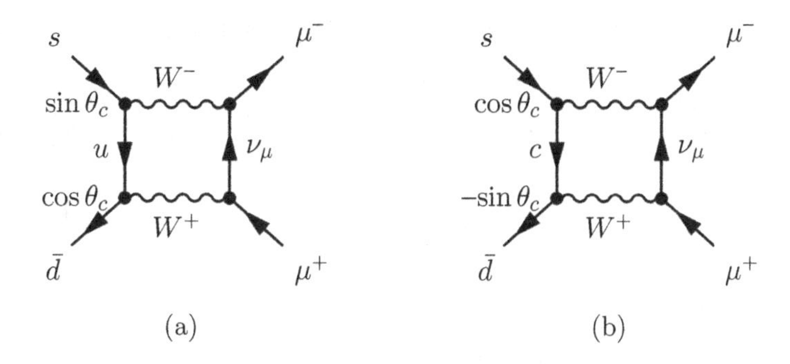

Fig. 9.5 The two box diagrams responsible for the K^0 decay into $\mu^+ + \mu^-$ at the lowest order. (a) A virtual u quark is exchanged. (b) A virtual c is exchanged.

The GIM Mechanism

In 1970, the particle physics community was puzzled by the very small branching ratio of the decay $K^0 \to \mu^+ + \mu^-$, about 9×10^{-9}, which was supposed to be described by the so-called box diagram shown in Fig. 9.5a. By symmetry with the leptonic sector where two families were already known, Glashow, Iliopoulos and Maïani (1970) postulated the existence of a fourth quark, the charm, such that another box diagram was possible with a c quark replacing u in the box (Fig. 9.5b). Since the c quark would couple to the s or d quark, it must be coupled to the linear combination of the d and s quarks orthogonal to the d', defining a new state s' such that

$$\begin{pmatrix} d' \\ s' \end{pmatrix} = \begin{pmatrix} \cos\theta_c & \sin\theta_c \\ -\sin\theta_c & \cos\theta_c \end{pmatrix} \begin{pmatrix} d \\ s \end{pmatrix}. \tag{9.19}$$

Hence, the W boson couples the c quark to s' and the u quark to d'. The u and d' quarks are respectively the counterparts of the ν_e and e^- in the leptonic sector, while c, s' play the role of ν_μ, μ^-. According to Eq. (9.19), the amplitude of the first diagram in Fig. 9.5a is proportional to $g_w^4 \cos\theta_c \sin\theta_c$, while the second diagram in Fig. 9.5b is proportional to $-g_w^4 \cos\theta_c \sin\theta_c$ (the power 4 is coming from the four vertices). If the u and c quarks had the same mass, these two diagrams would perfectly cancel, and at the lowest order, the decay would not be possible. To explain the very low value of the branching ratio, Glashow, Iliopoulos and Maïani estimated in 1970 that the c-quark had to have a mass between 1 and 3 GeV. Four years later, the charm quark was discovered as a constituent of the J/ψ meson[7] made of a pair $c\bar{c}$. Its mass is $m_c \simeq 1.3$ GeV! This suppression effect is now known as the GIM mechanism, with GIM standing for the names of the three physicists.

[7] One may be surprised by such a name for a particle. This meson was discovered independently by a team at the Stanford Linear Accelerator Center (SLAC) headed by Richter and a team at the Brookhaven National Laboratory (BNL) headed by Ting. Richter called it ψ, while Ting named it J. It is now known as the J/ψ.

The CKM Matrix

Nowadays, six quarks have been discovered. The Cabibbo–Kobayashi–Maskawa (CKM) matrix, V_{CKM}, is an extension of the matrix (9.19) to the three families, i.e.

$$\begin{pmatrix} d' \\ s' \\ b' \end{pmatrix} = \underbrace{\begin{pmatrix} V_{ud} & V_{us} & V_{ub} \\ V_{cd} & V_{cs} & V_{cb} \\ V_{td} & V_{ts} & V_{tb} \end{pmatrix}}_{V_{\text{CKM}}} \begin{pmatrix} d \\ s \\ b \end{pmatrix}. \tag{9.20}$$

The fields d', s' and b' are the fields of the weak eigenstates, while d, s and b are the fields of the mass eigenstates.[8] The d', s' and b' are thus the quarks coupled to the u, c and t quarks by a W exchange,

$$W^{\pm} \begin{pmatrix} u \\ d' \end{pmatrix} \begin{pmatrix} c \\ s' \end{pmatrix} \begin{pmatrix} t \\ b' \end{pmatrix}.$$

Because of the CKM matrix, an up-type quark can couple to any down-type quark that is mass eigenstate. The CKM matrix is unitary ($V^{\dagger}V = \mathbb{1}$), and hence, the following equalities hold:

$$\sum_k |V_{ik}|^2 = \sum_i |V_{ik}|^2 = 1, \qquad \sum_k V_{ik}V^*_{jk} = \sum_k V_{ki}V^*_{kj} = 0 \; (i \neq j). \tag{9.21}$$

They translate the fact that a down-type quark is necessarily coupled to an up-type quark and vice versa. Two down-type (or up-type) quarks cannot be coupled together by a W boson exchange. The charge rising current, which changes a down-type quark to an up-type quark, reads

$$j^{\mu}_{cc+} = \frac{g_w}{\sqrt{2}}(\bar{u}, \bar{c}, \bar{t})\gamma^{\mu}\frac{1}{2}(1-\gamma^5)V_{\text{CKM}} \begin{pmatrix} d \\ s \\ b \end{pmatrix}. \tag{9.22}$$

The charge lowering current is obtained with the Hermitian conjugate,

$$j^{\mu}_{cc-} = (j^{\mu}_{cc+})^{\dagger} = \frac{g_w}{\sqrt{2}}(\bar{d}, \bar{s}, \bar{b})V^{\dagger}_{\text{CKM}}\gamma^{\mu}\frac{1}{2}(1-\gamma^5) \begin{pmatrix} u \\ c \\ t \end{pmatrix}. \tag{9.23}$$

The vertex factors are then

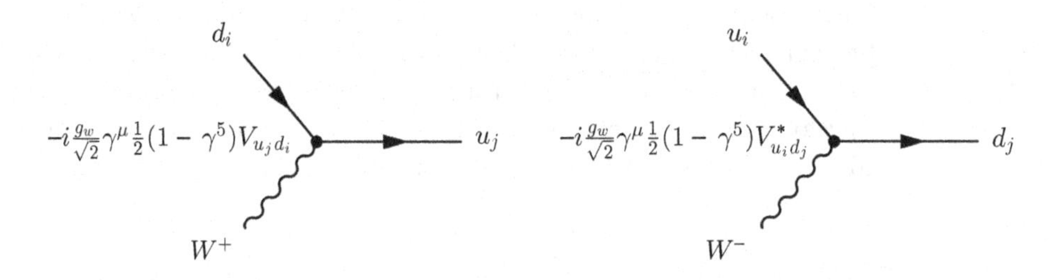

[8] See footnote 6 in this chapter.

where u_i is an up-type quark (of charge +2/3) $u_{i=1,2,3} = u, c, t$ and d_i a down-type quark (of charge $-1/3$) $d_{i=1,2,3} = d, s, b$. Looking at the expression of the charged current (9.23), we conclude that the complex conjugate of the CKM matrix is used in conjunction with the adjoint spinor of a down-type quark (or antiquark). Therefore, processes, such as $u \to d + W^+$, $\bar{d} \to \bar{u} + W^+$, $u + \bar{d} \to W^+$ or $W^- \to \bar{u} + d$, all have V_{ud}^* in the expression of their amplitude.

It is important to realise that, in the quark sector, the weak interaction mediated by a W boson not only depends on the coupling constant g_w (used in the leptonic sector) but also on the CKM coefficients. The CKM matrix modifies the strength of charged current interactions. Its coefficients are not predicted by the theory (as we shall see in Chapter 10) and, therefore, must be measured using various charged current processes. Since V_{CKM} is a 3×3 unitary matrix, it depends on four real parameters. Indeed, any 3×3 unitary matrix can be decomposed as the product of three rotations (three real parameters, the angles $\theta_{ij=12,13,23}$) plus one complex parameter interpreted as a phase $\delta \in [0, 2\pi]$. A common choice (but there are others) is

$$\begin{aligned} V_{\text{CKM}} &= \begin{pmatrix} 1 & 0 & 0 \\ 0 & c_{23} & s_{23} \\ 0 & -s_{23} & c_{23} \end{pmatrix} \begin{pmatrix} c_{13} & 0 & s_{13}e^{-i\delta} \\ 0 & 1 & 0 \\ -s_{13}e^{i\delta} & 0 & c_{13} \end{pmatrix} \begin{pmatrix} c_{12} & s_{12} & 0 \\ -s_{12} & c_{12} & 0 \\ 0 & 0 & 1 \end{pmatrix} \\ &= \begin{pmatrix} c_{12}c_{13} & s_{12}c_{13} & s_{13}e^{-i\delta} \\ -s_{12}c_{23} - c_{12}s_{23}s_{13}e^{i\delta} & c_{12}c_{23} - s_{12}s_{23}s_{13}e^{i\delta} & s_{23}c_{13} \\ s_{12}s_{23} - c_{12}c_{23}s_{13}e^{i\delta} & -c_{12}s_{23} - s_{12}c_{23}s_{13}e^{i\delta} & c_{23}c_{13} \end{pmatrix}, \end{aligned} \tag{9.24}$$

where $c_{ij} = \cos\theta_{ij}$ and $s_{ij} = \sin\theta_{ij}$. If the third generation of quarks did not mix with the first one, the angle θ_{12} would be exactly the Cabibbo angle, θ_c. The mixing is, however, very small, and hence $\theta_{12} \simeq \theta_c$. The CKM matrix elements are determined by a global fit that uses all available experimental measurements and imposes several constraints, such as the assumption of only three generations of quarks and the unitarity of the matrix (i.e., $|V_{ub}|^2 + |V_{cb}|^2 + |V_{tb}|^2 = 1$). The magnitudes of the elements are (Particle Data Group, 2022)

$$\left|V_{\text{CKM}}^{\text{exp}}\right| = \begin{pmatrix} 0.97401(11) & 0.22650(48) & 3.61(10) \times 10^{-3} \\ 0.22636(48) & 0.97320(11) & 0.04053(72) \\ 8.54(20) \times 10^{-3} & 0.03978(71) \times 10^{-3} & 0.999172(30) \end{pmatrix}, \tag{9.25}$$

and the parameters of Eq. (9.24) are

$$s_{12} = 0.22650(48), \quad s_{13} = 0.00361(10), \quad s_{23} = 0.04053(72), \quad \delta = 1.196(44). \tag{9.26}$$

(The uncertainties indicated in parenthesis have been made symmetric.) Note that the matrix is almost diagonal, implying that the dominant couplings are those within the same generation of quarks: $u \leftrightarrow d$, $c \leftrightarrow s$ and $t \leftrightarrow b$. The mixing between the first and third families is almost zero.

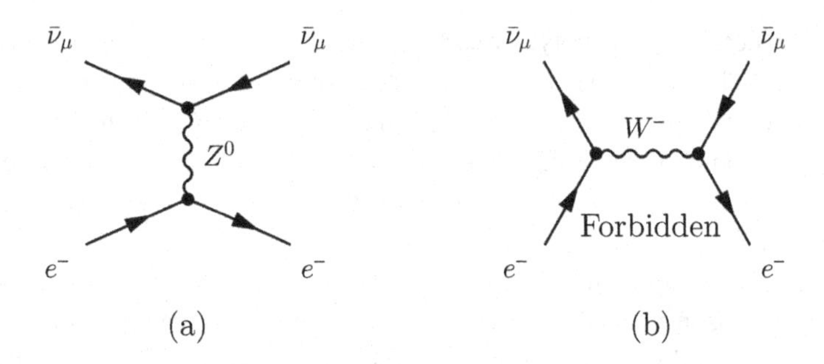

Fig. 9.6 (a) The Feynman diagram (at the lowest order) describing the reaction $\bar{\nu}_\mu e^- \to \bar{\nu}_\mu + e^-$ via a Z^0 boson exchange. (b) The exchange by a $W^\pm$ boson is forbidden in the Standard Model.

9.3.5 Neutral Currents

In the mid-1960s, Glashow (1961), Salam (1968) and Weinberg (1967) developed a theory to unify the weak interaction and electromagnetism. This theory, called *electroweak,* will be described in Chapter 10, but it predicted the existence of a neutral boson, the Z^0, that mediates the weak interaction without a change of the fermion electric charge. Such an interaction via the weak *neutral current* predicted the reactions

$$\bar{\nu}_\mu + e^- \to \bar{\nu}_\mu + e^-, \qquad \bar{\nu}_\mu + N \to \bar{\nu}_\mu + X, \tag{9.27}$$

where X denotes any hadron produced by the reaction. The first reaction is described by the Feynman diagram in Fig. 9.6a. Note that a W cannot be exchanged, the diagram in Fig. 9.6b being forbidden in the Standard Model since the W only couples leptons of the same generation. In 1973, the processes (9.27) were observed at CERN in the Gargamelle experiment (a bubble chamber of 4.8 m long and 2 m in diameter that can be seen now in the CERN garden!). The Z^0 boson necessarily couples a fermion with its anti-fermion. Therefore, a coupling such as $Z^0 \to e^- + \mu^-$ inducing a *Flavour Changing Neutral Current* (FCNC) is forbidden in the Standard Model (but actively searched for as a sign of new physics). The measurement of the cross section of the neutrino scattering on a nucleon in the reaction (9.27) showed that it was about a third of the corresponding charged current reaction, $\bar{\nu}_\mu + N \to \mu^+ + X$. It suggested that the structure of the neutral current is different than that of the charged current, as will be confirmed in Chapter 10. We shall see that not only the left-handed chirality state is involved in the neutral current but also the right-handed chirality state, even if the theory remains chiral (left- and right-handed components being treated differently).

About 10 years after the discovery of neutral currents, the Z^0 boson was observed for the first time at CERN in 1983, still in the UA1 and UA2 experiments few months after the discovery of the $W^\pm$. It was discovered in its leptonic decay

$$p + \bar{p} \to Z^0 + X \to \ell^+ + \ell^- + X,$$

where $\ell = e$ or μ. A decade later, it has been extensively studied at the LEP e^+e^- collider and its mass is now very precisely measured (Particle Data Group, 2022),

$$M_Z = 91.1876 \pm 0.0021 \text{ GeV}/c^2. \tag{9.28}$$

As for the W boson, its large mass explains the weakness of the weak interaction at low energy.

9.3.6 Failure of the Model

There is a major problem with the weak theory described so far: it is *not* renormalisable! Such a high order ($O(g_w^4)$) process, $\nu_\mu\bar{\nu}_\mu \to \nu_\mu\bar{\nu}_\mu$, corresponding to the diagram

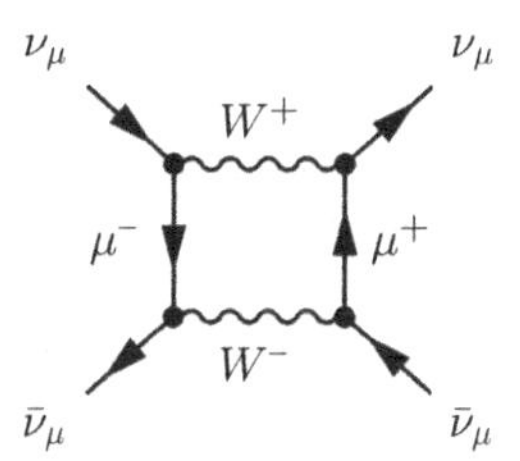

is divergent (because of the longitudinal polarisation contribution $q^\mu q^\nu/m_W^2$ in the propagator)! Even if this process cannot be detected, it is a problem for the theory! One can show that this problem is related to the fact that the W and Z bosons exposed in this chapter are *not* quanta of gauge fields. What is missing here is a gauge invariance to make the theory renormalisable. Chapter 10 is dedicated to this topic.

9.4 The Case of Neutrinos

9.4.1 Massless Neutrinos

Until the 1990s, the neutrinos were assumed to be massless. In these conditions, the solutions of the Dirac equation simplify. Looking back to the two coupled Eqs. (5.53), setting $m = 0$ and focusing on the $E = +|\boldsymbol{p}|$ solutions, the coupled equations read

$$\begin{cases} |\boldsymbol{p}|\, u_a - \boldsymbol{\sigma}\cdot\boldsymbol{p}\, u_b &= 0, \\ \boldsymbol{\sigma}\cdot\boldsymbol{p}\, u_a - |\boldsymbol{p}|\, u_b &= 0, \end{cases} \tag{9.29}$$

where u_a and u_b are the two-component spinors of the wave function $\psi(x)$. Let us define the following linear combinations:

$$u_L = (u_a - u_b)/2, \qquad u_R = (u_a + u_b)/2.$$

From Eqs. (9.29), it is easy to check that they satisfy the two independent equations

$$\frac{\boldsymbol{\sigma}\cdot\boldsymbol{p}}{2|\boldsymbol{p}|}u_L = -\frac{1}{2}u_L, \qquad \frac{\boldsymbol{\sigma}\cdot\boldsymbol{p}}{2|\boldsymbol{p}|}u_R = \frac{1}{2}u_R.$$

Since $\boldsymbol{\sigma} \cdot \boldsymbol{p}/(2|\boldsymbol{p}|)$ is the helicity projector (equivalent to chirality in the massless assumption), u_L and u_R are thus, respectively, the left- and right-handed 2-spinors. Let us define the two wave functions $\chi_{L/R}(x) = u_{L/R}e^{-ip\cdot x}$ such that the original wave function $\psi(x)$ can be expressed as

$$\psi(x) = \begin{pmatrix} u_a \\ u_b \end{pmatrix} e^{-ip\cdot x} = \begin{pmatrix} u_R + u_L \\ u_R - u_L \end{pmatrix} e^{-ip\cdot x} = \begin{pmatrix} \chi_R(x) + \chi_L(x) \\ \chi_R(x) - \chi_L(x) \end{pmatrix}.$$

Since $\psi(x)$ satisfies the Dirac equation $i\gamma^\mu \partial_\mu \psi(x) = 0$, it follows using the Dirac representation of the Dirac matrices

$$\begin{aligned} i\left(\partial_0 \chi_R(x) + \partial_0 \chi_L(x) + \sigma^i \partial_i \chi_R(x) - \sigma^i \partial_i \chi_L(x)\right) &= 0, \\ i\left(-\partial_0 \chi_R(x) + \partial_0 \chi_L(x) - \sigma^i \partial_i \chi_R(x) - \sigma^i \partial_i \chi_L(x)\right) &= 0, \end{aligned}$$

which decouple into

$$\begin{aligned} i\left(\partial_0 - \sigma^i \partial_i\right) \chi_L(x) &= 0, \\ i\left(\partial_0 + \sigma^i \partial_i\right) \chi_R(x) &= 0. \end{aligned}$$

These two equations, known as the Weyl equations,[9] independently describe the evolution of a massless left-handed particle and a massless right-handed particle. Thus, these can be considered as two independent particles. Until the late 1990s, neutrinos were considered massless. Therefore, ν_L and ν_R were two different particles. Moreover, only ν_L is sensitive to weak interactions. There was thus no way (even in principle) to detect a ν_R. Consequently, ν_R did not have any physical meaning and was then absent from the original Standard Model.

9.4.2 Solar and Atmospheric Neutrino Puzzles

Although the neutrino masses have never been measured so far (only upper limits exist), a long adventure of several decades has led to the conclusion that neutrinos are massive. It started in 1968 with an experiment called Homestake, located 1 478 m below the surface in a mine in South Dakota (USA). Homestake, led by R. Davis, aimed at detecting solar neutrinos. The sun produces a huge amount of neutrinos (whose typical energy extends up to 15 MeV for an average of about 0.5 MeV), about 2×10^{38} ν_e per second, via various thermonuclear reactions. The flux reaching the earth is large, about 6.5×10^{10} cm^{-2} s^{-1}, but the tiny interaction cross section (cf. the discussion in Section 1.4.4) imposes the use of large detectors to have a chance to detect a few events. Homestake was made of a steel tank of 6.1 m in diameter and 14.6 m long containing 133 tons of $^{35}_{17}$Cl. The reaction, $\nu_e + {}^{35}_{17}\text{Cl} \to {}^{35}_{18}\text{Ar} + e^-$, was used to capture solar neutrinos. Then, argon produced over an exposure time of the order of one month was extracted through chemical methods and

[9] Using another representation of the Dirac matrices, called the chiral representation, we would have found these equations more simply.

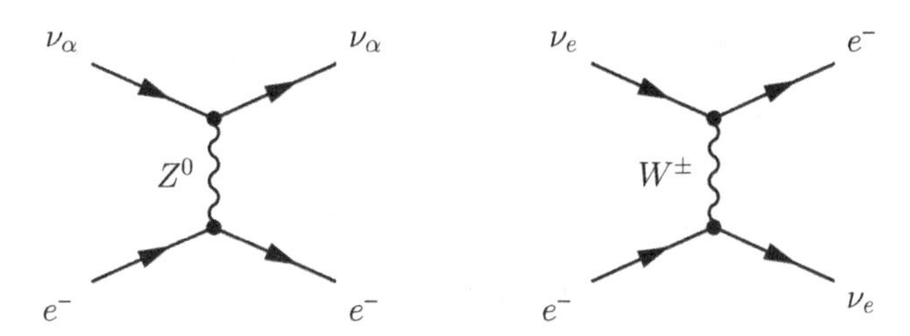

Fig. 9.7 Diagrams (at the lowest order) contributing to the elastic scattering of neutrinos on electrons. Left: for all flavours of neutrinos ($\alpha = e, \mu, \tau$). Right: valid only for ν_e.

counted (the Auger electron emitted by radioactive argon was detected). The experiment counted argon over a total of 25 years of observation. Surprisingly, the measured flux of solar neutrinos was about three times smaller than the expected flux. This puzzling deficit called the 'missing solar neutrinos problem' was further confirmed by other experiments[10] in the following decades. The missing solar neutrinos problem had two possible explanations at that time: either the standard solar model predicting the neutrino production from thermonuclear reactions was wrong, or neutrino physics was more complicated than initially foreseen.

The resolution of the puzzle came from two water Cherenkov experiments: Super-Kamiokande and SNO. Super-Kamiokande is an experiment located 1 000 m underground in the Kamioka mine in Japan. It consists of a tank (42 m in height by 39.3 m in diameter) filled with 50 000 tons of water.[11] The tank is equipped with 11 146 photomultipliers (PMTs) of 50-cm diameter in its inner part (see Section 1.4.4 to learn how a photomultiplier works). A neutrino entering the water may interact elastically with an electron $\nu_\alpha + e^- \rightarrow \nu_\alpha + e^-$ (cf. Fig. 9.7), producing a relativistic electron that generates Cherenkov light (see Section 1.4.2) detected by the PMTs. For energetic neutrinos (thus from boron production), the electron is boosted in the direction of the neutrino. Therefore, one can deduce the initial direction of the neutrino from the orientation of the Cherenkov light-cone and select only neutrinos originating from the sun. This allowed Super-Kamiokande to detect about 22 400 solar neutrino events by 2001 (the experiment started in 1996) and to measure the boron solar neutrino flux with an unprecedented accuracy of 3% (Super-Kamiokande Collaboration, 2001). The elastic scattering cross section of electrons (ES) is sensitive to all neutrino flavours, but ν_e dominates because the constructive interference between the two diagrams is only possible for ν_e in Fig. 9.7. The differences in cross sections imply that the flux measured by this reaction is $\phi_{\text{SK}}^{\text{ES}}(\nu_\alpha) = \phi(\nu_e) + 0.154[\phi(\nu_\mu) + \phi(\nu_\tau)]$. On the other hand, the SNO (Sudbury Neutrino Observatory) experiment located 2 090 m underground in the Creighton Mine near Sudbury, Canada, had the advantage of offering complementary detection channels using a vessel filled with 1 000 tons of heavy water (D_2O) equipped

[10] As the neutrino cross section increases with neutrino energy, the experiments are mainly sensitive to the high-energy spectrum of solar neutrinos, particularly those produced by the decay of boron that emits ν_e up to 15 MeV. It was the case of Homestake, where almost 80% of the rate was from that source. Gallium experiments (GALLEX, GNO, SAGE) access neutrinos of lower energy, thanks to a reduced detection threshold.

[11] Since 2020, gadolinium has been dissolved in the water tank.

with 9 456 PMTs of 20-cm diameter. The charged current reaction (CC) with deuterium, $\nu_e + d \to e^- + p + p$, is only possible for electron neutrinos mainly emitted from boron. In 2001, three months after the release of Super-Kamiokande's solar neutrino flux measurement, SNO published the first measurement based on the charged current reaction, $\phi_{\text{SNO}}^{\text{CC}}(\nu_e) = \phi(\nu_e)$ (SNO Collaboration, 2001). Therefore, it was possible to compare the two fluxes without relying on the standard solar model: if neutrinos from the sun are only ν_es, both fluxes must be equal. They measured, however, a difference of more than 3.3 standard deviations between both results, with $\phi_{\text{SK}}^{\text{ES}}(\nu_\alpha) \simeq 0.75\,\phi_{\text{SNO}}^{\text{CC}}(\nu_e)$. Moreover, SNO could also measure the flux based on the elastic scattering of electrons and found a consistent result (but with less accuracy) with Super-Kamiokande. Furthermore, the estimated total flux $\phi(\nu_e) + \phi(\nu_\mu) + \phi(\nu_\tau)$ was in agreement with predictions of the standard solar model. Therefore, the data were evidence of a non-electron active flavour component in the solar neutrino flux, i.e. $\phi(\nu_\mu) + \phi(\nu_\tau) > 0$, reaching the earth. In other words, although ν_es were emitted by the sun (given that fusion processes cannot produce ν_μ or ν_τ), a fraction reaching the earth are ν_μ or ν_τ. This is the phenomenon of neutrino oscillations explained in the following sections. One year later, the evidence was further strengthened by the measurement of the flux with the neutral current (NC) reaction, $\nu_\alpha + d \to \nu_\alpha + n + p$, which is equally sensitive to all neutrino flavours, $\phi_{\text{SNO}}^{\text{NC}}(\nu_\alpha) = \phi(\nu_e) + \phi(\nu_\mu) + \phi(\nu_\tau)$. It was consistent with predictions of the standard solar model (SNO Collaboration, 2002).

At the same time, another puzzle emerged. The atmospheric neutrino flux revealed a deficit compared with the expectations. Atmospheric neutrinos are created by the interactions of primary cosmic rays (see Section 1.3.1) with the nuclei of the atmosphere. Hadrons (pions and kaons) generated by those collisions mostly decay into muons, which in turn decay into neutrinos and electrons. As $\pi^+ \to \mu^+ + \nu_\mu$ and $\mu^+ \to e^+ + \nu_e + \bar{\nu}_\mu$ (and the charge conjugate reactions), if muons decay in flight before reaching the ground, typically for $E < 1$ GeV, one expects $\phi(\nu_\mu) + \phi(\bar{\nu}_\mu) \simeq 2[\phi(\nu_e) + \phi(\bar{\nu}_e)]$. For $E > 1$ GeV, the fraction should be larger, with some muons reaching the ground. During the 1980s and early 1990s, Kamiokande (the predecessor of Super-Kamiokande, about 20 times smaller) and IMB (another water Cherenkov experiment located in a mine close to Lake Erie in the USA) observed that the ratio ν_μ/ν_e was less than expected. However, at that time, people were more likely to question the predictions than to accept the neutrino oscillations hypothesis, and more striking evidence was needed. It came with Super-Kamiokande, which, in 1998, already had about four times more atmospheric neutrino events than Kamiokande. First, Super-Kamiokande confirmed the deficit with greater precision for both sub- and multi-GeV energy ranges. Second, the convincing evidence for oscillation came from the zenith angle distributions. The ν_μ data showed a deficit of events in the upward-going direction (so for neutrinos travelling through the Earth) with respect to the downward-going direction, with more than six standard deviations of statistical significance (Super-Kamiokande Collaboration, 1998). This asymmetry was not observed for ν_e events. Therefore, the rate of upward-going ν_μs travelling several hundred kilometres further than downward-going ones was reduced, whereas the rate of ν_e was not. The logical conclusion was that ν_es do not oscillate as far as the flight length is less than the diameter of the Earth, while ν_μs oscillate to other types of neutrinos, most likely to ν_τs that could not be detected by the experiment. Therefore, the disappearance of atmospheric neutrinos must be sensitive to the oscillation parameter governing the transition $\nu_\mu \leftrightarrow \nu_\tau$, while we saw that the solar neutrino disappearance must be sensitive to the oscillation parameter governing the transitions $\nu_e \rightarrow \nu_\mu$ or ν_τ.

This series of discoveries led to several Nobel prizes: R. Davis and M. Koshiba (Homestake and Kamiokande experiments) in 2002 and T. Kajita (Super-Kamiokande) and A. B. McDonald (SNO) in 2015.

9.4.3 The PMNS Matrix

Neutrino oscillations (i.e., flavour changing) are a consequence of the fact that the neutrino eigenstates of the weak interaction differ from the mass eigenstates of the free Hamiltonian. This scenario is then similar to the quark case of the previous sections. After the quark mixing, we now have the neutrino mixing! However, there is a historical and physical difference. Historically, quarks were first discovered through the strong interaction and electromagnetism. The usual labels u, d, etc. name the strong or equivalently the mass eigenstates.[12] Only later it was realised that the weak interaction acts on other states. In the case of neutrinos, as neutrinos are only sensitive to the weak interaction, the usual neutrinos ν_e, ν_μ and ν_τ are, by definition, those associated with the charged leptons that are used to detect them: they are necessarily the weak eigenstates. Only since the late 1990s, it has been realised that they differ from the mass eigenstates. The physical difference between quark oscillations and neutrino oscillations is that there is no way to detect mass eigenstates of neutrinos directly. Unlike quarks, the neutrino mass eigenstates are not sensitive to other interactions.[13]

As for the quarks, there is a unitary matrix similar to the CKM matrix, called PMNS for Pontecorvo–Maki–Nakagawa–Sakata, that relates the mass eigenstates described by the fields ν_1, ν_2 and ν_3 to the weak eigenstates described by the fields ν_e, ν_μ and ν_τ,

$$\begin{pmatrix} \nu_e \\ \nu_\mu \\ \nu_\tau \end{pmatrix} = \underbrace{\begin{pmatrix} V_{e1} & V_{e2} & V_{e3} \\ V_{\mu 1} & V_{\mu 2} & V_{\mu 3} \\ V_{\tau 1} & V_{\tau 2} & V_{\tau 3} \end{pmatrix}}_{V_{\text{PMNS}}} \begin{pmatrix} \nu_1 \\ \nu_2 \\ \nu_3 \end{pmatrix}. \tag{9.30}$$

Assuming that neutrinos are Dirac fermions,[14] the matrix V_{PMNS} has the same parametrisation as in Eq. (9.24) with mixing angles θ_{12}, θ_{13} and θ_{23} and a single phase $\delta \in [0, 2\pi]$. However, so far, the elements of the neutrino matrix are poorly known compared with the mixing quark matrix. It is, therefore, a very active field of research. Measurements from solar and atmospheric neutrinos and, more recently, from neutrinos produced artificially by accelerators and reactors allow constraining the elements of the matrix. For example, in 2014, the T2K experiment in Japan observed in the Super-Kamiokande detector electron neutrinos from a muon neutrino beam (produced at J-PARC in Tokai, 295 km

[12] Cf. footnote 6 in this chapter.

[13] Well, they must be sensitive to gravity. Needless to say that detecting mass eigenstate neutrinos via their gravitational interactions is hopeless (currently and for the foreseeable future).

[14] In Chapter 12, Section 12.3, another hypothesis, called Majorana fermions, is presented.

away from Kamioka, where Super-Kamiokande is installed). It was the first direct observation of the oscillation phenomenon yielding the appearance of electron neutrinos from a muon neutrino beam, with a statistical significance of more than seven standard deviations (T2K Collaboration, 2014). Those kinds of experiments, called *long baseline experiments*, provide today's constraints with rather good accuracy. In particular, using the appearance mode and by comparison of neutrino and anti-neutrino oscillation probabilities, a direct search for CP violation (due to the parameter δ in the PMNS matrix, as we shall see later) becomes possible.

As of today, assuming the unitary of the matrix and three active neutrinos, the absolute values of the elements of the matrix are estimated from a global fit of all measurements, which gives (Esteban et al., 2020)

$$|V_{\mathrm{PMNS}}| = \begin{pmatrix} 0.801-0.845 & 0.513-0.579 & 0.143-0.155 \\ 0.234-0.500 & 0.471-0.689 & 0.637-0.776 \\ 0.271-0.525 & 0.477-0.694 & 0.613-0.756 \end{pmatrix}. \tag{9.31}$$

The ranges of the elements cover three standard deviations of the parameters. Using the same parametrisation of the matrix as that of the CKM matrix in Eq. (9.24), the corresponding mixing angles and the phase are

$$\theta_{12} \simeq 31.3^{\circ} - 35.9^{\circ}, \quad \theta_{13} \simeq 8.2^{\circ} - 8.9^{\circ}, \quad \theta_{23} \simeq 40.2^{\circ} - 51.7^{\circ}, \quad \delta = 120^{\circ} - 369^{\circ}. \tag{9.32}$$

Compared to the CKM matrix in Eq. (9.25), the difference is striking: the PMNS matrix is not diagonal at all, and all elements have sizeable values. This raises many theoretical questions for understanding such a difference between the lepton and quark sectors.

One may wonder why the mixing is applied to the neutrinos and not to the charged leptons, which are considered with a definite mass. All charged leptons have the same quantum numbers, with the only difference being their masses. Thus, the flavour of a charged lepton is determined by its mass. However, the mass difference between charged leptons is very large. Consequently, the mass measurement uncertainties are small enough to discriminate the three masses, and moreover, the large difference in masses yields very distinct experimental signatures (fast decay of the τ, bremsstrahlung of the light e^-, etc.). Therefore, there is no ambiguity in the assignment of the nature of the particle. Hence, charged leptons with a definite flavour are, by definition, particles with definite mass. On the other hand, neutrinos are never directly detected: they are identified via the charged leptons produced by the weak interaction, which define the flavour of the neutrino. Therefore, flavour neutrinos are not required to have a definite mass, and the mixing implies that they are superpositions of neutrinos with definite masses.

9.4.4 Neutrino Oscillations Formalism

As the PMNS matrix in Eq. (9.30) gives the relation between the flavour fields ν_α ($\alpha = e, \mu, \nu$) and the mass fields ν_i ($i = 1, 2, 3$), a neutrino state $|\nu_\alpha\rangle$, which comes from the creation operator contained in the hermitian conjugate field $\nu_\alpha^\dagger$, satisfies the relation

$$|\nu_\alpha\rangle = \sum_{k=1}^{3} V_{\alpha k}^* |\nu_k\rangle , \tag{9.33}$$

with the normalisation $\langle \nu_\beta|\nu_\alpha\rangle = \delta_{\beta\alpha}$, $\langle \nu_i|\nu_k\rangle = \delta_{ik}$. Let us consider that a neutrino mass eigenstate, $|\nu_k\rangle$, with a mass m_k, is created in the spacetime point $(0, \mathbf{0})$. Its energy E_k and momentum $\boldsymbol{p}$ satisfy $E_k^2 = |\boldsymbol{p}|^2 + m_k^2$. As $|\nu_k\rangle$ is an eigenstate of the free Hamiltonian, the evolution of the state in spacetime is

$$|\nu_k(x,t)\rangle = e^{-i(E_k t - \boldsymbol{p}\cdot \boldsymbol{x})} |\nu_k\rangle .$$

Let us assume that the three mass eigenstates propagate in the same direction over a distance L for a time t. It follows for the state $|\nu_\alpha\rangle$ that

$$|\nu_\alpha(L,t)\rangle = \sum_k V_{\alpha k}^* e^{-i(E_k t - pL)} |\nu_k\rangle. \tag{9.34}$$

Since V_{PMNS} is a unitary matrix, $\left(V_{\text{PMNS}}^*\right)^{-1} = (V_{\text{PMNS}})^\mathsf{T}$, and hence, the inversion of Eq. (9.33) yields

$$\begin{pmatrix} |\nu_1\rangle \\ |\nu_2\rangle \\ |\nu_3\rangle \end{pmatrix} = \begin{pmatrix} V_{e1} & V_{\mu 1} & V_{\tau 1} \\ V_{e2} & V_{\mu 2} & V_{\tau 2} \\ V_{e3} & V_{\mu 3} & V_{\tau 3} \end{pmatrix} \begin{pmatrix} |\nu_e\rangle \\ |\nu_\mu\rangle \\ |\nu_\tau\rangle \end{pmatrix}$$

so that $|\nu_k\rangle = \sum_\beta V_{\beta k} |\nu_\beta\rangle$. Therefore, Eq. (9.34) reads

$$|\nu_\alpha(L,t)\rangle = \sum_{k=1}^{3} \sum_{\beta = e,\mu,\tau} V_{\alpha k}^* e^{-i(E_k t - pL)} V_{\beta k} |\nu_\beta\rangle,$$

which is a function of flavour neutrinos only. The probability of observing the transition $\nu_\alpha \to \nu_\beta$ is $|\langle \nu_\beta|\nu_\alpha(L,t)\rangle|^2$, i.e.

$$P_{\nu_\alpha\to\nu_\beta}(L,t) = \left|\sum_k V_{\alpha k}^* e^{-i(E_k t - pL)} V_{\beta k}\right|^2 = \sum_{k,j} V_{\alpha k}^* V_{\beta k} V_{\alpha j} V_{\beta j}^* e^{-i(E_k t - pL)} e^{i(E_j t - pL)}.$$

For relativistic neutrinos, $t \approx L$ and $E_k \approx E \approx p$. Therefore, the phase $E_k t - pL$ reads

$$E_k t - pL = (E_k - p)L = \frac{E_k^2 - p^2}{E_k + p} L = \frac{m_k^2}{2E} L,$$

with a similar expression for $E_j t - pL$. It follows that

$$P_{\nu_\alpha\to\nu_\beta}(L) = \sum_{k,j} V_{\alpha k}^* V_{\beta k} V_{\alpha j} V_{\beta j}^* \exp\left(-i\frac{\Delta m_{kj}^2}{2E} L\right), \tag{9.35}$$

where

$$\Delta m_{kj}^2 = m_k^2 - m_j^2. \tag{9.36}$$

Note that

$$\sum_{k,j} V_{\alpha k}^* V_{\beta k} V_{\alpha j} V_{\beta j}^* = \delta_{\alpha\beta} \tag{9.37}$$

because the unitarity of the PMNS matrix imposes $\sum_k V^*_{\alpha k} V_{\beta k} = \sum_j V_{\alpha j} V^*_{\beta j} = \delta_{\alpha\beta}$. Therefore, for $L = 0$, $P_{\nu_\alpha \to \nu_\beta}(0) = \delta_{\alpha\beta}$. In other words, there is no change in flavour without propagation. To observe a change in flavour between the place of creation of neutrinos and the place of their detection at a distance L, Eq. (9.35) tells us that three conditions have to be fulfilled:

- the corresponding PMNS matrix elements must not be 0,
- neutrinos must be massive, i.e. $m_k \neq 0$ or $m_j \neq 0$ in Eq. (9.36), and
- their masses must not be degenerate ($m_k \neq m_j$).

It is convenient to introduce the oscillation length,

$$L^{\rm osc}_{kj} = \frac{4\pi E}{\Delta m^2_{kj}}, \tag{9.38}$$

and decompose the Formula (9.35) as

$$P_{\nu_\alpha \to \nu_\beta}(L) = \sum_k |V_{\alpha k}|^2 |V_{\beta k}|^2 + 2\Re \left\{ \sum_{k>j} V^*_{\alpha k} V_{\beta k} V_{\alpha j} V^*_{\beta j} \exp\left(-i 2\pi \frac{L}{L^{\rm osc}_{kj}} \right) \right\}, \tag{9.39}$$

or, equivalently (see Problem 9.3),

$$P_{\nu_\alpha \to \nu_\beta}(L) = \delta_{\alpha\beta} - 4 \sum_{k>j} \Re \left\{ V^*_{\alpha k} V_{\beta k} V_{\alpha j} V^*_{\beta j} \right\} \sin^2 \left(\frac{\Delta m^2_{kj} L}{4E} \right) + 2 \sum_{k>j} \Im \left\{ V^*_{\alpha k} V_{\beta k} V_{\alpha j} V^*_{\beta j} \right\} \times \sin \left(\frac{\Delta m^2_{kj} L}{2E} \right). \tag{9.40}$$

The first term in Eq. (9.39) does not depend on L. Far from the production place of the neutrino, i.e. for $L \gg L^{\rm osc}_{kj}$, the probability approaches this constant value because of the finite resolution in the measurement of E and L. Indeed, one has to convolute the theoretical probability by the detector resolution. Denoting by $\phi(L/E)$ the resolution of the variable L/E, the average of the functions $\sin^2(\Delta m^2_{kj} L/(4E)) = [1 - \cos(\Delta m^2_{kj} L/(2E))]/2$ and $\sin(\Delta m^2_{kj} L/(2E))$ in Eq. (9.40) is calculated with

$$\langle f(L/E) \rangle = \int f(L/E) \phi(L/E) d(L/E).$$

One can analytically perform the calculation if ϕ is a Gaussian centred on $\langle L/E \rangle$ with a standard deviation $\sigma_{L/E}$. One finds for f a sine or a cosine function,

$$\left\langle f\left(\frac{\Delta m^2_{kj} L}{2E} \right) \right\rangle = f\left(\frac{\Delta m^2_{kj}}{2} \langle L/E \rangle \right) \exp\left[-\frac{1}{2} \left(\frac{\Delta m^2_{kj}}{2} \sigma_{L/E} \right)^2 \right].$$

Assuming that $\sigma_{L/E} \propto \langle L/E \rangle$, we observe that for very large L/E, the average above vanishes. Therefore, from Eq. (9.40), the average probability becomes

$$\langle P_{\nu_\alpha \to \nu_\beta} \rangle = \delta_{\alpha\beta} - 2 \sum_{k>j} \Re \left\{ V^*_{\alpha k} V_{\beta k} V_{\alpha j} V^*_{\beta j} \right\} = \sum_k |V_{\alpha k}|^2 |V_{\beta k}|^2.$$

Strictly speaking, even if there is a transition from one flavour to another (mixing), the average probability does not constitute an oscillation. The term 'oscillation' is reserved for the case where the probability depends on the distance (or equivalently, the time) as in Eq. (9.35) or (9.39). Solar neutrino experiments, for which L is the Sun–Earth distance, are only sensitive to the average probability. When the same flavour is observed as the one that is produced, we talk about survival probabilities. These are the quantities that disappearance experiments measure, as in atmospheric experiments. According to Eq. (9.40), for $\alpha = \beta$, $V^*_{\alpha k} V_{\beta k} V_{\alpha j} V^*_{\beta j} = |V_{\alpha k}|^2 |V_{\alpha j}|^2$ is real, and hence, the survival probability reads

$$P_{\nu_\alpha \to \nu_\alpha}(L) = 1 - 4 \sum_{k>j} |V_{\alpha k}|^2 |V_{\alpha j}|^2 \sin^2 \left(\frac{\Delta m^2_{kj} L}{4E} \right). \tag{9.41}$$

For very large L, as the average of the sine squared is about $1/2$ (in the Gaussian case), the survival probability approaches

$$\langle P_{\nu_\alpha \to \nu_\alpha} \rangle \simeq 1 - 2 \sum_{k>j} |V_{\alpha k}|^2 |V_{\alpha j}|^2. \tag{9.42}$$

Using standard units, the oscillation length in Eq. (9.38) and the argument in Eq. (9.40) or (9.41) of the squared sine read

$$L^{\rm osc}_{kj}(\text{km}) = 2.47 \frac{E(\text{GeV})}{\Delta m^2_{kj}(\text{eV}^2)}, \qquad \frac{\Delta m^2_{kj} L}{4E} = 1.27 \frac{\Delta m^2_{kj}(\text{eV}^2) L(\text{km})}{E(\text{GeV})}. \tag{9.43}$$

We shall see soon that Δm^2_{kj} is small, of the order of $O(10^{-3}\text{–}10^{-4})$ eV^2. Typical neutrino experiments trying to observe neutrino oscillations use either (anti-)neutrinos ($\bar{\nu}_e$) from nuclear plants (with typical energy of the order of a few MeV) or neutrinos from accelerators (a few GeV). Consequently, the oscillation length is necessarily large (macroscopic), ranging typically from 1 km (Double Chooz, Daya Bay, Reno experiments, etc.) to several hundreds of kilometres (T2K, OPERA, etc.). Those are the typical distances between the source of neutrinos and their detection to be able to observe the oscillation phenomenon. Since a given experiment is often sensitive to the influence of the mixing of two neutrinos and not three (such as the atmospheric neutrinos in Super-Kamiokande, which is not sensitive to ν_e), it is instructive to determine the oscillation probabilities for two neutrino flavours, ν_α and ν_β. In that case, the mixing matrix is reduced to a 2×2 unitary matrix that reads

$$V^{2\times 2}_{\rm PMNS} = \begin{pmatrix} \cos\theta & \sin\theta \\ -\sin\theta & \cos\theta \end{pmatrix}. \tag{9.44}$$

As $V^{2\times 2}_{\rm PMNS}$ is a real matrix, the general Formula (9.40) simplifies to

$$\begin{aligned} P_{\nu_\alpha \to \nu_\beta}(L) &= \delta_{\alpha\beta} - 4 V_{\alpha 2} V_{\beta 2} V_{\alpha 1} V_{\beta 1} \sin^2 \left(\frac{\Delta m^2 L}{4E} \right) \\ &= \begin{cases} 1 - \sin^2(2\theta) \sin^2 \left(\frac{\Delta m^2 L}{4E} \right), & \alpha = \beta, \\ \sin^2(2\theta) \sin^2 \left(\frac{\Delta m^2 L}{4E} \right), & \alpha \neq \beta. \end{cases} \end{aligned} \tag{9.45}$$

Global constraints on various sources of data (solar, atmospheric, accelerator neutrinos, etc.) give an estimate of Δm^2_{kj} and the three mixing angles. For example, atmospheric neutrinos in Super-Kamiokande showed that the two-flavour approximation is valid in

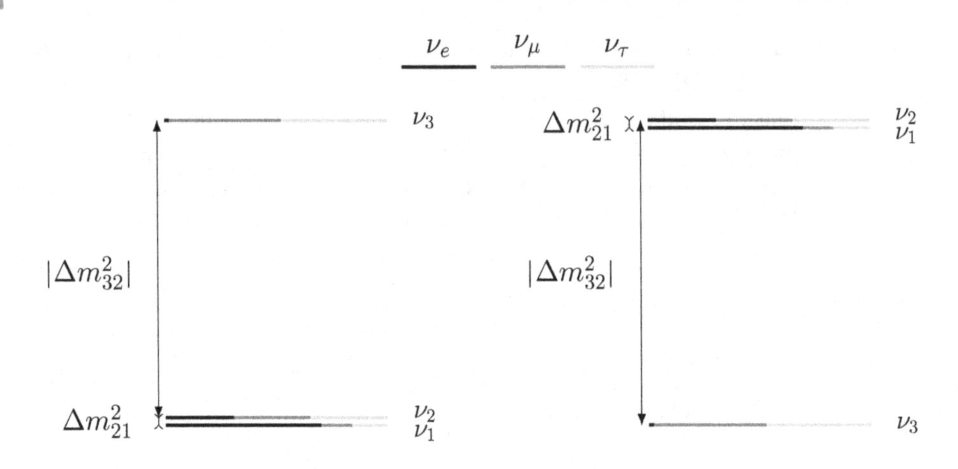

Fig. 9.8 The neutrino mass hierarchy. Left, normal; right, inverted. The probability to be measured as a given flavour eigenstate (at $L = 0$) is indicated by the grey scale.

this specific case (since there was no sensitivity to ν_e). In the original paper of Super-Kamiokande Collaboration (1998), Super-Kamiokande fitted the value of θ and Δm^2 in $P_{\nu_\mu \to \nu_\mu}(L)$ using Eq. (9.45) and found $\sin^2(2\theta) \simeq 1$ and $\Delta m^2 \simeq 2.2 \times 10^{-3}$ eV^2. Nowadays, the global fit in the three-flavour framework gives (Esteban et al., 2020)

$$\Delta m^2_{21} \simeq 7.42 \times 10^{-5} \text{ eV}^2, \qquad |\Delta m^2_{31}| \approx |\Delta m^2_{32}| \simeq 2.5 \times 10^{-3} \text{ eV}^2.$$

Atmospheric and solar neutrinos constrain mainly Δm^2_{31} and Δm^2_{21}, respectively. The three squared-mass differences are not independent since $\Delta m^2_{32} + \Delta m^2_{21} = \Delta m^2_{31}$ [see the definition in Eq. (9.36)]. As the values of Δm^2_{21} and $|\Delta m^2_{31}| \approx |\Delta m^2_{32}|$ differ by almost two orders of magnitude, the neutrino mass spectrum is not regularly spaced (see Fig. 9.8). Two of them, ν_1 and ν_2, are almost degenerate. Note that the signs of Δm^2_{31} and Δm^2_{32} are still unknown. Two scenarios, as shown in Fig. 9.8, are possible: the so-called *normal mass hierarchy*, which assumes, $m_1 < m_2 < m_3$; and the *inverted mass hierarchy*, with $m_3 < m_1 < m_2$.

9.4.5 Considerations on the Measurement of Neutrinos Masses and the Oscillation Phenomenon

The ν_k neutrinos have, by definition, a definite mass, but experiments detect ν_α. What mass can be measured for ν_α? By definition, $|\nu_k\rangle$ is an eigenstate of the mass operator

$$\hat{P}^\mu \hat{P}_\mu \, |\nu_k\rangle = m_k^2 \, |\nu_k\rangle,$$

where $\hat{P}^\mu$ is the 4-momentum operator. The state $|\nu_\alpha\rangle$, a linear combination of $|\nu_k\rangle$ in Eq. (9.33), is not a mass eigenstate since

$$\hat{P}^\mu \hat{P}_\mu \, |\nu_\alpha\rangle = \sum_k m_k^2 V^*_{\alpha k} \, |\nu_k\rangle.$$

Its average mass squared is, however,

$$\langle m_\alpha^2 \rangle = \langle \nu_\alpha | \hat{P}^\mu \hat{P}_\mu | \nu_\alpha \rangle = \sum_k m_k^2 V_{\alpha k}^* \langle \nu_\alpha | \nu_k \rangle = \sum_k m_k^2 |V_{\alpha k}|^2.$$

The same result would have been obtained by using the expression of $|\nu_\alpha(x,t)\rangle$ given by Eq. (9.34) at a time t. On average, we would measure $\langle m_\alpha^2 \rangle$, which does not depend on t or L (and so on the distance of the detector), but event by event, we would measure m_k^2 with a probability $|V_{\alpha k}|^2$. Now, let us imagine that we evaluate the neutrino mass by measuring the 4-momenta of the other particles of a reaction. The neutrino 4-momentum would be inferred from 4-momentum conservation. Let us assume an ideal detector able to measure perfectly the 4-momenta with perfect accuracy. For example, in the pion decay, $\pi^+ \to \mu^+ + \nu_\mu$, according to Eq. (3.32), the neutrino mass would be

$$m_\nu^2 = m_\pi^2 + m_\mu^2 - 2m_\pi \sqrt{m_\mu^2 + |\boldsymbol{p}_\mu^*|},$$

where $\boldsymbol{p}_\mu^*$ is the muon momentum in the pion rest frame (which would be perfectly known from the ideal measurements of $\boldsymbol{p}_\pi$ and $\boldsymbol{p}_\mu$). In quantum mechanics, the mass measurement forces the quantum superposition of neutrino mass eigenstates to collapse on the massive neutrino whose mass has been measured, i.e. $m_\nu^2 = m_k^2$. Therefore, there is no more oscillation after the measurement since after the collapse, the state is $|\nu_k(x,t)\rangle = |\nu_k\rangle\, e^{-i(E_k t - \boldsymbol{p}\cdot\boldsymbol{x})}$, and the probability of detecting the state $|\nu_\beta\rangle$ at t when $|\nu_\alpha\rangle$ was produced at $t = 0$ no longer depends on time, i.e.

$$P_{\nu_\alpha \to \nu_k \to \nu_\beta} = \left| \langle \nu_\beta | \nu_k(x,t) \rangle \right|^2 |\langle \nu_k | \nu_\alpha \rangle|^2 = |V_{\beta k}|^2 |V_{\alpha k}|^2.$$

When the measurement is accurate enough to determine m_k^2 with an uncertainty smaller than Δm_{ij}^2, the mass eigenstate ν_k is unambiguously determined, which prevents the oscillation pattern. What is the reason for the destruction of the oscillation pattern? It relies on the uncertainty principle. Coming back to the example of the pion decay, if the pion momentum is perfectly known, its position becomes unknown and, therefore, the place where the neutrino has been created is unknown too. Thus, the uncertainty on L exceeds the oscillation length L_{kj}^{osc}, which washes out the oscillation pattern. A rigorous proof of this explanation would require using the wave packet treatment instead of the naive plane-wave approximation.[15] As proven in Giunti and Kim (2007), this introduces the notion of a coherence length, L^{coh}, related to the size of the neutrino wave packets. When this size is larger than L_{kj}^{osc}, or when L is larger than L^{coh}, the oscillation is suppressed.

9.4.6 Lepton Number: Conserved or Not Conserved?

With the weak interaction (and electromagnetism), the individual lepton number is conserved at each vertex, as shown by Fig. 9.3, where the charged lepton is always associated with the neutrino of the same generation. However, because of neutrino oscillations, a neutrino produced with a given flavour can be detected after its propagation with another flavour. Hence, there is clearly no conservation of the individual lepton number for

[15] In addition, the wave packet treatment eliminates idealising assumptions like assuming that all ν_k have the same momentum during the propagation.

neutrinos. But what about charged leptons? Theoretically, the conclusion is the same. Indeed, let us consider the following diagram describing the muon decay $\mu^- \to e^- + \gamma$ (the photon is only there to conserve the 4-momentum since $m_\mu \neq m_e$):

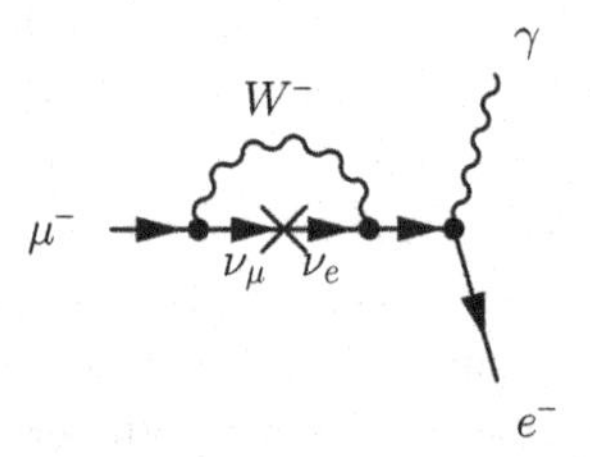

The muon neutrino produced at the first vertex may oscillate to an electron neutrino when it is absorbed at the second vertex, with a probability $P_{\nu_\mu \to \nu_e}$ given by Eq. (9.35). The cross in the diagram symbolises this neutrino mixing due to the oscillation. This reaction, theoretically allowed, leads to the muon and electron lepton number violations. However, the probability of such reaction is proportional to $(\Delta m^2_{ij}/m^2_W)^2$ (Bilenky and Giunti, 2001). Since $\Delta m^2 \lesssim 10^{-3}$ eV^2, this decay is virtually impossible to observe: a branching ratio of the order of 10^{-56} is expected! Hence, even if, strictly speaking, the individual lepton number for the charged lepton is also violated, it remains in all practical cases conserved. Note that this decay has been experimentally searched for without success. The current experimental limit for the branching ratio is (Particle Data Group, 2022)

$$\text{BR}(\mu^\pm \to e^\pm + \gamma) < 4.2 \times 10^{-13}$$

at the 90% confidence level. Therefore, it seems impossible to probe the theoretical value. In the modern Standard Model, which includes neutrino oscillation, the individual lepton number is thus not conserved. However, the total lepton number remains always conserved: when a lepton is present in the initial state (of any flavour), there must be a lepton in the final state (but not necessarily with the same flavour).

9.5 CP Violation

9.5.1 Introduction

We saw previously that the weak interaction violates the parity symmetry. It is easy to check that it also violates charge conjugation. Indeed, in Section 9.2.3, we observed that in the leptonic decay of the charged pion $\pi^- \to \ell^- + \bar{\nu}_\ell$, both the anti-neutrino and the charged lepton ($\ell^- = \mu^-$ or e^-) are in the right-handed helicity states, i.e. $\pi^- \to \ell^-_R + \bar{\nu}_{\ell R}$. As the charge conjugation transforms every particle in its antiparticle (and vice versa), the charge conjugate decay is $\pi^+ \to \ell^+_R + \nu_{\ell R}$, which is forbidden since only left-handed neutrinos interact. Hence, the weak interaction violates both the P and C symmetries. However, if in addition, we apply parity transformation, helicities are changed, yielding

$$\pi^- \to \ell^-_R + \bar{\nu}_{\ell R} \xrightarrow{\text{charge conjugation+parity}} \pi^+ \to \ell^+_L + \nu_{\ell L},$$

which is perfectly possible and gives a similar rate as the original reaction. Therefore, in this specific example, the weak interaction seems to preserve the combination of charge conjugation and parity (in any order), called the CP symmetry. However, we shall see that in some rare cases, CP is violated.

9.5.2 CP Transformation of Fields

If CP is violated by the weak interaction, it indicates that the weak interaction Lagrangian is not invariant under the CP transformation. Let us take a closer look at what this means in theoretical terms. We shall derive the weak interaction Lagrangian in the next chapter when the electroweak theory is presented, but at this stage, we can observe that, as for QED in Eq. (6.29) or QCD in Eq. (8.44), the interaction Lagrangian must couple the vector boson, carrier of the interaction, $W^{\pm}$, to the (charged) current, i.e.

$$\mathcal{L}_{\text{int}} = -j^{\mu}_{cc+} W^{+}_{\mu} - \left(j^{\mu}_{cc+} W^{+}_{\mu}\right)^{\dagger} = -j^{\mu}_{cc+} W^{+}_{\mu} - j^{\mu}_{cc-} W^{-}_{\mu}.$$

For quarks, j^{μ}_{cc+} has been given in Eq. (9.22). It is of the form

$$j^{\mu}_{cc+} = \frac{g_W}{\sqrt{2}} \bar{u}_i \gamma^{\mu} \frac{1-\gamma^5}{2} V_{ij} d_j, \tag{9.46}$$

with the up-type quark fields, $u_{i=1,2,3} = u, c, t$, the down-type quark fields, $d_{j=1,2,3} = d, s, b$, and V_{ij}, the CKM matrix coefficient. Therefore, the interaction Lagrangian contains terms (up to a factor) such as

$$\mathcal{L}_{\text{int}} \supset V_{ij} \bar{u}_i \gamma^{\mu} \frac{1-\gamma^5}{2} d_j W^{+}_{\mu} + V^{*}_{ij} \bar{d}_j \gamma^{\mu} \frac{1-\gamma^5}{2} u_i W^{-}_{\mu}.$$

Note that it is expressed with the mass eigenfields of quarks. For leptons, j^{μ}_{cc+} was given with the weak eigenfields in Eq. (9.15), yielding the terms in the interaction Lagrangian

$$\mathcal{L}_{\text{int}} \supset \bar{\nu}_{\ell} \gamma^{\mu} \frac{1-\gamma^5}{2} \ell W^{+}_{\mu} + \bar{\ell} \gamma^{\mu} \frac{1-\gamma^5}{2} \nu_{\ell} W^{-}_{\mu},$$

with $\ell = e, \mu, \tau$. However, if we express it with mass eigenfields, as $\nu_{\ell} = \sum_i V_{\ell i} \nu_i$, where $V_{\ell i}$ is the PMNS matrix coefficient, it follows that the interaction Lagrangian contains terms such as

$$\mathcal{L}_{\text{int}} \supset V^{*}_{\ell i} \bar{\nu}_i \gamma^{\mu} \frac{1-\gamma^5}{2} \ell W^{+}_{\mu} + V_{\ell i} \bar{\ell} \gamma^{\mu} \frac{1-\gamma^5}{2} \nu_i W^{-}_{\mu}.$$

Therefore, both for quarks and leptons, the interaction Lagrangian contains terms with the following form:

$$\mathcal{L}_{\text{int}} \supset V_{ij} \overline{\psi}_i \gamma^{\mu} \frac{1-\gamma^5}{2} \psi_j W_{\mu} + V^{*}_{ij} \overline{\psi}_j \gamma^{\mu} \frac{1-\gamma^5}{2} \psi_i W^{\dagger}_{\mu}, \tag{9.47}$$

where ψ_i and ψ_j are the mass eigenfields of fermions, V is a mixing matrix (either CKM or PMNS) and W_{μ} is either the W^{+}_{μ} or W^{-}_{μ} vector field. The parity and charge conjugation transformations were given in Chapter 5 for the scalar, fermion and vector fields. As W is a vector field like the photon, its party transformation is that of Eq. (5.134). For the charge conjugation, one has to take into account that, unlike the photon, the W is not its antiparticle, and therefore, charge conjugation must produce the antiparticle,[16] i.e.

$$\hat{C} W_\mu \hat{C}^{-1} = \eta_C W_\mu^\dagger, \tag{9.48}$$

where η_C is the intrinsic charge conjugation of the W. Globally, the CP transformation of W thus reads

$$\left(\hat{C}\hat{\mathcal{P}}\right) W_0(x) \left(\hat{C}\hat{\mathcal{P}}\right)^{-1} = -\eta_{\rm CP} W_0^\dagger(x'), \qquad \left(\hat{C}\hat{\mathcal{P}}\right) W_k(x) \left(\hat{C}\hat{\mathcal{P}}\right)^{-1} = \eta_{\rm CP} W_k^\dagger(x'), \tag{9.49}$$

where $x' = (x^0, -\boldsymbol{x})$ and the intrinsic CP phase of the field, $\eta_{\rm CP}$, is simply the product of the parity and charge conjugation phases, $\eta_{\rm CP} = \eta_C \eta_P$. Let us now calculate the CP transform of the first term in Eq. (9.47). Following the calculation that led to the result of Eq. (5.137) and introducing the chirality projectors (5.98), it is easy to check that

$$\hat{\mathcal{P}} \overline{\psi}_i(x) \gamma^\mu P_L \psi_j(x) \hat{\mathcal{P}}^{-1} = \psi_i^\dagger(x') \gamma^\mu \gamma^0 P_R \psi_j(x') = \begin{cases} \overline{\psi}_i(x') \gamma^0 P_R \psi_j(x'), & \mu = 0 \\ -\overline{\psi}_i(x') \gamma^k P_R \psi_j(x'), & \mu \neq 0. \end{cases} \tag{9.50}$$

Similarly, following the calculations leading to Eq. (5.142), we find (assuming that both ψ_i and ψ_j have the same intrinsic C phase) that

$$\hat{C} \overline{\psi}_i \gamma^\mu P_R \psi_j \hat{C}^{-1} = -\psi_j^\dagger (M^\mu)^{\rm T} \psi_i, \quad \text{with} \quad M^\mu = \hat{\rm C}^\dagger \gamma^0 \gamma^\mu P_R \hat{\rm C}. \tag{9.51}$$

Using the explicit expression of the charge conjugation matrix (5.84), we deduce that (see Problem 9.4)

$$\hat{C} \overline{\psi}_i \gamma^\mu P_R \psi_j \hat{C}^{-1} = -\overline{\psi}_j \gamma^\mu P_L \psi_i. \tag{9.52}$$

Therefore, it follows from Eqs. (9.50) and (9.52) that

$$\left(\hat{C}\hat{\mathcal{P}}\right) \overline{\psi}_i(x) \gamma^\mu P_L \psi_j(x) \left(\hat{C}\hat{\mathcal{P}}\right)^{-1} = \begin{cases} -\overline{\psi}_j(x') \gamma^0 P_R \psi_i(x'), & \mu = 0 \\ \overline{\psi}_j(x') \gamma^k P_R \psi_i(x'), & \mu \neq 0. \end{cases} \tag{9.53}$$

Given Eqs. (9.49) and (9.53), we conclude that the CP transform of the first term in Eq. (9.47) is

$$\left(\hat{C}\hat{\mathcal{P}}\right) V_{ij} \overline{\psi}_i \gamma^\mu \frac{1-\gamma^5}{2} \psi_j W_\mu \left(\hat{C}\hat{\mathcal{P}}\right)^{-1} = \eta_{\rm CP} V_{ij} \overline{\psi}_j \gamma^\mu \frac{1-\gamma^5}{2} \psi_i W_\mu^\dagger. \tag{9.54}$$

Assigning for the W boson, $\eta_{\rm CP} = 1$, we would obtain the second term, and thus an invariant Lagrangian, if the mixing matrix element was real valued ($V_{ij} = V_{ij}^*$). Therefore, the source of CP violation in the weak interaction is the consequence of the complex nature of the CKM and PMNS matrices.

9.5.3 CP Violation in the Neutral Kaon System

The first violation of the CP symmetry was observed in the $K^0 - \overline{K}^0$ system in 1964 by James Cronin and Val Fitch. The $K^0 \equiv |d\bar{s}\rangle$ and $\overline{K}^0 \equiv |\bar{d}s\rangle$ carry ± 1 unit of the strangeness

[16] From this perspective, the transformation is similar to the scalar field in Eq. (5.24).

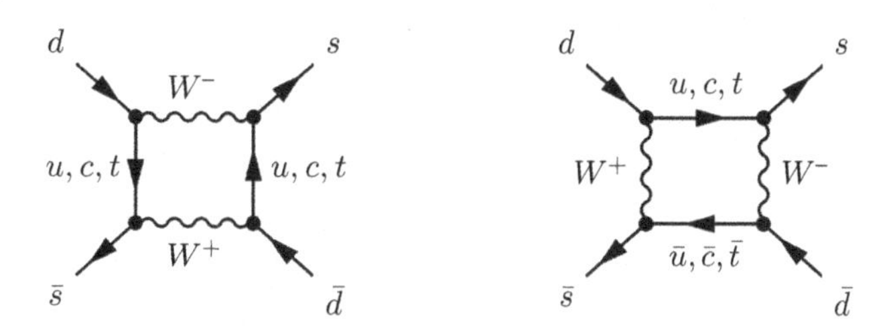

Fig. 9.9 Box diagram contributions to the mixing of K^0 and $\overline{K}^0$.

quantum number (+1 for K^0 and -1 for $\overline{K}^0$) and, thus, are strangeness eigenstates. They are very easily produced by the strong interaction in reaction, such as,

$$K^+ + n \to K^0 + p, \quad K^- + p \to \overline{K}^0 + n.$$

However, K^0 (or $\overline{K}^0$), an eigenstate of the strong interaction (and strangeness), can transform spontaneously in the vacuum to its antiparticle and vice versa, $K^0 \leftrightarrow \overline{K}^0$, thanks to weak couplings contributing to the two box diagrams shown in Fig. 9.9. Since K^0 and $\overline{K}^0$ are permanently oscillating, they cannot have a definite lifetime. As strangeness is conserved by the strong interaction and electromagnetism, kaons decay to lighter mesons that do not contain the s-quark (pions) by the weak interaction. If a K^0 is produced at $t = 0$, $|\psi(0)\rangle = |K^0\rangle$, its time evolution is thus governed by a Hamiltonian that includes the weak interaction, such as

$$|\psi(t)\rangle = a(t)\,|K^0\rangle + b(t)\,|\overline{K}^0\rangle + \sum_i c_i(t)\,|f_i\rangle,$$

where $|f_i\rangle$ are all possible decay products (excluding $|K^0\rangle$ and $|\overline{K}^0\rangle$). This Hamiltonian written in the basis that spans all decay products, K^0, and $\overline{K}^0$ would be Hermitian; however, if we restrict our discussion to the evolution in K^0 or $\overline{K}^0$ only and ignore the decay products, the 2×2 effective Hamiltonian H in the subspace $\{|K^0\rangle, |\overline{K}^0\rangle\}$ is no longer Hermitian. Without loss of generality, a complex matrix can be decomposed into

$$H = M - \frac{i}{2}\Gamma, \tag{9.55}$$

where $M = (H + H^\dagger)/2$ and $\Gamma = i(H - H^\dagger)$ are of the kind 2×2 Hermitian matrices. Note that if the weak interaction did not exist, K^0 and $\overline{K}^0$ would be the eigenstates of the Hamiltonian, H_0, which would correspond to the Hamiltonian of the strong interaction, electromagnetism and the free Hamiltonian. They would be stable, and M in Eq. (9.55) would be diagonal with $M = \mathrm{diag}(m_0, m_0)$, where m_0 would be the K^0 and $\overline{K}^0$ mass, while Γ would be zero. In Supplementary 9.1, some properties of the CPT symmetry are presented. In particular, applying Eq. (9.62) to M and Γ with $\psi_1 = \psi_2 = K^0$, we deduce

$$\langle \overline{K}^0 | M | \overline{K}^0 \rangle = \langle K^0 | M | K^0 \rangle, \qquad \langle \overline{K}^0 | \Gamma | \overline{K}^0 \rangle = \langle K^0 | \Gamma | K^0 \rangle. \tag{9.56}$$

In the absence of the weak interaction, K^0 and $\overline{K}^0$ would thus have the same well-defined mass. With the weak interaction, Eq. (9.56) imposes $M_{11} = M_{22}$ and $\Gamma_{11} = \Gamma_{22}$, and therefore, the effective Hamiltonian (9.55) takes the form

$$H = \begin{pmatrix} m & M_{12} \\ M_{12}^* & m \end{pmatrix} - \frac{i}{2}\begin{pmatrix} \gamma & \Gamma_{12} \\ \Gamma_{12}^* & \gamma \end{pmatrix}. \tag{9.57}$$

The diagonalisation of this 2×2 matrix is easy (and left as an exercise), giving the two complex eigenvalues

$$\lambda_\pm = m - \frac{i}{2}\gamma \pm \sqrt{\left(M_{12}^* - \frac{i}{2}\Gamma_{12}^*\right)\left(M_{12} - \frac{i}{2}\Gamma_{12}\right)} = m_\pm - \frac{i}{2}\gamma_\pm, \tag{9.58}$$

where $m_\pm = \Re(\lambda_\pm)$ and $\gamma_\pm = -2\Im(\lambda_\pm)$ are the well-defined masses and decay widths of the two eigenstates, their lifetimes being $1/\gamma_\pm$. Anticipating other neutral meson systems, let us generalise the rest of the development by denoting the neutral mesons M^0 and $\overline{M}^0$ and the mass eigenstates M_H and M_L, where M_H has the highest mass and M_L has the lowest. It is conventional to define the mass eigenstates with

$$|M_L\rangle = p\,|M^0\rangle + q\,|\overline{M}^0\rangle, \qquad |M_H\rangle = p\,|M^0\rangle - q\,|\overline{M}^0\rangle, \tag{9.59}$$

where $|p|^2 + |q|^2 = 1$. We shall see that depending on the characteristics of $|M_L\rangle$ and $|M_H\rangle$ in terms of CP behaviour, M_L may be associated with the eigenstate λ_+ or λ_- (and M_H with λ_- or λ_+). Solving the eigenvalues equation $H\,|M_L\rangle = \lambda_\pm\,|M_L\rangle$, i.e. $H\binom{p}{q} = \lambda_\pm\binom{p}{q}$ leads to

$$\begin{array}{l} |M_L\rangle = \lambda_+\,|M_L\rangle\,, \\ |M_H\rangle = \lambda_-\,|M_H\rangle\,, \end{array} \frac{q}{p} = \sqrt{\frac{M_{12}^* - \frac{i}{2}\Gamma_{12}^*}{M_{12} - \frac{i}{2}\Gamma_{12}}}; \qquad \begin{array}{l} |M_L\rangle = \lambda_-\,|M_L\rangle\,, \\ |M_H\rangle = \lambda_+\,|M_H\rangle\,, \end{array} \frac{q}{p} = -\sqrt{\frac{M_{12}^* - \frac{i}{2}\Gamma_{12}^*}{M_{12} - \frac{i}{2}\Gamma_{12}}}. \tag{9.60}$$

In the kaon system, the two eigenstates have very close mass, about 498 MeV/c^2, with the mass difference being only 3.5×10^{-12} MeV/c^2. Therefore, instead of labelling the states by their masses, we use their lifetimes, which differ by more than two orders of magnitude. The state with the largest lifetime is called *K long*, K_L, and the other *K short*, K_S, with

$$\tau_{K_L} \simeq 5 \times 10^{-8}\ \mathrm{s}, \qquad \tau_{K_S} \simeq 9 \times 10^{-11}\ \mathrm{s}.$$

It turns out that K_L is the heaviest, and therefore, we will use the linear combinations

$$|K_L\rangle = p\,|K^0\rangle - q\,|\overline{K}^0\rangle, \qquad |K_S\rangle = p\,|K^0\rangle + q\,|\overline{K}^0\rangle. \tag{9.61}$$

Supplement 9.1. CPT symmetry

If CPT is a good symmetry, any observable $\hat{O}$ ($\hat{O}$ being Hermitian) commutes with the CPT operator $\hat{S} = \hat{\mathcal{C}}\hat{\mathcal{P}}\hat{\mathcal{T}}$ transformation, i.e. $\hat{O}\hat{S} = \hat{S}\hat{O}$ (cf. Section 4.2). Note that the time reversal operator is an anti-unitary operator. Indeed, a transition between two states $|\psi_1\rangle \to |\psi_2\rangle$ becomes after the time reversal a transition $|\psi_2\rangle \to |\psi_1\rangle$. Therefore, $\langle(\hat{\mathcal{T}}\psi_1)|(\hat{\mathcal{T}}\psi_2)\rangle = \langle\psi_2|\psi_1\rangle$, which defines an anti-unitary operator [see Eq. (4.2)]. It follows that $\hat{S}$, including $\hat{\mathcal{T}}$ is an anti-unitary operator too. Now, let us denote by $|\overline{\psi}_1\rangle = \hat{S}\,|\psi_1\rangle$ and $|\overline{\psi}_2\rangle = \hat{S}\,|\psi_2\rangle$ the CPT transformation of states $|\psi_1\rangle$ and $|\psi_2\rangle$, respectively. Given that $\hat{O}^\dagger = \hat{O}$, the anti-unitary property leads to the equality

$$\langle\hat{S}\psi_1|\hat{S}\hat{O}\psi_2\rangle = \langle\hat{O}\psi_2|\psi_1\rangle = \langle\psi_1|\hat{O}|\psi_2\rangle^* = \langle\psi_2|\hat{O}|\psi_1\rangle.$$

Supplement 9.1. (cont.)

But, as $\hat{O}$ commutes with $\hat{S}$, we also have

$$\langle \hat{S}\psi_1 | \hat{S}\hat{O}b\psi_2 \rangle = \langle \hat{S}\psi_1 | \hat{O}\hat{S}\psi_2 \rangle = \langle \overline{\psi}_1 | \hat{O} | \overline{\psi}_2 \rangle.$$

Therefore,

$$\langle \overline{\psi}_1 | \hat{O} | \overline{\psi}_2 \rangle = \langle \psi_2 | \hat{O} | \psi_1 \rangle. \tag{9.62}$$

We have denoted the CPT transformation of states with a bar symbol, which suggests that they are the antiparticles of the original states. Indeed, consider the Fock operator of a spinless particle $a_{\boldsymbol{p}}$ and the Fock operator of its antiparticle $b_{\boldsymbol{p}}$. As the momentum $\boldsymbol{p}$ is a function of the time derivative, we expect up to a phase $\hat{\mathcal{T}} a_{\boldsymbol{p}} \hat{\mathcal{T}}^{-1} \propto a_{-\boldsymbol{p}}$, and hence,

$$\hat{S} a_{\boldsymbol{p}} \hat{S}^{-1} = \hat{C}\hat{\mathcal{P}}\hat{\mathcal{T}} a_{\boldsymbol{p}} \hat{\mathcal{T}}^{-1} \hat{\mathcal{P}}^{-1} \hat{C}^{-1} \propto \hat{C}\hat{\mathcal{P}} a_{-\boldsymbol{p}} \hat{\mathcal{P}}^{-1} \hat{C}^{-1} \propto \hat{C} a_{\boldsymbol{p}} \hat{C}^{-1} \propto b_{\boldsymbol{p}},$$

where the transformations (5.22) and (5.25) have been used. It shows that the particle state is transformed under CPT in the antiparticle state.

Let us now make the connection with CP violation. The K^0 is the ground state of $d\bar{s}$ mesons. Therefore, its parity is given by Eq. (4.17), with no orbital angular momentum ($l = 0$), yielding $\hat{\mathcal{P}} |K^0\rangle = -|K^0\rangle$. For the same reason, $\hat{\mathcal{P}} |\overline{K}^0\rangle = -|\overline{K}^0\rangle$. As K^0 and $\overline{K}^0$ are not the charge-conjugation eigenstates, their phase under this transformation is arbitrary. Let us choose $\hat{C} |K^0\rangle = -|\overline{K}^0\rangle$ and $\hat{C} |\overline{K}^0\rangle = -|K^0\rangle$ such that their CP phases are

$$\hat{C}\hat{\mathcal{P}} |K^0\rangle = |\overline{K}^0\rangle, \qquad \hat{C}\hat{\mathcal{P}} |\overline{K}^0\rangle = |K^0\rangle. \tag{9.63}$$

(The opposite convention would not change the measurable quantities.) Consequently, these two states are *not* the eigenstates of CP. However, it is straightforward to check that the linear combinations,

$$|K_1\rangle = \left(|K^0\rangle + |\overline{K}^0\rangle\right)/\sqrt{2}, \qquad |K_2\rangle = \left(|K^0\rangle - |\overline{K}^0\rangle\right)/\sqrt{2}, \tag{9.64}$$

are CP even and CP odd eigenstates, respectively. Kaons can easily decay in pions because of their large difference in masses ($\simeq 498\ \mathrm{MeV}/c^2$ versus $\simeq 140\ \mathrm{MeV}/c^2$). The decay channels in two pions ($\pi^+\pi^-$, $\pi^0\pi^0$) or in three pions ($\pi^+\pi^-\pi^0$, $\pi^0\pi^0\pi^0$) are kinematically possible. In Section 9.1.2, we saw that two- and three-pion systems resulting from the decay of a spin-0 particle (as K^0) have parities

$$\hat{\mathcal{P}} |2\pi\rangle = \eta_P(2\pi) |2\pi\rangle = +|2\pi\rangle, \qquad \hat{\mathcal{P}} |3\pi\rangle = \eta_P(3\pi) |3\pi\rangle = -|3\pi\rangle. \tag{9.65}$$

The charge conjugation of a system made of a scalar and its antiparticle has been given in Eq. (4.23). Therefore,

$$\hat{C} |2\pi\rangle = \eta_C(2\pi) |2\pi\rangle = (-1)^l |2\pi\rangle, \tag{9.66}$$

where l is the relative orbital angular momentum. Since both kaons and pions are spinless particles, angular momentum conservation imposes $l = 0$, leading to $\eta_C(\pi^+\pi^-) = \eta_C(\pi^0\pi^0) = +1$. It follows that the CP number of the two-pion system final state is

$$\eta_{CP}(2\pi) = \eta_C(2\pi) \times \eta_P(2\pi) = 1 \times 1 = 1. \tag{9.67}$$

The charge conjugation of the three-pion final state is a bit more tricky to determine. For $\pi^0\pi^0\pi^0$, as π^0 is an eigenstate of the charge conjugation with[17] $\eta_C(\pi^0) = 1$, $\eta_C(3\pi^0) = 1^3 = 1$. For $\pi^+\pi^-\pi^0$, we have $\eta_C(\pi^+\pi^-\pi^0) = \eta_C(\pi^+\pi^-) \times \eta_C(\pi^0) = (-1)^{l_1} \times 1 = (-1)^{l_1}$, where l_1 is the angular momentum number between the two charged pions, and the result of Eq. (9.66) has been used. The difference of mass between K^0 and three pions is only $\simeq$80 MeV. Hence, in the decay $K^0 \to \pi^+ + \pi^- + \pi^0$, $l_1 = 0$ is much more likely to occur than $l_1 > 0$ (only quantum tunnelling would allow the latter case). We thus conclude that in most cases, $\eta_C(3\pi) = 1$, and therefore,

$$\eta_{CP}(3\pi) = \eta_C(3\pi) \times \eta_P(3\pi) = 1 \times -1 = -1. \tag{9.68}$$

Let us recap. Three kinds of kaons have been distinguished:

- K^0 and $\overline{K}^0$, produced by the strong interaction.
- K_S and K_L, the physical eigenstates of the total Hamiltonian that includes both weak and strong interactions, with well-defined masses and lifetimes.
- K_1 and K_2, the CP eigenstates with $\eta_{CP}(K_1) = 1$ and $\eta_{CP}(K_2) = -1$.

If CP were conserved, the two-pion final states (CP even) must result from the decay of a CP-even state, i.e. $K_1 \to 2\pi$. Similarly, $K_2 \to 3\pi$. Since the phase space in the two-pion final states is much larger, we expect K_1 to have a much smaller lifetime. Therefore, we can identify the states K_S and K_L as

$$K_S = K_1 \to 2\pi, \qquad K_L = K_2 \to 3\pi, \qquad \text{if CP is conserved.}$$

In 1964, Christenson, Cronin, Fitch and Turlay checked this conclusion by realising the following experiment at Brookhaven National Laboratory (USA) (Christenson et al., 1964). By dumping a proton beam (30 GeV) into a beryllium target, they produced a K^0 beam, a mixture of the two physical states K_L and K_S. About 18 m after the production point, they placed spectrometers to measure the momentum of the charged pions, decay products of the $K_{L,S}$ (the π^0 contribution was deduced from energy–momentum conservation). As K_S is a short-lived meson ($c\tau = 2.7$ cm), it decays promptly in the beam, and therefore, the only mesons flying over 18 m are K_L. Among the 22 700 events analysed, 45 corresponded to a 2π decay (Christenson et al., 2013). Hence, K_L could decay in 0.2% of cases in 2π and 99.8% in 3π. They concluded that CP is violated by the weak interaction at the level of 0.2% in the kaon system. In 1980, Cronin and Fitch were awarded the Nobel prize for their discovery.

9.5.4 Classification of CP Violation Types

There are several interpretations of the observed CP violation in the kaon system. The so-called *indirect CP violation*, often called *CP violation from mixing*, assumes that the mass eigenstates K_L and K_S are actually defined by a linear combination of K_1 and K_2, i.e.

[17] π^0 decays into two photons, and we saw in Section 5.4.4 that $\eta_C(\gamma) = -1$. Conclusion: $\eta_C(\pi^0) = \eta(\gamma)^2 = 1$.

$$|K_L\rangle = \frac{1}{\sqrt{1+|\epsilon|^2}} (|K_2\rangle + \epsilon\, |K_1\rangle) = \frac{1}{\sqrt{2(1+|\epsilon|^2)}} \left[(1+\epsilon)\,|K^0\rangle - (1-\epsilon)\,|\overline{K}^0\rangle\right], \tag{9.69a}$$

$$|K_S\rangle = \frac{1}{\sqrt{1+|\epsilon|^2}} (|K_1\rangle + \epsilon\, |K_2\rangle) = \frac{1}{\sqrt{2(1+|\epsilon|^2)}} \left[(1+\epsilon)\,|K^0\rangle + (1-\epsilon)\,|\overline{K}^0\rangle\right]. \tag{9.69b}$$

The observation of the decay $K_L \to 2\pi$ is then possible because of its $\epsilon\,|K_1\rangle$ component. The comparison between Eq. (9.61) and Eqs. (9.69a), (9.69b) leads to the identification

$$p = \frac{1+\epsilon}{\sqrt{2(1+|\epsilon|^2)}}, \qquad q = \frac{1-\epsilon}{\sqrt{2(1+|\epsilon|^2)}}, \qquad \frac{q}{p} = \frac{1-\epsilon}{1+\epsilon}. \tag{9.70}$$

Since $\epsilon \simeq 0.3 \times 10^{-3}$, $|K_L\rangle$ is mostly $|K_2\rangle = (|K^0\rangle - |\overline{K}^0\rangle)/\sqrt{2}$ and $|K_S\rangle$ is mostly $|K_1\rangle = (|K^0\rangle + |\overline{K}^0\rangle)/\sqrt{2}$. In the approximation where CP violation is neglected, $\epsilon = 0$, $q/p = 1 > 0$, and thus, according to Eq. (9.60), the solution with the plus sign must be chosen, i.e.

$$\frac{q}{p} = +\sqrt{\frac{M_{12}^* - \frac{i}{2}\Gamma_{12}^*}{M_{12} - \frac{i}{2}\Gamma_{12}}}. \tag{9.71}$$

Notice that $q/p = 1$ would imply that M_{12} and Γ_{12} are real numbers. Now, as K_L and K_S evolve with times with their respective phase $\exp[-i(m_{L,S} - i\Gamma_{L,S}/2)]$, it follows from Eqs. (9.69a) and (9.69b) that the probabilities for finding a K^0 or $\overline{K}^0$ at t are (see Problem 9.5)

$$P_{K^0\to K^0}(t) = P_{\overline{K}^0\to \overline{K}^0}(t) = \frac{1}{4}\left[e^{-\Gamma_L t} + e^{-\Gamma_S t} + 2e^{-(\Gamma_L+\Gamma_S)t/2}\cos(\Delta m\, t)\right], \tag{9.72a}$$

$$P_{K^0\to \overline{K}^0}(t) = \frac{1}{4}\left[e^{-\Gamma_L t} + e^{-\Gamma_S t} - 2e^{-(\Gamma_L+\Gamma_S)t/2}\cos(\Delta m\, t)\right]\left|\frac{q}{p}\right|^2, \tag{9.72b}$$

$$P_{\overline{K}^0\to K^0}(t) = \frac{1}{4}\left[e^{-\Gamma_L t} + e^{-\Gamma_S t} - 2e^{-(\Gamma_L+\Gamma_S)t/2}\cos(\Delta m\, t)\right]\left|\frac{p}{q}\right|^2, \tag{9.72c}$$

where $\Delta m = m_L - m_s > 0$. Therefore, as long as $|q/p| \neq 1$, indirect CP violation generates differences between the probabilities

$$P_{K^0\to \overline{K}^0}(t) \neq P_{\overline{K}^0\to K^0}(t). \tag{9.73}$$

Since

$$\left|\frac{q}{p}\right|^2 = \left|\frac{1-\epsilon}{1+\epsilon}\right|^2 = \frac{1+|\epsilon|^2 - 2\Re(\epsilon)}{1+|\epsilon|^2 + 2\Re(\epsilon)}, \tag{9.74}$$

the constraint $|q/p| \neq 1$ requires $\Re(\epsilon) \neq 0$. The non-zero value of the real part of the mixing parameter ϵ thus represents the magnitude of indirect CP violation. It turns out to be the main source of CP violation in the kaon system.

The second kind of CP violation is the so-called *direct CP violation*. In this scenario, $K_L = K_2$ and $K_S = K_1$ are the CP eigenstates, but the CP symmetry is directly violated in the decay process itself: $K_2 \to 2\pi$ and $K_1 \to 3\pi$ decays are then possible. It can be quantified by observing a difference between the amplitudes

$$\left|\frac{A_{\overline{K}^0\to \bar{f}}}{A_{K^0\to f}}\right| \neq 1, \tag{9.75}$$

where $\bar{f}$ is the CP conjugate of f. This source of CP violation was observed for the first time in 1988 (in the NA31 experiment at CERN and later in NA48 still at CERN and kTeV at Fermilab, USA) in the kaon decay into two pions by measuring

$$\eta_{+-} = \frac{A_{K_L\to\pi^+\pi^-}}{A_{K_s\to\pi^+\pi^-}} = \epsilon + \epsilon', \qquad \eta_{00} = \frac{A_{K_L\to\pi^0\pi^0}}{A_{K_S\to\pi^0\pi^0}} = \epsilon - 2\epsilon'.$$

Note that as the two-pion system is a CP eigenstate, there is no distinction between f and $\bar{f}$. Without giving details, the parameter ϵ is present because the amplitudes above are related to $A_{K^0\to 2\pi}$ and $A_{\overline{K}^0\to 2\pi}$ by the combinations in Eqs. (9.69a) and (9.69b). The parameter ϵ' is thus representative of direct CP violation. Note that if $\epsilon' = 0$, $\eta_{+-} = \eta_{00}$. The present measurements observe, however, a very small difference (Particle Data Group, 2022):

$$|\eta_{+-}| = (2.232 \pm 0.011) \times 10^{-3}, \qquad |\eta_{00}| = (2.220 \pm 0.011) \times 10^{-3}.$$

This indicates a minor contribution of the direct violation in the kaon systems where

$$|\epsilon| = (2.228 \pm 0.011) \times 10^{-3}, \qquad \Re(\epsilon'/\epsilon) = (1.66 \pm 0.23) \times 10^{-3}. \tag{9.76}$$

Finally, the presence of these two types of CP violation opens the possibility of a third type, usually called *CP violation in interferences*. When a common CP eigenstate f is accessible to both K^0 and $\overline{K}^0$, the interferences between the amplitudes of the decay $K^0 \to f$ and the same decay with mixing, i.e. $K^0 \to \overline{K}^0 \to f$, generate CP violation characterised by the quantity

$$\lambda_f = \frac{q}{p}\frac{A_{\overline{K}^0\to f}}{A_{K^0\to f}}. \tag{9.77}$$

Even in the absence of indirect CP violation, i.e. $|q/p| = 1$, and the absence of direct CP violation, i.e. $|A_{\overline{K}^0\to f}/A_{K^0\to f}| = 1$ (here $\bar{f} = f$), there could be CP violation if $\Im(\lambda_f) \neq 0$, as illustrated in Problem 9.6. This quantity turns out to be related to $\Im(\epsilon)$. CP violation in interferences has also been observed in $K \to 2\pi$ decays.

In summary, the three types of CP violation have been observed in the kaon system. They are characterised by $\Re(\epsilon) = (1.66 \pm 0.02) \times 10^{-3}$ (indirect), $\Re(\epsilon') = (2.5 \pm 0.4) \times 10^{-6}$ (direct) and $\Im(\epsilon) = (1.57 \pm 0.02) \times 10^{-3}$ (interferences).

9.5.5 CP Violation in the Quark Sector

The box diagrams in Fig. 9.9 and the effective Hamiltonian (9.57) are easily applicable to other neutral meson systems: $B^0 - \overline{B}^0$ ($B^0 = d\bar{b}$), $B^0_s - \overline{B}^0_s$ ($B^0_s = s\bar{b}$) and $D^0 - \overline{D}^0$ ($D^0 = c\bar{u}$), replacing, for instance, the strange quark by a b for $B^0 - \overline{B}^0$. For B^0_q mesons ($q = d, s$), the eigenstates of the effective Hamiltonian have almost the same lifetimes but significant mass differences. Therefore, they are labelled by their mass, i.e. B_H and B_L, with H and L standing for heavy and light. This difference with the kaons is mainly due to the large mass of B mesons, about 5.3 GeV$/c^2$. Neutral kaons have a small mass, and therefore, their decays are dominated by a single CP final state. Due to the phase space, we saw that the kaon for which the 2π channel is dominant (K_S) is necessarily short-lived, while the other is long-lived. This argument no longer holds for B mesons since many CP final states are possible for both eigenstates because of the large phase space available. Therefore, B_H and B_L necessarily have a similar lifetime. For B_q^0 mesons, various asymmetry measurements show that $|q/p|$ is very close to 1. Moreover, a very small value of ϵ is expected in the Standard Model for B mesons. Therefore, CP violation due to the mixing is negligible for Bs. Nevertheless, in 2001, CP violation of the B meson decay (so direct CP violation) was discovered by the Belle experiment at KEK in Japan and the Babar experiment at SLAC (USA). CP violation by interferences has also been observed. Finally, in 2019, the LHCb collaboration at CERN announced the discovery of CP violation in the charm system by analysing the difference in decay rates between D^0 and $\overline{D}^0$, decaying into the K^+K^- or $\pi^+\pi^-$ pairs.

We showed that CP violation is due to the phase present in the mixing matrix. In Appendix A.4.7, about the U(N) group, we show that an $N \times N$ unitary matrix depends on N^2 real parameters. If the matrix was real, it would correspond to a rotation matrix depending on $N(N-1)/2$ angles (i.e., the dimension of SO(N)). Therefore, in general, there are $N(N+1)/2$ remaining phases allowing complex elements of the matrix. Nevertheless, looking at the charge current (9.46), we observe that a phase φ in the element $V_{ij} = |V_{ij}|e^{i\varphi}$ can be absorbed by a redefinition of the phase of the quark fields,

$$\bar{u}_i|V_{ij}|e^{i\varphi}d_j = \overline{e^{-i\varphi}u_i}|V_{ij}|d_j = \bar{u}_i|V_{ij}|\left(e^{i\varphi}d_j\right),$$

i.e. $u_i \to e^{i\varphi}u_i$ or $d_j \to e^{-i\varphi}d_j$. With N quark families, we would thus expect to absorb $2N$ phases. However, if both up- and down-type quarks are changed by the same amount, it leaves the CKM matrix invariant. Therefore, with N quark families,

$$N_\varphi = N(N+1)/2 - (2N-1) = (N-1)(N-2)/2$$

physical phases remain. We realise that $N_\varphi > 0$ if $N > 2$. In other words, there must be at least three quark families to generate CP violation in the quark sector. In 1973, when Kobayashi and Maskawa extended the Cabibbo model to three families, there was no experimental signature of a third family! In a sense, the discovery of CP violation in 1964 predicted the existence of a third family.

The parametrisation of the 3×3 CKM matrix in Eq. (9.24) results from a particular choice of the quark field phases. Another choice would lead to another representation. It is, therefore, convenient to define a quantity representative of CP violation that would be independent of that choice. This quantity is based on a geometric representation exploiting the unitary of the CKM matrix,

$$V_{i1}V_{j1}^* + V_{i2}V_{j2}^* + V_{i3}V_{j3}^* = 0, \qquad V_{1i}V_{1j}^* + V_{2i}V_{2j}^* + V_{3i}V_{3j}^* = 0, \qquad (i \neq j). \quad (9.78)$$

Since the matrix elements are complex valued (otherwise, CP symmetry would be conserved), these six relations ($i,j \in [1,3]$), a sum of three terms, must form six *unitarity triangles* in the complex plane. A change of the phase of one of the quark fields simply rotates the triangle depending on the quark field without changing its area. Moreover, as shown in Problem 9.7, the six triangles have the same non-zero area, $|J|/2$, where

$$J = \Im(V_{i\alpha}^* V_{j\alpha} V_{i\beta} V_{j\beta}^*) = \Im(V_{\alpha i}^* V_{\alpha j} V_{\beta i} V_{\beta j}^*), \quad (9.79)$$

for all integers $i,j,\alpha,\beta \in [1,3]$, with $i \neq j$ and $\alpha \neq \beta$. As the area is independent of the quark phases, so is J, called the *Jarlskog invariant*. Using the standard parametrisation (9.24), one easily finds that with $i = \beta = 1$ and $j = \alpha = 2$,

$$J = \Im(V_{12}^* V_{22} V_{11} V_{21}^*) = c_{12}c_{13}^2 c_{23}s_{12}s_{13}s_{23}\sin(\delta). \tag{9.80}$$

The Jarlskog invariant necessarily appears in the calculation of any CP violating observable, as CP violation requires non-trivial triangles (i.e with a non-zero area) in the complex plane. Using the fitted values of the CKM matrix parameters from Eq. (9.26) yields

$$J^{\text{exp}} = 3.00(12) \times 10^{-5}. \tag{9.81}$$

It is a small number, indicating that CP violation in the quark sector is small.[18]

9.5.6 CP Violation in the Leptonic Sector

The existence of the PMNS matrix revealed by the discovery of the neutrino oscillation phenomenon opens the possibility of CP violation in the leptonic sector. According to the oscillation formula (9.39), we observe that when $\alpha \neq \beta$, $P_{\nu_\alpha \to \nu_\beta}$ can be different from $P_{\nu_\beta \to \nu_\alpha}$. Indeed, in Eq. (9.39),

$$\Re\left[V_{\alpha k}^* V_{\beta k} V_{\alpha j} V_{\beta j}^* \exp\left(-i2\pi \frac{L}{L_{kj}^{\text{osc}}}\right)\right] \neq \Re\left[V_{\beta k}^* V_{\alpha k} V_{\beta j} V_{\alpha j}^* \exp\left(-i2\pi \frac{L}{L_{kj}^{\text{osc}}}\right)\right]$$

as long as the elements of PMNS are complex valued. Now, as the CPT transformation of the reaction $\nu_\beta \to \nu_\alpha$ is $\bar{\nu}_\alpha \to \bar{\nu}_\beta$,[19] assuming the conservation of the CPT symmetry, we would expect $P_{\nu_\beta \to \nu_\alpha} = P_{\bar{\nu}_\alpha \to \bar{\nu}_\beta}$. Therefore, a complex PMNS matrix implying $P_{\nu_\alpha \to \nu_\beta} \neq P_{\nu_\beta \to \nu_\alpha}$ also implies $P_{\nu_\alpha \to \nu_\beta} \neq P_{\bar{\nu}_\alpha \to \bar{\nu}_\beta}$. Since the reaction $\bar{\nu}_\alpha \to \bar{\nu}_\beta$ is just the CP transformation of the reaction $\nu_\alpha \to \nu_\beta$, experiments looking for CP violation in the leptonic sector compare these two probabilities. Long baseline neutrino experiments, such as T2K in Japan or NOvA in the USA, using a ν_μ beam or a $\bar{\nu}_\mu$ beam flying over 295 km and 810 km, respectively, look for the appearance of ν_e or $\bar{\nu}_e$. At the time of writing (2022), the experimental situation was not conclusive. While the NOvA data disfavour combinations of oscillation parameters that give rise to a large asymmetry in the rates of ν_e and $\bar{\nu}_e$ appearance (NOvA Collaboration, 2022), the T2K data tend to favour an asymmetry (T2K Collaboration, 2021). Therefore, the CP phase δ quoted in Eq. (9.32) is compatible with the CP conserving scenario $\sin(\delta) = 0$ as well as the maximal CP violation scenario $\sin(\delta) = \pm 1$. The leptonic Jarlskog invariant similar to Eq. (9.80) is then written as

$$J = J^{\max}\sin(\delta), \quad \text{with} \quad J^{\max} = c_{12}c_{13}^2 c_{23}s_{12}s_{13}s_{23}. \tag{9.82}$$

[18] It does not mean that all CP violating observables are of the order of J. Indeed, usually, one exploits ratios of CP violating quantities with CP conserving ones that may be small too. For instance, in kaon mixing, $\epsilon \simeq 10^{-3}$, not 10^{-5}.

[19] The only interacting (anti-)neutrinos are left- (right-)handed, so the CPT transformation of $\nu_\beta \to \nu_\alpha \equiv \nu_{\beta_L} \to \nu_{\alpha L}$ reads

$$\nu_{\beta_L} \to \nu_{\alpha L} \underset{\text{C}}{\Rightarrow} \bar{\nu}_{\beta L} \to \bar{\nu}_{\alpha L} \underset{\text{P}}{\Rightarrow} \bar{\nu}_{\beta R} \to \bar{\nu}_{\alpha R} \underset{\text{T}}{\Rightarrow} \bar{\nu}_{\alpha R} \to \bar{\nu}_{\beta R} \equiv \bar{\nu}_\alpha \to \bar{\nu}_\beta.$$

We also observe that the CP transformation of $\nu_\beta \to \nu_\alpha \equiv \nu_{\beta_L} \to \nu_{\alpha L}$ is $\bar{\nu}_{\beta R} \to \bar{\nu}_{\alpha R} \equiv \bar{\nu}_\beta \to \bar{\nu}_\alpha$.

A global fit of the data from various experiments yields (Esteban et al., 2020)

$$J^{\max \exp} = 0.0332(8). \tag{9.83}$$

Depending on the actual value of $\sin(\delta)$ (still unknown), the status of the leptonic CP violation could be larger or smaller than the CP violation in the quark sector. The next generation of long baseline experiments, DUNE (USA) and Hyper-Kamiokande (Japan), expected to start in the late 2020s, should reveal the presence of CP violation in the lepton sector or constrain the phase δ in the leptonic mixing matrix to be smaller than about 10°.

Problems

9.1 Using the energy–momentum conservation in the reaction $\mathrm{Sm}^* \to \mathrm{Sm}+\gamma$, check that if the difference in the energy level between Sm^* and Sm is E_0, then $\Delta E = E_0 - E_\gamma = E_0^2/(2mc^2)$, where m is the mass of Sm (and Sm^*). Use $mc^2 = 141.51$ GeV and $E_0 = 963$ keV.

9.2 In this problem, we neglect the masses of the muon decay products but not the muon mass (except when clearly justified), and we consider unpolarised muons. In the Standard Model, almost 100% of the muon decays correspond to $\mu^- \to e^- + \bar{\nu}_e + \nu_\mu$.

1. Using the labels $\mu^-(p) \to e^-(k) + \bar{\nu}_e(k') + \nu_\mu(p')$ for the 4-momenta, determine the amplitude of the process and show that the spin-averaged amplitude squared is $|\overline{\mathcal{M}}|^2 = 64\, G_F^2\, (p.k')(p'.k)$.
2. Show that the decay width can be expressed as

$$\frac{\mathrm{d}\Gamma_\mu}{\mathrm{d}k^0\,\mathrm{d}k'^0} = \frac{G_F^2}{2\pi^3} m_\mu\, k'^0 (m_\mu - 2k'^0)\, \theta(m_\mu - k^0 - k'^0), \tag{9.84}$$

where $\theta(x)$ is the Heaviside function and m_μ is the muon mass.

3. Explain why $k^0 \le \frac{m_\mu}{2}$ and $\frac{m_\mu}{2} - k^0 \le k'^0 \le \frac{m_\mu}{2}$. Conclude that the muon lifetime is given in natural units by

$$\tau_\mu = \frac{192\pi^3}{G_F^2 m_\mu^5}. \tag{9.85}$$

9.3 Using Eq. (9.37), check that Eq. (9.39) can be written as Eq. (9.40).

9.4 Using $(M^\mu)^\intercal = (M^\mu)^{\dagger *}$, calculate $(M^\mu)^\intercal$ with $M^\mu = \hat{\mathrm{C}}^\dagger \gamma^0 \gamma^\mu P_R \hat{\mathrm{C}}$ from Eq. (9.51). *Hint: in the Dirac representation, γ^2 is a pure imaginary matrix, with all other matrices being real.* Deduce the expression of the charge conjugate transformation given by Eq. (9.52).

9.5 We denote by $\lambda_L = m_L - i\Gamma_L/2$ and $\lambda_S = m_S - i\Gamma_S/2$ the eigenvalues of the effective Hamiltonian (9.57) corresponding to K_L and K_S eigenstates, respectively. Invert Eqs. (9.69a) and (9.69b) where $|K_L\rangle = p\,|K^0\rangle - q\,|\overline{K}^0\rangle$ and $|K_S\rangle = p\,|K^0\rangle + q\,|\overline{K}^0\rangle$ and show that at time t, $|K^0\rangle$ and $|\overline{K}^0\rangle$ evolve according to

$$|K^0(t)\rangle = g_+(t)\,|K^0\rangle - \frac{q}{p} g_-(t)\,|\overline{K}^0\rangle = \frac{1}{2}\left(e^{-i\lambda_L t} + e^{-i\lambda_S t}\right)|K^0\rangle - \frac{q}{p}\frac{1}{2}\left(e^{-i\lambda_L t} - e^{-i\lambda_S t}\right)|\overline{K}^0\rangle,$$

$$|\overline{K}^0(t)\rangle = g_+(t)\,|\overline{K}^0\rangle - \frac{p}{q} g_-(t)\,|K^0\rangle = \frac{1}{2}\left(e^{-i\lambda_L t} + e^{-i\lambda_S t}\right)|\overline{K}^0\rangle - \frac{p}{q}\frac{1}{2}\left(e^{-i\lambda_L t} - e^{-i\lambda_S t}\right)|K^0\rangle.$$

Show that $|g_\pm(t)|^2 = \frac{1}{2} e^{-\Gamma t}[\cosh(\Delta\Gamma t/2) \pm \cos(\Delta m t)]$, where $\Gamma = (\Gamma_L + \Gamma_s)/2$, $\Delta\Gamma = \Gamma_L - \Gamma_S$, $\Delta m = m_L - m_S > 0$, and deduce the probabilities in Eqs. (9.72a)–(9.72c).

9.6 This problem uses results and notations from Problem 9.5. We denote by $A_{K^0\to f} = \langle f|T|K^0\rangle$ and $A_{\overline{K}^0\to f} = \langle f|T|\overline{K}^0\rangle$ the amplitudes of the transition $K^0/\overline{K}^0 \to f$ (the symbol T is equivalent to the S-matrix in the presence of a transition and restricted to the neutral kaon subsystem). Show that

$$\left|A_{K^0\to f}(t)\right|^2 = |A_{K^0\to f}|^2 \left(|g_+(t)|^2 + |\lambda_f|^2|g_-(t)|^2 - 2\Re\{\lambda_f g_+^*(t)g_-(t)\}\right),$$

$$\left|A_{\overline{K}^0\to f}(t)\right|^2 = |A_{K^0\to f}|^2 \left(|g_-(t)|^2 + |\lambda_f|^2|g_+(t)|^2 - 2\Re\{\lambda_f g_+(t)g_-^*(t)\}\right)\left|\frac{p}{q}\right|^2,$$

where λ_f is given in Eq. (9.77). Check that $g_+^*(t)g_-(t) = -\frac{1}{2}e^{-\Gamma t}[\sinh(\Delta\Gamma t/2) + i\sin(\Delta m t)]$. When there are no direct CP violation ($|A_{\overline{K}^0\to f}/A_{K^0\to f}|^2 = 1$) and indirect violation ($|q/p|^2 = 1$), show that the asymmetry takes the form

$$\frac{|A_{\overline{K}^0\to f}(t)|^2 - |A_{K^0\to f}(t)|^2}{|A_{\overline{K}^0\to f}(t)|^2 + |A_{K^0\to f}(t)|^2} = \frac{\Im(\lambda_f)\sin(\Delta m t)}{\cosh(\Delta\Gamma t/2) + \Re(\lambda_f)\sinh(\Delta\Gamma t/2)}.$$

9.7 We recall that the area of a triangle ABC is $|\boldsymbol{AB} \times \boldsymbol{AC}|/2$. Starting from the first equation of Eq. (9.78) and denoting $\boldsymbol{AB} = (\Re(V_{i1}V_{j1}^*), \Im(V_{i1}V_{j1}^*), 0)^\intercal$ and $\boldsymbol{AC} = (\Re(V_{i2}V_{j2}^*), \Im(V_{i2}V_{j2}^*), 0)^\intercal$, show that the area of ABC is $|\Im(V_{i1}^*V_{j1}V_{i2}V_{j2}^*)|/2$. To prove that all triangles have the same area, show from the first equation of Eq. (9.78) that

$$\begin{aligned}\Im(V_{i1}^*V_{j1}V_{i2}V_{j2}^*) &= -\Im(V_{i1}^*V_{j1}V_{i3}V_{j3}^*) = -\Im(V_{i2}^*V_{j2}V_{i1}V_{j1}^*) = \Im(V_{i2}^*V_{j2}V_{i3}V_{j3}^*)\\ &= -\Im(V_{i3}^*V_{j3}V_{i2}V_{j2}^*) = \Im(V_{i3}^*V_{j3}V_{i1}V_{j1}^*).\end{aligned}$$

Establish a similar relation starting from the second equation of Eq. (9.78) and conclude that the area of all unitary triangles is related to the Jarlskog invariant (9.79).

9.8 Indicate whether the following reactions are allowed or not. If not, specify the reason. When allowed, which interaction is involved? The useful particle quantum numbers are $I(J^P)$: $p = \frac{1}{2}(\frac{1}{2}^+)$, $n = \frac{1}{2}(\frac{1}{2}^+)$, $\pi^\pm = 1(0^-)$; $I(J^{PC})$: $\pi^0 = 1(0^{-+})$, $\gamma = 1^{--}$.

$$\begin{array}{lll} 1.\ n \to p + e^+ + \nu_e, & 2.\ p + \pi^- \to n + \pi^0, & 3.\ p \to n + e^+ + \nu_e,\\ 4.\ \pi^0 \to \gamma\gamma, & 5.\ p \to e^+ + \nu_e, & 6.\ \pi^0 \to \gamma. \end{array}$$

9.9 Electron–neutrino scattering.
Let us suppose that the energy of particles is large enough to neglect their mass but remains negligible with respect to the weak boson mass. The two following reactions are considered:

$$(1)\ \nu_\mu + e^- \to \nu_e + \mu^-, \quad (2)\ \bar{\nu}_e + e^- \to \bar{\nu}_\mu + \mu^-.$$

1. Draw the Feynman diagrams and deduce the two amplitudes. Use the labels: $\nu_\mu(k) + e^-(p) \to \nu_e(k') + \mu^-(p')$ and $\bar{\nu}_e(k) + e^-(p) \to \bar{\nu}_\mu(k') + \mu^-(p')$.
2. With the help of formulas (9.8) and (9.9), show that

$$\overline{|\mathcal{M}_1|^2} = 64G_F^2\frac{s^2}{4} \text{ and } \overline{|\mathcal{M}_2|^2} = 64G_F^2\frac{t^2}{4}.$$

3. Conclude that

$$\frac{\sigma(\nu_\mu + e^- \to \nu_e + \mu^-)}{\sigma(\bar{\nu}_e + e^- \to \bar{\nu}_\mu + \mu^-)} = 3.$$

10 Electroweak Interaction

At this stage of the book, we have all the pieces to build the Standard Model of particle physics. In this chapter, we shall see how gauge theories underlie all elementary interactions (except gravity). Surprisingly, this necessitates encompassing electromagnetism and the weak interaction into a unified theory called the electroweak interaction theory.

10.1 Weak Isospin

In Chapter 9, we learned that the charged current only couples left-handed particles. We can form doublets of the fields that are the weak eigenstates,

$$\begin{pmatrix}\nu_e\\ e\end{pmatrix}_L,\ \begin{pmatrix}\nu_\mu\\ \mu\end{pmatrix}_L,\ \begin{pmatrix}\nu_\tau\\ \tau\end{pmatrix}_L,\ \begin{pmatrix}u\\ d'\end{pmatrix}_L,\ \begin{pmatrix}c\\ s'\end{pmatrix}_L,\ \begin{pmatrix}t\\ b'\end{pmatrix}_L,$$

such that the doublets remain unchanged by the charged weak interaction mediated by a $W^\pm$ boson. Note that the down-type quarks here are the eigenstates of the weak interaction, denoted by the prime symbol. This structure recalls the early days of the strong interaction, where neutron and proton were considered in a strong isospin doublet (see Section 8.1.1, p. 249). As SU(2) was the symmetry group of the strong isospin, it is then tempting to consider that $\mathrm{SU}(2)_L$ is the symmetry group describing the weak interactions. The subscript L is added to emphasise that this group acts non-trivially only on the left component of the chiral fields. Let us define a weak isospin quantum number denoted with the letter T instead of the letter I to clearly distinguish the weak isospin T from the strong isospin I. Particles now carry a new quantum number, the weak isospin T and its projection T_3, so

$$T=\frac{1}{2},\ \begin{matrix}T_3=+\frac{1}{2}\rightarrow\\ T_3=-\frac{1}{2}\rightarrow\end{matrix}\quad \begin{pmatrix}\nu_e\\ e\end{pmatrix}_L,\ \begin{pmatrix}\nu_\mu\\ \mu\end{pmatrix}_L,\ \begin{pmatrix}\nu_\tau\\ \tau\end{pmatrix}_L,\ \begin{pmatrix}u\\ d'\end{pmatrix}_L,\ \begin{pmatrix}c\\ s'\end{pmatrix}_L,\ \begin{pmatrix}t\\ b'\end{pmatrix}_L. \tag{10.1}$$

In terms of quantum numbers, we could have used the mass eigenstate instead of the weak eigenstate since a linear combination of fields with the same isospin preserves the isospin value. Given that the right-handed components are not sensitive to the charged current of the weak interaction, they belong to a weak isospin singlet, i.e.

$$T=T_3=0,\quad \nu_{eR}(?),\ \nu_{\mu R}(?),\ \nu_{\tau R}(?),\ e_R,\ \mu_R,\ \tau_R,\ u_R,\ c_R,\ t_R,\ d_R,\ s_R,\ b_R. \tag{10.2}$$

A question mark has been added to the neutrinos since, so far, the existence of their right-handed chirality components remains hypothetical, even if we know now that neutrinos are massive particles. For convenience, let us denote the weak isospin doublets and singlets of leptons and quarks by

$$L_l^i = \begin{pmatrix} \nu^i \\ e^i \end{pmatrix}_L, \quad L_q^i = \begin{pmatrix} u^i \\ d'^i \end{pmatrix}_L, \quad \nu_R^i, \; e_R^i, \; u_R^i, \; d_R'^i, \tag{10.3}$$

where $i = 1, 2, 3$ denotes the family index ($e^1 = e$, $e^2 = \mu$, $e^3 = \tau$ and similarly for the other fermions). We will use the generic notation

$$L^i = \begin{pmatrix} \psi_L^i \\ \psi_L'^i \end{pmatrix}, \quad \psi_R^i, \; \psi_R'^i, \tag{10.4}$$

for any weak isodoublet (leptons ℓ or quark q) and the two corresponding weak isosinglets. For fermion fields, we expect the free Lagrangian of the theory to be based on the usual Dirac Lagrangian. Since the mass term of a Dirac field ψ can be decomposed into

$$m\bar{\psi}\psi = m\bar{\psi}(P_L + P_R)(P_L + P_R)\psi = m\bar{\psi}(P_L\psi_L + P_R\psi_R) = m(\overline{\psi_R}\psi_L + \overline{\psi_L}\psi_R),$$

through the introduction of the chirality projectors P_L and P_R (given that $P_L P_R = 0$), the free Lagrangian reads

$$\mathcal{L} = \sum_{q,\ell}\sum_{i=1}^{3} \overline{L^i}(i\partial\!\!\!/)L^i + \overline{\psi_R^i}(i\partial\!\!\!/)\psi_R^i + \overline{\psi_R'^i}(i\partial\!\!\!/)\psi_R'^i - m_i(\overline{\psi_R^i}\psi_L^i + \overline{\psi_L^i}\psi_R^i) - m_i'(\overline{\psi_R'^i}\psi_L'^i + \overline{\psi_L'^i}\psi_R'^i).$$

However, under an $\mathrm{SU}(2)_L$ gauge transformation, only the left fields are transformed. Therefore, this Lagrangian with a mass term combining both the left- and right-handed fields is not invariant, since the transformation of the left-handed fields cannot be compensated by the (non-)transformation of the right-handed ones. The free Lagrangian is thus not gauge invariant! Hence, either we give up the isodoublet model, or we find another mechanism to generate the particle masses. The latter scenario will be the path to follow, implemented via the so-called *Higgs mechanism* that we shall see in the next chapter. Therefore, at this stage, let us consider that *all fermions are massless*. The Lagrangian of the free theory is then

$$\mathcal{L}_{\text{free}}^{\text{EW}} = \sum_{q,\ell}\sum_{i=1}^{3} \overline{L^i} i\partial\!\!\!/ L^i + \overline{\psi_R^i} i\partial\!\!\!/ \psi_R^i + \overline{\psi_R'^i} i\partial\!\!\!/ \psi_R'^i = \sum_f \overline{L} i\partial\!\!\!/ L + \overline{\psi_R} i\partial\!\!\!/ \psi_R + \overline{\psi_R'} i\partial\!\!\!/ \psi_R', \tag{10.5}$$

where, to simplify the notation, the double summation is now replaced by a simple summation over fermions f, and the generation indexes are dropped. A doublet is rotated in the weak isospin space according to

$$L \xrightarrow{\mathrm{SU}(2)_L} e^{-i\alpha\cdot T} L, \tag{10.6}$$

where the three generators T_a ($a = 1, 2, 3$) of the $\mathrm{SU}(2)_L$ group are related to the Pauli matrices

$$T_a = \frac{1}{2}\sigma_a, \quad \text{with} \quad \sigma_1 = \begin{pmatrix} 0 & 1 \\ 1 & 0 \end{pmatrix}, \quad \sigma_2 = \begin{pmatrix} 0 & -i \\ i & 0 \end{pmatrix}, \quad \sigma_3 = \begin{pmatrix} 1 & 0 \\ 0 & -1 \end{pmatrix}, \tag{10.7}$$

in exactly the same way as for the strong isospin (and the usual spin). The Lagrangian (10.5) is invariant under a global $\mathrm{SU}(2)_L$ transformation (the singlet components being unchanged). Gauge theories are based on local invariance, i.e. when the transformation varies in each point of spacetime,

$$L \xrightarrow{\mathrm{SU}(2)_L} e^{-ig_w \boldsymbol{\alpha}(x)\cdot\frac{\boldsymbol{\sigma}}{2}} L. \tag{10.8}$$

The weak coupling constant associated with the $\mathrm{SU}(2)_L$ group, g_w, has been explicitly factorised. The formalism developed in Section 8.2.4 in the context of $\mathrm{SU}(3)_c$, the gauge group of QCD, is straightforward to transpose to $\mathrm{SU}(2)_L$. Imposing local gauge invariance generates gauge bosons via the covariant derivative analogous to Eq. (8.43),

$$\mathrm{D}_\mu = \partial_\mu + ig_w \frac{\sigma_a}{2} W^a_\mu, \tag{10.9}$$

where the summation over $a = 1$ to 3 is implicit. There are as many gauge bosons (denoted above by W_1, W_2 and W_3) as the number of generators (for $\mathrm{SU}(3)_c$, we have eight gluons). Moreover, these gauge bosons belong to the so-called adjoint representation (see Appendix A), as the eight gluons belong to the **8** representation. Hence, in the case of $\mathrm{SU}(2)_L$, the three weak bosons belong to the **3** representation (cf. Appendix B). Supplement 10.1 gives some details about this statement. The three gauge bosons give rise to the interaction term

$$\mathcal{L}_{\mathrm{int}} = \sum_f -g_w j^\mu_a W^a_\mu, \tag{10.10}$$

where the three conserved currents ($a = 1, 2, 3$) are

$$j^\mu_a = \bar{L}\gamma^\mu \frac{\sigma_a}{2} L. \tag{10.11}$$

As an example, let us express the three currents for the lepton doublet of the first family. They read

$$j^\mu_a = (\overline{\nu_{eL}}, \overline{e_L})\gamma^\mu \frac{\sigma_a}{2}\begin{pmatrix}\nu_{eL}\\ e_L\end{pmatrix}\left\{\begin{array}{l} j^\mu_1 = \left(\overline{\nu_{eL}}\gamma^\mu e_L + \overline{e_L}\gamma^\mu \nu_{eL}\right)/2,\\ j^\mu_2 = \left(-i\overline{\nu_{eL}}\gamma^\mu e_L + i\overline{e_L}\gamma^\mu \nu_{eL}\right)/2,\\ j^\mu_3 = \left(\overline{\nu_{eL}}\gamma^\mu \nu_{eL} - \overline{e_L}\gamma^\mu e_L\right)/2,\end{array}\right. \tag{10.12}$$

using the expressions (10.7) of the Pauli matrices. Since the charged current of the weak interaction transforms a member of the doublet to the other member,

$$j^\mu_{cc+} = \overline{\nu_e}\gamma^\mu \frac{1}{2}(1-\gamma^5)e = \overline{\nu_{eL}}\gamma^\mu e_L, \qquad j^\mu_{cc-} = \left(j^\mu_{cc+}\right)^\dagger = \overline{e_L}\gamma^\mu \nu_{eL},$$

we remark that the following linear combinations of j^μ_1 and j^μ_2,

$$j^\mu_{cc\pm} = j^\mu_1 \pm i j^\mu_2 = \bar{L}\gamma^\mu \frac{\sigma_\pm}{2} L, \tag{10.13}$$

match the charged current, the matrices $\sigma_\pm$ being

$$\sigma_\pm = \sigma_1 \pm i\sigma_2. \tag{10.14}$$

This combination is valid for any doublet. Hence, the interaction term (10.10) can be written as

$$\begin{aligned}\mathcal{L}_{\mathrm{int}} &= \textstyle\sum_f -g_w \left(j^\mu_1 W^1_\mu + j^\mu_2 W^2_\mu + j^\mu_3 W^3_\mu\right)\\ &= \textstyle\sum_f -g_w \left(\frac{1}{2}[j^\mu_{cc+} + j^\mu_{cc-}]W^1_\mu - i\frac{1}{2}[j^\mu_{cc+} - j^\mu_{cc-}]W^2_\mu + j^\mu_3 W^3_\mu\right)\\ &= \textstyle\sum_f -g_w \left(\frac{1}{2}[W^1_\mu - iW^2_\mu]j^\mu_{cc+} + \frac{1}{2}[W^1_\mu + iW^2_\mu]j^\mu_{cc-} + j^\mu_3 W^3_\mu\right).\end{aligned}$$

The charge current must be coupled to the physical charged bosons $W^{\pm}$. Therefore, the linear combinations $W^1_\mu \mp iW^2_\mu$ must be identified with $W^{\pm}$, up to a normalisation factor. Defining the normalised fields of $W^{\pm}$ by[1]

$$\boxed{W^{\pm}_\mu = \frac{1}{\sqrt{2}}\left[W^1_\mu \mp iW^2_\mu\right]}, \tag{10.15}$$

the interaction term reads

$$\mathcal{L}_{\text{int}} = \sum_f -\frac{g_w}{\sqrt{2}}\left(W^+_\mu j^\mu_{cc+} + W^-_\mu j^\mu_{cc-}\right) + \sum_f -g_w j^\mu_3 W^3_\mu. \tag{10.16}$$

We observe that the square root in the definition of the coupling $g_w/\sqrt{2}$ entering the vertex factors of the weak interaction given in Fig. 9.3 is simply a consequence of the combination of the two gauge fields W^1 and W^2.

Now, we know that the weak interaction also involves a neutral current with transitions of the kind of those described by j^μ_3 in Eq. (10.12). It is then tempting to identify the neutral boson W^3 with Z^0. However, this cannot be correct. The first reason is that in this model the neutral current would have exactly the same $V - A$ structure as the charged current since the three bosons would be members of the weak isospin triplet. However, we saw in Chapter 9, Section 9.3.5, that experimentally it is not the case. Another reason is that the three bosons should have the same mass, which is not correct. Therefore, $SU(2)_L$ cannot be the gauge group of the weak interaction!

10.2 Weak Hypercharge

In 1961, long before the weak neutral current was discovered, Glashow and later, in 1964, Salam and Ward suggested extending the gauge group of the weak interaction from $SU(2)_L$ to $SU(2)_L \times U(1)_Y$. Their initial motivations were to unify the weak interaction and electromagnetism, as we shall see. The subscript Y in $U(1)_Y$ stands for the *weak hypercharge*, which is *not* the strong hypercharge introduced in Section 8.1.4. The names weak isospin and weak hypercharge obviously come from the analogy with the strong interaction where we had a similar mathematical structure: SU(2) for the strong isospin, and even if it was not mentioned in the QCD chapter, the conservation of the strong hypercharge (a simple number) is necessarily related to the symmetry group U(1). How did Glashow et al. manage to unify the weak interaction and electromagnetism, which seem so different (a very short range chiral theory for the former, and an infinite range parity conserved theory for the latter)? Consider first the electromagnetic current of an electron,

$$j^\mu_{\text{em}} = q\bar{e}\gamma^\mu e = \frac{1}{2}(-1)\left(\bar{e}\gamma^\mu(1-\gamma^5)e + \bar{e}\gamma^\mu(1+\gamma^5)e\right) = -\overline{e_L}\gamma^\mu e_L - \overline{e_R}\gamma^\mu e_R.$$

[1] It is not surprising: we have already seen in Chapter 5 that charged particles are necessarily a superposition of two real fields. Here, charged bosons of spin 1 are necessarily a superposition of two real vector fields.

The left and right components appear symmetrical. Now, comparing this expression to j_3^μ in (10.12), we observe that

$$j_{\rm em}^\mu - j_3^\mu = -\frac{1}{2}\left(\overline{\nu_{eL}}\gamma^\mu \nu_{eL} + \overline{e_L}\gamma^\mu e_L\right) - \overline{e_R}\gamma^\mu e_R = -\frac{1}{2}\overline{L_\ell^1}\gamma^\mu L_\ell^1 - \overline{e_R^1}\gamma^\mu e_R^1 + 0 \times \overline{\nu_R^1}\gamma^\mu \nu_R^1, \quad (10.17)$$

where the compact notation of weak isodoublets and isosinglets has been reintroduced. The difference between those two currents can be interpreted as a third current acting on both the doublet and the two singlets of the theory, with the coefficients $-1/2$ for the doublet, -1 for the electron singlet and 0 for the neutrino singlet. Now, consider the action of the hypercharge group $\mathrm{U}(1)_Y$ on the doublet and singlets. By definition of U(1), it generates a simple change of phase, and therefore, if $\mathrm{U}(1)_Y$ is a symmetry, the two members of a doublet are necessarily treated on the same footing

$$L_\ell^i \xrightarrow{\mathrm{U}(1)_Y} \left[e^{-i\alpha(x)\frac{Y}{2}}\right] L_\ell^i = e^{-i\alpha(x)\frac{y_\ell}{2}} L_\ell^i,$$

where a factor 1/2 was introduced to agree with the convention.[2] Both the left-handed electron and left-handed neutrino carry the same weak hypercharge, $YL_\ell^i = y_\ell L_\ell^i$. According to the current (10.17), $y_\ell/2$ is identified with the coefficient[3] $-1/2$, i.e. $y_\ell = -1$. Similarly, with the isosinglets,

$$e_R^i \xrightarrow{\mathrm{U}(1)_Y} e^{-i\alpha(x)\frac{y_e}{2}} e_R^i, \quad \nu_R^i \xrightarrow{\mathrm{U}(1)_Y} e^{-i\alpha(x)\frac{y_\nu}{2}} \nu_R^i,$$

we identify $y_e = -2$ and $y_\nu = 0$. By construction, the current generated by the weak hypercharge satisfies

$$j_{\rm em}^\mu - j_3^\mu = \frac{1}{2} j_Y^\mu, \quad (10.18)$$

which can be translated in terms of the generators of each type of current, Q the electrical charge generator for the electromagnetism, T_3 the generator associated with j_3^μ and Y the generator of $\mathrm{U}(1)_Y$, as

$$\boxed{Q = T_3 + \frac{Y}{2}.} \quad (10.19)$$

This relation mimics the Gell-Mann–Nishijima relation (8.8) used in the strong interaction. Glashow indeed proposed that it also holds for the weak quantities. Note that despite their formal similarities, Eqs. (10.19) and (8.8) are, however, intrinsically different. Equation (8.8) applies only to quarks, whereas (10.19) is valid for all elementary fermions. The weak isospin distinguishes the chirality of particles, whereas the strong isospin does not.

With Eqs. (10.19), (10.1) and (10.2), we can easily determine the quantum numbers of all elementary fermions. They are listed in Table 10.1. A word of caution about neutrinos. Their right-handed component of chirality has zero quantum numbers: they have no electric charge, have $y_\nu = 0$ and as they belong to an isosinglet, they have $T = 0$. In other words, the right-handed neutrinos (or left-handed anti-neutrinos), if they exist, do not interact

[2] This factor 1/2 allows having the same Gell-Mann–Nishijima relation among the three generators as that of the strong interaction.

[3] It is obviously similar to the QED group U(1), for which a transformation due to the generator Q, i.e. $e^{-ie\alpha(x)Q}\psi$, generates an electromagnetic current $q\bar{\psi}\gamma^\mu\psi$, where $Q\psi = q\psi$.

Table 10.1 Quantum numbers of weak isospin and hypercharge for quarks and leptons.

Fields	Q	T	T_3	Y
$\nu_{eL}, \nu_{\mu L}, \nu_{\tau L}$	0	1/2	1/2	−1
e_L, μ_L, τ_L	−1	1/2	−1/2	−1
$\nu_{eR}, \nu_{\mu R}, \nu_{\tau R}$	0	0	0	0
e_R, μ_R, τ_R	−1	0	0	−2
u_L, c_L, t_L	2/3	1/2	1/2	1/3
d'_L, s'_L, b'_L	−1/3	1/2	−1/2	1/3
u_R, c_R, t_R	2/3	0	0	4/3
d'_R, s'_R, b'_R	−1/3	0	0	−2/3

except by gravity as they are finally massive.[4] They are said to be *sterile*, i.e. having no interaction under the gauge group.

10.3 Electroweak Unification

With the introduction of the weak hypercharge, the gauge group is $\mathrm{SU}(2)_L \times \mathrm{U}(1)_Y$. Let us clearly show that it describes both the weak interaction and electromagnetism. The generators of the respective groups are $T_a = \frac{\sigma_a}{2}$ and $\frac{Y}{2}$. The local transformation of the weak isodoublets and the isosinglets are then

$$L \xrightarrow{\mathrm{SU}(2)_L\times\mathrm{U}(1)_Y} e^{-ig_w\boldsymbol{\alpha}(x)\cdot\frac{\boldsymbol{\sigma}}{2}-ig\beta(x)\frac{Y}{2}}L,\ \psi_R \xrightarrow{\mathrm{SU}(2)_L\times\mathrm{U}(1)_Y} e^{-ig\beta(x)\frac{Y}{2}}\psi_R,\ \psi'_R \xrightarrow{\mathrm{SU}(2)_L\times\mathrm{U}(1)_Y} e^{-ig\beta(x)\frac{Y}{2}}\psi'_R,$$

where L is a doublet field (L^i_ℓ or L^i_q), ψ_R is a singlet field of up-type fermions (ν^i_R, u^i_R), ψ'_R a singlet field of down-type fermions (e^i_R, d'^i_R) [cf. the notations in Eq. (10.4)] and $\boldsymbol{\alpha}(x), \beta(x)$ are the local parameters of the transformation. The coupling constants g_w and g are associated with the $\mathrm{SU}(2)_L$ and $\mathrm{U}(1)_Y$ transformations, respectively. As for QED or QCD, the theory is invariant under a gauge transformation if an interaction term generated by the covariant derivative is added to the free Lagrangian (10.5) :

$$\begin{aligned} L: \mathrm{D}_\mu &= \partial_\mu + ig_w\tfrac{\sigma_a}{2}W^a_\mu + ig\tfrac{Y}{2}B_\mu, \\ \psi_R,\ \psi'_R: \mathrm{D}_\mu &= \partial_\mu + ig\tfrac{Y}{2}B_\mu, \end{aligned} \tag{10.20}$$

where W^a_μ are the three gauge bosons of $\mathrm{SU}(2)_L$ ($a = 1, 2, 3$) and B_μ is the gauge boson of $\mathrm{U}(1)_Y$. Note that the field B_μ is not the electromagnetic field. It does not couple to the electric charge but with the weak hypercharge. Introducing the covariant derivative into the free Lagrangian $\mathcal{L}^{\mathrm{EW}}_{\mathrm{free}}$ in Eq. (10.5) then yields

$$\mathcal{L}_f = \sum_f \overline{L}i\not{D}L + \overline{\psi_R}i\not{D}\psi_R + \overline{\psi'_R}i\not{D}\psi'_R = \mathcal{L}^{\mathrm{EW}}_{\mathrm{free}} + \mathcal{L}_{\mathrm{int}}, \tag{10.21}$$

[4] As massive particles, they might also interact with the Higgs field, as we shall see later.

Supplement 10.1. Adjoint representation and W gauge bosons

The generators T_a of the adjoint representation are determined by the matrices whose elements are given by $(T_a)_{ij} = -if_{aij}$, the constants f_{aij} being the structure constant of the group (see Appendix A and the example of QCD in Section 8.2.4). For $SU(2)_L$, $a = 1, 2, 3$ and $f_{aij} = \epsilon_{aij}$ because of the commutators of two Pauli matrices. Therefore, T_3, whose elements are $-i\epsilon_{3ij}$, is given in the adjoint representation by the matrix

$$T_3 = \begin{pmatrix} 0 & -i & 0 \\ i & 0 & 0 \\ 0 & 0 & 0 \end{pmatrix}.$$

In this representation, we have for $W^1 = \begin{pmatrix} 1 \\ 0 \\ 0 \end{pmatrix}$, $W^2 = \begin{pmatrix} 0 \\ 1 \\ 0 \end{pmatrix}$, $W^3 = \begin{pmatrix} 0 \\ 0 \\ 1 \end{pmatrix}$,

$$T_3 W^1 = iW^2, \quad T_3 W^2 = -iW^1, \quad T_3 W^3 = 0W^3.$$

Hence, according to the definition (10.15) of the charged W boson, it follows that

$$T_3 W^+ = -W^+, \quad T_3 W^- = +W^-, \quad T_3 W^3 = 0W^3.$$

Conclusion: the physical bosons $W^\pm$ are the eigenstates of the T_3 operator and form, with the W^3 boson, the weak isospin triplet

$$\begin{pmatrix} W^- \\ W^0 \\ W^+ \end{pmatrix}. \tag{10.22}$$

The charge generator, Q, satisfies the relation $Q = T_3 + \frac{Y}{2}$, but W bosons carrying no hypercharge, Q is reduced to T_3. Consequently,

$$QW^\pm = T^3 W^\pm = \mp W^\pm, \quad QW^3 = 0W^3.$$

It might be counterintuitive to find a positive charge for the W^- field (and the same remark for W^+), but it is actually explained by the expression of the charged field in terms of the Fock operators. The W^k field expression, with $k = 1$ or 2, is similar to that of the photon (but with three polarisation states),

$$W^{k\,\mu}(x) = \int \frac{d^3\boldsymbol{p}}{(2\pi)^3 2E_p} \sum_\lambda \epsilon^\mu(p,\lambda)\alpha^k_{p,\lambda} e^{-ip\cdot x} + \epsilon^{*\mu}(p,\lambda)\alpha^{k\,\dagger}_{p,\lambda} e^{+ip\cdot x},$$

the Fock operators, $\alpha^k_{p,\lambda}$, destroying a quantum of the field W^k, while $\alpha^{k\,\dagger}_{p,\lambda}$ creates one. Hence, according to the definition (10.15) of the charged fields, it yields the charged field expansion,

$$W^{-\,\mu}(x) = \int \frac{d^3\boldsymbol{p}}{(2\pi)^3 2E_p} \sum_\lambda \epsilon^\mu(p,\lambda)\alpha^-_{p,\lambda} e^{-ip\cdot x} + \epsilon^{*\mu}(p,\lambda)\alpha^{+\,\dagger}_{p,\lambda} e^{+ip\cdot x},$$

with the Fock operators $\alpha^\pm_{p,\lambda} = (\alpha^1_{p,\lambda} \mp i\alpha^2_{p,\lambda})/\sqrt{2}$. We thus see that the field $W^{-\,\mu}$ destroys a W^- boson or creates a W^+. Hence, the action of $W^{-\,\mu}$ is to increase the charge of the system by $+1$ unit and therefore, we do have $QW^- = +W^-$. Similarly, the action of the $W^{+\,\mu}$ is the opposite of that of the $W^{-\,\mu}$. We could

Supplement 10.1. (cont.)

have anticipated this result since, in the Lagrangian (10.16), the field W^- is coupled to j_{cc-}, which describes a charge-decreasing transition. The charge balance in the Lagrangian is thus respected.

the interaction term between the fermions and the gauge bosons being explicitly

$$\mathcal{L}_{\rm int} = -\sum_f g_w \overline{L}\gamma^\mu \frac{\sigma_a}{2} L\, W_\mu^a + g\,\overline{L}\gamma^\mu \frac{Y}{2} L\, B_\mu + g\,\overline{\psi_R}\gamma^\mu \frac{Y}{2}\psi_R\, B_\mu + g\,\overline{\psi'_R}\gamma^\mu \frac{Y}{2}\psi'_R\, B_\mu .$$

We saw in the previous section that W_μ^1 and W_μ^2 mix together to form the two charged $W^\pm$ bosons. Using Eqs. (10.14) and (10.15), $\mathcal{L}_{\rm int}$ then reads

$$\begin{aligned}\mathcal{L}_{\rm int} = -\sum_f \Bigg[& \frac{g_w}{\sqrt{2}} \overline{L}\gamma^\mu \frac{\sigma_+}{2} L\, W_\mu^+ + \frac{g_w}{\sqrt{2}} \overline{L}\gamma^\mu \frac{\sigma_-}{2} L\, W_\mu^- \\ & + g_w \overline{L}\gamma^\mu \frac{\sigma_3}{2} L\, W_\mu^3 + g\,\overline{L}\gamma^\mu \frac{Y}{2} L\, B_\mu \\ & + g\,\overline{\psi_R}\gamma^\mu \frac{Y}{2}\psi_R\, B_\mu + g\,\overline{\psi'_R}\gamma^\mu \frac{Y}{2}\psi'_R\, B_\mu \Bigg].\end{aligned} \tag{10.23}$$

Let us focus on the neutral components described by the last two lines of the Lagrangian above. Following Glashow, Salam[5] and Weinberg, we want to identify the electromagnetic field. We already know that it cannot be W_μ^3 (since the latter couples to neutrinos) or B_μ. But let us assume that the photon field A_μ is a linear combination of the two neutral fields W_μ^3 and B_μ. The orthogonal combination to A^μ is also a neutral field that must contribute to the weak interaction. We are going to identify it with the Z boson, but at the time when the model was built, the weak neutral current had not been discovered. It was then a strong prediction of the theory. We thus define

$$\boxed{\begin{pmatrix} A_\mu \\ Z_\mu \end{pmatrix} = \begin{pmatrix} \cos\theta_w & \sin\theta_w \\ -\sin\theta_w & \cos\theta_w \end{pmatrix} \begin{pmatrix} B_\mu \\ W_\mu^3 \end{pmatrix}}, \tag{10.24}$$

where θ_w is the Weinberg angle describing the mixing of the original gauge fields. Inverting Eq. (10.24), the second line of Eq. (10.23) then reads

$$\begin{aligned} g_w \overline{L}\gamma^\mu \tfrac{\sigma_3}{2} L\, W_\mu^3 + g\,\overline{L}\gamma^\mu \tfrac{Y}{2} L\, B_\mu = \; & \overline{L}\gamma^\mu \left(g_w \cos\theta_w \tfrac{\sigma_3}{2} - g\sin\theta_w \tfrac{Y}{2}\right) L\, Z_\mu \\ & + \overline{L}\gamma^\mu \left(g_w \sin\theta_w \tfrac{\sigma_3}{2} + g\cos\theta_w \tfrac{Y}{2}\right) L\, A_\mu , \end{aligned} \tag{10.25}$$

while the last line of Eq. (10.23) reads

$$\begin{aligned} g\,\overline{\psi_R}\gamma^\mu \tfrac{Y}{2}\psi_R\, B_\mu + g\,\overline{\psi'_R}\gamma^\mu \tfrac{Y}{2}\psi'_R\, B_\mu = & -g\sin\theta_w \left(\overline{\psi_R}\gamma^\mu \tfrac{Y}{2}\psi_R + \overline{\psi'_R}\gamma^\mu \tfrac{Y}{2}\psi'_R\right) Z_\mu \\ & + g\cos\theta_w \left(\overline{\psi_R}\gamma^\mu \tfrac{Y}{2}\psi_R + \overline{\psi'_R}\gamma^\mu \tfrac{Y}{2}\psi'_R\right) A_\mu . \end{aligned} \tag{10.26}$$

[5] And the unfortunate Ward, who was the fourth for the electroweak theory Nobel prize, which only rewards three persons.

The terms in Eqs. (10.25) and (10.26) that are coupled with the electromagnetic field are

$$g_w \sin\theta_w \overline{L}\gamma^\mu T_3 L + g\cos\theta_w \left(\overline{L}\gamma^\mu \frac{Y}{2} L + \overline{\psi_R}\gamma^\mu \frac{Y}{2}\psi_R + \overline{\psi'_R}\gamma^\mu \frac{Y}{2}\psi'_R\right), \tag{10.27}$$

with $T_3 = \sigma_3/2$. Let us examine the term in the brackets. Inserting $Y/2 = Q - T_3$ and noting that, by definition, the isosinglets have $T_3 = 0$ and thus $Y/2 = Q$, it follows that

$$\begin{aligned}\overline{L}\gamma^\mu \frac{Y}{2} L + \overline{\psi_R}\gamma^\mu \frac{Y}{2}\psi_R + \overline{\psi'_R}\gamma^\mu \frac{Y}{2}\psi'_R &= -\overline{L}\gamma^\mu T_3 L + \overline{\psi_L}\gamma^\mu Q\psi_L + \overline{\psi'_L}\gamma^\mu Q\psi'_L \\ &\quad + \overline{\psi_R}\gamma^\mu Q\psi_R + \overline{\psi'_R}\gamma^\mu Q\psi'_R,\end{aligned}$$

where the isodoublet L has been decomposed in each of its components [see the notations in Eq. (10.4)] in the terms proportional to Q. Since (check it)

$$\overline{\psi_L}\gamma^\mu Q\psi_L + \overline{\psi_R}\gamma^\mu Q\psi_R = \overline{\psi}\gamma^\mu Q(P_L + P_R)\psi = \overline{\psi}\gamma^\mu Q\psi$$

(and similarly for ψ'), Eq. (10.27) simplifies to

$$(g_w \sin\theta_w - g\cos\theta_w)\overline{L}\gamma^\mu T_3 L + g\cos\theta_w \left(\overline{\psi}\gamma^\mu Q\psi + \overline{\psi'}\gamma^\mu Q\psi'\right). \tag{10.28}$$

This term is coupled with the electromagnetic field A_μ. Hence, we wish to identify it with the electromagnetic current

$$e j^\mu_{\text{em}} = e\left(\overline{\psi}\gamma^\mu Q\psi + \overline{\psi'}\gamma^\mu Q\psi'\right)$$

(the operator Q giving the charge in units of the elementary charge e). A comparison of Eq. (10.28) to the expression above requires the equalities

$$\boxed{g_w \sin\theta_w = g\cos\theta_w} \quad \text{and} \quad \boxed{e = g\cos\theta_w = g_w \sin\theta_w.} \tag{10.29}$$

Combined with the constraint $1 = \cos^2\theta_w + \sin^2\theta_w = e^2/g_w^2 + e^2/g^2$, this implies

$$\boxed{e = \frac{g_w g}{\sqrt{g_w^2 + g^2}}} \quad \text{or} \quad \boxed{\cos\theta_w = \frac{g_w}{\sqrt{g_w^2 + g^2}}, \qquad \sin\theta_w = \frac{g}{\sqrt{g_w^2 + g^2}}}. \tag{10.30}$$

Eq. (10.30) is a major step towards the unification of electromagnetism and weak interaction. The coupling constant of electromagnetism e is now expressed as a function of the two coupling constants g_w and g of the electroweak interaction.

So far, we have focused on the coupling with the electromagnetic field. However, there are terms in Eqs. (10.25) and (10.26) that are coupled with the Z_μ field. Proceeding as before, they finally read (see Problem 10.1)

$$\frac{g_w}{\cos\theta_w}\left[\overline{L}\gamma^\mu T_3 L - \sin^2\theta_w \left(\overline{\psi}\gamma^\mu Q\psi + \overline{\psi'}\gamma^\mu Q\psi'\right)\right]. \tag{10.31}$$

Since the Z boson is neutral, Eq. (10.31) is called the weak neutral current (up to a constant). Dropping the coupling constant of the Z boson,

$$g_{\text{nc}} = g_w/\cos\theta_w, \tag{10.32}$$

the weak neutral current is thus defined as

$$j^\mu_{\text{nc}} = \overline{L}\gamma^\mu T_3 L - \sin^2\theta_w \left(\overline{\psi}\gamma^\mu Q\psi + \overline{\psi'}\gamma^\mu Q\psi'\right) = j^\mu_3 - \sin^2\theta_w j^\mu_{\text{em}}. \tag{10.33}$$

We can now come back to the Lagrangian (10.23) describing the whole electroweak interaction with fermions. Inserting all the previous steps, it reads in terms of the physical gauge fields,

$$\mathcal{L}_{\text{int}}^{\text{EW}} = -\sum_f \left\{ \frac{g_w}{\sqrt{2}} j^\mu_{cc+} W^+_\mu + \frac{g_w}{\sqrt{2}} j^\mu_{cc-} W^-_\mu + e j^\mu_{\text{em}} A_\mu + \frac{g_w}{\cos\theta_w} j^\mu_{\text{nc}} Z_\mu \right\} \tag{10.34}$$

$$= -\sum_f \left\{ \frac{g_w}{\sqrt{2}} \overline{L} \gamma^\mu T_+ L W^+_\mu + \frac{g_w}{\sqrt{2}} \overline{L} \gamma^\mu T_- L W^-_\mu \right. \tag{10.35a}$$

$$+ e \left(\overline{\psi} \gamma^\mu Q \psi + \overline{\psi'} \gamma^\mu Q \psi' \right) A_\mu \tag{10.35b}$$

$$\left. + \frac{g_w}{\cos\theta_w} \left[\overline{L} \gamma^\mu T_3 L - \sin^2\theta_w \left(\overline{\psi} \gamma^\mu Q \psi + \overline{\psi'} \gamma^\mu Q \psi' \right) \right] Z_\mu \right\}, \tag{10.35c}$$

with

$$T_+ = \frac{\sigma_+}{2} = \begin{pmatrix} 0 & 1 \\ 0 & 0 \end{pmatrix}, \quad T_- = \frac{\sigma_-}{2} = \begin{pmatrix} 0 & 0 \\ 1 & 0 \end{pmatrix}, \quad T_3 = \frac{\sigma_3}{2} = \begin{pmatrix} \frac{1}{2} & 0 \\ 0 & -\frac{1}{2} \end{pmatrix}, \tag{10.36}$$

and a given generation of quarks or leptons is defined by the fields

$$L = \begin{pmatrix} \psi_L \\ \psi'_L \end{pmatrix}, \quad \psi_R, \ \psi'_R \quad \text{with} \quad \psi = \psi_L + \psi_R, \quad \psi' = \psi'_L + \psi'_R. \tag{10.37}$$

The couple (ψ, ψ') stands for (ν_e, e^-), (ν_μ, μ^-), (ν_τ, τ^-), (u, d'), (c, s') or (t, b').

Let us recap: the four physical gauge bosons of the electroweak interaction, A_μ, Z_μ and $W^\pm_\mu$, are now consistently coupled to the matter fields with electromagnetism and the weak interaction. The couplings to all gauge bosons involve only three free parameters, g_w, e and the angle θ_w.

10.4 Feynman Rules of Gauge Boson—Fermion Interactions

10.4.1 Interactions with $W^\pm$

The Lagrangian (10.35) allows us to deduce the Feynman rules corresponding to the interaction between the gauge fields and the matter fields. The line (10.35a) describes the interaction of $W^\pm$ with the fermions. The T_+ matrix mixes the two components of a weak isospin doublet, inducing the desired charged-current transition: $\psi_L \to \psi'_L$ or $\psi'_L \to \psi_L$. Only the left-handed chirality fields are involved. The vertex factor is $-i\frac{g_w}{\sqrt{2}}\gamma^\mu$ in terms of left-handed fields or $-i\frac{g_w}{\sqrt{2}}\gamma^\mu\frac{1}{2}(1-\gamma^5)$ when we include the projection on the left-handed components, i.e. $\psi = P_L\psi$. The Feynman diagram is then

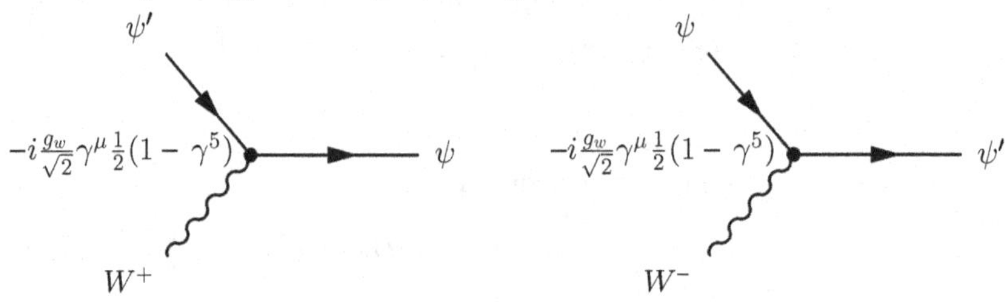

We have recovered what we found in Chapter 9. Keep in mind that for down-type quarks (ψ'), one should use the V_{CKM} matrix element if the mass eigenstates are used (instead of the weak eigenstates).

10.4.2 Interactions with γ

The line (10.35b) specifies the interaction with photons. As expected, this is that of the usual QED. Left-handed and right-handed chiralities are treated on an equal footing. The coupling constant is the charge of the particle $q = eQ$. The diagram is thus

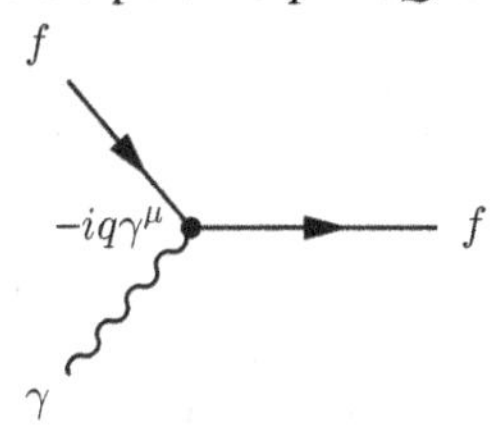

where f is any charged fermions (ψ or ψ').

10.4.3 Interactions with Z^0

Line (10.35c) gives the coupling of matter with the Z^0 boson. Note that both the left-handed chirality (through the isodoublet) and the right-handed chirality (through the field ψ and ψ') are involved. But unlike for the photon, the strength of the coupling differs for the two chiralities. Let us expand line (10.35c) using T_3 in Eq. (10.36). It yields

$$\begin{aligned}\mathcal{L}^{\text{EW}}_{\text{int } Z^0} &= -\sum_f \frac{g_w}{\cos\theta_w}\left[\tfrac{1}{2}\overline{\psi_L}\gamma^\mu\psi_L - \tfrac{1}{2}\overline{\psi'_L}\gamma^\mu\psi'_L - \sin^2\theta_w\left(\overline{\psi}\gamma^\mu Q\psi + \overline{\psi'}\gamma^\mu Q\psi'\right)\right]Z_\mu \\ &= -\sum_f \frac{g_w}{\cos\theta_w}\left[\overline{\psi}\gamma^\mu\tfrac{1}{2}\left(\tfrac{1}{2} - 2\sin^2\theta_w Q - \tfrac{1}{2}\gamma^5\right)\psi\right. \\ &\left.+\overline{\psi'}\gamma^\mu\tfrac{1}{2}\left(-\tfrac{1}{2} - 2\sin^2\theta_w Q + \tfrac{1}{2}\gamma^5\right)\psi'\right]Z_\mu.\end{aligned}$$

Defining the coefficients

$$\begin{aligned} c_V^f &= \tfrac{1}{2} - 2\sin^2\theta_w Q, & c_A^f &= \tfrac{1}{2}, \\ c_V^{f'} &= -\tfrac{1}{2} - 2\sin^2\theta_w Q, & c_A^{f'} &= -\tfrac{1}{2}, \end{aligned} \tag{10.38}$$

it follows that the Lagrangian describing the interaction with the Z^0 boson is

$$\mathcal{L}^{\text{EW}}_{\text{int } Z^0} = -\sum_f \frac{g_w}{\cos\theta_w}\left[\overline{\psi}\gamma^\mu\tfrac{1}{2}\left(c_V^f - c_A^f\gamma^5\right)\psi + \overline{\psi'}\gamma^\mu\tfrac{1}{2}\left(c_V^{f'} - c_A^{f'}\gamma^5\right)\psi'\right]Z_\mu. \tag{10.39}$$

The coefficients (10.38) for each fermion are listed in Table 10.2.

Table 10.2 Coefficients involved in the coupling of fermions with the Z^0 boson.

f	c_V	c_A
ν_e, ν_μ, ν_τ	$1/2$	$1/2$
e^-, μ^-, τ^-	$-1/2 + 2\sin^2\theta_w \simeq -0.04$	$-1/2$
u, c, t	$1/2 - \frac{4}{3}\sin^2\theta_w \simeq +0.19$	$1/2$
d, s, b	$-1/2 + \frac{2}{3}\sin^2\theta_w \simeq -0.35$	$-1/2$

Note that neutrinos are the fermions that couple only to the Z boson via their left-handed chirality components (since $c_V = c_A = 1/2$). This originates from the fact that ν_R has all its gauge numbers equal to zero, i.e. $Q = T_3 = Y = 0$, and hence, has no coupling to a gauge field. In contrast, both left-handed and right-handed components of charged leptons are involved, as

$$c_V - c_A\gamma^5 = (c_V - c_A)P_R + (c_V + c_A)P_L \tag{10.40}$$

differs from P_L as long as $c_V \neq c_A$. It is also important to realise that, unlike the interaction with $W^{\pm}$, there is no flavour changing of fermions with Z^0. The Z^0 boson couples to the same fermion: for instance, $Z^0 \to u\bar{u}$, while for the charged weak boson, we have $W^+ \to u\bar{d}$. The structure of the Lagrangian (10.39) forbids couplings such as $Z^0 \to u\bar{c}$ or $Z^0 \to d\bar{s}$ as there is no term such as $\overline{\psi_1}\left(c_V - c_A\gamma^5\right)\psi_2$ with $\psi_1 \neq \psi_2$. One may wonder whether the weak or mass eigenstates couple to Z^0. Even if the fields ψ and ψ' are the weak eigenstates in Eq. (10.39), it turns out that the Z^0 boson couples in exactly the same way to the mass eigenstates. In other words, both bases lead to the same coupling. This claim is proved in Supplement 10.2.

The Feynman rules are now easy to deduce from the Lagrangian (10.39). They read

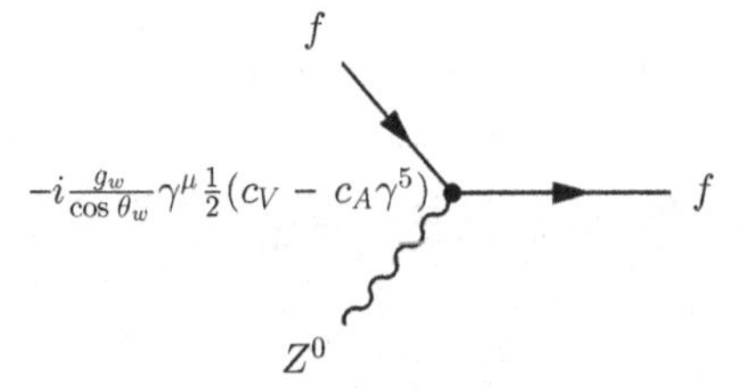

The fermion f can be leptons or quarks, up or down type, weak or mass eigenstate as you wish. The coefficients c_V and c_A are listed in Table 10.2.

10.5 Consequences of Gauge Field Transformations

Along with QCD, based on the gauge symmetry group $SU(3)_c$, the electroweak theory based on the gauge symmetry group $SU(2)_L \times U(1)_Y$ is another example of a non-Abelian theory, also called a Yang–Mills theory. Originally in 1954, Yang (the theorist who hypothesised parity violation in the weak interaction) and Robert Mills developed the theoretical framework of a non-Abelian theory in attempting to describe the strong interaction. However, their theory predicted a massless charged boson in conflict with the experimental facts and was soon forgotten. It was only 10 years later, when theorists like Sheldon Glashow worked on the unification of electromagnetism and weak interaction, that Yang–Mills theories were fully appreciated.

Let us recall what we found for the gauge field transformation in QCD. The gluon field transformation is given in Eq. (8.47) and reproduced here,

$$G'^a_\mu = G^a_\mu + g_s\alpha^b(x)f_{abc}\,G^c_\mu + \partial_\mu\alpha^a(x),$$

Supplement 10.2. Coupling of the Z^0 boson to the mass eigenstates

In the Lagrangian (10.39), the fields coupled to the Z^0 are the weak eigenstates, for instance, $\psi = u, c, t$ or $\psi' = d', s', b'$ in the quark sector. The up-type quarks are, by convention, both the weak and the mass eigenstates, the distinction between the two kinds of states being carried only by the down-type quarks. The weak eigenstates d', s', b' are connected to the mass eigenstates by the V_{CKM} mixing matrix (9.20), with, for instance, $d' = \sum_{i=d,s,b} V_{ui} q_i$, where the sloppy notation $q_d = d, q_s = s$ and $q_b = b$ is used. Since the coefficients c_V and c_A, in Eq. (10.38), only depend on the electric charge of the quarks, all down-type quarks d', s', b' have the same coefficients. Therefore, defining the matrix

$$\Gamma^\mu = \gamma^\mu \frac{1}{2} \left(c_V - c_A \gamma^5 \right),$$

the Lagrangian (10.39) reads

$$\begin{aligned} \mathcal{L} &= -\sum_{f'=d',s',b'} \frac{g_W}{\cos\theta_W} \left[\overline{f'} \Gamma^\mu f' \right] Z_\mu \\ &= -\sum_{k=u,c,t} \frac{g_W}{\cos\theta_W} \left[(\textstyle\sum_{j=d,s,b} \overline{V_{kj} q_j}) \Gamma^\mu (\sum_{i=d,s,b} V_{ki} q_i) \right] Z_\mu \\ &= -\frac{g_W}{\cos\theta_W} \sum_{i,j=d,s,b} \left[\overline{q_j} \Gamma^\mu q_i \right] Z_\mu \sum_{k=u,c,t} V_{ki} V_{kj}^*. \end{aligned}$$

As the CKM matrix is a unitary matrix satisfying the relations (9.21), it follows that

$$\begin{aligned} \mathcal{L} &= -\frac{g_W}{\cos\theta_W} \Big(\sum_{i \neq j=d,s,b} \left[\overline{q_j} \Gamma^\mu q_i \right] Z_\mu \underbrace{\sum_{k=u,c,t} V_{ki} V_{kj}^*}_{=0} + \sum_{i=d,s,b} \left[\overline{q_i} \Gamma^\mu q_i \right] Z_\mu \underbrace{\sum_{k=u,c,t} V_{ki} V_{ki}^*}_{=1} \Big) \\ &= -\sum_{i=d,s,b} \frac{g_W}{\cos\theta_W} \left[\overline{q_i} \Gamma^\mu q_i \right] Z_\mu \end{aligned}$$

Conclusion: the Lagrangian coupling the down-type quarks to the Z^0 can equivalently be written as

$$\mathcal{L} = -\sum_{f'=d',s',b'} \frac{g_W}{\cos\theta_W} \left[\overline{f'} \gamma^\mu \frac{1}{2} (c_V - c_A \gamma^5) f' \right] Z_\mu = -\sum_{f=d,s,b} \frac{g_W}{\cos\theta_W} \left[\overline{f} \gamma^\mu \frac{1}{2} (c_V - c_A \gamma^5) f \right] Z_\mu.$$

The weak neutral current is then diagonal in flavours, coupling the weak eigenstates or mass eigenstates. The same reasoning with the up-type components in the lepton sector[a] would have yielded a similar conclusion.

[a] For historical reasons, unfortunately the quark and lepton sectors follow different conventions. In the former, the weak isospin components $-1/2$ differ from the mass eigenstates, while in the latter, this holds for the $+1/2$ components.

where f_{abc} are the structure constants verifying

$$\left[\frac{\lambda_a}{2}, \frac{\lambda_b}{2} \right] = i f_{abc} \frac{\lambda_c}{2}.$$

The field transformation ensures that the QCD Lagrangian,

$$\mathcal{L} = -\frac{1}{4} G^a_{\mu\nu} G^{\mu\nu}_a,$$

based on the tensor,

$$G^a_{\mu\nu} = \partial_\mu G^a_\nu - \partial_\nu G^a_\mu - g_s f_{abc} G^b_\mu G^c_\nu,$$

is gauge invariant. We can easily transpose our findings to the electroweak gauge symmetry, $SU(2)_L \times U(1)_Y$. The $SU(2)_L$ generators obey the Lie algebra relations

$$\left[\frac{\sigma_a}{2}, \frac{\sigma_b}{2}\right] = i\epsilon_{abc}\frac{\sigma_c}{2},$$

where $\sigma_{a,b,c}$ are the Pauli matrices and ϵ_{abc} is the antisymmetric tensor. The integers a, b, c take the values 1, 2 or 3. By analogy with QCD, we infer that the transformation of the three gauge fields W^1, W^2 and W^3 reads

$$W'^a_\mu = W^a_\mu + g_w\alpha^b(x)\epsilon_{abc}\, W^c_\mu + \partial_\mu\alpha^a(x) \tag{10.41}$$

with an implicit summation over the repeated indices b and c. The transformation above ensures that the Lagrangian

$$\mathcal{L} = -\frac{1}{4}F^a_{\mu\nu}F^{\mu\nu}_a,$$

based on the tensor

$$F^a_{\mu\nu} = W^a_{\mu\nu} - g_w\epsilon_{abc}W^b_\mu W^c_\nu \quad \text{with} \quad W^a_{\mu\nu} = \partial_\mu W^a_\nu - \partial_\nu W^a_\mu, \tag{10.42}$$

is gauge invariant under $SU(2)_L$. For $U(1)_Y$, the situation is much simpler since the group is Abelian and mathematically equivalent to the usual U(1) of QED. Consequently, we have the gauge field transformation

$$B'_\mu = B_\mu + \partial_\mu\alpha(x). \tag{10.43}$$

The Lagrangian, $-\frac{1}{4}B_{\mu\nu}B^{\mu\nu}$, built with the tensor

$$B_{\mu\nu} = \partial_\mu B_\nu - \partial_\nu B_\mu, \tag{10.44}$$

is then $U(1)_Y$ invariant. The Lagrangian of the electroweak gauge fields finally reads

$$\mathcal{L}^{\text{EW}}_{\text{gauge}} = -\frac{1}{4}F^a_{\mu\nu}F^{\mu\nu}_a - \frac{1}{4}B_{\mu\nu}B^{\mu\nu}. \tag{10.45}$$

It is invariant by construction under the gauge transformations $SU(2)_L \times U(1)_Y$, given the field transformations (10.41) and (10.43). The next step is to express this Lagrangian in terms of the physical fields $W^\pm_\mu$, A_μ and Z_μ, using the definitions (10.15) and (10.24). This lengthy (and tedious) calculation is presented in Supplements 10.3 and 10.4. The resulting Lagrangian is

$$\mathcal{L}^{\text{EW}}_{\text{gauge}} = \mathcal{L}^{\text{EW}}_{\text{gauge free}} + \mathcal{L}^{\text{EW}}_{\text{trilinear}} + \mathcal{L}^{\text{EW}}_{\text{quartic}}, \tag{10.46}$$

with

$$\mathcal{L}^{\text{EW}}_{\text{gauge free}} = -\frac{1}{4}\left[(W^-_{\mu\nu})^\dagger W^{-\mu\nu} + (W^+_{\mu\nu})^\dagger W^{+\mu\nu} + Z_{\mu\nu}Z^{\mu\nu} + A_{\mu\nu}A^{\mu\nu}\right], \tag{10.47}$$

and

$$\begin{aligned}\mathcal{L}^{\text{EW}}_{\text{trilinear}} = -ig_w\Big[&(\cos\theta_w Z^\mu + \sin\theta_w A^\mu)(W^-_{\mu\nu}W^{+\nu} - W^+_{\mu\nu}W^{-\nu}) \\ &+(\cos\theta_w Z_{\mu\nu} + \sin\theta_w A_{\mu\nu})W^{+\mu}W^{-\nu}\Big],\end{aligned} \tag{10.48}$$

$$
\begin{aligned}
\mathcal{L}^{\mathrm{EW}}_{\text{quartic}} = -\frac{g_w^2}{2}\Big[& 2\cos^2\theta_w (W^+_\mu W^{-\mu} Z_\nu Z^\nu - W^+_\mu W^{-\nu} Z_\nu Z^\mu) \\
& +2\sin^2\theta_w (W^+_\mu W^{-\mu} A_\nu A^\nu - W^+_\mu W^{-\nu} A_\nu A^\mu) \\
& +2\cos\theta_w \sin\theta_w (2W^+_\mu W^{-\mu} Z_\nu A^\nu - W^+_\mu W^{-\nu} Z_\nu A^\mu - W^+_\mu W^{-\nu} A_\nu Z^\mu) \\
& -W^+_\mu W^{+\mu} W^-_\nu W^{-\nu} + W^+_\mu W^{-\mu} W^-_\nu W^{+\nu}\Big]. \qquad (10.49)
\end{aligned}
$$

The Lagrangian (10.46) specifies the interaction of the four electroweak gauge bosons. Equation (10.47) just describes the propagation of free fields. We shall see in the next section how it will provide the propagators of the massive fields $W^\pm_\mu$, Z_μ and of the massless photon A_μ. The Lagrangian (10.48) describes the interactions between three gauge fields due to the *trilinear couplings*, while Eq. (10.49) corresponds to the interactions between four gauge fields due to *quartic couplings*. Those trilinear and quartic couplings are possible because of the non-Abelian nature of the theory. They are analogous to the 3-gluon and 4-gluon couplings.

10.6 Feynman Rules for the Self-interactions of Gauge Bosons

10.6.1 Trilinear Gauge Field Couplings

- **Coupling $Z^0 W^+ W^-$:**
  It is given by the subpart of the Lagrangian (10.48),

$$\mathcal{L}_{Z^0W^+W^-} = -ig_w \cos\theta_w \left[Z^\mu W^-_{\mu\nu} W^{+\nu} - Z^\mu W^+_{\mu\nu} W^{-\nu} + Z_{\mu\nu} W^{+\mu} W^{-\nu}\right].$$

Let us denote the 4-momenta of W^+, W^- and Z^0 respectively by k_1, k_2 and k_3. Given the correspondence between spacetime derivatives and momentum via the Fourier transformation $\partial_\mu \to -ik_\mu$, the tensor $W^-_{\mu\nu}$ reads in momentum space

$$iW^-_{\mu\nu} = i(\partial_\mu W^-_\nu - \partial_\nu W^-_\mu) \to k_{2\mu} W^-_\nu - k_{2\nu} W^-_\mu$$

and similarly for the other tensors, so that, after re-arrangement of the dummy Lorentz indices, $\mathcal{L}_{Z^0W^+W^-}$ reads in momentum space

$$\mathcal{L}_{Z^0W^+W^-} = g_w \cos\theta_w \left[(k_1 - k_2)_\mu\, g_{\lambda\nu} + (k_2 - k_3)_\nu\, g_{\mu\lambda} + (k_3 - k_1)_\lambda\, g_{\mu\nu}\right] Z^\mu W^{-\lambda} W^{+\nu}.$$

The vertex factor is then straightforward to determine from the Lagrangian, resulting from multiplication by i times the number of permutations (here, there is just one single possibility) after stripping the Lagrangian of all fields. This gives

$W^-(\lambda, k_2)$

$ig_w \cos\theta_w \left[(k_1 - k_2)_\mu\, g_{\lambda\nu} + (k_2 - k_3)_\nu\, g_{\mu\lambda} + (k_3 - k_1)_\lambda\, g_{\mu\nu}\right]$ $Z^0(\mu, k_3)$

$W^+(\nu, k_1)$

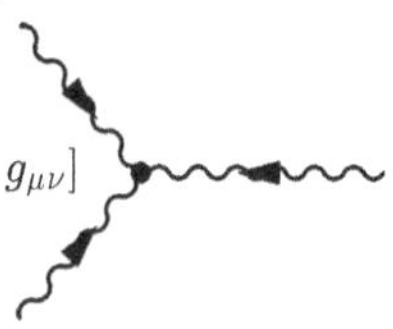

The 3-momenta are oriented towards the vertex (cf. footnote 20 in Chapter 9).

Supplement 10.3. Expression of the gauge field Lagrangian as a function of the physical bosons: $F^1_{\mu\nu}F^{1\mu\nu} + F^2_{\mu\nu}F^{2\mu\nu}$

The Lagrangian of the gauge fields in Eq. (10.45) is expressed with the field tensors $F^a_{\mu\nu}$ and $B_{\mu\nu}$, which are functions of the gauge fields W^1_μ, W^2_μ, W^3_μ and B_μ. Since the charged fields $W^\pm$ are a function of $W^{1,2}$, they must result from the term $F^1_{\mu\nu}F^{1\mu\nu} + F^2_{\mu\nu}F^{2\mu\nu}$. The properties of the antisymmetric tensor in the definition (10.42) of $F^a_{\mu\nu}$ imply

$$\begin{aligned} F^1_{\mu\nu} &= W^1_{\mu\nu} - g_w\epsilon_{123}W^2_\mu W^3_\nu - g_w\epsilon_{132}W^3_\mu W^2_\nu = W^1_{\mu\nu} - g_w(W^2_\mu W^3_\nu - W^3_\mu W^2_\nu), \\ F^2_{\mu\nu} &= W^2_{\mu\nu} - g_w\epsilon_{213}W^1_\mu W^3_\nu - g_w\epsilon_{231}W^3_\mu W^1_\nu = W^2_{\mu\nu} + g_w(W^1_\mu W^3_\nu - W^3_\mu W^1_\nu). \end{aligned}$$

Playing with dummy indices, it follows that

$$\begin{aligned} F^1_{\mu\nu}F^{1\mu\nu} + F^2_{\mu\nu}F^{2\mu\nu} = \; & W^1_{\mu\nu}W^{1\mu\nu} + W^2_{\mu\nu}W^{2\mu\nu} + 4g_w W^{3\mu}(W^1_{\mu\nu}W^{2\nu} - W^2_{\mu\nu}W^{1\nu}) \\ & +2g_w^2\left[(W^1_\mu W^{1\mu} + W^2_\mu W^{2\mu})W^3_\nu W^{3\nu} - (W^1_\mu W^{1\nu} + W^2_\mu W^{2\nu})W^3_\nu W^{3\mu}\right]. \end{aligned}$$

Given the definition of $W^\pm$ fields (10.15), it is then straightforward to show that

$$\begin{aligned} & W^1_{\mu\nu}W^{1\mu\nu} + W^2_{\mu\nu}W^{2\mu\nu} = 2W^+_{\mu\nu}W^{-\mu\nu}, \quad W^1_{\mu\nu}W^{2\nu} - W^2_{\mu\nu}W^{1\nu} = i(W^-_{\mu\nu}W^{+\nu} - W^+_{\mu\nu}W^{-\nu}); \\ & W^1_\mu W^{1\mu} + W^2_\mu W^{2\mu} = 2W^+_\mu W^{-\mu}, \qquad W^1_\mu W^{1\nu} + W^2_\mu W^{2\nu} = W^+_\mu W^{-\nu} + W^-_\mu W^{+\nu}. \end{aligned}$$

This leads to the expression

$$\begin{aligned} F^1_{\mu\nu}F^{1\mu\nu} + F^2_{\mu\nu}F^{2\mu\nu} = \; & 2W^+_{\mu\nu}W^{-\mu\nu} + 4ig_w W^{3\mu}(W^-_{\mu\nu}W^{+\nu} - W^+_{\mu\nu}W^{-\nu}) \\ & +2g_w^2\left[2W^+_\mu W^{-\mu}W^3_\nu W^{3\nu} - (W^+_\mu W^{-\nu} + W^-_\mu W^{+\nu})W^3_\nu W^{3\mu}\right]. \end{aligned}$$

The last step is to express the field W^3_μ as a function of Z_μ and A_μ. According to the definition (10.24), $W^{3\mu} = \cos\theta_w Z^\mu + \sin\theta_w A^\mu$. This yields the formulas:

$$\begin{aligned} W^3_\nu W^{3\nu} &= \cos^2\theta_w Z_\nu Z^\nu + \sin^2\theta_w A_\nu A^\nu + 2\cos\theta_w \sin\theta_w Z_\nu A^\nu; \\ W^3_\nu W^{3\mu} &= \cos^2\theta_w Z_\nu Z^\mu + \sin^2\theta_w A_\nu A^\mu + \cos\theta_w \sin\theta_w (Z_\nu A^\mu + A\nu Z^\mu). \end{aligned}$$

Moreover, by changing the dummy indices or their position, we deduce the two following formulas:

$$\begin{aligned} (W^+_\mu W^{-\nu} + W^-_\mu W^{+\nu})W^3_\nu W^{3\mu} &= 2W^+_\mu W^{-\nu}W^3_\nu W^{3\mu}; \\ 2W^+_{\mu\nu}W^{-\mu\nu} &= (W^-_{\mu\nu})^\dagger W^{-\mu\nu} + (W^+_{\mu\nu})^\dagger W^{+\mu\nu}. \end{aligned}$$

Gathering all this finally yields

$$\begin{aligned} F^1_{\mu\nu}F^{1\mu\nu} + F^2_{\mu\nu}F^{2\mu\nu} = \; & (W^-_{\mu\nu})^\dagger W^{-\mu\nu} + (W^+_{\mu\nu})^\dagger W^{+\mu\nu} \\ & +4ig_w(\cos\theta_w Z^\mu + \sin\theta_w A^\mu)(W^-_{\mu\nu}W^{+\nu} - W^+_{\mu\nu}W^{-\nu}) \\ & +2g_w^2\Big[2\cos^2\theta_w W^+_\mu W^{-\mu}Z_\nu Z^\nu + 2\sin^2\theta_w W^+_\mu W^{-\mu}A_\nu A^\nu \\ & \qquad +4\cos\theta_w\sin\theta_w W^+_\mu W^{-\mu}Z_\nu A^\nu\Big] \\ & -2g_w^2\Big[2\cos^2\theta_w W^+_\mu W^{-\nu}Z_\nu Z^\mu + 2\sin^2\theta_w W^+_\mu W^{-\nu}A_\nu A^\mu \\ & \qquad +2\cos\theta_w\sin\theta_w W^+_\mu W^{-\nu}(Z_\nu A^\mu + A_\nu Z^\mu)\Big]. \end{aligned}$$

- **Coupling $\gamma W^+ W^-$:**
  The procedure is similar to the previous one. From Eq. (10.48), one reads

$$\begin{aligned}\mathcal{L}_{\gamma W^+W^-} &= -ig_w \sin\theta_w \left[A^\mu W^-_{\mu\nu} W^{+\nu} - A^\mu W^+_{\mu\nu} W^{-\nu}) + A_{\mu\nu} W^{+\mu} W^{-\nu}\right] \\ &= g_w \sin\theta_w \left[(k_1 - k_2)_\mu \, g_{\lambda\nu} + (k_2 - k_3)_\nu \, g_{\nu\lambda} + (k_3 - k_1)_\lambda \, g_{\mu\nu}\right] A^\mu W^{-\lambda}.\end{aligned}$$

Supplement 10.4. Expression of the gauge field Lagrangian as a function of the physical bosons: $F^3_{\mu\nu} F^{3\mu\nu} + B_{\mu\nu} B^{\mu\nu}$

The remaining term of the Lagrangian (10.45), $F^3_{\mu\nu} F^{3\mu\nu} + B_{\mu\nu} B^{\mu\nu}$, gives rise, among other things, to the kinetic term of the Z^0 and the photon. As

$$\begin{aligned}F^3_{\mu\nu} = W^3_{\mu\nu} - g_w \epsilon_{312} W^1_\mu W^2_\nu - g_w \epsilon_{321} W^2_\mu W^1_\nu &= W^3_{\mu\nu} - g_w (W^1_\mu W^2_\nu - W^2_\mu W^1_\nu) \\ &= W^3_{\mu\nu} + ig_w (W^+_\mu W^-_\nu - W^-_\mu W^+_\nu),\end{aligned}$$

it follows that

$$F^3_{\mu\nu} F^{3\mu\nu} = W^3_{\mu\nu} W^{3\mu\nu} + 4ig_w W^3_{\mu\nu} W^{+\mu} W^{-\nu} - 2g_w^2 (W^+_\mu W^{+\mu} W^-_\nu W^{-\nu} - W^+_\mu W^{-\mu} W^-_\nu W^{+\nu}).$$

Moreover, according to the expression (10.24) of the Z and A fields, we deduce

$$\begin{aligned}W^3_{\mu\nu} W^{3\mu\nu} + B_{\mu\nu} B^{\mu\nu} &= (\cos\theta_w Z_{\mu\nu} + \sin\theta_w A_{\mu\nu})(\cos\theta_w Z^{\mu\nu} + \sin\theta_w A^{\mu\nu}) \\ &\quad + (-\sin\theta_w Z_{\mu\nu} + \cos\theta_w A_{\mu\nu})(-\sin\theta_w Z^{\mu\nu} + \cos\theta_w A^{\mu\nu}) \\ &= Z_{\mu\nu} Z^{\mu\nu} + A_{\mu\nu} A^{\mu\nu}.\end{aligned}$$

Therefore, globally

$$\begin{aligned}F^3_{\mu\nu} F^{3\mu\nu} + B_{\mu\nu} B^{\mu\nu} = \; & Z_{\mu\nu} Z^{\mu\nu} + A_{\mu\nu} A^{\mu\nu} + 4ig_w (\cos\theta_w Z_{\mu\nu} + \sin\theta_w A_{\mu\nu}) W^{+\mu} W^{-\nu} \\ & -2g_w^2 (W^+_\mu W^{+\mu} W^-_\nu W^{-\nu} - W^+_\mu W^{-\mu} W^-_\nu W^{+\nu}) \qquad (10.50)\end{aligned}$$

Finally, gathering Eq. (10.50) from Supplement 10.3 and Eq. (10.50) yields the expressions (10.47)–(10.49).

Recalling the relation $e = g_w \sin\theta_w$, the vertex factor is then

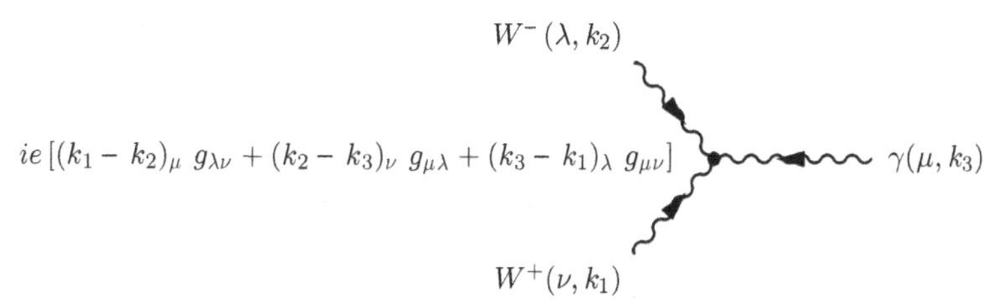

10.6.2 Quartic Gauge Field Couplings

- **Coupling $Z^0 Z^0 W^+ W^-$:**
  This coupling results from the subpart of the Lagrangian (10.49),

$$\begin{aligned}\mathcal{L}_{Z^0Z^0W^+W^-} &= -g_w^2 \cos^2\theta_w \left[W^+_\mu W^{-\mu} Z_\nu Z^\nu - W^+_\mu W^{-\nu} Z_\nu Z^\mu\right] \\ &= -g_w^2 \cos^2\theta_w \tfrac{1}{2} \left[2g_{\mu\nu} g_{\alpha'\beta'} W^{+\mu} W^{-\nu} Z^{\alpha'} Z^{\beta'} - 2g_{\alpha'\nu} g_{\mu\beta'} W^{+\mu} W^{-\nu} Z^{\alpha'} Z^{\beta'}\right],\end{aligned}$$

where the dummy Lorentz indices have been renamed to ease the determination of the vertex factor. Here, there is a subtlety due to the two undistinguishable Z^0 particles, which can be generated from $2! = 2$ different field configurations (this is why above the factor 2 has been explicitly factored out). Indeed, a Feynman diagram with the labels $W^{+\mu}, W^{-\nu}, Z^\alpha, Z^\beta$ can result from the field configuration $(\alpha', \beta') = (\alpha, \beta)$ or $(\alpha', \beta') = (\beta, \alpha)$. Hence, $2g_{\mu\nu}g_{\alpha'\beta'} = g_{\mu\nu}g_{\alpha\beta} + g_{\mu\nu}g_{\beta\alpha}$ and $2g_{\alpha'\nu}g_{\mu\beta'} = g_{\alpha\nu}g_{\mu\beta} + g_{\beta\nu}g_{\mu\alpha}$. Therefore, the vertex factor for $Z^0Z^0W^+W^-$ is

$$\begin{aligned} &i \times 2! \times \left(-g_w^2 \cos^2\theta_w \tfrac{1}{2}\right)\left[g_{\mu\nu}(g_{\alpha\beta} + g_{\beta\alpha}) - (g_{\alpha\nu}g_{\mu\beta} + g_{\beta\nu}g_{\mu\alpha})\right] = \\ &-ig_w^2 \cos^2\theta_w \left[2g_{\mu\nu}g_{\alpha\beta} - g_{\alpha\nu}g_{\mu\beta} - g_{\beta\nu}g_{\mu\alpha}\right], \end{aligned}$$

which is represented below

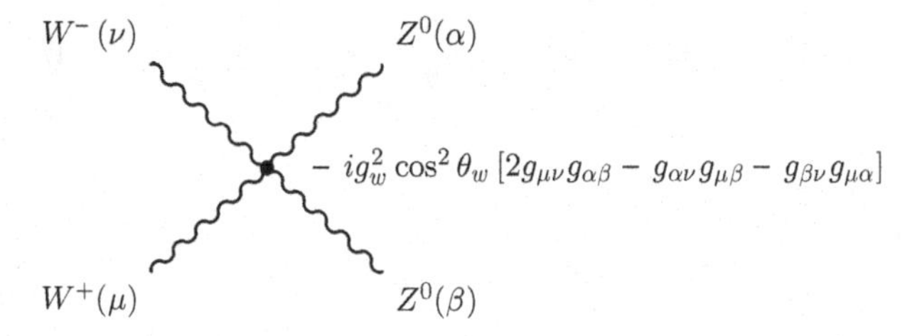

- **Coupling $\gamma\gamma W^+W^-$:**
It is given by the subpart of the Lagrangian (10.49),

$$\mathcal{L}_{\gamma\gamma W^+W^-} = -g_w^2 \sin^2\theta_w \left[W_\mu^+ W^{-\mu} A_\nu A^\nu - W_\mu^+ W^{-\nu} A_\nu A^\mu\right].$$

No need to do the calculation since the Lagrangian is similar to that of $Z^0Z^0W^+W^-$, notwithstanding the change $g_w^2 \cos^2\theta_w \to g_w^2 \sin^2\theta_w = e^2$. Hence, we obtain the vertex factor

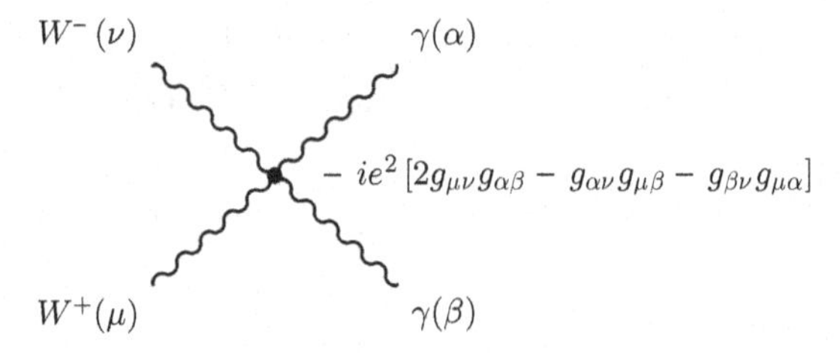

- **Coupling $\gamma Z^0 W^+W^-$:**
It is given by the subpart of the Lagrangian (10.49),

$$\begin{aligned} \mathcal{L}_{\gamma Z^0 W^+W^-} &= -g_w^2 \cos\theta_w \sin\theta_w \left[2W_\mu^+ W^{-\mu} Z_\nu A^\nu - W_\mu^+ W^{-\nu} Z_\nu A^\mu - W_\mu^+ W^{-\nu} A_\nu Z^\mu\right] \\ &= -eg_w \cos\theta_w \left[2W_\mu^+ W^{-\mu} Z_\nu A^\nu - W_\mu^+ W^{-\nu} Z_\nu A^\mu - W_\mu^+ W^{-\nu} A_\nu Z^\mu\right]. \end{aligned} \tag{10.51}$$

The reader will easily be convinced that the vertex factor is (Problem 10.2)

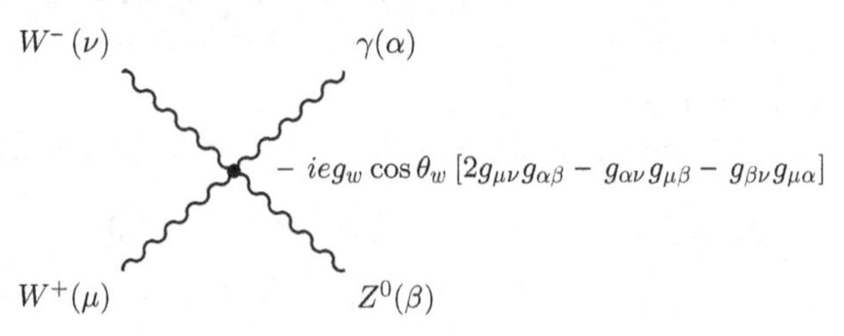

- **Coupling $W^+W^-W^+W^-$:**
  It is given by the subpart of the Lagrangian (10.49),

$$\mathcal{L}_{W^+W^-W^+W^-} = \frac{g_w^2}{2}\left[W_\mu^+ W^{+\mu} W_\nu^- W^{-\nu} - W_\mu^+ W^{-\mu} W_\nu^- W^{+\nu}\right]. \tag{10.52}$$

Here again, we cannot distinguish the two W^+ and two W^-. As an exercise, the reader can easily determine the vertex factor (Problem 10.3). It is

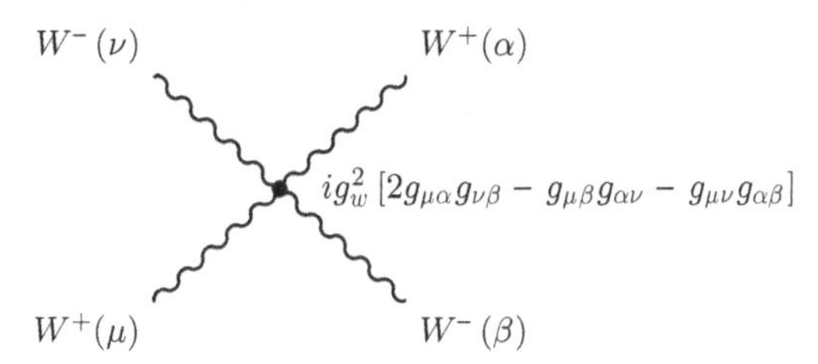

10.7 Conclusions

The theory defined by the Lagrangian

$$\mathcal{L} = \mathcal{L}_{\text{free}}^{\text{EW}} + \mathcal{L}_{\text{int}}^{\text{EW}} + \mathcal{L}_{\text{gauge}}^{\text{EW}},$$

where $\mathcal{L}_{\text{free}}^{\text{EW}}$, $\mathcal{L}_{\text{int}}^{\text{EW}}$ and $\mathcal{L}_{\text{gauge}}^{\text{EW}}$ are given by Eqs. (10.5), (10.35) and (10.46), respectively, describes the free fermions of the three generations, the free gauge bosons γ, $W^\pm$ and Z^0, the interactions of the fermions with the gauge bosons and the interactions between the gauge bosons. It encompasses both QED and the weak interaction. Due to its non-Abelian algebra, the electroweak interaction includes couplings of gauge bosons to each other. The existence of those couplings has been extensively checked by experiments. For instance, Fig. 10.1 shows the measurement of the W pair production at the LEP e^+e^- collider as a function of the CM energy. The data are consistent with the Standard Model prediction (the solid line), which includes the graph at lowest order as shown in Fig. 10.2. The upper dashed curve in Fig. 10.1 ignored all trilinear couplings. Therefore, only graph (a) in Fig. 10.2 contributes to the cross section at lowest order. The lower dashed curve excludes the trilinear coupling ZWW [graph (c) in Fig. 10.2] but includes γWW [graph (b)]. Data require both trilinear couplings to be correctly described.

Nevertheless, the electroweak theory presented so far has a serious issue: none of the particles (fermions and bosons) involved in this theory has a mass. A mass term as

$$m\overline{\psi}\psi = m\overline{\psi_L}\psi_R + m\overline{\psi_R}\psi_L$$

mixes components having different weak isospin and weak hypercharge and so cannot conserve T_3 and Y. So, gauge invariance would imply that all particles are massless (!). The next chapter explains how to introduce particle masses in the theory while still preserving gauge symmetry.

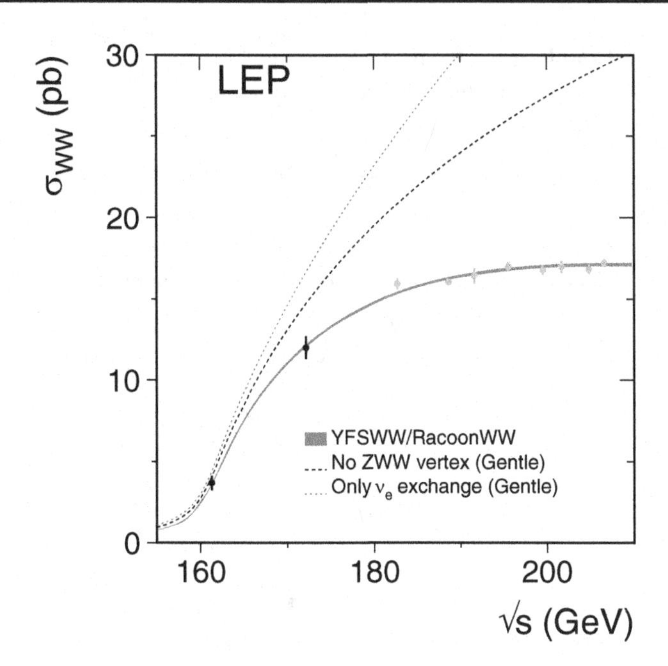

Fig. 10.1 Measurement of the W pair production at LEP and the comparison with the Standard Model prediction (solid line). The dashed curves correspond to predictions if the trilinear couplings ZWW and γWW are ignored. Figure reprinted with permission from Elsevier (2013).

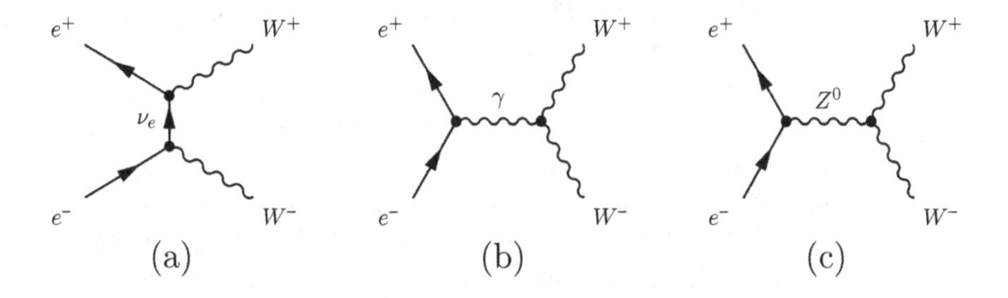

Fig. 10.2 Diagrams contributing at lowest order to the W pair production at the e^+e^- collider. Only (b) and (c) involve a trilinear coupling of gauge bosons.

Problems

10.1 Starting with the sum of Eqs. (10.25) and (10.26), check the result (10.31) using Eq. (10.30).

10.2 From Eq. (10.51), check the vertex factor of $\gamma Z^0 W^+ W^-$ given on p. 390.

10.3 Starting from Eq. (10.52), express the Lagrangian of the interaction of four W as a function of the field labels $W^{+\alpha'}$, $W^{+\mu'}$, $W^{-\beta'}$, $W^{-\nu'}$. List the different combinations leading to a

Feynman diagram with the labels $W^{+\alpha}$, $W^{+\mu}$, $W^{-\beta}$, $W^{-\nu}$ and deduce the vertex factor,

$$ig_w^2\left[2g_{\mu\alpha}g_{\nu\beta} - g_{\mu\beta}g_{\alpha\nu} - g_{\mu\nu}g_{\alpha\beta}\right].$$

10.4 Determination of the number of light neutrinos.

At the LEP (e^+e^- collider), the total decay width Γ_z of the Z^0 boson has been measured accurately using the Z-lineshape. In addition, with the reactions $e^+e^- \to \ell^+\ell^-$ (ℓ being a charged leptons) and $e^+e^- \to$ jets at $\sqrt{s} = M_Z$, it was possible to estimate the branching ratio of $Z^0 \to \ell^+\ell^-$ and $Z^0 \to$ hadrons denoted respectively by BR(ℓ) and BR(h).

1. Assuming lepton universality and neglecting all fermion masses, explain why the invisible decay width Γ_{inv} can be measured as $\Gamma_{\text{inv}} = \Gamma_z[1 - \text{BR}(h) - 3\text{BR}(\ell)]$. The measurement gave $\Gamma_{\text{inv}} = 499.0 \pm 1.5$ MeV.
2. How are the number of light neutrinos with masses smaller than $M_Z/2$ and Γ_{inv} related? To reduce systematic errors, it is more accurate to use

$$N_\nu = \frac{\Gamma_{\text{inv}}}{\Gamma_\ell} \times \frac{\Gamma_\ell^{th}}{\Gamma_\nu^{th}},$$

where Γ_ℓ^{th} and Γ_ν^{th} are the theoretical partial decay width of the Z into charged leptons or into neutrinos, respectively. The measured value at the LEP is $\Gamma_\ell = 83.984 \pm 86$ MeV.

3. We wish now to calculate Γ_ℓ^{th} and Γ_ν^{th}.
   1. Determine the amplitude of $Z^0 \to f\bar{f}$ where f is any fermion. Use ϵ^μ for the Z^0 polarisation vector and p_1 and p_2 for the 4-momenta of f and $\bar{f}$, respectively.
   2. Show that the spin-averaged amplitude squared $\overline{|\mathcal{M}|^2}$ can be expressed as:

$$\overline{|\mathcal{M}|^2} = \frac{g_w^2}{12\cos^2\theta_w}\left(-g_{\mu\nu} + \frac{(p_1+p_2)_\mu(p_1+p_2)_\nu}{m_Z^2}\right) \times \sum_{s_1,s_2}\left[\bar{u}(p_1)\gamma^\mu(c_V - c_A\gamma^5)v(p_2)\right]\left[\bar{u}(p_1)\gamma^\nu(c_V - c_A\gamma^5)v(p_2)\right]^*.$$

   Hint: use for averaging the polarisation of massive vector bosons,

$$\sum_\lambda \epsilon_\mu^*(\lambda, q)\epsilon_\nu(\lambda, q) = -g_{\mu\nu} + \frac{q_\mu q_\nu}{m_Z^2}.$$

   3. Neglecting fermion masses, check that $(p_1 + p_2)_\mu\left[\bar{u}(p_1)\gamma^\mu(c_V - c_A\gamma^5)v(p_2)\right] = 0$, and conclude that

$$\overline{|\mathcal{M}|^2} = \frac{g_w^2}{12\cos^2\theta_w}\left(-g_{\mu\nu}\right)(c_V^2 + c_A^2)Tr\left[\gamma^\mu \not{p}_1\gamma^\nu \not{p}_2\right].$$

   4. Finally, show that

$$\Gamma(Z^0 \to f\bar{f}) = \frac{g_w^2}{48\pi\cos^2\theta_w} m_Z(c_V^2 + c_A^2).$$

4. Using the numerical value $\sin^2\theta_w = 0.23$ and Table 10.2, calculate the number of light neutrinos.
5. Additional question: calculate the Z^0 lifetime, using $m_Z = 91.19$ GeV and $g_w^2 = 0.426$.

10.5 Following a similar approach to exercise 10.4, calculate the decay width of the W boson. Use the numerical values $g_w^2 = 0.426$ and $m_W = 80.38$ GeV.

11 Electroweak Symmetry Breaking

The electroweak theory described in the previous chapter assumed that all particles are massless since a mass term would not respect the $SU(2)_L \times U(1)_Y$ gauge symmetry. However, experimentally, it is well known that most particles have a mass. This chapter explains how we can reconcile massive particles with gauge symmetry. The notion of spontaneous symmetry breaking is introduced, and the Brout–Englert–Higgs mechanism is presented. The rest of the chapter is devoted to the experimental discovery of the Higgs boson and its properties.

11.1 Spontaneous Symmetry Breaking

11.1.1 A Simple Example with a U(1) Symmetry

As a starting point, let us recall the Lagrangian describing a free massive scalar boson,

$$\mathcal{L} = \partial_\mu \phi^\dagger \, \partial^\mu \phi - \mu^2 \phi^\dagger \phi,$$

with

$$\phi = (\phi_1 + i\phi_2)/\sqrt{2},$$

and ϕ_1 and ϕ_2 being two real scalar fields. If the scalar field ϕ describes a particle that is its own antiparticle, then $\phi^\dagger = \phi$ and $\phi_2 = 0$. Applying the Euler–Lagrange equation in Eq. (6.13) to the field $\phi^\dagger$ leads to the well-known Klein–Gordon equation (see p. 170). Now, let us consider the Lagrangian

$$\mathcal{L} = \partial_\mu \phi^\dagger \, \partial^\mu \phi - V(\phi), \tag{11.1}$$

where V is a scalar potential having the form[1]

$$V(\phi) = \mu^2 \phi^\dagger \phi + \lambda (\phi^\dagger \phi)^2 = \mu^2 |\phi|^2 + \lambda |\phi|^4. \tag{11.2}$$

For $\lambda = 0$ and $\mu^2 > 0$, the Klein–Gordon Lagrangian is recovered. The Lagrangian (11.1) is invariant under a U(1) transformation, $\phi \to \phi' = e^{-i\alpha}\phi$. By definition, the ground state of the theory corresponds to the minimum energy state, an eigenstate of the Hamiltonian operator. For a real field, we saw that the Hamiltonian (density) was

[1] The form is appropriate to generate spontaneous symmetry breaking and is constrained by the gauge invariance and the renormalisability of the theory.

defined by Eq. (6.8). For a complex field depending on two scalar fields, one should instead use

$$\mathcal{H} = \frac{\partial \mathcal{L}}{\partial \overset{\circ}{\phi}} \overset{\circ}{\phi} + \overset{\circ}{\phi}{}^{\dagger} \frac{\partial \mathcal{L}}{\partial \overset{\circ}{\phi}{}^{\dagger}} - \mathcal{L}.$$

This yields with the Lagrangian (11.1) (see Problem 11.1),

$$\mathcal{H} = |\overset{\circ}{\phi}|^2 + |\nabla \phi|^2 + \mu^2 |\phi|^2 + \lambda |\phi|^4.$$

The first two terms are positive and vanish if ϕ is a constant field not depending on space-time. Therefore, the minimum of $\mathcal{H}$ is reached for a constant field ϕ_0 that minimises the last two terms, i.e. the potential $V(\phi_0)$. The potential is bounded from below if $\lambda \geq 0$; otherwise, an infinite value of ϕ would yield infinite negative energy, and therefore there would be no ground state, which is unphysical. Two situations have to be considered:

- **Case** $\mu^2 \geq 0$.

The minimum is reached by the trivial constant field $\phi_0 = 0$. The ground state is unique and invariant under the U(1) symmetry. The expansion of the Lagrangian in terms of the real scalar fields ϕ_1 and ϕ_2 yields

$$\mathcal{L} = \sum_{k=1}^{2} \frac{1}{2} \left(\partial_\mu \phi_k \partial^\mu \phi_k - \mu^2 \phi_k^2 \right) + \frac{\lambda}{4} \left(\phi_1^2 + \phi_2^2 \right)^2,$$

showing that it describes two scalar bosons with the same mass, μ. They interact via the quartic coupling λ.

- **Case** $\mu^2 < 0$.

The extrema of the potential $V(\phi) = \frac{\mu^2}{2}(\phi_1^2 + \phi_2^2) + \frac{\lambda}{4}(\phi_1^2 + \phi_2^2)^2$ satisfy

$$\begin{cases} \partial V/\partial \phi_1 = \mu^2 \phi_1 + \lambda(\phi_1^2 + \phi_2^2)\phi_1 = 0, \\ \partial V/\partial \phi_2 = \mu^2 \phi_2 + \lambda(\phi_1^2 + \phi_2^2)\phi_2 = 0. \end{cases}$$

The solution $\phi_1 = \phi_2 = 0$ implies $V = 0$. As $\mu^2 < 0$, this defines a maximum. The other solution implies

$$(\phi_1^2 + \phi_2^2) = 2|\phi|^2 = \frac{-\mu^2}{\lambda} \equiv v^2 > 0, \quad \text{i.e.} \quad |\phi| = \sqrt{\frac{-\mu^2}{2\lambda}} = \frac{v}{\sqrt{2}}. \tag{11.3}$$

Therefore, all constant fields satisfying $|\phi| = v/\sqrt{2}$ correspond to the minimum energy. There is degeneracy of the ground states, with all fields lying on a circle in the complex plane $(\Im(\phi), \Re(\phi))$ of radius $v/\sqrt{2}$ are possible ground states. This is illustrated in Fig. 11.1, where the potential (11.2) has the famous shape of a Mexican hat. Those ground states respect the initial U(1) symmetry in the sense that a change of the phase of fields does not change the circle representing the solutions. It is not surprising since we started initially from a Lagrangian and thus a Hamiltonian invariant under a change of phase of the field ϕ. However, as soon as a particular (arbitrary) solution is chosen as *the* ground state, this ground state is no longer invariant under a U(1) transformation since it is transformed into another state on the circle. This situation where the ground state has less symmetry

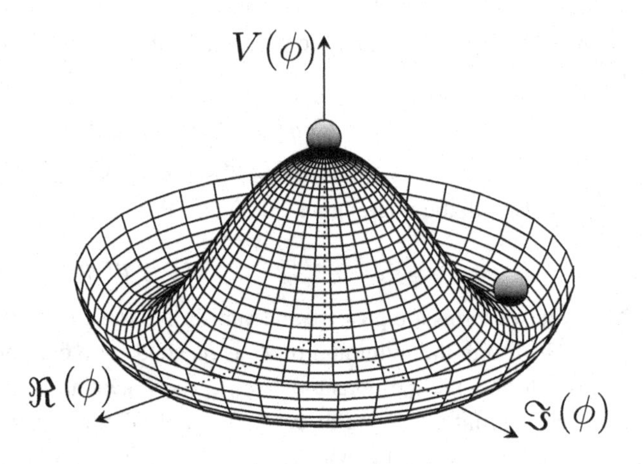

Fig. 11.1 The potential $V(\phi)$ in Eq. (11.2) when $\mu^2 < 0$. The ball at the top has the maximum potential energy, and the corresponding state is unstable. The one in the gutter symbolises the degenerated ground states.

than the original Lagrangian is called a *spontaneous symmetry breaking*. Typically, this happens in a ferromagnet where, above the critical Curie temperature T_c, the electron spins forming tiny magnetic dipoles cannot maintain a fixed direction because of thermal fluctuations. When $T < T_c$, there are enough dipoles stably oriented in the same random direction to create a spontaneous and permanent magnetic field: the material acquires a magnetisation, with a random orientation in space that breaks the initial rotational symmetry. Any direction can be spontaneously chosen by the state minimising the energy.

So far, the field ϕ has been considered a classical field, and we simply established that the ground state corresponds to a field satisfying $|\phi| = v/\sqrt{2}$. In quantum field theory, it is assumed that the *average values* of quantum fields correspond to the classical values. When $\mu^2 \geq 0$, we then expect $\langle 0|\phi|0\rangle = 0$, which is what we are used to when expanding the free field ϕ in terms of creation and annihilation operators,

$$\phi(x) \propto \int dk \left(a_k e^{-ik\cdot x} + a_k^\dagger e^{+ik\cdot x} \right),$$

with $a_k |0\rangle = 0$. The quantity $\langle 0|\phi|0\rangle$ is the *vacuum expectation value* of the field, often abbreviated *v.e.v*. For $\mu^2 < 0$, the correspondence principle between the classical and quantum fields leads to

$$|\langle 0|\phi|0\rangle| = \frac{v}{\sqrt{2}}. \tag{11.4}$$

This situation is new: the field does not vanish in the vacuum, with its v.e.v. being non-zero. One could say that, on average, there is *something* in the vacuum.[2] This new ground state no longer corresponds to an empty vacuum but a uniform distribution of the field ϕ. We note that as for the classical fields, there is again a manifold of degenerate ground states Ω since the condition (11.4) implies

$$\langle \Omega|\phi|\Omega\rangle = \frac{v}{\sqrt{2}} e^{i\Omega}.$$

[2] Section 11.9.3 tries to specify what is meant by *something* in the vacuum.

The phase $e^{i\Omega}$ has no incidence on the energy spectrum since the initial Lagrangian and thus the Hamiltonian are invariant under a change of phase due to the U(1) symmetry. In quantum field theory, particles are considered to be excitations of the vacuum. In order to have a perturbative quantum field theory, let us consider small excitations about the true vacuum. Without loss of generality, we expand ϕ about the vacuum state $|\Omega = 0\rangle$, where the v.e.v. $\langle 0|\phi|0\rangle$ is real, using the convenient parametrisation

$$\phi(x) = \frac{1}{\sqrt{2}}\left[v + h(x)\right] e^{i\theta(x)}. \tag{11.5}$$

The fields $h(x)$ and $\theta(x)$ are two real scalar fields like $\phi_1(x)$ and $\phi_2(x)$. They correspond to the radial (h) and azimuthal (θ) excitations. Note that the azimuthal excitations are in the direction where the potential does not change and, therefore, at a zero energy cost. Assuming $\langle 0|h|0\rangle = \langle 0|\theta|0\rangle = 0$ yields the correct v.e.v. for ϕ since

$$\langle 0|\phi(x)|0\rangle = \frac{v}{\sqrt{2}}\langle 0|1 + i\theta(x) + \cdots|0\rangle + \frac{1}{\sqrt{2}}\langle 0|h(x)(1 + i\theta(x) + \cdots)|0\rangle = \frac{v}{\sqrt{2}}.$$

Again, if the field $\theta(x)$ is shifted by a constant value Ω, we would find $\langle 0|\phi(x)|0\rangle = \frac{v}{\sqrt{2}}e^{i\Omega}$, namely another ground state. In other words, we find that the shift of $\theta(x)$, i.e. the azimuthal excitation, costs no energy and, therefore, we can anticipate that necessarily $\theta(x)$ is a massless field. Let us check this explicitly. The insertion of the field (11.5) into the original Lagrangian (11.1) yields (see Problem 11.2)

$$\mathcal{L} = \frac{1}{2}\left(\partial_\mu h \partial^\mu h - \left[\sqrt{2}|\mu|\right]^2 h^2\right) + \frac{1}{2}\partial_\mu\tilde{\theta}\partial^\mu\tilde{\theta} + \lambda\frac{v^4}{4} + \cdots, \tag{11.6}$$

where $\tilde{\theta}(x) = v\theta(x)$. The dots correspond to trilinear and quartic interactions of h, and interactions between h and $\tilde{\theta}$. Note that the constant term $\lambda v^4/4$ has no impact on the equations of motions and hence can be ignored. We observe that the field h is a massive field with a mass $\sqrt{2}|\mu|$, whereas $\tilde{\theta}$ is massless since there are no quadratic terms in $\tilde{\theta}$. The spectrum of the theory with spontaneous symmetry breaking (one massive boson and one massless boson) is drastically different from that of the theory with $\mu^2 > 0$ (two massive bosons with the same mass). The massless spin-0 boson is called a *Goldstone*[3] boson. It appears when the Lagrangian is invariant under a global gauge transformation (U(1) in this subsection), but the vacuum does not respect that symmetry due to a non-zero vacuum expectation value of the field.

11.1.2 Spontaneous Breaking of the Global $SU(2)_L \times U(1)_Y$ Symmetry

This time, the field ϕ must be sensitive to both $SU(2)_L$ and $U(1)_Y$. Therefore, ϕ must be extended to a weak isospin doublet with non-zero weak hypercharge. Weinberg, in 1967,

[3] Jeffrey Goldstone is an English physicist who proved a theorem that states that in a theory where there is a spontaneous breaking of continuous symmetry, massless scalar bosons appear. His name was given to such bosons, as well as to the theorem.

considered the following doublet:

$$\Phi = \begin{pmatrix} \phi^+ \\ \phi^0 \end{pmatrix} = \begin{pmatrix} \frac{1}{\sqrt{2}}(\phi_1 + i\phi_2) \\ \frac{1}{\sqrt{2}}(\phi_3 + i\phi_4) \end{pmatrix}, \tag{11.7}$$

where $\phi_{1,2,3,4}$ are four real scalar fields. As usual, the most positively charged component is the upper one. The complex field ϕ^+ has $T_3 = +1/2$, while ϕ^0 has $T_3 = -1/2$. Note that since ϕ^0 is a complex scalar field, the neutral particle described by ϕ^0 would be different from its antiparticle. Assuming that the Gell-Mann–Nishijima relation (10.19) is satisfied, the hypercharge[4] of the fields ϕ^+ and ϕ^0 is $Y\Phi = 2(Q - T_3)\Phi$, i.e.

$$\boxed{y_\Phi = 1.} \tag{11.8}$$

We still consider the Lagrangian

$$\mathcal{L} = \partial_\mu \Phi^\dagger \, \partial^\mu \Phi - V(\Phi), \tag{11.9}$$

where V now reads

$$\begin{aligned} V(\Phi) &= \mu^2 \phi^\dagger \Phi + \lambda(\Phi^\dagger \Phi)^2, \qquad \mu^2 < 0, \\ &= \frac{\mu^2}{2}(\phi_1^2 + \phi_2^2 + \phi_3^2 + \phi_4^2) + \frac{\lambda}{4}(\phi_1^2 + \phi_2^2 + \phi_3^2 + \phi_4^2)^2. \end{aligned} \tag{11.10}$$

This time, the potential V lies in a four-dimensional space (the four independent real scalar fields). The Mexican hat of Fig. 11.1 is thus an illustration in two of these four dimensions.

The Lagrangian (11.9) is invariant under the global transformation $\mathrm{SU}(2)_L \times \mathrm{U}(1)_Y$. Indeed, with Φ carrying both T_3 and Y numbers, its transformation under the gauge group is

$$\Phi \to \Phi' = U e^{-i\alpha \frac{Y}{2}} \Phi, \quad \text{with} \quad U = e^{-i\boldsymbol{\alpha} \cdot \frac{\boldsymbol{\sigma}}{2}},$$

and, therefore,

$$\partial_\mu \Phi'^\dagger \, \partial^\mu \Phi' = \partial_\mu \Phi^\dagger \, \underbrace{U^\dagger U}_{=\mathbb{1}} \, \partial^\mu \Phi = \partial_\mu \Phi^\dagger \, \partial^\mu \Phi,$$

and $\Phi'^\dagger \Phi' = \Phi^\dagger \Phi$. As in the previous section, the ground state (with minimal energy) is given by the classical fields minimising the potential. Equation (11.3) is now replaced with

$$(\phi_1^2 + \phi_2^2 + \phi_3^2 + \phi_4^2) = \frac{-\mu^2}{\lambda} \equiv v^2, \quad \text{i.e.} \quad \Phi^\dagger \Phi = \frac{-\mu^2}{2\lambda} = \frac{v^2}{2}. \tag{11.11}$$

There is again an infinity of ground states. We could arbitrarily choose one of them as *the* ground state. For instance, $\phi_1 = \phi_2 = \phi_4 = 0$ and $\phi_3 = v$. Then, the doublet field corresponding to that ground state is

$$\Phi_{\text{ground state}} = \begin{pmatrix} 0 \\ \frac{v}{\sqrt{2}} \end{pmatrix}. \tag{11.12}$$

[4] Our argument is a bit sloppy. In fact, the Higgs hypercharge must be 1 to keep the Yukawa Lagrangian invariant under a $\mathrm{U}(1)_Y$ transformation (see Section 11.5.1). Consequently, because of the Gell-Mann–Nishijima relation, the upper and lower components of the doublet are positively charged and neutral, respectively. Hence the notation used here.

Why this particular choice? As the charged component $\phi^+ = 0$, the ground state (11.12) is electrically neutral. This choice is consistent with our universe, where the vacuum is electrically neutral. Moreover, we shall see later that this choice preserves the $U(1)_{em}$ gauge invariance of electromagnetism and keeps the photon massless. Note, however, that the ground state violates both the $SU(2)_L$ and $U(1)_Y$ transformations. None of the four generators of the group ($\sigma_{i=[1,3]}$ and Y) keeps the ground state invariant. Indeed, if it were invariant, we would have

$$e^{-i\alpha\Lambda}\Phi_{\text{ground state}} = \Phi_{\text{ground state}},$$

where Λ is $\frac{1}{2}\sigma_{i=1,2,3}$ or $Y/2$. It would imply that $\Lambda\Phi_{\text{ground state}} = 0$. However,

$$\begin{cases} \sigma_1\Phi_{\text{ground state}} = \begin{pmatrix} 0 & 1 \\ 1 & 0 \end{pmatrix}\begin{pmatrix} 0 \\ \frac{v}{\sqrt{2}} \end{pmatrix} = \begin{pmatrix} \frac{v}{\sqrt{2}} \\ 0 \end{pmatrix} \neq 0, \\ \sigma_2\Phi_{\text{ground state}} = \begin{pmatrix} 0 & -i \\ i & 0 \end{pmatrix}\begin{pmatrix} 0 \\ \frac{v}{\sqrt{2}} \end{pmatrix} = \begin{pmatrix} -i\frac{v}{\sqrt{2}} \\ 0 \end{pmatrix} \neq 0, \\ \sigma_3\Phi_{\text{ground state}} = \begin{pmatrix} 1 & 0 \\ 0 & -1 \end{pmatrix}\begin{pmatrix} 0 \\ \frac{v}{\sqrt{2}} \end{pmatrix} = \begin{pmatrix} 0 \\ -\frac{v}{\sqrt{2}} \end{pmatrix} \neq 0, \\ Y\Phi_{\text{ground state}} = \begin{pmatrix} 1 & 0 \\ 0 & 1 \end{pmatrix}\begin{pmatrix} 0 \\ \frac{v}{\sqrt{2}} \end{pmatrix} = \begin{pmatrix} 0 \\ \frac{v}{\sqrt{2}} \end{pmatrix} \neq 0. \end{cases}$$

(Recall that $y_\Phi = 1$.) Nevertheless, note that the generator $Q = T_3 + \frac{Y}{2} = \frac{\sigma_3}{2} + \frac{Y}{2}$ keeps the ground state invariant as

$$(\sigma_3 + Y)\Phi_{\text{ground state}} = \left(\begin{pmatrix} 1 & 0 \\ 0 & -1 \end{pmatrix} + \begin{pmatrix} 1 & 0 \\ 0 & 1 \end{pmatrix}\right)\begin{pmatrix} 0 \\ \frac{v}{\sqrt{2}} \end{pmatrix} = 0.$$

Therefore, even if the ground state does not respect the initial $SU(2)_L \times U(1)_Y$ symmetry of the electroweak Lagrangian, it respects the subgroup of symmetry: $U(1)_{em}$ generated by Q. The vacuum remains neutral. The result of the spontaneous symmetry breaking is summarised by

$$\boxed{SU(2)_L \times U(1)_Y \xrightarrow{\text{spontaneous symmetry breaking}} U(1)_{em}.} \tag{11.13}$$

Now, let us come back to the quantum field and leave the classical field. The quantum field version of the classical field minimum (11.11) is

$$\langle 0|\Phi^\dagger\Phi|0\rangle = \frac{-\mu^2}{2\lambda} = \frac{v^2}{2}.$$

Once again, the vacuum is not *empty* since the field ϕ has a non-zero v.e.v. To identify the particle spectrum of the theory, we still consider small perturbations about the ground state (11.12). The generalisation of the parametrisation (11.5) to the doublet (11.7) is straightforward, yielding

$$\Phi(x) = \frac{1}{\sqrt{2}}e^{i\theta^a(x)\sigma_a}\begin{pmatrix} 0 \\ v + h(x) \end{pmatrix},$$

where a runs from 1 to 3 and the four fields θ_1, θ_2, θ_3 and h are four real scalar fields. For small excitations (keeping only the first order), it follows that

$$\Phi \simeq \frac{1}{\sqrt{2}}(1 + i\theta^a\sigma_a)\begin{pmatrix} 0 \\ v + h \end{pmatrix} = \frac{1}{\sqrt{2}}\begin{pmatrix} 1 + i\theta^3 & i\theta^1 - \theta^2 \\ i\theta^1 + \theta^2 & 1 - i\theta^3 \end{pmatrix}\begin{pmatrix} 0 \\ v + h \end{pmatrix}$$
$$\simeq \begin{pmatrix} 0 \\ \frac{v}{\sqrt{2}} \end{pmatrix} + \begin{pmatrix} -\frac{v}{\sqrt{2}}\theta^2 + i\frac{v}{\sqrt{2}}\theta^1 \\ \frac{h}{\sqrt{2}} - i\frac{v}{\sqrt{2}}\theta^3 \end{pmatrix}.$$

Inserting the expression above into the Lagrangian (11.9), one can show that

$$\mathcal{L} = \frac{1}{2}\left(\partial_\mu h \partial^\mu h - \left[\sqrt{2}|\mu|\right]^2 h^2\right) + \frac{1}{2}\left(\partial_\mu\tilde{\theta}^1\partial^\mu\tilde{\theta}^1 + \partial_\mu\tilde{\theta}^2\partial^\mu\tilde{\theta}^2 + \partial_\mu\tilde{\theta}^3\partial^\mu\tilde{\theta}^3\right) + \cdots,$$

with $\tilde{\theta}^i = v\theta^i$, and where the dots do not contain quadratic terms in $\tilde{\theta}^i$. Consequently, the spontaneous symmetry breaking generates a massive field h with $m_h = \sqrt{2}|\mu|$ and three massless Goldstone bosons. It is worth mentioning that if the symmetry $\mathrm{SU}(2)_L \times \mathrm{U}(1)_Y$ were fully broken, we would have four massless Goldstone bosons. This is due to the Goldstone theorem that states that there are as many Goldstone bosons as the number of broken generators. However, in the present case, we have seen that $\mathrm{U}(1)_{\mathrm{em}}$ remains unbroken. Therefore, only $4 - 1 = 3$ massless Goldstone bosons appear. Since none of the massless quanta (i.e., particles) associated with these Goldstone fields has been discovered, it would be a serious issue, except that these Goldstone bosons are non-physical particles. The field h is commonly called the Higgs field, referring to Peter Higgs (Higgs, 1964), one of the physicists who along with Englert and Brout (1964) and Guralnik, Hagen and Kibble (1964) independently proposed a mechanism to eliminate the three non-physical massless Goldstone bosons. This mechanism is presented in the next section.

11.2 The Brout–Englert–Higgs Mechanism

In a theory that contains gauge fields, we are going to establish that the Goldstone boson can be eliminated, showing that they do not have a physical existence. The electroweak theory presented in Chapter 10 is based on the local gauge symmetry $\mathrm{SU}(2)_L \times \mathrm{U}(1)_Y$. The Lagrangian of scalar fields introduced in Eq. (11.9) is invariant under the global symmetry $\mathrm{SU}(2)_L \times \mathrm{U}(1)_Y$. Imposing local invariance requires substituting ∂_μ with the covariant derivative D_μ, which depends on the gauge fields. Since the scalar doublet ϕ in Eq. (11.7) carries both the weak isospin and the weak hypercharge, the covariant derivative to be applied on the doublet is

$$\mathrm{D}_\mu = \partial_\mu + ig_w\frac{\sigma_i}{2}W^i_\mu + ig\frac{Y}{2}B_\mu, \tag{11.14}$$

where W^1, W^2, W^3 and B are the gauge bosons already introduced in the previous chapter, which transform under the gauge group as in Eqs. (10.41) and (10.43). Then, the Lagrangian (11.9) becomes

$$\mathcal{L}^{\mathrm{EW}}_\Phi = (\mathrm{D}_\mu\Phi)^\dagger\,\mathrm{D}^\mu\Phi - V(\Phi), \quad \text{with} \quad V(\Phi) = \mu^2\Phi^\dagger\Phi + \lambda(\Phi^\dagger\Phi)^2, \quad \mu^2 < 0,\ \lambda > 0. \tag{11.15}$$

The introduction of the gauge bosons also requires adding to the theory the Lagrangian $\mathcal{L}^{\text{EW}}_{\text{gauge}}$ from Eq. (10.46). As $\mu^2 < 0$ and $\lambda > 0$, the minimum of the potential is obtained for a doublet with a non-zero v.e.v. We make the choice[5]

$$\langle 0|\Phi|0\rangle = \begin{pmatrix} 0 \\ \frac{v}{\sqrt{2}} \end{pmatrix}, \quad \text{with} \quad v^2 = -\mu^2/\lambda. \tag{11.16}$$

As in the previous section, this choice eliminates the charged component of the doublet. Note that the v.e.v. does not depend on spacetime: physics is assumed to be invariant under translations. Therefore, v must be a constant. To identify the physical particle spectrum (remember, particles are excitation from a ground state, which is the vacuum), we still use a parametrisation about the minimum. The scalar doublet ϕ then reads

$$\Phi(x) = \frac{1}{\sqrt{2}} e^{i\theta^a(x)\sigma_a} \begin{pmatrix} 0 \\ v + h(x) \end{pmatrix}. \tag{11.17}$$

By construction, the Lagrangian $\mathcal{L}^{\text{EW}}_{\Phi} + \mathcal{L}^{\text{EW}}_{\text{gauge}}$ [defined in (10.46)] is gauge invariant under an $\text{SU}(2)_L \times \text{U}(1)_Y$ transformation, regardless of the chosen vacuum. With a gauge theory, we have, however, the freedom to perform a local transformation of the scalar doublet (11.17) that eliminates the Goldstone bosons. Indeed, we can do the SU(2) transformation

$$\Phi(x) \to \Phi'(x) = U_\alpha(x)\Phi(x) = \frac{1}{\sqrt{2}} U_\alpha(x) e^{i\theta^a(x)\sigma_a} \begin{pmatrix} 0 \\ v + h(x) \end{pmatrix}, \text{ with } U_\alpha(x) = e^{-i\alpha^a(x)\frac{\sigma_a}{2}}, \tag{11.18}$$

such that the α_as are chosen in each spacetime point to satisfy the equation

$$U_\alpha(x) e^{i\theta^a(x)\sigma_a} = \mathbb{1}.$$

As $e^{i\theta^a(x)\sigma_a} = U_{-2\theta}(x)$ is also a matrix of SU(2), and all elements of the group have an inverse, there is always a solution to the previous equation. This particular choice of gauge is called the *unitary gauge*. It is appropriate to reveal the particle spectrum of the theory, keeping only the physical degrees of freedom. In Supplement 11.1, the consequences of this choice are discussed. In the unitary gauge, the scalar doublet takes the simple form

$$\boxed{\Phi(x) = \frac{1}{\sqrt{2}} \begin{pmatrix} 0 \\ v + h(x) \end{pmatrix},} \tag{11.19}$$

satisfying

$$\langle 0|\Phi^\dagger\Phi|0\rangle = \frac{-\mu^2}{2\lambda} = \frac{v^2}{2}, \quad \text{with} \quad \langle 0|h(x)|0\rangle = 0. \tag{11.20}$$

The doublet Φ is usually called the Higgs doublet and $h(x)$ is called the Higgs field, whose quanta represent the excitations of the vacuum associated with the Higgs boson. In many textbooks, the breaking of the electroweak symmetry by a non-zero v.e.v. of the Higgs doublet is qualified as a 'spontaneous symmetry breaking of the local gauge symmetry'. This is an abuse of terminology as, strictly speaking, it has been shown by Elitzur (1975)

[5] All the other fields of the theory, gauge bosons and later fermions, which carry a spin, must have a zero v.e.v., i.e. they cannot be present in the vacuum on average. Otherwise, as spin is sensitive to rotation, the vacuum would have a non-zero angular momentum, and the invariance of space under rotation would be broken. The only fields that can have a non-zero v.e.v. are scalar fields.

that a local gauge symmetry cannot be spontaneously broken.[6] The $U(1)_{em}$ gauge symmetry remains, however, unbroken and, as a consequence of Noether's theorem, the electric charge is conserved.

11.3 Generation of Boson Masses

In Lagrangians, field masses are generated by quadratic terms in the fields when kinetic terms are also present. The kinetic terms are brought by $\mathcal{L}^{EW}_{gauge}$ in Eq. (10.46). The quadratic terms must then be read from $\mathcal{L}^{EW}_{\Phi}$ in Eq. (11.15). Using the expression of the Pauli matrices and recalling that the Higgs hypercharge satisfies $Y\phi = \phi$, the covariant derivative (11.14) in $\mathcal{L}^{EW}_{\Phi}$ is explicitly

$$\mathrm{D}_\mu \Phi = \left(\partial_\mu + i g_w \frac{\sigma_i}{2} W^i_\mu + i g \frac{Y}{2} B_\mu\right)\Phi = \begin{pmatrix} \partial_\mu + i\frac{g_w}{2}W^3_\mu + i\frac{g}{2}B_\mu & i\frac{g_w}{2}(W^1_\mu - iW^2_\mu) \\ i\frac{g_w}{2}(W^1_\mu + iW^2_\mu) & \partial_\mu - i\frac{g_w}{2}W^3_\mu + i\frac{g}{2}B_\mu \end{pmatrix}\Phi.$$

With the insertion of the explicit expression (11.19) of the field Φ and the definition of $W^\pm$ from $W^{1,2}$ [Eq. (10.15)], it follows that

$$\mathrm{D}_\mu \Phi = \begin{pmatrix} i\frac{g_w}{2}W^+_\mu(v+h) \\ \frac{1}{\sqrt{2}}\left[\partial_\mu h - i(v+h)(\frac{g_w}{2}W^3_\mu - \frac{g}{2}B_\mu)\right] \end{pmatrix}.$$

As W^3 and B are two real fields while $\left(W^\pm\right)^\dagger = W^\mp$, $(\mathrm{D}_\mu\Phi)^\dagger$ reads

$$(\mathrm{D}_\mu\Phi)^\dagger = \left(-i\frac{g_w}{2}W^-_\mu(v+h),\ \ \frac{1}{\sqrt{2}}\left(\partial_\mu h + i(v+h)\left(\frac{g_w}{2}W^3_\mu - \frac{g}{2}B_\mu\right)\right)\right).$$

It follows that the product $(\mathrm{D}_\mu\Phi)^\dagger\,\mathrm{D}^\mu\Phi$ yields

$$\begin{aligned}(\mathrm{D}_\mu\Phi)^\dagger\,\mathrm{D}^\mu\Phi &= \frac{g_w^2}{4}(v+h)^2 W^-_\mu W^{+\mu} + \frac{1}{2}\partial_\mu h\partial^\mu h + \frac{1}{8}(v+h)^2(g_w^2+g^2)\left(\frac{g_w}{\sqrt{g_w^2+g^2}}W^3_\mu \right.\\ &\left. - \frac{g}{\sqrt{g_w^2+g^2}}B_\mu\right)\left(\frac{g_w}{\sqrt{g_w^2+g^2}}W^{3\mu} - \frac{g}{\sqrt{g_w^2+g^2}}B^\mu\right) \\ &= \frac{g_w^2}{4}(v+h)^2 W^-_\mu W^{+\mu} + \frac{1}{2}\partial_\mu h\partial^\mu h + \frac{1}{8}(v+h)^2\frac{g_w^2}{\cos^2\theta_w}Z_\mu Z^\mu,\end{aligned} \tag{11.21}$$

where the expression (10.24) of Z_μ and the definition of the Weinberg angle (10.30) have been used. Now, as $\Phi^\dagger\Phi = (v+h)^2/2$ and using $\lambda = -\mu^2/v^2$ from Eq. (11.20), the Higgs potential reads explicitly

$$V(\Phi) = \mu^2\Phi^\dagger\Phi + \lambda(\Phi^\dagger\Phi)^2 = \frac{\mu^2 v^2}{2} - \mu^2 h^2 - \frac{\mu^2}{v}h^3 - \frac{\mu^2}{4v^2}h^4. \tag{11.22}$$

[6] More specifically, what is broken is the global part of the group made of the set of all constant elements $g(x)$ of $SU(2)_L \times U(1)_Y$, excluding the set of all $g(x)$ such that $g(x)$ approaches the identity at spatial infinity ($|x| \to \infty$); see Nair (2005, chapter 12). Global gauge symmetries, by definition, act trivially (i.e., do not change) on all observables (mass of particles, for example). An interesting philosophical discussion about these subtle notions concerning the spontaneous breaking of global and local theories can be found in Smeenk (2006), Friederich (2013) and Strocchi (2019).

The first term does not depend on any field and has, therefore, no bearing on the Euler-Lagrange equation. We can omit it. Therefore, combining Eqs. (11.21) and (11.22), the Lagrangian $\mathcal{L}_\Phi^{\text{EW}}$ reduces to

$$\begin{aligned}\mathcal{L}_\Phi^{\text{EW}} =& \frac{1}{2}(\partial_\mu h \partial^\mu h + 2\mu^2 h^2) + \frac{g_w^2 v^2}{4} W_\mu^- W^{+\mu} + \frac{g_w^2 v^2}{8\cos^2\theta_w} Z_\mu Z^\mu \\ &+ \frac{g_w^2}{4} h^2 W_\mu^- W^{+\mu} + \frac{g_w^2 v}{2} h W_\mu^- W^{+\mu} + \frac{g_w^2 v}{4\cos^2\theta_w} h Z_\mu Z^\mu + \frac{g_w^2}{8\cos^2\theta_w} h^2 Z_\mu Z^\mu \\ &+ \frac{\mu^2}{v} h^3 + \frac{\mu^2}{4v^2} h^4.\end{aligned} \tag{11.23}$$

The Lagrangian combining the gauge fields and the Higgs field is

$$\mathcal{L}_\Phi^{\text{EW}} + \mathcal{L}_{\text{gauge}}^{\text{EW}} = \mathcal{L}_\Phi^{\text{EW}} + \mathcal{L}_{\text{gauge free}}^{\text{EW}} + \mathcal{L}_{\text{trilinear}}^{\text{EW}} + \mathcal{L}_{\text{quartic}}^{\text{EW}},$$

where these Lagrangians have been defined in Eqs. (10.47)–(10.49) and (11.23). Expanding $\mathcal{L}_\Phi^{\text{EW}} + \mathcal{L}_{\text{gauge free}}^{\text{EW}}$, the Lagrangian $\mathcal{L}_\Phi^{\text{EW}} + \mathcal{L}_{\text{gauge}}^{\text{EW}}$ reads after rearrangement using $W_\mu^- W^{+\mu} = \frac{1}{2}[(W_\mu^+)^\dagger W^{+\mu} + (W_\mu^-)^\dagger W^{-\mu}]$,

$$\begin{aligned}\mathcal{L}_\Phi^{\text{EW}} + \mathcal{L}_{\text{gauge}}^{\text{EW}} =& \frac{1}{2}(\partial_\mu h \partial^\mu h + 2\mu^2 h^2) \\ &- \frac{1}{4}(W_{\mu\nu}^-)^\dagger W^{-\mu\nu} + \frac{1}{2}\left(\frac{g_w v}{2}\right)^2 (W_\mu^-)^\dagger W^{-\mu} \\ &- \frac{1}{4}(W_{\mu\nu}^+)^\dagger W^{+\mu\nu} + \frac{1}{2}\left(\frac{g_w v}{2}\right)^2 (W_\mu^+)^\dagger W^{+\mu} \\ &- \frac{1}{4} Z_{\mu\nu} Z^{\mu\nu} + \frac{1}{2}\left(\frac{g_w v}{2\cos\theta_w}\right)^2 Z_\mu Z^\mu \\ &- \frac{1}{4} A_{\mu\nu} A^{\mu\nu} \\ &+ \frac{g_w^2 v}{2} h W_\mu^- W^{+\mu} + \frac{g_w^2}{4} h^2 W_\mu^- W^{+\mu} + \frac{g_w^2 v}{4\cos^2\theta_w} h Z_\mu Z^\mu + \frac{g_w^2}{8\cos^2\theta_w} h^2 Z_\mu Z^\mu \\ &+ \frac{\mu^2}{v} h^3 + \frac{\mu^2}{4v^2} h^4 \\ &+ \mathcal{L}_{\text{trilinear}}^{\text{EW}} + \mathcal{L}_{\text{quartic}}^{\text{EW}}.\end{aligned} \tag{11.24}$$

We can now derive the Euler–Lagrange equations for the different fields. For the Higgs field (the first line), this leads as before to the Klein–Gordon equation (recalling $\mu^2 < 0$), where the Higgs boson mass is

$$\boxed{m_H = \sqrt{2}|\mu| = \sqrt{2\lambda}\, v.} \tag{11.25}$$

The Euler–Lagrange equation for $X = W^-$, W^+ or Z [the second, third and fourth lines of (11.24), respectively] leads to the Proca equation describing a massive particle of spin 1, i.e. in the absence of interactions,

$$(\square^2 + m_X^2) X^\mu = 0.$$

As for the Klein–Gordon equation, the mass is read from the quadratic term. Therefore, we deduce

$$\boxed{m_{W^\pm} = \frac{g_w v}{2},} \tag{11.26}$$

and

$$\boxed{m_Z = \frac{g_w v}{2\cos\theta_w} = \frac{m_W}{\cos\theta_w}.} \tag{11.27}$$

Note that there is no $A_\mu A^\mu$ term [see the fifth line of (11.24)], meaning that the photon is, as expected, massless

$$\boxed{m_\gamma = 0.} \tag{11.28}$$

Interestingly, Eq. (11.26) gives access to the numerical value of v. Indeed, since g_w and m_W are related to the Fermi constant by Eq. (9.17), one finds using the experimental value of G_F,

$$v^2 = \frac{4m_W^2}{g_w^2} = \frac{1}{\sqrt{2}G_F} \simeq 246 \text{ GeV}. \tag{11.29}$$

The vacuum expectation value of the Higgs doublet is thus

$$\langle 0|\Phi^\dagger\Phi|0\rangle = \frac{v^2}{2} \simeq (174 \text{ GeV})^2. \tag{11.30}$$

It sets the scale of the electroweak symmetry breaking. This energy corresponds to a temperature of roughly $T_c = 2 \times 10^{15}$ K. In the big-bang model of the evolution of the Universe, this phase transition $SU(2)_L \times U(1)_Y \to U(1)_{em}$, where the electroweak interaction splits into the weak force and electromagnetism, occurs at 10^{-11} s when the temperature is about T_c.

We have accomplished a major step: the Higgs mechanism inducing the electroweak symmetry breaking yields the massive gauge bosons of the weak interaction and keeps a massless gauge boson for electromagnetism. Before the symmetry breaking, we had four massless vector bosons (W^1, W^2, W^3 and B) and four scalar fields coming from the Higgs doublet. The Lagrangian $\mathcal{L}_\Phi^{EW} + \mathcal{L}_{gauge}^{EW}$ given by Eqs. (11.15) and (10.46) is also gauge invariant. In terms of degrees of freedom, this represents 4×2 polarisations + 4 scalars = 12. However, with the introduction of the Higgs doublet in the unitary gauge, we fixed the gauge. The Lagrangian $\mathcal{L}_\Phi^{EW} + \mathcal{L}_{gauge}^{EW}$ given by Eq. (11.24) now has mass terms of gauge bosons that would break the gauge invariance of the initial formulation of the Lagrangian. This is what is meant by the *electroweak symmetry breaking*. We then have three massive vector bosons W^+, W^- and Z, one massless vector boson (γ), and one massive scalar boson (H). This gives a total of 3×3 polarisations (recall that $\partial_\mu X^\mu = 0$ in the Procca equation) + 2 polarisations (γ) + 1 scalar = 12 degrees of freedom. Therefore, there is no loss of degrees of freedom, as should be the case, since we have just chosen a particular gauge to express the Lagrangian. The gauge choice shifted the three massless Goldstone modes into the three gauge fields W^+, W^- and Z via their longitudinal polarisations. What is remarkable about this way of giving a mass to the three weak bosons is that at no point we have given up the initial gauge invariance. The underlying gauge symmetry is actually only hidden.

Supplement 11.1. Considerations about the choice of the gauge

In order to eliminate the Goldstone bosons, we chose in Section 11.2 an appropriate gauge, called unitary gauge, that reveals the interpretation of the particle spectrum of the theory as shown in Section 11.3. The fact that the Goldstone bosons can be eliminated just by a judicious choice of the gauge shows they are not physical particles. When a theory is invariant under a gauge symmetry, the physics must not depend on a particular choice of gauge. In other words, we could have decided not to perform the gauge transformation (11.18) and keep the Goldstone bosons. In this case, these unphysical fields would have to be included in the Feynman diagrams with their specific propagators. The gauge bosons themselves would have their propagators modified, but overall, all these modifications would cancel so that the physics would, fortunately, remain the same. The proof of this statement is, however, not trivial at all.

11.4 Higgs Couplings to Bosons

The penultimate line of the Lagrangian (11.24) and the previous one show that the Higgs boson couples to the massive gauge fields via trilinear and quartic terms. The trilinear ones read

$$\mathcal{L}^{\mathrm{EW}}_{hWW} = \frac{g_w^2 v}{2} hW_\mu^- W^{+\mu} = g_w m_W \, hW_\mu^- W^{+\mu}, \tag{11.31}$$

$$\mathcal{L}^{\mathrm{EW}}_{hZZ} = \frac{g_w^2 v}{4\cos^2\theta_w} hZ_\mu Z^\mu = \frac{1}{2}\frac{g_w}{\cos\theta_w} m_Z \, hZ_\mu Z^\mu. \tag{11.32}$$

When proceeding the same as in Section 10.6 (in particular, pay attention to the two identical Z bosons), the vertex factors are given by the diagrams (a) and (b) in Fig. 11.2. The Higgs couplings to gauge bosons can be expressed more symmetrically. Replacing g_w with $2m_w/v$, or equivalently with $2(\sqrt{2}G_F)^{1/2}\, m_W$, gives the couplings

$$c_W = g_w m_w = \frac{2}{v} m_W^2 = 2\left(\sqrt{2}G_F\right)^{1/2} m_W^2$$

$$c_Z = \frac{g_W}{\cos\theta_W} m_Z = \frac{2}{v} m_Z^2 = 2\left(\sqrt{2}G_F\right)^{1/2} m_Z^2,$$

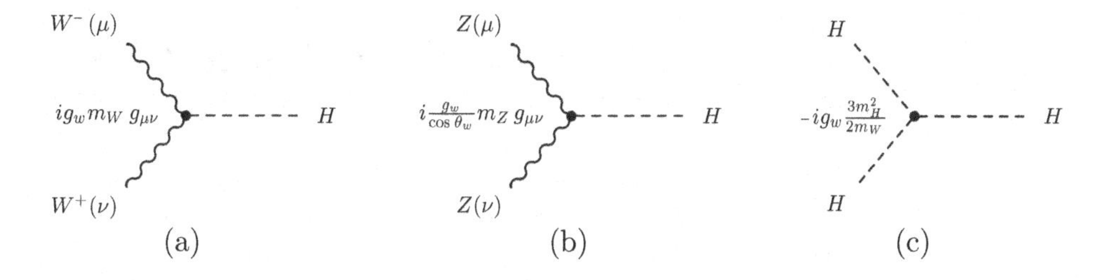

Fig. 11.2 Trilinear couplings with a Higgs boson. (a) The coupling of H with two W bosons. (b) The coupling of H with two Z bosons. (c) The coupling of three Higgs bosons.

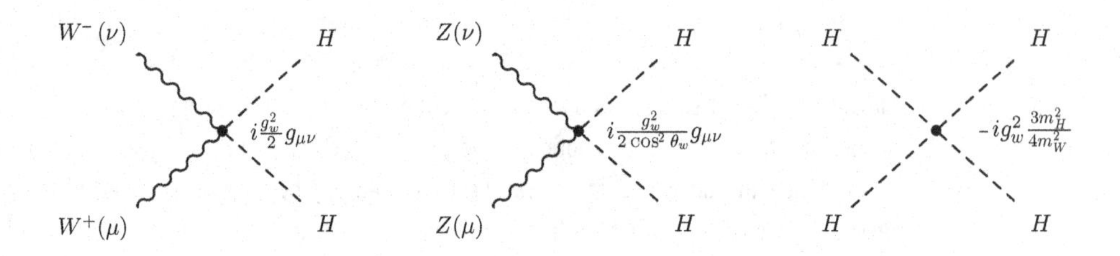

Fig. 11.3 Quartic couplings with a Higgs boson.

where Eq. (11.27) has been used. Therefore, both couplings have a similar form, i.e.

$$c_V = \frac{2}{v}m_V^2 = 2\left(\sqrt{2}G_F\right)^{1/2} m_V^2, \quad V = W, Z. \tag{11.33}$$

They are quadratic in the mass of the gauge bosons. The other trilinear coupling in the Lagrangian (11.24) is the self-coupling of the Higgs boson given by

$$\mathcal{L}_{hhh}^{\mathrm{EW}} = \frac{\mu^2}{v}h^3 = -\lambda v h^3 = -\frac{m_H^2}{2v}h^3 = -g_w\frac{m_H^2}{4m_W}h^3. \tag{11.34}$$

This term corresponds to $3! = 6$ possible permutations of the three identical Higgs, and hence, the vertex factor is six times larger, giving diagram (c) in Fig. 11.2.

Now, let us consider the quartic couplings with a Higgs boson in the Lagrangian (11.24). They read

$$\mathcal{L}_{hhWW}^{\mathrm{EW}} = \frac{g_w^2}{4}h^2 W_\mu^- W^{+\mu}, \tag{11.35}$$

$$\mathcal{L}_{hhZZ}^{\mathrm{EW}} = \frac{g_w^2}{8\cos^2\theta_w}h^2 Z_\mu Z^\mu, \tag{11.36}$$

$$\mathcal{L}_{hhhh}^{\mathrm{EW}} = \frac{\mu^2}{4v^2}h^4 = -\frac{\lambda}{4}h^4 = -g_w^2\frac{m_H^2}{32m_W^2}h^4. \tag{11.37}$$

In Eq. (11.35), there are two possible permutations of H, yielding a vertex factor proportional to $2 \times g_w^2/4$. In Eq. (11.36), we have two permutations for the Higgs and two for the Z, and so the vertex factor is proportional to $2\times2\times g_w^2/(8\cos^2\theta_w)$. Finally, in (11.37), there are $4! = 24$ permutations of Higgs, giving the factor $24 \times (-g_w^2 m_H^2)/(32m_W^2)$. Figure 11.3 summarises these vertex factors.

11.5 Generation of the Fermion Masses

The Higgs mechanism gives rise to mass terms of the weak gauge bosons while keeping the photon massless. It fixes two issues at once by restoring massive bosons and eliminating the unphysical Goldstone bosons. We saw that, technically, this was accomplished through the coupling of the Higgs doublet to the covariant derivative D_μ. However, such a mechanism cannot generate the fermion masses since the fermion fields do not show up in the definition of D_μ. Another mechanism must be at work.

11.5.1 Masses without Fermion Mixing

In this section, for pedagogical reasons, the generation of the fermion masses is exposed by assuming that the weak eigenstates correspond to the mass eigenstates. It does not change the philosophy of the mechanism. In the next section, we do away with this simplification.

In Section 10.1, we explained why a fermion mass term is not invariant under the electroweak gauge transformation: the mass couples the two different chirality states: the left-handed field belonging to an isospin doublet, while the right-handed ones are singlets (which are unaffected under an $SU(2)_L$ transformation). We have to find a way to couple left-handed states to right-handed ones without explicitly breaking the gauge invariance. Well, it turns out that the Higgs doublet in Eq. (11.7) can do this job. For simplicity, let us consider, for the moment, only the first generation of leptons. We couple a doublet of fermions to a fermion singlet and the Higgs doublet via an arbitrary coupling constant g_e (a real number), such that

$$\mathcal{L} = -g_e\overline{L_e}\Phi\, e_R + \text{h.c.} = -g_e(\overline{\nu_{eL}}, \overline{e_L})\begin{pmatrix}\phi^+\\ \phi^0\end{pmatrix} e_R + \text{h.c.} = -g_e(\overline{\nu_{eL}}\phi^+ + \overline{e_L}\phi^0)e_R + \text{h.c.},$$

where $L_e = \binom{\nu_{eL}}{e_L}$. As the Hermitian conjugate of the adjoint fermion field is $\overline{\psi}^\dagger = \gamma^0\psi$, it follows that

$$\mathcal{L} = -g_e\overline{L_e}\Phi\, e_R - g_e\overline{e_R}\Phi^\dagger L_e. \tag{11.38}$$

Let us check the invariance of this Lagrangian under an $SU(2)_L \times U(1)_Y$ transformation. Denoting the eigenvalues of the hypercharge generator Y by y_{L_e}, y_Φ and y_{e_R} for the corresponding doublet and singlet fields, the transformation reads

$$\begin{aligned}\mathcal{L}' \to \mathcal{L}' &= -g_e\left(\overline{e^{-i\alpha^a(x)\frac{\sigma_a}{2}-i\alpha(x)\frac{y_{Le}}{2}}L_e}\right)\left(e^{-i\alpha^a(x)\frac{\sigma_a}{2}-i\alpha(x)\frac{y_\Phi}{2}}\Phi\right)\left(e^{-i\alpha(x)\frac{y_{eR}}{2}}e_R\right) + \text{h.c.}\\ &= -g_e\overline{L_e}\left(e^{-i\alpha^a(x)\frac{\sigma_a}{2}}\right)^\dagger e^{-i\alpha^a(x)\frac{\sigma_a}{2}}e^{i\frac{\alpha(x)}{2}(y_{Le}-y_\Phi-y_{eR})}\Phi\, e_R + \text{h.c.}\\ &= -g_e\overline{L_e}\Phi\, e_R\, e^{i\frac{\alpha(x)}{2}(y_{Le}-y_\Phi-y_{eR})} + \text{h.c.}\end{aligned}$$

The Lagrangian would be invariant if $y_\Phi = y_{L_e} - y_{e_R}$. According to Table 10.1, where the fermion hypercharge values are listed, $y_{L_e} = 1$ and $y_{e_R} = -2$. Therefore, $y_\Phi = +1$, in order to ensure the gauge invariance of the Lagrangian. This justifies the choice made in the definition (11.8), and this implies with the Gell-Mann–Nishijima relation (10.19) that the upper component of the Higgs doublet is positively charged and the lower one is neutral. Note that for all down-type fermions, either lepton $\ell = e, \mu, \tau$ or quark $q = d, s, b$, $y_\Phi = 1 = y_{L_\ell} - y_{\ell_R} = y_{L_q} - y_{q_R}$ in Table 10.1. We shall address the case of the up-type fermions in a moment. Now, in the unitary gauge, $\phi^+ \to 0$ and $\phi^0 \to (v+h)/\sqrt{2}$, and therefore, the Lagrangian (11.38) now reads

$$\begin{aligned}\mathcal{L} &= -g_e(\overline{\nu_{eL}}, \overline{e_L})\begin{pmatrix}0\\ \frac{1}{\sqrt{2}}(v+h)\end{pmatrix} e_R - g_e\overline{e_R}\left(0, \frac{1}{\sqrt{2}}(v+h)\right)\begin{pmatrix}\nu_{eL}\\ e_L\end{pmatrix}\\ &= -\frac{g_e}{\sqrt{2}}\left(\overline{e_L}e_R + \overline{e_R}e_L\right)(v+h).\end{aligned}$$

As $\overline{e_L}e_R + \overline{e_R}e_L = \bar{e}e$, the Lagrangian is simply

$$\mathcal{L} = -m_e\bar{e}e - \frac{m_e}{v}\bar{e}eh,$$

where

$$m_e = \frac{g_e}{\sqrt{2}}v.$$

With the quadratic term $-m_e\bar{e}e$, we have just created a mass term for the electron! We also conclude that the Higgs boson couples to the electrons with a coupling proportional to the electron mass. A coupling between a scalar and two spin-1/2 fermions (Dirac field) is called a *Yukawa* coupling, referring to the Yukawa model of the strong interaction where the two states of the nucleon were coupled to the pion, a scalar boson (see the introductory part of this book in Section 1.2.3).

We can easily generalise the Lagrangian (11.38) to the other down-type fermions. Let us introduce the notation

$$L_\ell^1 = \begin{pmatrix} \nu_{eL} = \ell_L^1 \\ e_L = \ell_L'^1 \end{pmatrix}, \begin{matrix} \nu_{eR} = \ell_R^1 \\ e_R = \ell_R'^1 \end{matrix} \quad L_\ell^2 = \begin{pmatrix} \nu_{\mu L} = \ell_L^2 \\ \mu_L = \ell_L'^2 \end{pmatrix}, \begin{matrix} \nu_{\mu R} = \ell_R^2 \\ \mu_R = \ell_R'^2 \end{matrix} \quad L_\ell^3 = \begin{pmatrix} \nu_{\tau L} = \ell_L^3 \\ \tau_L = \ell_L'^3 \end{pmatrix}, \begin{matrix} \nu_{\tau R} = \ell_R^3 \\ \tau_R = \ell_R'^3 \end{matrix}$$

$$L_q^1 = \begin{pmatrix} u_L = q_L^1 \\ d_L = q_L'^1 \end{pmatrix}, \begin{matrix} u_R = q_R^1 \\ d_R = q_R'^1 \end{matrix} \quad L_q^2 = \begin{pmatrix} c_L = q_L^2 \\ s_L = q_L'^2 \end{pmatrix}, \begin{matrix} c_R = q_R^2 \\ s_R = q_R'^2 \end{matrix} \quad L_q^3 = \begin{pmatrix} t_L = q_L^3 \\ b_L = q_L'^3 \end{pmatrix}, \begin{matrix} t_R = q_R^3 \\ b_R = q_R'^3 \end{matrix} \tag{11.39}$$

The generalisation reads

$$\mathcal{L}_{\text{Yuk}} = \sum_i -g_{\ell'}^i \overline{L_\ell^i} \Phi \, \ell_R'^i - g_{\ell'}^i \overline{\ell_R'^i} \Phi^\dagger L_\ell^i - g_{q'}^i \overline{L_q^i} \Phi \, q_R'^i - g_{q'}^i \overline{q_R'^i} \Phi^\dagger L_{q'}^i, \tag{11.40}$$

where $g_{\ell'}^i$ and $g_{q'}^i$ are add hoc couplings. The spontaneous symmetry breaking generates the mass terms and the coupling of the fermions to the Higgs. Then, in the unitary gauge, the Lagrangian takes the form

$$\mathcal{L}_{\text{Yuk}} = \sum_i -m_{\ell'}^i \overline{\ell'^i}\ell' - \frac{m_{\ell'}^i}{v}\overline{\ell'^i}\ell'^i h - m_{q'}^i \overline{q'^i}q' - \frac{m_{q'}^i}{v}\overline{q'^i}q'^i h, \quad \text{with} \quad m_{\ell'}^i = \frac{g_{\ell'}^i}{\sqrt{2}}v, \quad m_{q'}^i = \frac{g_{q'}^i}{\sqrt{2}}v.$$

How can we generate the mass of the up-type fermions? We observe that down-type fermions get their mass because the vacuum expectation value, v, of the Higgs isodoublet is located in the lower isospin coordinate ($T_3 = -1/2$). Therefore, if we manage to couple the fermion isodoublet and isosinglet to a scalar isodoublet with v in $T_3 = +1/2$ coordinate, then we would generate masses of up-type fermions through a similar mechanism. The question is thus to identify an object such as $\begin{pmatrix} (v+h)/\sqrt{2} \\ 0 \end{pmatrix}$. In Appendix B, Section B.4, we mention that the fundamental representation of SU(2), $\mathcal{D}^{1/2}$, and the complex conjugate representation, $\mathcal{D}^{*1/2}$, are two equivalent representations, and consequently, the anti-doublet of the doublet $\begin{pmatrix} \alpha \\ \beta \end{pmatrix}$ can be described in $\mathcal{D}^{1/2}$, provided the following transformation with the matrix (B.17), i.e.

$$i\sigma_2 \begin{pmatrix} \alpha^* \\ \beta^* \end{pmatrix} = \begin{pmatrix} 0 & 1 \\ -1 & 0 \end{pmatrix} \begin{pmatrix} \alpha^* \\ \beta^* \end{pmatrix} = \begin{pmatrix} \beta^* \\ -\alpha^* \end{pmatrix}.$$

The upper and lower components of the doublet are thus interchanged (modulo a change of sign). This leads to the charge conjugate formula of the isospin formula mentioned in Supplement 8.1. Hence, let us consider the charge conjugate of the Higgs iso-doublet,

$$\Phi^c = i\sigma_2 \Phi^* = \begin{pmatrix} 0 & 1 \\ -1 & 0 \end{pmatrix} \begin{pmatrix} (\phi^+)^* \\ \phi^{0*} \end{pmatrix} = \begin{pmatrix} \phi^{0*} \\ -\phi^- \end{pmatrix}. \tag{11.41}$$

The quantum numbers of Φ^c are by construction opposite to those of ϕ. In particular,

$$y_{\Phi^c} = -y_\Phi = -1. \tag{11.42}$$

This can be easily checked using the Gell-Mann–Nishijima relation (10.19). In the unitary gauge, after spontaneous symmetry breaking, since $\phi^{0*} = (v+h)/\sqrt{2}$, the charge conjugate doublet has the desired form

$$\Phi^c = \begin{pmatrix} \frac{1}{\sqrt{2}}(v+h) \\ 0 \end{pmatrix}.$$

The generalisation of the Lagrangian (11.40) to all fermions thus reads [still using notations (11.39)]

$$\begin{aligned} \mathcal{L}^{\text{EW}}_{\text{Yuk}} = \sum_i \Big\{ & -g^i_{\ell'} \overline{L^i_\ell} \Phi\, \ell'^i_R - g^i_{\ell'} \overline{\ell'^i_R} \Phi^\dagger L^i_\ell - g^i_\ell \overline{L^i_\ell} \Phi^c\, \ell^i_R - g^i_\ell \overline{\ell^i_R} \Phi^{c\dagger} L^i_\ell \\ & -g^i_{q'} \overline{L^i_q} \Phi\, q'^i_R - g^i_{q'} \overline{q'^i_R} \Phi^\dagger L^i_q - g^i_q \overline{L^i_q} \Phi^c\, q^i_R - g^i_q \overline{q^i_R} \Phi^{c\dagger} L^i_q \Big\}, \end{aligned} \tag{11.43}$$

which after spontaneous symmetry breaking becomes

$$\begin{aligned} \mathcal{L}^{\text{EW}}_{\text{Yuk}} = \sum_i \Big\{ & -m^i_{\ell'} \overline{\ell'^i} \ell' - \frac{m^i_{\ell'}}{v} \overline{\ell'^i} \ell'^i h - m^i_\ell \overline{\ell^i} \ell^i - \frac{m^i_\ell}{v} \overline{\ell^i} \ell^i h \\ & -m^i_{q'} \overline{q'^i} f' - \frac{m^i_{q'}}{v} \overline{q'^i} q'^i h - m^i_q \overline{q^i} q^i - \frac{m^i_q}{v} \overline{q^i} q^i h \Big\}, \\ \text{with} \quad & m^i_\ell = \frac{g^i_\ell}{\sqrt{2}} v, \quad m^i_{\ell'} = \frac{g^i_{\ell'}}{\sqrt{2}} v, \quad m^i_q = \frac{g^i_q}{\sqrt{2}} v, \quad m^i_{q'} = \frac{g^i_{q'}}{\sqrt{2}} v, \end{aligned} \tag{11.44}$$

Let us stress again that the Lagrangian (11.43) respects the $\mathrm{SU}(2)_L \times \mathrm{U}(1)_Y$ symmetry.[7] The symmetry is, however, hidden in Eq. (11.44) by using the unitary gauge where the Higgs doublet is expanded around the ground state.

The coupling of fermions to the Higgs field generates their mass. Note, however, that the masses are not predicted at all: the Yukawa couplings are free parameters, and there are as many couplings as fermion species. In contrast, an important prediction of the model is the coupling of the Higgs boson to fermions, which is necessarily proportional to the fermion masses since, according to Eq. (11.44), the coupling constant is [using Eq. (11.26)]

$$-\frac{m_f}{v} = -g_w \frac{m_f}{2m_W} = -(\sqrt{2}G_F)^{1/2} m_f. \tag{11.45}$$

This yields the Feynman diagram

[7] We have again $y_{\Phi^c} = -1 = y_{L^i_f} - y_{f^i_R}$ in Table 10.1, where $f = \ell$ or q. It ensures the gauge invariance of the Lagrangian part involving up-type fermions.

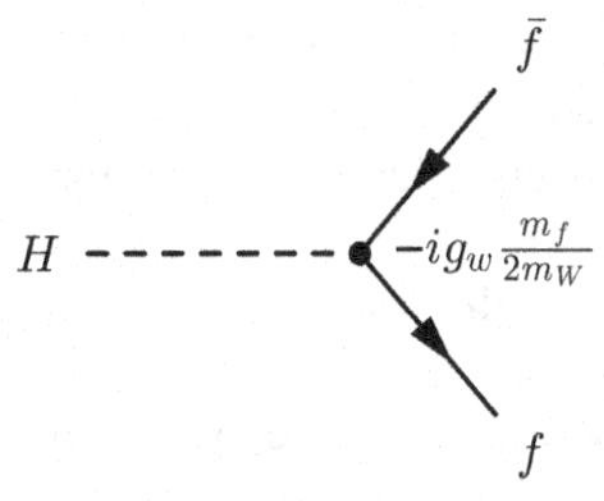

The coupling to fermions is thus linear in the fermion masses, whereas we saw in Eq. (11.33) that it is quadratic in the gauge boson masses.

11.5.2 Masses of Quarks with Quark Mixing

So far, we did not pay attention to the mixing of fermions. Recall that this mixing results from the distinction between the (electro-)weak eigenstates and the mass eigenstates. Using just one Yukawa coupling per fermion in the previous section led to the equality of the two kinds of eigenstates.[8] When the one-particle states produced by the Fock operators of the fields are the eigenstates of the $SU(2)_L \times U(1)_Y$ gauge group, the doublets and singlets remain stable under a rotation in the gauge group space. They transform in a definite way under the symmetry group, i.e. they belong to representations of the symmetry group. Let us denote these gauge fields producing such eigenstates with the tilde symbols (we are going to see how they are related to the fields used so far). In the quark sector, we introduce thc notation

$$\tilde{L}_q^i = \begin{pmatrix} \tilde{u}_L^i \\ \tilde{d}_L^i \end{pmatrix}, \; \tilde{u}_R^i, \; \tilde{d}_R^i,$$

with $\tilde{u}^{i=1,2,3} = \tilde{u}, \tilde{c}, \tilde{t}$ and $\tilde{d}^{i=1,2,3} = \tilde{d}, \tilde{s}, \tilde{b}$. The most general gauge-invariant Yukawa Lagrangian that can be built with these fields is

$$\mathcal{L}_{\text{Yuk}}^{\text{EW}}(q) = \sum_{j,k=1}^{3} -g_{jk}^d \overline{\tilde{L}_q^j} \Phi \, \tilde{d}_R^k - g_{jk}^u \overline{\tilde{L}_q^j} \Phi^c \, \tilde{u}_R^k + \text{h.c.}, \tag{11.46}$$

where g_{jk}^d and g_{jk}^u are 18 Yukawa couplings. The Lagrangian after spontaneous symmetry breaking and after the expansion of the quark doublets reads

$$\mathcal{L}_{\text{Yuk}}^{\text{EW}}(q) = \sum_{j,k=1}^{3} -\frac{g_{jk}^d}{\sqrt{2}} \overline{\tilde{d}_L^j} \tilde{d}_R^k (v+h) - \frac{g_{jk}^u}{\sqrt{2}} \overline{\tilde{u}_L^j} \tilde{u}_R^k (v+h) + \text{h.c.}$$

Defining

$$m_{jk}^u = g_{jk}^u \frac{v}{\sqrt{2}}, \quad m_{jk}^d = g_{jk}^d \frac{v}{\sqrt{2}}, \tag{11.47}$$

so that

$$\mathcal{L}_{\text{Yuk}}^{\text{EW}}(q) = \sum_{j,k=1}^{3} -m_{jk}^d \overline{\tilde{d}_L^j} \tilde{d}_R^k - m_{jk}^u \overline{\tilde{u}_L^j} \tilde{u}_R^k - \frac{m_{jk}^d}{v} \overline{\tilde{d}_L^j} \tilde{d}_R^k h - \frac{m_{jk}^u}{v} \overline{\tilde{u}_L^j} \tilde{u}_R^k h + \text{h.c.},$$

[8] See, however, the footnote 6 of Chapter 9 of this book, for the notion of quark mass eigenstate.

leads to the Lagrangian in matrix notation

$$\mathcal{L}^{\text{EW}}_{\text{Yuk}}(q) = -\overline{\tilde{\boldsymbol{u}}_L} M^u \tilde{\boldsymbol{u}}_R - \overline{\tilde{\boldsymbol{d}}_L} M^d \tilde{\boldsymbol{d}}_R - \frac{1}{v}\overline{\tilde{\boldsymbol{u}}_L} M^u \tilde{\boldsymbol{u}}_R\, h - \frac{1}{v}\overline{\tilde{\boldsymbol{d}}_L} M^d \tilde{\boldsymbol{d}}_R\, h + \text{h.c.}, \tag{11.48}$$

where

$$\tilde{\boldsymbol{u}} = \begin{pmatrix} \tilde{u}^1 \\ \tilde{u}^2 \\ \tilde{u}^3 \end{pmatrix}, \quad \tilde{\boldsymbol{d}} = \begin{pmatrix} \tilde{d}^1 \\ \tilde{d}^2 \\ \tilde{d}^3 \end{pmatrix}, \quad M^u = \begin{pmatrix} m^u_{11} & m^u_{12} & m^u_{13} \\ m^u_{21} & m^u_{22} & m^u_{23} \\ m^u_{31} & m^u_{32} & m^u_{33} \end{pmatrix}, \quad M^d = \begin{pmatrix} m^d_{11} & m^d_{12} & m^d_{13} \\ m^d_{21} & m^d_{22} & m^d_{23} \\ m^d_{31} & m^d_{32} & m^d_{33} \end{pmatrix}. \tag{11.49}$$

The mass eigenstates are the ones that would be unmixed in the Lagrangian above, i.e. those having a definite mass by definition. A general $N \times N$ complex matrix can be diagonalised with the aid of two $N \times N$ unitary matrices.[9] Let us introduce U^u_L and U^u_R, the two unitary matrices for M^u, and U^d_L and U^d_R the two unitary matrices for M^d, such that

$$U^u_L M^u U^{u\dagger}_R = \begin{pmatrix} m_u & & \\ & m_c & \\ & & m_t \end{pmatrix} = M^u_{\text{diag}}, \quad U^d_L M^d U^{d\dagger}_R = \begin{pmatrix} m_d & & \\ & m_s & \\ & & m_b \end{pmatrix} = M^d_{\text{diag}}.$$

Therefore, the mass eigenstates are

$$\boldsymbol{u_L} = \begin{pmatrix} u_L \\ c_L \\ t_L \end{pmatrix} = U^u_L \tilde{\boldsymbol{u}}_L, \quad \boldsymbol{u_R} = \begin{pmatrix} u_R \\ c_R \\ t_R \end{pmatrix} = U^u_R \tilde{\boldsymbol{u}}_R, \quad \boldsymbol{d_L} = \begin{pmatrix} d_L \\ s_L \\ b_L \end{pmatrix} = U^d_L \tilde{\boldsymbol{d}}_L, \quad \boldsymbol{d_R} = \begin{pmatrix} d_R \\ s_R \\ b_R \end{pmatrix} = U^d_R \tilde{\boldsymbol{d}}_R. \tag{11.50}$$

In the Lagrangian (11.48), terms such as $\overline{\tilde{\boldsymbol{u}}_L} M^u \tilde{\boldsymbol{u}}_R$ also read $\overline{\boldsymbol{u_L}} M^u_{\text{diag}} \boldsymbol{u_R}$. Recalling that $\overline{u_L} u_R + \overline{u_R} u_L = \bar{u}u$, it follows the following expression of the Lagrangian (11.48):

$$\begin{aligned} \mathcal{L}^{\text{EW}}_{\text{Yuk}}(q) &= -\overline{\boldsymbol{u_L}} M^u_{\text{diag}} \boldsymbol{u_R} - \overline{\boldsymbol{d_L}} M^d_{\text{diag}} \boldsymbol{d_R} - \frac{1}{v}\overline{\boldsymbol{u_L}} M^u_{\text{diag}} \boldsymbol{u_R}\, h - \frac{1}{v}\overline{\boldsymbol{d_L}} M^d_{\text{diag}} \boldsymbol{d_R}\, h + \text{h.c.} &(11.51)\\ &= \sum_{q=u,d,c,s,t,b} -m_q \bar{q}q - \frac{m_q}{v}\bar{q}qh. &(11.52) \end{aligned}$$

Even in the presence of mixing, the Higgs boson coupling to the mass eigenstates is diagonal, i.e. the same flavour is produced in processes such as $H \to q\bar{q}$, with a coupling given by Eq. (11.45).

11.5.3 Connection with the CKM Matrix

In the previous section, we have introduced the possibility of quark mixing. The CKM matrix is related to the unitary matrices already introduced. To identify the CKM matrix, let us express the weak charged current using the mass eigenstates. According to its definition (10.13), for all quarks, we have

$$j^\mu_{cc+} = \sum_{i=1}^{3} \overline{\tilde{L}^i_q} \gamma^\mu \frac{\sigma_+}{2} \tilde{L}^i_q = \sum_{i=1}^{3} \overline{\tilde{u}^i_L} \gamma^\mu \tilde{d}^i_L = \overline{\tilde{\boldsymbol{u}}_L} \gamma^\mu \tilde{\boldsymbol{d}}_L = \overline{\tilde{\boldsymbol{u}}_L} U^{u\dagger}_L U^u_L \gamma^\mu U^{d\dagger}_L U^d_L \tilde{\boldsymbol{d}}_L = \overline{\boldsymbol{u_L}} U^u_L \gamma^\mu U^{d\dagger}_L \boldsymbol{d_L}.$$

[9] For real matrices or Hermitian matrices, only one unitary matrix is needed.

Therefore, in the mass eigenstate basis, the charged current reads

$$j^{\mu}_{cc+} = \overline{\boldsymbol{u}_L}\gamma^{\mu}U^u_L U^{d\dagger}_L \boldsymbol{d}_L.$$

It does not separately depend on the unitary matrices U^u_L and U^d_L but on their product. Comparing this expression to Eq. (9.22), we identify the CKM mixing matrix with

$$V_{\text{CKM}} = U^u_L U^{d\dagger}_L. \tag{11.53}$$

The down-type quark fields with the prime symbols, for example, in Eq. (10.1), are thus simply

$$\boldsymbol{d}'_L = V_{\text{CKM}}\,\boldsymbol{d}_L = U^u_L U^{d\dagger}_L \boldsymbol{d}_L = U^u_L \tilde{\boldsymbol{d}}_L. \tag{11.54}$$

Note that due to the definition (11.53), the CKM matrix is clearly a 3×3 unitary matrix since U^u_L and U^d_L are both unitary. In terms of the three up-type quark mass eigenstates u, c, t, the three quarks doublets are thus

$$\begin{pmatrix} u \\ d' \end{pmatrix}, \quad \begin{pmatrix} c \\ s' \end{pmatrix}, \quad \begin{pmatrix} t \\ b' \end{pmatrix}, \tag{11.55}$$

where the primed quarks are given by the linear combination (11.54). It is customary to leave the three up-type quarks of charge $Q = 2/3$ unmixed and to attribute the mixing exclusively to the quarks with $Q = -1/3$. Obviously, we could have chosen another convention, i.e. instead define $\boldsymbol{u}'_L = V^{\dagger}_{\text{CKM}}\boldsymbol{u}_L$. In this case, the three quark doublets would have been

$$\begin{pmatrix} u' \\ d \end{pmatrix}, \quad \begin{pmatrix} c' \\ s \end{pmatrix}, \quad \begin{pmatrix} t' \\ b \end{pmatrix}.$$

It is worth noting that the electromagnetic current and the weak neutral current are not affected by the change of basis (11.54): they are diagonal in all bases. For example, the electromagnetic current of up-type quarks is

$$j^{\mu} = \frac{2}{3}\overline{\boldsymbol{u}}\gamma^{\mu}\boldsymbol{u} = \frac{2}{3}\left(\overline{\boldsymbol{u}_L}\gamma^{\mu}\boldsymbol{u}_L + \overline{\boldsymbol{u}_R}\gamma^{\mu}\boldsymbol{u}_R\right).$$

On the other hand, inserting Eq. (11.50) yields

$$j^{\mu} = \frac{2}{3}\left(\overline{\tilde{\boldsymbol{u}}_L}U^{u\dagger}_L\gamma^{\mu}U^u_L\tilde{\boldsymbol{u}}_L + \overline{\tilde{\boldsymbol{u}}_R}U^{u\dagger}_R\gamma^{\mu}U^u_R\tilde{\boldsymbol{u}}_R\right) = \frac{2}{3}\left(\overline{\tilde{\boldsymbol{u}}_L}\gamma^{\mu}\tilde{\boldsymbol{u}}_L + \overline{\tilde{\boldsymbol{u}}_R}\gamma^{\mu}\tilde{\boldsymbol{u}}_R\right).$$

Thus, the expression has the same form in both bases. A similar conclusion is obtained for the neutral current.

11.5.4 Masses of Charged Leptons with Lepton Mixing

The generation of the mass of the charged leptons follows exactly the same principle as for the quarks. The lepton field eigenstates of the gauge group are

$$\tilde{L}^i_{\ell} = \begin{pmatrix} \tilde{\nu}^i_L \\ \tilde{e}^i_L \end{pmatrix}, \quad \tilde{\nu}^i_R, \quad \tilde{e}^i_R.$$

with $\tilde{\nu}^{i=1,2,3} = \tilde{\nu}_e, \tilde{\nu}_{\mu}, \tilde{\nu}_{\tau}$ and $\tilde{e}^{i=1,2,3} = \tilde{e}, \tilde{\mu}, \tilde{\tau}$. To underline the parallelism with the quark sector, we keep for the moment the right-handed neutrino fields, even if we have shown that they are not charged under the gauge group. In the original formulation of the Standard Model, they were simply ignored. We will come back to this in Section 11.5.5 and Chapter 12. The most general gauge-invariant Yukawa Lagrangian that can be written with these fields is then

$$\mathcal{L}_{\text{Yuk}}^{\text{EW}}(\ell) = \sum_{j,k=1}^{3} -g_{jk}^{e} \overline{\tilde{L}_{\ell}^{j}} \Phi\, \tilde{e}_{R}^{k} - g_{jk}^{\nu} \overline{\tilde{L}_{\ell}^{j}} \Phi^{c}\, \tilde{\nu}_{R}^{k} + \text{h.c.}, \tag{11.56}$$

with the 18 Yukawa couplings g_{jk}^{e} and g_{jk}^{ν}. Introducing the same notations as for quarks, $m_{jk}^{\nu} = g_{jk}^{\nu} \frac{v}{\sqrt{2}}$, $m_{jk}^{e} = g_{jk}^{e} \frac{v}{\sqrt{2}}$, and

$$\tilde{\boldsymbol{\nu}} = \begin{pmatrix} \tilde{\nu}^1 \\ \tilde{\nu}^2 \\ \tilde{\nu}^3 \end{pmatrix}, \quad \tilde{\boldsymbol{e}} = \begin{pmatrix} \tilde{e}^1 \\ \tilde{e}^2 \\ \tilde{e}^3 \end{pmatrix}, \quad M^{\nu} = \begin{pmatrix} m_{11}^{\nu} & m_{12}^{\nu} & m_{13}^{\nu} \\ m_{21}^{\nu} & m_{22}^{\nu} & m_{23}^{\nu} \\ m_{31}^{\nu} & m_{32}^{\nu} & m_{33}^{\nu} \end{pmatrix}, \quad M^{e} = \begin{pmatrix} m_{11}^{e} & m_{12}^{e} & m_{13}^{e} \\ m_{21}^{e} & m_{22}^{e} & m_{23}^{e} \\ m_{31}^{e} & m_{32}^{e} & m_{33}^{e} \end{pmatrix}, \tag{11.57}$$

the diagonalisation of the Lagrangian after spontaneous symmetry breaking in the unitary gauge yields

$$\begin{aligned} \mathcal{L}_{\text{Yuk}}^{\text{EW}}(\ell) &= -\overline{\boldsymbol{n}_L} M_{\text{diag}}^{\nu} \boldsymbol{n}_R - \overline{\boldsymbol{e}_L} M_{\text{diag}}^{e} \boldsymbol{e}_R - \frac{1}{v} \overline{\boldsymbol{n}_L} M_{\text{diag}}^{\nu} \boldsymbol{n}_R\, h - \frac{1}{v} \overline{\boldsymbol{e}_L} M_{\text{diag}}^{e} \boldsymbol{e}_R\, h + \text{h.c.} \\ &= -\overline{\boldsymbol{n}} M_{\text{diag}}^{\nu} \boldsymbol{n} - \overline{\boldsymbol{e}} M_{\text{diag}}^{e} \boldsymbol{e} - \frac{1}{v} \overline{\boldsymbol{n}} M_{\text{diag}}^{\nu} \boldsymbol{n}\, h - \frac{1}{v} \overline{\boldsymbol{e}} M_{\text{diag}}^{e} \boldsymbol{e}\, h, \end{aligned} \tag{11.58}$$

where the lepton mass eigenstates have been introduced:

$$\boldsymbol{n}_L = \begin{pmatrix} \nu_{1L} \\ \nu_{2L} \\ \nu_{3L} \end{pmatrix} = U_L^{\nu}\, \tilde{\boldsymbol{\nu}}_L, \quad \boldsymbol{n}_R = \begin{pmatrix} \nu_{1R} \\ \nu_{2R} \\ \nu_{3R} \end{pmatrix} = U_R^{\nu}\, \tilde{\boldsymbol{\nu}}_R, \quad \boldsymbol{e}_L = \begin{pmatrix} e_L \\ \mu_L \\ \tau_L \end{pmatrix} = U_L^{e}\, \tilde{\boldsymbol{e}}_L, \quad \boldsymbol{e}_R = \begin{pmatrix} e_R \\ \mu_R \\ \tau_R \end{pmatrix} = U_R^{e}\, \tilde{\boldsymbol{e}}_R,$$

with U_L^{ν}, U_R^{ν}, U_L^{e} and U_R^{e} being the four matrices ensuring the diagonalisation of M^{ν} and M^{e}, i.e.

$$M_{\text{diag}}^{\nu} = \begin{pmatrix} m_{\nu_1} & & \\ & m_{\nu_2} & \\ & & m_{\nu_3} \end{pmatrix}, \quad M_{\text{diag}}^{e} = \begin{pmatrix} m_e & & \\ & m_{\mu} & \\ & & m_{\tau} \end{pmatrix}.$$

Here again, by construction, the coupling of the charged leptons to the Higgs boson is proportional to their masses.

The usual neutrino fields $\nu_e, \nu_{\mu}, \nu_{\tau}$, called the flavour neutrinos, are associated with the mass eigenstates of the charged leptons in the weak interaction. As the charged current coupled to the W boson, the current reads for leptons

$$j_{cc+}^{\mu} = \sum_{i=1}^{3} \overline{\tilde{L}_{\ell}^{i}} \gamma^{\mu} \frac{\sigma_+}{2} \tilde{L}_{\ell}^{i} = \overline{\tilde{\boldsymbol{\nu}}_L} \gamma^{\mu} \tilde{\boldsymbol{e}}_L = \overline{\boldsymbol{n}_L} U_L^{\nu} \gamma^{\mu} U_L^{e\dagger} \boldsymbol{e}_L = \overline{U_L^{e} U_L^{\nu\dagger} \boldsymbol{n}_L} \gamma^{\mu} \boldsymbol{e}_L.$$

The PMNS unitary mixing matrix in the lepton sector is then identified with

$$V_{\text{PMNS}} = U_L^{e} U_L^{\nu\dagger}, \tag{11.59}$$

while the flavour neutrino fields are

$$\boldsymbol{\nu}_L = \begin{pmatrix} \nu_{eL} \\ \nu_{\mu L} \\ \nu_{\tau L} \end{pmatrix} = V_{\text{PMNS}}\, \boldsymbol{n}_L = U_L^e U_L^{\nu\dagger} \boldsymbol{n}_L = U_L^e \tilde{\boldsymbol{\nu}}_L. \tag{11.60}$$

The only characteristic that distinguishes the three charged leptons is their mass. So, by definition, the flavour of a charged lepton is identified by its mass (which governs its decay channel, etc.). On the other hand, neutrinos can only be detected indirectly by identifying the charged leptons produced in weak interactions. The neutrino flavour is then, by definition, the flavour of the detected charged leptons, implying that flavour neutrinos are not required to have a definite mass. In terms of the mass (and flavour) eigenstates e, μ, τ, the three leptons doublets are thus $\binom{\nu_e}{e}, \binom{\nu_\mu}{\mu}, \binom{\nu_\tau}{\tau}$.

In the original Standard Model without ν_R, neutrinos are massless, and the PMNS matrix is not needed. Consequently, the matrices U_L^ν and U_R^ν are also superfluous. Therefore, the charged current simply takes the form

$$j^\mu_{cc+} = \overline{\tilde{\boldsymbol{\nu}}_L} \gamma^\mu \tilde{\boldsymbol{e}}_L = \overline{\tilde{\boldsymbol{\nu}}_L} \gamma^\mu U_L^{e\dagger} U_L^e \tilde{\boldsymbol{e}}_L = \overline{\tilde{\boldsymbol{\nu}}_L} U_L^{e\dagger} \gamma^\mu \boldsymbol{e}_L = \overline{U_L^e \tilde{\boldsymbol{\nu}}_L} \gamma^\mu \boldsymbol{e}_L.$$

Flavour neutrinos are simply defined as the linear combination $\boldsymbol{\nu}_L = U_L^e \tilde{\boldsymbol{\nu}}_L$. One can then absorb the coefficients of the U_L^e matrix into a redefinition of the phase of $\tilde{\nu}_L$ fields to match $\boldsymbol{\nu}_L = \tilde{\boldsymbol{\nu}}_L$.

11.5.5 Masses of Neutrinos: At the Edge of the Standard Model

The experimental evidence is now clear: neutrinos do have a mass even if their masses have not been measured yet. The developments exposed in Section 11.5.4, and in particular, the Lagrangian (11.58), lead to the generation of the neutrino masses via a Dirac mass term similar to that of charged leptons, i.e. by assuming that $\nu = \nu_L + \nu_R$, with $\nu_L = P_L \nu$ and $\nu_R = P_R \nu$ (P_L and P_R are the usual chirality projectors). Indeed, Eq. (11.58) is of the form

$$\mathcal{L} = -m\overline{\nu}\nu = -m\,(\overline{\nu_L}\nu_R + \overline{\nu_R}\nu_L). \tag{11.61}$$

In this model, the two chiral components are two aspects of the same Dirac particle, with ν_L being the active component under the gauge group, whereas ν_R is sterile (inactive).

However, note that people are not fully satisfied by the explanation of the origin of neutrino mass coming from a Dirac mass term only. First, in such a scenario, the Yukawa parameter g_{ν_e} would be lower than 10^{-11} (given that the upper limit of the neutrino mass is of the order of 1 eV), while the Yukawa parameter of the charged lepton of the same generation, the electron, member of the same isodoublet, is about five orders of magnitude larger. Second, the necessary introduction of the fields ν_R, carrying no gauge quantum numbers (sterile) and thus not detectable, seems a bit arbitrary. Third, we shall see in Chapter 12 that in the presence of a sterile ν_R, neutrinos offer other possibilities than a Yukawa coupling with the Higgs field to generate their mass.

11.6 Discovery of the Higgs Boson

11.6.1 Context

On 4 July 2012, the ATLAS and CMS experiments at CERN announced the discovery of a new particle, a boson, with a mass of about 125 GeV/c^2. The discovery resulted from the analysis of two years of data-taking at the LHC and appeared in the plots of the invariant mass as a new resonance. In Section 1.5.4, the criteria used in high energy physics to claim a discovery were explained. In short, the probability under the background-only hypothesis of observing the actual number of events must be less than 2.87×10^{-7}. This thus requires a large amount of data, explaining why it took more than two years at the LHC. The importance of this discovery was emphasised one year later, on 8 October 2013, when François Englert and Peter Higgs were awarded the Nobel prize.

According to our current knowledge, the particle discovered has the properties of the long-sought Higgs boson of the Standard Model. The Higgs boson is the phenomenological footprint of the Higgs mechanism, with the Higgs bosons appearing as 'vacuum excitations' of the Higgs field. Since masses play the role of coupling strengths between particles and the Higgs field, the coupling of the Higgs boson to any other particle is predicted to be a function of the particle's mass in the field equations. A phenomenological confirmation of the Higgs mechanism, thus, requires finding the Higgs boson, measuring its quantum numbers (spin, electrical charge, etc.) and finally verifying the dependence of the Higgs coupling with the mass of the particle considered. Once a value for the hypothetical Higgs-boson mass is assumed, predictions for production and decay rates are possible. Based on these predictions, experimental searches of the Higgs boson have been performed, and it took 25 years after the formulation of the Higgs mechanism until a significant mass range could be probed with the start of the operation of the Large Electron Positron Collider (LEP) at CERN in 1989. The search was carried on at the Tevatron proton–anti-proton collider from 2002 to 2011 at Fermilab before the LHC reached sufficient performance, leading to the discovery of the Higgs boson in 2012.

11.6.2 LEP Contributions

The LEP e^+e^- collider operated between 1989 and 2000 with two phases: the LEP1 phase operated at the CM energy $\sqrt{s} = 91$ GeV until 1995, when the LEP2 phase took over during five years at $\sqrt{s}$ varying from 136 to 209 GeV. At e^+e^- colliders, the importance of the various Higgs production mechanisms strongly depends on the CM energy of the collider. Figure 11.4 illustrates the lowest order Feynman diagrams of the main production channels. Up to LEP2 energies, the production is dominated by the ZH production via the 'Higgs-strahlung' reaction, $e^+e^- \to ZH$. The reach for the Higgs mass is then typically of the order of $\sqrt{s} - m_Z - \epsilon$, where ϵ is a few GeV. At future e^+e^- colliders with higher CM energy, the fusion of W or Z bosons production mode would become predominant, with the corresponding cross section scaling as $\ln(s)$.

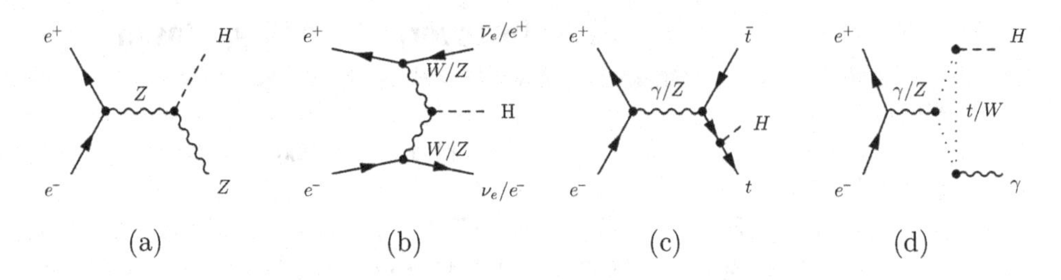

Fig. 11.4 Feynman diagrams (at the lowest order) for the main Higgs-boson production channels at e^+e^- colliders: (a) Higgs-strahlung, (b) vector–boson fusion, (c) top-quark associated production, (d) $H\gamma$ production.

At LEP1, the final state with the largest sensitivity was the reaction $e^+e^- \rightarrow Z \rightarrow Z^*H \rightarrow \nu\bar{\nu}H$, where the symbol Z^* denotes an off-shell Z boson. For m_H larger than 15 GeV, the decay into a pair of b quarks was dominant. The topology was characterised by the presence of missing energy (due to the neutrinos) and the hadronic activity from the Higgs decay, producing two b-jets.[10] At LEP1, the four experiments (ALEPH, DELPHI, OPAL and L3) excluded a Higgs boson with a mass below approximately 64 GeV.

With the increase of the CM energy, the Z in the final state $e^+e^- \rightarrow Z^* \rightarrow ZH$ can be on-shell at LEP2. In contrast to LEP1, the hadronic decays of the Z boson can be used, benefitting from the mass constraint of two jets to be close to m_Z. The most sensitive final state is then $e^+e^- \rightarrow Z^* \rightarrow ZH \rightarrow q\bar{q}\, b\bar{b}$. Even if a slight excess of events was observed by the ALEPH experiment for $m_H = 114$ GeV, this excess was not confirmed by the other LEP experiments, and therefore no striking hints for the production of a Higgs boson were observed at LEP2. Finally, the combination of the four experiments (Barate et al., 2003) excluded, with a 95% confidence level, the Higgs-boson mass hypotheses below 114.4 GeV.

11.6.3 Tevatron Contributions

The Tevatron was a proton–anti-proton collider installed at Fermilab near Chicago, which produced its first accelerated beam in 1983 and ceased operations in 2011. Its CM energy gradually increased from 1.6 TeV to finally reaching 1.96 TeV in 2001. One of its main contributions to high energy physics is the discovery of the top quark by the two experiments CDF and D0 operating at the Tevatron in 1995.

Proton–anti-proton collisions at the Tevatron are dominated by the huge total inelastic cross section due to processes mediated by the strong force. It is more than 10 orders of magnitude higher than the expected cross section for Higgs-boson production as presented in Fig. 11.5. In addition to these large cross sections, the large instantaneous luminosity

[10] The b-jets are jets compatible with the hadronisation of a b quark and thus contain a B hadron. B hadrons are characterised by a high mass that explains why they statistically produce more particles when they decay. In addition, they are long-living particles (because of the CKM coefficients). These two properties justify why b-jets consist of a large number of tracks and why they result from a secondary vertex (a vertex displaced with respect to the main vertex of the collision). They also often contain a charged lepton because of the semi-leptonic decays of the B or C hadrons.

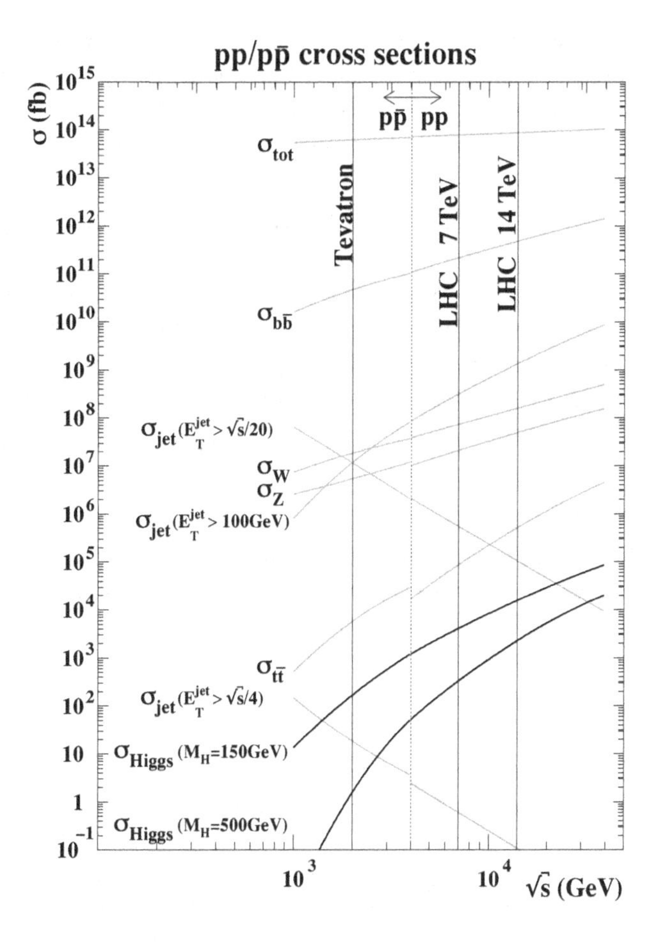

Fig. 11.5 Main signal and background cross sections at hadron colliders (proton–anti-proton for the Tevatron and proton–proton for the LHC). Reprinted with permission from Elsevier (2013).

of hadron colliders implies that not all interactions can be recorded. In order to limit the volume of data recorded, an online selection (called *trigger* in many high energy physics experiments) has to be applied mainly based on the presence of photons, charged leptons or large energy deposits in calorimeters. Hence, even if Higgs-boson topologies with fully hadronic final states lead to the highest production rates, they cannot be exploited. More details will be given in Section 11.6.6 in the context of the LHC where the problem is similar.

At the Tevatron, the delivered luminosity and the Higgs-production cross section were sufficient to be sensitive, at a 95% confidence level, to a Higgs boson with a mass between 90 and 190 GeV/c^2. Over the years, exploiting many decay channels and production modes, the results of the searches of the Higgs boson by the two experiments CDF and D0 were combined in an attempt to reveal the presence of the Higgs boson in the Tevatron data. Unfortunately, only limits were set. For instance, at the date of the Higgs discovery by the LHC, the Tevatron combination excluded the presence of the Higgs boson in the mass range 149–182 GeV$/c^2$ (CDF and D0 Collaboration, 2013). Interestingly, the largest deviation from the background-only hypothesis was observed for a mass of about 120 GeV$/c^2$, very close to 125 GeV$/c^2$ given the mass resolution, with a significance of 3 standard deviations, not enough to claim a discovery.

11.6.4 Constraints from the Self-consistency of the Theory

In Section 11.3, we saw that the mass of the Higgs-boson is a free parameter of the Standard Model. However, indirect constraints on its mass can be derived. The Higgs boson contributes to the W and Z vacuum polarisation through loop effects. The relation between the Z mass and the W mass given in Eq. (11.27) is only true at the tree level. When loops are included,

$$\rho \equiv \frac{m_W^2}{m_Z^2 \cos^2\theta_w} = \frac{1}{1-\Delta\rho}. \tag{11.62}$$

The value $\Delta\rho = 0$ corresponds to the absence of loops since, according to Eq. (11.27), $\rho = 1$ at the tree level. The quantity $\Delta\rho$ receives one-loop corrections that depend quadratically on the top mass (this is by far the dominant contribution) and on the logarithm of the Higgs-boson mass.[11] Those corrections are represented in the diagrams below for the W mass.

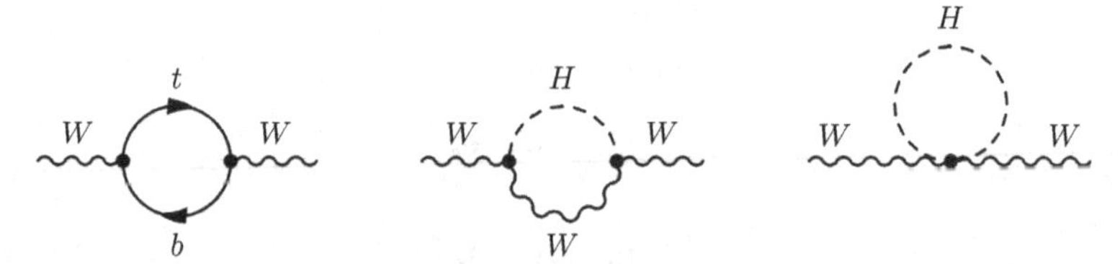

This rather loose sensitivity is exploited to constrain the Higgs mass assuming the validity of the Standard Model. Just before the Higgs discovery, a global fit to precision electroweak data gave in March 2012 (LEP Electroweak Working Group, 2012) $m_H = 94^{+29}_{-24}$ GeV or $m_H < 152$ GeV at a 95% confidence level.

11.6.5 Production of the Higgs Boson at the LHC

The total production cross section at the LHC is about 20 times larger than that at the Tevatron (see Fig. 11.5). Consequently, with an integrated luminosity of about 10 fb^{-1} collected by the two Tevatron experiments, it was expected that the LHC experiments would take over the previous searches with less than 1 fb^{-1}, which was realised in 2011. Indeed, by the end of 2011, with less than 5 fb^{-1} collected at a CM energy $\sqrt{s} = 7$ TeV, the full mass range for the Higgs mass above 130 GeV was excluded. The next 5 fb^{-1} collected in 2012 at $\sqrt{s} = 8$ TeV led to the discovery of the Higgs boson.

In proton–proton collisions at the LHC, the four main Higgs-boson production modes are shown in Fig. 11.6. They are gluon–gluon fusion (ggF), vector–boson fusion (VBF), Higgs-strahlung often called VH (for V = W or Z) and associated production with a top

[11] With two- and three-loop corrections, the dependence is also quadratic in the Higgs-boson mass, but for a light Higgs boson, those corrections remain negligible compared to the one-loop corrections (see Djouadi, 2008).

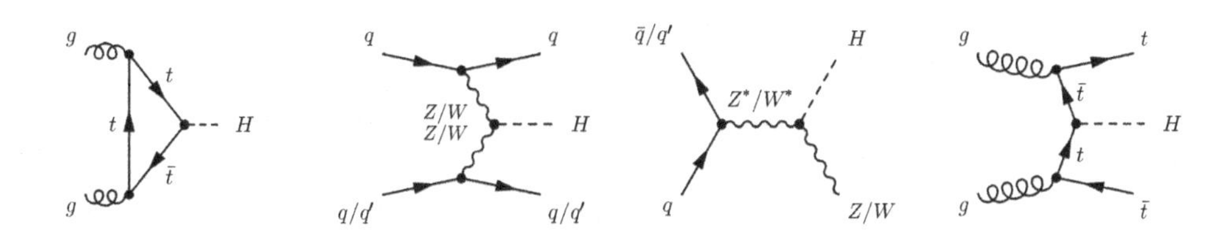

Fig. 11.6 Feynman diagrams of the main Higgs production modes at hadron colliders. From left to right: gluon–gluon fusion (ggF), vector–boson fusion (VBF), Higgs-strahlung (VH) and $t\bar{t}H$.

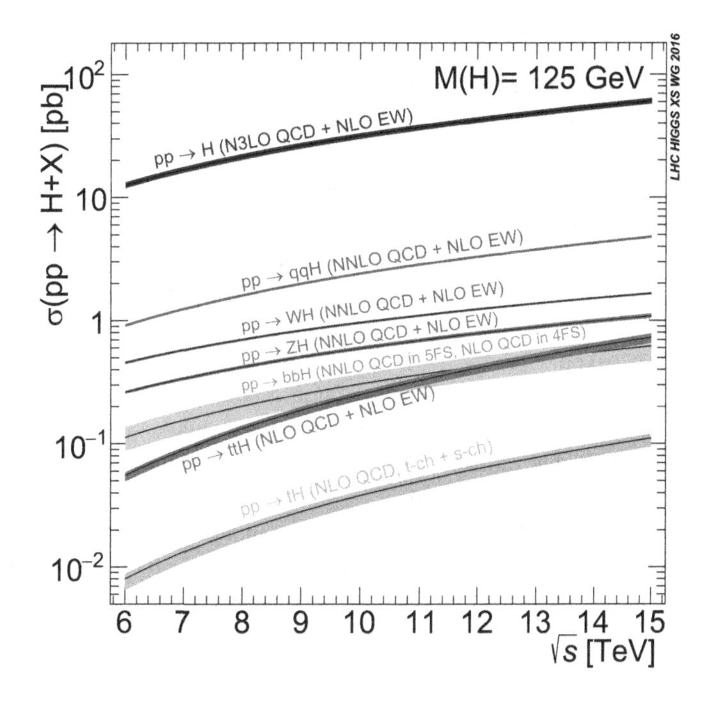

Fig. 11.7 Cross sections of the main Higgs production modes as a function of the CM energy. Source: LHC Higgs Cross Section Working Group (2017). CERN Yellow Reports: Monographs, CERN 2017-002, is licensed under Creative Commons CC BY 4.0.

quark ($t\bar{t}H$). Their expected cross sections for a Higgs boson with a mass of 125 GeV$/c^2$ are presented in Fig. 11.7 as a function of the CM energy.

The main production mode is the gluon–gluon fusion process, in which two gluons of the protons interact to produce a Higgs boson via a loop of heavy quarks. Note that the loop is required since the Higgs boson does not couple directly to the massless gluons. As the Higgs coupling to fermions is proportional to their mass and the top quark is by far the most massive one, the process is dominated at the leading order by a top-quark loop. Although the loop involves an extra factor α_s^2, the ggF cross section remains the largest one because of the large parton density function for gluons (see Fig. 7.8) and the large coupling of the Higgs boson to the top quarks.

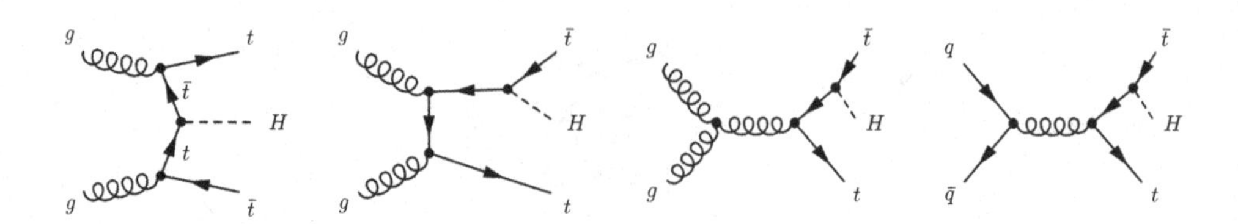

Fig. 11.8 Feynman diagrams contributing to the $t\bar{t}H$ production mode.

For the other production modes, additional particles are produced in association with the Higgs boson. The second most likely production mode is VBF. In this process, two quarks from the protons radiate massive vector bosons, which interact to produce a Higgs boson. At the LHC, the W fusion cross section is approximately three times higher than that of Z fusion because of the different couplings of the quarks to the W and Z bosons. The experimental signature of that production mode is cleaner than the gluon–gluon fusion. Indeed, since that process is purely electroweak and there is no colour exchange between the quarks, the QCD activity is concentrated around the outgoing quarks. Consequently, analyses sensitive to this production mode typically require two forward jets with large pseudorapidity separation and large di-jet invariant mass, with no jet in-between.

In the case of the VH production mode also called Higgs-strahlung, a valence quark interacts with an antiquark from the sea-quarks to produce a massive vector boson (W or Z) that radiates a Higgs boson. The WH cross section is approximately twice as high as the ZH cross section.

The last of the four main production modes corresponds to the associated production of a Higgs boson with a heavy quark, either a t- or b-quark, with the process being referred to as the $t\bar{t}H$ or $b\bar{b}H$ production mode, respectively. Those processes can be initiated either by two incoming gluons, or a $q\bar{q}$ initial state, as presented in Fig. 11.8. They have similar cross sections at $\sqrt{s} = 13$ TeV. Since no loop is involved, unlike the gluon–gluon fusion, the $t\bar{t}H$ production mode enables one to directly probe the Higgs coupling to the top quark. It is interesting to directly measure this coupling because the mass of the top quark is so large that the corresponding Yukawa coupling ($y_t = \sqrt{2}m_t/v$) must be of the order of unity, which is rather atypical. Therefore, a large variety of models beyond the Standard Model predict a modified top-Higgs coupling as compared with the other fermion couplings.

11.6.6 Decay Channels of the Higgs Boson

Since the Higgs boson couples to every massive particle, it can decay in many different channels. It can even couple indirectly, via boson or fermion loops, to photons and gluons. At the LHC, final states containing energetic leptons or photons are often chosen because hadronic channels suffer from a large background due to all of the QCD processes involved in proton–proton collisions (see Fig. 11.5). In addition, the detector resolution of variables

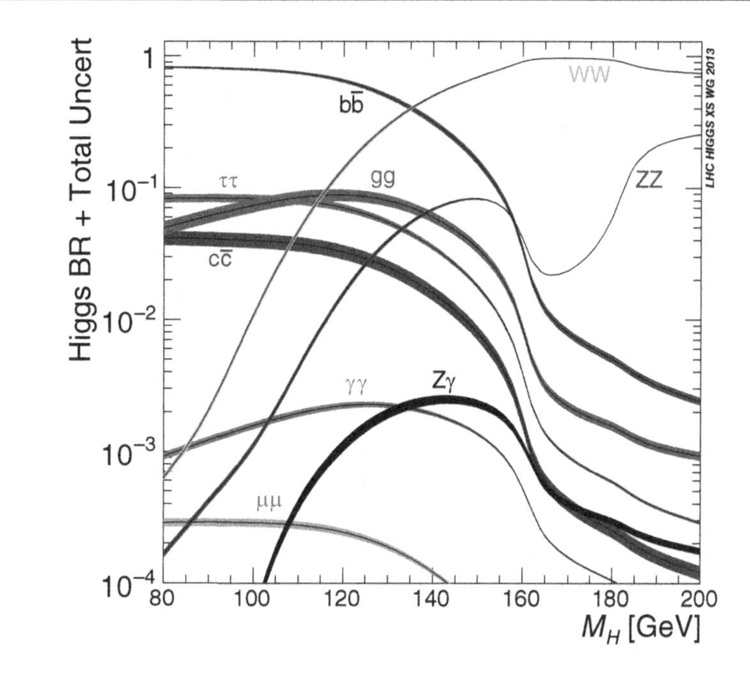

Fig. 11.9 Branching ratios of the Higgs boson. The width of the bands reflects the total uncertainties. Source: LHC Higgs Cross-Section Working Group (2013). CERN Yellow Reports: Monographs, CERN 2013-004, is licensed under Creative Commons CC BY 3.0.

associated with leptons and photons is also usually much better than those related to quarks or gluons.

The branching ratios depend on the Higgs mass and are presented in Fig. 11.9. Given that the Higgs mass is about 125 GeV, the highest branching ratio corresponds to the decay $H \to b\bar{b}$ (57.7%). This decay mode suffers from a high QCD background, which can be partially reduced using dedicated algorithms to identify jets originating from b quarks. However, the QCD production of $b\bar{b}$ pairs represents a very important irreducible background that strongly limits the sensitivity of a search of the $H \to b\bar{b}$ decay targeting the gluon–gluon fusion production mode. Therefore, searches for the $H \to b\bar{b}$ decay mostly focus on the VH production mode (see Section 11.6.5). The presence of leptons and/or missing transverse energy[12] in the final state is then used to select events with $Z \to \ell\ell$, $Z \to \nu\nu$ or $W \to \ell\nu$.

The decay $H \to \tau\bar{\tau}$ represents the other main decay of the Higgs boson into fermions accessible at the LHC, with a branching ratio of 6.4%. However, the resolution on the invariant mass of the Higgs boson is limited in that channel. Indeed, since neutrinos are involved in the decay of the τ leptons, the 4-momenta of the τs cannot be accurately

[12] At hadron colliders, the missing transverse energy is defined as the imbalance in the transverse momentum with respect to the beam axis of all reconstructed particles in the event. It is often used to infer the presence of neutrinos (see Section 1.4.4).

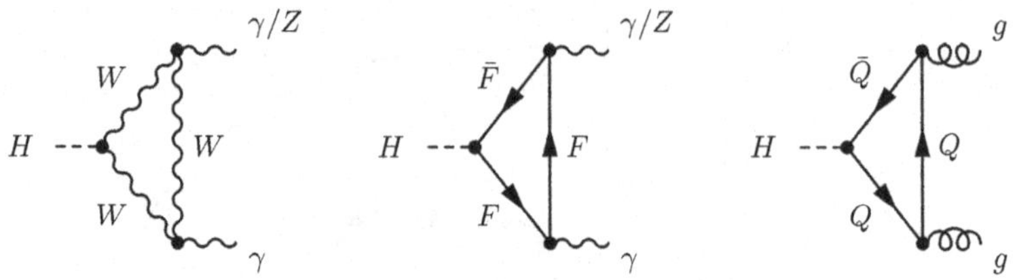

Fig. 11.10 Feynman diagrams of the loop-induced decays $H \to g + g$ and $H \to \gamma + \gamma/Z$. The symbole F denotes a heavy fermion, while Q is a heavy quark.

evaluated, and the reconstructed invariant mass has to take into account the contributions from the missing transverse energy. With the high statistics acquired at the LHC, other decays of the Higgs boson into lighter fermions have been recently accessible, although they suffer either from a very high QCD background ($H \to c\bar{c}$, BR $= 2.7\%$) or from a very low branching ratio ($H \to \mu^-\mu^+$, BR $= 2.2 \times 10^{-4}$).

The Higgs boson also couples to massive electroweak bosons. Therefore, it decays into ZZ (BR $= 2.7\%$) or W^+W^- (BR $= 21.6\%$).[13] Since it turns out that $m_H < 2m_{Z,W}$, only one of the electroweak bosons produced in this decay can be on-shell. Despite its very low branching ratio (1.3×10^{-4}), the decay mode $H \to ZZ \to 4\ell$, where ℓ represents an electron or a muon, was one of the main discovery channels of the Higgs boson in 2012 (ATLAS Collaboration, 2012; CMS Collaboration, 2012b). It benefits from an excellent 4ℓ-invariant mass resolution and a large signal-over-background ratio of about $50\% - 100\%$. For the $H \to WW$ decay channel, the searches mainly focused on the $H \to WW \to 2\ell 2\nu$ decay channel. It has, however, a poor resolution of the Higgs boson mass due to the presence of neutrinos, which can only be measured with low precision using missing transverse energy.

Finally, the Higgs boson can also decay into bosons via loops of fermions (mostly top quarks) or W bosons. Although the Higgs boson does not couple to massless particles as gluons or photons at the tree level, an effective coupling can be generated at the loop level, as shown in Fig. 11.10. This leads to the decays $H \to gg$, $H \to \gamma\gamma$ and $H \to Z\gamma$. Because these decays are loop induced, they are suppressed by two extra powers of the coupling constant (α_s or α) with respect to the tree-level decays described previously. Nevertheless, this suppression is mitigated, given the large coupling of the Higgs boson to massive virtual particles (namely the top quark and the W boson). This is, for instance, the case for the $H \to gg$ decay that has a branching ratio of 8.6%. However, the sensitivity of that decay mode is very low because of its large, irreducible QCD background. In contrast, the decays $H \to \gamma\gamma$ (BR $= 2.3 \times 10^{-3}$) and $H \to Z\gamma$ (BR $= 1.5 \times 10^{-3}$) have

[13] It might be surprising that the branching ratio in WW is larger than in ZZ since we saw in Eq. (11.33) that the coupling with gauge bosons is proportional to the square of the mass of the boson. Naively, we would expect BR(WW) < BR(ZZ). However, two effects contribute to the opposite. First, as $m_H \simeq 125$ GeV$/c^2$, one of the two gauge bosons produced by the Higgs decay must be off-shell. The W being lighter than the Z, it is a bit less off-shell than the W, and its branching ratio is thus less weakened. Second, because the two Zs are indistinguishable, the branching ratio in ZZ is weakened by an extra factor 1/2 (statistics due to permutations).

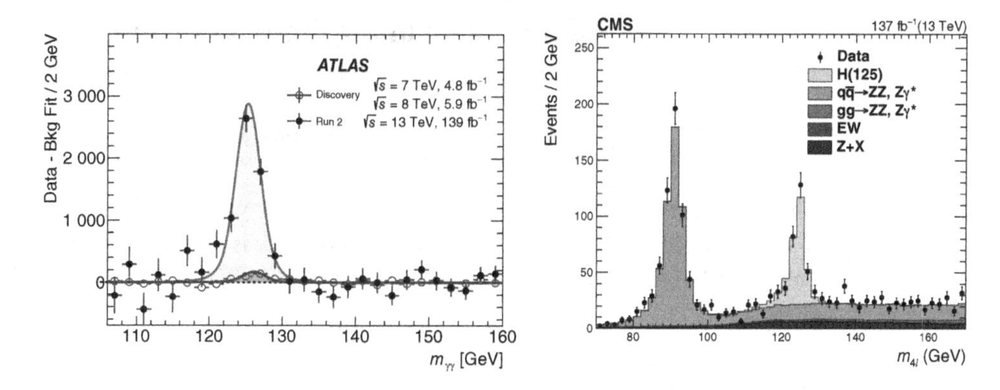

Fig. 11.11 Left: Distribution of the di-photon invariant mass spectrum observed by the ATLAS experiment from 2015 to 2018. Source: ATLAS Collaboration (2022a). The image is licensed under Creative Commons CC BY 4.0. Right: Distribution of the 4ℓ-invariant mass spectrum observed in the CMS experiment from 2016 to 2018. See the text for details. Left source: ATLAS Collaboration (2022a). Right source: CMS Collaboration (2021). The image is licensed under Creative Commons CC BY 4.0.

enough sensitivity to be probed at the LHC. The very high resolution of the di-photon invariant mass helped make the $H \to \gamma\gamma$ decay channel one of the main discovery channels of the Higgs boson in 2012, despite its high background from QCD $\gamma\gamma$ production.

In conclusion, the two main discovery channels at the LHC were $H \to \gamma\gamma$ and $H \to ZZ^* \to 4\ell$. The invariant masses of both the final states are shown in Fig. 11.11, with data collected by the ATLAS and CMS experiments at the LHC from 2015 to 2018 at a CM energy of $\sqrt{s} = 13$ TeV. Similar plots were produced in 2012 by both collaborations for the discovery announcement but with a much smaller dataset (about 30 times fewer statistics). The residuals of the di-photon invariant mass spectrum (i.e., the data from the ATLAS experiment from which the expected background has been subtracted) are represented with the solid points in the left plot of Fig. 11.11. The curve in the solid line with the shaded area corresponds to the expected Higgs boson signal. One can appreciate the improvement since the time of the discovery, with the data represented by the open circles and the signal model fitting with the lower curve. In the right plot of Fig. 11.11, the data from the CMS experiment (the solid points with error bars) are compared with the Standard Model Higgs boson expectation for the $H \to ZZ^* \to 4\ell$ decay channel. The Higgs signal predicted for $m_H = 125$ GeV$/c^2$ is depicted by the solid histogram centred at 125 GeV$/c^2$. The other plain histograms represent the estimated background processes. The large peak for a 4ℓ-invariant mass around 91 GeV$/c^2$ is due to the Z^0 boson. Although the Z^0 boson does not couple directly to four leptons, it can decay into two leptons, where one of the two leptons is off-shell and emits a photon that splits into two new charged leptons. This radiative process then allows the reaction $Z^0 \to 4\ell$. The statistical significance of both plots is now far above the 5 standard deviations threshold required to claim a discovery.

11.7 Measurement of the Higgs Boson Properties

11.7.1 Mass and Quantum Numbers

Thanks to the two high mass resolution channels, $H \to \gamma\gamma$ and $H \to ZZ^* \to 4\ell$, the Higgs mass measured by the ATLAS and CMS experiments is now known with an accuracy of one per mille. A global combination of the results from Particle Data Group (2022) gives

$$m_H = 125.25 \pm 0.17 \ \text{GeV}/c^2. \tag{11.63}$$

According to Eq. (11.25), the value of μ or equivalently λv^2 in the Higgs scalar potential is thus known.

The observation of the decay of the Higgs boson into two photons also provides several useful indications. The Higgs boson is electrically neutral, and its spin is necessarily an integer: 0, 1 or 2. It is thus a boson. We also note that photons being C-odd eigenstates of the charge conjugation, if C-conservation is assumed in Higgs decays, the observed Higgs boson is C-even. As photons are massless, the Landau–Yang theorem (Landau, 1948; Yang, 1950) excludes that the Higgs boson is a vector boson. Both ATLAS and CMS have investigated several models with $J = 0, 1, 2$ and an odd or even parity. The comparison of the angular distribution of the Higgs decay products ($\gamma\gamma, ZZ^* \to 4\ell, WW^* \to 2\nu 2\ell$) expected by those models to the observed distributions shows that the data are compatible with $J^{PC} = 0^{++}$, i.e. the quantum numbers of the Standard Model Higgs boson. Alternative numbers are excluded with a confidence level of 99.9% (Particle Data Group, 2022).

11.7.2 Couplings

One of the strong features of the Higgs boson in the Standard Model is the dependence of the coupling to massive particles on the mass of the particles. More precisely, we saw that the coupling to massive fermions f is linear in the mass [Eq. (11.45)], while the coupling to massive gauge boson $V = W$ or Z is quadratic [Eq. (11.33)], i.e.

$$c_f^{\text{SM}} = \frac{m_f}{v}, \qquad c_V^{\text{SM}} = \frac{2}{v} m_V^2.$$

In other words, the masses of fermions and bosons divided by v are expected to scale respectively as $m_f/v = c_f^{\text{SM}}$ and $m_V/v = \sqrt{c_V^{\text{SM}}/(2v)}$. It is convenient to introduce coupling modifiers defined as $\kappa_f = c_f^{\text{meas}}/c_f^{\text{SM}}$ and $\kappa_V = c_V^{\text{meas}}/c_V^{\text{SM}}$ to parametrise possible deviations of the couplings with respect to the Standard Model expectations. Experiments typically measure the product of a production cross section by a branching ratio, from which the products of coupling modifiers can be constrained. Measurements in perfect agreement with the Standard Model would correspond to $\kappa_f = \kappa_V = 1$. However, in the general case,

$$\begin{aligned} c_f^{\text{meas}} &= \kappa_f c_f^{\text{SM}} = \kappa_f \frac{m_f}{v}, \\ \sqrt{\frac{c_V^{\text{meas}}}{2v}} &= \sqrt{\kappa_V}\sqrt{\frac{c_V^{\text{SM}}}{2v}} = \sqrt{\kappa_V}\frac{m_V}{v}. \end{aligned}$$

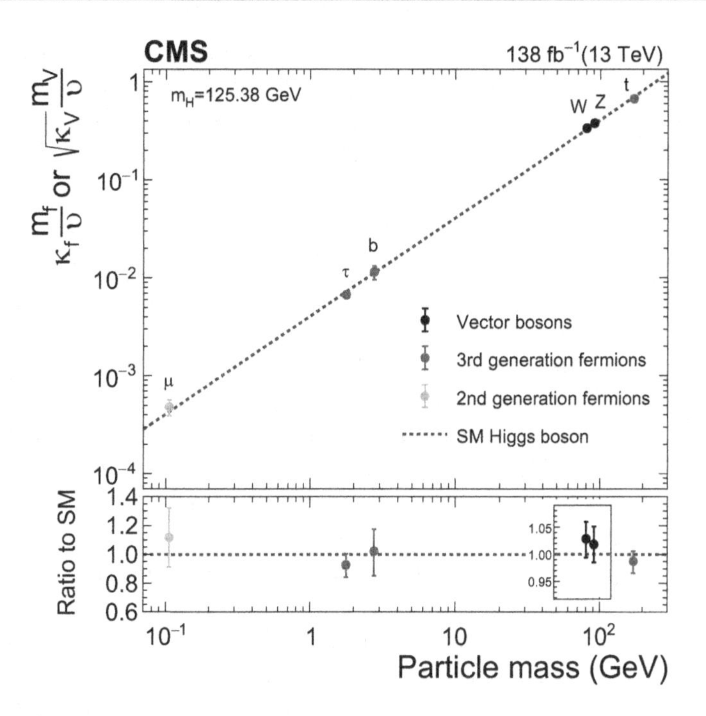

Fig. 11.12 Best-fit values of the parametrised Higgs couplings (see text) as a function of particle masses. The dashed line indicates the predicted dependence on the particle mass in the case of the Standard Model Higgs boson. The lower panel shows the ratios of the measured coupling modifiers values to their Standard Model predictions. Source: CMS Collaboration (2022): CMS Collaboration, 2022, Nature, **607**, 60–68, is licensed under Creative Commons CC BY.

In particular, plotting the measured quantities c_f^{meas} and $\sqrt{c_V^{\text{meas}}/2v}$ as a function of the mass of the particles should yield a perfect linear dependence in the case of the Standard Model (with a slope $1/v$), whereas deviations would break this linear dependence. Figure 11.12 presents the results from the CMS experiment (CMS Collaboration, 2022) with the data collected from 2016 to 2018 at a CM energy $\sqrt{s} = 13$ GeV. It is worth noticing that the coupling of the Higgs boson to fermions of the second generation (namely muons here) has started to be measured. The coupling to charm quarks should be the next one, as limits have already been obtained (experiments are currently sensitive to couplings about ten times larger than the Standard Model one). So far, all the measured Higgs bosons couplings are in very good agreement with the Standard Model expectations.

11.8 Prospects for Higgs Physics with Future Colliders

With data collected at the LHC last years, an extensive set of combined measurements from ATLAS and CMS of the Higgs-boson production and its decay rates has been published, for instance, in CMS Collaboration (2022) and ATLAS Collaboration (2022b), yielding many constraints on the Higgs couplings to vector bosons and fermions. The five production processes, ggF, VBF, WH, ZH and ttH and the main decay channels, $H \to ZZ^*$, WW^*, $\gamma\gamma$, $\tau^+\tau^-$, $b\bar{b}$ and $\mu^+\mu^-$ have been probed (sometimes with limited statistical power as for $H \to \mu^+\mu^-$). So far, the particle discovered in 2012 by the ATLAS and CMS Collaborations at the LHC has properties consistent with the Higgs boson of the Standard Model within the experimental uncertainties. However, more data will be needed to provide more stringent constraints on the couplings of the various production and decay modes, in particular for the second generation of fermions (probing the couplings of the first generation seems inaccessible at the LHC).

The next big challenge of the coming decades will be the measurement of the self-couplings of the Higgs field at the heart of the Higgs mechanism. The trilinear self-coupling of the Higgs boson in Eq. (11.34) would give an independent measurement of the parameter λ (besides the measurement of the Higgs mass itself). Since the LHC is expected to run until the late 2030s, accumulating up to 3 000 fb^{-1}, the determination of the trilinear coupling seems in the reach of the LHC, in particular, by exploiting events containing the production of two Higgs bosons (the Higgs boson produced in Fig. 11.6 is then off-shell and decays into two Higgs bosons). A measurement of λ at a relative accuracy of about 50% at 1 standard deviation is envisaged. On the other hand, the measurement of the quartic coupling seems unrealistic if this coupling is that of the Standard Model.

Beyond 2040, several projects are under consideration for the next generation of colliders. The Future Circular Collider (FCC) at CERN (Future Circular Collider Study, Vol. 2, 2019) is likely to be the successor of the LHC. If approved, this huge collider of almost 100 km circumference will collide, in a first stage called FCC-*ee*, electron–positron beams with CM energy ranging from 91 to 365 GeV. At the FCC-*ee*, the Higgs production channels are those already displayed in Fig. 11.4 in the context of the LEP collider and are largely dominated by the Higgs-strahlung $e^+e^- \to HZ$. As the initial state is perfectly known (unlike the situation at LHC, where the hard process is a function of the parton density functions), the cross-section σ_{HZ} can be precisely determined, with the presence of the Higgs boson signalled by the recoil of the Z boson, where the Z decays into two charged leptons, i.e.

$$m_H^2 = m_{\text{recoil}}^2 = s + m_Z^2 - 2\sqrt{s}(E_\ell^+ + E_\ell^-).$$

The coupling of the Higgs boson to the Z, g_{HZZ}, will then be measured with excellent accuracy, of the order of 0.2% (Future Circular Collider Study, Vol. 1, 2019). Therefore, the Higgs-boson width can be inferred by counting the number of $e^+e^- \to HZ$ events where the Higgs-boson decays into a pair of two Z bosons ($H \to ZZ^*$), as this number is proportional to

$$\sigma_{HZ} \times \frac{\Gamma(H \to ZZ)}{\Gamma_H} \propto \frac{g_{HZZ}^4}{\Gamma_H}.$$

Accuracy on the width of about 1.3% would then be reached (Future Circular Collider Study, Vol. 1, 2019). The Higgs trilinear self-coupling can also be probed at FCC-*ee*, mostly from indirect measurements via loops of Higgs bosons as shown in the two diagrams (there are others) below:

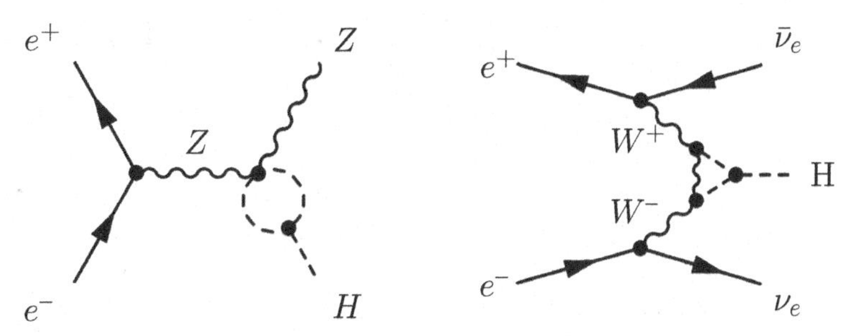

The relative accuracy would be about 44% which, combined with results from the LHC at high luminosity, can be slightly improved.

11.9 Consequences of the Higgs Boson Discovery

11.9.1 Is Such a Mass Natural?

The existence of the Higgs boson as a fundamental scalar particle raises questions. As for any fundamental particle in the Standard Model, the physical mass of a particle m_{phys}, which is measured, corresponds to the pole in the propagator of the particle. However, as soon as higher orders in perturbation theory are considered, the pole mass is not only the mass parameter used in the Lagrangian m_0 (the *bare* mass), but it receives additional corrections δm such as $m_{\text{phys}} = m_0 + \delta m$ [see a quantum field theory course giving more details about the renormalisation procedure, for instance, Peskin and Schroeder (1995)]. These corrections depend on the scale of validity of the Standard Model, Λ, i.e. $\delta m = \delta m(\Lambda)$. It turns out that these corrections to m^2 behave as Λ^2 in the case of the Higgs boson (due to its spin 0), while they behave as $\log(\Lambda)$ for elementary fermions. Therefore, they can drive the Higgs mass to large values if the Standard Model is expected to be valid at very high energy (several hundreds of TeV, for instance). However, we now know that the Higgs mass is about 125 GeV$/c^2$. How to explain such a rather low value compared with Λ? One solution would be to assume that new physics must appear at an energy scale slightly above the electroweak symmetry breaking, let us say about $\Lambda = 1$ TeV. This scenario is not confirmed by the searches performed at the LHC for the moment: there are no signs of the supersymmetry theory nor of the composite Higgs models, which are two examples of theories avoiding quadratic corrections. The other solution assumes that the Standard Model is actually valid at a very high energy scale, but one has to adjust with extreme precision the initial parameter m_0, such as $m_0 + \delta m$ remains of the order of 125 GeV$/c^2$. It requires an extreme fine-tuning of the fundamental parameters of the theory, which is not natural at all, and hence, is not philosophically satisfactory. Indeed, the observable properties of a theory are expected to be stable against minute variations of its fundamental parameters. These two alternatives suggest that the presence of this Higgs boson with such a low mass is probably the sign of something new, but so far, nobody can tell what!

11.9.2 Mass in Modern Physics

Let us first recall how mass is understood in the theory of special relativity. In contrast to Newtonian mechanics, the mass of the system is not a measure of the amount of matter. Quoting Einstein (1905), 'the mass of a body is a measure of the energy contained in it', meaning that the mass is equivalent to the rest energy of a body. As recalled in Chapter 2, the mass is a 4-scalar and thus does not depend on the frame. In Newtonian mechanics, the mass of a body is also a measure of its inertia (and the source of gravitational force). This is no longer the case in special relativity. Indeed, inertia is the tendency of an object to resist any change in its motion, and thus any change in its velocity (i.e., the acceleration). If the inertia only depended on a single number, the mass, we would still have in special relativity the Newtonian relationship between the force and the (usual) acceleration: $\boldsymbol{F} = \frac{\mathrm{d}\boldsymbol{p}}{\mathrm{d}t} = m\boldsymbol{a}$. However, the relation reads in special relativity

$$\boldsymbol{F} = \frac{\mathrm{d}\boldsymbol{p}}{\mathrm{d}t} = \frac{\mathrm{d}(\gamma m \boldsymbol{v})}{\mathrm{d}t} = m\left(\frac{\mathrm{d}\gamma}{\mathrm{d}t}\boldsymbol{v} + \gamma\boldsymbol{a}\right) = m\left(\frac{\gamma^3}{c^2}(\boldsymbol{v}\cdot\boldsymbol{a})\boldsymbol{v} + \gamma\boldsymbol{a}\right).$$

The first term containing $\boldsymbol{v}\cdot\boldsymbol{a}$ implies that the resistance of a body to the force accelerating it depends not only on the mass but also on the angle between the force and the velocity. It should be noticed that both in special relativity and Newtonian mechanics, the mass is an intrinsic property of the particle.

Now, in the light of this chapter, how is the mass of an elementary particle understood? We have seen that it results from the interaction of the particle with the Higgs field. Is it so surprising that a field can create massive particles from particles initially massless? Actually, it is even pretty common. A proton (or any light hadron) is made of valence quarks that can be considered massless compared with the proton mass. Consequently, the proton mass is largely dominated by the QCD potential responsible for the cohesion of hadrons, and hence the mass is largely dominated by the interactions of the hadron constituents with the gluon field. Again, the mass results from an interaction with a field. Most of the mass of ordinary matter is actually pure energy of moving quarks and gluons. Similarly, the interaction of massless particles (gauge bosons and fermions as well) with the Higgs field generates the mass of the particles. Therefore, it is important to realise that this implies that the mass is no longer an intrinsic property of the particle. If the scalar field vacuum expectation value vanishes (as in the early times of the Universe when the electroweak symmetry was not broken), the elementary particles would still be massless. The intrinsic property is now transferred to the coupling to the Higgs field.[14]

11.9.3 Vacuum and Higgs Field

In quantum field theory, the vacuum is a complicated notion. It is not empty at all because of quantum fluctuations – a source of virtual particles that emerge and quickly disappear. We have learned that the Higgs field permeates the Universe in such a way that its vacuum expectation value is different from zero (at any time, anywhere, because the Higgs field

[14] At least for the moment, there is still no theory explaining the values of the Yukawa couplings of fermions. On the other hand, for gauge bosons, the couplings to the Higgs field are a direct consequence of gauge symmetry.

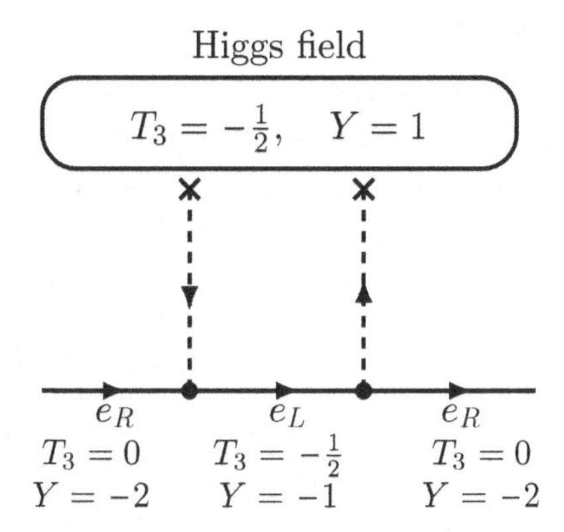

Fig. 11.13 Interaction of the Higgs field condensate (the vacuum expectation value) with a fermion, here, an electron.

is static and infinitely long-lived in the absence of a known dynamic process), allowing elementary particles to acquire a mass. Therefore, the vacuum is not empty, even in the absence of quantum fluctuations (in other words, setting $\hbar \to 0$ cancels the quantum fluctuations but not the Higgs field).

The Higgs field carries non-zero weak isospin and hypercharge quantum numbers: $Y = 1$ and $T_3 = -1/2$. The vacuum is then an unlimited source of these quantum numbers, which are constantly exchanged with the particles moving in, as shown in Fig. 11.13. Only quantum numbers are exchanged, not momentum or energy. Changing T_3 and Y implies changing the nature of the particle: a left-handed state becomes a right-handed one and vice versa, which for massless particles are two independent states and thus two independent particles. The electric charge of the particle remains, however, unchanged since the Higgs field is neutral $Q = T_3 + Y/2 = -1/2 + 1/2 = 0$.

In the jargon of condensed matter physics, such a vacuum is called a condensate: it is a coherent state with an undefined number of particles in the same state, here $T_3 = -1/2$, $Y = 1$, and described by a non-zero field corresponding to the minimal energy. The Universe is no longer symmetric since there are preferred weak charges. However, the electroweak symmetry is only hidden: there is still conservation of the weak hypercharges and isospin via the Higgs condensate, even if, for us, this appears as a mass term that seems to violate the electroweak symmetry. The effect of the condensate is to couple the two chiralities of fermions (the probability to flip the chirality is given by the Yukawa coupling), generating an effective mass even if fermions are originally massless. Recall that the Hamiltonian of a free Dirac particle does not commute with the chirality (see Section 5.3.5). Therefore, the mass eigenstate (i.e., a Hamiltonian eigenstate) is necessarily made of a superposition of left- and right-handed chiralities, precisely what the interaction with the Higgs condensate produces. Consequently, a physical fermion observed with a well-defined mass cannot have well-defined weak charges.

As the Higgs field permeates the entire Universe, it is legitimate to wonder whether it can be considered as an absolute reference frame (a kind of ether). The answer is no. The Higgs field changes some of the underlying properties of the vacuum, i.e. the empty space itself. There is no meaning to asking whether you are moving relative to empty space. The Higgs field does not fill the empty space with *something*: it is a *part of the empty space* itself. Moreover, the vacuum expectation value of the Higgs field is constant everywhere. All elementary particles interact with it in the same way, no matter how they are moving in space and where or when they are moving in space.[15] The Higgs field is thus as unmeasurable as space itself. The only way to interpret the presence of the Higgs field is via its excitations, namely via the observation of Higgs bosons.

Problems

11.1 Check that $\mathcal{L} = \mathring{\phi}^\dagger \mathring{\phi} + \partial_i \phi^\dagger \partial^i \phi - V(\phi)$ leads to the Hamiltonian density $\mathcal{H} = |\mathring{\phi}|^2 + |\nabla\phi|^2 + \mu^2|\phi|^2 + \lambda|\phi|^4$, where ϕ is a complex scalar field.

11.2 Using the definition $v = \sqrt{-\mu^2/\lambda}$, show that with the parametrisation of ϕ given in Eq. (11.5), the potential $V(\phi)$ in Eq. (11.2) reads

$$V(\phi) = -\frac{\lambda}{4}v^4 + \lambda v^2 h^2 + \lambda v h^3 + \frac{\lambda}{4}h^4.$$

Then deduce the expression (11.6) of the Lagrangian (11.1).

11.3 In Section 11.1.2, we saw that the spontaneous symmetry breaking of the $\mathrm{SU}(2)_L \times \mathrm{U}(1)_Y$ symmetry results from the introduction of a weak isospin doublet Φ of hypercharge $Y = 1$ into the Lagrangian $\mathcal{L}_\Phi = (\mathrm{D}_\mu\Phi)^\dagger(\mathrm{D}^\mu\Phi) - V(\Phi)$, with

$$\mathrm{D}_\mu\Phi = \left(\partial_\mu + ig_w\frac{\sigma_i}{2}W_\mu^i + ig\frac{Y}{2}B_\mu\right)\Phi,$$

and where the spontaneous symmetry breaking occurs when $\Phi \longrightarrow \langle\Phi\rangle = \frac{1}{\sqrt{2}}\binom{0}{v}$. This problem identifies the conditions to ensure that photons stay massless.

1. Check that $(\mathrm{D}_\mu\Phi)^\dagger = \partial_\mu\Phi^\dagger - ig_w\Phi^\dagger\frac{\sigma_i}{2}W_\mu^i - i\frac{g}{2}B_\mu\Phi^\dagger$.
2. Develop the kinetic term $(\mathrm{D}_\mu\Phi)^\dagger(\mathrm{D}^\mu\Phi)$ and isolate all terms $\propto A_\mu A^\mu$ contributing to the photon mass term. Check that the photon mass is given by

$$m_\gamma^2 = \frac{v^2}{4}\left(g^2\cos^2\theta_w + g_w^2\sin^2\theta_w - 2g_w g\sin\theta_w\cos\theta_w\right).$$

Conclude that the photon is indeed massless.

3. Let us imagine that the Φ doublet has an unknown hypercharge Y. Justify that the photon mass can be obtained from the previous formula using $g \to gY$, and show that $m_\gamma^2 = v^2e^2(Y-1)^2/4$. Conclude that the photon is massless if, and only if, the non-zero vacuum expectation value of the field Φ is carried by the electrically neutral component, i.e., if and only if, the Higgs boson is neutral.

11.4 We study the Higgs boson decaying into a pair of fermions: $H \to f\bar{f}$. For labelling, use p_1 for f and p_2 for $\bar{f}$. The mass m_f of the fermion is not neglected.

1. Draw the Feynman diagram and determine its amplitude.

[15] In many attempts to explain the Higgs mechanism to non-experts, the Higgs field is presented as a source of friction, justifying why particles cannot move at the speed of light, and hence, accounting for their apparent mass. A limitation of this analogy is that for particles at rest, there is no friction at all.

2. Show that the squared average amplitude is $\overline{|\mathcal{M}|^2} = 4\sqrt{2}m_f^2 G_F(p_1 \cdot p_2 - m_f^2)\mathcal{N}_c(f)$, where $\mathcal{N}_c(f)$ is 3 for a quark and 1 for a lepton.
3. Conclude that

$$\Gamma(H \to f\bar{f}) = \frac{G_F}{4\pi\sqrt{2}} M_H\, m_f^2 \left(1 - \frac{4m_f^2}{M_H^2}\right)^{\frac{3}{2}} \mathcal{N}_c(f).$$

12 The Standard Model and Beyond

This concluding chapter recaps what has been learned in the previous chapters about the Standard Model. This model is highly successful in describing particle physics phenomena. Some of its successes are briefly underlined. However, as with any model, it has also its weaknesses that are mentioned. The most important open questions in particle physics are addressed in the second part of the chapter.

12.1 The Standard Model

12.1.1 Gathering All Pieces Together

Let us summarise the Standard Model of particle physics exposed in the previous chapters. The Standard Model is a model based on quantum field theory coupled with gauge theory to describe the interactions. Its gauge group is

$$\boxed{\mathrm{SU}(3)_c \times \mathrm{SU}(2)_L \times \mathrm{U}(1)_Y.} \tag{12.1}$$

Three groups appear in this product. The colour group of QCD is $\mathrm{SU}(3)_c$. It describes the strong interaction by an exchange of colour quantum numbers. The electroweak interaction is based on the gauge group $\mathrm{SU}(2)_L \times \mathrm{U}(1)_Y$. A unified framework developed by Glashow (1961), Weinberg (1967) and Salam (1968) describes this interaction by an exchange of weak isospin quantum numbers for $\mathrm{SU}(2)_L$ and weak hypercharge for $\mathrm{U}(1)_Y$, encompassing electromagnetism and the weak interaction. After spontaneous symmetry breaking due to the presence of the Higgs field with a non-zero vacuum expectation value, the weak interaction and electromagnetism appear as two distinct interactions, the carriers of the weak force ($W^\pm$, Z^0) being massive, while that of electromagnetism (γ) is massless. The gauge group $\mathrm{SU}(3)_c$ is assumed to remain unbroken, with its gauge boson (gluons) being massless. Therefore, the result of the symmetry breaking is

$$\boxed{\mathrm{SU}(3)_c \times \mathrm{SU}(2)_L \times \mathrm{U}(1)_Y \xrightarrow{\text{s.s.b.}} \mathrm{SU}(3)_c \times \mathrm{U}(1)_{\text{em}}.}$$

The gauge fields are vector fields (spin 1). They comprise the eight gluons (g), the three weak bosons (W^+, W^-, Z^0) and the photon (γ). The field symbols are indicated below in brackets:

$$g\ (G_\mu^{a=[1,8]}), \quad W^+\ (W_\mu^+), \quad W^-\ (W_\mu^+), \quad Z^0\ (Z_\mu), \quad \gamma\ (A_\mu).$$

The eight gluons form an octet of $\mathrm{SU}(3)_c$.

Table 12.1 Field content of the Standard Model with their representation under $SU(3)_c \times SU(2)_L \times U(1)_Y$.

Fields	Name	$SU(3)_c \times SU(2)_L \times U(1)_Y$
G	Gluons	$(\mathbf{8}, \mathbf{1}, 0)$
W^i	Gauge bosons of $SU(2)_L$	$(\mathbf{1}, \mathbf{3}, 0)$
B	Gauge boson of $U(1)_Y$	$(\mathbf{1}, \mathbf{1}, 0)$
L_ℓ	Left-handed leptons	$(\mathbf{1}, \mathbf{2}, -1)$
$\nu_{\ell R}$	Right-handed neutrinos	$(\mathbf{1}, \mathbf{1}, 0)$
ℓ_R	Right-handed charged leptons	$(\mathbf{1}, \mathbf{1}, -2)$
L_q	Left-handed quarks	$(\mathbf{3}, \mathbf{2}, 1/3)$
u_R	Right-handed up-quark type	$(\mathbf{3}, \mathbf{1}, 4/3)$
d_R	Right-handed down-quark type	$(\mathbf{3}, \mathbf{1}, -2/3)$
Φ	Higgs doublet	$(\mathbf{1}, \mathbf{2}, 1)$

The matter fields are described by Dirac fields as spin-1/2 fermions. Their representation depends on the gauge group. In terms of the $SU(2)_L$ representation, the left-handed chiral components of the fermion fields form doublets of weak isospin, while the right-handed chiral components are singlets:

$$L = \begin{pmatrix} \psi_L \\ \psi'_L \end{pmatrix}, \begin{matrix} \psi_R \\ \psi'_R \end{matrix} \quad \left\{ \begin{matrix} \begin{pmatrix} \nu_{eL} \\ e_L \end{pmatrix}, \begin{matrix} \nu_{eR} \\ e_R \end{matrix} & \begin{pmatrix} \nu_{\mu L} \\ \mu_L \end{pmatrix}, \begin{matrix} \nu_{\mu R} \\ \mu_R \end{matrix} & \begin{pmatrix} \nu_{\tau L} \\ \tau_L \end{pmatrix}, \begin{matrix} \nu_{\tau R} \\ \tau_R \end{matrix} \\ \begin{pmatrix} u_L \\ d_L \end{pmatrix}, \begin{matrix} u_R \\ d_R \end{matrix} & \begin{pmatrix} c_L \\ s_L \end{pmatrix}, \begin{matrix} c_R \\ s_R \end{matrix} & \begin{pmatrix} t_L \\ b_L \end{pmatrix}, \begin{matrix} t_R \\ b_R. \end{matrix} \end{matrix} \right. \tag{12.2}$$

Right-handed neutrinos were absent from the original formulation of the Standard Model. Their existence is not yet proven but seems reasonable since neutrino oscillation has shown that they are massive. For the $SU(3)_c$ representations, quark fields are a member of a triplet of colour:

$$q_f = \begin{pmatrix} f_1 \\ f_2 \\ f_3 \end{pmatrix} \quad \left\{ \begin{pmatrix} u_1 \\ u_2 \\ u_3 \end{pmatrix} \begin{pmatrix} d_1 \\ d_2 \\ d_3 \end{pmatrix} \begin{pmatrix} s_1 \\ s_2 \\ s_3 \end{pmatrix} \begin{pmatrix} c_1 \\ c_2 \\ c_3 \end{pmatrix} \begin{pmatrix} b_1 \\ b_2 \\ b_3 \end{pmatrix} \begin{pmatrix} t_1 \\ t_2 \\ t_3 \end{pmatrix} \right. . \tag{12.3}$$

The index 1, 2 or 3 refers to the colour index.

Finally, the last particle of the Standard Model is the Higgs boson, the only scalar of the theory. It is a member of an $SU(2)_L$ doublet and carries a hypercharge $y_\Phi = 1$, which reads after spontaneous symmetry breaking in the unitary gauge

$$\Phi = \begin{pmatrix} \phi^+ \\ \phi^0 \end{pmatrix} \xrightarrow{\text{s.s.b.}} \Phi = \begin{pmatrix} 0 \\ \frac{1}{\sqrt{2}}(v+h) \end{pmatrix}. \tag{12.4}$$

Table 12.1 summarises how the fields of the Standard Model transform under the gauge group. For $U(1)_Y$, the number indicated corresponds to the hypercharge value. For $SU(3)_c$ and $SU(2)_L$, the number in bold is the representation labelled by its dimension (so, **1** is the trivial representation, making no change).

The Lagrangian of the Standard Model results from the sum of several Lagrangians:

$$\mathcal{L}^{SM} = \underbrace{\mathcal{L}^{QCD}_{\text{free}} + \mathcal{L}^{QCD}_{\text{int}} + \mathcal{L}^{QCD}_{\text{gauge}}}_{\mathcal{L}^{QCD}} + \underbrace{\mathcal{L}^{\text{EW}}_{\text{free}} + \mathcal{L}^{\text{EW}}_{\text{int}} + \mathcal{L}^{\text{EW}}_{\text{gauge}} + \mathcal{L}^{\text{EW}}_{\Phi} + \mathcal{L}^{\text{EW}}_{\text{Yuk}}}_{\mathcal{L}^{\text{EW}}}, \tag{12.5}$$

with

$$\mathcal{L}^{QCD}_{\text{free}} = \sum_{f=u,d,s,c,b,t} \bar{q}_f(i\gamma^\mu \partial_\mu - m)q_f, \tag{12.6a}$$

$$\mathcal{L}^{QCD}_{\text{int}} = \sum_{f=u,d,s,c,b,t} -g_s\, \bar{q}_f \gamma^\mu \frac{\lambda_a}{2} q_f\, G^a_\mu, \tag{12.6b}$$

$$\mathcal{L}^{QCD}_{\text{gauge}} = -\frac{1}{4} G^{\mu\nu}_a G^a_{\mu\nu}, \tag{12.6c}$$

with the expression of $G^{\mu\nu}_a$ being given in Eq. (8.56). Given this expression, it is the QCD Lagrangian (12.6c) that causes the coupling of gluons to each other.

In the electroweak Lagrangian part,

$$\mathcal{L}^{\text{EW}}_{\text{free}} = \sum_f \overline{L} i \partial\!\!\!/ L + \overline{\psi_R} i \partial\!\!\!/ \psi_R + \overline{\psi'_R} i \partial\!\!\!/ \psi'_R, \tag{12.7a}$$

where the sum over f is understood as a sum over the representation given in Eq. (12.2), and hence, there are six contributions for the leptons and quarks of the three generations. The interaction term reads

$$\begin{aligned}\mathcal{L}^{\text{EW}}_{\text{int}} = -\sum_f \Bigg\{ & \frac{g_w}{\sqrt{2}} \overline{\psi_L} \gamma^\mu V \psi'_L\, W^+_\mu + \frac{g_w}{\sqrt{2}} \overline{\psi'_L} \gamma^\mu V^\dagger \psi_L\, W^-_\mu \\ & + e\left(\overline{\psi}\gamma^\mu Q \psi + \overline{\psi'}\gamma^\mu Q \psi'\right) A_\mu \\ & + \frac{g_w}{\cos\theta_w} \left[\overline{\psi}\gamma^\mu \frac{1}{2}\left(c^f_V - c^f_A \gamma^5\right)\psi + \overline{\psi'}\gamma^\mu \frac{1}{2}\left(c^{f'}_V - c^{f'}_A \gamma^5\right)\psi' \right] Z_\mu \Bigg\},\end{aligned} \tag{12.7b}$$

where $V = V_{\text{CKM}}$ for quarks and $V = \mathbb{1}$ for leptons since we use the neutrino eigenstates of the weak interaction as matter fields in the representation (12.2). For the charged current interaction, the coupling with the $W^\pm$ bosons only involves the left-handed chiral components, the source of the maximal violation of parity. The coefficients of the neutral current c_V and c_A are listed in Table 10.2. As $c_V \neq c_A$ for all particles except neutrinos, parity is also violated by the neutral current but not maximally, i.e. the right-handed chiral component is also coupled to the Z boson (but not as much as the left-handed one). The electroweak gauge field Lagrangian reads

$$\mathcal{L}^{\text{EW}}_{\text{gauge}} = -\frac{1}{4} F^a_{\mu\nu} F^{\mu\nu}_a - \frac{1}{4} B_{\mu\nu} B^{\mu\nu}. \tag{12.7c}$$

Its explicit expression as a function of the physical vector fields ($W^\pm_\mu$, Z_μ and A_μ) results from the sum of Eqs. (10.47)–(10.49). In the scalar sector of the theory, the Lagrangian that includes the scalar potential inducing the spontaneous symmetry breaking is

$$\mathcal{L}^{\text{EW}}_{\Phi} = (\text{D}_\mu \Phi)^\dagger\, \text{D}^\mu \Phi - \left(\mu^2 \Phi^\dagger \Phi + \lambda (\Phi^\dagger \Phi)^2\right) \quad (\text{with } \mu^2 < 0,\ \lambda > 0). \tag{12.8}$$

It is described in detail in Eq. (11.23). The Lagrangian with the Yukawa couplings restoring the fermion masses simply reads after spontaneous symmetry breaking and diagonalisation,

$$\mathcal{L}_{\text{Yuk}}^{\text{EW}} = \sum_f -m_f \overline{\psi}\psi \left(1 + \frac{h}{v}\right) + \sum_{f'} -m_{f'} \overline{\psi}'\psi' \left(1 + \frac{h}{v}\right). \tag{12.9}$$

Finally, the Higgs mechanism restoring the particle masses also ensures that the Standard Model is a renormalisable theory. This was shown by t'Hooft and Veltman, who were awarded the Nobel prize in 1999 for their work on developing a method of renormalising non-Abelian gauge theories.

12.1.2 Summary of Conserved Quantities by the Interactions

Throughout our journey among the interactions, we have encountered several quantities that are conserved or partially conserved by the interactions: the electric charge, the baryon number, the lepton number, parity, etc. Table 12.2 summarises those quantities. Note that one may define a fermion number defined as $F = B + L$, where B and L are the baryon and lepton numbers, respectively. The fermion number is always conserved since both B and L

Table 12.2 Quantities conserved (□) or violated (□) by the different interactions.

Interaction	QED	QCD	Weak
Baryon number	□	□	□
Lepton number	□	□	□
Total isospin I	□†	□	□
Isospin third component I_3	□	□	□
Lepton flavour (l_e, l_μ, l_τ)	□	□	□ and □*
Quark flavour (strangeness, charm number, etc.)	□	□	□
Parity P	□	□	□
Charge conjugation C	□	□	□
CP	□	□	□
Time reversal T	□	□	□
CPT	□	□	□

†Obvious in $\pi^0 \to \gamma\gamma$. Note, however, that I_3 is always conserved by QED since QED current never changes the flavour of the fermions. This can be shown, for example, with the isospin doublet $q = \binom{u}{d}$ by expressing the electromagnetic Lagrangian of the u quark as a function of q, i.e. $\mathcal{L} = \frac{2}{3}e\bar{u}\gamma^\mu u A_\mu = \frac{e}{3}\bar{q}\gamma^\mu(1+\sigma_3)q$, where σ_3 is the third Pauli matrix. If the isospin were conserved, then the Lagrangian would be invariant under the isospin symmetry and thus under the transformation $q \to q' = Uq$, with U being the unitary SU(2) matrix given by $U = e^{-i(\alpha_1\sigma_1+\alpha_2\sigma_2+\alpha_3\sigma_3)}$ (α_i are real numbers). Under this transformation, $\mathcal{L} \to \mathcal{L}' = \frac{e}{3}\bar{q}\gamma^\mu(1 + U^\dagger\sigma_3 U)q$. Generally, however, $U^\dagger\sigma_3 U \neq \sigma_3$ because the Pauli matrices do not commute with each other. So, isospin is not conserved by QED. However, if U just rotates I_3, i.e. $U = e^{-i(\alpha_3\sigma_3)}$, the Lagrangian remains invariant, implying that I_3 is conserved.
*As soon as a neutrino is involved, mixing is possible. See Section 9.4.6.

are individually conserved. The last line of Table 12.2 concerns the CPT symmetry based on the product (in any order) of parity, charge conjugation and time reversal. As all interactions are described by quantum field theory, a general theorem mainly based on Lorentz invariance states that CPT is conserved by quantum field theory. We have already observed in Eq. (9.56) that a consequence of CPT symmetry is the equality of the masses and the lifetimes of a particle and its antiparticle. This prediction has been tested for several particles. One of the most stringent tests comes from kaons (Particle Data Group, 2022) where

$$\frac{\left|m(K^0) - m(\overline{K}^0)\right|}{\frac{1}{2}\left[m(K^0) + m(\overline{K}^0)\right]} < 6 \times 10^{-19}\ \mathrm{GeV}/c^2.$$

Since CP is violated by the weak interaction, T must also be violated, which has been confirmed by the BaBar experiment (USA) in 2012 with the comparison of the rates of a transition in B meson decays with its time-reversed transition.

12.1.3 Successes of the Standard Model

So far, there is almost no significant and persistent sign of deviations between experimental measurements and the predictions of the Standard Model. The adverb *almost* is used because we shall see in the next sections a few hints of possible discrepancies. Many successes have already been reported in the previous chapters: the impressively accurate prediction of the electron anomalous magnetic moment, the prediction of the charm quark (the GIM mechanism), of the Z^0 and $W^\pm$ bosons, of multi-jet events with the emission of gluons, etc. In the last decade of the twentieth century, at the LEP and SLC e^+e^- colliders, very accurate measurements of the hadronic cross section at various energies around the Z^0 mass were obtained, $e^+ + e^- \to Z^0 \to$ hadrons. It determined the so-called Z^0 lineshape, i.e. the shape of the cross section as a function of energy, providing a precise measurement of the Z^0 mass and its total decay width. It gave access to many predictions and measurements, such as the number of light neutrinos, the precise determination of the Weinberg angle and the prediction of the top quark mass. They are briefly mentioned in what follows.

Number of Light Neutrinos

The number of light neutrinos with a mass lower than half of the Z^0 mass is inferred from the measurement of $\Gamma_{\rm inv}$, the invisible partial decay width, i.e. when the Z^0 decays in channels that cannot be detected. The invisible partial decay width is determined by the subtraction of measured quantities, i.e.

$$\Gamma_{\rm inv} = \Gamma_{\rm tot} - \Gamma_{ee} - \Gamma_{\mu\mu} - \Gamma_{\tau\tau} - \Gamma_{\rm had}, \tag{12.10}$$

where $\Gamma_{\rm had}$ is the partial decay width in hadronic channels, $e^+e^- \to Z^0 \to$ hadrons and Γ_{ee}, the partial decay width $e^+e^- \to Z^0 \to e^+e^-$ (and similarly for the other charged leptons). Assuming that only neutrinos contribute to $\Gamma_{\rm inv}$, the number of neutrinos is then

$$N_\nu = \frac{\Gamma_{\rm inv}}{\Gamma_{\nu\nu}} = \frac{\Gamma_{\rm inv}}{\Gamma_{\ell\ell}}\left(\frac{\Gamma_{\ell\ell}}{\Gamma_{\nu\nu}}\right)_{\rm SM}. \tag{12.11}$$

The ratio $\left(\frac{\Gamma_{\ell\ell}}{\Gamma_{\nu\nu}}\right)_{\rm SM}$ is taken from the prediction of the Standard Model (see Problem 10.4 for the evaluation of the relevant quantities). Systematic errors are reduced in the calculation of the ratio compared with $\Gamma_{\nu\nu}$ by itself. The ratio $\Gamma_{\rm inv}/\Gamma_{\ell\ell}$ is taken from measurements of the cross section and branching ratios by assuming lepton universality of the neutral current. More specifically, at the peak of the resonance, the hadronic cross section and the fermion production cross section are

$$\sigma^0_{\rm had} = \frac{12\pi}{m_Z^2}\frac{\Gamma_{ee}\Gamma_{\rm had}}{\Gamma^2_{\rm tot}}, \qquad \sigma^0_{ff} = \frac{12\pi}{m_Z^2}\frac{\Gamma_{ee}\Gamma_{ff}}{\Gamma^2_{\rm tot}}. \tag{12.12}$$

Assuming lepton universality, $\Gamma_{ee} = \Gamma_{\mu\mu} = \Gamma_{\ell\ell}$, while $\Gamma_{\tau\tau} = \Gamma_{\ell\ell}(1+\delta_\tau)$ because of the large tau mass ($\delta_\tau \simeq -0.23\%$). From Eqs. (12.10) and (12.12), it follows that (Problem 12.1)

$$\frac{\Gamma_{\rm inv}}{\Gamma_{\nu\nu}} = \frac{\Gamma_{\rm tot}-\Gamma_{\rm had}}{\Gamma_{\ell\ell}} - 3 - \delta_\tau = \sqrt{\frac{12\pi}{m_Z^2\sigma^0_{\rm had}}\frac{\Gamma_{\rm had}}{\Gamma_{\ell\ell}}} - \frac{\Gamma_{\rm had}}{\Gamma_{\ell\ell}} - 3 - \delta_\tau. \tag{12.13}$$

The value of $\Gamma_{\ell\ell}$ can be inferred from the measured cross sections of σ^0_{ee}, $\sigma^0_{\mu\mu}$ and $\sigma^0_{\tau\tau}$ in Eq. (12.12). The accuracy of the measurement of the cross sections is therefore crucial for a good evaluation of the above ratio. Using the data collected at the LEP collider, the quantities m_Z, $\Gamma_{\rm tot}$, $\sigma^0_{\rm had}$ and $\Gamma_{\ell\ell}$ were evaluated, and the number of light neutrinos was determined by the LEP and SLD Collaboration (2006) and recently updated by Janot and Jadach (2020) to be

$$N_\nu = 2.9963 \pm 0.0074.$$

This number is compatible with the three neutrinos of the Standard Model and excludes other light-active neutrinos (i.e., with $m_\nu < m_Z/2$). Therefore, in the context of the Standard Model, it is reasonable to assume that there are only three generations of fermions.

The Top Mass and the Weinberg Angle

In the same spirit as the measurement of the number of light neutrinos, the e^+e^- colliders provided strong constraints on the top mass before its discovery. These constraints are mostly based on precise measurements of the coefficients c_V and c_A involved in the Z^0 couplings to fermions, which are related to the Weinberg angle in the Standard Model via Eq. (10.38). In Eq. (10.40), we saw that the chiral left- and right-handed couplings of the Z^0 to fermions are not equal. Therefore, even if the electron and positron beams are unpolarised, the Z^0 boson exhibits a net polarisation along the beam axis that is transmitted to its decay products. The produced fermions in $e^+ + e^- \to Z^0 \to f + \bar{f}$ thus have a net helicity generating an asymmetry of their angular distribution. Two asymmetries have been extensively studied: the forward–backward and the left–right asymmetries. The forward–backward asymmetry is defined by

$$A^f_{FB} = \frac{\sigma_F - \sigma_B}{\sigma_F + \sigma_B}, \tag{12.14}$$

where σ_F (σ_B) is the cross section of forward (backwards) fermions, i.e. with $\cos\theta \geq 0$ ($\cos\theta < 0$), with the angle θ being the angle between the incoming electron and the outgoing fermion. At the Stanford Linear Collider (SLC) where the electron beam was polarised, the left–right asymmetry was also measured,

$$A_{LR}^f = \frac{\sigma_L - \sigma_R}{\sigma_L + \sigma_R}, \tag{12.15}$$

where σ_L (σ_R) is the cross section for a left- (right-)handed polarised electron. Let us take, as an example, the reaction $e^+ + e^- \to Z^0 \to \mu^+ + \mu^-$, where the energy of the collision is very close to $\sqrt{s} \simeq m_Z$. With the help of Problem 12.2, it is not difficult to show that the amplitudes squared are

$$\begin{aligned} |\mathcal{M}_{LL}|^2 &= s^2|C|^2 \left(c_L^e c_L^\mu\right)^2 (1+\cos\theta)^2, \quad |\mathcal{M}_{RR}|^2 = s^2|C|^2 \left(c_R^e c_R^\mu\right)^2 (1+\cos\theta)^2, \\ |\mathcal{M}_{LR}|^2 &= s^2|C|^2 \left(c_L^e c_R^\mu\right)^2 (1-\cos\theta)^2, \quad |\mathcal{M}_{RL}|^2 = s^2|C|^2 \left(c_R^e c_L^\mu\right)^2 (1-\cos\theta)^2, \end{aligned} \tag{12.16}$$

with

$$c_R = c_V - c_A, \quad c_L = c_V + c_A, \quad C = \frac{\sqrt{2}G_F m_Z^2}{s - m_Z^2 + i m_Z \Gamma_Z}. \tag{12.17}$$

The first subscript, L or R in Eq. (12.16), refers to the helicity of the electron, while the second refers to that of the muon. Assuming an unpolarised positron beam and summing over the polarisation of the muon yield the differential cross section (see Problem 12.3)

$$\frac{d\sigma}{d\cos\theta} = \frac{3}{8}\sigma_{\text{tot}} \left[(1 - P_e \mathcal{A}_e)(1+\cos^2\theta) + 2(\mathcal{A}_e - P_e)\mathcal{A}_\mu \cos\theta\right], \tag{12.18}$$

where P_e represents the electron polarisation, i.e. $P_e = 1$ for a 100% right-handed beam and $P_e = -1$ for a 100% left-handed beam, while $A_{f=e,\mu}$ is the asymmetry parameter defined by

$$\mathcal{A}_f = \frac{\left(c_L^f\right)^2 - \left(c_R^f\right)^2}{\left(c_L^f\right)^2 + \left(c_R^f\right)^2} = 2\frac{c_V^f/c_A^f}{1 + \left(c_V^f/c_A^f\right)^2}. \tag{12.19}$$

The quantity σ_{tot} in Eq. (12.18) is the total cross section integrated over all angles for an unpolarised beam (i.e., $P_e = 0$). The measurement of $\mathcal{A}_f$ gives access to c_V^f/c_A^f and, therefore, according to Eq. (10.38) (see Problem 12.4), to

$$\sin^2\theta_w = \frac{1 - c_V^f/c_A^f}{4|Q|}, \tag{12.20}$$

where Q is the electric charge of the fermion. Equation (12.18) is straightforward to generalise to a final state $f\bar{f}$ replacing $\mathcal{A}_\mu$ with $\mathcal{A}_f$. Its integration yields (Problem 12.5)

$$\begin{aligned} \sigma_F &= \int_0^1 \left.\frac{d\sigma}{d\cos\theta}\right|_{P_e=0} = \frac{3}{8}\sigma_{\text{tot}}\left(\frac{4}{3} + \mathcal{A}_e\mathcal{A}_f\right), \quad \sigma_R = \int_{-1}^1 \left.\frac{d\sigma}{d\cos\theta}\right|_{P_e=1} = \sigma_{\text{tot}}\,(1 - \mathcal{A}_e), \\ \sigma_B &= \int_{-1}^0 \left.\frac{d\sigma}{d\cos\theta}\right|_{P_e=0} = \frac{3}{8}\sigma_{\text{tot}}\left(\frac{4}{3} - \mathcal{A}_e\mathcal{A}_f\right), \quad \sigma_L = \int_{-1}^1 \left.\frac{d\sigma}{d\cos\theta}\right|_{P_e=-1} = \sigma_{\text{tot}}\,(1 + \mathcal{A}_e). \end{aligned} \tag{12.21}$$

Therefore, the forward–backward (12.14) and left–right (12.15) asymmetries read

$$A_{FB}^f = \frac{3}{4}\mathcal{A}_e\mathcal{A}_f, \quad A_{LR}^f = \mathcal{A}_e. \tag{12.22}$$

The measurements of the asymmetries in various channels ($f = e, \mu, \tau$ and quarks) give an estimation of $\mathcal{A}_f$ and, thus, of $\sin^2\theta_W$.

In Section 11.6.4, we have already explained, in the context of the Higgs boson mass constraints, that radiative corrections of the $W^\pm$ and Z^0 boson masses are a quadratic function of the top mass depending on m_t^2/m_W^2 and m_t^2/m_Z^2, respectively, while they are logarithmic in the Higgs boson mass. The two Feynman diagrams below show the main contributions with a virtual top in the fermion loop:

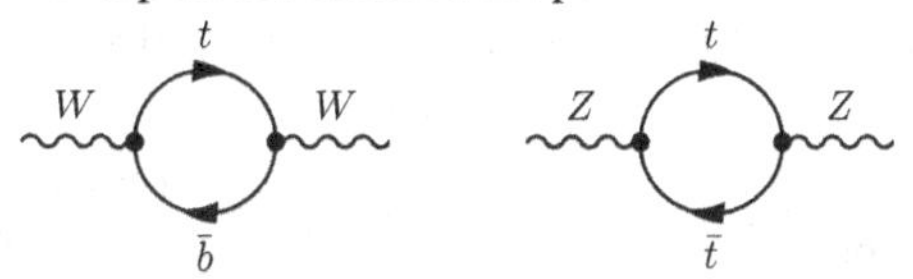

The top mass enters in the calculation of radiative corrections to a large number of electroweak observables obtained at LEP and SLC (from which the asymmetries described before are critical). Using the mass measurements of the W boson from the Tevatron experiments, a global fit of the top mass assuming the validity of the Standard Model was made with LEP and SLC data before the top quark discovery. The predicted value was (Campagnari and Franklin, 1997)

$$m_t = 178 \pm 8^{+17}_{-20}\ \mathrm{GeV}/c^2,$$

where the main uncertainty was due to the systematic error related to the unknown mass at that time of the Higgs boson (a range from 60 GeV/c^2 to 1 TeV/c^2 was used). In 1995, the CDF and D0 Collaborations discovered the top quark and measured (CDF Collaboration, 1995; D0 Collaboration, 1995)

$$m_t^{\mathrm{CDF}} = 176 \pm 13\ \mathrm{GeV}/c^2, \quad m_t^{\mathrm{D0}} = 199 \pm 30\ \mathrm{GeV}/c^2,$$

where the statistical and systematic errors have been added in quadrature. The up-to-date experimental value dominated by measurements at the LHC and a recent global fit of m_t are now (Particle Data Group, 2022)

$$m_t^{\mathrm{meas.}} = 172.89 \pm 0.59\ \mathrm{GeV}/c^2, \quad m_t^{\mathrm{SM}} = 176.3 \pm 1.9\ \mathrm{GeV}/c^2.$$

The predicted value is only 1.7 standard deviations above the average measurement, which remains impressive. It illustrates the consistency of the Standard Model.

Finally, it is worth mentioning a particularity of the top quark (beyond its very large mass). Its total decay width is measured to be $1.42^{+0.19}_{-0.15}$ GeV/c^2 (Particle Data Group, 2022). It is very large compared with other quarks because, due to the its large mass, the number of accessible decay channels is large (including the decay into an on-shell W boson, i.e. $t \to W^+ + b$, which is not possible for the other quarks). Therefore, the top lifetime is about 5×10^{-25} s, corresponding to a distance of only 0.1 fm before decaying. Given that the hadronisation process is governed by the QCD colour potential, which becomes very large for distances greater than 1 fm (see Section 8.5.4), the typical scale of distances of the hadronisation is 10 times larger than the mean free path of the top quark. Therefore, the top quark decays before 'hadronising'. It is the only quark that cannot be a constituent of hadrons.

12.1.4 Weaknesses of the Theory

Despite its predictive power, the Standard Model has several intrinsic weaknesses. Three of them are listed below.

A 'good' theory is viewed by many as a theory derived from first principles with few parameters adjusted to match the experimental observables. In its original formulation, where the neutrinos were considered massless, the free parameters of the Standard Model were: the six quark masses, the three charged lepton masses, the three coupling constants g_w, g, g_s, the Higgs potential parameters μ and λ, the four independent $V_{\rm CKM}$ constants (three angles and a phase) and the so-called θ angle of QCD. This last parameter is part of a term proportional to $\mathcal{L}_\theta \propto \theta\, \epsilon^{\mu\nu\rho\sigma} G^a_{\mu\nu}\, G_{a\rho\sigma}$ of the QCD Lagrangian, not mentioned so far, but nevertheless allowed by the gauge invariance under $SU(3)_c$. Such a term would generate CP violation in the QCD sector. As it has never been observed in the reactions induced by the strong interaction, the parameter θ is set to 0 in the Standard Model (the current limit on θ is about 2×10^{-10}). Therefore, the original Standard Model needed 19 free parameters. Now, since the neutrino sector appears richer than expected, there are at least seven new parameters: the three neutrino masses and the four parameters of the $V_{\rm PMNS}$ mixing matrix.[1] This increases the number of free parameters to 26. It is commonly admitted to be too many for a fundamental theory. The Standard Model is likely to be an effective theory of a more general theory.

Another weakness comes from the Higgs sector. So far, the Higgs potential does not come from a dynamical mechanism. It is a pure ad hoc postulate. The Yukawa couplings of fermions do not result from a theory predicting their values but are simply adjusted to match the fermion masses.

Finally, a major weakness is related to the connection between quantum field theory and general relativity. In general relativity, the vacuum energy density ρ is proportional to the cosmological constant Λ,

$$\Lambda = \kappa\rho, \quad \text{with} \quad \kappa = \frac{8\pi G_N}{c^2} = 1.86 \times 10^{-27} \text{cm gr}^{-1},$$

which couples to gravity via the Einstein field equations

$$R_{\mu\nu} - \frac{1}{2}\eta_{\mu\nu}R = \kappa T_{\mu\nu} + \Lambda\eta_{\mu\nu},$$

where $R_{\mu\nu}$ (the Ricci tensor) and R (the scalar curvature) refer to the curvature of spacetime, $\eta_{\mu\nu}$ is the metric tensor of spacetime and $T_{\mu\nu}$ is the energy–momentum tensor [see a general relativity textbook, e.g. Misner et al. (1973)]. The observations show that Λ is very small but not strictly zero,[2]

$$\Lambda_{\rm meas} \simeq 10^{-56}\ \text{cm}^{-2}.$$

How can we evaluate the energy density predicted by the Standard Model? In Supplement 6.1, we have already mentioned that fields generate an energy density potentially

[1] If neutrinos are Majorana particles, two additional phases would be present in the mixing matrix.

[2] In 1998, it was discovered that the expansion of the universe is accelerating, implying the possibility of a positive value for the cosmological constant. The actual value of $\Lambda_{\rm meas}$ is, however, model dependent (see Section 12.2.3).

extremely large (the zero-point energy of fields) that we promptly forgot by using the normal ordering of operators. In the Higgs sector, we have another source of energy density. Using $\mu^2 = -\lambda v^2$, the vacuum energy density coming from the Higgs potential is

$$\rho_H = V(\Phi) = \mu^2 \begin{pmatrix} 0 & \frac{v}{\sqrt{2}} \end{pmatrix} \begin{pmatrix} 0 \\ \frac{v}{\sqrt{2}} \end{pmatrix} + \lambda \left[\begin{pmatrix} 0 & \frac{v}{\sqrt{2}} \end{pmatrix} \begin{pmatrix} 0 \\ \frac{v}{\sqrt{2}} \end{pmatrix} \right]^2 = -\lambda \frac{v^4}{4} = -\frac{m_H^2 v^2}{8}.$$

Given the Higgs mass value (11.63) and the vacuum expectation value (11.29), $\rho_H \simeq -1.2 \times 10^8\ \mathrm{GeV}^4$. Therefore, after the conversion of ρ_H to the appropriate units (gr cm^{-3}), the vacuum energy density due to the Higgs potential contributes to the cosmological constant,

$$\Lambda_H = (1.86 \times 10^{-27} \mathrm{cm\ gr}^{-1}) \underbrace{\frac{-1.2 \times 10^8\ \mathrm{GeV}^4}{(1.97 \times 10^{-14}\ \mathrm{GeV} \cdot \mathrm{cm})^3}}_{\hbar c} \times \underbrace{1.7827 \times 10^{-24}}_{\mathrm{gr\ GeV}^{-1}}$$
$$= -5.2 \times 10^{-2} \mathrm{cm}^{-2}.$$

Not only is the wrong sign compared with the measured value, but the cosmological constant due to the Higgs potential is about 54 orders of magnitude larger. One might argue that Λ_H could become close to Λ_{meas} by adding a constant to the potential or because of the contribution of the zero-point energy of the other fields; however, this would require an extreme fine-tuning of the parameters, which is not satisfactory.

12.2 Possible Cracks in the Model

12.2.1 The Muon Magnetic Moment

In Chapter 6, we saw that the anomalous magnetic moment of the electron predicted by the theory, $a_e = (g_e - 2)/2$, is in agreement with the measured value with exceptional accuracy. The electron is obviously not the only particle with an anomalous magnetic moment: any elementary fermion is expected to have a g-factor slightly deviating from 2. The most precisely studied lepton is the electron, but the muon can also be explored with extreme precision. Moreover, the muon turns out to be a lepton more sensitive to hypothetical physics beyond the Standard Model. Indeed, if muons were massless, helicity would be conserved by all gauge bosons of the interactions and helicity flips would be forbidden. As muons have a mass about 200 times larger than electrons, helicity flips are allowed. This implies that the anomalous magnetic moment becomes sensitive to quantum fluctuations due to heavier particles or contributions from higher energy scales $\delta a_\mu \propto m_\mu^2/M$, where M may be the mass of a heavier particle belonging to the Standard Model or not (see Jegerlehner and Nyffeler, 2009). Therefore, the measurement of $a_\mu = (g_\mu - 2)/2$ can be considered a possible portal towards new physics.

Experimentally, the muon lifetime is large enough to allow the confinement of a muon on a circular orbit due to the presence of a strong magnetic field. The Larmor precession of the direction of the muon spin is sensitive to the value of a_μ. The first measurement of the anomalous magnetic moment of the muon was performed at Columbia University in 1960 [see Jegerlehner and Nyffeler (2009) for a review of all measurements]. The result $a_\mu = 1.22(8) \times 10^{-3}$ with a precision of about 5% showed no difference with the electron. The measurement has been performed more recently with a much higher accuracy at CERN and at BNL. The world average from Particle Data Group (2022) is

$$a_\mu^{\mathrm{exp}} = 1.16592091(63) \times 10^{-3}.$$

On the theory side, in QED, the value of $a_\ell = (g_\ell - 2)/2$ is universal for all leptons at the lowest order $(= \alpha/(2\pi))$. However, starting at 2-loops, higher order corrections include contributions from lepton loops in which different types of leptons can circulate, and results depend on the corresponding mass ratio m_μ/m_ℓ. These mass-dependent contributions explain that lepton universality is broken $a_e \neq a_\mu \neq a_\tau$. Overall, the QED, hadronic and electroweak contributions predict an anomalous gyromagnetic ratio of the muon (Particle Data Group, 2022)

$$a_\mu^{\mathrm{th}} = a_\mu^{\mathrm{QED}} + a_\mu^{\mathrm{EW}} + a_\mu^{\mathrm{had}} = 1.16591846(47) \times 10^{-3}.$$

Therefore, there is an overall discrepancy of 3.1 standard deviations between a_μ^{th} and a_μ^{exp}. It is significant but not large enough to be totally conclusive (the convention requires a statistical significance of 5σ before claiming a discovery). This might be a sign of new physics. However, the hadronic contribution entering the calculation of a_μ gives rise to the main uncertainty and might not be so well controlled. Two methods are used: one relies on a data-driven method based on data from $e^+e^- \to$ hadrons and the other on lattice QCD calculations that conflict with the data-driven evaluations at a level up to 3σ (Jegerlehner and Nyffeler, 2009; Particle Data Group, 2022). In conclusion, it is too early to claim that the Standard Model is being challenged.

12.2.2 Matter–Antimatter Asymmetry

Is There Really an Asymmetry?

Particles produced by colliders come in pairs: baryon and lepton numbers are conserved by the interactions of the Standard Model, and this is, so far, verified experimentally. These conservation laws imply that matter appears or disappears only as particle–antiparticle pairs. Therefore, it is legitimate to suppose that what is observed at the microscopic scale must also be true at the macroscopic scale, and hence, the Universe must contain equal amounts of matter and antimatter. However, if we look around us, we do not observe any bodies of antimatter. The solar wind, for example, is a probe for antimatter within the solar system. No spectacular annihilation between matter and antimatter is, however, observed. Moreover, the analysis of the cosmic ray flux does not show evidence of anti-nuclei reaching the earth beyond the expected secondary flux produced by collisions of particles with the interstellar medium. As cosmic rays are charged particles moving in the magnetic field of the Galaxy, they probably provide a sample of the material content of the Galaxy. On a wider scale, the diffuse γ-ray background, to which the matter–antimatter annihilations would contribute, is used to set limits on the presence of antimatter. Here again, the result is negative, and if domains of antimatter exist in the Universe, they are separated from us on scales certainly larger than the Virgo cluster (about 10 Mpc). Therefore, it is reasonable to conclude that the Universe is not (matter–antimatter) symmetric and contains little, if any, antimatter. Furthermore, the density of baryons n_B compared with the density of photons[3] n_γ is extremely small (Particle Data Group, 2022, Big-Bang nucleosynthesis review),

$$\eta = \frac{n_B}{n_\gamma} = \frac{N_B}{N_\gamma} = 6.1 \times 10^{-10}, \tag{12.23}$$

where N_B and N_γ are the total number of baryons and photons, respectively. In the Universe, as in the present time $N_{\overline{B}} \simeq 0$, $\Delta N_B/N_\gamma = (N_B - N_{\overline{B}})/N_\gamma \simeq \eta$. In the early Universe, cosmological models assume that $N_B \simeq N_{\overline{B}} \simeq N_\gamma$. As particles–antiparticles are created by pairs, the difference in the total number of baryons and anti-baryons, $\Delta N_B = N_B - N_{\overline{B}}$, is conserved, and neglecting the few extra photons emitted, for example, by stars, N_γ also remains constant. Therefore, Eq. (12.23) implies that there was a tiny excess of one baryon for every $1/\eta \simeq 10^9$ baryon–anti-baryon pairs (or for every 10^9 photons) before the annihilation of those pairs with the cooling down of the Universe due to its expansion.[4] The presence of matter in the Universe would be the result of this excess. The asymmetry between baryons and anti-baryons, i.e. the presence of a non-zero baryon number $B \neq 0$, must have arisen in the early evolution of the Universe before the primordial nucleosynthesis but after the inflation phase[5] (if any); otherwise, inflation would have diluted the baryon number by an enormous factor, such that B would be zero after this phase. Therefore, if after inflation at $t \simeq 0$, $B = 0$, a mechanism must be at work to generate a net baryon number before the nucleosynthesis epoch. Even without the inflation scenario (it is not yet firmly established, but nevertheless, commonly admitted among cosmologists), one might expect the Big-Bang to produce equal numbers of particles and antiparticles. Hence, the conclusion still holds: one needs a mechanism to move from $B = 0$ to $B \neq 0$.

Sakharov Conditions

In 1967, Sakharov stated three conditions for *baryogenesis*, i.e. the generation of an asymmetry between matter and antimatter: (i) baryon number violation; (ii) departure from thermal equilibrium and (iii) C and CP violation. Condition (i) is obvious otherwise if $B = 0$ at $t \simeq 0$, $B = 0$ for all $t > 0$. Now imagine that we have a process with baryon number violation: we start with an initial state i with $B = 0$ and get a final state f with $B \neq 0$. If $i \to f$ is in thermal equilibrium, then by definition, the rate for the inverse process satisfies $\Gamma(f \to i) = \Gamma(i \to f)$. Therefore, no net baryon asymmetry can be produced, hence condition (ii). Note that the expansion of the Universe satisfies condition (ii). However, the rate of expansion is likely to be too slow compared with the rate of B-violating processes.

[3] This ratio is obtained by two independent methods: the first uses the primordial nucleosynthesis model validated by the observation of the light element abundances, while the second is based on the measurements of cosmic microwave background temperature fluctuations. Both methods give consistent results.

[4] The Universe expansion breaks the equilibrium $p + \bar{p} \leftrightarrow \gamma + \gamma$, the photon energy becoming too small to produce a proton-antiproton pair.

[5] Inflation is a theory that supposes an exponential expansion of space in the early Universe. This theory offers the advantage of explaining the isotropy of the Universe, its flatness, etc.

For condition (iii), if C were an exact symmetry, the rate $\Gamma(i \to f)$ (still with $B = 0$ for i and $B \neq 0$ for f) would be equal to the rate with the antiparticles $\Gamma(\bar{i} \to \bar{f})$. Since the baryon number due to f is opposite to that of $\bar{f}$, the net baryon number B due to these reactions would vanish. Now, it is more subtle to see why CP must be violated too. Let us assume an initial state containing particles with momenta $\boldsymbol{p_i}$ and helicities λ_i and a final state containing particles with $\boldsymbol{p_f}$ and λ_f. If CP is conserved, the amplitude of the transition would be equal to the CP conjugate amplitude, i.e.

$$\mathcal{A}[i(\boldsymbol{p_i}, \lambda_i) \to f(\boldsymbol{p_f}, \lambda_f)] = \mathcal{A}[\bar{i}(-\boldsymbol{p_i}, -\lambda_i) \to \bar{f}(-\boldsymbol{p_f}, -\lambda_f)]. \tag{12.24}$$

Even though it is possible to create a baryon asymmetry in a certain region of the phase space where $\mathcal{A}(i \to f) = \mathcal{A}(\bar{i} \to \bar{f})$ (i.e., with the same momenta and helicities in both amplitudes), since the total rate requires to integrate the amplitude squared over all possible momenta and to sum/average over all possible spins, Eq. (12.24) results in a vanishing asymmetry. Therefore, both C and CP must be violated.

Baryon Number Violation

As the proton is the lightest baryon, one expects that, eventually, the proton can decay in scenarios where the baryon number is not conserved. In the Standard Model, we saw that the baryon number is conserved to all orders of perturbative theory (so, at each vertex of Feynman diagrams). However, a possible non-perturbative process in the Standard Model called *sphaleron* induces both violations of the baryon and lepton number (but the difference $B - L$ remains conserved) at a very large energy scale or, equivalently, a very high temperature, so in the early times of the Universe. [Such topics are treated in advanced quantum field theory courses, for instance, by Shifman (2012) or Weinberg (1995, volume II) and the more pedagogical presentation of Riotto (1998).] Moreover, many extensions of the Standard Model based on a gauge group that encompasses $SU(3)_c \times SU(2)_L \times U(1)_Y$ predict the non-conservation of baryon number. Those models are envisaged in the context of the grand unified theory (GUT), where the three gauge interactions are unified, the evolution of the running coupling constants with the energy scale reaching a common value at very high energy (typically at about 10^{15}–10^{16} GeV). Therefore, many experiments have searched for B-violating transitions. In general, these experiments are also involved in neutrino physics, as they share the same constraint: having a huge amount of matter to expect to see at least one decaying proton or to capture neutrinos.[6] So far, the results are negative: proton decay has never been observed, and only limits on its lifetime can be evaluated. The current best limits are set by the two experiments, SNO+ and Super-Kamiokande, already described in Section 9.4.2 in the context of neutrino oscillations. SNO+ sets a limit (SNO+ Collaboration, 2019) independent of the proton decay channel

$$\tau_p > 3.6 \times 10^{29} \text{ yr},$$

at the 90% confidence level. After a hypothetical proton decay, the remaining nucleus would be left in an excited state, and its de-excitation would emit γ-rays that could be

[6] Super-Kamiokande is one of them. The name KamiokaNDE actually stands for Kamioka Nucleon Decay Experiment, with Kamioka being the nearby village where the experiment is located.

detected by the technology used in SNO+. Therefore, this approach is model independent, as it puts no requirements on the particles produced in the decay. On the other hand, Super-Kamiokande tries to identify the decay products of the proton in the several decay channels, among which $p \to e^+\pi^0$ and $p \to \mu^+\pi^0$ (Super-Kamiokande Collaboration, 2020) and obtains in the electronic channel

$$\tau_p/\mathrm{BR}(p \to e^+\pi^0) > 2.4 \times 10^{34} \text{ yr}.$$

These results already exclude several extensions of the standard model to larger gauge groups. The simplest SU(5) predicts the proton decays in leptonic channels with a typical proton lifetime of about 10^{32} years. It is already ruled out.

Need for More CP Violation?

As described in Section 9.5, charge conjugation and the CP symmetry are violated by the weak interactions of the Standard Model. It is, therefore, legitimate to wonder whether CP violation is large enough to explain the matter-antimatter asymmetry. A fair answer is 'we don't know'. A naive estimation is obtained using the Jarlskog invariant, J, introduced in Section 9.5.5. We saw, there, that CP-violating processes always involve this invariant. In the quark sector, it is rather well measured [Eq. (9.81)]. As CP violation manifests if up-type fermions are not degenerate in mass as well as down-type fermions (otherwise, if two fermions are degenerate, the mixing angle between the two corresponding generations in the CKM or PMNS matrix becomes meaningless), the CP-violating parameter in the quark sector is

$$A_{\mathrm{CP}} = (m_c^2 - m_u^2)(m_t^2 - m_u^2)(m_t^2 - m_c^2)(m_s^2 - m_d^2)(m_b^2 - m_d^2)(m_b^2 - m_s^2)J.$$

Since A_{CP} has dimension twelve (in mass or energy), a dimensionless estimator of the magnitude of CP violation is A_{CP}/M^{12}, where M is a typical energy scale. At the high temperature needed for effective processes violating the baryon number (sphaleron) in the Standard Model, the scale is typically set by the electroweak phase transition of the order of $O(100 \text{ GeV})$. Inserting the measured value of J^{exp} from Eq. (9.81) and the quark masses from Table 1.1, one obtains $A_{\mathrm{CP}}/M^{12} \sim O(10^{-20})$. This is about 10 orders of magnitude smaller than the dimensionless quantity η in Eq. (12.23), characterising the matter-antimatter asymmetry. Similarly, we can play that game in the leptonic sector where J is approximately given by Eq. (9.83) in the maximal CP violation scenario. As the lepton masses are much smaller than quark masses, in particular for neutrinos, we expect an even smaller number for A_{CP}/M^{12} in the leptonic sector. Therefore, this naive approach concludes that the Standard Model cannot accommodate such an asymmetry between matter and anti-matter. However, the naive approach can be easily questioned, as explained in Cline (2006). With more sophisticated approaches, the answer to the question 'can the asymmetry be explained by the Standard Model?' is less conclusive [see, e.g., Gavela et al. (1994), for a negative answer and Servant (2014), for a less affirmative answer].

In conclusion, the asymmetry between matter and anti-matter is still an unresolved puzzle in modern physics, even if the satisfaction of the three Sakharov conditions is possible within the Standard Model or with a minimal extension of the model.

12.2.3 Dark Matter

Is Dark Matter Needed?

If the asymmetry between matter and anti-matter is likely to be a signal of physics beyond the Standard Model, another astrophysics observation provides a possible indication of the need for new physics. In a stable system of objects mutually bound by gravity, the virial theorem in mechanics states that the total kinetic energy of its components equals minus one-half their gravitational potential energy. The mass M of a galaxy can then be inferred by measuring the rotation velocity of its gas and stars, assuming circular Keplerian orbits. For spiral galaxies, as most of their mass is assumed to be concentrated in the luminous central bulge of the galaxy, one can evaluate the velocity v of a peripheral star, subject of the centripetal acceleration from Newton's laws, using the previous evaluation of the mass,

$$\frac{mv^2}{2} \simeq G_N \frac{mM}{r^2}.$$

The velocity should then be proportional to $r^{-1/2}$. A direct measurement of the star velocity shows, however, that this is not the case: the tangential velocities remain roughly constant with the distance r. One can perform these measurements on many stars or a wider scale with galaxies belonging to a cluster, varying the distance from the centre, to obtain the rotation curves, i.e. a plot of the orbital velocities as a function of r. The conclusion is invariably the same: objects rotate faster than expected. Therefore, two possibilities can be envisaged: either gravity does not behave as expected on a large scale, or there is a non-luminous contribution to the mass of galaxies. The first possibility implies a change in the gravitation theory. The modified Newtonian dynamics (MOND) proposed by Milgrom (1983) is one of the most popular alternative theories reproducing the rotation curves within galaxies. The second possibility involves another kind of matter, non-visible, that barely interacts electromagnetically (rotation curves have also been measured in radio astronomy) and, therefore, called dark matter.[7] Although there is still no consensus among physicists as to whether the modified gravity hypothesis or the dark matter approach is the right one, the prevailing opinion is rather in favour of the latter. This opinion is strengthened by the standard model of cosmology, ΛCDM, which depends only on six free parameters able to fit various observables, among which the anisotropy (and the polarisation) of the cosmic microwave background. In the ΛCDM model, the dark matter (non-baryonic matter) accounts for about 26.4% of the mass energy of the Universe, compared to only 4.9 % for the ordinary baryonic matter forming nuclei, the large fraction missing (about 69%) being the dark energy, i.e. a manifestation of the cosmological constant accounting for the acceleration of the Universe expansion (Planck Collaboration, 2020).

[7] One may wonder whether known 'standard' astrophysical objects, such as brown dwarfs and black holes (possibly primordial black holes), could be the source of dark matter. The answer is rather not. See, for instance, Carr et al. (2021) for a recent detailed review of the contribution of black holes to dark matter, where they estimate a contribution of about 10% (but with many theoretical assumptions).

Searches for Dark Matter

The hypothesis of dark matter has direct implications on particle physics: What particles could be a dark matter candidate? The simplest assumption about dark matter is that it has negligible interactions with other particles and that the dark matter particles are non-relativistic in the current stage of the evolution of the universe. They are, therefore, described as 'cold'.[8]

Within the Standard Model, the only possible candidates of dark matter particles are neutrinos, which barely interact but nevertheless have a mass. They are influenced by gravity and may induce gravitational effects, given that they are the most abundant (known) elementary particles in the Universe. However, their mass is so tiny (global cosmological measurements assuming the ΛCDM model constrain the sum of the three neutrino masses to be below approximately $0.1\ \mathrm{eV}/c^2$) that in the early Universe, they were not cold enough to explain the large-scale structure formations. It is estimated that standard neutrinos can account for at least 0.5% and at most 1.6% of the non-baryonic matter content of the Universe (Particle Data Group, 2022). Therefore, the Standard Model cannot accommodate the estimated amount of dark matter. In the literature, there are many possible extensions of the model proposing weakly interacting massive particles (WIMP). Supersymmetric Standard Model, familiarly called SUSY, is one of the most popular. In SUSY, every Standard Model particle has a supersymmetric partner with half a unit larger or smaller spin: fermions have scalar partners, while gauge bosons have fermionic partners. The Higgs sector is enriched with two Higgs doublets and their respective fermionic partners. Particles and super-partners belong to super-multiplets. Therefore, if supersymmetry, the transformation relating fermions to bosons and vice-versa, were an unbroken local symmetry, the super-partners would have the same mass as the standard particles. This contradicts the observations (nobody has ever seen a scalar electron) implying that SUSY is a broken symmetry. The initial motivation of SUSY models is to avoid the fine-tuning of the Higgs boson mass mentioned in Section 11.9.1. If the symmetry were exact, the quantum corrections to the mass coming from the super-partners would perfectly compensate those from standard particles. As supersymmetry is broken, the compensation is only approximate if the super-particle masses are not too different (typically within a few TeV's) from those of particles (otherwise, the fine-tuning problem would be reintroduced). An interesting by-product of SUSY models is that they provide dark matter candidates. More specifically, linear combinations of the neutral super-parters of the Higgs sector and those of the neutral gauge bosons [both have the same quantum numbers after the electroweak symmetry breaking, thus for $\mathrm{SU}(3)_c \times \mathrm{U}(1)_{\mathrm{em}}$] generate physical particles, called neutralinos. Depending on the SUSY models, the lightest of them, generically called LSP for the *lightest supersymmetric particle*, is stable and thus is a dark matter candidate.

[8] A scenario with cold dark matter favours the formation of structures with small galaxies forming first and galaxy clusters at a later stage. Indeed, let us assume that, in the early Universe, dark matter particles could travel over a mean length due to random motions. This distance must be smaller than the typical size of the primordial density fluctuations, otherwise, this would have washed out those fluctuations, which are supposed to seed the later structure formation by gradually aggregating particles (from dark matter and ordinary matter). Consequently, this constrains the minimum scale for later structure formation. Observations (counting the galaxies and looking at their size) are consistent with the cold dark matter scenario.

Supersymmetric particles, including the LSP, are extensively sought after at colliders, either through their direct production in collisions or through quantum correction effects in loops. So far, no SUSY particles have been found, and limits of the order of 1 TeV/c^2 are set for their masses. This places significant constraints on SUSY parameter space. On the other hand, dedicated experiments try to detect WIMPs without relying on a specific theoretical model for the dark matter candidate. They aim to observe elastic or inelastic scatters of galactic dark matter particles with atomic nuclei. The recoil of a nucleus (for elastic scatterings) is detected from scintillation light produced in scintillating devices or using cryogenic detectors (bolometers) encapsulating a crystal (often germanium) cooled to low temperature. The recoil of a nucleus is then detected by the temperature increase due to the lattice vibration. At the time of writing, here again, there has been no confirmed direct detection of dark matter.

12.3 The Intriguing Case of Neutrinos

Neutrinos are paradoxical: they are the most abundant elementary particles in nature but are the least well-known. In this section, some hypothetical specificities of neutrino physics that are not part of the Standard Model corpus are emphasised.

12.3.1 Majorana Fermions

Neutrinos are the only neutral elementary particles of the Standard Model. For such objects, we do not need the four degrees of freedom of a Dirac field to describe them (recall that a single Dirac field describes both a fermion and the anti-fermion with their two possible helicities, hence the four degrees of freedom for a given momentum). Indeed, starting from the Dirac equation and multiplying from the left by the two chirality projectors, one obtains

$$\begin{aligned} i\gamma^\mu \partial_\mu \psi_L &= m\psi_R, \\ i\gamma^\mu \partial_\mu \psi_R &= m\psi_L. \end{aligned} \tag{12.25}$$

Now, taking the complex conjugate of these equations and multiplying by the charge conjugate matrix $\hat{C}$ given by Eq. (5.84) yields

$$\begin{aligned} -i\hat{C}\gamma^{\mu *}\partial_\mu \psi_L^* &= m\hat{C}\psi_R^* = m\left(\psi_R\right)^c, \\ -i\hat{C}\gamma^{\mu *}\partial_\mu \psi_R^* &= m\hat{C}\psi_L^* = m\left(\psi_L\right)^c, \end{aligned}$$

where ψ^c is the charge conjugate field. Given the $\hat{C}$ matrix property (5.83), which states that $\hat{C}\gamma^{\mu *} = -\gamma^\mu \hat{C}$, it follows that

$$\begin{aligned} i\gamma^\mu \partial_\mu \left(\psi_L\right)^c &= m\left(\psi_R\right)^c, \\ i\gamma^\mu \partial_\mu \left(\psi_R\right)^c &= m\left(\psi_L\right)^c. \end{aligned} \tag{12.26}$$

Equations (12.26) with the charge conjugate fields have exactly the same structure as those with the fields in Eq. (12.25). However, notice that applying the chirality left-handed projector on $(\psi_L)^c$ yields

$$P_L(\psi_L)^c = P_L\hat{C}\psi_L^* = \hat{C}P_R\psi_L^* = \hat{C}(P_R\psi_L)^* = 0,$$

which shows that $(\psi_L)^c$ is a right-handed field. Similarly, we would find that $(\psi_R)^c$ is a left-handed field. Therefore, comparing Eq. (12.26) with (12.25), it is tempting to identify

$$\psi_R \equiv (\psi_L)^c, \quad \psi_L \equiv (\psi_R)^c, \tag{12.27}$$

implying that ψ_L and ψ_R would no longer be independent. As $((\psi_L)^c)^c = \psi_L$ [Eq. (5.87)], Eq. (12.26) would read

$$\begin{aligned} i\gamma^\mu\partial_\mu(\psi_L)^c &= m\psi_L, \\ i\gamma^\mu\partial_\mu\psi_L &= m(\psi_L)^c. \end{aligned} \tag{12.28}$$

Therefore, just ψ_L is needed to describe the particle. Moreover, these two equations are no longer independent. Indeed, taking the charge conjugate of the second equation and using the property of the $\hat{C}$ matrix (5.83), we recover the first equation.[9] Hence, while we initially needed two coupled equations to describe a particle corresponding to the ψ field, the identifications in Eq. (12.27) restrict the description of the particle to just one single equation. The initial four degrees of freedom of the Dirac field are reduced to only two for a field satisfying

$$\psi = \psi_L + (\psi_L)^c, \tag{12.29}$$

implying that

$$\psi = \psi^c. \tag{12.30}$$

Equation (12.30), a simple consequence of Eq. (12.27), is called the Majorana condition. The Majorana field (12.29) describes a particle that is its anti-particle,[10] as clearly shown by Eq. (12.30). Therefore, a Majorana field has just two degrees of freedom (for a given momentum, the fermion being its anti-fermion has two possible helicities). Clearly, only neutral fermions can satisfy this condition. It is, therefore, appropriate for neutrinos. Note that for a Majorana field defined in Eq. (12.29), as $P_L\psi = \psi_L$ and $P_R\psi = (\psi_L)^c$, a Lorentz invariant term as $\overline{\psi}\psi$ can be decomposed into

$$\overline{\psi}\psi = \left(\overline{P_L\psi} + \overline{P_R\psi}\right)(P_L\psi + P_R\psi) = \overline{P_L\psi}P_R\psi + \overline{P_R\psi}P_L\psi = \overline{\psi_L}(\psi_L)^c + \overline{(\psi_L)^c}\psi_L. \tag{12.31}$$

The quantisation of the Majorana field is very similar to the Dirac one, except that we have only one type of operator since there is no distinction between the particle and the anti-particles, i.e.

$$\psi(x) = \int \frac{d^3\boldsymbol{p}}{(2\pi)^3 2E_p} \sum_{r=1,2} \left[c_{p,r}u_r(p)e^{-ipx} + c^\dagger_{p,r}v_r(p)e^{+ipx}\right]. \tag{12.32}$$

As $v_r(p) = \hat{C}u_R^*(p) = i\gamma^2 u_R^*(p)$ (like Dirac fields), we do have $\psi^c = \psi$.

[9] Furthermore, the insertion of the second equation in the first gives $(\partial^\mu\partial_\mu + m^2)\psi_L = 0$.

[10] Ettore Majorana was an Italian physicist who discovered this possibility to describe a neutrino in the 1930s.

12.3.2 The Different Neutrino Masses

In this section, only one generation of neutrinos is considered for simplicity. This avoids the complication of the mixing of neutrino flavours, but the conclusions remain valid with three generations.

Dirac Mass

In Section 11.5.5, we studied a simple way to generate neutrino masses. We introduced a sterile (i.e., not charged under the gauge group) right-handed field ν_R to form a Dirac field ν (with four degrees of freedom) when associated with the active field ν_L, i.e.

$$\nu = \nu_L + \nu_R.$$

After the spontaneous symmetry breaking of the electroweak theory, following the same procedure with the Higgs mechanism as for other elementary fermions of the Standard Model, it gave rise to the Dirac mass term (11.61) reproduced below

$$\mathcal{L}_D^\nu = -m_D\overline{\nu}\nu = -m_D\left(\overline{\nu_L}\nu_R + \overline{\nu_R}\nu_L\right). \tag{12.33}$$

The parameter m_D is considered as a real positive number to be interpreted as a mass (one can always redefine the phase of ν_L appropriately to satisfy this condition). In addition to ν_L and ν_R, we can consider their charge conjugate fields $(\nu_L)^c$ and $(\nu_R)^c$ without adding any extra degrees of freedom to the theory. Note that ν_R being a sterile right-handed field, $(\nu_R)^c$ has the opposite chirality, and hence, it represents a sterile left-handed field. Therefore, the charge conjugate field of the Dirac neutrino (i.e., the anti-neutrino),

$$\nu^c = (\nu_L)^c + (\nu_R)^c,$$

is made with a sterile left-handed component and an active right-handed one. They can be used to form another Dirac mass term

$$\mathcal{L}_D^{\prime\nu} = -m_D\overline{\nu^c}\nu^c = -m_D\left(\overline{(\nu_R)^c}\,(\nu_L)^c + \overline{(\nu_L)^c}\,(\nu_R)^c\right). \tag{12.34}$$

However, it turns out that both Lagrangians (12.33) and (12.34) are the same. Indeed, a term like $\overline{(\nu_R)^c}\,(\nu_L)^c$ reads

$$\overline{(\nu_R)^c}\,(\nu_L)^c = \left(i\gamma^2\nu_R^*\right)^{\mathsf{T}*}\gamma^0 i\gamma^2\nu_L^* = \nu_R^{\mathsf{T}} i\gamma^2\gamma^0 i\gamma^2\nu_L^* = -\nu_R^{\mathsf{T}}\gamma^0\nu_L^* = -\nu_R^{\mathsf{T}}\gamma^0\nu_L^{\dagger\mathsf{T}} = -\nu^{\mathsf{T}}\gamma^0 P_L\nu^{\dagger\mathsf{T}}.$$

An expression as $\psi^{\mathsf{T}}\Gamma\psi^{\dagger\mathsf{T}}$ above, where Γ is any combination of γ matrices, can be re-arranged as follows, given the commutation rules of the fermion field (6.16):

$$\psi^{\mathsf{T}}\Gamma\psi^{\dagger\mathsf{T}} = \sum_{i,j}\psi_i\Gamma_{ij}\psi_j^\dagger = \sum_{i,j}-\psi_j^\dagger\Gamma_{ij}\psi_i = -\sum_{i,j}\psi_j^\dagger\left(\Gamma^{\mathsf{T}}\right)_{ji}\psi_i = -\psi^\dagger\Gamma^{\mathsf{T}}\psi,$$

where the contribution from the commutator is ignored because the normal ordering is implicitly used.[11] Applying this re-ordering with $\Gamma = \gamma^0 P_L$, we deduce $\nu^{\mathsf{T}}\gamma^0 P_L\nu^{\dagger\mathsf{T}} = -\nu^\dagger P_L\gamma^0\nu = -\nu_L^\dagger\gamma^0\nu_R$. Therefore,

[11] Interestingly, the term due to the commutator is $\delta^{(3)}(0)\mathrm{Tr}(\Gamma)$. The $\delta^{(3)}(0)$ can always be considered as a limit of a function $\delta^{(3)}(0) = \lim_{\boldsymbol{x}\to\boldsymbol{0}}\Delta(\boldsymbol{x})$. As this function is going to be multiplied by $\mathrm{Tr}(\Gamma)$, where $\Gamma = \gamma^0 P_L$ for which $\mathrm{Tr}(\gamma^0 P_L) = 0$ (because $\gamma^0 P_L = (\gamma^0 - \gamma^0\gamma^5)/2$ and γ^0 is a traceless matrix, while $\gamma^0\gamma^5$ is a product of an odd number of γ matrices which is traceless too), we keep a finite value.

$$\overline{(\nu_R)^c}\,(\nu_L)^c = \overline{\nu_L}\nu_R.$$

Similarly, we would find that $\overline{(\nu_L)^c}\,(\nu_R)^c = \overline{\nu_R}\nu_L$. Finally, since both Lagrangians (12.33) and (12.34) are identical, we can gather them into the Dirac mass term Lagrangian

$$\mathcal{L}_D^\nu = \frac{1}{2}\left(\mathcal{L}_D^\nu + \mathcal{L}_D^{\prime\nu}\right) = -\frac{1}{2}m_D\left(\overline{\nu_L}\nu_R + \overline{(\nu_R)^c}\,(\nu_L)^c + \overline{\nu_R}\nu_L + \overline{(\nu_L)^c}\,(\nu_R)^c\right).$$

Introducing the vectorial notation $\boldsymbol{N_L}$ for left-handed fields, $\boldsymbol{N_R}$ for right-handed fields, and the matrix M_D as follows:

$$\boldsymbol{N_L} = \begin{pmatrix} \nu_L \\ (\nu_R)^c \end{pmatrix}, \quad \boldsymbol{N_R} \equiv (\boldsymbol{N_L})^c = \begin{pmatrix} (\nu_L)^c \\ \nu_R \end{pmatrix}, \tag{12.35}$$

$$M_D = \begin{pmatrix} 0 & m_D \\ m_D & 0 \end{pmatrix}, \tag{12.36}$$

$\mathcal{L}_D^\nu$ above can be expressed as

$$\mathcal{L}_D^\nu = -\frac{1}{2}\left(\overline{\boldsymbol{N_L}}M_D\boldsymbol{N_R} + \overline{\boldsymbol{N_R}}M_D\boldsymbol{N_L}\right). \tag{12.37}$$

Note that the Lagrangian (12.37) is just a reformulation of the Lagrangian (12.33). It has simply the advantage of being expressed as a function of the four chiral fields of the neutrino sector (for one generation): ν_L, ν_R, $(\nu_L)^c = \hat{C}\nu_L^*$ and $(\nu_R)^c = \hat{C}\nu_R^*$, each field having only two degrees of freedom. The expression of the Dirac mass term (12.37) is not specific to neutrinos. Any massive Dirac particle can be described this way, but Eq. (12.37) is more appropriate when it is going to be combined with the Majorana mass term (only possible for neutrinos) in the following sections.

Majorana Mass

In the previous section, we introduced the sterile field ν_R to give mass to a neutrino. One may, however, wonder whether it is possible to generate a mass without this field originally not present in the Standard Model. The answer is positive, assuming that neutrinos are of Majorana type, as we are going to see now. Since $(\nu_L)^c$ is a chiral right-handed field, it can play the role of ν_R.[12] One should notice that a Yukawa term such as $\overline{L_\ell}\Phi^c\,(\nu_L)^c$, inspired by the Lagrangian (11.46), where ν_R is replaced by $(\nu_L)^c$, gives rise to a mass term very similar to the Lagrangian (11.61) after spontaneous symmetry breaking, i.e.

$$-\frac{1}{2}\overline{L_\ell}\Phi^c\,(\nu_L)^c \xrightarrow{\text{S.S.B.}} \mathcal{L} = -\frac{1}{2}m_L\left(\overline{(\nu_L)^c}\nu_L + \overline{\nu_L}\,(\nu_L)^c\right). \tag{12.38}$$

The numerical factor, $1/2$, accounts for the fact that ν_L and $(\nu_L)^c$ are not independent [the reader can check that the application of the Euler–Lagrange equation produces the Majorana field Eq. (12.28)]. Unfortunately, the Lagrangian (12.38), even before the spontaneous symmetry breaking, does not respect gauge symmetry. Indeed, $(\nu_L)^c$ carries

[12] Moreover, the charge-conjugate field $(\nu_L)^c$ has the property to transform as ν_L under a Lorentz transformation, which implies that $\overline{\nu_L}\,(\nu_L)^c$ is a Lorentz scalar that can thus be used in a Lagrangian.

quantum numbers opposite to those of ν_L, and consequently, Eq. (12.38) is not neutral in hypercharge,

$$\left.\begin{aligned} y_{\overline{L_\ell}} &= -y_{L_\ell} = -1, \\ y_{\Phi^c} &= -y_{\Phi} = -1, \\ y_{\nu_L^c} &= -y_{\nu_L} = -1, \end{aligned}\right\} \; y_{\overline{L_\ell}} + y_{\Phi^c} + y_{\nu_L^c} \neq 0,$$

nor in weak isospin, as the sum of three isospins 1/2 cannot produce a total isospin of zero. This implies that Φ^c should be a triplet of isospin (the combination $\frac{1}{2} + \mathbf{1} + \frac{1}{2}$ can yield $\mathbf{0}$) with $Y = 2$, which is not the case. Therefore, considering this term would impose a major change to the Standard Model.[13] We will ignore this possibility in the rest of this section.

Nevertheless, if ν_L is charged under the gauge symmetry, ν_R is not since it is sterile. Therefore, a Lagrangian similar to Eq. (12.38) based on ν_R,

$$\mathcal{L}_M^\nu = -\frac{1}{2} m_R \left(\overline{(\nu_R)^c} \nu_R + \overline{\nu_R} (\nu_R)^c \right) = -\frac{1}{2} \left(\overline{\boldsymbol{N}_L} M_R \boldsymbol{N}_R + \overline{\boldsymbol{N}_R} M_R \boldsymbol{N}_L \right), \tag{12.39}$$

respects the gauge symmetry by construction. We have introduced the matrix

$$M_R = \begin{pmatrix} 0 & 0 \\ 0 & m_R \end{pmatrix}, \tag{12.40}$$

where m_R is a positive real number (with a proper phase definition of ν_R if needed). Note that there is no need here for the spontaneous symmetry breaking (and thus a Higgs field) since ν_R does not carry any gauge quantum numbers.

Determination of Massive Neutrinos

The previous developments show that to introduce massive neutrinos (via a Dirac or Majorana term) without violating the gauge symmetry [and thus $m_L = 0$ in Eq. (12.38)], we need the additional sterile field ν_R. In principle, all terms allowed by a theory must be taken into account. This implies that with ν_L and ν_R, we also have to consider the terms involving their charge conjugates. Therefore, the most general Lagrangian with massive neutrinos is the sum of the two Lagrangians (12.37) and (12.39),

$$\begin{aligned} \mathcal{L}_{D+M}^\nu = \mathcal{L}_D^\nu + \mathcal{L}_M^\nu &= -\frac{1}{2} \left(\overline{\boldsymbol{N}_L} M \boldsymbol{N}_R + \overline{\boldsymbol{N}_R} M \boldsymbol{N}_L \right) \\ &= -\frac{1}{2} \left(\overline{\boldsymbol{N}_L} M (\boldsymbol{N}_L)^c + \overline{(\boldsymbol{N}_L)^c} M \boldsymbol{N}_L \right), \\ &= -\frac{1}{2} \overline{(\boldsymbol{N}_L)^c} M \boldsymbol{N}_L + \text{h.c.} \end{aligned} \tag{12.41}$$

with the matrix

$$M = M_D + M_R = \begin{pmatrix} 0 & m_D \\ m_D & m_R \end{pmatrix}. \tag{12.42}$$

[13] Using operators of higher dimension, i.e. a product of at least four fields in the Lagrangian, it is possible to respect the gauge invariance of the Standard Model with ν_L and ν_L^c. However, such a Lagrangian is not renormalisable. Hence, the theory would have to be considered an effective theory valid only at low energy. See, for example, Giunti and Kim (2007, p. 206).

Note that Lagrangian (12.41) is only possible for neutrino particles. Indeed, with a charged particle like the electron, the Majorana mass term with $\overline{e_R}\,(e_R)^c$ is forbidden by gauge symmetry since e_R carries gauge quantum numbers (this term would not even be neutral in terms of electric charge!).

We can reformulate the Lagrangian (12.41) by specifying the components of $(\boldsymbol{N_L})^c$. Denoting by ψ one of the two components of $\boldsymbol{N_L}$, i.e. $\psi = \nu_L$ or $(\nu_R)^c$, it follows, using the definition of the charge conjugate (5.86), that

$$\overline{(\psi)^c} = \left(\hat{C}\overline{\psi}^{\mathsf{T}}\right)^{\dagger}\gamma^0 = \left(\gamma^{0\mathsf{T}}\psi^{\mathsf{T}\dagger}\right)^{\dagger}\hat{C}^{\dagger}\gamma^0 = \psi^{\mathsf{T}}\gamma^{0\mathsf{T}}\hat{C}^{\dagger}\gamma^0 = -\psi^{\mathsf{T}}\hat{C}^{\dagger}, \tag{12.43}$$

where in the penultimate equality, $\gamma^{0\mathsf{T}}\hat{C}^{\dagger} = -\hat{C}^{\dagger}\gamma^{0\dagger} = -\hat{C}^{\dagger}\gamma^0$ has been deduced, using the property (5.83), i.e. $\left(\hat{C}\gamma^{0*}\right)^{\dagger} = \left(-\gamma^0\hat{C}\right)^{\dagger}$. Consequently, the Lagrangian (12.41) reads

$$\mathcal{L}^{\nu}_{D+M} = \frac{1}{2}\boldsymbol{N_L}^{\mathsf{T}}\hat{C}^{\dagger}M\,\boldsymbol{N_L} + \text{h.c.}, \tag{12.44}$$

where the matrix $\hat{C}$ is implicitly applied to each component of $\boldsymbol{N_L}^{\mathsf{T}}$. Compared to Eq. (12.41), this expression has the advantage of depending only on $\boldsymbol{N_L}$. In the Lagrangian (12.44), the fields ν_L and $(\nu_R)^c$ from the vector $\boldsymbol{N_L}$ [Eq. (12.35)] are clearly mixed by the mass matrix M. To determine the neutrino masses, we have to diagonalise this Lagrangian. Since M is a 2×2 real and symmetric matrix, it is easy to find its eigenvalues,

$$\mu_{1,2} = \frac{m_R}{2}\left(1 \mp \sqrt{1 + \left(\frac{2m_D}{m_R}\right)^2}\right), \tag{12.45}$$

while its eigenvectors are

$$\check{\boldsymbol{n}}_L = \begin{pmatrix} \check{\nu}_{1L} \\ \check{\nu}_{2L} \end{pmatrix} = \begin{pmatrix} \cos\theta & -\sin\theta \\ \sin\theta & \cos\theta \end{pmatrix}\begin{pmatrix} \nu_L \\ (\nu_R)^c \end{pmatrix} = U\boldsymbol{N_L}, \tag{12.46}$$

where

$$\tan 2\theta = \frac{2m_D}{m_R}. \tag{12.47}$$

Note that $\mu_1 \leq 0$ and $\mu_2 \geq 0$ in Eq. (12.45). If we want to find the masses of physical neutrinos (i.e., having only positive numbers), we can appropriately define the phase of the fields $\check{\nu}_i$ as follows. Since from Eq. (12.46), $\boldsymbol{N_L} = U^{\mathsf{T}}\check{\boldsymbol{n}}_L$, the Lagrangian (12.44) now reads

$$\mathcal{L}^{\nu}_{D+M} = \frac{1}{2}\check{\boldsymbol{n}}_L{}^{\mathsf{T}}U\hat{C}^{\dagger}M\,U^{\mathsf{T}}\check{\boldsymbol{n}}_L + \text{h.c.} = \frac{1}{2}\check{\boldsymbol{n}}_L{}^{\mathsf{T}}\hat{C}^{\dagger}\;UM\,U^{\mathsf{T}}\check{\boldsymbol{n}}_L + \text{h.c.},$$

where $UM\,U^{\mathsf{T}} = \text{diag}(\mu_1, \mu_2)$ is a diagonal matrix. Explicitly, it reads

$$\mathcal{L}^{\nu}_{D+M} = \frac{1}{2}\begin{pmatrix} \check{\nu}_{1L} \\ \check{\nu}_{2L} \end{pmatrix}^{\mathsf{T}}\hat{C}^{\dagger}\begin{pmatrix} \mu_1 & 0 \\ 0 & \mu_2 \end{pmatrix}\begin{pmatrix} \check{\nu}_{1L} \\ \check{\nu}_{2L} \end{pmatrix} + \text{h.c.} = \frac{1}{2}\begin{pmatrix} i\check{\nu}_{1L} \\ \check{\nu}_{2L} \end{pmatrix}^{\mathsf{T}}\hat{C}^{\dagger}\begin{pmatrix} -\mu_1 & 0 \\ 0 & \mu_2 \end{pmatrix}\begin{pmatrix} i\check{\nu}_{1L} \\ \check{\nu}_{2L} \end{pmatrix} + \text{h.c.}$$

Hence, we can define the physical neutrino mass eigenstates as

$$\boldsymbol{n}_L = \begin{pmatrix} \nu_{1L} \\ \nu_{2L} \end{pmatrix} = \begin{pmatrix} i\check{\nu}_{1L} \\ \check{\nu}_{2L} \end{pmatrix} = V\boldsymbol{N_L} = V\begin{pmatrix} \nu_L \\ (\nu_R)^c \end{pmatrix}, \quad \text{with} \quad V = \begin{pmatrix} i\cos\theta & -i\sin\theta \\ \sin\theta & \cos\theta \end{pmatrix}, \tag{12.48}$$

The Lagrangian $\mathcal{L}_{D+M}^{\nu}$ now reads

$$\mathcal{L}_{D+M}^{\nu} = \frac{1}{2}\begin{pmatrix}\nu_{1L}\\ \nu_{2L}\end{pmatrix}^{\mathsf{T}} \hat{C}^{\dagger} \begin{pmatrix} m_1 & 0 \\ 0 & m_2 \end{pmatrix} \begin{pmatrix}\nu_{1L}\\ \nu_{2L}\end{pmatrix} + \text{h.c.} = -\frac{1}{2}\overline{(\boldsymbol{n}_L)^c} \begin{pmatrix} m_1 & 0 \\ 0 & m_2 \end{pmatrix} \boldsymbol{n}_L + \text{h.c.}, \tag{12.49}$$

where the masses m_1 and m_2 are defined by

$$\begin{aligned} m_1 &= -\mu_1 = \frac{m_R}{2}\left(\sqrt{1+\left(\frac{2m_D}{m_R}\right)^2} - 1\right) = \sqrt{\frac{m_R^2}{4} + m_D^2} - \frac{m_R}{2}, \\ m_2 &= \mu_2 = \frac{m_R}{2}\left(1 + \sqrt{1+\left(\frac{2m_D}{m_R}\right)^2}\right) = \sqrt{\frac{m_R^2}{4} + m_D^2} + \frac{m_R}{2}. \end{aligned} \tag{12.50}$$

Introducing the massive fields

$$\nu_i = \nu_{i_L} + \left(\nu_{i_L}\right)^c, \quad \text{with} \quad i = 1, 2, \tag{12.51}$$

the Lagrangian (12.49) can be written more conventionally with the property (12.31), i.e.

$$\mathcal{L}_{D+M}^{\nu} = -\frac{1}{2}\left(m_1\, \overline{\left(\nu_{1_L}\right)^c}\nu_{1_L} + m_2\, \overline{\left(\nu_{2_L}\right)^c}\nu_{2_L}\right) + \text{h.c.} = -\frac{1}{2}\left(m_1\, \overline{\nu_1}\nu_1 + m_2\, \overline{\nu_2}\nu_2\right). \tag{12.52}$$

The definition (12.51) leading to the Lagrangian (12.52) implies that the two massive neutrinos are Majorana particles.

It might be surprising that, starting from one single flavour neutrino, we end up with two massive neutrinos. Actually, in terms of degrees of freedom, the balance is correct: we started with the four chiral fields ν_L, ν_R, $(\nu_L)^c$ and $(\nu_R)^c$. Only the two fields ν_L and ν_R are independent, and hence we had 2×2 degrees of freedom. We ended up with two Majorana fields and thus again 2×2 degrees of freedom.

Starting from the general Lagrangian (12.41), we conclude that two massive Majorana neutrinos are produced. We would obviously reach the same conclusion if the Dirac part of the Lagrangian is absent, i.e. for $m_D = 0$. Now, what happens when the Majorana part is removed, setting $m_R = 0$? We expect to recover a single massive Dirac neutrino. Indeed, according to Eqs. (12.47) and (12.50), $m_R = 0$ implies

$$\theta = \frac{\pi}{4}, \qquad m_1 = m_2 = m_D.$$

With the definition of the mass eigenstates (12.48), it follows that

$$\begin{cases} \nu_{1_L} = \dfrac{i}{\sqrt{2}} \quad \left[\nu_L - (\nu_R)^c\right] \\ \nu_{2_L} = \dfrac{1}{\sqrt{2}} \quad \left[\nu_L + (\nu_R)^c\right] \end{cases} \quad \text{and} \quad \begin{cases} \left(\nu_{1_L}\right)^c = -\dfrac{i}{\sqrt{2}} \quad \left[(\nu_L)^c - \nu_R\right] \\ \left(\nu_{2_L}\right)^c = \dfrac{1}{\sqrt{2}} \quad \left[(\nu_L)^c + \nu_R\right] \end{cases}$$

Since the two massive fields have the same mass, m_D, we can combine them to form the neutrino Dirac field of that mass. Inverting the system of equations above, we indeed deduce

$$\begin{cases} \nu_L = \dfrac{1}{\sqrt{2}} \quad \left[\nu_{2_L} - i\nu_{1_L}\right] \\ \nu_R = \dfrac{1}{\sqrt{2}} \quad \left[\left(\nu_{2_L}\right)^c - i\left(\nu_{2_L}\right)^c\right] \end{cases}$$

and therefore, $\nu = \nu_L + \nu_R = (\nu_2 - i\nu_1)/\sqrt{2}$. The massive Dirac field is thus equivalent to the two massive Majorana fields, as expected.

12.3.3 See-Saw Mechanism

One interesting case is obtained when $m_R \gg m_D$. According to (12.50), the masses become

$$m_1 \simeq \frac{m_D^2}{m_R} \simeq 0, \quad m_2 \simeq m_R.$$

The neutrino, ν_1, would be naturally extremely light. In this approximation, according to Eq. (12.47), the mixing angle $\theta \simeq 0$. Therefore, with the definition (12.48)

$$\nu_{1_L} \simeq i\nu_L, \quad \nu_{2_L} \simeq (\nu_R)^c.$$

In this scenario, called the *see-saw mechanism*, the active neutrino is naturally light, whereas the sterile neutrino is very massive. That could explain why the observed neutrinos are so light. Note that the assumption $m_R \gg m_D$ is not unjustified. Indeed, the Dirac mass m_D is generated by the spontaneous symmetry breaking of the electroweak theory. Therefore, we expect m_D to be of the order of the mass of other fermions of the Standard Model, so typically in the range from 1 MeV$/c^2$ to 100 GeV$/c^2$, proportional to the vacuum expectation value of the Higgs field. On the other hand, m_R does not result from such a mechanism and can be arbitrarily chosen since ν_R is a singlet of the gauge group. So, it can be potentially as high as the limit of validity of the theory, let us say, the Planck scale or a few orders of magnitude below.

The see-saw mechanism (i.e., the heavier the sterile neutrino is, the lighter the active neutrino is) remains valid (but much more complex) when the three generations of neutrinos are considered. The reader can consult the reference Giunti and Kim (2007) for the details.

12.3.4 Experimental Manifestation of a Majorana Neutrino

The nature of neutrinos, Dirac fermions or Majorana fermions, is not yet determined. In this section, we show how observations can decide.

Non-conservation of the Lepton Number

The lepton number (L) conservation is associated with the invariance of the Lagrangian of the Standard Model under a global U(1) phase transformation of the lepton fields[14], $\psi \to e^{i\alpha}\psi$, with $\psi = \ell_L, \ell_R, \nu_L, \nu_R$. However, the Majorana mass term described by

[14] Baryon number conservation is also associated with a similar global U(1) phase transformation but acting on the quark fields (see Section 8.1.2). Electric charge conservation is associated with a local U(1) phase transformation but acting on all fields (see Chapter 6).

the Lagrangian (12.49) breaks this invariance. Indeed, under a U(1) transformation of the lepton fields with a phase α, the term

$$\frac{1}{2}m_i\nu_{i_L}^{\mathsf{T}}\hat{C}^{\dagger}\nu_{i_L} \to e^{i2\alpha}\frac{1}{2}m_i\nu_{i_L}^{\mathsf{T}}\hat{C}^{\dagger}\nu_{i_L},$$

which shows the absence of lepton number conservation. This is expected since a Majorana neutrino is its own antiparticle, implying that it would have both $L = 1$ and $L = -1$.

Neutrinoless Double-Beta Decay

Regardless of the nature of the neutrino (Dirac or Majorana), only the left-handed chiral component interacts. In this respect, no difference is expected between these two kinds of neutrinos. The remarks of the previous paragraph show, however, that the manifestation of Majorana neutrinos must be preferentially searched for via the lepton number violation generated by the specific Majorana mass term. The golden channel is double-beta decay. In the Standard Model, the decay of a nucleus with mass number A and atomic number Z in $(Z,A) \to (Z+2,A) + 2e^- + 2\overline{\nu}$ emits two electrons and two anti-neutrinos (see the left-hand side of Fig. 12.1) such that the lepton number is conserved ($L = 0$ since the initial state does not contain leptons). This rare decay is possible in some isotopes where simple beta decay is kinematically forbidden or strongly suppressed by nuclear selection rules. However, if neutrinos are Majorana particles, the internal line of the right-hand diagram of Fig. 12.1 can represent on one side a neutrino and on the other side an anti-neutrino. This leads to a violation of the lepton number conservation by two units, with $L = 2$ in the final state. This process is called the neutrinoless double beta decay since no neutrino is observed in the final state $(Z,A) \to (Z+2,A) + 2e^-$. Note that if it exists, it should be extremely rare. Indeed, in the right-hand diagram of Fig. 12.1, the left-handed chiral component ν_L of the Majorana neutrino is involved in the upper vertex, whereas it is the right-handed component $(\nu_L)^c$ in the lower vertex with the anti-neutrino symbol. Therefore, if the Majorana neutrino helicity is negative (left-handed), the configuration of the lower vertex requires to project a left-handed helicity onto a right-handed chirality component, which is suppressed by a factor m_ν/E_ν (see Section 9.2.3).

The neutrinoless double beta decay is often referred to by the abbreviation $0\nu\beta\beta$ (or less often $2\beta0\nu$) in the literature, while the Standard Model process $(Z,A) \to (Z+2,A)+2e^-+$

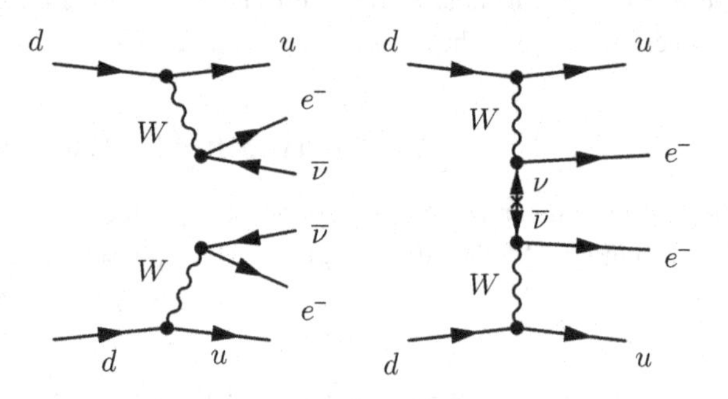

Fig. 12.1 The lowest order diagram of the double-beta decay in the Standard Model (left) and with Majorana neutrinos (right).

$2\overline{\nu}$ is referred to by $2\nu\beta\beta$. The normal process has been observed with several isotopes (^{100}Mo, ^{124}Xe, etc.). The typical half-lives $\tau_{1/2}^{2\nu\beta\beta}$ range from about 10^{19} to 10^{22} years (see Particle Data Group, 2022). It is clearly a rare process. On the other hand, neutrinoless double beta decay has never been observed. Only limits have been set. The best limit is obtained with the transition ${}^{76}_{32}\text{Ge} \to {}^{76}_{34}\text{Se} + 2e^-$, where $\tau_{1/2}^{0\nu\beta\beta} > 1.8 \times 10^{26}$ years.

Disappearance of the Active Neutrino

The charged current interaction couples the left-handed component of the neutrino to the left-handed component of the associated charged lepton via a W boson. The corresponding Lagrangian for one family of neutrinos is simply

$$\mathcal{L} = -\frac{g_w}{\sqrt{2}} \overline{\ell_L} \gamma^\mu \nu_L W_\mu + \text{h.c.}$$

However, according to the definition (12.48),

$$\begin{pmatrix} \nu_L \\ (\nu_R)^c \end{pmatrix} = V^\dagger \begin{pmatrix} \nu_{1_L} \\ \nu_{2_L} \end{pmatrix}.$$

As $\nu_L = V_{11}^* \nu_{1_L} + V_{21}^* \nu_{2_L}$, the previous Lagrangian becomes

$$\mathcal{L} = -\frac{g_w}{\sqrt{2}} \left(V_{11}^* \overline{\ell_L} \gamma^\mu \nu_{1_L} + V_{21}^* \overline{\ell_L} \gamma^\mu \nu_{2_L} \right) W_\mu + \text{h.c.},$$

implying that the charge current interaction would create a superposition of massive neutrinos. These massive neutrinos will propagate at different velocities [their phase $\exp(-iE_i t)$ with $E_i^2 = m_i^2 + p^2$ evolves differently], and therefore, when neutrinos are detected by a charged current interaction, a different linear superposition of ν_L and ν_R may be formed. We would thus have a phenomenon very similar to the oscillations described in Section 9.4.4 between the three neutrino flavours, except that this time, the oscillation is between the active neutrino ν_L and the sterile ν_R. The oscillation frequency would depend again on the squared mass difference

$$\Delta m^2 = m_2^2 - m_1^2 = m_R \sqrt{m_R^2 + 4m_D^2},$$

the mathematics being the same as in Section 9.4.4. But, experimentally, there is an important difference: only ν_L can be detected (ν_R being sterile). Therefore, the oscillation would be interpreted as a disappearance of the active neutrino, which would not be compensated by the appearance of a detectable neutrino (unlike the usual oscillation in Section 9.4.4). Experimentally, it is very challenging to detect the 'disappearance' of neutrinos because it supposes that the initial neutrino flux is very well controlled, which is usually not the case (there are large uncertainties on the nuclear cross sections). Moreover, if we suppose a see-saw mechanism, the active neutrino being already a mass-eigenstate ($\nu_{1_L} \simeq i\nu_L$), it would not oscillate at all, and thus, no disappearance would be expected.

The search for sterile neutrino(s) is an active experimental field. The reader can consult, for instance, the recent review from Dasgupta and Kopp (2021). Over the last decades, several 'anomalies' have been reported (from reactor and accelerator short baseline experiments, and with experiments using galium), but they remain insufficiently conclusive to claim the discovery of a sterile neutrino. Furthermore, the minimal sterile neutrino models are now excluded by global fits of neutrino data.

Problems

12.1 Check Eq. (12.13).

12.2 The reaction $e^-(k) + e^+(k') \to \mu^-(p) + \mu^+(p')$ can occur via the exchange of a γ or Z^0 boson. The fermion 4-momenta are denoted k, k', p, p' and the fermion masses are neglected. In this problem, the electron and positron beams have the same energy and collide back to back with total energy $\sqrt{s}$. For generality, the Z^0 total decay width Γ_Z is not neglected in the Z^0 propagator,

$$Z_{\text{propagator}} = i\frac{-g^{\mu\nu} + q^\mu q^\nu/m_Z^2}{q^2 - m_Z^2 + im_Z\Gamma_Z}. \tag{12.53}$$

1. At the lowest order, give the expression of the amplitude $\mathcal{M} = \mathcal{M}_\gamma + \mathcal{M}_Z$, where $\mathcal{M}_\gamma$ and $\mathcal{M}_Z$ are the amplitudes corresponding to the exchange of a γ and Z^0 boson, respectively.
2. Show that the term $q^\mu q^\nu/M_Z^2$ from the Z^0 propagator does not contribute to the amplitude.
3. We are interested in the polarised cross section, and the variables

$$c_R = c_V - c_A, \quad c_L = c_V + c_A, \quad \chi = \frac{\sqrt{2}G_F m_Z^2}{s - m_Z^2 + im_Z\Gamma_Z}\frac{s}{4\pi\alpha}, \tag{12.54}$$

are introduced (c_V/c_A are the coefficients of the Z^0 coupling, see Table 10.2, G_F and α, the Fermi coupling constant and the fine-structure constant, respectively). The reduced amplitudes are defined by:

$$A_{RR} = \left[\overline{v_L}(k')\gamma^\mu u_R(k)\right]\left[\overline{u_R}(p)\gamma_\mu v_L(p')\right], \quad A_{RL} = \left[\overline{v_L}(k')\gamma^\mu u_R(k)\right]\left[\overline{u_L}(p)\gamma_\mu v_R(p')\right],$$
$$A_{LR} = \left[\overline{v_R}(k')\gamma^\mu u_L(k)\right]\left[\overline{u_R}(p)\gamma_\mu v_L(p')\right], \quad A_{LL} = \left[\overline{v_R}(k')\gamma^\mu u_L(k)\right]\left[\overline{u_L}(p)\gamma_\mu v_R(p')\right],$$

where the subscript R or L refers to the helicity of the initial/final fermions. (Thus, a right-handed e^- and a left-handed μ^- correspond to A_{RL}.) Show that

$$\mathcal{M}_\gamma = \frac{4\pi\alpha}{s}\left(A_{LL} + A_{LR} + A_{RL} + A_{RR}\right), \tag{12.55}$$
$$\mathcal{M}_Z = \frac{4\pi\alpha}{s}\chi\left(c_L^e c_L^\mu A_{LL} + c_L^e c_R^\mu A_{LR} + c_R^e c_L^\mu A_{RL} + c_R^e c_R^\mu A_{RR}\right). \tag{12.56}$$

4. Using the helicity spinor eigenstates in Eq. (5.74), show that

$$A_{LL} = A_{RR} = -s(1 + \cos\theta), \quad A_{LR} = A_{RL} = s(1 - \cos\theta), \tag{12.57}$$

where θ is the angle between the electron and the muon and the e^- momentum is in the $+z$-axis direction.

5. Show that the differential cross section $\frac{d\sigma}{d\cos\theta}(e^-_L e^+_R \to \mu^-_L \mu^+_R)$ is

$$\frac{d\sigma}{d\cos\theta}(e^-_L e^+_R \to \mu^-_L \mu^+_R) = \frac{\pi\alpha^2}{2s}(1+\cos\theta)^2|1+\chi c^e_L c^\mu_L|^2, \tag{12.58}$$

and determine at which energy it is twice the contribution of the corresponding pure QED process. Use the constant values from Table 1, and Table 10.2, and neglect $\Gamma_Z \ll m_Z$.

6. Express now the differential cross section: $\frac{d\sigma}{d\cos\theta}(e^-_L e^+_R \to \mu^-_R \mu^+_L)$. Which value of the angle θ cancels the cross section? What is the physical interpretation of this result?

7. Experimentally, the polarisation of the beams may be controlled but not necessarily that of the final state. Give the expressions of $\frac{d\sigma}{d\cos\theta}(e^-_L e^+_R \to \mu^-\mu^+)$ and $\frac{d\sigma}{d\cos\theta}(e^-_R e^+_L \to \mu^-\mu^+)$.

12.3 This problem uses notations and results from Problem 12.2. We are now interested in the reaction $e^- + e^+ \to \mu^- + \mu^+$ for $\sqrt{s} \simeq m_Z$. Therefore, the γ exchange contribution can be safely neglected.

1. Using notations (12.17), check Eq. (12.16).
2. Let P_e be the electron polarisation, such as $P_e = 1$ for an electron beam 100% right-handed and $P_e = -1$ for a beam 100% left-handed. The positron beam is considered unpolarised, and the polarisation of outgoing particles is not measured. Show that the spin-averaged amplitude squared is

$$|\overline{\mathcal{M}}(Pe)|^2 = \frac{s^2|C|^2}{4}\Big[(c^e_L)^2+(c^e_R)^2\Big]\Big[(c^\mu_L)^2+(c^\mu_R)^2\Big]\Big[(1-P_e\mathcal{A}_e)(1+\cos^2\theta)+2(\mathcal{A}_e-P_e)\mathcal{A}_\mu\cos\theta\Big],$$

where the quantity $\mathcal{A}_{f=e,\mu}$ is defined in Eq. (12.19).

3. Deduce the expression of the corresponding cross section and show that it can be expressed as in Eq. (12.18).

12.4 Check Eq. (12.20).

12.5 Check Eq. (12.21).

Appendix A Elements of Group Theory for Particle Physics

Group theory has become an important ingredient of elementary particle physics, in particular, for the modelling of gauge interactions. This appendix gives an informal (i.e., not rigorous) introduction adapted for students in particle physics. It is mostly focused on matrix Lie groups and does not replace a real course with its mathematical foundations. Many textbooks treat this topic at various levels and from various perspectives. Among others, the reader can consult references Bump (2004), Hall (2015) and Helgason (1978) for a mathematician approach and Tung (1985) and Georgi (1999) for a physicist one.

A.1 Definition of a Group

A group G is a set of elements a, b, c, etc., with a composition law (whose symbol below is the multiplication) satisfying

- Closure: if $(a, b) \in G$ then $c = a \times b \in G$.
- Associativity: $a \times (b \times c) = (a \times b) \times c$.
- Identity: $\exists\, e \in G$ such as $\forall\, a \in G,\ e \times a = a \times e = a$.
- Inverse: $\forall\, a \in G,\ \exists\, a^{-1} \in G$ such as $a \times a^{-1} = a^{-1} \times a = e$.

If the group law is commutative, i.e. $\forall\, (a, b) \in G,\ a \times b = b \times a$, then the group is said *Abelian*. If the number of elements in the group is finite, the group is called a *finite* group. Otherwise, the group is *infinite*. A group can be *compact*. A rigorous definition of this term would take us beyond this simple summary, but for matrix groups (whose elements are matrices), it implies that all elements M of G are such that the components M_{ij} satisfy $|M_{ij}| \leq C$, where C is a constant.

An intuitive example of groups is the set of rotation transformations in $\mathbb{R}^3$. The combination of two rotations is clearly a rotation. We can perform a rotation and do the inverse rotation, define the identity rotation (i.e., a rotation with zero angle(s)), etc.

In the context of the Standard Model of particle physics, we encounter the notion of groups made from the *direct product* of other groups. Let G and H be two groups with the respective composition law $\circ$ and $\star$. The direct product $G \times H$ is a group whose set is the Cartesian product of G by H, namely the set of ordered pairs[1] (g, h), where g is an element of G and h is an element of H. The composition law of $G \times H$ denoted $\times$ below is then defined such as

[1] An ordered pair means that (g, h) is a different element than (h, g), both belonging to two different sets.

$$(g_1, h_1) \times (g_2, h_2) = (g_1 \circ g_2, h_1 \star h_2).$$

This ensures that $G \times H$ is a group: its identity element is the pair of the identity elements of G and H, (e, e), and the inverse of (g, h) is $(g, h)^{-1} \equiv (g^{-1}, h^{-1})$.

A.2 Representations

A.2.1 Definitions

Groups are easier to understand by studying the effects of their elements on objects. For instance, with rotation in $\mathbb{R}^3$, it is easy to visualise, i.e. *represent* what a rotation is when acting on a vector of $\mathbb{R}^3$. A *representation* $\mathcal{D}$ makes abstract transformations corresponding to a group G (such as rotations) more concrete by describing its elements $g \in G$ with linear operators $\mathcal{D}(g)$ acting on a vector space[2] V. The set of linear transformations on a vector space V itself forms a group denoted by $\mathrm{GL}(V)$, with GL standing for General Linear. For example, when rotations are characterised by their action on vectors in $\mathbb{R}^3$, they are represented by 3×3 matrices that form a representation of rotations. The name given to the representation is often that of the objects on which it acts. In the latter case, one would speak of a vector representation of rotations. We can also characterise rotations by their actions on 2-ispinors, leading to 2×2 transformation matrices that form the spinor representation of rotations. The same group is described here, the rotation transformations, but viewed from two different representations.

So, a representation is a way of visualising group elements by considering their action on a particular choice of vectors of a vector space V (which is, in most cases in physics, $\mathbb{R}^n, \mathbb{C}^n$ or a Hilbert space when manipulating quantum mechanics objects). The *dimension* of a representation is the dimension of the space on which it acts. Mathematically, a representation is a *group homomorphism* from the group G to the linear group $\mathrm{GL}(V)$. A function f is a group homomorphism from a group G to another group G' when it preserves the algebraic structure, i.e. if for all elements g_1, g_2 in G,

$$f{:}G \to G', \quad f(g_1 \times g_2) = f(g_1) \times f(g_2).$$

Beware that the multiplication symbol on the left-hand side is the composition law of G, whereas on the right-hand side, it is that of G'. A representation is thus a mapping from G to $\mathrm{GL}(V)$ that reflects the structure of the group, i.e.

$$\mathcal{D}{:}G \to \mathrm{GL}(V), \quad \mathcal{D}(g_1 \times g_2) = \mathcal{D}(g_1) \times \mathcal{D}(g_2), \quad \forall g_1, g_2 \in G. \tag{A.1}$$

Note that setting $g_2 = e$ in the equation above yields

$$\mathcal{D}(e) = I, \tag{A.2}$$

[2] The properties of vector spaces must not be confused with those of groups. They both share an associative composition law which, for vector spaces, is called the addition (if u and v are two vectors of V, then $u + v = v + u \in V$). However, vector spaces have another defined operation, the scalar multiplication, such that if $\alpha \in \mathbb{F}$ (in practice, the field $\mathbb{F}$ is $\mathbb{R}$ or $\mathbb{C}$), $\alpha \times u \in V$. There are other properties like the distributivity of the scalar multiplication, etc.

where I is the identity element in GL(V), while setting $g = g_2 = g_1^{-1}$ gives

$$(\mathcal{D}(g))^{-1} = \mathcal{D}(g^{-1}). \tag{A.3}$$

A group homomorphism f from G to G' might be bijective, i.e. its inverse f^{-1} is also a group homomorphism from G' to G. In that case, the function f is a *group isomorphism.* It ensures a one-to-one mapping. Clearly, a representation is a group homomorphism but not necessarily an isomorphism. For example, the trivial representation, such as $\mathcal{D}(g) = I$ for all g in G, is only a homomorphism. In passing, if the only element of g satisfying $\mathcal{D}(g) = I$ is $g = e$, then the representation is said to be *faithful* (injective).

When an orthonormal basis $\{\hat{\boldsymbol{e}}_i\}$ of the vector space V is chosen (assuming here a finite dimension), it becomes convenient to view representations both as linear operators and as matrices. A linear operator transforms a vector into another vector of V; thus,

$$\mathcal{D}(g)\hat{\boldsymbol{e}}_j = \sum_i \hat{\boldsymbol{e}}_i \left(\mathcal{D}(g)\right)_{ij}.$$

(I use here a common abuse of notation denoting the matrix on the right-hand side of the equation by the same symbol as the representation itself on the left-hand side.) The group homomorphism (A.1) then becomes a simple multiplication of matrices

$$(\mathcal{D}(g_1 \times g_2))_{ij} = (\mathcal{D}(g_1))_{ik} \left(\mathcal{D}(g_2)\right)_{kj}, \tag{A.4}$$

where the summation over k is implicit.

A representation of a matrix group [like SO(3) for rotations in three dimensions] that is given by the matrices of the group itself is called the *standard representation*. The vector representation of rotations is thus the standard representation of SO(3).

A.2.2 Equivalent Representations

Two representations can be *equivalent* (or *conjugate*). It means that if $\mathcal{D}$ acts on a vector space V and $\mathcal{D}'$ on V' (which can be V), then there is a vector space isomorphism φ from V to V' (which preserves the addition and scalar multiplication), such as for all elements g of the group G,

$$\mathcal{D}'(g) = \varphi \mathcal{D}(g) \varphi^{-1}.$$

Once bases of V and V' are chosen, two representations, $\mathcal{D} \in$ GL(V) and $\mathcal{D}' \in$ GL(V'), are equivalent when there is a fixed matrix $M{:}V \to V'$, such that

$$\mathcal{D}'(g) = M\mathcal{D}(g)M^{-1}, \tag{A.5}$$

where $\mathcal{D}'(g)$ and $\mathcal{D}(g)$ are the matrices representing $\mathcal{D}$ and $\mathcal{D}'$. The Formula (A.5) is just a matrix similarity transformation, and when $V = V'$, the representations are related to one another by a change of basis.

A.2.3 Irreducible Representations

We are mostly interested in *irreducible representations*, i.e. those are not *reducible*. A reducible representation leaves invariant a proper subspace of the vector space V on which it acts. A representation is irreducible if the only invariant subspace of V is $\{0\}$ and V itself. Mathematically, when a representation $\mathcal{D}_0 \in \mathrm{GL}(V_0)$ is *completely reducible*, it is equivalent to another representation $\mathcal{D} \in \mathrm{GL}(V)$, which is block diagonal, meaning that for all g of G, there is a fixed matrix $M{:}V_0 \to V$, such that

$$\mathcal{D}(g) = M\mathcal{D}_0(g)M^{-1} = \begin{pmatrix} \mathcal{D}_1(g) & \mathbb{0} \\ \mathbb{0} & \mathcal{D}_2(g) \end{pmatrix}, \quad \forall g \in G.$$

With the block-diagonal decomposition of $\mathcal{D}$, the representation $\mathcal{D}$ is said to be the *direct sum* of the representations $\mathcal{D}_1$ and $\mathcal{D}_2$, with each acting on their specific subspaces V_1 and V_2 of V, and denoted as

$$\mathcal{D} = \mathcal{D}_1 \oplus \mathcal{D}_2.$$

The original vector space V is thus the direct sum[3] of its subspaces, i.e.

$$V = V_1 \oplus V_2,$$

where the dimension of V is just the sum of the dimensions of V_1 and V_2 (assuming finite dimensional vector spaces). Hence, if $\boldsymbol{v_1} \in V_1$ and $\boldsymbol{v_2} \in V_2$, $\boldsymbol{v} = \boldsymbol{v_1} + \boldsymbol{v_2}$ is in V and

$$(\mathcal{D}_1(g) \oplus \mathcal{D}_2(g))(\boldsymbol{v_1} + \boldsymbol{v_2}) = \mathcal{D}_1(g)\boldsymbol{v_1} + \mathcal{D}_2(g)\boldsymbol{v_2} \in V.$$

(Here, I am referring again to the representative matrices.)

When a representation $\mathcal{D} \in \mathrm{GL}(V)$ is reducible but not completely reducible, there is a proper subspace that is invariant, but the supplementary subspace is not. The representative matrix then takes the form

$$\mathcal{D}(g) = \begin{pmatrix} \mathcal{D}_1(g) & \vdots & \mathcal{D}'(g) \\ \mathbb{0} & \vdots & \end{pmatrix}, \quad \forall g \in G.$$

Here, the proper space of $\mathcal{D}_1(g)$ is invariant, but that of $\mathcal{D}'(g)$ is not.

Finally, the following lemma and corollaries are important to study the irreducibility of representations (a concrete example of their use is shown in Section B.4).

Lemma A.1. (Schur's lemma) *Let $\mathcal{D}$ and $\mathcal{D}'$ be two irreducible representations on their respective space V and V'. Suppose φ, a linear operator from V to V', such that for all elements of the group G,*

$$\varphi\mathcal{D}(g) = \mathcal{D}'(g)\varphi, \quad \forall g \in G.$$

[3] The direct sum of two vector spaces V_1 and V_2 over a field $\mathbb{F}$ forms a vector space whose set corresponds to the Cartesian product of V_1 by V_2: V is thus the set of all ordered pairs $(\boldsymbol{v_1}, \boldsymbol{v_2})$, where $\boldsymbol{v_1} \in V_1$ and $\boldsymbol{v_2} \in V_2$. In the context of vector spaces, the ordered pairs $(\boldsymbol{v_1}, \boldsymbol{v_2})$ are commonly denoted as $\boldsymbol{v_1} + \boldsymbol{v_2}$. In addition to the Cartesian product construction, the direct sum specifies the algebraic structure: how to add two elements of V and how to multiply with scalars. Let $\boldsymbol{u_1}, \boldsymbol{v_1}$ be in V_1 and $\boldsymbol{u_2}, \boldsymbol{v_2}$ be in V_2. The Cartesian product of V_1 by V_2, $V = V_1 \times V_2$, tells us that $(\boldsymbol{u_1}, \boldsymbol{u_2}) \in V$ and $(\boldsymbol{v_1}, \boldsymbol{v_2}) \in V$. The algebraic structure required by the direct sum tells us that $(\boldsymbol{u_1}, \boldsymbol{u_2}) + (\boldsymbol{v_1}, \boldsymbol{v_2}) = (\boldsymbol{u_1} + \boldsymbol{v_1}, \boldsymbol{u_2} + \boldsymbol{v_2}) \in V$ and $\alpha \times (\boldsymbol{u_1}, \boldsymbol{u_2}) = (\alpha \times \boldsymbol{u_1}, \alpha \times \boldsymbol{u_2}) \in V$, with α being a scalar of the field $\mathbb{F}$. .

Then, either φ is a vector space isomorphism (bijective), and $\mathcal{D}(g)$ and $\mathcal{D}'(g)$ are two equivalent representations, or $\varphi = 0$.

Corollary A.1.1. *Any linear operator φ from V to V that commutes with an irreducible complex representation $\mathcal{D}(g)$ for all $g \in G$ is proportional to the identity operator.*

Corollary A.1.2. *If all operators φ that commute with a representation $\mathcal{D}(g)$ for all $g \in G$ are proportional to the identity operator, then $\mathcal{D}(g)$ is irreducible.*

The last corollary is just the reciprocal of the previous one. If $\mathcal{D}(g)$ were not irreducible, the projector on a non-trivial invariant vector subspace would be an operator, not proportional to the identity, that commutes with $\mathcal{D}(g)$.

A.2.4 Tensor Product Representations

We can form another kind of representation from the combination of two (or more) representations: the *tensor product representation*. Let $\mathcal{D}_1$ and $\mathcal{D}_2$ be two representations acting on their respective vector spaces V_1 and V_2. The tensor product representation (often called the direct product) is denoted as

$$\mathcal{D} = \mathcal{D}_1 \otimes \mathcal{D}_2.$$

It acts on the tensor product space[4] $V_1 \otimes V_2$ in the following way: if $\boldsymbol{u}$ and $\boldsymbol{v}$ are two vectors of respectively V_1 and V_2, then

$$\mathcal{D}(\boldsymbol{u} \otimes \boldsymbol{v}) = (\mathcal{D}_1 \boldsymbol{u}) \otimes (\mathcal{D}_2 \boldsymbol{v}) \in V_1 \otimes V_2.$$

Let $\{\hat{\boldsymbol{e}}^1_i\}$ be a basis of V_1 and $\{\hat{\boldsymbol{e}}^2_j\}$ a basis of V_2. And let us denote, in their respective basis, the representative matrices of the representations by the same symbols so that

$$\mathcal{D}_1 \hat{\boldsymbol{e}}^1_{i'} = \sum_i (\mathcal{D}_1)_{ii'} \hat{\boldsymbol{e}}^1_i, \quad \mathcal{D}_2 \hat{\boldsymbol{e}}^2_{j'} = \sum_j (\mathcal{D}_2)_{jj'} \hat{\boldsymbol{e}}^2_j.$$

Then, writing $\boldsymbol{u} = \sum_{i'} u_{i'} \hat{\boldsymbol{e}}^1_{i'}$ and $\boldsymbol{v} = \sum_{j'} v_{j'} \hat{\boldsymbol{e}}^2_{j'}$, it follows that

$$(\mathcal{D}_1 \boldsymbol{u}) \otimes (\mathcal{D}_2 \boldsymbol{v}) = \sum_{ii'jj'} (\mathcal{D}_1)_{ii'} (\mathcal{D}_2)_{jj'} \, u_{i'} v_{j'} \, \hat{\boldsymbol{e}}^1_i \otimes \hat{\boldsymbol{e}}^2_j.$$

We conclude that in the basis $\{\hat{\boldsymbol{e}}^1_i \otimes \hat{\boldsymbol{e}}^2_j\}$, the representative matrix of $\mathcal{D} = \mathcal{D}_1 \otimes \mathcal{D}_2$ is

$$\mathcal{D}_{ij,i'j'} = (\mathcal{D}_1)_{ii'} (\mathcal{D}_2)_{jj'}, \tag{A.6}$$

where the notation $ij, i'j'$ on the left-hand side denotes the component corresponding to the couple $(\hat{\boldsymbol{e}}^1_i \otimes \hat{\boldsymbol{e}}^2_j), (\hat{\boldsymbol{e}}^1_{i'} \otimes \hat{\boldsymbol{e}}^2_{j'})$. In the notation borrowed from quantum mechanics, it would be written as

$$\mathcal{D}_{ij,i'j'} = \langle (\hat{\boldsymbol{e}}^1_i \otimes \hat{\boldsymbol{e}}^2_j) | \mathcal{D} | (\hat{\boldsymbol{e}}^1_{i'} \otimes \hat{\boldsymbol{e}}^2_{j'}) \rangle = \langle \hat{\boldsymbol{e}}^1_i | \mathcal{D}_1 | \hat{\boldsymbol{e}}^1_{i'} \rangle \, \langle \hat{\boldsymbol{e}}^2_j | \mathcal{D}_2 | \hat{\boldsymbol{e}}^2_{j'} \rangle.$$

[4] We recall that if $\{\hat{\boldsymbol{e}}^1{}_{i=1,n}\}$ is a basis of V_1 and $\{\hat{\boldsymbol{e}}^2{}_{j=1,m}\}$ is a basis of V_2, then $\{\hat{\boldsymbol{e}}^1_i \otimes \hat{\boldsymbol{e}}^2_j\}$ is a basis of $W = V_1 \otimes V_2$, meaning that any elements of W can be written as $\sum_{ij} \alpha^{ij} \hat{\boldsymbol{e}}^1_i \otimes \hat{\boldsymbol{e}}^2_j$. The dimension of W is thus $\dim V_1 \times \dim V_2 = n \times m$. Note that unlike the direct sum $W = V_1 \oplus V_2$, where any element of W can be written from an element of V_1 and V_2, i.e. $w = \boldsymbol{v_1} + \boldsymbol{v_2}$, it is not the case for $W = V_1 \otimes V_2$. There are not necessarily two vectors, $\boldsymbol{v_1}$ and $\boldsymbol{v_2}$, such as $\boldsymbol{w} = \boldsymbol{v_1} \otimes \boldsymbol{v_2}$. Another difference is the scalar multiplication: while in $V_1 \oplus V_2$, we have $\alpha(\boldsymbol{v_1} + \boldsymbol{v_2}) = \alpha \boldsymbol{v_1} + \alpha \boldsymbol{v_2}$, in $V_1 \otimes V_2$, we have $\alpha(\boldsymbol{v_1} \otimes \boldsymbol{v_2}) = (\alpha \boldsymbol{v_1}) \otimes \boldsymbol{v_2} = \boldsymbol{v_1} \otimes (\alpha \boldsymbol{v_2})$.

Note that, in general, the tensor representation $\mathcal{D}_1 \otimes \mathcal{D}_2$ is not irreducible. For instance, the reader must be familiar with the decomposition of a tensor representation in the context of the addition of angular momenta. Denoting by $\mathcal{D}^{2j+1}$ the spin-j representation (see Appendix B), with $2j+1$ being the dimension of the representation, we have, for SU(2),

$$\mathcal{D}^{2j+1} \otimes \mathcal{D}^{2j'+1} = \bigoplus_{n=|j-j'|}^{j+j} \mathcal{D}^{2n+1}.$$

A.2.5 Conjugate, Contragredient and Unitary Representations

- **Conjugate representations.**

Once a representative matrix $\mathcal{D}$ of a representation of a group G is chosen, one can easily take its complex conjugate $\mathcal{D}^*$. It is then straightforward to check that $\mathcal{D}^*$ constitutes a representation since, for all elements g_1 and g_2 in G,

$$\mathcal{D}^*(g_1 \times g_2) = \mathcal{D}^*(g_1)\mathcal{D}^*(g_2).$$

Conjugate representations play an important role in particle physics since the elementary antiparticles belong to them.

- **Contragredient representations.**

Another possible construction of a representation from $\mathcal{D}$ relies on its transpose. The *contragredient representation* is defined as

$$\overline{\mathcal{D}}(g) = \mathcal{D}^{-1^{\mathsf{T}}}(g) = \mathcal{D}^{\mathsf{T}}(g^{-1}), \tag{A.7}$$

where the last equality is a consequence of Eq. (A.3).

- **Contragredient conjugate representations.**

In the same spirit as before, one can build the *contragredient conjugate representation* defined as

$$\overline{\mathcal{D}}^*(g) = \mathcal{D}^{-1^{\mathsf{T}*}}(g) = \mathcal{D}^{-1^{\dagger}}(g) = \mathcal{D}^{\dagger}(g^{-1}). \tag{A.8}$$

- **Unitary representations.**

Unitary representations are of primary importance, especially in quantum mechanics. A representation of a group G acting on a vector space V is called unitary if for all element g in G, the operator $\mathcal{D}(g)$ is unitary, i.e. $\mathcal{D}(g))^{\dagger}\mathcal{D}(g) = \mathcal{D}(g)\mathcal{D}(g))^{\dagger} = I$, where I is the identity operator. In quantum mechanics, V is a Hilbert space endowed with the Hermitian scalar product. If ψ, and φ are two states of V, and ψ' and φ' are their transformation by $\mathcal{D}(g)$, then

$$\langle\psi'|\varphi'\rangle = \langle\psi(\mathcal{D}(g))^{\dagger}\mathcal{D}(g)|\varphi\rangle = \langle\psi|\varphi\rangle.$$

Hence, the physics of the system is invariant (the probabilities are conserved), and therefore, the transformation represented by $\mathcal{D}(g)$ is a symmetry. Note that anti-unitary operators[5] can also represent symmetries (Wigner's theorem).

Unitary representations satisfy the following important properties [proofs can be found in Bump (2004)].

Theorem A.2. *Any representation of a finite or compact group on a Hilbert space is equivalent to a unitary representation.*

Theorem A.3. *Any unitary representation that is reducible is completely reducible.*

Therefore, if $\mathcal{D}$ is a reducible representation of a finite or compact group on a Hilbert space, then it is completely reducible.

Finally, note that for unitary representations whose representative matrices satisfy $\mathcal{D}^\dagger = \mathcal{D}^{-1}$, $\overline{\mathcal{D}}(g) = \mathcal{D}^{\dagger\,\mathsf{T}}(g) = \mathcal{D}^*(g)$, the conjugate and contragredient representations are the same. Obviously, the contragredient conjugate representation $\overline{\mathcal{D}}^*(g)$ is $\mathcal{D}(g)$.

A.3 Lie Groups

A.3.1 Definition, Generators

In physics, Lie groups are the most important groups because they give rise to conserved currents via the Noether theorem. The elements of a finite Lie group can be labelled by a finite set of continuous real parameters. These parameters then play the role of the (local) coordinates on a manifold corresponding to the complete parameters space. The set of real numbers, $\mathbb{R}$, obviously forms a Lie group, with the label being the number itself (so, a Lie group of one dimension). A more useful example is $\mathrm{GL}(n, \mathbb{F})$. It is set of the non-degenerate linear transformations on an n-dimensional vector space whose field $\mathbb{F}$ is, in practice, $\mathbb{C}$ or $\mathbb{R}$ [obviously, $\mathrm{GL}(n, \mathbb{C})$ contains $\mathrm{GL}(n, \mathbb{R})$]. This group is important because, in particle physics, the transformations are (almost) always linear: quantum mechanics is based on the linear algebra, gauge transformations are linear, so are Lorentz transformations, etc. We saw that linear transformations are represented by $n \times n$ matrices acting on the vector space, with the group composition law being the usual matrix multiplication for that representation. Since any element of the group must have an inverse, the matrices are invertible with non-zero determinants. Hence, $\mathrm{GL}(n, \mathbb{F})$ is the set of non-singular $n \times n$ matrices with elements from the field $\mathbb{F}$ (i.e., real or complex). Its dimension is n^2 when $\mathbb{F} = \mathbb{R}$ and $2n^2$ when $\mathbb{F} = \mathbb{C}$ if the Lie algebra (see below) is considered a real algebra or n^2 if it is a complex algebra. The group $\mathrm{GL}(n, \mathbb{F})$ is non-Abelian since matrix multiplication is generally non-commutative. Particle physics, being mostly interested by real or complex linear transformations, all Lie groups of interest are necessarily (closed) subgroups[6] of $\mathrm{GL}(n, \mathbb{F})$

[5] If U is an anti-unitary operator, $U^\dagger U = UU^\dagger = I$, but $\langle U\psi|U\varphi\rangle = \langle\psi|\varphi\rangle^*$, and thus, $U\lambda\,|\psi\rangle = \lambda^* U\,|\psi\rangle$, with $\lambda \in \mathbb{C}$.

[6] A closed subgroup of $\mathrm{GL}(n, \mathbb{R})$ or $\mathrm{GL}(n, \mathbb{C})$ is called a matrix Lie group.

since all possible linear transformations belong to this group. Therefore, in what follows, we will restrict ourselves to continuous groups of finite-size matrices.

The manifold of Lie groups has an important property: it is k-differentiable. An appropriate definition of differentiable manifolds requires a high level of mathematics in differential geometry (the interested reader can consult Dubrovin et al., 1984). Fortunately, most symmetry groups of interest to particle physicists can be described by real or complex matrices. Qualitatively, the consequence of a k-differentiable manifold is that an element of that group can be reached from an element infinitesimally close. In particular, one can expand a group element g near the identity element represented by the matrix $\mathbb{1}$, such that

$$g(\mathrm{d}\alpha) = \mathbb{1} - i\,\mathrm{d}\alpha\,\hat{G}, \quad \mathrm{d}\alpha \in \mathbb{R}, \tag{A.9}$$

where we suppose here that the Lie group depends on a single real continuous infinitesimal parameter $\mathrm{d}\alpha$. The symbol $\hat{G}$ is called the *infinitesimal generator* of the symmetry group (represented by a matrix in our case). The factor $-i$ is conventional among physicists because, with unitary transformations used in quantum mechanics, the generator is Hermitian (as shown in Section 4.2). Let us assume that under the group composition[7]

$$g(\alpha_1)g(\alpha_2) = g(\alpha_1 + \alpha_2), \quad \forall \alpha_1, \alpha_2 \in \mathbb{R}.$$

Defining the finite parameter as $\alpha = N\,\mathrm{d}\alpha$, with N being an integer, it follows that

$$g(\alpha) = (g(\mathrm{d}\alpha))^N = \left(\mathbb{1} - i\frac{\alpha}{N}\hat{G}\right)^N.$$

Hence, taking the limit $N \to \infty$, we conclude that a finite element can be obtained from the infinitesimal generator via

$$g(\alpha) = \exp(-i\alpha\hat{G}) \equiv \sum_{n=0}^{\infty} \frac{(-i\alpha)^n}{n!}(\hat{G})^n, \quad \alpha \in \mathbb{R}, \tag{A.10}$$

with α being the real continuous parameter of the Lie group. By definition of Lie groups, for all $\alpha \in \mathbb{R}$, $g(\alpha)$ is an element of the Lie group. Conversely, once an element $g(\alpha)$ of the group is known, we can recover the generator either with the identification of the terms in the infinitesimal development (A.9) or by differentiation,

$$\hat{G} = i\frac{\partial g(\alpha)}{\partial \alpha}\bigg|_{\alpha=0}. \tag{A.11}$$

Note that the 'exponentiation' (A.10) does not mean that all elements of the group can be obtained. It would imply that any element can be reached from the identity element. It is not always the case. For instance, GL(n, $\mathbb{R}$) is represented by matrices with a non-zero determinant. Since the parameters of this Lie group can vary continuously, the value of the determinant varies continuously. Starting from the identity, which has a positive determinant, if we want to reach an element with a negative determinant, we would have to go through a matrix with a vanishing determinant, which does not belong to the group. When continuous variations can take us to every group element, the group is said *connected*.

[7] Strictly speaking, one should define a continuous homomorphism from $\mathbb{R}$ to G called a *one-parameter subgroup* of the Lie group G. The reader can consult Bump (2004), Hall (2015) and Helgason (1978) for an appropriate definition beyond the simple matrix Lie groups.

When it is not possible, the group has several *components*. In the case of $GL(n, \mathbb{R})$, there are two components: matrices with positive or negative determinants.

A.3.2 Lie Algebra

An n-dimensional Lie group depends on n continuous parameters $\{\alpha_{a=1,\cdots,n}\}$, and there are as many generators $\hat{G}_a$ as the number of parameters. The group elements of the group component connected to the identity can then be reconstructed via the exponential map,

$$g(\{\alpha_a\}) = \exp\Big(- i\sum_{a=1}^{n} \alpha_a \hat{G}_a\Big), \quad \alpha_a \in \mathbb{R}. \tag{A.12}$$

The closure property of the group requires that the product (in the sense of the composition law) of two elements is also an element. Therefore, with two elements, $g(\{\alpha_a\})$ and $g(\{\alpha'_a\})$, there must be a set of parameters $\{\alpha''_a\}$ such as

$$\exp\Big(- i\sum_{a} \alpha''_a \hat{G}_a\Big) = \exp\Big(- i\sum_{a} \alpha'_a \hat{G}_a\Big)\exp\Big(- i\sum_{a} \alpha_a \hat{G}_a\Big). \tag{A.13}$$

The product of two exponentials of matrices is given by the Baker–Campbell–Hausdorff formula:

$$e^X e^Y = e^Z, \quad \text{with} \quad Z = X + Y + \frac{1}{2}[X,Y] + \frac{1}{12}\big[X,[X,Y]\big] + \frac{1}{12}\big[[X,Y],Y\big] + \cdots, \tag{A.14}$$

where the dots include combinations of the commutator $[X, Y]$ of higher orders. Formula (A.13) then imposes

$$-i\sum_a \alpha''_a \hat{G}_a = -i\sum_a (\alpha'_a + \alpha_a)\hat{G}_a - \frac{1}{2}\sum_{a,b} \alpha'_a \alpha_b [\hat{G}_a, \hat{G}_b] + \cdots$$

We conclude that Formula (A.13) is satisfied only if the commutator between two generators is a linear combination of generators, i.e.

$$[\hat{G}_a, \hat{G}_b] = i\sum_{c=1}^{n} f_{abc}\hat{G}_c, \tag{A.15}$$

where f_{abc} are the real constants called the *structure constants* of the group. The anti-symmetry of the commutator imposes

$$f_{abc} = -f_{bac}. \tag{A.16}$$

In addition, we can exploit the Jacobi identity of commutators,

$$\big[\hat{G}_a, [\hat{G}_b, \hat{G}_c]\big] + \big[\hat{G}_b, [\hat{G}_c, \hat{G}_a]\big] + \big[\hat{G}_c, [\hat{G}_a, \hat{G}_b]\big] = 0, \tag{A.17}$$

to deduce another equality satisfied by the structure constants. Indeed, according to the commutators (A.15),

$$\big[\hat{G}_a, [\hat{G}_b, \hat{G}_c]\big] = i\sum_d f_{bcd}[\hat{G}_a, \hat{G}_d] = -\sum_{d,e} f_{bcd} f_{ade} \hat{G}_e.$$

Hence, the Jacobi identity (A.17) is equivalent to

$$\sum_{d,e}(f_{bcd}f_{ade} + f_{cad}f_{bde} + f_{abd}f_{cde})\hat{G}_e = 0.$$

But, as all $\hat{G}_e$s are independent, it imposes that for all a, b, c, e,

$$\sum_{d=1}^{n} f_{bcd}f_{ade} + f_{cad}f_{bde} + f_{abd}f_{cde} = 0. \tag{A.18}$$

The set of generators $\hat{G}_{a=1,\cdots,n}$ can be seen as a basis of an n-dimensional vector space (generators can be added, multiplied by scalars, etc.), called the *Lie algebra*, spanned by the linear combinations $X = \sum_a \alpha_a \hat{G}_a$. The Lie algebra $\mathfrak{g}$ of a group G is first an *algebra*, that is, a vector space over a field $\mathbb{F}$ endowed with a bilinear product operation denoted $\star$ below, i.e.

$$\text{algebra:} \begin{cases} (\alpha_1 X_1 + \alpha_2 X_2) \star X_3 & = \alpha_1 X_1 \star X_3 + \alpha_2 X_2 \star X_3 \in \mathfrak{g}, \\ X_1 \star (\alpha_1 X_2 + \alpha_2 X_3) & = \alpha_1 X_1 \star X_2 + \alpha_2 X_1 \star X_3 \in \mathfrak{g}, \end{cases} \tag{A.19}$$

valid for all vectors, Xs, of the vector space and all scalars, αs, in $\mathbb{F}$ (in our case $\mathbb{C}$ or $\mathbb{R}$). The Lie algebra has the following additional properties:

$$\text{Lie algebra:} \begin{cases} X_1 \star X_2 = -X_2 \star X_1, \\ (X_1 \star X_2) \star X_3 + (X_2 \star X_3) \star X_1 + (X_3 \star X_1) \star X_2 = 0. \end{cases} \tag{A.20}$$

The bilinear product operation of a Lie algebra is usually called the *Lie bracket* and conventionally denoted $[.,.]$ instead of $\star$. For matrix Lie groups, the Lie bracket is simply the commutator of matrices, which satisfies all Lie algebra properties. The good point is that there is a theorem that states

Theorem A.4 (Ado's theorem). *Every finite-dimensional real or complex Lie algebra is (isomorphic to) a Lie algebra of square matrices, where the Lie bracket is the commutator of matrices.*

It is customary to refer to the Lie algebra of a group by using lower case Gothic characters corresponding to the group name. For example, $\mathfrak{su}(2)$ for SU(2).

Mathematicians usually use the following exponential map with the infinitesimal generator:

$$X \in \mathfrak{g} \mapsto \exp(X) \in G. \tag{A.21}$$

With the physicist's definition of the generators (A.12), since $\exp(X) = \exp(-i(iX))$, the generators $\hat{G} = iX$, and thus, the algebra generated by $\hat{G}$ is $i\mathfrak{g}$. In summary,

$$\text{Generators } \hat{G} \text{ for physics: } \hat{G} = iX \in i\mathfrak{g}. \tag{A.22}$$

A.3.3 Representation of Lie Algebra

As for representations of a group G, which are homomorphisms from G to $\mathrm{GL}(V)$, with V being a finite-dimensional real or complex vector space, representations of a Lie algebra

$\mathfrak{g}$ are Lie algebra homomorphisms from $\mathfrak{g}$ to $\mathfrak{gl}(V)$, where for matrix Lie groups, $\mathfrak{gl}(V)$ denotes the space of all linear operators from V to itself [i.e., End(V)]. The Lie algebra homomorphism must preserve the commutation relations of the algebra elements [and not the composition law as in representations of groups in Eq. (A.1)], implying that if X and X' are two elements of the algebra, then the representation d satisfies

$$d([X,X']) = [d(X), d(X')]. \tag{A.23}$$

Then, given the representation d of the Lie algebra $\mathfrak{g}$ acting on the vector space V, a representation $\mathcal{D}$ of the group G on V can be determined with the exponential map procedure:

$$X \in \mathfrak{g}, \quad X \overset{d}{\mapsto} d(X), \qquad g \in G, \quad g = e^X \overset{\mathcal{D}}{\mapsto} \mathcal{D}(g) = e^{d(X)}. \tag{A.24}$$

This is the recipe to easily find the representations of the group G. Note, however, that two different groups can share the same Lie algebra, as we shall see in the next section. Therefore, the 'exponentiation' procedure may not only give a representation of the Lie group of interest, but a representation of a larger group. An example is shown with SO(3) and SU(2) in Sections B.2 and B.4. Note, also, that if $d(X)$ and $d'(X)$ are two equivalent representations of the Lie algebra, i.e. $d'(X) = Md(X)M^{-1}$, then the expansion of the exponential series implies $e^{d'(X)} = Me^{d(X)}M^{-1}$, and therefore, the two representations of the group are also equivalent.

There is a particular representation of the Lie algebra called the *adjoint representation*. It acts directly on the vector space of the Lie algebra (the one whose basis is given by the generators) and corresponds to matrices whose elements are given by the structure constants. More specifically, consider the matrix T_a with elements

$$(T_a)_{bc} = -if_{abc}. \tag{A.25}$$

It is easy to show that the T_a's matrices then satisfy the commutation rule (A.15), i.e.

$$[T_a, T_b] = i\sum_c f_{abc}T_c. \tag{A.26}$$

Therefore, they constitute a representation of the Lie algebra.

A.4 Short Description of the Main Matrix Lie Groups and Their Algebra

A.4.1 Preamble about Matrix Lie Groups

Let G be a matrix Lie group (real or complex) and $\mathfrak{g}$ its Lie algebra. By definition of an algebra (A.19), if X is a matrix in $\mathfrak{g}$, then for all real numbers t, the matrix tX is in $\mathfrak{g}$. The Lie algebra $\mathfrak{g}$ is thus the set of all matrices X, such that $\exp(tX)$ is in G for all real numbers t (see Hall, 2015), and the Lie bracket is simply the commutator of matrices. We will use this definition extensively in what follows.

A.4.2 General Linear Groups GL(n, $\mathbb{R}$) and GL(n, $\mathbb{C}$)

Let us consider the group $\mathrm{GL}(n, \mathbb{F})$, the set of $n \times n$ invertible matrices with coefficients in $\mathbb{F} = \mathbb{R}$ or $\mathbb{C}$,

$$\mathrm{GL}(n, \mathbb{F}) = \{M \in M_n(\mathbb{F}) \mid \det(M) \neq 0\}, \tag{A.27}$$

where $M_n(\mathbb{F})$ is the space of all $n \times n$ matrices with entries in $\mathbb{F}$. Its Lie algebra using the exponential map (A.21) is simply

$$\mathfrak{gl}(n, \mathbb{F}) = \{X \in M_n(\mathbb{F})\}. \tag{A.28}$$

Let us check it with $\mathrm{GL}(n, \mathbb{R})$. If X is in $M_n(\mathbb{R})$, then $\exp(tX)$ is an $n \times n$ matrix, invertible (even if X is not) and real for all t, so that by definition $X \in \mathfrak{gl}(n, \mathbb{R})$. Conversely, if $\exp(tX)$ is a real $n \times n$ real matrix, then, for all t,

$$X = \frac{\mathrm{d}}{\mathrm{d}t} \exp(tX)\bigg|_{t=0}$$

is an $n \times n$ real matrix too. The dimension of $\mathrm{GL}(n, \mathbb{R})$ is n^2, and if $\mathfrak{gl}(n, \mathbb{C})$ is considered a real algebra (even if the matrices are complex), the dimension of $\mathrm{GL}(n, \mathbb{C})$ is double.

A.4.3 Special Linear Group SL(n, $\mathbb{C}$)

The special linear group $\mathrm{SL}(n, \mathbb{C})$ is a subgroup of $\mathrm{GL}(n, \mathbb{C})$ restricted to matrices with determinant one,

$$\mathrm{SL}(n, \mathbb{F}) = \{M \in \mathrm{GL}(n, \mathbb{C}) \mid \det(M) = 1\}. \tag{A.29}$$

- **Lie algebra.**

Let us find the matrices X such as $M = \exp(tX)$ is in $\mathrm{SL}(n, \mathbb{C})$ for all real t. First, consider the following theorem:

Property A.5. If X is an $n \times n$ real or complex matrix, then

$$\det(e^X) = e^{\mathrm{Tr}(X)}, \tag{A.30}$$

where det and Tr denote the determinant and the trace of matrices respectively.
Consequently, the Lie algebra of $\mathfrak{sl}(n, \mathbb{C})$ is

$$\mathfrak{sl}(n, \mathbb{C}) = \{X \in M_n(\mathbb{C}) \mid \mathrm{Tr}(\mathrm{X}) = 0\}. \tag{A.31}$$

Given the property, $\mathrm{Tr}(AB) = \mathrm{Tr}(BA)$, the trace of a commutator is necessarily zero. Therefore, $\mathfrak{sl}(n, \mathbb{C})$ is closed under the matrix commutator. The dimension of $\mathfrak{sl}(n, \mathbb{C})$ depends on whether we consider a real or complex Lie algebra (i.e., real or complex linear combinations of the generators). In the former case, the dimension is $2n^2 - 2$, whereas in the latter, it is $n^2 - 1$. If not specified, $\mathfrak{sl}(n, \mathbb{C})$ is usually considered a complex Lie algebra.

Let us give more details about $\mathfrak{sl}(2, \mathbb{C})$, the 2×2 complex traceless matrices,

$$\mathfrak{sl}(2, \mathbb{C}) = \left\{ \begin{pmatrix} \gamma & \beta \\ \delta & -\gamma \end{pmatrix} \mid \gamma, \beta, \delta \in \mathbb{C} \right\}.$$

This algebra plays an important role in the context of the Lorentz group (see Appendix D). As $\mathfrak{sl}(2,\mathbb{C})$ is a three-dimensional vector space, a possible basis is given by the three matrices

$$H = \begin{pmatrix} 1 & 0 \\ 0 & -1 \end{pmatrix}, \quad X_+ = \begin{pmatrix} 0 & 1 \\ 0 & 0 \end{pmatrix}, \quad X_- = \begin{pmatrix} 0 & 0 \\ 1 & 0 \end{pmatrix}.$$

However, if $\mathfrak{sl}(2,\mathbb{C})$ is treated as a real Lie algebra, only real linear combinations are allowed, and thus one has to add the three matrices iH, iX_+ and iX_- to constitute a six-dimensional vector space.

A.4.4 Orthogonal Group O(n)

The orthogonal group $\mathrm{O}(n)$ contains all $n \times n$ orthogonal real matrices, i.e. matrices verifying $MM^\intercal = M^\intercal M = \mathbb{1}$. Their determinant is thus necessarily $\det(M) = \pm 1 \neq 0$, and thus $\mathrm{O}(n)$ is a subgroup of $\mathrm{GL}(n,\mathbb{R})$:

$$\mathrm{O}(n) = \{M \in \mathrm{GL}(n,\mathbb{R}) \mid MM^\intercal = M^\intercal M = \mathbb{1}\}. \tag{A.32}$$

Its dimension is thus $n(n-1)/2$. Note that if $\boldsymbol{x}$ is a real vector $\boldsymbol{x} = (x_1, x_2, \ldots, x_n)^\intercal$, then the quadratic form

$$q(\boldsymbol{x}) = \boldsymbol{x}^\intercal \boldsymbol{x} = x_1^2 + x_2^2 + \cdots + x_n^2$$

is preserved when $\boldsymbol{x}$ is transformed by a matrix of $\mathrm{O}(n)$. The check is straightforward: for $\boldsymbol{x}' = M\boldsymbol{x}$, $q(\boldsymbol{x}') = \boldsymbol{x}^\intercal M^\intercal M \boldsymbol{x} = q(\boldsymbol{x})$.

- **Lie algebra.**

Let us find the matrices X, such as $M = \exp(tX)$ is in $\mathrm{O}(n)$ for all real t. By definition of $\mathrm{O}(n)$, the matrix M must satisfy $M^\intercal = M^{-1}$. Using the development of the exponential, it is clear that

$$M^\intercal = \sum_n \frac{t^n}{n!}(X^n)^\intercal = \exp(tX^\intercal).$$

In addition, $M^{-1} = \exp(-tX)$ since $\exp(tX)\exp(-tX) = \mathbb{1}$ according to Baker–Campbell–Hausdorff Formula (A.14). Hence, the matrix X is such that for all t,

$$\sum_n \frac{t^n}{n!}(X^\intercal)^n = \sum_n \frac{t^n}{n!}(-X)^n.$$

The polynomials in t on both sides of the equation must be equal. This implies that X is a skew-symmetric matrix, i.e satisfying $X + X^\intercal = 0$. Conversely, if X is skew-symmetric, then $M = \exp(X)$ is such that $M^\intercal = \exp(X^\intercal) = \exp(-X) = M^{-1}$, and so $M \in \mathrm{O}(n)$. Hence, let us define

$$\mathfrak{o}(n) = \{X \in M_n(\mathbb{R}) \mid X + X^\intercal = 0\}. \tag{A.33}$$

It is easy to check that $\mathfrak{o}(n)$ is closed under the Lie bracket (the matrix commutator) since if $X, X' \in \mathfrak{o}(n)$, $[X,X']^\intercal = X'^\intercal X^\intercal - X^\intercal X'^\intercal = X'X - XX' = -[X,X']$, and thus $[X,X'] \in \mathfrak{o}(n)$. Therefore, $\mathfrak{o}(n)$ has the property of a Lie algebra.

A.4.5 Special Orthogonal Group SO (n)

The special orthogonal group SO(n) is a subgroup of O(n) restricted to matrices with determinant 1,

$$\mathrm{SO}(n) = \{M \in \mathrm{O}(n) \mid \det(M) = 1\}. \tag{A.34}$$

The elements of SO(n) correspond to rotations in an n-dimensional space. The dimension of SO(n) is $n(n-1)/2$.

- **Lie algebra.**

The Lie algebra of SO(n) is actually the same as that of O(n). This is not surprising because O(n) is a group with two components, and the component containing the identity is just SO(n) (remember that the Lie algebra only concerns transformations close to the identity). This can easily be checked: as matrices of $\mathfrak{o}(n)$ satisfy $X + X^{\mathsf{T}} = 0$, they are traceless [$\mathrm{Tr}(X^{\mathsf{T}}) = \mathrm{Tr}(X)$ for a square matrix], and so are matrices of $\mathfrak{so}(n)$, i.e.

$$\mathfrak{so}(n) = \mathfrak{o}(n) = \{X \in M_n(\mathbb{R}) \mid X + X^{\mathsf{T}} = 0, \mathrm{Tr}(X) = 0\}. \tag{A.35}$$

The property (A.30) implies that the elements of SO(n) have determinant 1.

It is worth noting that for physics [see identity (A.22)], the Lie algebra generated by the generators $\hat{G}$ corresponds to matrices belonging to

$$\mathfrak{so}(n) = \{X \in iM_n(\mathbb{R}) \mid X + X^{\mathsf{T}} = 0, \mathrm{Tr}(X) = 0\}.$$

The matrix iX is thus real, with $X^* = -X$. This implies that $X^\dagger = (X^*)^{\mathsf{T}} = -X^{\mathsf{T}}$. On the other hand, since $X + X^{\mathsf{T}} = 0$, we conclude that the matrix X is necessarily Hermitian, and therefore,

$$\mathfrak{so}(n) = \{X \in iM_n(\mathbb{R}) \mid X^\dagger = X, \mathrm{Tr}(X) = 0\}. \tag{A.36}$$

The generators can then describe observables, i.e. physical quantities.

- **Generators.**

The role played by rotations in physics is of primary importance. Therefore, this section gives some details about SO(3), with the group of rotations in the three-dimensional space. An expression of the matrix representing a rotation of an angle α about an axis defined by the unit vector $\hat{\boldsymbol{n}}$ can be easily deduced from the Rodrigues formula,

$$\boldsymbol{x} \xrightarrow{R_{\hat{\boldsymbol{n}}}(\alpha)} \boldsymbol{x}' = R_{\hat{\boldsymbol{n}}}(\alpha)\boldsymbol{x} = \cos\alpha\,\boldsymbol{x} + (1-\cos\alpha)(\boldsymbol{x}\cdot\hat{\boldsymbol{n}})\hat{\boldsymbol{n}} + \sin\alpha(\hat{\boldsymbol{n}} \times \boldsymbol{x}). \tag{A.37}$$

For instance, the three matrices corresponding to rotations about the x, y and z axes are

$$R_x(\alpha) = \begin{pmatrix} 1 & 0 & 0 \\ 0 & \cos\alpha & -\sin\alpha \\ 0 & \sin\alpha & \cos\alpha \end{pmatrix}, R_y(\alpha) = \begin{pmatrix} \cos\alpha & 0 & \sin\alpha \\ 0 & 1 & 0 \\ -\sin\alpha & 0 & \cos\alpha \end{pmatrix}, R_z(\alpha) = \begin{pmatrix} \cos\alpha & -\sin\alpha & 0 \\ \sin\alpha & \cos\alpha & 0 \\ 0 & 0 & 1 \end{pmatrix}. \tag{A.38}$$

Note that we consider active transformations here, namely transformations with a fixed frame in which vectors are transformed (see the discussion in Section 2.3.4). A rotation requires three (continuous) real parameters, for instance, the two angles $\theta \in [0, \pi]$ and $\varphi \in [0, 2\pi]$ defining $\hat{\boldsymbol{n}}$, and the angle α restricted to $[0, \pi]$ since $R_{\hat{\boldsymbol{n}}}(\alpha) = R_{-\hat{\boldsymbol{n}}}(2\pi - \alpha)$.

As three-dimensional rotations depend on three parameters, SO(3) has necessarily three generators, whose matrix expression can be easily deduced from the matrices above. Using the exponential map for physics in Eq. (A.12), we have, for instance, $R_x(\alpha) = \exp(-i\alpha J_1)$, with J_1 being the (Hermitian) generator of a rotation about the x-axis given by

$$J_1 = i\,\frac{\partial}{\partial\alpha} R_x(\alpha)\bigg|_{\alpha=0} = \begin{pmatrix} 0 & 0 & 0 \\ 0 & 0 & -i \\ 0 & i & 0 \end{pmatrix}. \tag{A.39}$$

Similarly, the infinitesimal generators of rotations about the two other axes are deduced from the matrices $R_y(\alpha)$ and $R_z(\alpha)$ in Eq. (A.38). They read

$$J_2 = \begin{pmatrix} 0 & 0 & i \\ 0 & 0 & 0 \\ -i & 0 & 0 \end{pmatrix}, \quad J_3 = \begin{pmatrix} 0 & -i & 0 \\ i & 0 & 0 \\ 0 & 0 & 0 \end{pmatrix}. \tag{A.40}$$

As expected, the generators are represented by Hermitian traceless matrices. The three matrices of the generators satisfy the commutation relation

$$[J_i, J_j] = i\epsilon_{ijk} J_k, \tag{A.41}$$

where ϵ_{ijk} is the spatial antisymmetric tensor, and the sum over k is implicit. Note also that those matrices are simply defined by

$$(J_k)_{ij} = -i\epsilon_{ijk}. \tag{A.42}$$

Hence, we have just found the adjoint representation of the Lie algebra $\mathfrak{so}(3)$, with the spatial antisymmetric tensor playing the role of the structure constants of the Lie group. Note that this result is obtained immediately from the linearisation of the Rodrigues Formula (A.37): $\boldsymbol{x}' \simeq \boldsymbol{x} + \alpha\,\hat{\boldsymbol{n}} \times \boldsymbol{x}$. Given that the ith component of the vector $(\hat{\boldsymbol{n}} \times \boldsymbol{x})$ is $\epsilon_{ijk}\hat{n}_j x_k$ (with an implicit summation over j and k), it follows that

$$x'_i = x_i - i\alpha\,(i\epsilon_{ijk})\hat{n}_j x_k = x_i - i\alpha\,(-i\epsilon_{ikj})\hat{n}_j x_k = x_i - i\alpha\,(J_j)_{ik}\,\hat{n}_j x_k = x_i - i\alpha\,(\hat{\boldsymbol{n}}\cdot\boldsymbol{J})_{ik}\,x_k.$$

The formula above just corresponds to the linearisation of a rotation about the $\hat{\boldsymbol{n}}$ axis with an angle α whose general expression is

$$R_{\hat{\boldsymbol{n}}}(\alpha) = \exp(-i\alpha\,\hat{\boldsymbol{n}}\cdot\boldsymbol{J}) \quad \text{with} \quad \boldsymbol{J} = \begin{pmatrix} J_1 \\ J_2 \\ J_3 \end{pmatrix}, \tag{A.43}$$

or equivalently

$$R(\boldsymbol{\alpha}) = \exp(-i\boldsymbol{\alpha}\cdot\boldsymbol{J}), \tag{A.44}$$

where the length of the vector $\boldsymbol{\alpha}$ is the rotation angle $\alpha \in [0, \pi]$, and its direction coincides with the axis of rotation.

A.4.6 Generalised Orthogonal Groups O(*n*, *m*) and SO(*n*, *m*)

Let $\boldsymbol{x}$ be a real vector in $\mathbb{R}^{n+m}$, $\boldsymbol{x} = (x_1, x_2, \ldots, x_n, x_{n+1}, \ldots, x_{n+m})^{\mathsf{T}}$. By definition, the generalised orthogonal group $\mathrm{O}(n, m)$ preserves the quadratic form

$$q(\boldsymbol{x}) = \sum_{i=1}^{n} x_i^2 - \sum_{i=1}^{m} x_{n+i}^2.$$

Defining the $(n+m) \times (n+m)$ matrix $I_{n,m}$ by

$$I_{n,m} = \begin{pmatrix} \mathbb{1}_n & 0 \\ 0 & -\mathbb{1}_m \end{pmatrix},$$

the group $\mathrm{O}(n, m)$ is thus

$$\mathrm{O}(n, m) = \{M \in M_{n+m}(\mathbb{R}) \mid M^{\mathsf{T}} I_{n,m} M = I_{n,m}\}. \tag{A.45}$$

Matrices M of $\mathrm{O}(n, m)$ then have a determinant $\det(M) = \pm 1$. It is thus a subgroup of $\mathrm{GL}(n+m, \mathbb{R})$ and has the dimension $(n+m)(n+m-1)/2$. The special generalised orthogonal group $\mathrm{SO}(n, m)$ is the subgroup of $\mathrm{O}(n, m)$ with $\det(M) = 1$,

$$\mathrm{SO}(n, m) = \{M \in \mathrm{O}(n, m) \mid \det(M) = 1\}. \tag{A.46}$$

The Lie algebra of $\mathrm{O}(n, m)$ or $\mathrm{SO}(n, m)$ [both are the same $\mathfrak{o}(n, m) = \mathfrak{so}(n, m)$] is described in detail in Section D.1 devoted to the Lorentz group.

A.4.7 Unitary Groups U(*n*) and SU(*n*)

The set of all $n \times n$ unitary matrices is the unitary group $\mathrm{U}(n)$. Recall that a matrix M is unitary if

$$M^\dagger M = M M^\dagger = \mathbb{1}.$$

The determinant of matrices of $\mathrm{U}(n)$ verifies $|\det(M)| = 1$. Hence, $\mathrm{U}(n)$ is a subgroup of $\mathrm{GL}(n, \mathbb{C})$,

$$\mathrm{U}(n) = \{M \in \mathrm{GL}(n, \mathbb{C}) \mid M^\dagger M = M M^\dagger = \mathbb{1}\}. \tag{A.47}$$

$\mathrm{U}(n)$ is a real Lie group[8] of dimension n^2. Indeed, a matrix of $n \times n$ has $2n^2$ real parameters. The unitary constraint implies that $M^\dagger M$ is Hermitian. Therefore, the diagonal of $M^\dagger M$ is real, which removes n parameters, and the elements of the lower triangle are the complex conjugate of those of the upper triangle, removing $(2n^2 - 2n)/2$ real parameters. Overall, this leaves n^2 real parameters. By construction, $\mathrm{U}(n)$ preserves the Hermitian product

$$\langle z, z' \rangle = \sum_{i=1}^{n} z_i^* z_i',$$

with $z = (z_1, \ldots, z_n)^{\mathsf{T}}$ and $z' = (z_1', \ldots, z_n')^{\mathsf{T}}$. For $n = 1$, the matrix M is reduced to a simple complex number, and $\mathrm{U}(1)$ thus describes the set of complex numbers with absolute

[8] A real Lie group is a Lie group whose Lie algebra is a real Lie algebra.

value 1, i.e. $e^{i\theta}$. It is thus the group of phase transformations. Note that it is Abelian and, according to Eq. (A.15), the structure constant is necessarily zero.

The subset of unitary matrices with determinant 1 is the *special unitary group* SU(n),

$$\mathrm{SU}(n) = \{M \in \mathrm{GL}(n, \mathbb{C}) \mid M^\dagger M = MM^\dagger = \mathbb{1},\ \det(M) = 1\}. \tag{A.48}$$

It shall be described in detail for SU(2) and SU(3) in Appendices B and C, respectively. The dimension of the real Lie group SU(n) is $n^2 - 1$, one dimension less than U(n) due to the determinant constraint.

- **Lie algebra.**

Following the same approach as for O(n), consider a matrix $M = \exp(tX) \in \mathrm{U}(n)$, for all real t. The constraint of unitarity $M^\dagger = M^{-1}$ translates into

$$\sum_n \frac{t^n}{n!}(X^\dagger)^n = \sum_n \frac{t^n}{n!}(-X)^n.$$

Since the equality holds for all t, we conclude that the matrix X must be anti-Hermitian: $X^\dagger = -X$. Conversely, if X is anti-Hermitian, then $M^\dagger = \exp(tX^\dagger) = \exp(-tX) = M^{-1}$, and thus M is in U(n). The Lie algebra of U(n) is thus

$$\mathfrak{u}(n) = \{X \in M_n(\mathbb{C}) \mid X^\dagger = -X\}. \tag{A.49}$$

It is immediate to check that $\mathfrak{u}(n)$ is closed under the Lie bracket (the matrix commutator). For SU(n), the additional constraint $\det(M) = 1$ imposes, for all t,

$$e^{t\,\mathrm{Tr}(X)} = 1,$$

where the equality (A.30) is used. Therefore, necessarily, $\mathrm{Tr}(X) = 0$, and thus,

$$\mathfrak{su}(n) = \{X \in M_n(\mathbb{C}) \mid X^\dagger = -X,\ \mathrm{Tr}(X) = 0\}. \tag{A.50}$$

For physics, due to the definition of generators in Eq. (A.12), the Lie algebra spanned are obviously, respectively,

$$\mathfrak{u}(n) = \{X \in M_n(\mathbb{C}) \mid X^\dagger = X\}. \tag{A.51}$$

$$\mathfrak{su}(n) = \{X \in M_n(\mathbb{C}) \mid X^\dagger = X,\ \mathrm{Tr}(X) = 0\}. \tag{A.52}$$

Note that the constraints on the matrices X are the same in $\mathfrak{su}(n)$ in Eq. (A.52) and in $\mathfrak{so}(n)$ in Eq. (A.36), even if in the latter case, iX is real, while in the former, it is not necessary. We shall see in Appendix B what it means.

Appendix B Special Unitary Group SU(2)

B.1 Generators of SU(2)

The SU(2) group is the set of 2×2 unitary (complex) matrices with unit determinant, while its real Lie algebra $\mathfrak{su}(2)$ is the set of 2×2 (complex) Hermitian traceless matrices (see Section A.4.7). Therefore, a matrix X of $\mathfrak{su}(2)$ takes the form

$$X = \begin{pmatrix} a & b - ic \\ b + ic & -a \end{pmatrix} = a \begin{pmatrix} 1 & 0 \\ 0 & -1 \end{pmatrix} + b \begin{pmatrix} 0 & 1 \\ 1 & 0 \end{pmatrix} + c \begin{pmatrix} 0 & -i \\ i & 0 \end{pmatrix},$$

where a, b and c are three real numbers. A basis of Hermitian traceless matrices is then given by the three Pauli matrices,

$$\sigma_1 = \begin{pmatrix} 0 & 1 \\ 1 & 0 \end{pmatrix}, \quad \sigma_2 = \begin{pmatrix} 0 & -i \\ i & 0 \end{pmatrix}, \quad \sigma_3 = \begin{pmatrix} 1 & 0 \\ 0 & -1 \end{pmatrix}. \tag{B.1}$$

It is convenient to use $\{\sigma_i/2\}$ as the basis of the Lie algebra since the generators obey the commutation rules

$$\left[\frac{\sigma_i}{2}, \frac{\sigma_j}{2}\right] = i\epsilon_{ijk}\frac{\sigma_k}{2}. \tag{B.2}$$

By the exponential map, an element of SU(2) then takes the form

$$M = \exp\left(-i\boldsymbol{\alpha} \cdot \frac{\boldsymbol{\sigma}}{2}\right), \quad \text{with} \quad \boldsymbol{\alpha} = \begin{pmatrix} \alpha_1 \\ \alpha_2 \\ \alpha_3 \end{pmatrix} \in \mathbb{R}^3, \quad \boldsymbol{\sigma} = \begin{pmatrix} \sigma_1 \\ \sigma_2 \\ \sigma_3 \end{pmatrix}.$$

Denoting the unit vector $\hat{\boldsymbol{n}} = \boldsymbol{\alpha}/\alpha$, with $\alpha = |\boldsymbol{\alpha}|$, M reads

$$\exp\left(- i\alpha\hat{\boldsymbol{n}} \cdot \frac{\boldsymbol{\sigma}}{2}\right) = \sum_{p=0}^{\infty} \frac{(-i)^{2p}}{(2p)!}\left(\frac{\alpha}{2}\right)^{2p} (\hat{\boldsymbol{n}} \cdot \boldsymbol{\sigma})^{2p} + \frac{(-i)^{2p+1}}{(2p+1)!}\left(\frac{\alpha}{2}\right)^{2p+1} (\hat{\boldsymbol{n}} \cdot \boldsymbol{\sigma})^{2p+1}.$$

It is easy to check that $(\hat{\boldsymbol{n}} \cdot \boldsymbol{\sigma})^2 = \mathbb{1}$. Consequently,

$$M = \sum_{p=0}^{\infty} \frac{(-1)^p}{(2p)!}\left(\frac{\alpha}{2}\right)^{2p} \mathbb{1} - i\frac{(-1)^p}{(2p+1)!}\left(\frac{\alpha}{2}\right)^{2p+1} \hat{\boldsymbol{n}} \cdot \boldsymbol{\sigma},$$

i.e.

$$M = \exp\left(- i\alpha\hat{\boldsymbol{n}} \cdot \frac{\boldsymbol{\sigma}}{2}\right) = \cos\frac{\alpha}{2}\,\mathbb{1} - i\sin\frac{\alpha}{2}\,\hat{\boldsymbol{n}} \cdot \boldsymbol{\sigma} = \cos\frac{\alpha}{2}\,\mathbb{1} - i\frac{1}{\alpha}\sin\frac{\alpha}{2}\,\boldsymbol{\alpha} \cdot \boldsymbol{\sigma}. \tag{B.3}$$

The matrix $M \equiv M_{\hat{\boldsymbol{n}}}(\alpha)$ depends on the two parameters α and $\hat{\boldsymbol{n}}$. In principle, note that α can take any value between 0 and 4π. However, as $M_{\hat{\boldsymbol{n}}}(\alpha) = M_{-\hat{\boldsymbol{n}}}(4\pi - \alpha)$, we can restrict the range of α to $[0, 2\pi]$.

B.2 Isomorphism and Homomorphism between SU(2) and SO(3)

In Section A.4.7, we noted that both Lie algebras $\mathfrak{su}(n)$ and $\mathfrak{so}(n)$ are *almost* the same. To better appreciate this fact, the notion of *isomorphism* presented in Section A.2 must be first completed.

Two groups, G and G', are said to be *isomorphic* if they are not distinguishable in their mathematical structure. It means that there is a group isomorphism f that preserves the algebraic structure, ensuring a perfect one-to-one mapping between G and G',

$$f : G \to G', \quad f(g_1 \times g_2) = f(g_1) \times f(g_2), \quad \forall (g_1, g_2) \in G,$$
$$f^{-1}(g_1' \times g_2') = f^{-1}(g_1') \times f^{-1}(g_2'), \quad \forall (g_1', g_2') \in G'.$$

Two isomorphic groups are usually denoted $G \cong G'$. What about the Lie algebra? One can logically define a *Lie algebra homomorphism* from an algebra $\mathfrak{g}$ to an algebra $\mathfrak{g}'$ by a function f, such that

$$f : \mathfrak{g} \to \mathfrak{g}', \quad f([x_1, x_2]) = [f(x_1), f(x_2)], \quad \forall x_1, x_2 \in \mathfrak{g}.$$

If f is bijective,

$$f^{-1}([x_1', x_2']) = [f^{-1}(x_1'), f^{-1}(x_2')], \quad \forall x_1', x_2' \in \mathfrak{g}'.$$

Then f is a *Lie algebra isomorphism*, and both algebras are isomorphic, $\mathfrak{g} \cong \mathfrak{g}'$.

Let us come back to SU(2) and SO(3). Their Lie algebras are *almost* the same because $\mathfrak{so}(3)$ and $\mathfrak{su}(2)$ are isomorphic. A basis of $\mathfrak{so}(3)$ is given by the generators J_1, J_2 and J_3 in Eqs. (A.39) and (A.40), which verify the commutation rules (A.41). Similarly, a basis of $\mathfrak{su}(2)$ is given by the Pauli matrices (B.1), which obey the same commutation rules (B.2). They are isomorphic because the simple linear application that sends the $\frac{\sigma_i}{2}$s onto J_is, i.e.

$$f : \mathfrak{su}(2) \to \mathfrak{so}(3), \quad f\Big(\alpha_1 \frac{\sigma_1}{2} + \alpha_2 \frac{\sigma_2}{2} + \alpha_3 \frac{\sigma_3}{2}\Big) = \alpha_1 J_1 + \alpha_2 J_2 + \alpha_3 J_3, \quad \forall \alpha_1, \alpha_2, \alpha_3 \in \mathbb{R},$$

is clearly a Lie algebra isomorphism. Does it mean that both groups SU(2) and SO(3) are isomorphic? Actually, it does not. We know that every element of SU(2) can be written as the matrix $M(\boldsymbol{\alpha}) = \exp(-i\boldsymbol{\alpha} \cdot \boldsymbol{\sigma}/2)$ (B.3), while every element of SO(3) can be written as the matrix $R(\boldsymbol{\alpha}) = \exp(-i\boldsymbol{\alpha} \cdot \boldsymbol{J})$ (A.44). In both cases, the matrices depend on $\boldsymbol{\alpha}$. Hence, let us define the function

$$f : \mathrm{SU}(2) \to \mathrm{SO}(3), \quad f(M(\boldsymbol{\alpha})) = R(\boldsymbol{\alpha}).$$

In SU(2), we saw that $\alpha = |\boldsymbol{\alpha}|$ can be restricted in an interval of 2π width. In terms of topology, as $\boldsymbol{\alpha} = \alpha\hat{\boldsymbol{n}}$, with $\hat{\boldsymbol{n}}$ being a unit vector, elements of SU(2) thus lie in $\mathbb{R}^3$ within a sphere of radius 2π. Now, for all elements of SU(2), the element $f(M(\boldsymbol{\alpha}))$ exists in SO(3). The function f is thus a group homomorphism. However, it is only surjective and not bijective. Indeed, for instance, both elements of SU(2), $M(\boldsymbol{\alpha})$ and $M(-\boldsymbol{\alpha})$, with $\alpha = \pi$, are distinct since, according to Eq. (B.3), $M(-\boldsymbol{\alpha}) = -M(\boldsymbol{\alpha})\,|_{\alpha=\pi}$. However, the two corresponding elements in SO(3) are identical, as $R(-\boldsymbol{\alpha}) = R(\boldsymbol{\alpha})\,|_{\alpha=\pi}$ from the Rodrigues Formula (A.37). Another way of seeing the surjective nature of f is to compare the topology in $\mathbb{R}^3$ of both groups. Elements of SU(2) lie within a sphere of radius 2π, while those of

SO(3) are in a smaller sphere of radius π. Actually, the group homomorphism f is a two-to-one map of SU(2) to SO(3) since for a single element of SU(2), with $\alpha \in [0, 2\pi]$, there are two elements of SO(3): $R_{\hat{n}}(\alpha)$ and $R_{-\hat{n}}(2\pi - \alpha)$. The group SU(2) is said to be the *universal covering group*[1] of order two of the group SO(3).

B.3 Representations of the Lie Algebra $\mathfrak{su}(2)$

The commutation rules of the Lie algebras have been given in Eqs. (B.2) for $\mathfrak{su}(2)$ and (A.41) for $\mathfrak{so}(3)$, i.e.

$$[J_1, J_2] = iJ_3, \quad [J_2, J_3] = iJ_1, \quad [J_3, J_1] = iJ_2, \tag{B.4}$$

where for $\mathfrak{su}(2)$, we used $\boldsymbol{J} = \boldsymbol{\sigma}/2$. The reader must be familiar with them since they are those of the angular momentum encountered in quantum mechanics. The determination of the eigenvectors and eigenvalues is treated in any quantum mechanics textbook (see, e.g., Cohen-Tannoudji et al., 1977; Griffiths, 1995). A summary of the procedure is just recalled here. First, introduce the operator[2]

$$\boldsymbol{J}^2 = J_1^2 + J_2^2 + J_3^2. \tag{B.5}$$

The commutation rules (B.4) imply that $\boldsymbol{J}^2$ commutes with all J_i. Hence, there are common eigenstates between $\boldsymbol{J}^2$ and, for example, J_3, which we denote $|j, m\rangle$, such that

$$\boldsymbol{J}^2 \, |j, m\rangle = j(j+1) \, |j, m\rangle, \quad J_3 \, |j, m\rangle = m \, |j, m\rangle. \tag{B.6}$$

A priori, $j(j+1)$, the eigenvalues of $\boldsymbol{J}^2$ and, m, the eigenvalues of J_3 can be any real numbers since $\boldsymbol{J}^2$ and J_3 are Hermitian. However, j is necessarily positive because of the positivity of the squared norm $\| J \, |j, m\rangle \, \|^2 = j(j+1)$, where we assume $|j, m\rangle$ normalised. The second step is to introduce the ladder operators

$$J_\pm = J_1 \pm iJ_2. \tag{B.7}$$

[1] Loosely speaking, it means the following: imagine all possible continuous paths in the space parameters of a matrix Lie group G that start form a point x_0 and end up in the same point x_0. If for all those closed paths, we can continuously change their parameters, the paths staying still closed, such that at the end, they all coincide to the trivial path $\{x_0\}$, the group is said to be *simply connected*. SO(3) is, however, not simply connected. The space parameters form a sphere of radius π. Consider two diametrically opposed points on the surface of the sphere. They correspond to the same element. Now, visualise a path going through those points. Even if it is a closed path, there is no way to deform it to a single point. Roughly speaking, a *universal covering group* of G is a simply connected matrix Lie group G' together with a Lie group homomorphism $G \to G'$, such that the associated Lie algebra homomorphism $\mathfrak{g} \to \mathfrak{g}'$ is a Lie algebra isomorphism. In the case of SO(3), it is SU(2). In order to verify that SU(2) is simply connected, it is better to use the following parametrisation of SU(2) matrix: as $M^\dagger = M^{-1}$, with $\det(M) = 1$, the matrix M reads $M = \begin{pmatrix} a & -c^* \\ c & a^* \end{pmatrix}$, with $|a|^2 + |c|^2 = 1$. Decomposing $a = u_0 - iu_3$ and $c = u_2 - iu_1$, it leads to $u_0^2 + u_1^2 + u_2^2 + u_3^2 = 1$. Hence, all elements of SU(2) lie on the surface of a sphere of unit radius in $\mathbb{R}^4$ that is simply connected.

[2] Such operators, polynomials in generators, which commute with all generators of the irreducible representations, are called Casimir operators.

Since $J_\pm^\dagger = J_\mp$ and $J_\mp J_\pm = \boldsymbol{J}^2 - J_3^2 \mp J_3$, the squared norm of $J_\pm |j,m\rangle$ reads

$$\| J_\pm |j,m\rangle \|^2 = j(j+1) - m(m \pm 1) = (j \mp m)(j \pm m + 1) \geq 0.$$

This constraint with $j \geq 0$ leads to

$$-j \leq m \leq j. \tag{B.8}$$

Moreover, it implies $J_+ |j,m\rangle = 0$ if and only if $m = j$, and $J_- |j,m\rangle = 0$ if and only if $m = -j$. Now, using the commutation rules (B.4) and as $J_\pm$ commutes with $\boldsymbol{J}^2$, it follows that

$$\begin{aligned} &\forall m < j, \quad \boldsymbol{J}^2 (J_+ |j,m\rangle) = j(j+1) (J_+ |j,m\rangle), \quad J_3 (J_+ |j,m\rangle) = (m+1) (J_+ |j,m\rangle), \\ &\forall m > -j, \quad \boldsymbol{J}^2 (J_- |j,m\rangle) = j(j+1) (J_- |j,m\rangle), \quad J_3 (J_- |j,m\rangle) = (m-1) (J_- |j,m\rangle). \end{aligned}$$

Therefore, $J_+ |j,m\rangle$ is proportional[3] to $|j,m+1\rangle$ and $J_- |j,m\rangle$ to $|j,m-1\rangle$. We can obviously repeat the procedure

$$\begin{aligned} |j,m\rangle &\xrightarrow{J_+} |j,m+1\rangle \xrightarrow{J_+} |j,m+2\rangle \xrightarrow{J_+} \text{etc.} \\ |j,m\rangle &\xrightarrow{J_-} |j,m-1\rangle \xrightarrow{J_-} |j,m-2\rangle \xrightarrow{J_-} \text{etc.} \end{aligned}$$

But the eigenvalues of J_3 are bound by the inequality (B.8), and hence, the procedure can only stop after k steps when $m+k = j$ or $m-k = -j$. This implies that $j-m$ and $j+m$ are positive integers, and hence, this requires that j and m are simultaneously integers or half integers,

$$j = 0, \ \frac{1}{2}, \ 1, \ \frac{3}{2}, \ 2, \ldots; \qquad m = -j, \ -(j-1), \ldots, j-1, \ j. \tag{B.9}$$

Hence, for each j, there are $2j+1$ possible eigenstates of $\boldsymbol{J}^2$ and J_3. Let us now consider the *highest weight* eigenstate, normalised with a unit norm, $|j,j\rangle$, such that $J_+ |j,j\rangle = 0$. The successive actions of the J_- operator on $|j,j\rangle$ allow us to determine the so-called spin-j irreducible representation of the $\mathfrak{su}(2)$ algebra via the relations

$$J_\pm |j,m\rangle = \sqrt{j(j+1) - m(m \pm 1)}\, |j,m \pm 1\rangle. \tag{B.10}$$

The representations of the generators J_1, J_2 and J_3 are then obtained on the basis

$$\{|j,j\rangle, |j,j-1\rangle, \ldots, |j,-j\rangle\},$$

given that $J_1 = (J_- + J_+)/2$ and $J_2 = i(J_- - J_+)/2$. For instance, the three lowest dimensional representations are listed below.

- One-dimensional representation: $j = 0$. This is the trivial representation, with a single state $|0,0\rangle$ for which $J_1^0 = J_2^0 = J_3^0 = 0$.
- Two-dimensional representation: $j = 1/2$, with the two states $|1/2, \pm 1/2\rangle$:

$$J_-^{1/2} = \begin{pmatrix} 0 & 0 \\ 1 & 0 \end{pmatrix}, \quad J_+^{1/2} = \begin{pmatrix} 0 & 1 \\ 0 & 0 \end{pmatrix},$$

$$J_1^{1/2} = \frac{1}{2}\begin{pmatrix} 0 & 1 \\ 1 & 0 \end{pmatrix}, \quad J_2^{1/2} = \frac{1}{2}\begin{pmatrix} 0 & -i \\ i & 0 \end{pmatrix}, \quad J_3^{1/2} = \frac{1}{2}\begin{pmatrix} 1 & 0 \\ 0 & -1 \end{pmatrix}.$$

We recover the generators $J_i^{1/2} = \sigma_i/2$, those of the standard representation of $\mathfrak{su}(2)$.

[3] We assume non-degenerate eigenvalues.

- Three-dimensional representation: $j = 1$, with the three states $|1,-1\rangle, |1,0\rangle$ and $|1,1\rangle$:

$$J_-^1 = \begin{pmatrix} 0 & 0 & 0 \\ \sqrt{2} & 0 & 0 \\ 0 & \sqrt{2} & 0 \end{pmatrix}, \quad J_+^1 = \begin{pmatrix} 0 & \sqrt{2} & 0 \\ 0 & 0 & \sqrt{2} \\ 0 & 0 & 0 \end{pmatrix},$$

$$J_1^1 = \frac{1}{\sqrt{2}}\begin{pmatrix} 0 & 1 & 0 \\ 1 & 0 & 1 \\ 0 & 1 & 0 \end{pmatrix}, \quad J_2^1 = \frac{1}{\sqrt{2}}\begin{pmatrix} 0 & -i & 0 \\ i & 0 & -i \\ 0 & i & 0 \end{pmatrix}, \quad J_3^1 = \begin{pmatrix} 1 & 0 & 0 \\ 0 & 0 & 0 \\ 0 & 0 & -1 \end{pmatrix}.$$

At first glance, it may seem surprising to get different expressions of the generators than those of the standard representation in Eqs. (A.39) and (A.40) of $\mathfrak{so}(3)$. By construction, we have obtained a representation in which J_3^1 is diagonal. The generators obtained in Section A.4.5 for $\mathfrak{so}(3)$ have been expressed in the Cartesian basis $\hat{\boldsymbol{x}}, \hat{\boldsymbol{y}}, \hat{\boldsymbol{z}}$. Both representations are just equivalent via the similarity transformation (A.5) $J_i = S J_i^1 S^{-1}$, with

$$S = \frac{1}{\sqrt{2}}\begin{pmatrix} -1 & 0 & 1 \\ -i & 0 & -i \\ 0 & \sqrt{2} & 0 \end{pmatrix}, \quad S^{-1} = S^\dagger = \frac{1}{\sqrt{2}}\begin{pmatrix} -1 & i & 0 \\ 0 & 0 & \sqrt{2} \\ 1 & i & 0 \end{pmatrix}.$$

Hence $|1,1\rangle = -(\hat{\boldsymbol{x}} + i\hat{\boldsymbol{y}})/\sqrt{2}$, $|1,0\rangle = \hat{\boldsymbol{z}}$ and $|1,-1\rangle = (\hat{\boldsymbol{x}} - i\hat{\boldsymbol{y}})/\sqrt{2}$. The three-dimensional representation is thus equivalent to the adjoint representation.
- For the general spin-j representation, the J_-^j and J_+^j matrices are given by

$$J_-^j = \begin{pmatrix} 0 & \cdots & \cdots & \cdots & 0 \\ c_1^j & \ddots & & & \vdots \\ 0 & c_2^j & \ddots & & \vdots \\ \vdots & \ddots & \ddots & \ddots & \vdots \\ 0 & \cdots & 0 & c_{2j}^j & 0 \end{pmatrix}_{|j,j\rangle \ldots |j,-j\rangle}, \quad J_+^j = \begin{pmatrix} 0 & c_1^j & 0 & \cdots & 0 \\ \vdots & \ddots & c_2^j & \ddots & \vdots \\ \vdots & & \ddots & \ddots & 0 \\ \vdots & & & \ddots & c_{2j}^j \\ 0 & \cdots & \cdots & \cdots & 0 \end{pmatrix}_{|j,j\rangle \ldots |j,-j\rangle}. \tag{B.11}$$

The coefficients $c_{k=1,\ldots,2j}^j$ are deduced from Eq. (B.10), i.e.

$$c_k^j = \sqrt{k(2j-k+1)} = \left\{\sqrt{2j}, \sqrt{4j-2}, \sqrt{6j-6}, \ldots, \sqrt{6j-6}, \sqrt{4j-2}, \sqrt{2j}\right\}, \tag{B.12}$$

while J_3^j has the diagonal form

$$J_3^j = \begin{pmatrix} j & 0 & \cdots & \cdots & 0 \\ 0 & j-1 & \ddots & & \vdots \\ \vdots & \ddots & \ddots & \ddots & \vdots \\ \vdots & & \ddots & \ddots & 0 \\ 0 & \cdots & \cdots & 0 & -j \end{pmatrix}_{|j,j\rangle \ldots |j,-j\rangle}. \tag{B.13}$$

B.4 Representations of the SU(2) Group

In Section B.1, we saw that the elements $U(\boldsymbol{\alpha})$ of the SU(2) group are obtained by the 'exponentiation' procedure

$$U(\boldsymbol{\alpha}) = \exp\left(-i\boldsymbol{\alpha}\cdot\frac{\boldsymbol{\sigma}}{2}\right) = \exp\left(-i\alpha\hat{\boldsymbol{n}}\cdot\frac{\boldsymbol{\sigma}}{2}\right), \quad \alpha \in [0, 2\pi]. \tag{B.14}$$

Therefore, starting from a representation of the $\mathfrak{su}(2)$ Lie algebra with the generators $\boldsymbol{J}^j$ of the previous section, a representation of SU(2) is given by Eq. (A.24), i.e.

$$\mathcal{D}^j(U(\boldsymbol{\alpha})) = \exp\left(-i\boldsymbol{\alpha}\cdot\boldsymbol{J}^j\right) = \exp\left(-i\alpha\hat{\boldsymbol{n}}\cdot\boldsymbol{J}^j\right), \quad \alpha \in [0, 2\pi]. \tag{B.15}$$

Since the dimension of the representation of $\mathfrak{su}(2)$ is $2j+1$, the dimension of $\mathcal{D}^j$ is $2j+1$ too. This representation acts on the Hilbert space generated by the basis $\{|j,m\rangle\}$, with $m = -j,\ldots,j$. It is called the representation of spin j. Indeed, given $\boldsymbol{J}^2 = (J_1^j)^2 + (J_2^j)^2 + (J_3^j)^2$ and $[\boldsymbol{J}^2, J_i^j] = 0$, it follows that

$$\boldsymbol{J}^2\left(\mathcal{D}^j(U(\boldsymbol{\alpha}))\,|j,m\rangle\right) = \mathcal{D}^j(U(\boldsymbol{\alpha}))\boldsymbol{J}^2\,|j,m\rangle = j(j+1)\mathcal{D}^j(U(\boldsymbol{\alpha}))\,|j,m\rangle,$$

showing that $\mathcal{D}^j(U(\boldsymbol{\alpha})\,|j,m\rangle$ has spin j.

Let us show that the spin j representation is irreducible. The representation $\mathcal{D}^j$ of SU(2) is unitary by construction. According to theorem A.3, if $\mathcal{D}^j$ were reducible, then it would be completely reducible. In the appropriate basis, the representative matrices must then be block diagonal. Imagine a diagonal operator A that commutes with $\mathcal{D}$, i.e. $A_{kk}\mathcal{D}_{kl} = \mathcal{D}_{kl}A_{ll}$, (implicit summation). Then its diagonal elements corresponding to a block are necessarily all equals. If $\mathcal{D}^j$ is not irreducible, according to Schur's corollary A.1.2, A must not be the identity. If A commutes with the representative matrices, it commutes necessarily with the generators and thus with J_3^j. In the basis $\{|j,j\rangle, |j,j-1\rangle, \ldots |j,-j\rangle\}$, where J_3^j is diagonal, $J_3^j = \text{Diag}(j, j-1, \ldots, -j)$, A is a priori not necessarily diagonal. However, as $AJ_3^j = J_3^jA$ and J_3^j is diagonal, the elements A_{kl} of A satisfy $A_{kl}(J_3^j)_{ll} = (J_3^j)_{kk}A_{kl}$, i.e. $A_{kl}(j-l+1) = (j-k+1)A_{kl}$. Then, $A_{kl}(k-l) = 0$, and thus either $l = k$ or $A_{kl} = 0$. The matrix A is finally diagonal in that basis too. But A commuting with all generators commutes with J_+^j, which implies

$$(J_+^j)_{kl}A_{ll} = A_{kk}(J_+^j)_{kl}.$$

Hence, either $(J_+^j)_{kl} = 0$ or $A_{ll} = A_{kk}$. However, according to the expression (B.11) of J_+^j, $(J_+^j)_{k\,(k+1)}$ is not zero for all k, and hence, $A_{kk} = A_{(k+1)\,(k+1)}$. The matrix A is thus proportional to the identity, which contradicts the initial assumption that $\mathcal{D}^j$ is reducible. Therefore, as announced, $\mathcal{D}^j$ is irreducible.

It is interesting to note one specificity of the spin-1/2 representation. The fundamental representation $\mathcal{D}^{1/2}$ of SU(2),

$$\mathcal{D}^{\frac{1}{2}}(U(\boldsymbol{\alpha})) = \exp\left(-i\boldsymbol{\alpha}\cdot\frac{\boldsymbol{\sigma}}{2}\right), \tag{B.16}$$

and the contragredient representation $\overline{\mathcal{D}}^{1/2}$ (which is also the complex conjugate representation $\mathcal{D}^{*1/2}$ since the matrices are unitary) turn out to be two equivalent representations. Indeed, the matrix

$$C = i\sigma_2 = \begin{pmatrix} 0 & 1 \\ -1 & 0 \end{pmatrix} \tag{B.17}$$

is such that for any U matrix in the representation $\mathcal{D}^{1/2}$,

$$CUC^{-1} = \begin{pmatrix} 0 & 1 \\ -1 & 0 \end{pmatrix} \begin{pmatrix} a & b \\ c & d \end{pmatrix} \begin{pmatrix} 0 & -1 \\ 1 & 0 \end{pmatrix} = \begin{pmatrix} d & -c \\ -b & a \end{pmatrix}.$$

But

$$U^* = (U^{-1})^\mathsf{T} = \left(\frac{1}{\det(U)} \begin{pmatrix} d & -b \\ -c & a \end{pmatrix} \right)^\mathsf{T} = \begin{pmatrix} d & -c \\ -b & a \end{pmatrix}.$$

In conclusion, $\mathcal{D}^{1/2}$ and $\mathcal{D}^{*1/2} = \overline{\mathcal{D}}^{1/2}$ are two equivalent representations. It has interesting consequences: the transformation of spin-1/2 anti-fermions by U^* can then be described by transformations with U. Therefore, the same representation can be used for fermions and anti-fermions (e.g., see the discussion in Supplement 8.1). Two-component objets that transform according to the representation $\mathcal{D}^{1/2}$ are called *Pauli spinors*.

- **Remarks about SO(3) representations.**

As SU(2) and SO(3) share the same Lie algebra $\mathfrak{su}(2) \cong \mathfrak{so}(3)$, one may wonder whether the spin-j representations $\mathcal{D}^j$ found above are also valid representations of SO(3). Clearly, $\mathcal{D}^1$ is a representation of SO(3) since the representative matrices are elements of SO(3). What about $\mathcal{D}^{1/2}$? Consider the product of two rotations about the z-axis by an angle π. We obviously have $R_{\hat{z}}(\pi)R_{\hat{z}}(\pi) = R_{\hat{z}}(0)$, i.e. the identity element, whereas using Eq. (B.3)

$$\mathcal{D}^{1/2}(U(\pi\hat{z}))\mathcal{D}^{1/2}(U(\pi\hat{z})) = \left(\exp(-i\pi\frac{\sigma_3}{2})\right)^2 = (-i\mathbb{1})^2 = -\mathbb{1} = -\mathcal{D}^{1/2}(U(0\hat{z})).$$

Consequently, $\mathcal{D}^{1/2}$ is not a true representation of SO(3). It is a *projective representation*, a representation whose composition law differs only by a phase. It occurs because SO(3) is not simply connected. All spin-j, where j is half-integer, are projective representations as

$$\mathcal{D}^j(U(2\pi\hat{z})) = \exp(-i2\pi J_3^j) = \mathrm{Diag}((-1)^{2(j)}, (-1)^{2(j-1)}, \ldots, (-1)^{2(-j)}) = (-1)^{2j}\mathbb{1}.$$

Note that in the context of quantum mechanics, states are defined up to a phase. Therefore, projective representations are also appropriate to describe symmetry transformations.

Appendix C Special Unitary Group SU(3) and SU(N)

C.1 The SU(3) Group and Its Lie Algebra $\mathfrak{su}(3)$

The SU(3) group is the set of 3×3 unitary (complex) matrices with unit determinant, while its real Lie algebra $\mathfrak{su}(3)$ is the set of 3×3 (complex) Hermitian traceless matrices (see Section A.4.7). The dimension of $\mathfrak{su}(3)$ and thus of SU(3) is $3^2 - 1 = 8$. As explained in Section 8.1.4, a basis of $\mathfrak{su}(3)$ can be easily deduced from the three Pauli matrices, the basis of $\mathfrak{su}(2)$. A convenient choice is based on the Gell-Mann matrices,

$$
\lambda_1 = \begin{pmatrix} 0 & 1 & 0 \\ 1 & 0 & 0 \\ 0 & 0 & 0 \end{pmatrix}, \quad \lambda_2 = \begin{pmatrix} 0 & -i & 0 \\ i & 0 & 0 \\ 0 & 0 & 0 \end{pmatrix}, \quad \lambda_3 = \begin{pmatrix} 1 & 0 & 0 \\ 0 & -1 & 0 \\ 0 & 0 & 0 \end{pmatrix}, \quad \lambda_4 = \begin{pmatrix} 0 & 0 & 1 \\ 0 & 0 & 0 \\ 1 & 0 & 0 \end{pmatrix},
$$
$$
\lambda_5 = \begin{pmatrix} 0 & 0 & -i \\ 0 & 0 & 0 \\ i & 0 & 0 \end{pmatrix}, \quad \lambda_6 = \begin{pmatrix} 0 & 0 & 0 \\ 0 & 0 & 1 \\ 0 & 1 & 0 \end{pmatrix}, \quad \lambda_7 = \begin{pmatrix} 0 & 0 & 0 \\ 0 & 0 & -i \\ 0 & i & 0 \end{pmatrix}, \quad \lambda_8 = \begin{pmatrix} \frac{1}{\sqrt{3}} & 0 & 0 \\ 0 & \frac{1}{\sqrt{3}} & 0 \\ 0 & 0 & \frac{-2}{\sqrt{3}} \end{pmatrix},
$$

all satisfying $\mathrm{Tr}(\lambda_a \lambda_b) = 2\delta_{ab}$, as the Pauli matrices. The infinitesimal generators of SU(3) are then $T_a = \lambda_a/2$, and a matrix M of SU(3) is thus obtained with

$$
M = \exp\left(-i\boldsymbol{\alpha} \cdot \frac{\boldsymbol{\lambda_a}}{2}\right), \quad \text{with} \quad \boldsymbol{\alpha} = \begin{pmatrix} \alpha_1 \\ \vdots \\ \alpha_8 \end{pmatrix} \in \mathbb{R}^8, \quad \boldsymbol{\lambda} = \begin{pmatrix} \lambda_1 \\ \vdots \\ \lambda_8 \end{pmatrix}.
$$

Note that, by construction, T_1, T_2 and T_3 generate an SU(2) subgroup of SU(3); let us call it $\mathrm{SU}(2)_T$, since their actions on the subspace of dimension 2 spanned by the vectors $\boldsymbol{e_1} = (1,0,0)^\mathsf{T}$ and $\boldsymbol{e_2} = (0,1,0)^\mathsf{T}$ are governed by the 2×2 Pauli matrices. Similarly, T_6, T_7, U_3 and, T_4, T_5, V_3, with

$$
U_3 = \frac{\sqrt{3}}{2}T_8 - \frac{T_3}{2} = \begin{pmatrix} 0 & 0 & 0 \\ 0 & \frac{1}{2} & 0 \\ 0 & 0 & -\frac{1}{2} \end{pmatrix}, \quad V_3 = \frac{\sqrt{3}}{2}T_8 + \frac{T_3}{2} = \begin{pmatrix} \frac{1}{2} & 0 & 0 \\ 0 & 0 & 0 \\ 0 & 0 & -\frac{1}{2} \end{pmatrix}, \tag{C.1}
$$

generate two other SU(2) subgroups, $\mathrm{SU}(2)_U$ and $\mathrm{SU}(2)_V$, respectively, with subspaces spanned by $\boldsymbol{e_2}$ and $\boldsymbol{e_3} = (0,0,1)^\mathsf{T}$ for the former, and $\boldsymbol{e_1}$ and $\boldsymbol{e_3}$ for the latter. The generators obey the commutation rules

$$
[T_a, T_b] = if_{abc}T_c,
$$

where f_{abc} are the fully antisymmetric structure constants (i.e., $f_{bac} = f_{cba} = f_{acb} = -f_{abc}$), whose non-zero values are

$$f_{123} = 1, \quad f_{147} = f_{246} = f_{257} = f_{345} = -f_{156} = -f_{367} = \frac{1}{2}, \quad f_{458} = f_{678} = \frac{\sqrt{3}}{2}.$$

C.2 Irreducible Representations

The generators T_3 and T_8 are already diagonal. Therefore, we can find common eigenvectors of these two operators. Instead of T_8, it is conventional to introduce the generator

$$Y = \frac{2}{\sqrt{3}} T_8 = \begin{pmatrix} \frac{1}{3} & 0 & 0 \\ 0 & \frac{1}{3} & 0 \\ 0 & 0 & -\frac{2}{3} \end{pmatrix} \tag{C.2}$$

and find the eigenvectors $|m, y\rangle$ of both T_3 and Y, with the respective eigenvalues m and y. Moreover, because of $\mathrm{SU}(2)_T$, $\mathrm{SU}(2)_U$ and $\mathrm{SU}(2)_V$ subgroups already mentioned, it is natural to introduce the corresponding SU(2) ladder operators

$$T_\pm = T_1 \pm iT_2\,, \quad U_\pm = T_6 \pm iT_7\,, \quad V_\pm = T_4 \pm iT_5, \tag{C.3}$$

to build the representations of the $\mathfrak{su}(2)$ sub-algebras with the weight method followed in Section B.3. Given the matrix expression of $T_\pm$, $U_\pm$ and $V_\pm$ (deduced from their definition above), it is easy to check the commutators

$$\begin{aligned} &[Y, T_\pm] = 0, && [T_3, T_\pm] = \pm T_\pm, \\ &[Y, U_\pm] = \pm U_\pm, && [T_3, U_\pm] = \mp\tfrac{1}{2} U_\pm, \\ &[Y, V_\pm] = \pm V_\pm, && [T_3, V_\pm] = \pm\tfrac{1}{2} V_\pm, \end{aligned} \tag{C.4}$$

and

$$\begin{aligned} &[T_3, Y] = [T_3, U_3] = [T_3, V_3] = [Y, U_3] = [Y, V_3] = 0, \\ &[T_+, U_+] = [T_+, V_+] = [U_+, V_+] = 0. \\ &[T_-, U_-] = -V_-, \; [T_-, V_-] = [U_-, V_-] = 0. \end{aligned} \tag{C.5}$$

Let us specify the action of the ladder operators on the eigenvector $|y, m\rangle$. As an example, the commutator $[Y, U_+] = U_+$ in Eq. (C.4) implies that $Y(U_+ |m, y\rangle) = yU_+ |m, y\rangle + U_+ |m, y\rangle$. Thus, $U_+ |m, y\rangle$ has the eigenvalue $y + 1$ of Y. Similarly, $[T_3, U_+] = -\frac{1}{2} U_+$ implies that $T_3 |m, y\rangle$ has the eigenvalue $m - 1/2$ of T_3. Hence, U_+ increases y by 1 and decreases m by $1/2$. The action of all ladder operators is summarised in Fig. 8.5, where in the context of SU(3) flavour, m, T_3 and $T_\pm$ are denoted by i_3, I_3 and $I_\pm$, respectively, with the letter I standing for the isospin.

Now that we have identified the actions of the ladder operators, we can generalise the method used in SU(2) to find the irreducible representations of SU(3), starting from the highest weight eigenvector, i.e. the one satisfying

$$T_+ |m_0, y_0\rangle = U_+ |m_0, y_0\rangle = V_+ |m_0, y_0\rangle = 0.$$

Since U_3 or V_3 commute with T_3 and Y [see Eq. (C.5)], the eigenvector $|y_0, m_0\rangle$ is also an eigenvector of U_3 and V_3, with the eigenvalues

$$u_0 = \frac{3}{4}y_0 - \frac{m_0}{2}, \quad v_0 = \frac{3}{4}y_0 + \frac{m_0}{2}, \tag{C.6}$$

respectively [see Eqs. (C.1) and (C.2)]. On the basis of the study of SU(2), we know that, necessarily, m_0, u_0 and v_0, the eigenvalues of T_3, U_3 and V_3, respectively, are integers or half-integers [remember that T_+, T_-, T_3 generate $\mathrm{SU}(2)_T$, U_+, U_-, U_3 generate $\mathrm{SU}(2)_U$ and V_+, V_-, V_3 generate $\mathrm{SU}(2)_V$]. So, let us introduce two integers, such that $p = 2m_0$ and $q = 2u_0$, and therefore $v_0 = (p+q)/2$ [see Eq. (C.6)]. An irreducible representation of SU(3) is thus characterised by two independent integers p and q, whereas in SU(2), only one integer (the spin j) is needed. Therefore, irreducible representations of SU(3) are denoted $\mathcal{D}(p,q)$, for which the highest weight has

$$m_0 = \frac{p}{2}, \quad y_0 = \frac{p}{3} + \frac{2q}{3}. \tag{C.7}$$

The other eigenvectors of this representation are generated by the successive actions (in any order) of T_-, U_- and V_- on $|m_0, y_0\rangle$. The weight diagram of a general irreducible representation, where the eigenvectors are represented in the (m,y) plane, then has the shape of a hexagon, as depicted in Fig. C.1. When $p = 0$ or $q = 0$, the shape is a triangle, and when both $p = q = 0$, the trivial representation is reduced to a point. Unlike SU(2), in SU(3), the eigenvectors may be degenerate, i.e. there are multiple eigenvectors for a given weight (m,y). In such a case, they can be distinguished by the 'square of the total spin' along the T, U or V direction. Let us take the T direction. As an example in Fig.

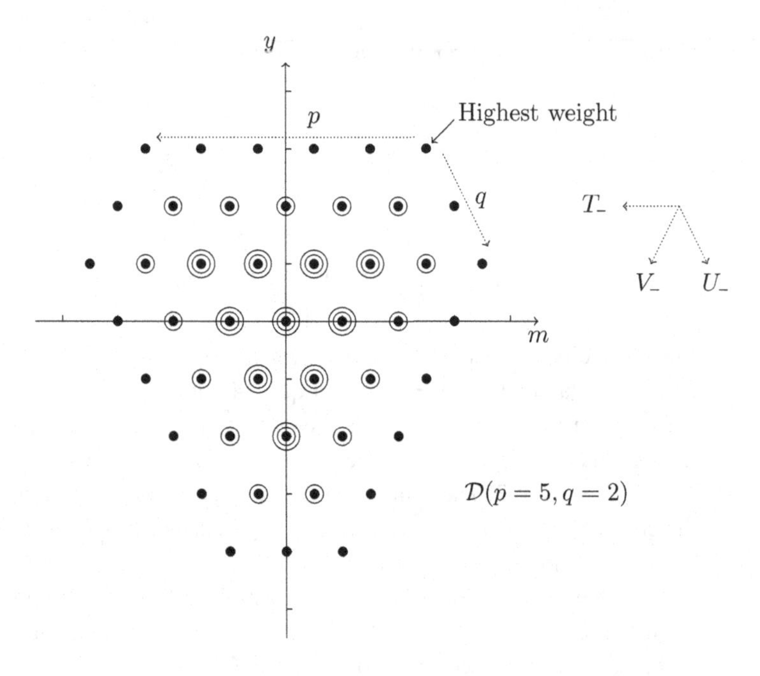

Fig. C.1 Weight diagram of the representation $\mathcal{D}(p,q)$, with $p = 5, q = 2$. The highest weight is located at $m = 5/2$ and $y = 3$. The action of the lowering ladder operators T_-, U_-, V_- is visualised with the direction of the three dotted arrows on the right-hand side. The multiplicity of eigenvectors for a given weight (m,y) is represented by the open circles.

C.1, among the 18 states located at $y = 1$, 8 belongs to a total spin $S = 4$, 6 to $S = 3$ and 4 to $S = 2$. The degeneracy is related to the number of non-equivalent paths that can be followed from the highest weight to reach the weight of interest by applying only T_- and U_- since $V_- = -\left[T_-, U_-\right]$ in Eq. (C.5). In Fig. C.1, the highest weight of the representation $\mathcal{D}(5,2)$ has the weight $(y = 3, m = 5/2)$. The weight $(y = 2, m = 2)$ is reached by U_-T_- and T_-U_-, hence the multiplicity 2. Similarly, to reach $(y = 1, m = 3/2)$, among the four possible paths $P_1 = T_-^2 U_-^2$, $P_2 = T_-U_-T_-U_-$, $P_3 = U_-T_-U_-T_-$ and $P_4 = U_-^2 T_-^2$, one can realise that only three are independent by using the commutator $[T_-, U_-]$ $(P_4 = P_1 - 2P_2 + 2P_3)$. Overall, for a general representation, the states in the external layer, i.e. on the boundary of the weight diagram (describing a hexagon, a triangle, or a single point), are non-degenerate. When one moves by one layer at a time inside the diagram, the multiplicity is increased by one unit at each layer, until a triangular shape or a single point is reached. The multiplicity remains constant in the subsequent moves. The dimension of the representation corresponds to the total number of states. One can show that

$$\dim[\mathcal{D}(p,q)] = (p+1)(q+1)\left(1 + \frac{p+q}{2}\right). \tag{C.8}$$

The first irreducible representations with the lowest dimensions are as follows:

- One-dimensional representation $\mathcal{D}(0,0)$. As $p = q = 0$, $m_0 = y_0 = 0$. This is the trivial representation **1** with a single state $|0,0\rangle$. The weight diagram is reduced to a single point.
- Three-dimensional representation $\mathcal{D}(1,0)$. The highest weight state is $|m_0 = \frac{1}{2}, y_0 = \frac{1}{3}\rangle$ [Eq. (C.7)]. The two other states obtained by the action of T_- and V_- are $|-\frac{1}{2}, \frac{1}{3}\rangle$ and $|0, -\frac{2}{3}\rangle$. This representation is the fundamental representation **3**, with a triangular weight diagram (see Fig. 8.9).
- Three-dimensional representation $\mathcal{D}(0,1)$. The highest weight state is $|0, \frac{2}{3}\rangle$. The two other states obtained by the action of U_- and V_- are $|-\frac{1}{2}, -\frac{1}{3}\rangle$ and $|\frac{1}{2}, -\frac{1}{3}\rangle$. This representation is the conjugate representation $\bar{\mathbf{3}}$ with a triangular weight diagram (Fig. 8.9).
- Eight-dimensional representation $\mathcal{D}(1,1)$. The highest weight state is $|\frac{1}{2}, 1\rangle$. This representation is the adjoint representation **8**. It has the shape of a hexagon (see, e.g., Fig. 8.4).

C.3 SU(3) Tensor Methods

The tensor methods provide an alternative for building the SU(3) irreducible representations, in particular, with its application of Young tableaux (presented in the next section). Let us consider the SU(3) fundamental representation **3**. The 3×3 matrices U in this representation act on states $|\psi\rangle$ with three components ψ^i $(i = 1,2,3)$ on the standard basis, such that it transforms $|\psi\rangle$ into $|\psi'\rangle$ according to the component transformations

$$\psi'^i = U^i{}_{i'}\, \psi^{i'},$$

where $U^{i}{}_{i'}$ is the component of U in line i and column i'. We can form a tensorial product of states such as $|\psi_1\rangle \otimes |\psi_2\rangle$. Its components $T^{i_1 i_2} = \psi_1^{i_1} \psi_2^{i_2}$ forms a tensor of rank-2 that transforms as

$$T'^{i_1 i_2} = U^{i_1}_{\ i'_1} U^{i_2}_{\ i'_2} T^{i'_1 i'_2}.$$

Unlike SU(2), where the fundamental representation $\mathcal{D}^{1/2} \equiv \mathbf{2}$ and the conjugate representation $\mathcal{D}^{*1/2} = \bar{\mathbf{2}}$ are equivalent (see Section B.4), it is not the case with SU(3). Therefore, if we now denote the components of the state in $\bar{\mathbf{3}}$ by ψ_j, it transforms according to

$$\psi'_j = (U^*)_j^{\ j'} \psi_{j'} = \psi_{j'} (U^\dagger)^{j'}_{\ j}$$

(the change of the $\psi_{j'}$ position in the last equality is just there to ease the interpretation of the result in terms of matrix multiplications), while the tensor components of $|\psi_1\rangle \otimes |\psi_2\rangle$ transform as

$$T'_{j_1 j_2} = (U^*)_{j_1}^{\ j'_1} (U^*)_{j_2}^{\ j'_2} T_{j'_1 j'_2} = T_{j'_1 j'_2} (U^\dagger)^{j'_1}_{\ j_1} (U^\dagger)^{j'_2}_{\ j_2}.$$

The generalisation to tensors of rank p in the upper components and q in the lower components, $T^{i_1,\ldots,i_p}_{j_1,\ldots,j_q}$, is then

$$T'^{i_1 \cdots i_2}_{j_1 \cdots j_q} = U^{i_1}_{\ i'_1} \cdots U^{i_p}_{\ i'_p} (U^*)_{j_1}^{\ j'_1} \cdots (U^*)_{j_q}^{\ j'_q} T^{i'_1 \cdots i'_p}_{j'_1 \cdots j'_q} = U^{i_1}_{\ i'_1} \cdots U^{i_p}_{\ i'_p} T^{i'_1 \cdots i'_p}_{j'_1 \cdots j'_q} (U^\dagger)^{j'_1}_{\ j_1} \cdots (U^\dagger)^{j'_q}_{\ j_q}.$$

Any tensor can be decomposed into symmetric and antisymmetric tensors (with respect to its indices). In the simple rank-2 example $T^{i_1 i_2}$, it suffices to consider the linear combinations $T^{i_1 i_2} \pm T^{i_2 i_1}$. Consider now the following linear combinations:

$$T'^{i_1 i_2} \pm T'^{i_2 i_1} = U^{i_1}_{\ i'_1} U^{i_2}_{\ i'_2} T^{i'_1 i'_2} \pm U^{i_2}_{\ i'_1} U^{i_1}_{\ i'_2} T^{i'_1 i'_2} = U^{i_1}_{\ i'_1} U^{i_2}_{\ i'_2} (T^{i'_1 i'_2} \pm T^{i'_2 i'_1}),$$

where for obtaining the last step, the dummy indices have been redefined in the second term. This relation shows that the symmetry of a tensor is preserved by the transformation. Hence, we have a representation for symmetric rank-2 tensors and another for antisymmetric rank-2 tensors. The generalisation to rank-p tensors with fully symmetric or fully antisymmetric representations is straightforward. An important property in SU(3), also true in SU(N) and admitted here, is that

Traceless tensors $T^{i_1 \ldots i_p}_{j_1 \ldots j_q}$ that are fully symmetric in the upper components and fully symmetric in the lower components carry the irreducible representation $\mathcal{D}(p,q)$.

Traceless tensors are tensors satisfying $\delta^{j_1}_{i_1} T^{i_1 \ldots i_p}_{j_1 \ldots j_q} = 0$. From a tensor with lower indices, note that one can always define a tensor with upper indices by the contraction with the rank-3 antisymmetric Levi–Civita tensor ϵ^{ijk} [defined in Eq. (2.53) and in more detail in Appendix G], i.e.

$$t^{i_1 \ldots i_p \, k_1 l_1 \ldots k_q l_q} = \epsilon^{j_1 k_1 l_1} \cdots \epsilon^{j_q k_q l_q} T^{i_1 \ldots i_p}_{j_1 \ldots j_q}. \tag{C.9}$$

Each lower index, j, is converted into two upper indices (k,l). By construction, t is antisymmetric in each exchange $k \leftrightarrow l$ but symmetric in the exchange of two (k,l) pairs, provided that T is fully symmetric in the interchange of its lower indices. Therefore, a tensor $T^{i_1 \ldots i_p}_{j_1 \ldots j_q}$ carrying the irreducible representation $\mathcal{D}(p,q)$ leads to tensor $t^{i_1 \ldots i_p \, k_1 l_1 \ldots k_q l_q}$, which is fully symmetric under the interchange of i indices, antisymmetric under the interchange of any $k \leftrightarrow l$ and symmetric under the interchange of two (k,l) pairs.

C.4 SU(*N*) and Young Tableaux

The Young tableaux[1] is a graphical method to build irreducible representations of all SU(N) groups by symmetrising a tensor $T^{i_1\ldots i_p}_{j_1\ldots j_q}$. A box is used to stand for each index in the states. The fundamental representation of SU(N) depends only on one single index since a state in the fundamental representation is a vector of N components. Therefore, the fundamental representation is represented by a single box, i.e. $\mathbf{N} = \square$. All other representations are represented by diagrams having several connected boxes extending over several columns and/or several rows. All rows are left justified, and all columns top justified. A fully symmetric tensor of rank k is represented by a row of k boxes, while a fully antisymmetric tensor is a column of k boxes. For SU(3), the symmetry properties of the tensor $t^{i_1\ldots i_p\, k_1 l_1 \ldots k_q l_q}$ in Eq. (C.9) carrying the irreducible representation $\mathcal{D}(p,q)$ lead to the Young diagram

$$\mathcal{D}(p,q):\quad \begin{array}{|c|c|c|c|c|c|}\hline k_1 & \cdots & k_q & i_1 & \cdots & i_p \\ \hline l_1 & \cdots & l_q \\ \cline{1-3}\end{array}$$

with p boxes $i_1 \ldots i_p$ and q columns $k_1 l_1 \ldots k_q l_q$.

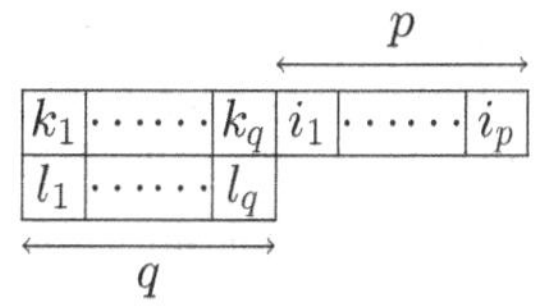

A general tensor with, for instance, two indices and no particular symmetry, t^{ij}, can be decomposed into a symmetric tensor $t^{ij} + t^{ji}$ represented by the diagram $\square\!\square$ and an antisymmetric tensor $t^{ij} - t^{ji}$ represented by the diagram $\begin{array}{|c|}\hline \, \\ \hline \, \\ \hline\end{array}$. In SU(2), the fundamental representation $\mathbf{2}$ and the conjugate representation $\bar{\mathbf{2}}$ are equivalent. Hence, both are represented by a single box. In SU(3), they are not equivalent and $\bar{\mathbf{3}}$ is represented by a column of two boxes since, according to Eq. (C.9), $t^{k_1 l_1} = \epsilon^{j_1 k_1 l_1}\, T_{j_1}$ is fully antisymmetric under the interchange $k_1 \leftrightarrow l_1$. Similarly, the conjugate representation in SU(N) is represented by a column of $N-1$ boxes. [Eq. (C.9) has to be extended with the rank-N Levi–Civita tensor.] Here are some examples of simple Young diagrams in SU(2) and SU(3):

$$\begin{array}{l} \mathrm{SU}(2){:}\,\mathbf{2} = \mathcal{D}^{\frac{1}{2}} \equiv \square, \quad \bar{\mathbf{2}} = \mathcal{D}^{*\frac{1}{2}} \equiv \square, \quad \mathbf{3} = \mathcal{D}^{1} \equiv \square\!\square; \\ \mathrm{SU}(3){:}\,\mathbf{3} = \mathcal{D}(1,0) \equiv \square, \quad \bar{\mathbf{3}} = \mathcal{D}(0,1) \equiv \begin{array}{|c|}\hline \, \\ \hline \, \\ \hline\end{array}, \quad \mathbf{8} = \mathcal{D}(1,1) \equiv \begin{array}{|c|c|}\hline \, & \, \\ \hline \, \\ \cline{1-1}\end{array}. \end{array} \tag{C.10}$$

The dimension of a diagram (and thus of the representation) is given by $d = A/B$. To determine the integer A, proceed as follows. Put the value N for boxes in the diagonal, $N+1$ for boxes immediately above the diagonal, $N-1$ immediately below, $N+2$ above the diagonal labelled with $N+1$, etc. (see the left-hand side diagram of Fig. C.2). The value of A is the product of all those numbers. The integer B also results from the product of numbers associated with each box. To determine the value of a box, starting from the right-hand end of the row containing the box of interest, draw a horizontal line up to the box, then turn downward until the vertical line leaves the diagram (see the right-hand side diagram of Fig. C.2). The value associated with the box is the number of boxes crossed. The procedure is illustrated in Fig. C.2 for the box of interest marked with the bullet. The dotted line crossed four boxes. One has to repeat the procedure for all boxes of the diagram

[1] Tableaux is a French word meaning tables (plural). I will mainly use the word diagram instead of tableaux.

N	$N+1$	$N+2$	$N+3$
$N-1$	N	$N+1$	
$N-2$	$N-1$	N	
$N-3$			

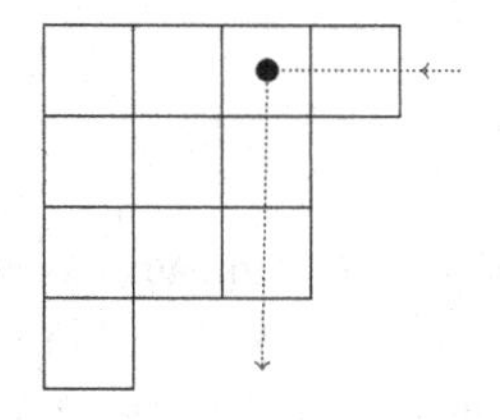

Fig. C.2 The calculation of the dimension of a representation with Young diagrams. See the text.

and multiply all the obtained numbers to get the value of B. Let us take the example in SU(2) of the representation □□. Factor A is $2 \times 3 = 6$, while B is $1 \times 2 = 2$. Hence, $d = 6/2 = 3$. Note that no useful diagram can have more than N boxes in any column since it would give a factor $A = 0$ and thus a dimension of zero. As an exercise, check the dimensions of the other representations given in Eq. (C.10). We observe that a single column of N boxes has dimension 1 since $d = (N!)/(N!)$. It represents the singlet of the representation and is usually represented by the symbol •. For instance, in SU(2), $\mathbf{1} \equiv \begin{smallmatrix}\square\\\square\end{smallmatrix} =$ •. Consequently, a diagram containing such a column necessarily has the same dimension as the same diagram without the column. In other words, pursuing the example of SU(2), the diagrams $\begin{smallmatrix}\square\square\square\square\\\square\square\end{smallmatrix}$ and $\begin{smallmatrix}\square\square\square\\\square\end{smallmatrix}$ are both equivalent to □□, and it is simpler to keep the last diagram to represent the representation.

Young diagrams provide an easy way to perform tensor products of representations and to decompose the result in irreducible representations. The recipe (without proof) is the following. Put the bigger diagram on the left and the other on the right (it is more convenient but not mandatory). In the diagram on the right, insert the letter a in the boxes of the first row, the letter b in the second row and so on. Then, distribute the a-boxes to the left diagram. You can attach an a-box only to the right or to the bottom of a box from the left diagram. There are several rules to only keep acceptable diagrams. *Rule 1: the number of boxes in any given row is not smaller (but can be equal) than the number in any of the lower rows. Rule 2: valid diagrams have no more than one a-box per column.* Let us take, first, the example in SU(3) of $\bar{\mathbf{3}} \otimes \mathbf{3}$. The distribution of a-boxes is simple since

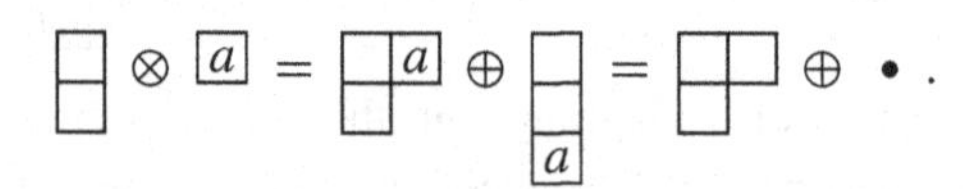

Note that $\begin{smallmatrix}\square\\\square\square\end{smallmatrix}$ is forbidden by rule 1. The calculation of the dimension of $\begin{smallmatrix}\square\square\\\square\end{smallmatrix}$ gives $2 \times 3 \times 4/(1 \times 3 \times 1) = 8$. We thus recover the well-known result $\bar{\mathbf{3}} \otimes \mathbf{3} = \mathbf{8} \oplus \mathbf{1}$. Now consider the more complicated case of $\mathbf{8} \otimes \mathbf{8}$ in SU(3). The distribution of a-boxes gives

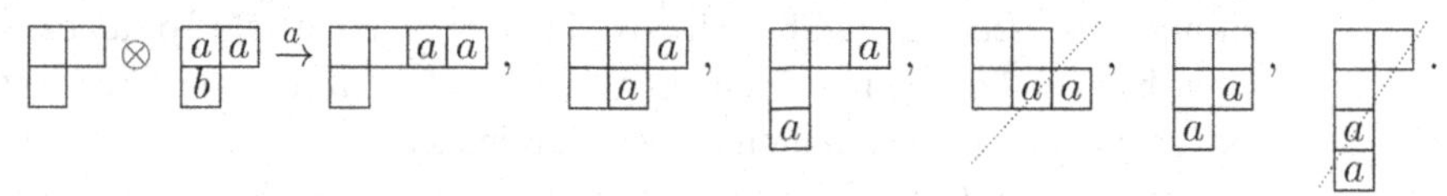

The first discarded diagram above violates rule 1, while the second violates rule 2. Note that the first column of the latter diagram is also too long [in SU(3), the maximum column length is 3]. The next step is to distribute the b-boxes to the valid diagrams obtained after the first step. Use the same rules as for the a-boxes with one further constraint. *Rule 3:*

count the number of a- and b-boxes by following a path starting from the top rightmost box and going along the first row towards the left, then moving to the rightmost box of the second row, towards the left, etc. (see the picture below). The number of a-boxes must be greater than or equal to the number of b-boxes at any step of the path. If not, throw away the diagram.

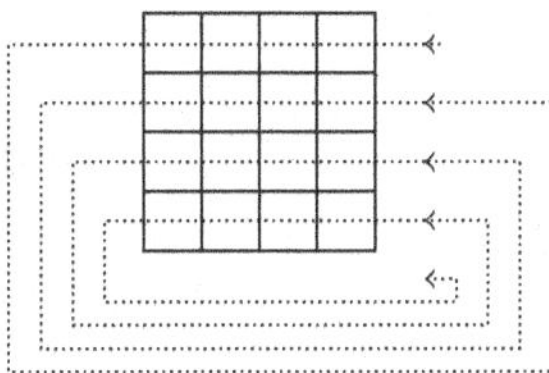

If there are also c-, d-, e-, etc. boxes, then at each step, the number of a-boxes must be greater than or equal to the number of b-boxes that must be greater than or equal to the number of c-boxes, etc. Here are some examples:

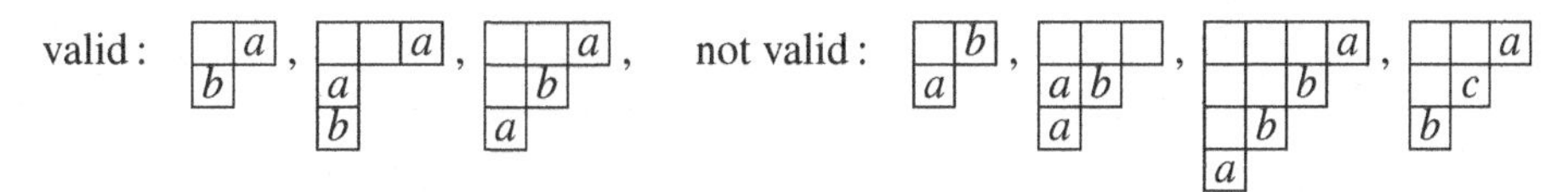

Continuing our example of $\mathbf{8} \otimes \mathbf{8}$, the distribution of b-boxes gives

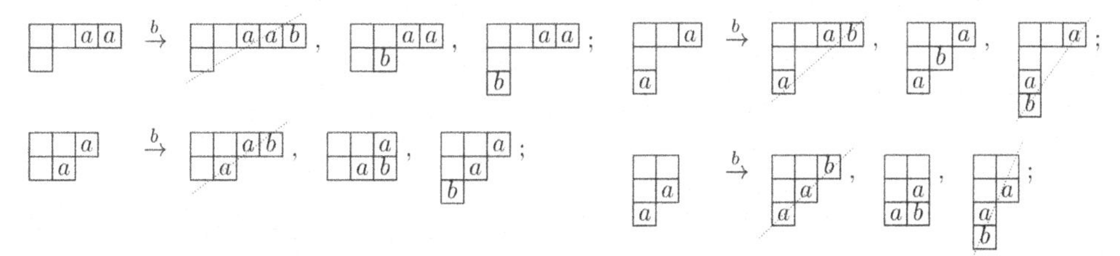

In the first line, the first two discarded diagrams are excluded by rule 3. The last diagram has a column length greater than $N = 3$ and thus should not be considered for SU(3) representations. In the second line, the three diagrams are excluded for the same reasons as in the first line. Therefore, we keep

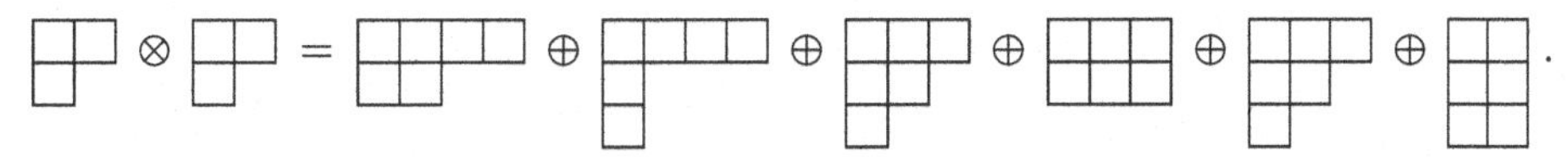

Several diagrams can be simplified since they contain columns of three boxes. After simplification, the result finally reads

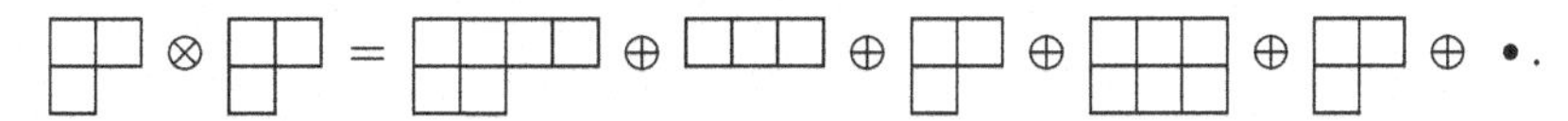

The calculation of the dimension of the diagrams tells us that $\equiv \mathbf{27}$, $\equiv \mathbf{10}$, $\equiv \mathbf{8}$ and $\bullet \equiv \mathbf{1}$. What about ? The calculation of the dimension gives 10. Actually, it corresponds to the conjugate representation $\overline{\mathbf{10}}$. In SU(3), we already know that the tensorial product of $\mathbf{3}$ with its conjugate representation $\bar{\mathbf{3}}$ generates a singlet since $\bar{\mathbf{3}} \otimes \mathbf{3} = \mathbf{8} \oplus \mathbf{1}$. It is also true in SU($N$) for the product of any representation with its conjugate. Therefore, given the Young diagram of a representation, its product with the conjugate representation must be able to yield, among other things, the singlet, which is represented by a column diagram of length N or equivalently multi-columns of length N. This gives us the recipe to find the diagram of the conjugate representation: complete the diagram of the representation by boxes to produce columns of N boxes. The added boxes correspond to the conjugate representation after a rotation of 180° (to get a valid diagram satisfying rule 1) as illustrated below in SU(4) for **36** and $\overline{\mathbf{36}}$ (the grey area).

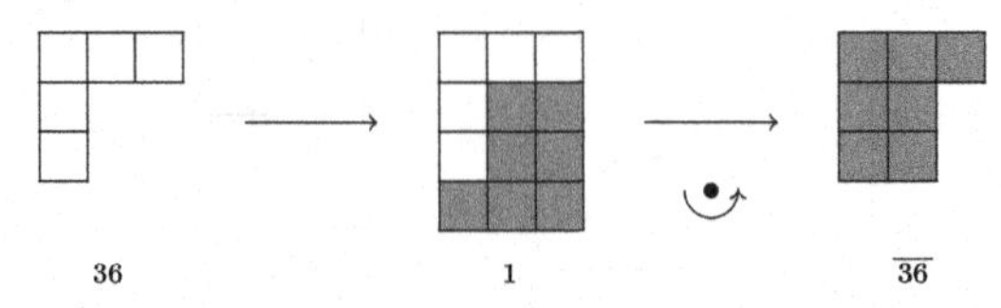

The same procedure shows easily that in SU(3), the conjugate representation of $\mathbf{10} \equiv$ ☐☐☐ is given by $\overline{\mathbf{10}} \equiv$ ⊞⊞. Hence, we conclude

$$\mathbf{8} \otimes \mathbf{8} = \mathbf{27} \oplus \mathbf{10} \oplus \overline{\mathbf{10}} \oplus \mathbf{8} \oplus \mathbf{8} \oplus \mathbf{1}.$$

Appendix D Lorentz and Poincaré Groups

D.1 Lorentz Group

D.1.1 Generalities

In Section 2.2, we saw that a Lorentz transformation, $\Lambda^{\mu}{}_{\nu}$, must satisfy Eq. (2.14) resulting from the conservation of the scalar product in the Minkowski space,

$$g_{\rho\sigma}\Lambda^{\rho}{}_{\mu}\Lambda^{\sigma}{}_{\nu} = g_{\mu\nu}, \tag{D.1}$$

where $g_{\mu\nu}$ is the metric tensor of the Minkowski space. We then deduced that the product of two Lorentz transformations is a Lorentz transformation, that the inverse of a Lorentz transformation is defined and constitutes a Lorentz transformation, and that the identity matrix is also a Lorentz transformation. These properties are those of a group, i.e. $O(1,3)$ when the Minkowski metric $(+,-,-,-)$ is used. The restricted Lorentz group whose elements are the proper and orthochronous Lorentz transformations (i.e., $\det\Lambda = 1$ and $\Lambda^{0}{}_{0} \geq 1$) is denoted by $SO^{+}(1,3)$. It is generated by ordinary spatial rotations whose group is isomorphic to SO(3) and Lorentz boosts (which do not constitute a group). A rotation needs three parameters to be fully defined (for instance, the two angles defining the axis of rotation and the angle of rotation about the axis). Similarly, a boost requires three parameters (the value of the boost β and its direction). Therefore, six parameters are needed for Lorentz transformations of the restricted Lorentz group. The restricted Lorentz group is a Lie group since the six parameters can vary continuously, and therefore, its matrix elements can be built by 'exponentiating' linear combinations of its six infinitesimal generators.

D.1.2 Generators of the Restricted Lorentz Group

Lorentz transformations corresponding to an active rotation of angle α leave the time component of 4-vectors unchanged, while their spatial components are affected by the rotation matrices $R_x(\alpha)$, $R_y(\alpha)$ and $R_z(\alpha)$ given in Eq. (A.38). The corresponding matrices are then

$$\Lambda_a(\alpha) = \begin{pmatrix} 1 & 0 & 0 & 0 \\ 0 & & & \\ 0 & & R_a(\alpha) & \\ 0 & & & \end{pmatrix},$$

where the axis $a = x, y$ or z. Following the same approach as in Section A.4.5, i.e. $J_a = i\frac{\partial}{\partial\alpha}\Lambda_a(\alpha)\big|_{\alpha=0}$, we obtain the three infinitesimal generators of rotations

$$J_1 = \begin{pmatrix} 0 & 0 & 0 & 0 \\ 0 & 0 & 0 & 0 \\ 0 & 0 & 0 & -i \\ 0 & 0 & i & 0 \end{pmatrix}, \quad J_2 = \begin{pmatrix} 0 & 0 & 0 & 0 \\ 0 & 0 & 0 & i \\ 0 & 0 & 0 & 0 \\ 0 & -i & 0 & 0 \end{pmatrix}, \quad J_3 = \begin{pmatrix} 0 & 0 & 0 & 0 \\ 0 & 0 & -i & 0 \\ 0 & i & 0 & 0 \\ 0 & 0 & 0 & 0 \end{pmatrix}. \tag{D.2}$$

By construction, J_1, J_2 and J_3 satisfy the usual commutation relations of the Lie algebra of the rotation group [and of SU(2)], i.e.

$$[J_i, J_j] = i\epsilon_{ijk}J_k, \tag{D.3}$$

with ϵ_{ijk} being the spatial antisymmetric tensor and where the sum over k is implicit. The three generators of the Lorentz boost can be obtained similary. Consider, for example, a boost along the x-axis direction

$$\Lambda = \begin{pmatrix} \gamma & \beta\gamma & 0 & 0 \\ \beta\gamma & \gamma & 0 & 0 \\ 0 & 0 & 1 & 0 \\ 0 & 0 & 0 & 1 \end{pmatrix} = \begin{pmatrix} \cosh\zeta & \sinh\zeta & 0 & 0 \\ \sinh\zeta & \cosh\zeta & 0 & 0 \\ 0 & 0 & 1 & 0 \\ 0 & 0 & 0 & 1 \end{pmatrix},$$

with ζ, being the rapidity [see Eq. (2.44)]. Here again, the active viewpoint is followed, and therefore, the transformation is equivalent to that in Eq. (2.21) where the particle is boosted from its rest frame to the lab frame. The corresponding generator is thus given by

$$K_1 = i\frac{\partial}{\partial\zeta}\begin{pmatrix} \cosh\zeta & \sinh\zeta & 0 & 0 \\ \sinh\zeta & \cosh\zeta & 0 & 0 \\ 0 & 0 & 1 & 0 \\ 0 & 0 & 0 & 1 \end{pmatrix}\Bigg|_{\zeta=0} = \begin{pmatrix} 0 & i & 0 & 0 \\ i & 0 & 0 & 0 \\ 0 & 0 & 0 & 0 \\ 0 & 0 & 0 & 0 \end{pmatrix}. \tag{D.4}$$

Similarly, for a boost along the y- or z-axis, the generators read

$$K_2 = \begin{pmatrix} 0 & 0 & i & 0 \\ 0 & 0 & 0 & 0 \\ i & 0 & 0 & 0 \\ 0 & 0 & 0 & 0 \end{pmatrix}, \quad K_3 = \begin{pmatrix} 0 & 0 & 0 & i \\ 0 & 0 & 0 & 0 \\ 0 & 0 & 0 & 0 \\ i & 0 & 0 & 0 \end{pmatrix}. \tag{D.5}$$

We thus deduce the following commutation relations (with an implicit summation over k):

$$[K_i, K_j] = -i\epsilon_{ijk}J_k, \quad [K_i, J_j] = i\epsilon_{ijk}K_k. \tag{D.6}$$

In addition to Eq. (D.3), the relations above define the Lie algebra of the restricted Lorentz group. Note that the first commutation in relation (D.6) shows that the product of two boosts is generally equivalent to a rotation (except when the two boosts are performed in the same direction). Therefore, even if pure rotations constitute a group, the pure boost transformations do not.

It is convenient to gather the six Lorentz generators into a single tensor $J_{\rho\sigma}$. Consider an infinitesimal Lorentz transformation $\Lambda \simeq \mathbb{1} + \omega$ (where ω is the matrix with the infinitesimal parameters). The component of the infinitesimal Lorentz transformation then reads[1]

$$\Lambda^{\mu}_{\ \nu} = \delta^{\mu}_{\nu} + \omega^{\mu}_{\ \nu}, \tag{D.7}$$

where δ^{μ}_{ν} denotes the Kronecker symbol (2.8), i.e. $\delta^{\mu}_{\nu} = g_{\nu\alpha}g^{\alpha\mu} = g_{\nu}^{\ \mu} = g^{\mu}_{\ \nu}$ (the latter equality is due to the symmetry of the metric tensor as $g_{\nu\alpha}g^{\alpha\mu} = g_{\alpha\nu}g^{\mu\alpha}$). Since any Lorentz transformation satisfies Eq. (D.1), it reads, for infinitesimal transformations,

$$g_{\mu\nu} = g_{\rho\sigma}(\delta^{\rho}_{\mu} + \omega^{\rho}_{\ \mu})(\delta^{\sigma}_{\nu} + \omega^{\sigma}_{\ \nu}) = g_{\rho\sigma}\delta^{\rho}_{\mu}\delta^{\sigma}_{\nu} + g_{\rho\sigma}\delta^{\rho}_{\mu}\omega^{\sigma}_{\ \nu} + g_{\rho\sigma}\omega^{\rho}_{\ \mu}\delta^{\sigma}_{\nu}.$$

where only the first order has been kept. The only non-zero terms are those for which the indices of the Kronecker symbol are the same. It follows that

$$g_{\mu\nu} = g_{\mu\nu} + g_{\mu\sigma}\omega^{\sigma}_{\ \nu} + g_{\rho\nu}\omega^{\rho}_{\ \mu} = g_{\mu\nu} + \omega_{\mu\nu} + \omega_{\nu\mu}.$$

Therefore, ω is antisymmetric, i.e. $\omega_{\mu\nu} = -\omega_{\nu\mu}$, confirming that only six independent parameters (and not 16) are needed to describe a Lorentz transformation. The infinitesimal Lorentz transformation thus writes

$$\Lambda^{\mu}_{\ \nu} = \delta^{\mu}_{\nu} + \omega^{\mu}_{\ \nu} = \delta^{\mu}_{\nu} + g^{\mu\sigma}\delta^{\rho}_{\nu}\,\omega_{\sigma\rho} = \delta^{\mu}_{\nu} + \frac{1}{2}(g^{\mu\sigma}\delta^{\rho}_{\nu}\,\omega_{\sigma\rho} - g^{\mu\sigma}\delta^{\rho}_{\nu}\,\omega_{\rho\sigma}).$$

As the indices ρ and σ are dummy indices, we can swap them in the second term, leading to

$$\Lambda^{\mu}_{\ \nu} = \delta^{\mu}_{\nu} + \frac{1}{2}\omega_{\sigma\rho}(g^{\mu\sigma}\delta^{\rho}_{\nu} - g^{\mu\rho}\delta^{\sigma}_{\nu}) = \delta^{\mu}_{\nu} - \frac{i}{2}\omega_{\sigma\rho}(ig^{\mu\sigma}\delta^{\rho}_{\nu} - ig^{\mu\rho}\delta^{\sigma}_{\nu}). \tag{D.8}$$

Defining the anti-symmetric tensor[2] (for indices σ and ρ),

$$(J^{\sigma\rho})^{\mu}_{\ \nu} = ig^{\mu\sigma}\delta^{\rho}_{\nu} - ig^{\mu\rho}\delta^{\sigma}_{\nu}, \tag{D.9}$$

the infinitesimal Lorentz transformation finally reads

$$\Lambda^{\mu}_{\ \nu} = \delta^{\mu}_{\nu} - \frac{i}{2}\omega_{\sigma\rho}\,(J^{\sigma\rho})^{\mu}_{\ \nu}, \tag{D.10}$$

or in terms of matrices,

$$\Lambda = \mathbb{1} - \frac{i}{2}\omega_{\sigma\rho}J^{\sigma\rho}. \tag{D.11}$$

Therefore, a finite Lorentz transformation of the restricted Lorentz group corresponds to the matrix

$$\Lambda = \exp\left(-\frac{i}{2}\omega_{\sigma\rho}J^{\sigma\rho}\right). \tag{D.12}$$

[1] Notice that $x'^{\mu} = \Lambda^{\mu}_{\ \nu}x^{\nu} = (\delta^{\mu}_{\nu} + \omega^{\mu}_{\ \nu})x^{\nu} = (g_{\nu\alpha}g^{\alpha\mu} + g_{\nu\alpha}\omega^{\mu\alpha})x^{\nu} = (g^{\mu\alpha} + \omega^{\mu\alpha})x_{\alpha}$.

[2] Alternatively, since $\omega^{\mu}_{\ \nu} = g^{\mu\sigma}\delta^{\rho}_{\nu}\,\omega_{\sigma\rho} = g_{\nu\rho}\delta^{\mu}_{\sigma}\omega^{\sigma\rho}$, one can define the tensor $\left(J_{\sigma\rho}\right)^{\mu}_{\ \nu} = ig_{\nu\rho}\delta^{\mu}_{\sigma} - ig_{\nu\sigma}\delta^{\mu}_{\rho}$. It leads to the equalities

$$(J_{0i})^{\mu}_{\ \nu} = -\left(J^{0i}\right)^{\mu}_{\ \nu} = g_{i\alpha}\left(J^{0\alpha}\right)^{\mu}_{\ \nu}, \quad \left(J_{ij}\right)^{\mu}_{\ \nu} = \left(J^{ij}\right)^{\mu}_{\ \nu} = g_{i\alpha}g_{j\beta}\left(J^{\alpha\beta}\right)^{\mu}_{\ \nu},$$

which ensures that $\omega_{\sigma\rho}J^{\sigma\rho} = \omega^{\sigma\rho}J_{\sigma\rho}$.

The tensor $J^{\sigma\rho}$ includes the six generators that read, with Eq. (D.9),

$$J^{01} = \begin{pmatrix} 0 & i & 0 & 0 \\ i & 0 & 0 & 0 \\ 0 & 0 & 0 & 0 \\ 0 & 0 & 0 & 0 \end{pmatrix}, \quad J^{02} = \begin{pmatrix} 0 & 0 & i & 0 \\ 0 & 0 & 0 & 0 \\ i & 0 & 0 & 0 \\ 0 & 0 & 0 & 0 \end{pmatrix}, \quad J^{03} = \begin{pmatrix} 0 & 0 & 0 & i \\ 0 & 0 & 0 & 0 \\ 0 & 0 & 0 & 0 \\ i & 0 & 0 & 0 \end{pmatrix},$$

$$J^{12} = \begin{pmatrix} 0 & 0 & 0 & 0 \\ 0 & 0 & -i & 0 \\ 0 & i & 0 & 0 \\ 0 & 0 & 0 & 0 \end{pmatrix}, \quad J^{13} = \begin{pmatrix} 0 & 0 & 0 & 0 \\ 0 & 0 & 0 & -i \\ 0 & 0 & 0 & 0 \\ 0 & i & 0 & 0 \end{pmatrix}, \quad J^{23} = \begin{pmatrix} 0 & 0 & 0 & 0 \\ 0 & 0 & 0 & 0 \\ 0 & 0 & 0 & -i \\ 0 & 0 & i & 0 \end{pmatrix}, \tag{D.13}$$

with the symmetry properties

$$J^{\sigma\rho} = -J^{\rho\sigma}, (J^{00} = J^{11} = J^{22} = J^{33} = 0). \tag{D.14}$$

Comparing Eq. (D.13) with Eqs. (D.2), (D.4) and (D.5), it follows the identification of the generators J_i and K_i:

$$\begin{aligned} J^{01} &= K_1, \quad J^{02} = K_2, \quad J^{03} = K_3, \\ J^{12} &= J_3, \quad J^{13} = -J_2, \quad J^{23} = J_1, \end{aligned} \tag{D.15}$$

or equivalently

$$J^{0i} = K_i, \quad J^{ij} = \epsilon_{ijk} J_k. \tag{D.16}$$

The equation above can be easily inverted to find the expression of K_i and J_i from the tensor $J^{\mu\nu}$. It yields

$$K_i = J^{0i}, \quad J_i = \frac{1}{2}\epsilon_{ijk} J^{jk}. \tag{D.17}$$

The commutation rules of the restricted Lorentz group (D.3) and (D.6) can elegantly be grouped into one single relation, using the $J^{\sigma\rho}$ tensor, i.e.

$$[J^{\mu\nu}, J^{\sigma\rho}] = i(g^{\nu\sigma} J^{\mu\rho} - g^{\mu\sigma} J^{\nu\rho} + g^{\mu\rho} J^{\nu\sigma} - g^{\nu\rho} J^{\mu\sigma}). \tag{D.18}$$

This relation fully defines the $\mathfrak{so}^+(1,3)$ Lie algebra. Note that the derivation of the expression of $J^{\mu\nu}$ has only exploited the definition (D.1) of a general Lorentz transformation. Hence, Eq. (D.18) also defines the $\mathfrak{so}(1,3)$ Lie algebra.

D.1.3 Homomorphism of SL(2, $\mathbb{C}$) into SO$^+$(1, 3)

In Section B.2, we showed that there is a group homomorphism corresponding to a two-to-one mapping of SU(2) to SO(3), with SU(2) being the universal covering group of SO(3), both sharing the same Lie algebra. The situation of the Lorentz group SO$^+(1,3)$ turns out to be very similar, but this time the universal covering group is SL(2, $\mathbb{C}$). In order to reveal the homomorphism, let us characterise Lorentz transformations with Hermitian matrices.

Let $\mathcal{H}_2$ be the vector space of 2×2 (complex) Hermitian matrices over $\mathbb{R}$. Any 2×2 Hermitian matrices X can then be written as

$$X = \begin{pmatrix} x^0 + x^3 & x^1 - ix^2 \\ x^1 + ix^2 & x^0 - x^3 \end{pmatrix}, \tag{D.19}$$

where the x^is are real numbers. A basis of $\mathcal{H}_2$ is $\mathcal{B} = \{\sigma_0, \sigma_1, \sigma_2, \sigma_3\}$, where $\sigma_0 = \mathbb{1}$ and $\sigma_{i>0}$ are the Pauli matrices. In this basis, the (real) components of X are then

$$[X]_{\mathcal{B}} = \begin{pmatrix} x^0 \\ x^1 \\ x^2 \\ x^3 \end{pmatrix}, \quad X = [X]^{\mu}_{\mathcal{B}}\sigma_\mu = x^\mu \sigma_\mu. \tag{D.20}$$

Given the following properties of σ_μ matrices:

$$\mathrm{Tr}(\sigma_\mu \sigma_\nu) = 2\delta_{\mu\nu}, \quad \sigma_\mu^2 = \mathbb{1},$$

the component $[X]^{\mu}_{\mathcal{B}}$ is[3]

$$[X]^{\mu}_{\mathcal{B}} = x^\mu = \frac{1}{2}\mathrm{Tr}(X\sigma_\mu). \tag{D.21}$$

As expected, for $X = \sigma_\nu$, we do have $[\sigma_\nu]^{\mu}_{\mathcal{B}} = \delta_{\mu\nu}$. Note that in Eq. (D.19), $\det(X) = (x^0)^2 - (x^1)^2 - (x^2)^2 - (x^3)^2$. Therefore, if we interpret $[X]_{\mathcal{B}}$ as a 4-vector x, $\det(X) = \|x\|^2$. Let us define the action f_M of a matrix M of $\mathrm{SL}(2, \mathbb{C})$ on X that

$$X \xrightarrow{f_M} X' = f_M(X) = MXM^\dagger. \tag{D.22}$$

The matrix X' also belongs to the set of 2×2 Hermitian matrices and its coordinates $[X']_{\mathcal{B}} = x'$ satisfy $\|x'\|^2 = \det(X') = \det(X) = \|x\|^2$, as $\det(M) = 1$. Consequently, f_M performs a linear mapping from $\mathcal{H}_2$ into itself and preserves the norm of 4-vectors. In the basis $\mathcal{B}$, f_M is then represented by a 4×4 matrix acting on $[X]_{\mathcal{B}}$, denoted $f(M)$, such that

$$f(M): \quad [X']_{\mathcal{B}} = f(M)[X]_{\mathcal{B}} \Leftrightarrow x' = f(M)x. \tag{D.23}$$

Since $\|x'\|^2 = \|x\|^2$, $f(M)$ preserves the Minkowski norm, and thus, $f(M)$ is an element of $\mathrm{O}(1,3)$. Therefore, the mapping performed by f is

$$\begin{aligned} f: \mathrm{SL}(2,\mathbb{C}) &\rightarrow \mathrm{O}(1,3) \\ M &\mapsto f(M). \end{aligned} \tag{D.24}$$

In order to be a homomorphism, it must satisfy $f(M_1M_2) = f(M_1)f(M_2)$ for all matrices M_1 and M_2 in $\mathrm{SL}(2,\mathbb{C})$. Equations (D.22) and (D.23) imply that for all M_1, M_2 and X,

$$\begin{cases} f(M_1M_2)[X]_{\mathcal{B}} & = [f_{M_1M_2}(X)]_{\mathcal{B}} = [M_1M_2XM_2^\dagger M_1^\dagger]_{\mathcal{B}}, \\ f(M_1)f(M_2)[X]_{\mathcal{B}} & = f(M_1)[f_{M_2}(X)]_{\mathcal{B}} = f(M_1)[M_2XM_2^\dagger]_{\mathcal{B}} = [M_1M_2XM_2^\dagger M_1^\dagger]_{\mathcal{B}}. \end{cases}$$

This shows that $f(M_1M_2) = f(M_1)f(M_2)$. We can go further and show that f is actually a homomorphism from $\mathrm{SL}(2,\mathbb{C})$ to $\mathrm{SO}^+(1,3)$. If we admit that $\mathrm{SL}(2,\mathbb{C})$ is simply connected (and it is!), the determinant of $f(M)$ must vary continuously as a function of the continuous parameters of M. But with $f(M)$ being in $\mathrm{O}(1,3)$, the only possible values of the determinant are ± 1. Taking $M = \mathbb{1}$, $f_{\mathbb{1}}(X) = X$, and thus, $f(\mathbb{1}) = \mathbb{1}$. Hence, the image of $\mathrm{SL}(2,\mathbb{C})$ by f includes the identity matrix, which has determinant 1. The determinant of $f(M)$, being a continuous function, cannot jump from 1 to -1. So, the homomorphism is at most

[3] Do not worry about the apparent inconsistent position of the μ index in Eq. (D.21). Pauli matrices are not 4-vectors. So quantity such as $X = x^\mu \sigma_\mu$ is not a (Lorentz) scalar, even if the Einstein notation is used.

from SL(2, $\mathbb{C}$) to SO(1, 3). Moreover, since $[X']_{\mathcal{B}} = f(M)[X]_{\mathcal{B}}$, and given Eq. (D.21), it follows that

$$[X']^{\mu}_{\mathcal{B}} = \frac{1}{2}\mathrm{Tr}(X'\sigma_{\mu}) = \frac{1}{2}\mathrm{Tr}(MXM^{\dagger}\sigma_{\mu}) = \frac{1}{2}\mathrm{Tr}(M\sigma_{\nu}M^{\dagger}\sigma_{\mu})[X]^{\nu}_{\mathcal{B}},$$

where X has been replaced by $[X]^{\nu}_{\mathcal{B}}\sigma_{\nu}$ with Eq. (D.20) to move to the last equality. We thus conclude

$$f(M)^{\mu}{}_{\nu} = \frac{1}{2}\mathrm{Tr}(M\sigma_{\nu}M^{\dagger}\sigma_{\mu}). \tag{D.25}$$

Therefore, $f(M)^{0}{}_{0} = \mathrm{Tr}(MM^{\dagger})/2$, and writing M of SL(2, $\mathbb{C}$) as $M = \begin{pmatrix} \alpha & \beta \\ \gamma & \delta \end{pmatrix}$ yields $f(M)^{0}{}_{0} = (|\alpha|^2 + |\beta|^2 + |\gamma|^2 + |\delta|^2)/2$. Therefore, $f(M)^{0}{}_{0} > 0$, and, as announced, f is a homomorphism from SL(2, $\mathbb{C}$) to $\mathrm{SO}^{+}(1,3)$.

Note that according to Eq. (D.25), $f(-M) = f(M)$, i.e. two matrices of SL(2, $\mathbb{C}$) correspond to the same restricted Lorentz transformation. Consequently, we have only a two-to-one homomorphism, and some finite-dimensional representations of SL(2, $\mathbb{C}$) are not true representations but only projective representations of $\mathrm{SO}^{+}(1,3)$. For matrices of SL(2, $\mathbb{C}$) close to the identity, the two groups look the same, explaining why they share the same Lie algebra (i.e., isomorphic algebra), as we shall see in the next section.

D.1.4 Finite-Dimensional Representations

In Section D.1.2., we saw that any Lorentz transformation could be represented by the 4×4 matrix (D.12), which constitutes the standard four-dimensional representation of the Lorentz group that acts on 4-vectors. Note that the argument of the exponential in Eq. (D.12) reads

$$-\frac{i}{2}\omega_{\sigma\rho}J^{\sigma\rho} = -i\left(\sum_{j>i}\omega_{ij}J^{ij} + \sum_{i}\omega_{0i}J^{0i}\right) = -i\left(\sum_{j>i,k}\omega_{ij}\epsilon_{ijk}J_k + \sum_{i}\omega_{0i}K_i\right).$$

Let us introduce the variables α_i and ζ_i such that

$$\omega_{ij} = \epsilon_{ijk}\alpha_k, \quad \omega_{0i} = \zeta_i. \tag{D.26}$$

This parametrisation corresponds to a rotation of angle $|\boldsymbol{\alpha}|$ about the axis $\hat{\boldsymbol{\alpha}} = \boldsymbol{\alpha}/|\boldsymbol{\alpha}|$ and a boost of rapidity $|\boldsymbol{\zeta}|$ in the direction $\hat{\boldsymbol{\zeta}} = \boldsymbol{\zeta}/|\boldsymbol{\zeta}|$. The four-dimensional standard representation (D.12) of the Lorentz group then takes the form

$$\mathcal{D}(\Lambda) = \exp\left(-\frac{i}{2}\omega_{\sigma\rho}J^{\sigma\rho}\right) = \exp\left(-i\boldsymbol{\alpha}\cdot\boldsymbol{J} - i\boldsymbol{\zeta}\cdot\boldsymbol{K}\right), \tag{D.27}$$

where $\boldsymbol{J}$ and $\boldsymbol{K}$ are defined as vectors of matrices, i.e.

$$\boldsymbol{J} = \begin{pmatrix} J_1 \\ J_2 \\ J_3 \end{pmatrix}, \quad \boldsymbol{K} = \begin{pmatrix} K_1 \\ K_2 \\ K_3 \end{pmatrix}.$$

Their components are specified in Eqs. (D.2), (D.4) and (D.5). The representation (D.27) is called the vector representation since it acts on 4-vectors. In order to present other finite-dimensional representations, it is convenient to introduce the following complex linear combinations of the Lorentz generators:

$$M_i = \frac{1}{2}(J_i + iK_i), \quad N_i = \frac{1}{2}(J_i - iK_i), \tag{D.28}$$

or equivalently,

$$J_i = M_i + N_i, \quad K_i = -i\,(M_i - N_i). \tag{D.29}$$

Indeed, given the commutation relations (D.3) and (D.6), it is straightforward to show that

$$[M_i, M_j] = i\epsilon_{ijk}M_k, \quad [N_i, N_j] = i\epsilon_{ijk}N_k, \quad [M_i, N_j] = 0. \tag{D.30}$$

Thus, M_i and N_i satisfy the same commutation rules as the generators of SU(2). There is, however, a subtlety. If the original Lie algebra $\mathfrak{so}(1,3)$, spanned by the generators J_i and K_i, is real (the rotation and boost parameters being real numbers), the one spanned by J_i and iK_i and thus by M_i and N_i becomes complex (still with the same six generators). Let us denote it by $\mathfrak{so}(1,3)_{\mathbb{C}}$. Mathematically, we do a *complexification*[4] of the Lie algebra $\mathfrak{so}(1,3) \hookrightarrow \mathfrak{so}(1,3)_{\mathbb{C}}$. Therefore, M_i and N_i both satisfy the $\mathfrak{su}(2)_{\mathbb{C}}$ complex Lie algebra. In other words, we have found

$$\mathfrak{so}(1,3)_{\mathbb{C}} \cong \mathfrak{su}(2)_{\mathbb{C}} \oplus \mathfrak{su}(2)_{\mathbb{C}}. \tag{D.32}$$

This decomposition is important because we have already determined all finite-dimensional real irreducible representations of $\mathfrak{su}(2)$ in Section B.3. The finite-dimensional complex irreducible representations of $\mathfrak{su}(2)_{\mathbb{C}}$ are thus the same as $\mathfrak{su}(2)$ but with complex linear combinations of the generators. Note that any matrix of $\mathfrak{su}(2)_{\mathbb{C}}$ can be written as $X+iX'$, where X and X' are matrices of $\mathfrak{su}(2)$. As a matrix of $\mathfrak{su}(2)$ writes $x_1\sigma_1+x_2\sigma_2+x_3\sigma_3$ with x_1, x_2 and x_3 three real numbers, a matrix of $\mathfrak{su}(2)_{\mathbb{C}}$ writes $z_1\sigma_1 + z_2\sigma_2 + z_3\sigma_3$ with three complex numbers, i.e.

$$\left\{ \begin{pmatrix} z_3 & z_1 - iz_2 \\ z_1 + iz_2 & -z_3 \end{pmatrix} \middle| \; z_1, z_2, z_3 \in \mathbb{C} \right\} = \left\{ \begin{pmatrix} \gamma & \beta \\ \delta & -\gamma \end{pmatrix} \middle| \; \gamma, \beta, \delta \in \mathbb{C} \right\}.$$

We thus conclude that $\mathfrak{su}(2)_{\mathbb{C}}$ is isomorphic to $\mathfrak{sl}(2,\mathbb{C})$, the complex vector space of complex traceless matrices. In Appendix B, Section B.2, we have seen that $\mathfrak{su}(2)$ and $\mathfrak{so}(3)$ are isomorphic. We now realise that the complex representations of $\mathfrak{su}(2)$ [and $\mathfrak{so}(3)$] are thus those of $\mathfrak{sl}(2,\mathbb{C})$. Therefore, the decomposition (D.32) is also equivalent to

$$\mathfrak{so}(1,3)_{\mathbb{C}} \cong \mathfrak{sl}(2,\mathbb{C}) \oplus \mathfrak{sl}(2,\mathbb{C}). \tag{D.33}$$

[4] The mathematical details of a complexification can be found, for example, in Hall (2015, chapter 3). The important points are as follows. (i) When $\mathfrak{g}$ is a real Lie algebra of a matrix Lie group, then $\mathfrak{g}_{\mathbb{C}} = \mathfrak{g} \otimes \mathbb{C}$, the complexified of $\mathfrak{g}$, is isomorphic to the set of matrices in $M_n(\mathbb{C})$ made up of $X + iX'$, where X and X' are in $\mathfrak{g}$, i.e.

$$\mathfrak{g}_{\mathbb{C}} \cong \mathfrak{g} \oplus i\mathfrak{g}. \tag{D.31}$$

(ii) Every finite-dimensional representation $d(X)$ of a Lie algebra $\mathfrak{g}$ extends uniquely to $\mathfrak{g}_{\mathbb{C}}$ via $d(X + iX') \equiv d(X) + id(X')$, with X, X' in $\mathfrak{g}$. The complex representations of $\mathfrak{g}$ are thus representations of $\mathfrak{g}_{\mathbb{C}}$.

In conclusion, in order to find the finite-dimensional representations of $\mathfrak{so}(1,3)$, one has to find the finite-dimensional representations of $\mathfrak{su}(2)_{\mathbb{C}} \oplus \mathfrak{su}(2)_{\mathbb{C}}$ or $\mathfrak{sl}(2,\mathbb{C}) \oplus \mathfrak{sl}(2,\mathbb{C})$. The role played by $\boldsymbol{M}$ and $\boldsymbol{N}$ in Eqs. (D.30) for $\mathfrak{su}(2)_{\mathbb{C}}$ or $\mathfrak{sl}(2,\mathbb{C})$ is thus that of $\boldsymbol{J}$ for $\mathfrak{su}(2)$ in Section B.3. Then, we can label the representations with two independent integers or half-integers, m and n, equivalent to a spin number, each defining a representation of dimensions $2m+1$ and $2n+1$, whose matrices are given by Eqs. (B.11)–(B.13), where j has to be replaced by m or n. The difficulty is thus to make explicit the representations of $\mathfrak{su}(2)_{\mathbb{C}} \oplus \mathfrak{su}(2)_{\mathbb{C}}$. In this regard, the following property is useful: if G and G' are two matrix Lie groups with Lie algebras $\mathfrak{g}$ and $\mathfrak{g}'$, respectively, and $\mathcal{D}$ and $\mathcal{D}'$ are the corresponding representations of G and G', the representation $\mathcal{D} \otimes \mathcal{D}'$ of $G \times G'$ is associated with the representation of $\mathfrak{g} \oplus \mathfrak{g}'$ denoted by $d \otimes d'$ given by

$$(d \otimes d')(X, X') = d(X) \otimes I + I \otimes d'(X')$$

for all X in $\mathfrak{g}$ and all X' in $\mathfrak{g}'$ [justifications can be found, for instance, in Hall (2015, chapter 4)]. Hence, if in the decomposition (D.32), the representation of the first $\mathfrak{su}(2)_{\mathbb{C}}$ is the spin-m representation with the generators M_i^m and that of the second $\mathfrak{su}(2)_{\mathbb{C}}$ is the spin-n representation with the generators N_i^n, then Eq. (D.29) must be transformed into

$$J_i^{m,n} = M_i^m \otimes \mathbb{1}_{2n+1} + \mathbb{1}_{2m+1} \otimes N_i^n\,, \quad K_i^{m,n} = -i\left(M_i^m \otimes \mathbb{1}_{2n+1} - \mathbb{1}_{2m+1} \otimes N_i^n\right). \quad \text{(D.34)}$$

The representations of the generators of $\mathfrak{so}(1,3)_{\mathbb{C}}$ and thus those of $\mathfrak{so}(1,3)$, which are the same but restricted to the real linear combinations, denoted above by $J_i^{m,n}$ and $K_i^{m,n}$, have dimension $(2m+1) \times (2n+1)$. The representations of the Lorentz group $\mathrm{SO}^+(1,3)$ (connected to the identity) are then given by [see Eq. (D.27)]

$$\mathcal{D}^{(m,n)}(\Lambda) = \exp\left(-i\boldsymbol{\alpha} \cdot \boldsymbol{J}^{m,n} - i\boldsymbol{\zeta} \cdot \boldsymbol{K}^{m,n}\right), \quad \text{(D.35)}$$

with $\boldsymbol{\alpha}$ being the rotation vector and $\boldsymbol{\zeta}$ the rapidity vector. In the literature, the representations of the Lorentz group are often simply denoted by (m,n). As the linear combination of the generators reads

$$-i\boldsymbol{\alpha} \cdot \boldsymbol{J}^{m,n} - i\boldsymbol{\zeta} \cdot \boldsymbol{K}^{m,n} = -i(\boldsymbol{\alpha} - i\boldsymbol{\zeta}) \cdot \boldsymbol{M}^m \otimes \mathbb{1}_{2n+1} - i\mathbb{1}_{2m+1} \otimes (\boldsymbol{\alpha} + i\boldsymbol{\zeta}) \cdot \boldsymbol{N}^n,$$

note that the representations are thus equivalent to

$$\mathcal{D}^{(m,n)}(\Lambda) = \exp\left(-i(\boldsymbol{\alpha} - i\boldsymbol{\zeta}) \cdot \boldsymbol{M}^m\right) \otimes \exp\left(-i(\boldsymbol{\alpha} + i\boldsymbol{\zeta}) \cdot \boldsymbol{N}^n\right), \quad \text{(D.36)}$$

which can also be written as

$$\mathcal{D}^{(m,n)}(\Lambda) = \mathcal{D}^{(m,0)}(\Lambda) \otimes \mathcal{D}^{(0,n)}(\Lambda), \quad \text{(D.37)}$$

since $\boldsymbol{M}^{\mathbf{0}} = 0$ or $\boldsymbol{N}^{\mathbf{0}} = 0$ (i.e., the trivial representations) in the argument of the exponential leads to 1. In conclusion, any (irreducible) representation can be built from $\mathcal{D}^{(m,0)}$ and $\mathcal{D}^{(0,n)}$.

D.1.5 Effect of Parity on the Representations

Parity transformation reverts the spatial coordinates of vectors but keeps pseudo-vectors unchanged. Therefore, the angular momentum must not be changed, whereas the boost is reverted. Its effect on the generators is thus expected to be

$$\boldsymbol{J}^{m,n} \xrightarrow{\text{parity}} \boldsymbol{J}^{m,n}, \quad \boldsymbol{K}^{m,n} \xrightarrow{\text{parity}} -\boldsymbol{K}^{m,n}. \tag{D.38}$$

We can easily check this with the standard representation for which the elements of the group are matrices of SO$(1,3)$. In this representation, the generators J_i and K_i are given by Eqs. (D.2), (D.4) and (D.5), while the parity transformation is

$$\hat{P} = \begin{pmatrix} 1 & 0 & 0 & 0 \\ 0 & -1 & 0 & 0 \\ 0 & 0 & -1 & 0 \\ 0 & 0 & 0 & -1 \end{pmatrix}.$$

It is then straightforward to establish that $\hat{P}J_i\hat{P}^{-1} = J_i$ and $\hat{P}K_i\hat{P}^{-1} = -K_i$. Therefore, as parity transformation changes $\boldsymbol{J}^{m,n}$ and $\boldsymbol{K}^{m,n}$ according to Eq. (D.38), Eq. (D.35) is changed into

$$\mathcal{D}^{(m,n)}(\Lambda) \xrightarrow{\text{parity}} \exp\left(-i\boldsymbol{\alpha}\cdot\boldsymbol{J}^{m,n} + i\boldsymbol{\zeta}\cdot\boldsymbol{K}^{m,n}\right),$$

which implies after the insertion of Eq. (D.34),

$$\begin{aligned} \mathcal{D}^{(m,0)}(\Lambda) = \exp\left(-i(\boldsymbol{\alpha} - i\boldsymbol{\zeta})\cdot\boldsymbol{M}^m\right) &\xrightarrow{\text{parity}} \exp\left(-i(\boldsymbol{\alpha} + i\boldsymbol{\zeta})\cdot\boldsymbol{M}^m\right) = \mathcal{D}^{(0,m)}(\Lambda), \\ \mathcal{D}^{(0,n)}(\Lambda) = \exp\left(-i(\boldsymbol{\alpha} + i\boldsymbol{\zeta})\cdot\boldsymbol{N}^n\right) &\xrightarrow{\text{parity}} \exp\left(-i(\boldsymbol{\alpha} - i\boldsymbol{\zeta})\cdot\boldsymbol{N}^n\right) = \mathcal{D}^{(n,0)}(\Lambda). \end{aligned}$$

The role of m and n is thus interchanged.

D.1.6 Explicit Construction of the Most Important Finite-Dimensional Representations

Equation (D.36) or equivalently (D.35) gives us the recipe to construct any (irreducible) representation of dimension $(2m+1)\times(2n+1)$ of the Lorentz group. Obviously, the trivial representation of dimension 1 is $\mathcal{D}^{(0,0)} = 1$, obtained by $\boldsymbol{M}^0 = \boldsymbol{N}^0 = 0$.

Two-Dimensional Representations

• The first representation is obtained with $m = 1/2, n = 0$, for which $\boldsymbol{M}^{1/2} = \boldsymbol{\sigma}/2$ and $\boldsymbol{N}^0 = 0$. Equation (D.34) then yields

$$\boldsymbol{J}^{\frac{1}{2},0} = \frac{\boldsymbol{\sigma}}{2}, \quad \boldsymbol{K}^{\frac{1}{2},0} = -i\frac{\boldsymbol{\sigma}}{2}.$$

Therefore, using Eq. (D.35), the representation $\mathcal{D}^{(1/2,0)}$ takes the form

$$\mathcal{D}^{(\frac{1}{2},0)}(\Lambda) = \exp\left(-i(\boldsymbol{\alpha} - i\boldsymbol{\zeta})\cdot\frac{\boldsymbol{\sigma}}{2}\right). \tag{D.39}$$

This representation is the standard representation of SL(2, $\mathbb{C}$) since $(\boldsymbol{\alpha}-i\boldsymbol{\zeta})\cdot\boldsymbol{\sigma}/2=\sum z_i\sigma_i$, with $z_i \in \mathbb{C}$, corresponds to a general matrix of $\mathfrak{sl}(2,\mathbb{C})$. Note the difference with respect to the standard representation of SU(2) in Eq. (B.16), where the $\boldsymbol{\sigma}$ matrices are multiplied by the real numbers $\boldsymbol{\alpha}$. The representation $\mathcal{D}^{(1/2,0)}$ acts on two-dimensional objects in $\mathbb{C}^2$ called the *Weyl spinors*. Let us denote a the Weyl spinor χ_L, with the label L standing for 'left' for reasons explained in a few pages. For a pure rotation of angle α about the axis $\hat{\boldsymbol{n}}$ or a pure boost ζ in the direction $\hat{\boldsymbol{n}}$, the spinor χ_L is transformed with Eq. (D.39) into

$$\chi_L \xrightarrow{\Lambda=R(\alpha,\hat{\boldsymbol{n}})} \chi_L' = \exp\left(-i\alpha\hat{\boldsymbol{n}}\cdot\frac{\boldsymbol{\sigma}}{2}\right)\chi_L, \quad \chi_L \xrightarrow{\Lambda=B(\zeta,\hat{\boldsymbol{n}})} \chi_L' = \exp\left(-\zeta\hat{\boldsymbol{n}}\cdot\frac{\boldsymbol{\sigma}}{2}\right)\chi_L. \quad \text{(D.40)}$$

• The second two-dimensional representation is obtained, with $m=0, n=1/2$, i.e. $\boldsymbol{M}^{0}=0$ and $\boldsymbol{N}^{1/2}=\boldsymbol{\sigma}/2$, yielding

$$\boldsymbol{J}^{0,\frac{1}{2}}=\frac{\boldsymbol{\sigma}}{2}, \quad \boldsymbol{K}^{0,\frac{1}{2}}=i\frac{\boldsymbol{\sigma}}{2}.$$

The representation $\mathcal{D}^{(0,1/2)}$ thus reads

$$\mathcal{D}^{(0,\frac{1}{2})}(\Lambda)=\exp\left(-i(\boldsymbol{\alpha}+i\boldsymbol{\zeta})\cdot\frac{\boldsymbol{\sigma}}{2}\right). \quad \text{(D.41)}$$

It is also a standard representation of SL(2, $\mathbb{C}$) since the argument of the exponential is still of the form $\sum z_i\sigma_i$, with $z_i \in \mathbb{C}$. Actually, $\mathcal{D}^{(0,1/2)}$ is just the contragredient conjugate representation of $\mathcal{D}^{(1/2,0)}$. Indeed, as $\sigma_i^\dagger=\sigma_i$, according to the definition (A.8) of the contragredient conjugate representation,

$$\overline{\mathcal{D}^{(\frac{1}{2},0)}}^{*}(\Lambda)=\left(\mathcal{D}^{(\frac{1}{2},0)}\right)^{\dagger}(-\Lambda)=\exp\left(+i((-\boldsymbol{\alpha})+i(-\boldsymbol{\zeta}))\cdot\frac{\boldsymbol{\sigma}}{2}\right)=\mathcal{D}^{(0,\frac{1}{2})}(\Lambda).$$

However, both representations are not equivalent. Representation $\mathcal{D}^{(0,1/2)}$ also acts on Weyl spinors, but in order to distinguish[5] them from those transformed by $\mathcal{D}^{(1/2,0)}$, we denote them by χ_R, with R standing for 'right'. The effects of a pure rotation or a pure boost are

$$\chi_R \xrightarrow{\Lambda=R(\alpha,\hat{\boldsymbol{n}})} \chi_R' = \exp\left(-i\alpha\hat{\boldsymbol{n}}\cdot\frac{\boldsymbol{\sigma}}{2}\right)\chi_R, \quad \chi_R \xrightarrow{\Lambda=B(\zeta,\hat{\boldsymbol{n}})} \chi_R' = \exp\left(+\zeta\hat{\boldsymbol{n}}\cdot\frac{\boldsymbol{\sigma}}{2}\right)\chi_R. \quad \text{(D.42)}$$

Both kinds of spinors transform in the same way under rotations, namely as spin$-1/2$ Pauli spinors, see Eq. (B.16), but differ for pure boost transformations.

Four-Dimensional Representations

• Setting $m=n=1/2$ gives the four-dimensional representation, with $\boldsymbol{M}^{1/2}=\boldsymbol{N}^{1/2}=\boldsymbol{\sigma}/2$,

$$\mathcal{D}^{(\frac{1}{2},\frac{1}{2})}(\Lambda)=\mathcal{D}^{(\frac{1}{2},0)}(\Lambda)\otimes\mathcal{D}^{(0,\frac{1}{2})}(\Lambda)=\exp\left(-i(\boldsymbol{\alpha}-i\boldsymbol{\zeta})\cdot\frac{\boldsymbol{\sigma}}{2}\right)\otimes\exp\left(-i(\boldsymbol{\alpha}+i\boldsymbol{\zeta})\cdot\frac{\boldsymbol{\sigma}}{2}\right). \quad \text{(D.43)}$$

It acts on $\mathbb{C}^2\otimes\mathbb{C}^2$, which is isomorphic to $\mathbb{C}^4$. This representation is actually equivalent to the standard representation of $\mathrm{SO}^{+}(1,3)$. To prove it, let us check that the generators

[5] In the literature, the components of Weyl spinors are often labelled with or without dotted indices depending on whether they are transformed with $\mathcal{D}^{(0,1/2)}$ or $\mathcal{D}^{(1/2,0)}$.

obtained from Eq. (D.34) are equivalent to the generators of the standard representation in Eqs. (D.2), (D.4) and (D.5). Equation (D.34) yields

$$J_i^{\frac{1}{2},\frac{1}{2}} = \frac{\sigma_i}{2} \otimes \mathbb{1} + \mathbb{1} \otimes \frac{\sigma_i}{2}, \quad K_i^{\frac{1}{2},\frac{1}{2}} = -i\left(\frac{\sigma_i}{2} \otimes \mathbb{1} - \mathbb{1} \otimes \frac{\sigma_i}{2}\right).$$

Expanding the tensor product, it follows the matrices expressions in the tensorial basis

$$J_1^{\frac{1}{2},\frac{1}{2}} = \frac{1}{2}\begin{pmatrix} 0 & 1 & 1 & 0 \\ 1 & 0 & 0 & 1 \\ 1 & 0 & 0 & 1 \\ 0 & 1 & 1 & 0 \end{pmatrix}, \quad J_2^{\frac{1}{2},\frac{1}{2}} = \frac{1}{2}\begin{pmatrix} 0 & -i & -i & 0 \\ i & 0 & 0 & -i \\ i & 0 & 0 & -i \\ 0 & i & i & 0 \end{pmatrix}, \quad J_3^{\frac{1}{2},\frac{1}{2}} = \begin{pmatrix} 1 & 0 & 0 & 0 \\ 0 & 0 & 0 & 0 \\ 0 & 0 & 0 & 0 \\ 0 & 0 & 0 & -1 \end{pmatrix},$$

$$K_1^{\frac{1}{2},\frac{1}{2}} = \frac{1}{2}\begin{pmatrix} 0 & i & -i & 0 \\ i & 0 & 0 & -i \\ -i & 0 & 0 & i \\ 0 & -i & i & 0 \end{pmatrix}, \quad K_2^{\frac{1}{2},\frac{1}{2}} = \frac{1}{2}\begin{pmatrix} 0 & 1 & -1 & 0 \\ -1 & 0 & 0 & -1 \\ 1 & 0 & 0 & 1 \\ 0 & 1 & -1 & 0 \end{pmatrix}, \quad K_3^{\frac{1}{2},\frac{1}{2}} = \begin{pmatrix} 0 & 0 & 0 & 0 \\ 0 & -i & 0 & 0 \\ 0 & 0 & i & 0 \\ 0 & 0 & 0 & 0 \end{pmatrix}.$$

They are related to the generators J_i and K_i in Eqs. (D.2), (D.4) and (D.5) by the change of basis $J_i = SJ_i^{\frac{1}{2},\frac{1}{2}}S^{-1}$, $K_i = SK_i^{\frac{1}{2},\frac{1}{2}}S^{-1}$ with

$$S = \frac{1}{\sqrt{2}}\begin{pmatrix} 0 & 1 & -1 & 0 \\ 1 & 0 & 0 & -1 \\ i & 0 & 0 & i \\ 0 & -1 & -1 & 0 \end{pmatrix}, \quad S^{-1} = S^{\dagger} = \frac{1}{\sqrt{2}}\begin{pmatrix} 0 & 1 & -i & 0 \\ 1 & 0 & 0 & -1 \\ -1 & 0 & 0 & -1 \\ 0 & -1 & -i & 0 \end{pmatrix}.$$

The representations of the generators are thus equivalent as announced.

• Another important four-dimensional representation can be constructed via the direct sum

$$\mathcal{D}^{\text{Dirac}}(\Lambda) = \mathcal{D}^{(\frac{1}{2},0)}(\Lambda) \oplus \mathcal{D}^{(0,\frac{1}{2})}(\Lambda), \tag{D.44}$$

with its 4×4 representative matrix being

$$\mathcal{D}^{\text{Dirac}}(\Lambda) = \begin{pmatrix} \exp\left(-i(\boldsymbol{\alpha} - i\boldsymbol{\zeta}) \cdot \frac{\boldsymbol{\sigma}}{2}\right) & \mathbb{0} \\ \mathbb{0} & \exp\left(-i(\boldsymbol{\alpha} + i\boldsymbol{\zeta}) \cdot \frac{\boldsymbol{\sigma}}{2}\right) \end{pmatrix}. \tag{D.45}$$

It acts on objects called the *Dirac bi-spinors* made of the two previous Weyl spinors,

$$\psi = \begin{pmatrix} \chi_L \\ \chi_R \end{pmatrix}.$$

We have seen in Section D.1.5 that the parity interchanges the representations $\mathcal{D}^{m,0}$ and $\mathcal{D}^{0,m}$. Hence, under parity, χ_L and χ_R are swapped, which can be represented by the action on ψ of the 4×4 matrix

$$\hat{P} = \begin{pmatrix} \mathbb{0} & \mathbb{1} \\ \mathbb{1} & \mathbb{0} \end{pmatrix}. \tag{D.46}$$

A theory where parity is conserved cannot be described by χ_L or χ_R only.[6] That is why the two Weyl spinors are stacked together to form a Dirac spinor. In Section 4.3.3, we

[6] Well, this is not always correct. Neutral particles can be described by Weyl spinors only, as presented in Chapter 12 with Majorana neutrinos.

realised that a massive particle of mass m could be an eigenstate of parity only if it is at rest. As

$$\begin{pmatrix} \mathbb{0} & \mathbb{1} \\ \mathbb{1} & \mathbb{0} \end{pmatrix} \begin{pmatrix} \chi_L(\mathbf{0}) \\ \chi_R(\mathbf{0}) \end{pmatrix} = \begin{pmatrix} \chi_R(\mathbf{0}) \\ \chi_L(\mathbf{0}) \end{pmatrix} = \eta \begin{pmatrix} \chi_L(\mathbf{0}) \\ \chi_R(\mathbf{0}) \end{pmatrix},$$

necessarily when $\boldsymbol{p} = \mathbf{0}$, $\chi_L(\mathbf{0}) = \chi_R(\mathbf{0})$ (up to the phase η that can be absorbed in the definition of the parity operator). For a particle with a momentum $\boldsymbol{p} = p\hat{\boldsymbol{n}}$, the Dirac spinor is then deduced using a boost of rapidity ζ such that

$$\begin{pmatrix} \chi_L(\boldsymbol{p}) \\ \chi_R(\boldsymbol{p}) \end{pmatrix} = \begin{pmatrix} \exp\left(-\zeta\hat{\boldsymbol{n}}\cdot\frac{\boldsymbol{\sigma}}{2}\right) & \mathbb{0} \\ \mathbb{0} & \exp\left(\zeta\hat{\boldsymbol{n}}\cdot\frac{\boldsymbol{\sigma}}{2}\right) \end{pmatrix} \begin{pmatrix} \chi_R(\mathbf{0}) \\ \chi_R(\mathbf{0}) \end{pmatrix},$$

where $p = m\sinh\zeta$ and $E = p^0 = m\cosh\zeta$ [see Eq. (2.44)]. Eliminating $\chi_R(\mathbf{0})$, this system of equations is equivalent to

$$\begin{pmatrix} -\mathbb{1} & \exp\left(-\zeta\hat{\boldsymbol{n}}\cdot\boldsymbol{\sigma}\right) \\ \exp\left(\zeta\hat{\boldsymbol{n}}\cdot\boldsymbol{\sigma}\right) & -\mathbb{1} \end{pmatrix} \begin{pmatrix} \chi_L(\boldsymbol{p}) \\ \chi_R(\boldsymbol{p}) \end{pmatrix} = 0.$$

The exponential can be easily calculated following the procedure that has led to the Formula (B.3). It reads

$$\exp\left(\pm\zeta\hat{\boldsymbol{n}}\cdot\boldsymbol{\sigma}\right) = \cosh\zeta\ \mathbb{1} \pm \sinh\zeta\ \hat{\boldsymbol{n}}\cdot\boldsymbol{\sigma} = \frac{1}{m}\left(p^0\,\mathbb{1} \pm \boldsymbol{p}\cdot\boldsymbol{\sigma}\right).$$

It follows that

$$\begin{pmatrix} -m\,\mathbb{1} & p^0\,\mathbb{1} - \boldsymbol{p}\cdot\boldsymbol{\sigma} \\ p^0\,\mathbb{1} + \boldsymbol{p}\cdot\boldsymbol{\sigma} & -m\,\mathbb{1} \end{pmatrix} \begin{pmatrix} \chi_L(\boldsymbol{p}) \\ \chi_R(\boldsymbol{p}) \end{pmatrix} = 0,$$

which after the introduction of matrices

$$\gamma^0 = \begin{pmatrix} \mathbb{0} & \mathbb{1} \\ \mathbb{1} & \mathbb{0} \end{pmatrix}, \quad \boldsymbol{\gamma} = \begin{pmatrix} \mathbb{0} & \boldsymbol{\sigma} \\ -\boldsymbol{\sigma} & \mathbb{0} \end{pmatrix}, \tag{D.47}$$

simply reads

$$(p^0\gamma^0 - \boldsymbol{p}\cdot\boldsymbol{\gamma} - m)\psi(\boldsymbol{p}) = 0, \quad \text{with} \quad \psi(\boldsymbol{p}) = \begin{pmatrix} \chi_L(\boldsymbol{p}) \\ \chi_R(\boldsymbol{p}) \end{pmatrix}.$$

This equation is just the Dirac equation in the momentum space [Eq. (5.48)]. The Dirac spinor we introduced is, as expected, a solution of the Dirac equation. Notice that the γ matrices in Eq. (D.47) satisfy the Clifford algebra in Eq. (5.29) (the check is straightforward), and, therefore, they are Dirac matrices. They are, however, given in another basis than those of the Dirac representation in Eq. (5.30). This representation is the Weyl representation, also called the chiral representation. They are related by $\gamma^\mu_{\text{Dirac}} = U\gamma^\mu_{\text{Weyl}}U^{-1}$, with $U = \left(\begin{smallmatrix} \mathbb{1} & \mathbb{1} \\ -\mathbb{1} & \mathbb{1} \end{smallmatrix}\right)/\sqrt{2}$ and $U^{-1} = U^\intercal$. Finally, in the Weyl representation, the chirality matrix $\gamma^5 = i\gamma^0\gamma^1\gamma^2\gamma^3$ is diagonal and reads

$$\gamma^5 = \begin{pmatrix} -\mathbb{1} & \mathbb{0} \\ \mathbb{0} & \mathbb{1} \end{pmatrix}.$$

Consequently,

$$\gamma^5 \begin{pmatrix} \chi_L(\boldsymbol{p}) \\ \chi_R(\boldsymbol{p}) \end{pmatrix} = \begin{pmatrix} -\chi_L(\boldsymbol{p}) \\ \chi_R(\boldsymbol{p}) \end{pmatrix}.$$

Therefore, if we decompose ψ into

$$\psi = \psi_L + \psi_R = \begin{pmatrix} \chi_L(\boldsymbol{p}) \\ 0 \end{pmatrix} + \begin{pmatrix} 0 \\ \chi_R(\boldsymbol{p}) \end{pmatrix},$$

we have ψ_L and ψ_R, two eigenstates of γ^5 with respective eigenvalues -1 and $+1$. This justifies the denomination 'Left' and 'Right' of the Weyl spinors as they are related to the eigenstates of the chirality operator γ^5 (see Section 5.3.5 for more details about chirality).

D.1.7 Field Representations

Fields depend explicitly on spacetime coordinates and are thus affected by Lorentz transformations. Consider, for example, a scalar field $\varphi(x)$. The 4-vector x is changed by the Lorentz transformation to another 4-vector x', with its four components given by $x'^\mu = \Lambda^\mu{}_\nu x^\nu$. A scalar field is, by definition, Lorentz invariant, and hence, the transformed field φ' must satisfy[7]

$$\varphi'(x') = \varphi(x),$$

or equivalently

$$\varphi'(x) = \varphi(\Lambda^{-1}x).$$

Consider the infinitesimal transformation in Eq. (D.7). If $x = x^\nu e_\nu$ (where e_ν is a vector of the basis of 4-vector positions),

$$\Lambda^{-1}x = \left(\Lambda^{-1}\right)^\mu{}_\nu x^\nu e_\mu = \left(\delta^\mu_\nu - \omega^\mu{}_\nu\right) x^\nu e_\mu = (x^\mu - \omega^\mu{}_\nu x^\nu) e_\mu = x - \omega^\mu{}_\nu x^\nu e_\mu. \tag{D.48}$$

Therefore, keeping only terms up to the first order in $\omega^\mu{}_\nu$, the transformed field becomes

$$\varphi'(x) = \varphi(x) - \sum_{\mu=0}^{3} \omega^\mu{}_\nu x^\nu \frac{\partial}{\partial x^\mu}\varphi(x) = \varphi(x) - \omega^\mu{}_\nu x^\nu \partial_\mu \varphi(x).$$

We thus conclude

$$\varphi'(x) = (1 - \omega^{\mu\nu} x_\nu \partial_\mu)\varphi(x) = \left(1 - \frac{1}{2}\omega^{\mu\nu}(x_\nu \partial_\mu - x_\mu \partial_\nu)\right)\varphi(x),$$

where the antisymmetry property of $\omega^{\mu\nu}$ has been used. In order to ease the identification of the differential form of generators, we adopt a similar form as for the vector representation in Eq. (D.11), i.e.

$$\varphi'(x) = \left(1 - \frac{i}{2}\omega^{\mu\nu}(ix_\mu \partial_\nu - ix_\nu \partial_\mu)\right)\varphi(x).$$

Therefore, the generators simply read

$$L_{\mu\nu} = i(x_\mu \partial_\nu - x_\nu \partial_\mu). \tag{D.49}$$

One can easily check the consistency of this result. For instance, according to Eq. (D.15), we expect L^{12} to correspond to the third component of the (orbital) angular momentum.

[7] The functional form of the field changes from φ to φ', but its value at the transformed spacetime point remains unchanged: this is the Lorentz scalar behaviour.

Since $L^{12} = i(x^1\partial^2 - x^2\partial^1)$, and given the 3-momentum operators in Eq. (5.3), we have $L^{12} = xp_y - yp_x$, which is, as expected, the third component of the orbital angular momentum. Other components would be found similarly. Since both $J^{\mu\nu}$ in Eq. (D.9) and $L^{\mu\nu}$ in (D.49) generate Lorentz transformations, they share the Lie Algebra given by the commutation rules (D.18) (the reader is invited to check this). In other words, $L_{\mu\nu}$ in Eq. (D.49) is just the representation of the generators $J_{\mu\nu}$ with differential operators.

For a finite Lorentz transformation, one has to 'exponentiate' the infinitesimal transformation, leading to the application of the differential operator

$$\mathcal{D}^{\text{diff}}(\Lambda) = \exp\left(-\frac{i}{2}\omega_{\sigma\rho}L^{\sigma\rho}\right). \tag{D.50}$$

Hence, a field with one single component, such as the scalar field, is transformed into

$$\varphi'(x) = \varphi(\Lambda^{-1}x) = \mathcal{D}^{\text{diff}}(\Lambda)\varphi(x). \tag{D.51}$$

Fields can have several components like (Lorentz) vector fields with four spacetime components or Dirac fields with four spinor components. In Sections D.1.4 and D.1.6, we saw how multi-component objects like vectors or spinors transform under Lorentz transformation via finite-dimensional representations $\mathcal{D}^{(m,n)}(\Lambda)$. It introduced generators $J_{\mu\nu}$, or equivalently, $\boldsymbol{J}$ and $\boldsymbol{K}$, which are matrices acting on the components of the object. In order to avoid confusion with what follows, let us rename the generators $J_{\mu\nu}$ of Section D.1.4 in $\mathcal{S}_{\mu\nu}$. The main formula of the finite-dimensional representations is reproduced below [see Eq. (D.27) and its extension to the representation with dimension $(2m+1)(2n+1)$]:

$$\mathcal{D}^{(m,n)}(\Lambda) = \exp\left(-i\boldsymbol{\alpha}\cdot\boldsymbol{J}^{m,n} - i\boldsymbol{\zeta}\cdot\boldsymbol{K}^{m,n}\right) = \exp\left(-\frac{i}{2}\omega_{\sigma\rho}\left(\mathcal{S}^{m,n}\right)^{\sigma\rho}\right), \tag{D.52}$$

where $(\mathcal{S}^{m,n})^{\sigma\rho}$ are given by Eq. (D.16), i.e.

$$(\mathcal{S}^{m,n})^{0i} = K_i^{m,n}, \quad (\mathcal{S}^{m,n})^{ij} = \epsilon_{ijk}J_k^{m,n}, \tag{D.53}$$

and $\omega_{\sigma\rho}$ by Eq. (D.26).

A field depending on spacetime coordinates with several components is sensitive to both kinds of transformations (those affecting spacetime coordinates and those affecting the field components). Hence, when spacetime coordinates x^μ are transformed with

$$x'^\mu = \Lambda^\mu{}_\nu x^\nu, \tag{D.54}$$

the components $\varphi^a(x)$ of the field are transformed into

$$\varphi'^a(x) = \left(\mathcal{D}^{(m,n)}(\Lambda)\right)^a{}_b\, \mathcal{D}^{\text{diff}}(\Lambda)\, \varphi^b(x), \tag{D.55}$$

i.e.

$$\varphi'^a(x) = \exp\left(-\frac{i}{2}\omega_{\sigma\rho}\left(L^{\sigma\rho}\,\mathbb{1} + (\mathcal{S}^{m,n})^{\sigma\rho}\right)\right)^a{}_b \varphi^b(x) = \exp\left(-\frac{i}{2}\omega_{\sigma\rho}\left(\mathcal{J}^{m,n}\right)^{\sigma\rho}\right)^a{}_b \varphi^b(x), \tag{D.56}$$

where the following operator has been introduced:

$$(\mathcal{J}^{m,n})^{\sigma\rho} = L^{\sigma\rho}\,\mathbb{1} + (\mathcal{S}^{m,n})^{\sigma\rho}. \tag{D.57}$$

There are many indices in the previous formulas. Let us see again what these objects are: $(\mathcal{S}^{m,n})^{\sigma\rho}$ is the matrix that corresponds to the representation (m,n). The indices

$\sigma, \rho = \{0, 1, 2, 3\}$ via Eq. (D.53) give us the correspondence with the initial matrices $\boldsymbol{J}^{m,n}$ and $\boldsymbol{K}^{m,n}$. The operator $L^{\sigma\rho}$ is not a matrix. It is a differential operator. The six real numbers $\omega_{\sigma\rho}$ (six because $\omega_{\sigma\rho}$ is antisymmetric) specify the parameters of the Lorentz transformation. Finally, indices a and b label the different components of the field. They may vary from 0 or 1 to an arbitrary number. For instance, 1–4 for a Dirac field and 0–3 for a vector field.

D.2 Poincaré Group

Lorentz invariance leaves the norm of 4-vectors, x^2, constant under a change of inertial frame. However, physics imposes that the speed of light is constant in any inertial frame, implying that the infinitesimal element,

$$\mathrm{d}s^2 = (c\,\mathrm{d}t)^2 - (\mathrm{d}x^2 + \mathrm{d}y^2 + \mathrm{d}z^2), \tag{D.58}$$

is constant too. Lorentz transformations, $\mathrm{d}x'^{\mu} = \Lambda^{\mu}_{\ \nu}\,\mathrm{d}x^{\nu}$, obviously satisfy this constraint ($\mathrm{d}x^{\mu}$ is a 4-vector), but also spacetime translations, $x'^{\mu} = x^{\mu} + a^{\mu}$, with a^{μ} being a constant 4-vector since in such a case, $\mathrm{d}x'^{\mu} = \mathrm{d}x^{\mu}$. The Poincaré group is the set of transformations,

$$x' = T(\Lambda, a)x = \Lambda x + a, \tag{D.59}$$

that leaves $\mathrm{d}s^2$ constant. Including Lorentz transformations and spacetime translations, it depends on $6 + 4 = 10$ parameters. It is easy to check that it constitutes a group with the properties

$$T(\Lambda', a')T(\Lambda, a) = T(\Lambda'\Lambda, a' + \Lambda' a), \quad e = T(\mathbb{1}, 0), \quad T^{-1}(\Lambda, a) = T(\Lambda^{-1}, -\Lambda^{-1}a). \tag{D.60}$$

We have already found the Lie algebra (and thus the generators) of the Lorentz group. One can now determine the additional contribution of spacetime translations by looking at an infinitesimal Poincaré transformation of the scalar field, for instance. Combining Eqs. (D.59) and (D.60), it reads

$$\varphi'(x) = \varphi(T^{-1}(\Lambda, a)x) = \varphi(\Lambda^{-1}x - \Lambda^{-1}a).$$

The quantity $\Lambda^{-1}x$ has already been calculated in Eq. (D.48), $\Lambda^{-1}x = x^{\mu}e_{\mu} - \omega^{\mu}_{\ \nu}x^{\nu}e_{\mu}$. It led to the generators $L_{\mu\nu}$ (D.49) of Lorentz transformations. Similarly, up to the first order in $\omega^{\mu}_{\ \nu}$ and a^{μ}, $\Lambda^{-1}a \simeq a^{\mu}e_{\mu}$, yielding

$$\varphi'(x) \simeq \varphi(x^{\mu}e_{\mu} - a^{\mu}e_{\mu} - \omega^{\mu}_{\ \nu}x^{\nu}e_{\mu}) \simeq \left(1 - ia^{\mu}(-i\partial_{\mu}) - \frac{i}{2}\omega^{\mu\nu}L_{\mu\nu}\right)\varphi(x).$$

This leads to the identification of the four well-known generators of translations, i.e. the energy–momentum operator of quantum mechanics,

$$P_{\mu} = -i\partial_{\mu}. \tag{D.61}$$

With the 10 generators of the Poincaré group, it is then easy to check the following commutation rules:

$$[P_{\mu} \cdot p_{\nu}] = 0, \quad [L_{\mu\nu}, P_{\rho}] = i(g_{\nu\rho}P_{\mu} - g_{\mu\rho}P_{\nu}). \tag{D.62}$$

This completes that of $J_{\mu\nu}$ given in Eq. (D.18).

The classification of the (irreducible) representations of the Poincaré group supposes quantities that are conserved to label the representation. Such quantities must be associated with (Casimir) operators that commute with the generators of the representation of interest. Those operators for the Poincaré group are

$$\begin{aligned} P^2 &= P^\mu P_\mu, \\ W^2 &= W^\lambda W_\lambda, \end{aligned} \tag{D.63}$$

where W^λ is the Pauli–Lubanski 4-vector operator,

$$W^\lambda = \frac{1}{2}\epsilon^{\lambda\mu\nu\rho} J_{\mu\nu} P_\rho. \tag{D.64}$$

One can check that P^2 and W^2 commute with P^μ and $J^{\mu\nu}$ and commute with each other. Moreover, P and W satisfy $P^\mu W_\mu = 0$. Let us consider a one-particle state, $|p,\sigma\rangle$, with a 4-momentum p and other quantum numbers denoted by σ. By definition,

$$P^\mu \, |p,\sigma\rangle = p^\mu \, |p,\sigma\rangle .$$

Therefore, $P^2\,|p,\sigma\rangle = p^2\,|p,\sigma\rangle = m^2$, where m is the particle mass. The particle mass is thus unchanged under the Poincaré symmetry. To determine the eigenvalue of W^2, let us consider a massive particle $m > 0$ at rest, $p_0 = (m,\mathbf{0})$. Given the properties of the Levi–Civita tensor, the components of W are

$$\begin{aligned} W^0\,|p_0,\sigma\rangle &= \frac{1}{2}\epsilon^{0\mu\nu\rho} J_{\mu\nu} P_\rho\,|p_0,\sigma\rangle = \frac{1}{2}\epsilon^{0\mu\nu 0} J_{\mu\nu}\, m\,|p_0,\sigma\rangle = 0, \\ W^i\,|p_0,\sigma\rangle &= \frac{1}{2}\epsilon^{i\mu\nu 0} J_{\mu\nu}\, m\,|p_0,\sigma\rangle = \frac{1}{2}\epsilon^{ijk0} J_{jk}\, m\,|p_0,\sigma\rangle = \frac{1}{2}\epsilon^{ijk0}\epsilon_{jkl} J_l\, m\,|p_0,\sigma\rangle . \end{aligned}$$

The last equality for $W^i\,|p_0,\sigma\rangle$ results from the definition (D.16) of J_{jk}. As the only non-zero term in $\epsilon^{ijk0}\epsilon_{jkl}$ requires $l = i$, with $\epsilon^{ijk0}\epsilon_{jki} = \epsilon^{ijk0}\epsilon_{ijk} = 2$ (both permutations, jk and kj, must be taken into account), it follows that

$$W^i\,|p_0,\sigma\rangle = mJ_i\,|p_0,\sigma\rangle .$$

Therefore,

$$W^2\,|p_0,\sigma\rangle = ((W^0)^2 - (\mathbf{W})^2)\,|p_0,\sigma\rangle = -m^2(\boldsymbol{J})^2\,|p_0,\sigma\rangle = -m^2 J(J+1)\,|p_0,\sigma\rangle .$$

The other quantum number that labels the representations of the Poincaré group is thus the angular momentum in the frame where the particle is at rest, in other words, its spin. A massive particle is thus defined by two numbers, its mass and its spin. For massless particles, a rest frame does not exist, but one can show that the helicity quantum number plays the role of the spin to label the representation (see the discussion on p. 110).

Appendix E Calculating with the Dirac Delta Distribution

The Dirac delta distribution is defined by

$$\int_{-\infty}^{+\infty} f(x)\delta(x-x_0)\ \mathrm{d}x = f(x_0) \tag{E.1}$$

for all smooth functions f (i.e., that have derivatives to all orders) on $\mathbb{R}$ with a compact support. The delta distribution can be composed with a continuously differentiable function $g(x)$ by

$$\delta(g(x)) = \sum_i \frac{\delta(x-x_i)}{\left|\frac{\partial g}{\partial x}(x_i)\right|}, \tag{E.2}$$

where x_i are the real roots of the g function ($g(x_i) = 0$). If g has no roots, then $\delta(g(x)) = 0$. Note that Eq. (E.2) is a generalisation of the well-known properties

$$\delta(\alpha x) = \frac{\delta(x)}{|\alpha|}, \quad \delta(-x) = \delta(x). \tag{E.3}$$

The combination of Eqs. (E.1) and (E.2) implies that

$$\int_{-\infty}^{+\infty} \delta(g(x))\ \mathrm{d}x = \sum_i \frac{1}{\left|\frac{\partial g}{\partial x}(x_i)\right|}. \tag{E.4}$$

Applying Eq. (E.4) to $g(p^0) = p^2 - m^2 = (p^0)^2 - |\boldsymbol{p}|^2 - m^2 = (p^0)^2 - E^2 = (p^0 - E)(p^0 + E)$ yields

$$\int_{-\infty}^{+\infty} \delta(p^2 - m^2)\ \mathrm{d}p^0 = \frac{1}{E} = 2\int_{0}^{+\infty} \delta(p^2 - m^2)\ \mathrm{d}p^0,$$

with $\delta(p^2 - m^2)$ being an even function of p^0. Introducing the Heaviside step function, θ, one gets the following useful identity:

$$\int_{-\infty}^{+\infty} \delta(p^2 - m^2)\theta(p^0)\ \mathrm{d}p^0 = \frac{1}{2E}, \tag{E.5}$$

which becomes in four dimensions

$$\int_{-\infty}^{+\infty} \mathrm{d}^4p\ \delta(p^2 - m^2)\theta(p^0) = \int_{-\infty}^{+\infty} \frac{\mathrm{d}^3\boldsymbol{p}}{2E}. \tag{E.6}$$

Another interesting application of Eq. (E.2) is with $g(x) = f(x) - f(x_0)$. The root of $g(x)$ is obviously $x = x_0$, and as $\frac{\partial g}{\partial x} = \frac{\partial f}{\partial x}$, we obtain

$$\delta(f(x) - f(x_0)) = \frac{\delta(x - x_0)}{\left|\frac{\partial f}{\partial x}(x_0)\right|}. \tag{E.7}$$

Finally, one can get a very useful expression of the delta distribution via the Fourier transform,

$$\tilde{f}(\xi) = \int_{-\infty}^{+\infty} f(x)e^{-i\xi x}\,\mathrm{d}x \Leftrightarrow f(x) = \frac{1}{2\pi}\int_{-\infty}^{+\infty} \tilde{f}(\xi)e^{i\xi x}\,\mathrm{d}\xi. \tag{E.8}$$

Applying it to the delta distribution yields

$$\tilde{\delta}(\xi) = \int_{-\infty}^{+\infty} \delta(x)e^{-i\xi x}\,\mathrm{d}x = 1,$$

implying that

$$\boxed{\delta(x) = \frac{1}{2\pi}\int_{-\infty}^{+\infty} e^{i\xi x}\,\mathrm{d}\xi \quad \text{and also} \quad \delta^{(n)}(x) = \frac{1}{(2\pi)^n}\int e^{i\xi\cdot x}\,\mathrm{d}^n\xi.} \tag{E.9}$$

Appendix F Contour Integration in the Complex Plane

This appendix is an informal and practical introduction to contour integration in the complex plane. Many integrals appearing in particle physics (propagators, for instance) can be calculated with this technique. No proof is given here, and many mathematical aspects are ignored. The reader can consult a mathematics textbook such as Boas (2005) for a more rigorous approach.

Let C be a curve connecting two complex points, z_A and z_B, and $f(z)$ be a complex function with real and imaginary parts u and v, respectively. Decomposing z as $x + iy$, the integral of the complex function $f(z)$ over the curve C is

$$I = \int_{C:z_A \to z_B} f(z)\,\mathrm{d}z = \int_C u(x,y)\,\mathrm{d}x - v(x,y)\,\mathrm{d}y + i\int_C v(x,y)\,\mathrm{d}x + u(x,y)\,\mathrm{d}y. \qquad \text{(F.1)}$$

When the curve C is parametrised as a function of a real variable, t, Eq. (F.1) becomes an integration over a single variable, yielding

$$I = \int_{t_A}^{t_B} f(z(t))\frac{\mathrm{d}z(t)}{\mathrm{d}t}\,\mathrm{d}t.$$

For instance, if C is the unit circle in the complex plane, $z = e^{i\theta}$, then C is naturally parametrised with θ.

From now on, we will be interested in *analytic* functions. By definition, a function f is analytic if the derivatives of its real and imaginary parts satisfy

$$\frac{\partial u(x,y)}{\partial x} = \frac{\partial v(x,y)}{\partial y}, \quad \frac{\partial u(x,y)}{\partial y} = -\frac{\partial v(x,y)}{\partial x}.$$

An interesting property of analytic functions is that their derivatives of all orders are defined.

When the curve C is closed (usually called a contour) and the function f is analytic on and inside C, according to Cauchy's theorem,

$$I = \oint_C f(z)\,\mathrm{d}z = 0, \qquad \text{(F.2)}$$

where the integral symbol $\oint$ is reserved for closed contours. One can then show that for any point z_0 inside C, the nth-order derivative of f in that point is given by

$$f^{(n)}(z_0) = \frac{n!}{2\pi i}\oint_C \frac{f(\zeta)}{(\zeta - z_0)^{n+1}}\,\mathrm{d}\zeta, \qquad \text{(F.3)}$$

where the closed contour is run counterclockwise in the complex plane. Remarkably, Eq. (F.3) does not depend on the contour. Any closed contour enclosing z_0 satisfies Eq. (F.3). The formula is valid for all n, including for $n = 0$ corresponding to the function f itself.

This particular case is known as Cauchy's integral formula. An important consequence of Eq. (F.3) is the Laurent series, a kind of Taylor series for analytic functions. If C_1 and C_2 are two circles centred at z_0, and $f(z)$ is analytic in the domain between the two circles, then the Laurent series expanded about z_0 reads

$$f(z) = \sum_{n=-\infty}^{+\infty} a_n(z-z_0)^n = a_0+a_1(z-z_0)+a_2(z-z_0)^2+\cdots+\frac{a_{-1}}{z-z_0}+\frac{a_{-2}}{(z-z_0)^2}+\cdots, \quad \text{(F.4)}$$

with

$$a_n = \frac{1}{2\pi i}\oint_C \frac{f(\zeta)}{(\zeta - z_0)^{n+1}}\,\mathrm{d}\zeta. \quad \text{(F.5)}$$

Note that if f is also analytic inside the two circles, the Laurent series reduces to the Taylor series. Indeed, for $n < 0$, $a_n = 0$, since the integrand in Eq. (F.5) would be analytic in this case, and for $n \geq 0$, $a_n = f^{(n)}(z_0)/n!$ according to Eq. (F.3). The pole z_0 is of order m if m is the smallest positive integer for which $(z - z_0)^m f(z)$ is analytic at z_0.

Laurent series is a powerful tool to calculate integrals over a real interval, I. Instead of calculating $\int_I f(x)\,\mathrm{d}x$, one considers its complex version $\int_C f(z)\,\mathrm{d}z$, where C is a closed contour that contains the real interval I. The coefficient a_{-1} of the Laurent series plays a special role, since it is

$$a_{-1} = \frac{1}{2\pi i}\oint_C f(\zeta)\,\mathrm{d}\zeta \equiv \mathrm{Res}(f, z_0).$$

Hence, if a_{-1} is known, so is the integral. This important coefficient is called the *residue* of $f(z)$ at $z = z_0$, denoted $\mathrm{Res}(f, z_0)$. It leads us to the residue theorem stating that

$$\oint_C f(z)\,\mathrm{d}z = 2\pi i \sum_k \mathrm{Res}(f, z_k), \quad \text{(F.6)}$$

where the z_ks are the poles contained inside the contour C, and the integral around C is in the counterclockwise direction (otherwise, there is a minus sign). With a pole of order 1, the residue is found with

$$\mathrm{Res}(f, z_k) = \lim_{z\to z_k}(z - z_k)f(z). \quad \text{(F.7)}$$

This is a particular case of the general formula for a pole of order m,

$$\mathrm{Res}(f, z_k) = \lim_{z\to z_k} \frac{1}{(m-1)!}\frac{d^{(m-1)}}{\mathrm{d}z^{(m-1)}}(z - z_k)^m f(z). \quad \text{(F.8)}$$

In summary, the calculation of the integral consists in defining a contour, finding the poles inside the contour, and calculating the corresponding residues.

As an example of a calculation, consider the real and complex integrals

$$I = \int_{-\infty}^{+\infty} \frac{e^{ix}}{x^2+\lambda^2}\,\mathrm{d}x, \quad I_z = \oint_C \frac{e^{iz}}{z^2+\lambda^2}\,\mathrm{d}z.$$

with λ being a non-zero real. In order to find the expression of I, we are going to calculate I_z with the residue theorem. The poles are found among the roots of

$$z^2 + \lambda^2 = (z - i\lambda)(z + i\lambda),$$

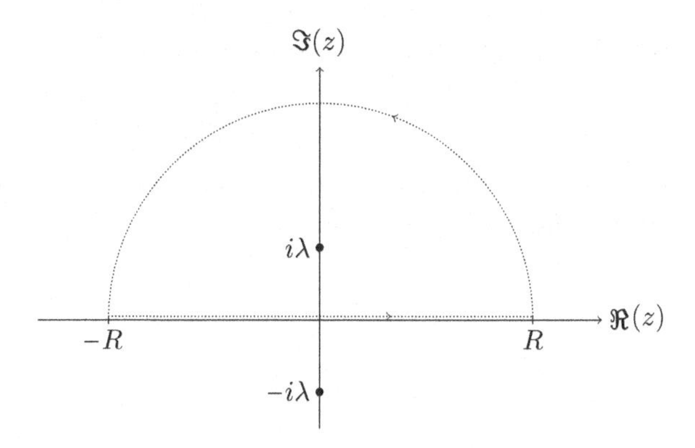

i.e. $z_1 = i\lambda$ and $z_2 = -i\lambda$. Let us consider first the case $\lambda > 0$. The figure above shows a possible contour with a dotted line. Along the real axis, z is obviously a real variable, x, varying from $-R$ to R. We will take the limit $R \to \infty$ at the end of the calculation. Along the semicircle of radius R, $z = Re^{i\theta} = R\cos\theta + iR\sin\theta$, with θ varying from 0 to π. Note that the exponential in I_z is $\exp(iR\cos\theta - R\sin\theta)$. The integral is thus convergent as with θ in $[0, \pi]$, $\sin\theta \geq 0$. In other words, only contours in the plane $\Im(z) \geq 0$ are possible. With our contour, the expression of I_z is

$$I_z = \lim_{R\to\infty} \int_{-R}^{R} \frac{e^{ix}}{x^2+\lambda^2}\, \mathrm{d}x + \int_0^{\pi} \frac{e^{iR\cos\theta - R\sin\theta}}{R^2 e^{2i\theta} + \lambda^2} iRe^{i\theta}\, \mathrm{d}\theta.$$

The second integral tends to 0 when taking the limit $R \to \infty$, while the first converges to I. The residue at $z_1 = i\lambda$ (the only singular point in the contour) is

$$\mathrm{Res}(f, z_1) = \lim_{z\to i\lambda} (z - i\lambda)\frac{e^{iz}}{z^2+\lambda^2} = \frac{e^{-\lambda}}{2i\lambda}.$$

Hence, for $\lambda > 0$, the residue theorem (F.6) leads us to conclude

$$\int_{-\infty}^{+\infty} \frac{e^{ix}}{x^2+\lambda^2}\, \mathrm{d}x = \pi \frac{e^{-\lambda}}{\lambda}.$$

When $\lambda < 0$, the singular point in the semicircle is $z_2 = -i\lambda$ and the residue is

$$\mathrm{Res}(f, z_2) = \lim_{z\to -i\lambda} (z + i\lambda)\frac{e^{iz}}{z^2+\lambda^2} = -\frac{e^{\lambda}}{2i\lambda}.$$

It follows that

$$\int_{-\infty}^{+\infty} \frac{e^{ix}}{x^2+\lambda^2}\, \mathrm{d}x = -\pi \frac{e^{\lambda}}{\lambda}.$$

Hence, for all $\lambda \neq 0$,

$$\int_{-\infty}^{+\infty} \frac{e^{ix}}{x^2+\lambda^2}\, \mathrm{d}x = \pi \frac{e^{-|\lambda|}}{|\lambda|}.$$

Another example, useful for the calculation of propagators, is to show the following identity of the step function $\theta(t)$ (with $\theta(t) = 0$ for $t < 0$ and 1 for $t > 0$):

$$\theta(t) = \lim_{\epsilon\to 0^+} \frac{i}{2\pi} \int_{-\infty}^{+\infty} \frac{e^{-ixt}}{x+i\epsilon}\, \mathrm{d}x, \tag{F.9}$$

where ϵ is a positive real number. Let us calculate

$$I = \oint_C \frac{e^{-izt}}{z + i\epsilon}\, dz,$$

In the complex plane, there is a singular point for $z_0 = -i\epsilon$. Depending on the sign of t we are going to use one of the contours drawn below.

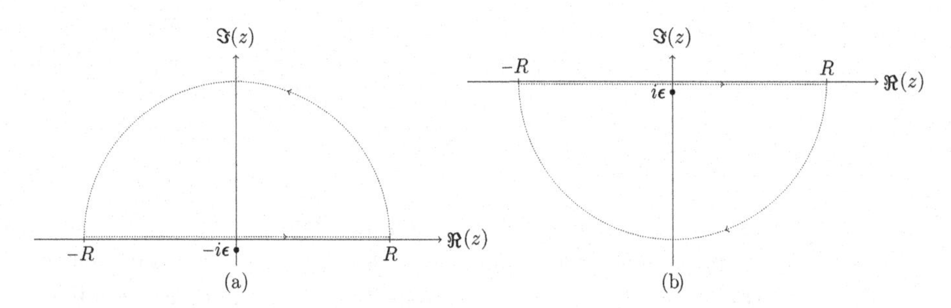

For $t < 0$, along the semicircle in (a), $z = Re^{i\theta}$, with $\theta \in [0^+, \pi]$. The integral along the circle tends to zero for large R since $\exp(-izt) = \exp(iz|t|) = \exp(iR|t|\cos\theta)\exp(-R|t|\sin\theta)$. As the singular point z_0 is not enclosed in the contour, $I = 0$. Therefore, the integral along the real axis is zero as well. For $t > 0$, we must use now the contour (b) because $\exp(-izt) = \exp(-iRt\cos\theta)\exp(R|t|\sin\theta)$ converges to zero only if $\sin\theta < 0$, i.e. for $\theta \in [0^-, -\pi]$. This time, z_0 is enclosed in the contour, and the application of the residue theorem leads to $I = -2\pi i e^{-\epsilon t}$. The negative sign at the front arises because the integral around C is in the clockwise direction. The integral along the real axis is thus $-2\pi i e^{-\epsilon t}$. In summary,

$$\int_{-\infty}^{+\infty} \frac{e^{-ixt}}{x + i\epsilon}\, dx = \begin{cases} 0, & \text{for } t < 0 \\ -2\pi i e^{-\epsilon t}, & \text{for } t > 0. \end{cases}$$

Its limit for $\epsilon \to 0^+$ is thus 0 for $t < 0$ and $-2\pi i$ otherwise, which justifies the identity (F.9).

Appendix G Useful Formulas with γ Matrices

Most of the following formulas are encountered in Chapters 5, 6 and 9.

Basics on γ^μ Matrices

$$\{\gamma^\mu, \gamma^\nu\} = \gamma^\mu \gamma^\nu + \gamma^\nu \gamma^\mu = 2g^{\mu\nu} \tag{G.1}$$

$$(\gamma^0)^2 = 1, \quad (\gamma^{k=1,2,3})^2 = -1 \tag{G.2}$$

$$\gamma^{0\dagger} = \gamma^0, \quad \gamma^{1\dagger} = -\gamma^1, \quad \gamma^{2\dagger} = -\gamma^2, \quad \gamma^{3\dagger} = -\gamma^3 \tag{G.3a}$$

$$\gamma^{\mu\dagger} = \gamma^0 \gamma^\mu \gamma^0 \tag{G.3b}$$

$$\gamma_\mu = g_{\mu\nu} \gamma^\nu \tag{G.4}$$

Equations (G.1) and (G.4) yield

$$\gamma^\mu \gamma_\mu = 4 \tag{G.5}$$

$$\gamma^\mu \gamma^\nu \gamma_\mu = -2\gamma^\nu \tag{G.6}$$

$$\gamma^\mu \gamma^\nu \gamma^\rho \gamma_\mu = 4g^{\nu\rho} \tag{G.7}$$

$$\gamma^\mu \gamma^\nu \gamma^\rho \gamma^\eta \gamma_\mu = -2\gamma^\eta \gamma^\rho \gamma^\nu \tag{G.8}$$

For any 4-vector,

$$\not{p} = p_\mu \gamma^\mu \tag{G.9}$$

$$\not{p}^2 = p^2, \tag{G.10}$$

where $\not{p}^2 = p_\mu p_\nu \gamma^\mu \gamma^\nu = p_\mu p_\nu (2g^{\mu\nu} - \gamma^\nu \gamma^\mu) = 2p^2 - \not{p}^2$ has been used to establish Eq. (G.10).

Basics on γ^5 Matrices

$$\gamma^5 = i\gamma^0 \gamma^1 \gamma^2 \gamma^3 \tag{G.11}$$

$$(\gamma^5)^2 = 1 \tag{G.12}$$

$$\gamma^{5\dagger} = \gamma^5 \tag{G.13}$$

$$\{\gamma^5, \gamma^\mu\} = \gamma^5 \gamma^\mu + \gamma^\mu \gamma^5 = 0. \tag{G.14}$$

Traces of γ Matrices

Given that, by construction, the γ matrices are traceless, the following properties can be easily deduced from the Clifford algebra (G.1):

$$\mathrm{Tr}(\mathbb{1}) = 4 \tag{G.15}$$
$$\mathrm{Tr}(\gamma^\mu) = 0 \tag{G.16}$$
$$\mathrm{Tr}(\gamma^5) = 0 \tag{G.17}$$
$$\mathrm{Tr}(\gamma^\mu\gamma^\nu) = 4g^{\mu\nu} \tag{G.18a}$$
$$\mathrm{Tr}(\not{p}\not{k}) = 4p\cdot k. \tag{G.18b}$$

Equality (G.17) is obtained by inserting $(\gamma^0)^2 = 1$, i.e. $\mathrm{Tr}(\gamma^\mu) = \mathrm{Tr}(\gamma^\mu\gamma^0\gamma^0) = -\mathrm{Tr}(\gamma^0\gamma^\mu\gamma^0) = -\mathrm{Tr}(\gamma^\mu\gamma^0\gamma^0)$, which leads to the result (G.17). For the trace of a larger number of γ^μ matrices, we have

$$\mathrm{Tr}(\text{odd nb of } \gamma^\mu) = 0 \tag{G.19}$$
$$\mathrm{Tr}(\gamma^\rho\gamma^\mu\gamma^\eta\gamma^\nu) = 4(g^{\rho\mu}g^{\eta\nu} - g^{\rho\eta}g^{\mu\nu} + g^{\rho\nu}g^{\mu\eta}) \tag{G.20a}$$
$$\mathrm{Tr}(\not{p}_1\not{p}_2\not{p}_3\not{p}_4) = 4(p_1\cdot p_2\, p_3\cdot p_4 - p_1\cdot p_3\, p_2\cdot p_4 + p_1\cdot p_4\, p_2\cdot p_3) \tag{G.20b}$$

The proofs of Eqs. (G.19) and (G.20) are given on p. 201. As γ^5 is made of four γ^μ matrices, Eq. (G.19) implies that the trace of the product of an odd number of γ^μ matrices with γ^5 is zero. For an even number of γ^μ matrices, here are a few useful formulas:

$$\mathrm{Tr}(\gamma^5\gamma^\mu\gamma^\nu) = 0 \tag{G.21}$$
$$\mathrm{Tr}(\gamma^5\gamma^\mu\gamma^\nu\gamma^\rho\gamma^\eta) = 4i\epsilon^{\mu\nu\rho\eta} \tag{G.22}$$
$$\mathrm{Tr}\left(\not{p}\gamma^\mu(1-\gamma^5)\not{p}'\gamma^\nu(1-\gamma^5)\right) = 8(p^\mu p'^\nu - p\cdot p' g^{\mu\nu} + p^\nu p'^\mu - ip_\rho p'_\eta \epsilon^{\rho\mu\eta\nu}). \tag{G.23}$$

To prove Eq. (G.21), insert $(\gamma^\sigma)^2 = s = \pm 1$ with $\sigma \neq \mu, \nu$, anti-commute three times γ^α towards the right and finally use the cyclicity of the trace, i.e.

$$\begin{aligned}\mathrm{Tr}(\gamma^5\gamma^\mu\gamma^\nu) &= s\mathrm{Tr}(\gamma^\alpha\gamma^\alpha\gamma^5\gamma^\mu\gamma^\nu) = -s\mathrm{Tr}(\gamma^\alpha\gamma^5\gamma^\mu\gamma^\nu\gamma^\alpha) = -s\mathrm{Tr}(\gamma^\alpha\gamma^\alpha\gamma^5\gamma^\mu\gamma^\nu)\\ &= -\mathrm{Tr}(\gamma^5\gamma^\mu\gamma^\nu).\end{aligned}$$

The proof of Eq. (G.23) is given on p. 338. The proof of Eq. (G.22) requires first to give some details about the Levi–Civita symbol, $\epsilon^{\mu\nu\rho\eta}$. This rank-4 antisymmetric tensor with the interchange of any two indices is defined by

$$\epsilon^{\mu\nu\rho\eta}\begin{cases}\epsilon^{\mu\nu\rho\eta} = \epsilon^{0123}, & \text{for an even permutation of } 0,1,2,3;\\ \epsilon^{\mu\nu\rho\eta} = -\epsilon^{0123}, & \text{for an odd permutation of } 0,1,2,3;\\ \epsilon^{\mu\nu\rho\eta} = 0, & \text{otherwise (when two indices are equal).}\end{cases} \tag{G.24}$$

The number of permutations corresponds to the number of interchanges of any two indices. The usual convention for the initialisation of the tensor is[1]

$$\epsilon_{0123} = -\epsilon^{0123} = 1. \tag{G.25}$$

[1] Some authors use the opposite convention leading to opposite signs in front of the symbols $\epsilon^{\mu\nu\rho\eta}$ in Eqs. (G.22) and (G.23).

(The usual rule for raising and lowering indices applies to the tensor $\epsilon^{\mu\nu\rho\eta}$.) The two previous equations imply that

$$\begin{aligned}
\epsilon^{\mu\nu\rho\eta} &= -\epsilon^{\nu\mu\rho\eta} = \epsilon^{\nu\rho\mu\eta} = -\epsilon^{\nu\rho\eta\mu}, \\
\epsilon^{\mu\nu\rho\eta} &= -\epsilon^{\nu\rho\eta\mu} = \epsilon^{\rho\eta\mu\nu} = -\epsilon^{\eta\mu\nu\rho}, \\
\epsilon^{\mu\nu\rho\eta}\epsilon_{\mu\nu\alpha\beta} &= -2(\delta^{\rho}_{\alpha}\delta^{\eta}_{\beta} - \delta^{\rho}_{\beta}\delta^{\eta}_{\alpha}).
\end{aligned} \tag{G.26}$$

Note that a cyclic permutation of indices, as in the second line of Eq. (G.26), implies three permutations of two indices and thus yields a minus sign. This is not the case for the spatial antisymmetric tensor ϵ_{ijk}, defined in Eq. (2.53), since in the latter case, only two permutations of two indices are needed to realise a cyclic permutation. The spatial antisymmetric tensor is a rank-3 Levi–Civita tensor. It is restricted to the three spatial coordinates with the initialisation, $\epsilon_{123} = 1$. By definition, $\epsilon^{ijk} = \epsilon_{ijk}$.

The proof of Eq. (G.22) requires first to note that the trace is necessarily proportional to $\epsilon^{\mu\nu\rho\eta}$. Indeed, if the indices are all different, the trace is totally antisymmetric with the interchange of two indices due to the anti-commutation rule of γ's matrices. If two or three indices are the same, since $(\gamma^{\mu})^2 = \pm 1$, we can apply the property (G.21), leading to a null trace. If the four indices are the same, $(\gamma^{\mu})^4 = 1$, and we apply property (G.17), giving again zero. Thus, as announced, the trace is necessarily proportional to $\epsilon^{\mu\nu\rho\eta}$. To set the value of the constant c in front of $\epsilon^{\mu\nu\rho\eta}$, let us choose $\mu = 0, \nu = 1, \rho = 2, \eta = 3$ and multiply by i: $\mathrm{Tr}(\gamma^5 i\gamma^0\gamma^1\gamma^2\gamma^3) = ic\epsilon^{0123} = -ic$ [with the convention in Eq. (G.25)]. As $\gamma^5 = i\gamma^0\gamma^1\gamma^2\gamma^3$ and $(\gamma^5)^2 = \mathbb{1}$ (identity matrix), we thus have $\mathrm{Tr}(\mathbb{1}) = -ic$, giving $c = 4i$. This leads to the result of Eq. (G.22).

Appendix H Fermi's Golden Rule and Time-Dependent Perturbation

H.1 Fermi's Golden Rule

H.1.1 Non-relativistic Quantum Mechanics Approach

Let us consider a free particle in a stationary state, satisfying the time-independent Schrödinger equation

$$H_0 \, |n\rangle = E_n \, |n\rangle, \tag{H.1}$$

where the eigenstates of H_0 are normalised such that $\langle m|n\rangle = \delta_{mn}$. We consider, for the moment, a non-relativistic wave function normalised to one particle in a box of volume $\mathcal{V}$, i.e.

$$\langle \boldsymbol{x}|n\rangle = \phi_n(\boldsymbol{x}) = \frac{1}{\sqrt{\mathcal{V}}} e^{i\boldsymbol{p}_n\cdot\boldsymbol{x}}, \quad \text{with} \quad \langle n|n\rangle = \int_{\mathcal{V}} \phi_n^*(\boldsymbol{x})\phi_n(\boldsymbol{x}) \, \mathrm{d}^3\boldsymbol{x} = 1. \tag{H.2}$$

The particle is subjected to interaction during a time T through a potential $V(t)$ that is small compared with the unperturbed Hamiltonian H_0. Moreover, during T, the potential is constant, not depending on time. Far from the region where the potential is non-negligible, the particle is considered free and described by a stationary state of H_0. Then, the potential takes the form

$$V(t) = \begin{cases} 0, & t < 0 \text{ or } t \geq T; \\ V, & t \in [0, T]. \end{cases}$$

The question is to determine the state of the particle, $|\psi(t)\rangle$, given that before the interaction, $|\psi(t)\rangle = |i\rangle$ for $t \leq 0$, $|i\rangle$ being a stationary state. The evolution of $|\psi(t)\rangle$ is given by

$$i\frac{\mathrm{d}}{\mathrm{d}t} \, |\psi(t)\rangle = (H_0 + V(t)) \, |\psi(t)\rangle, \tag{H.3}$$

and inserting the expansion of $|\psi(t)\rangle$ in the $|n\rangle$ basis, $|\psi(t)\rangle = \sum_n a_n(t) \, |n\rangle \, e^{-iE_nt}$, leads to

$$i\sum_n \frac{\mathrm{d}}{\mathrm{d}t} a_n(t) \, |n\rangle \, e^{-iE_nt} = \sum_n V(t) a_n(t) \, |n\rangle \, e^{-iE_nt}. \tag{H.4}$$

The probability that the particle is in a stationary state $|f\rangle$ at $t > T$ is $|\, \langle f|\psi(t)\rangle \,|^2 = |a_f(t)|^2$. Projecting Eq. (H.4) onto $\langle f|$ gives the evolution of $a_f(t)$,

$$\frac{\mathrm{d}}{\mathrm{d}t} a_f(t) = -i \sum_n a_n(t) V_{fn}(t) \, e^{i(E_f - E_n)t}, \tag{H.5}$$

where $V_{fn}(t)$ is defined by

$$V_{fn}(t) = \langle f|V(t)|n\rangle = \int_{\mathcal{V}} \phi_f^*(\boldsymbol{x})V(\boldsymbol{x},t)\phi_n(\boldsymbol{x})\ \mathrm{d}^3\boldsymbol{x}. \tag{H.6}$$

As the particle is in the eigenstate $|i\rangle$ for $t \le 0$, the coefficients a_n satisfy

$$a_i(t \le 0) = 1, \quad a_{n\neq i}(t \le 0) = 0. \tag{H.7}$$

Now, if we assume that the potential is small enough, the condition (H.7) must be reasonably satisfied[1] for $t \in [0, T]$, with coefficients $a_n(t)$ changing only slowly from their initial values because of the presence of the potential. Hence, Eq. (H.5) simplifies to

$$\frac{\mathrm{d}}{\mathrm{d}t}a_f(t) = -iV_{fi}(t)\ e^{i(E_f-E_i)t}, \quad \text{for } t \in [0, T], \tag{H.8}$$

whose solution is

$$a_f(t) = -i\int_0^t \mathrm{d}t'\ V_{fi}(t')\ e^{i(E_f-E_i)t'}, \quad \text{for } t \in [0, T]. \tag{H.9}$$

This expression is valid as long as $|a_f(t)| \ll 1$. With the assumption that, when $t \in [0, T]$, the potential does not depend on time, the coefficient at $t = T$, when the potential vanishes, is

$$a_f(T) = -iV_{fi}\int_0^T \mathrm{d}t'\ e^{i(E_f-E_i)t'} = -iV_{fi}\, e^{i(E_f-E_i)\frac{T}{2}} f_T(E_f - E_i), \tag{H.10}$$

where

$$f_T(E) = 2\frac{\sin\left(E\frac{T}{2}\right)}{E} = T\frac{\sin\left(E\frac{T}{2}\right)}{E\frac{T}{2}}. \tag{H.11}$$

Therefore, the transition rate, i.e. the probability per unit time of a transition from the state $|i\rangle$ to the state $|f\rangle \neq |i\rangle$, is

$$\Gamma_{i\to f} = \frac{|a_f(T)|^2}{T} = 2\pi|V_{fi}|^2\frac{f_T^2(E_f - E_i)}{2\pi T} = 2\pi|V_{fi}|^2\delta_T(E_f - E_i), \tag{H.12}$$

with

$$\delta_T(E_f - E_i) = \frac{f_T^2(E_f - E_i)}{2\pi T} = \frac{1}{\pi}\frac{\sin^2\left((E_f - E_i)\frac{T}{2}\right)}{(E_f - E_i)^2\frac{T}{2}}. \tag{H.13}$$

For large T, this function $\delta_T(E_f - E_i)$ is strongly peaked when $E_f \simeq E_i$. Hence, only final states having a substantial $|V_{fi}|$ and energy close enough to the initial energy can be reasonably reached. This effect is an illustration of the Heisenberg uncertainty principle $\Delta E\Delta T \simeq \hbar$. Asymptotically, for infinite T, one can show that [see, for instance, Cohen-Tannoudji et al. (1977, appendix II, vol. 2)]

$$\lim_{T\to+\infty} \delta_T(E_f - E_i) = \delta(E_f - E_i). \tag{H.14}$$

Then, Eq. (H.12) becomes Fermi's golden rule,

$$\boxed{\Gamma_{i\to f} = 2\pi|T_{fi}|^2\delta(E_f - E_i),} \tag{H.15}$$

[1] More rigorously, what is done here is an expansion to the first order of perturbations.

where T_{fi} is the transition matrix element

$$T_{fi} = T_{fi}^{(1)} + T_{fi}^{(2)} + \cdots . \tag{H.16}$$

The previous calculation was performed at the first order of the perturbation (as known as the Born approximation), where $T_{fi}^{(1)} = V_{fi} = \langle f|V|i\rangle$ just corresponds to the matrix element of the transition between the initial and the final states. We can improve the calculation of T_{fi} by moving to higher orders. The expression of $a_f^{(1)}(t)$ in Eq. (H.9) was obtained under the assumption that conditions (H.7) hold even for $t > 0$, whereas the expression of $\mathrm{d}a_f(t)/\mathrm{d}t$ in Eq. (H.5) is exact. We then proceed by successive iterations. For the second iteration, we can make the following approximation:

$$a_i^{(2)}(t) \simeq 1, \quad a_{n\neq i}^{(2)}(t) = a_{n\neq i}^{(1)}(t), \quad \text{for } t > 0, \tag{H.17}$$

where $a_{n\neq i}^{(1)}(t)$ is obtained from Eq. (H.9). Given that $|f\rangle$ does not play a particular role, Eq. (H.9) is also valid for any state $|n\rangle$ provided that $n \neq i$. Therefore, we obtain

$$\frac{\mathrm{d}}{\mathrm{d}t}a_f^{(2)}(t) \simeq -iV_{fi}(t)\, e^{i(E_f-E_i)t} - i\sum_{n\neq i}\left(-i\int_0^t \mathrm{d}t'\, V_{ni}(t')\, e^{i(E_n-E_i)t'}\right) V_{fn}(t)\, e^{i(E_f-E_n)t}. \tag{H.18}$$

But for $t \in [0, T]$, where V_{ni} does not depend on time,

$$-i\int_0^t \mathrm{d}t'\, V_{ni}(t')\, e^{i(E_n-E_i)t'} = V_{ni}\,\frac{e^{i(E_n-E_i)t} - 1}{E_i - E_n}.$$

Therefore, the integration of Eq. (H.18) yields

$$\begin{aligned} a_f^{(2)}(T) &= \int_0^T \mathrm{d}t \left(-iV_{fi}\, e^{i(E_f-E_i)t} - i\sum_{n\neq i}\left(V_{ni}\frac{e^{i(E_n-E_i)t} - 1}{E_i - E_n}\right) V_{fn}\, e^{i(E_f-E_n)t}\right) \\ &= -i\left(V_{fi} + \sum_{n\neq i}\frac{V_{fn}V_{ni}}{E_i - E_n}\right)\int_0^T \mathrm{d}t\, e^{i(E_f-E_i)t} + i\sum_{n\neq i}\frac{V_{fn}V_{ni}}{E_i - E_n}\int_0^T \mathrm{d}t\, e^{i(E_f-E_n)t}. \end{aligned}$$

Compare this expression to that of $a_f^{(1)}$ in Eq. (H.10). If there was only the term with the first integral above, following the argument of the first-order calculation, we would conclude that as long as T is large enough, the only important contribution arises from final states having $E_f \simeq E_i$. Therefore, we would identify the following second-order contribution:

$$T_{fi}^{(2)} = \sum_{n\neq i}\frac{V_{fn}V_{ni}}{E_i - E_n} = \sum_{n\neq i}\frac{\langle f|V|n\rangle\,\langle n|V|i\rangle}{E_i - E_n}. \tag{H.19}$$

Note that this term, as well as $T_{fi}^{(1)}$, leads to a linear probability with T, or equivalently, a constant transition rate. On the other hand, when E_n differs from E_i and E_f, the term of the second integral contributing to $a_f^{(2)}$ cannot converge to a delta distribution and oscillates faster as T increases. Therefore, this term does not have a linear behaviour at large T and necessarily becomes negligible. In conclusion, the expression (H.19) remains the only significant contribution at large T to $T_{fi}^{(2)}$. This second-order term includes two transitions to move from the state $|i\rangle$ to $|f\rangle$ via an intermediate state $|n\rangle$. There is no delta function $\delta(E_i - E_n)$ or $\delta(E_n - E_f)$, implying that energy conservation is not required for the transitions induced with the intermediate state. However, the energy is globally conserved since there is an overall $\delta(E_i - E_f)$ in Fermi's golden rule Formula (H.15).

H.1.2 Quantum Field Approach

Consider the first-order $S^{[n=1]}$ of the S-matrix (6.56) between a state $|i\rangle$ and $|f\rangle$,

$$\langle f|S^{[1]}|i\rangle = -i\int_{-\infty}^{+\infty} \mathrm{d}t\,\langle f|H_{\mathrm{int\,I}}|i\rangle.$$

Replacing the bounds of integration by 0 and T and introducing the Schrödinger representation of the Hamiltonian, H_{int} with $H_{\mathrm{int\,I}} = e^{itH_0}H_{\mathrm{int}}\,e^{-itH_0}$ [Eq. (6.43)] yield

$$\langle f|S^{[1]}|i\rangle = -i\int_0^T \mathrm{d}t\,\langle f|e^{itH_0}H_{\mathrm{int}}\,e^{-itH_0}|i\rangle = -i\int_0^T \mathrm{d}t\,e^{itE_f}\,\langle f|H_{\mathrm{int}}|i\rangle\,e^{-itE_i}.$$

If the Hamiltonian does not depend on time, as in the previous section, it follows that

$$\langle f|S^{[1]}|i\rangle = -i\,\langle f|H_{\mathrm{int}}|i\rangle\int_0^T \mathrm{d}t\,e^{it(E_f-E_i)} = -i\,\langle f|H_{\mathrm{int}}|i\rangle\,T\frac{\sin\left[(E_f-E_i)T/2\right]}{\left[(E_f-E_i)T/2\right]}e^{i(E_f-E_i)T/2}.$$

The transition rate at large time thus reads

$$\Gamma_{i\to f} = \lim_{T\to+\infty}\frac{|\,\langle f|S^{[1]}|i\rangle\,|^2}{T} = |\,\langle f|H_{\mathrm{int}}|i\rangle\,|^2\lim_{T\to+\infty} T\frac{\sin^2\left[(E_f-E_i)T/2\right]}{\left[(E_f-E_i)T/2\right]^2}.$$

The limit converges to $2\pi\,\delta(E_f-E_i)$ [see Eq. (H.14)], and hence, we recover Fermi's golden rule expression,

$$\Gamma_{i\to f} = 2\pi|\,\langle f|H_{\mathrm{int}}|i\rangle\,|^2\delta(E_f-E_i).$$

H.2 Transition Rate

Fermi's golden rule in Eq. (H.15) supposes that the final state $|f\rangle$ is precisely known. In practice, the final state is rather described by a set of final states belonging to a given phase-space element. Let us consider an infinitesimal phase-space element such that $|V_{fi}|$ remains constant for these dN states. The infinitesimal transition rate is then the sum of the probabilities of each states, yielding

$$\mathrm{d}\Gamma_{i\to f} = 2\pi\,|T_{fi}|^2\delta(E_f-E_i)\,\mathrm{d}N = 2\pi\,|T_{fi}|^2\delta(E_f-E_i)\,\frac{\mathrm{d}N}{\mathrm{d}E_f}\,\mathrm{d}E_f. \tag{H.20}$$

The integration of Eq. (H.20) yields

$$\Gamma_{i\to f} = 2\pi\,|T_{fi}|^2\left.\frac{\mathrm{d}N}{\mathrm{d}E_f}\right|_{E_i} = 2\pi\,|T_{fi}|^2\rho(E_i), \tag{H.21}$$

where $\rho(E_i) = \mathrm{d}N/\,\mathrm{d}E_f|_{E_i}$ is the density of states. Equation (H.21) is Fermi's golden rule for the total transition rate with the density of state ρ. As is well known from quantum mechanics, in the case of a single particle in a box of sides L, the components of its momentum are quantised as $p_{x,y,z} = \frac{2\pi}{L}n_{x,y,z}$. Therefore, the number of states in $\mathrm{d}^3\boldsymbol{p} = \mathrm{d}p_x\,\mathrm{d}p_y\,\mathrm{d}p_z$

is $\mathrm{d}N = \mathrm{d}n_x\,\mathrm{d}n_y\,\mathrm{d}n_z = \mathcal{V}\ \mathrm{d}^3\boldsymbol{p}/(2\pi)^3$, leading to the infinitesimal transition rate from Eq. (H.20),

$$\mathrm{d}\Gamma_{i\to f} = 2\pi\,|T_{fi}|^2\delta(E_f - E_i)\ \mathcal{V}\frac{\mathrm{d}^3\boldsymbol{p}}{(2\pi)^3},$$

with $\mathcal{V} = L^3$. The generalisation to an initial state with n_i particles and final states containing n_f is straightforward, yielding

$$\mathrm{d}\Gamma_{i\to f} = 2\pi\,|T_{fi}|^2\delta(E'_1 + \cdots + E'_{n_f} - E_1 - \cdots - E_{n_i})\prod_{k=1}^{n_f}\mathcal{V}\frac{\mathrm{d}^3\boldsymbol{p}'_k}{(2\pi)^3}, \tag{H.22}$$

where variables related to the final state are denoted by a prime symbol. To the first order of perturbation, generalising Eq. (H.6) and using Eq. (H.2), T_{fi} reads

$$\begin{aligned} T_{fi}^{(1)} &= V_{fi} = \langle f|V|i\rangle \\ &= \int_{\mathcal{V}} \phi_1'^{*}(\boldsymbol{x}'_1)\cdots\phi_{n_f}'^{*}(\boldsymbol{x}'_{n_f})\ V(\boldsymbol{X})\ \phi_1(\boldsymbol{x}_1)\cdots\phi_{n_i}(\boldsymbol{x}_{n_i})\ \mathrm{d}\boldsymbol{X} \\ &= \mathcal{V}^{-\frac{n_i+n_f}{2}}\int_{\mathcal{V}} e^{i\boldsymbol{p}_1\cdot\boldsymbol{x}_1}\ldots e^{i\boldsymbol{p}_{n_i}\cdot\boldsymbol{x}_{n_i}}\,e^{-i\boldsymbol{p}'_1\cdot\boldsymbol{x}'_1}\ldots e^{-i\boldsymbol{p}'_{n_f}\cdot\boldsymbol{x}'_{n_f}}\ V(\boldsymbol{X})\ \mathrm{d}\boldsymbol{X}, \end{aligned}$$

where the symbols $\boldsymbol{X} = \boldsymbol{x}'_1, ..., \boldsymbol{x}'_{n_f}, \boldsymbol{x}_1, ..., \boldsymbol{x}_{n_i}$ and $\mathrm{d}\boldsymbol{X} = \mathrm{d}^3\boldsymbol{x}'_1 ..., \mathrm{d}^3\boldsymbol{x}'_{n_f}, \mathrm{d}^3\boldsymbol{x}_1, ..., \mathrm{d}^3\boldsymbol{x}_{n_i}$ have been introduced. As physics is invariant under translation, $V(\boldsymbol{X})$ only depends on the differences in positions and not on the positions themselves. So, let us make the change of variables $\boldsymbol{r}_i = \boldsymbol{x}_i - \boldsymbol{x}_1$, with $i > 1$ and $\boldsymbol{r}' = \boldsymbol{x}'_i - \boldsymbol{x}_1$. This yields

$$V_{fi} = \mathcal{V}^{-\frac{n_i+n_f}{2}}\int_{\mathcal{V}} \mathrm{d}^3\boldsymbol{x}_1 e^{i(\boldsymbol{p}_1+\cdots+\boldsymbol{p}_{n_i}-\boldsymbol{p}'_1-\cdots-\boldsymbol{p}'_{n_f})\cdot\boldsymbol{x}_1}\,\hat{V}_{fi}, \tag{H.23}$$

with

$$\hat{V}_{fi} = \int_{\mathcal{V}} \mathrm{d}^3\boldsymbol{r}_2\cdots\mathrm{d}^3\boldsymbol{r}_{n_i}\,\mathrm{d}^3\boldsymbol{r}'_1\cdots\mathrm{d}^3\boldsymbol{r}'_{n_f}e^{i\boldsymbol{p}_2\cdot\boldsymbol{r}_1}\ldots e^{i\boldsymbol{p}_{n_i}\cdot\boldsymbol{r}_{n_i}}\,e^{-i\boldsymbol{p}'_1\cdot\boldsymbol{r}'_1}\ldots e^{-i\boldsymbol{p}'_{n_f}\cdot\boldsymbol{r}'_{n_f}}\ V(\boldsymbol{r}_2, ..., \boldsymbol{r}_{n_i}, \boldsymbol{r}'_1, ..., \boldsymbol{r}'_{n_f}).$$

For L large enough compared with the range of the interaction described by V, $\hat{V}$ does not depend on the volume $\mathcal{V} = L^3$. However, the integral in (H.23) does depend on L. Let us denote $\Delta\boldsymbol{p} = \boldsymbol{p}_1 + \cdots + \boldsymbol{p}_{n_i} - \boldsymbol{p}'_1 - \cdots - \boldsymbol{p}'_{n_f}$. The integral $\int_{\mathcal{V}} \mathrm{d}^3\boldsymbol{x}_1 e^{i\Delta\boldsymbol{p}\cdot\boldsymbol{x}_1}$ appearing in Eq. (H.23) is easy to calculate, yielding

$$\int_{\mathcal{V}} \mathrm{d}^3x_1 e^{i\Delta p\cdot x_1} = L\frac{\sin(\Delta p_x L/2)}{\Delta p_x L/2}\,L\frac{\sin(\Delta p_y L/2)}{\Delta p_y L/2}\,L\frac{\sin(\Delta p_z L/2)}{\Delta p_z L/2}.$$

Therefore, $|V_{fi}|^2$ or $|T_{fi}^{(1)}|^2$ in Eq. (H.22) reads

$$|V_{fi}|^2 = \mathcal{V}^{-(n_i+n_f)}L^3 L\frac{\sin^2(\Delta p_x L/2)}{(\Delta p_x L/2)^2}\,L\frac{\sin^2(\Delta p_y L/2)}{(\Delta p_y L/2)^2}\,L\frac{\sin^2(\Delta p_z L/2)}{(\Delta p_z L/2)^2}|\hat{V}_{fi}|^2.$$

We recognise the same functions as in Eq. (H.14) with T replaced with L so that for L large enough, $|V_{fi}|^2$ converges to

$$|V_{fi}|^2 = \mathcal{V}^{1-(n_i+n_f)}2\pi\,\delta(\Delta p_x)\ 2\pi\,\delta(\Delta p_y)\ 2\pi\,\delta(\Delta p_z)|\hat{V}_{fi}|^2. \tag{H.24}$$

Hence, inserting Eq. (H.24) into (H.22) yields the transition rate (with $\Delta E = E'_1 + \cdots + E'_{n_f} - E_1 - \cdots - E_{n_i}$)

$$\begin{aligned} \mathrm{d}\Gamma_{i\to f} &= 2\pi \mathcal{V}^{1-(n_i+n_f)}\, 2\pi\,\delta(\Delta p_x)\, 2\pi\,\delta(\Delta p_y)\, 2\pi\,\delta(\Delta p_z)\, |\hat{V}_{fi}|^2 \delta(\Delta E) \prod_{k=1}^{n_f} \mathcal{V} \frac{\mathrm{d}^3 \boldsymbol{p}'_k}{(2\pi)^3} \\ &= \mathcal{V}^{1-n_i} (2\pi)^4 \delta^{(4)}(p'_1 + \cdots + p'_{n_f} - p_1 - \cdots - p_{n_i})\, |\hat{V}_{fi}|^2 \prod_{k=1}^{n_f} \frac{\mathrm{d}^3 \boldsymbol{p}'_k}{(2\pi)^3}, \end{aligned} \tag{H.25}$$

where $\delta^{(4)}(p'_1 + \cdots) = \delta(E'_1 + \cdots)\delta(p'_{1_x} + \cdots)\delta(p'_{1_z} + \cdots)\delta(p'_{1_z} + \cdots)$.

In high energy physics, we are interested in transition rates valid in the relativistic regime. We saw in Chapter 5 that it is more appropriate to normalise the density of particles to $2E$ per unit volume. Each relativistic wave function then contributes to a multiplicative factor $2E$ in the matrix element squared. Let us define the new Lorentz invariant matrix element $\mathcal{M}_{fi}$ evaluated with the relativistic normalisation of wave functions,

$$|\mathcal{M}_{fi}|^2 = \prod_{k=1}^{n_i} 2E_k \prod_{k=1}^{n_f} 2E'_k\, |\hat{V}_{fi}|^2.$$

Choosing arbitrary phases, $\hat{V}_{fi}$ reads

$$\hat{V}_{fi} = \prod_{k=1}^{n_i} \frac{1}{\sqrt{2E_k}} \prod_{k=1}^{n_f} \frac{1}{\sqrt{2E'_k}}\, \mathcal{M}_{fi}. \tag{H.26}$$

Another way of realising that there are these $1/2E$ factors without relying on the wave functions themselves is as follows. When we established the Formula (H.22), we counted the number of states per particle in a box of volume $\mathcal{V}$ containing one particle: $\mathrm{d}N = \mathcal{V}\ \mathrm{d}^3\boldsymbol{p}/(2\pi)^3$. Given that there are finally $2E$ particles in that box, $\mathrm{d}N$ becomes

$$\mathrm{d}N = \mathcal{V} \frac{\mathrm{d}^3 \boldsymbol{p}}{2E(2\pi)^3}.$$

Finally, the transition rate in Eq. (H.25) expressed in terms of Lorentz invariant quantities reads

$$\mathrm{d}\Gamma_{i\to f} = \mathcal{V}^{1-n_i} (2\pi)^4 \delta^{(4)}(p'_1 + \cdots + p'_{n_f} - p_1 - \cdots - p_{n_i}) |\mathcal{M}_{fi}|^2 \prod_{k=1}^{n_i} \frac{1}{2E_k} \prod_{k=1}^{n_f} \frac{\mathrm{d}^3 \boldsymbol{p}'_k}{(2\pi)^3 2E'_k}. \tag{H.27}$$

References

Abada A., Abbrescia M., AbdusSalam S. S. et al., 2019a, *Eur. Phys. J.* C **79**, 474.

Abada A., Abbrescia M., AbdusSalam S. S. et al., 2019b, *Eur. Phys. J. Spec. Top.* **228**, 261–263.

Aharonov Y. and Bohm D., 1961, *Phys. Rev.* **122**, 1649.

Aitchison I. J. R. and Hey A. J. G., 2003a, *Gauge Theories in Particle Physics*, Vol. 1, IOP, London.

Aitchison I. J. R. and Hey A. J. G., 2003b, *Gauge Theories in Particle Physics*, Vol. 2, IOP, London.

Anderson C. D. and Neddermeyer S. H., 1937, *Phys. Rev.* **51**, 884.

Anderson C. D. and Neddermeyer S. H., 1938, *Phys. Rev.* **54**, 88.

Aoyama T., Hayakawa M., Kinoshita T., and Nio M., 2012, *Phys. Rev. Lett.* **109**, 111807.

Aoyama T., Hayakawa M., Kinoshita T., and Nio M., 2014, arXiv:1412.8284.

ATLAS Collaboration, 2022a, https://atlas.web.cern.ch/Atlas/GROUPS/PHYSICS/PAPERS/HIGG-2021-23/ giving additional materials to the publication ATLAS coll, 2022b.

ATLAS Collaboration, 2022b, *Nature* **607**, 52–59.

ATLAS Collaboration, Aad G. et al., 2012, *Phys. Lett.* B **716**, 1–29.

Atwood W. B., Bjorken J. D., Brodsky S. J. et al., 1982, *Lectures on Lepton Nucleon Scattering and Quantum Chromodynamics, Progress in Mathematical Physics*, Vol. 6, Birkhäuser, Basel.

Baker S. and Cousins R. D., 1984, *Nucl. Instrum. Meth.* **221**, 437.

Barate R. et al. (LEP Working Group for Higgs boson searches and ALEPH and DELPHI and L3 and OPAL Collaborations), 2003, *Phys. Lett.* B **565**, 61.

Barlow R. J., 1989, *Statistics: A Guide to the Use of Statistical Methods in the Physical Sciences*, John Wiley & Sons, New York.

Bethe H. A., 1930, *Ann. Phys.* **5**, 325.

Bethe H. A., 1932, *Z. Phys.* **76**, 293.

Bethke S., 2007, *Prog. Part. Nucl. Phys.* **58**, 351–386.

Bilenky S. M. and Giunti C., 2001, *Int. J. Mod. Phys.* A **16**, 3931–3949.

Bjorken J. D., 1969, *Phys. Rev.* **179**, 1547.

Bjorken J. D. and Paschos E. A., 1969, *Phys. Rev.* **185**, 1975.

Bloch F., 1933, *Z. Phys.* **81**, 363.

Boas M. L., 2005, *Mathematical Methods in the Physical Sciences*, 3rd ed., John Wiley & Sons, New York.

Bodek A., Breidenbach M., Dubin D. L. et al., 1979, *Phys. Rev.* D **20**, 1471.

Bohm G. and Zech G., 2010, *Introduction to Statistics and Data Analysis for Physicists*, Deutsches Elektronen-Synchrotron, Hamburg.

Breidenbach M., Friedman J. I., Kendall H. W. et al., 1969, *Phys. Rev. Lett.* **23**, 935–939.
Brun R. and Rademakers F., 1997, *Nucl. Inst. & Meth. Phys. Res.* A **389**.
Bump D., 2004, *Lie Groups*, Series *Graduate Texts in Mathematics*, Vol. 225, Springer-Verlag, New York.
Burkhardt H. and Pietrzyk B., 1995, *Phys. Lett.* B **356**, 398.
Buzhana P., Dolgosheina B., Filatov L. et al., 2003. *Nucl. Instrum. Methods* A **504**, 48.
Campagnari C. and Franklin M., 1997, *Rev. Mod. Phys.* **69**, 137–212.
Campbell J. M., Huston J. W., and Stirling W. J., 2007, *Rep. Prog. Phys.* **70**, 89.
Cardoso M., Cardoso N., and Bicodo P., 2010, *Phys. Rev.* D **81**, 034504.
Carr B., Kohri K., Sendouda Y., and Yokoyama J., 2021, *Rep. Prog. Phys.* **84**, 116902.
CDF Collaboration, Abe F. et al., 1995, *Phys. Rev. Lett.* **74**, 2626–2631.
CDF and D0 Collaborations, Aaltonen T. et al., 2013, *Phys. Rev.* D **88**, 052014.
Christenson J. H., Cronin J. W., Fitch V. L., and Turlay R., 1964, *Phys. Rev. Lett.* **13**, 138.
Cline J. M., 2006, *Lectures at Les Houches Summer School*, arXiv:hep-ph/0609145.
CMS Collaboration, 2022, *Nature.* **607**, 60–68.
CMS Collaboration, Chatrchyan S. et al., 2012a, *J. Instrument.* **7**, P10002.
CMS Collaboration, Chatrchyan S. et al., 2012b, *Phys. Lett.* B **716**, 30–61.
CMS Collaboration, Sirunyan A. M. et al., 2021, *Eur. Phys. J.* C **81**, 488.
Cohen-Tannoudji C., Diu B., and Laloë F., 1997, *Quantum Mechanics*, Wiley, New York, p. 1524.
Cowan G., Cranmer K., Gross E., and Vitells O., 2011, *Eur. Phys. J.* C **71**, 1554.
D0 Collaboration, Abachi S. et al., 1995, *Phys. Rev. Lett.* **74**, 2632–2637.
Dasgupta B. and Kopp J., 2021, *Phys. Rep.* **928**, 1–63.
Daya Bay Collaboration, An F. P. et al., 2012, *Phys. Rev. Lett.* **108**, 171803.
De Florian D., Sassot R., Stratmann M., and Vogelsang W., 2014, *Phys. Rev. Lett.* **113**, 012001.
Dittmaier S. and Schumacher M., 2013, *Prog. Part. Nucl. Phys.* **70**, 1–54.
Djouadi A., 2008, *Phys. Rep.* **457**, 1–216.
Dubrovin B. A., Fomenko A. T., and Novikov S. P., 1984, *Modern Geometry – Methods and Applications*, Springer-Verlag, New York.
Dürr S., Fodor Z., Frison J. et al., 2008, *Science* **322**, 1224.
Dydak F., Navarria F. L., Overseth O. E. et al., 1977, *Nucl. Phys.* B **118**, 1.
Einstein A., 1905, *Ann. Phys.* **18**, 639.
Elitzur S., 1975, *Phys. Rev.* D **12**, 3978.
Ellis J., Gaillard M. K., and Ross G. G., 1976, *Nucl. Phys.* B **111**, 253.
EMC Collaboration, Ashman J. et al., 1988, *Phys. Lett.* B **206**, 364.
Englert F. and Brout R., 1964, *Phys. Rev. Lett.* **13**, 32.
Esteban, I., Gonzalez-Garcia, M., Maltoni, M. et al., 2020, *J. High Energy Phys.* **2020**, 178. See also www.nu-fit.org.
Fabjan W. and Gianotti F., 2003, *Rev. Mod. Phys.* **75**, 1243.
Feldman G. J. and Cousins R. D., 1998, *Phys. Rev.* D**57**, 3873.
Feynman R. P., 1969, *Phys. Rev. Lett.* **23**, 1415.
Feynman R. P. and Gell-Mann M., 1958, *Phys. Rev.* **109**, 193.
Formaggio J. A. and Zeller G. P., 2012, *Rev. Mod. Phys.* **84**, 1307.

Friederich S., 2013, *Eur. J. Phil. Sci.*, **3**, 157.

Gaisser T. K., 1990, *Cosmic Rays and Particle Physics*, Cambridge University Press, Cambridge.

Gattringer C. and Lang C. B., 2010, *Quantum Chromodynamics on the Lattice, Lecture Notes in Physics Series*, Vol. 788, Springer, Berlin.

Gavela M. B., Hernández P., Orloff J., Pène O., 1994, *Mod. Phys. Lett.* A **9**, 795–810.

Gelis F., 2019, *Quantum Field Theory*, Cambridge University Press, Cambridge.

Georgi H., 1999, *Lie Algebras in Particle Physics: From Isospin to Unified Theories*, 2nd ed., Westview, Boulder, CO.

Giunti C. and Kim C. W., 2007, *Fundamentals of Neutrino Physics and Astrophysics*, Oxford University Press, Oxford.

Glashow S. L., 1961, *Nucl. Phys.* **22**, 579.

Glashow S. L., Iliopoulos J., and Maiani L., 1970, *Phys. Rev.* D **106**, 1285.

Goldhaber M., Grodzins L., and Sunyar A. W., 1958, *Phys. Rev.* **109**, 1015.

Greenberg O. W., 2015, *Phys. Today.* **68**, 33.

Grieder P. K. F., 2001, *Cosmic Rays at Earth*, Elsevier Science, Amsterdam.

Griffiths D. J., 1987, *Introduction to Elementary Particles*, Harper & Row, New York.

Griffiths D. J., 1995, *Introduction to Quantum Mechanics*, Prentice-Hall, Englewood Cliffs, NJ.

Grupen C. and Shwartz B., 2011, *Particle Detectors*, 2nd ed., Cambridge University Press, Cambridge.

Guralnik G. S., Hagen C. R., and Kibble T. W. B., 1964, *Phys. Rev. Lett.* **13**, 585.

Hall, B. C., 2015, *Lie Groups, Lie Algebras, and Representations: An Elementary Introduction, Graduate Texts in Mathematics Series,* Vol. 222, 2nd ed., Springer-Verlag, New York.

Halzen F. and Martin D., 1984, *Quarks & Leptons*, Wiley, New York.

Hanneke D., Fogwell S., and Gabrielse G., 2008, *Phys. Rev. Lett.* **100**, 120801.

Hanneke D., Fogwell Hoogerheide S., and Gabrielse G., 2011, *Phys. Rev.* A **83**, 052122.

Helgason S., 1978, *Differential Geometry, Lie Groups and Symmetric Spaces, Pure and Applied Mathematics Series*, Vol. 80, Academic Press, New York.

Higgs, P. W., 1964, *Phys. Rev. Lett.* **13**, 508.

Ho-Kim Q., Pham X. Y., 1998, *Elementary Particles and Their Interactions*, Springer-Verlag, Berlin.

Hughes E. B., Griffy T. A., Yearian M. R., and Hofstadter R., 1965, *Phys. Rev.* **477**, 1.

Hung P. Q. and Sakurai J. J., 1981, *Annu. Rev. Nucl. Part. Sci.* **31**, 375–438.

Jackson J. D., 1998, *Classical Electrodynamics*, 3rd ed., John Wiley & Sons, New York.

JADE Collaboration, Bartel W. et. al.,1985, *Z. Phys. C: Part. Fields* **26**, 507–513.

James F., 2006, *Statistical Methods in Experimental Physics*, 2nd ed., World Scientific, Singapore.

Janot P. and Jadach S., 2020, *Phys. Lett.* B **803**, 135319.

Jegerlehner F. and Nyffeler A., 2009, *Phys. Rept.* **477**, 1–110.

Jelley J. V., 1958, *Čerenkov Radiation and Its Applications*, Pergamon Press, London.

Kendall H. W., 1991, *Rev. Mod. Phys.* **63**, 597.

Kittel C., 2005, *Introduction to Solid State Physics*, 8th ed., John Wiley & Sons, New York.

Knapp A. W., 2001, *Representation Theory of Semisimple Groups: An Overview Based on Examples*, Princeton University Press, Princeton, NJ.

L3 Collaboration, Adeva B. et al., 1991, *Phys. Lett.* B **263**, 551.

Lattes C. M. G. et al., 1947a, *Nature.* **159**, 694.

Lattes C. M. G. et al., 1947b, *Nature.* **160**, 453, 486.

Landau L. D., 1948, *Dokl. Akad. Nauk USSR.* **60**, 207.

Leader E. and Lorcé C., 2014, *Phys. Rep.* **541**, 163–248.

Lee T. D. and Yang C. N., 1956, *Phys. Rev.* **104**, 254.

Leinweber D. B., 2003–2004, *Visualizations of Quantum Chromodynamics*, www.physics.adelaide.edu.au/theory/staff/leinweber/VisualQCD/Nobel/index.html

Leo W. R., 1994, *Techniques for Nuclear and Particle Physics Experiments*, Springer-Verlag, Berlin Heidelberg.

LEP Collaborations (ALEPH, DELPHI, L3, OPAL) and SLD Collaboration, the LEP Electroweak Working Group, and the SLD Electroweak and Heavy Flavour Groups, 2006, *Phys. Rep.* **427**, 257–454.

LEP Electroweak Working Group, 2012, http://lepewwg.web.cern.ch/LEPEWWG/

LEP Electroweak Working Group, ALEPH, DELPH, OPAL and L3 Collaborations, 2013, *Phys. Rep.* **532**, 119.

LHC Higgs Cross-Section Working Group, de Florian D. et al., 2017, *CERN Yellow Reports: Monographs*, CERN 2017-002. https://twiki.cern.ch/twiki/bin/view/LHCPhysics/LHCHXSWG

LHC Higgs Cross-Section Working Group, Heinemeyer S. et al., 2013, *CERN Yellow Reports: Monographs*, CERN 2013-004. https://twiki.cern.ch/twiki/bin/view/LHCPhysics/LHCHXSWG

Lohmann, W., Kopp, R., and Voss, R.,1985, Energy loss of muons in the energy range 1-10000 GeV, *CERN Yellow Reports: Monographs*, CERN 85-03.

Lyons L., 1986, *Statistics for Nuclear and Particle Physicists*, Cambridge University Press, NewYork.

Marshak R. E., 1992, *Curr. Sci.* **63**, 60–64.

Milgrom M.,1983, *Astrophys. J.*, **270**, 365–389,

Misner C. W., Thorne K. S., and Wheeler J. A., 1973, *Gravitation (Physics Series)*, W. H. Freeman, New York.

Nair V. P., 2005, *Quantum Field Theory, A Modern Perspective*, Springer, New York.

Noether E., 1918, *Math. Phys. Klass.*, 235–257.

NOvA Collaboration, Acero M. A. et al., 2022, *Phys. Rev.* D **106**, 032004.

Odom B., Hanneke D., D'Urso B., and Gabrielse G., 2006, *Phys. Rev. Lett.*, **97**, 030801.

OPAL Collaboration, Abbiendi G. et al., 2004, *Eur. Phys. J.* **C 33**, 173–212.

Particle Data Group, Workman R. L. et al., 2022, *Prog. Theor. Exp. Phys.* **2022**, 083C01.

Perkins D. H., 1982, *Introduction to High Energy Physics*, 3rd ed., Addison-Wesley, Boston, MA.

Peskin M. E. and Schroeder D. V., 1995, *An Introduction to Quantum Field Theory*, Westview, Boulder, CO.

Planck Collaboration, Aghanim N. et al., 2020, *Astron. Astrophys.* **641**, A6.

Povh B., Rith K., Scholz C., Zetsche F., and Lavelle M., 2008, *Particles and Nuclei*, Springer, Berlin.
Riotto A., 1998, *Theories of Baryogenesis*, arXiv:hep-ph/9807454.
Rossi B., 1952, *High Energy Particles*, Prentice-Hall, Inc., Englewood Cliffs, NJ.
Sachs R. G., 1962, *Phys. Rev.* **126**, 2256.
Salam A., 1968, *Conf. Proc.* C **680519**, 367.
Salam G., 2015, *PhD Lecture*, see for example, the lecture 3 given at the ICTP–SAIFR school on QCD and LHC physics, July 2015, Sao Paulo, Brazil: https://gsalam.web.cern.ch/gsalam/teaching/PhD-courses.html
Schmidt F. and Knapp J., 2005, *CORSIKA Shower Images*, www-zeuthen.desy.de/~jknapp/fs/showerimages.html.
Servant G., 2014, *Phys. Rev. Lett.* **113**, 171803.
Shafer J. B., Murray J. J., and Huwe D. O., 1963, *Phys. Rev. Lett.* **10**, 179.
Shifman M., 2012, *Advanced Topics in Quantum Field Theory: A Lecture Course*, Cambridge University Press, Cambridge.
Simon F., 2019, *Nucl. Instrum. Methods.* A **926**, 85.
Smeenk C., 2006, *Phil. Sci.* **73**, 487–499.
SNO Collaboration, Ahmad Q. R. et al., 2001, *Phys. Rev. Lett.* **87**, 7.
SNO Collaboration, Ahmad Q. R. et al., 2002, *Phys. Rev. Lett.* **89**, 1.
SNO+ Collaboration, Anderson M. et al., 2019, *Phys. Rev.* D **99**, 032008.
Street J. C. and Stevenson E. C., 1937, *Phys. Rev.* **52**, 1003.
Strocchi F., 2019, *Symmetry Breaking in the Standard Model*, Publications of the Scuola Normale Superiore, Edizioni della Normale Pisa.
Sudarshan E. C. G. and Marshak R. E., 1957, *Proceedings of the Padua-Venice Conference on Mesons and Newly Discovered Particles*, p. V–14. Reprinted in 'Development of the Theory of Weak Interactions', P. K. Kabir (ed.), Gordon and Breach, New York.
Super-Kamiokande Collaboration, Fukuda S. et al., 1998, *Phys. Rev. Lett.* **81**, 1562.
Super-Kamiokande Collaboration, Fukuda S. et al., 2001, *Phys. Rev. Lett.* **86**, 5651.
Super-Kamiokande Collaboration, Takenaka A. et al., 2020, *Phys. Rev.* D **102**, 112011.
T2K Collaboration, Abe K. et al., 2014, *Phys. Rev. Lett.* **112**, 061802.
T2K Collaboration, Abe K. et al., 2021, *Phys. Rev.* D **103**, 112008.
TASSO Collaboration, Brandelik R. et al., 1980, *Phys. Lett.* B **97**, 453.
Tung W. K., 1985, *Group Theory in Physics*, World Scientific, Singapore.
UA1 Collaboration, Arnison G. et al., 1983, *Phys. Lett.* B **122**, 103.
Weinberg S., 1995, *The Quantum Theory of Fields,* Vol. 1, Cambridge University Press, Cambridge.
Weinberg S., 1967, *Phys. Rev. Lett.* **19**, 1264.
Wiedemann H., 2007, *Particle Accelerator Physics*, 3rd ed., Springer, New York.
Wilson Kenneth G., 1974, *Phys. Rev.* D **10**, 8, 2445.
Wu C. S., Ambler E., Hayward R. W., Hoppes D. D., and Hudson R. P., 1957, *Phys. Rev.* D **105**, 1413.
Yang C. N., 1950, *Phys. Rev.* **77**, 242.
Yukawa, H., 1935, *Proc. Phys. Math. Soc. Japan.* **17**, 48.
ZEUS Collaboration, Abramowicz H. et al., 2016, *Phys. Lett.* B **757**, 468.

Index

Printed in the United States
by Baker & Taylor Publisher Services